FUNDAMENTAL AND APPLIED LASER PHYSICS

Proceedings of the Esfahan Symposium,
August 29 to September 5, 1971

FUNDAMENTAL AND APPLIED LASER PHYSICS

Proceedings of the Esfahan Symposium
August 29 to September 5, 1971

Edited by

MICHAEL S. FELD

ALI JAVAN

NORMAN A. KURNIT

Department of Physics
Massachusetts Institute of Technology

A WILEY-INTERSCIENCE PUBLICATION

JOHN WILEY & SONS, New York • London • Sydney • Toronto

Library of Congress Cataloging in Publication Data

Main entry under title:
Fundamental and applied laser physics.

   1.   Quantum electronics—Congresses. 2. Lasers—
Congresses. 3. Spectrum analysis—Congresses.
I. Feld, Michael S., 1940-   ed. II. Javan,
Ali, 1926-   ed. III. Kurnit, Norman A., 1939-   ed.
IV. Title.   *75715*

QC680.F86   535.5'8   73-392

ISBN 0-471-25701-X

Printed in the United States of America

10 9 8 7 6 5 4 3 2 1

Esfahan Symposium on

Fundamental and Applied

Laser Physics

Honorary Chairman

His Excellency, A. A. Hoveyda
Prime Minister of Iran

Symposium Director

A. Javan

International Symposium
Committee

A. Javan, Chairman
M. S. Feld, Secretary
M. Reza Amin
J. Ducuing
P. L. Kelley
N. A. Kurnit
V. S. Letokhov
F. Partovi
D. Ramachandra Rao
B. P. Stoicheff
A. Szöke

Organizing Committee

M. Reza Amin, Chairman
F. Partovi, Secretary
A. Javan
K. Jenab
P. L. Kelley
G. Motamedi

Symposium Coordinator

A. Banani

Symposium sponsored by
International Union of
Pure and Applied Physics

The symposium was held under the auspices of Arya-Mehr University of
Technology, Tehran, Iran, with the cooperation of Esfahan University
and the Massachusetts Institute of Technology. The symposium took
place on the campus of Esfahan University, Esfahan, Iran, August 29 –
September 5, 1971.

v

PREFACE

In the decade since the laser became a reality, it has found application in a variety of scientific disciplines and has had a major impact on both fundamental and applied areas of research. Its influence on technology has been considerable and promises to be even greater in the coming years. It may be, in fact, that many of the more important ideas and applications are yet to come.

In organizing the symposium, it seemed timely to bring together an international group of scientists to look back over the past advances and examine the areas in which progress remains to be made. In this spirit, participants were asked not only to review progress in their field and report on current developments, but to emphasize those areas which promise to have future impact on science and technology. A panel discussion devoted to these areas of increasing importance was held midway through the conference; the transcript of that discussion is presented by way of introduction to this volume.

It was neither desirable nor possible to cover in this symposium all aspects of laser physics; instead a few selected areas were chosen for discussion. These included some of the earlier developments in the field, such as nonlinear optics and light scattering, and some recent topics which are now beginning to see their major applications, such as precision spectroscopy; in addition, the areas covered included those which are just emerging, such as plasma production for controlled fusion and x-ray generation, and applications in chemistry. A number of participants were invited from closely related areas which have influenced developments in the field of quantum electronics. In this way it was hoped to broaden the discussions, both formal and informal, beyond the confines of specialized areas. How well this succeeded will perhaps be best judged in another 5 or 10 years when the predictions in these papers can be compared with actual developments.

Those of us present at the symposium came away with a sense of excitement in anticipation of these future possibilities, and with pleasant recollections of the stimulating and cordial atmosphere of Esfahan. We are pleased to share these feelings with our colleagues through this volume.

Michael S. Feld   Ali Javan   Norman A. Kurnit

ACKNOWLEDGMENTS

The Esfahan Symposium would not have been possible without the assistance
of the Imperial Government of Iran. We are especially grateful to His Imperial Majesty Shahanshah Arya-Mehr Pahlavi for his interest and support.
We also most appreciative of His Excellency, Prime Minister A. Abbas
Hoveyda, who served as Honorary Chairman and gave a memorable address
at the opening ceremonies.

We owe great thanks to Dr. Manouchehr Eqbal and the National Iranian
Oil Company for their enormous help and generous support of the symposium.
We thank, too, the other numerous groups in the United States and Iran for
their sponsorship and aid.

The extraordinary dedication and hard work of the organizing committee
was responsible for the success of the conference. In Iran, Dr. M. Reza
Amin, Chancellor of Arya-Mehr University, and Dr. Ghasem Motamedi,
Chancellor of Esfahan University, attended a myriad details both before and
during the conference. Each hosted a sumptuous conference banquet, as did
the Prime Minister and His Excellency, Dr. Gholam Reza Kianpour, the
Governor-General of Esfahan. The success of the conference was also due
in large part to the tireless efforts of the Secretary of the Organizing
Committee, Dr. Firooz Partovi of Arya-Mehr University--we owe him and
his wife, Farideh, a special debt of thanks. Dr. Paul Kelley of Lincoln
Laboratories also gave in extra measure of his time and expertise. And all
of the participants valued Dr. Amin Banani, the Symposium Coordinator,
who arranged unforgettable cultural events.

We particularly thank a number of students and colleagues at MIT who
helped so effectively with the preparation of this volume: M. M. Burns,
T. W. Ducas, G. M. Elchinger, R. E. Francke, J. Goldhar, H. P.
Grieneisen, M. A. Guerra, S. Hamadani, I. P. Herman, F. Keilmann,
M. J. Kelly, M. Ketabi, J. Leite, J. C. MacGillivray, A. T. Mattick,
R. E. McNair, J.-P. Monchalin, A. Sanchez, R. M. Osgood, Jr., R. L.
Sheffield, J. G. Small, and J. E. Thomas. Many of them also gave
generously and unfailingly of their time and effort during the symposium.

The secretarial assistance at the time of the conference of Frances
Aschheim, Kate French, Jill Ali-Akbarian, Ferial Ardalan, and Zahra
Taheri is gratefully acknowledged. Our special thanks go to Elizabeth
Kreilkamp and Frances Gedziun for their care and diligence in typing the
manuscript.

Finally, we thank the people of Esfahan, who welcomed us collectively
and individually with great warmth and friendship.

CONTENTS

THE FUTURE COURSE OF QUANTUM ELECTRONICS

Panel Discussion

Esfahan Symposium on Fundamental and Applied Laser Physics

September 2, 1971

PANEL DISCUSSION PARTICIPANTS

Professor Nicolaas Bloembergen
Department of Physics
Harvard University
Cambridge, Massachusetts, U.S.A.

Professor Pierre Jacquinot
Centre National de la Recherche
   Scientifique
Laboratoire Aimé Cotton
Orsay, France

Professor Ali Javan
Department of Physics
Massachusetts Institute of Technology
Cambridge, Massachusetts, U.S.A.

Dr. Raymond E. Kidder
Head, Q Division
Lawrence Radiation Laboratory
Livermore, California, U.S.A.

Professor Sergio P. S. Porto
Departments of Physics and Electrical
   Engineering
University of Southern California
Los Angeles, California, U.S.A.

Professor Alexandr M. Prokhorov
Director of Oscillation Laboratory
P. N. Lebedev Physical Institute
Academy of Sciences
Moscow, U.S.S.R.

Professor Arthur L. Schawlow
Physics Department
Stanford University
Stanford, California, U.S.A.

Professor Boris Stoicheff
Department of Physics
University of Toronto
Toronto, Ontario, Canada

Professor Charles H. Townes
Physics Department
University of California
Berkeley, California, U.S.A.

CONTRIBUTORS FROM AUDIENCE

Professor C. O. Alley
Department of Physics and Astronomy
University of Maryland
College Park, Maryland, U.S.A.

Dr. Michel Duguay
Member of Technical Staff
Bell Telephone Laboratory
Murray Hill, New Jersey, U.S.A.

Professor Pierre Giacomo
Assistant Director
Bureau International des Poids et
   Mesures
Sèvres, France

Dr. John Hall
Member of Technical Staff, JILA
University of Colorado & National
   Bureau of Standards
Boulder, Colorado, U.S.A.

Professor R. V. Khokhlov
Department of Physics
Moscow University
Moscow, U.S.S.R.

Professor Benjamin Lax
Director, National Magnet Laboratory
Massachusetts Institute of Technology
Cambridge, Massachusetts, U.S.A.

Professor Willie Low
The Hebrew University of Jerusalem
Department of Physics
Microwave Division
Jerusalem, Israel

Dr. Philip Mallozzi
Project Leader
Battelle Memorial Institute
Columbus, Ohio, U.S.A.

Dr. Chandra K. N. Patel
Director of Electronic Research
   Laboratory
Bell Telephone Laboratories
Holmdel, New Jersey, U.S.A.

Dr. Rolf Sinclair
Director of Atomic and Molecular
   Physics
National Science Foundation
Washington, D.C., U.S.A.

Professor Shaul Yatsiv
The Racah Institute of Physics
The Hebrew University of
   Jerusalem
Jerusalem, Israel

TEXT OF THE DISCUSSION

Schawlow. I'm supposed to be the moderator, so, panelists, let me tell you
the format we are going to use. First, we are going to ask each
of the distinguished panelists to speak for about three minutes,
not more. Each one should say something about some aspect of
lasers or their field of lasers or applications, scientific or
others; let's be as wild as possible and talk about the future!
After the panelists have had their say and perhaps rebutted each
other I hope we will have time for our even more distinguished
audience to correct the panelists and extend their remarks. I'm
not on the panel, but let me just take a minute to introduce the
topic and sort of set things in perspective. May we have the
first slide please. (Laughs) This appeared on the newsstands in
England last year: "Lasers in action--will the Death Ray find an
honest job?" (Loud laughter) We do not propose to answer that
question here (unless somebody wants to). We can think, however, not about applications, but about what is possible, what
we can do, because, remember, the laser is a prime example of
pure technology. One hears about pure science; there is also

pure technology as discussed by Daedalus in <u>New Scientist</u> a few
years ago. At that time he gave us an example of pure technology
in the making of a square phonograph record with a rectilinear
quadrilateral device that would play it so well that the corners
wouldn't be noticeable. (Mild laughter) Well, much of what goes
on in the laser field is the same thing. We try hard to find out
what we can do, and then let the business man worry about
whether or not he actually wants to do it.

We will start with Professor Bloembergen.

Comments by Professor N. Bloembergen

A reasonably safe prediction on future progress in a branch of physics consists always of pushing the values of important physical parameters to new
limits. In the case of laser physics one can think of pushing the frequency
to the uv and perhaps to the x-ray region. There are no known obstacles
to obtain laser light beams in the vacuum uv down to about 1200 Å wavelength by extension of familiar techniques of nonlinear optics. Clearly a
lot of materials research and development is required. One may reasonably
expect that within a decade high-resolution spectroscopy with tunable laser
sources may be done anywhere between 100 and 0.1 μ.
     The push toward the x-ray region will be more difficult. Perhaps the
pulsed x-ray emission from laser induced high-Z plasmas may be used to
pump an x-ray laser. A big payoff would be x-ray holography, making
visible the electronic structure of matter, in particular the structure of
biochemical molecules.
     Other physical parameters which may be pushed are the time duration
and peak power level of ultrashort light pulses. There seems to be no fundamental limitation why light pulses as short as 10^{-14} sec could not be
obtained. For flux densities exceeding 10^{12} W/cm^2, corresponding to light
field amplitudes of about 3×10^7 V/cm, all materials will ionize rapidly in
very short times, of the order of a picosecond or less. In a completely
ionized plasma the peak flux could be raised further. Short, powerful,
tunable pulses from picosecond dye lasers will clearly be important in analyzing excited states and intermediate products in photochemical reactions.
     Turning to shorter range predictions and more immediate developments,
I wish to point out that nonlinear spectroscopy is likely to grow enormously.
The exploration of this field has only just begun, and it is potentially much
richer than linear spectroscopy. One has frequency and spatial dispersion
in multidimensional spaces. Instead of the second-rank linear dielectric
tensor, one must worry about the symmetry properties of third- and fourth-

rank tensors. One could probe the band structure at points in highly absorbing regions, by observation of nonlinear processes in the transparent regions.

High-resolution absorption-dip spectroscopy is another example of nonlinear spectroscopy. Talks at this symposium have shown that one can now
do in the near and far infrared all the tricks that have been done in magnetic
resonance at microwave and radiofrequencies a few decades ago. One can
foresee important analytical and chemical applications of these techniques
with infrared lasers.

Comments by Professor A. M. Prokhorov

There are many things to speak about, but one of the fields, the far-infrared
and submillimeter region, will be developed during the next few years because this region is still not very popular; there are no available sources
and no available materials. I think it will be developed because, for solid-
state physics, this is a rather important region. The other field which I want
to mention is stimulated Raman scattering; we can use Raman components
to have a mode-locked laser with pulse duration about 10^{-14} sec. And this
would be very interesting and important because, at such short pulses, all
electronic nonlinearities will be involved. On the other hand, if you will investigate the interaction of this high-power radiation with matter you should
not bother about multiphoton absorption. There will be only a tunneling effect because the power will be very high and the tunneling time will be short,
comparable with the period of the light. This is not the only application.
For chemical applications it will also be important. But, it is difficult to
predict what will happen before actual experiments are made. One thing
which we can now foresee is that the electronic nonlinearity, with a time
constant of 10^{-15} seconds, is rather important when we discuss the self-
focusing process for picosecond and subpicosecond pulses. Subpicosecond
pulses will be developed in the coming years.

Comments by Professor S. P. S. Porto

I would like to mention three fields, three things that we are going to do
ourselves and we think are quite interesting. One is the question of the application of lasers to crystal structure. Natural crystal structure is now
studied through x-ray diffraction. It's very simple when one knows the
structure to calculate precisely how many phonons will be available and what
the polarizability tensor is, and so on. One can go this way or one can go
backwards by finding the phonons that exist by finding the polarizability tensor, and then go back and find the crystal structure. But where Raman

spectroscopy has an advantage over x-ray analysis is that x-ray crystallography measures the average configuration while in Raman spectroscopy you are making the analysis in a much shorter time domain, so that you can see variations in that average structure. For instance, if you have a crystal which is ferroelectric at one temperature or pyroelectric for some other temperature there will be domains existing after the transition has taken place which are inaccessible to the x-ray spectroscopist but which we can see. If those domains are of the order of a wavelength of light in size we can even get $k = 0$ phonons, and we can analyze these very precisely and know the nature of the distortion of the crystal structure. If the distortions are very small, much smaller than the wavelengths of light, we have essentially a breakdown of k conservation and broad features appear that will probably not be as important. The other aspect of this that I think is very important is to study the phase transition itself. As the crystal is heated or cooled, a phase transition occurs. The following picture is the best way to visualize what causes the phase transition. Imagine there is a phonon which becomes soft and decreases in frequency as the phase transition approaches. As the frequency decreases it crosses the frequency of another phonon. A coupled oscillator problem then occurs; the phase transition then becomes a reality. Looking for the microscopic mechanism of phase transitions is right now, in my opinion, the hottest subject for which one can find an application of the laser.

Comments by Professor C. H. Townes

Any review of the past development of quantum electronics is likely to make one enthusiastically optimistic about the future. Every two or three years there has been another revolution in what appears practical; over ten or fifteen years this characteristic has opened up the field to wide applications and a great range of fruitful developments. One can reasonably expect striking progress to continue for some time, for there is yet a considerable way to go before we reach any natural or physical limits in the useful characteristics and performance of lasers. Acceptance of this conclusion, and recognition that both optics and electronics have been remarkably useful as separate fields, makes the marriage between the two which is quantum electronics seem indeed a powerful combination. There appears every reason to expect that laser techniques will permeate much of our science and technology, and in time even get into the home. One cannot do justice to such a large and diffuse field in a few minutes. Hence I shall make only a few additional comments on very specific areas in which I myself am working. These also tempt me to reminisce a bit.

I first became interested in trying to make a molecular oscillator in

order to produce submillimeter and far-infrared oscillators for spectroscopic purposes. This original goal has not yet been fully achieved, although
many other striking successes have occurred. While the prospect for convenient scientific techniques in the far infrared still looks good, this region
is in fact one of the last to which quantum electronics is contributing, and
much remains to be achieved. Further development of the infrared and particularly of the far infrared is of great scientific importance, and probably
will occur rapidly over the next few years. One field for which it is especially important is astronomy. A very large fraction of the total radiation
from astronomical objects is in the infrared and far infrared; yet these
spectral regions have scarcely been tapped for astronomical purposes, and
they badly need development. Unfortunately, the atmosphere absorbs much
of the infrared spectral region rather severely, and has prevented anything
more than very rudimentary astronomy in the submillimeter region. But
with high-flying planes, balloons, and spacecraft, we can work above the
atmosphere. Furthermore, there is much good work to be done in the atmospheric windows where infrared is transmitted. Quantum electronics,
including coherent techniques and good amplifiers, is a key to further development of infrared techniques, and it should have an enormous impact on
astronomy. At the University of California, Berkeley, we are accordingly
attempting several technical developments.

One example of our efforts to properly utilize quantum electronics in
the infrared region is a long-baseline stellar interferometer for 10 μ
wavelength. By long, I mean in the realm of kilometers, and with angular
resolution far beyond any previously obtainable in the infrared. We are also
trying to develop techniques for infrared pictures by using up-conversion.
Sensitive detectors, utilizing Josephson effects and maser-type amplifiers,
represent other goals. Heterodyne techniques are being applied in the infrared for very high spectral resolution of line emission from astronomical
objects. Most of these goals will, I feel sure, be achieved in some way. A
few will be remarkably successful, others will be replaced by better ideas.
Regardless of what may be the success of these particular efforts, the rapid
development of quantum electronics resulting from the work of many laboratories will undoubtedly help open up an important and as yet very meagerly
explored branch of astronomy.

Comments by Professor A. Javan

I address my comments mainly to the applications of the nonlinearities of
atomic and molecular resonances in laser spectroscopy. The importance of
this field of study became evident in the early years following the advent of
gas lasers when a variety of nonlinear effects were applied to obtain enor-

mous reductions of the spectral line profiles in a host of atomic systems, enabling measurements of their line structures. The potential impact of this field has now been considerably enhanced with the recent advent of the realization of the necessary technology to mix and compare widely different frequencies of electromagnetic waves over a range extending from the microwave into the far and the near infrared, and soon with the aid of the already known methods, into the visible portion of spectrum. We can now look forward to the exploitation of these methods in the next decade of spectroscopic application and improvements in precision and accuracy of many orders of magnitude. In addition to the molecular spectra in the infrared and optical region, we must now go back to our simple elements and reexamine their spectra. Atomic hydrogen and ionized and neutral helium must be examined again. These will surely lead to a redetermination of the Rydberg constant with much improved accuracy. Other fundamental constants and processes will be determined with better precision. And very soon, we will also know an ultimate value for the speed of light.

A recent development has opened the way to observe the nonlinear resonances under extremely dilute gas conditions where pressure effects are at a minimum. An inspection shows that in special cases it may be possible to find the center frequencies of these resonances reproducibly to within accuracies exceeding one part in 10^{14}. This, together with our ability to determine the absolute frequencies in the optical and infrared region, will enable the realization of much improved atomic clocks, capable of better testing some fundamental laws of nature, possibly predictions of some general relativistic effects.

Since my time is up, I would just like to mention that we have been aware and have heard here in Esfahan of some vague speculations on the possibility of laser emission in the x-ray region. Let us remind ourselves that a practical way to achieve this has not yet been discussed, but I forecast that this is within the realm of possibility and will be discovered in the decades to come.

Comments by Professor B. P. Stoicheff

During my talk the other day you learned that only 10 years ago very few scientists were involved in Raman spectroscopy and that only a handful had heard of Brillouin scattering. With the development of the laser, this situation has rapidly changed, so much so that it is difficult to keep up with the literature, and as I demonstrated, it is impossible to review the advances in this field in one hour. We also learned from Dr. Lee that last year alone 30 ion lasers were sold specifically for experiments in Raman and Brillouin scattering. Clearly, there has been a revolution in the field of light scattering in the last 5 or 6 years--and it seems to me it is about time that many of us left it!

I can foresee that in a few years a similar revolution will occur in infrared spectroscopy and probably in the visible region--and I do feel sorry for the old timers, those spectroscopists working in these regions who have not kept up with the new techniques (which we have heard about from Brewer, Feld, Javan, and others) because there is no doubt that they will be swamped.

One of the most difficult regions for work in spectroscopy is the vacuum ultraviolet. Materials which transmit such radiation are almost nonexistent and those with even 50% reflectance are few. This is also a region where we need new ideas for laser sources. Some advance has already been made but this is just a beginning. New lasers in the vacuum uv will eventually lead to improved optical components, new materials, and much more spectroscopic research in this region.

One of the states of matter which we know very little about is the liquid phase. There have been many beautiful experiments carried out in light scattering with lasers particularly in the critical region. With further work we should be able to make some headway to a better understanding of liquids.

And one final remark. Of course, it would be one of the greatest discoveries to find another population somewhere in the universe. I believe that contact can probably only happen through the use of lasers and masers. This is not a new idea; some searches have already been made and I think we should continue to explore.

Comments by Professor P. Jacquinot

I would like to give here the point of view of a laboratory which is essentially interested in atomic spectroscopy and atomic structure. In such a laboratory there are some limitations in the use of lasers because many problems can be solved only by means of emission spectroscopy with conventional sources and spectrometers (including the recently developed Fourier spectrometers): for instance the study of spectra of the various neutral and ionized atomic species is still a major problem since only about 20% of these spectra are known and correctly analyzed. And of course the spectra emitted by celestial objects have to be observed with spectrometers. However, even in emission spectroscopy, lasers can be helpful, especially in the infrared regions, since some lines of interest for the understanding of spectra are too faint to be observed otherwise than in a laser; in our laboratory we have succeeded in obtaining about a hundred lasing lines of eight metallic elements, which have brought useful information in atomic spectroscopy.

But of course all the lines which can be observed in absorption could be studied much better with lasers than with conventional techniques. So far we have studied only specific cases where there is a coincidence of the line to be studied with the line emitted by a laser, and very high resolution has been obtained in isotope shifts and hyperfine structures of rare gases even by limiting ourselves to linear absorption. But of course this can become of more general use when tunable lasers are available. In addition, the use of saturated absorption is highly recommendable in many cases.

But there are also different methods of using lasers as sources, essentially resting upon the optical pumping produced by laser light, this pumping being detected by nonoptical means. For instance in a Rabi experiment with an atomic beam the magnetic moment of paramagnetic states can be reversed by the pumping in the C region and this can be used as a sensitive means to detect optical absorption transitions; this experiment is being developed for measurements of isotope shifts of radioactive elements. Another method which could be simpler and of more general use rests upon the recoil suffered by atoms of an atomic beam when they absorb photons of a laser beam; here preliminary experiments are just beginning.

My conclusion is that although a great part of atomic spectroscopy will still use spectrometers for the analysis of emission spectra, the use of lasers should become very important when tunable lasers are available with the following qualities: wide range, high stability, sharp lines, high power, and, in some cases, cw action.

Comments by Dr. R. E. Kidder

The area that I am most concerned with and interested in is possibly the farthest-off application that has been discussed so far. However, I think it is an exceedingly important one: the production of useful power from thermonuclear fusion.

In this context the laser has already provided us with a new approach to fusion which differs markedly from previous approaches by eliminating the troublesome "magnetic bottle." This doesn't necessarily mean that this particular approach will ultimately turn out to be the best one, but I think it indicates that already the laser is widening the scope of possibilities for controlled fusion. More generally, the laser beam's precise controllability in space and time may well play an essential role in the exceedingly difficult task of controlled fusion in ways that are now unforeseen.

One other application that I'd like to mention is the use of lasers in chemical processing, and particularly in isotope separation. For this, one needs favorable spectroscopy and chemistry, both of which should be available for separating such isotopes as those of carbon, nitrogen, and oxygen. The separation of deuterium from hydrogen has already been demonstrated in the laboratory experiments of Mayer, Kwok, Gross, and Spencer. (We didn't

attempt this at Livermore on grounds of insufficient challenge, Edward
Teller saying that one should be able to separate deuterium "with your bare
hands.") At the other end of the periodic table is uranium, where the chemistry is more severely limited and the spectroscopy is not yet known with
sufficient resolution. However, here too one may expect that eventually a
laser scheme for isotope separation will be discovered.

So these are two applications, controlled fusion and isotope separation,
that I think I can add to the list compiled by the other panelists, and predict
that lasers are likely to make a very big difference in both.

Schawlow. Let's now open the discussion to the floor. I'd like to begin by
just saying that we didn't mention the x-ray generation, which
sounds as if we may be very close to it, or somebody else may
be, by the brute force method of pumping in lots of laser power.
This, I believe, will happen fairly soon and I think the applications will be quite surprising ones, just as the applications of lasers were not all confined to communication. I think also that it
is a very exciting time now since we really are getting tunable lasers; some of them are even cw, but they all have the properties
that Professor Jacquinot asks for, and at least are now extending
into the visible and ultraviolet regions, which is right for doing
photochemistry, although clever people have ways of doing it in
the infrared. I think also for spectroscopic analysis of complex
spectra a laser is very useful because you excite just particular
states; thus, when we excite an iodine molecule with a broadband
lamp the fluorescence is extremely rich and complex. With a laser you get only two lines of $\Delta j = +1$ and -1 for each value of v.
It's a very much simpler spectrum--perhaps a hundred times
simpler, and similar simplifications may be obtained elsewhere.

Lax. I would like to comment on a remark made by Professor Prokhorov. Submillimeter spectroscopy is already here and far-infrared lasers are now being used in conjunction with high magnetic
fields to do resonant spectroscopy. The 337-μ HCN laser is a
good example. With a pulsed magnet available at any laboratory,
it can be used to extend resonant spectroscopic studies of the microwave type into the infrared.

Schawlow. This is only true if either the laser or the sample is tunable in a
magnetic field.

Lax. Another important advance is in the field of atomic spectroscopy, the use of two-photon studies using high power lasers. So we can now study transitions which are forbidden for single photons. I believe such studies will lead to a revolution in atomic spectroscopy, from the infrared all the way to the ultraviolet.

Low.   I would like to illustrate a remark made by Professor Bloembergen by an example. Not very much is known today about the band structures of various systems, particularly three-dimensional systems. Two-photon laser spectroscopy can now be used, in principle, to study the structure of such bands in the same way as is done in atomic systems. I think that this field is just opening up.

The next thing I have to say relates to comments made by Dr. Stoicheff. I think that Brillouin scattering will be important in the study of chemical reactions in the time domain of 10^{-7} to 10^{-10} sec. I also think it is important to utilize Brillouin scattering to generate acoustic waves in the region beyond 40 kMHz. For example, the 100 to 200 kMHz region would fall within the dispersion region of some important cases enabling the determination of their band structure.

Finally, I would like to challenge the Panel to comment on the use of lasers in nuclear physics.

Schawlow.   Would anyone in the Panel like to make some comments on Willie Low's challenge?

Townes.   One application in nuclear and particle physics is the generation of coherent polarized gamma rays by scattering a laser beam off of a high-energy beam of electrons. This is now a standard technique.

On another note, I might say I am very impressed with all the spectacularly beautiful work reported at this Conference, some of which I was not familiar with, and this keeps me quite optimistic about future developments, particularly when coupled with surprises which are bound to come along in the future.

Prokhorov.   I would like to comment on the application of lasers to isotope separation. We have recently performed an interesting experiment in our laboratories on BCl_3 (Boron trichloride) which has two isotopes. Using a CO_2 laser, we can selectively excite one isotopic species into a highly excited vibrational state from which the molecules dissociate. The vibrational quantum number of this state is about 30. To remove the isotope, we added hydrogen, but it reacted violently with the free radicals which were formed and produced enormous shockwaves.

Javan.   Have you published this work?

Prokhorov.   It has just been submitted for publication.

Yatsiv.   I would like to make a comment in the area of chemical lasers. Despite the fact that some advances have already been made, we are still in quite a poor position. The glow worm can convert chemical energy into light better than we can. I feel that in another few years we should be able to convert chemical energy into light much more efficiently than we do now. This is a field in which a lot of work is already being done. There is a big advantage in using liquids for this purpose because of energy dissipation problems. In the case of gaseous systems, the pressure and temperature rises because of the small heat capacity. One should do much better using liquids. Also, conversion of electronic transitions instead of infrared vibrational transitions has some advantages. Almost no one is working in this field now.

Schawlow.   I think that one of the important things needed is an efficient visible laser. The need for such a laser was mentioned in connection with thermonuclear fusion but I think that it would find all sorts of everyday uses also, like cutting metals, communications, photography, and, if the cost was low, perhaps even typewriter erasers! How does one get such a laser? I don't know. My guess is that a high-density gas is a more likely candidate for the active medium of such a device than is a liquid. The gas atoms would have large oscillator strengths and be pumped by a broad-band source, something like the gaseous analogue of a semiconductor laser.

Alley.   I would like to mention another application of lasers of high power densities to nuclear physics. I think one is now on the verge of being able to observe short pulses of powers sufficiently high to observe photon-photon interactions due to virtual pair production. Using fast time resolution, such experiments might just be feasible in the next few years. Another application which has not been mentioned so far is precision ranging over large distances using very short laser pulses. We are currently measuring the point-to-point distances of the earth-moon system with an accuracy of 30 cm, and we think that by

utilizing mode-locked lasers and taking atmospheric corrections into account, accuracies better than 3 cm will be achieved
within the next year or two.

Schawlow.   We may not know where we are going, but we will know where
the moon is.

Javan.   I think this business of photon-photon scattering due to vacuum
polarization effect will probably be more difficult than it appears because of the nonlinear properties of matter which
could dominate the scattering processes even at extremely low
pressures.

Prokhorov.   Yes, the cross sections are very small.

Townes.   I would like to add one more thought on the nuclear physics
business. Just to make an illustration -- if one is optimistic
and takes Keith Boyer's estimates of available power density,
then it may be possible to accelerate electrons to very useful
energies. Ray Chiao has written a paper on this fairly recently. With energies available at present, one can get into the region of maybe tens of hundreds of millions of electron volts.
With another few orders of magnitude one might get particles
of even higher energies. Such an energy source would be very
useful.

Schawlow.   I take it the acceleration is produced by radiation pressure.

Townes.   Yes.

Schawlow.   If we ever do get the illusive x-ray laser, one possible application would be the coherent excitation of nuclear states. Then
maybe we could do wave-echo (photon-echo) experiments in the
nuclear region.

Duguay.   Nobody has mentioned medical applications yet. Since most of
us are supported by taxpayers' money, perhaps we should
worry about it a little bit. All of the predictions made by this
distinguished Panel seem to be leading to new and sophisticated
industrial technology. We must remember however, that there
are undoubtedly some important medical applications as well.
For example, I would like to suggest that when picosecond
pulses of x rays are available, one can use them to range

through the body to see what is wrong inside when you feel bad. As you know, approximately half of all operations are done so that the doctor can see what is wrong at a predetermined depth inside of you. Also, another example is the powerful spectroscopic methods that Javan, Stoicheff, and other people have described in this Symposium. These methods could be useful in detecting the presence of bacteria and viruses, and thus, helpful in the medical field. So I would suggest that some of you experimenters put a drop of blood in your cells and see what you can do with it.

Schawlow. One other area, if you are going to be socially conscious, is the application of lasers to remote analysis of the atmosphere. We, of course, have read a number of reports about measuring the density of the atmosphere at high altitude by means of resonant scattering such as in the experiment of detecting the sodium atoms of extremely low densities high up in the atmosphere. Raman scattering can also be used for remote analysis of density and temperature. Now the people who work on pollution will probably say that there are already a number of other methods of getting the same information. But in fact, if you press them, you will find that they really don't have much information about the conditions of the atmosphere at high altitudes, and there aren't that many other methods available.

Patel. I would like to take issue on the subject of pollution. I think there are ways of detecting pollutants in the atmosphere. The problem is not one of detection but of prevention, and I don't think the laser is going to provide a way of doing anything to prevent pollution.

Schawlow. My technician, Ken Sherwood, has suggested the use of high-power lasers as after-burners for smoke stacks (laughter).

Sinclair. Since we are being socially conscious, I would like to point out that in this discussion no one has considered the problem of laser safety. This is an aspect of the use of lasers which I think is grossly under-discussed, under-thought and under-researched, and I would like to toss out the thought that more people should think about it.

Khokhlov. I would like to attract the attention of the audience to possible applications of ideas in laser physics to other fields, like neutrons, protons, and nuclei. It may be possible to get coherent emission of neutrons, protons, and other particles. It may also be possible to observe Raman-type interactions between particles. The idea of light scattering may be applicable in interactions between such particles. I cannot comment on a possible source of pumping. This is clearly a problem. One possibility may be application of high intensity light of the order of what is written there on the board (10^{21} W/cm^2), but there may also be other possibilities.

Schawlow.   Let me ask Professor Khokhlov a question about protons and neutrons. The fact that they are Fermi particles and they have a nonzero rest mass is of importance. Have you thought about these problems?

Javan.   Actually I had been thinking of making a comment on some far out ideas and I see that Professor Khokhlov has been thinking along similar lines. In fact, I was about to propose that instead of masers and lasers -- we should think of "basers" to describe boson amplification by stimulated emission effect.

Schawlow.   But Professor Khokhlov suggests that fermions can be used.

Javan.   So maybe I was not quite following what he has been suggesting. I don't think anything can be done with fermions but if we stick to bosons, we might have a chance. So let us still think of a baser.

Giacomo.   Some people have suggested using lasers as primary standards of time and length measurement. I would like to insist that it is essential that primary standards in different laboratories or different countries agree among each other at least within one part in 10^{11} or 10^{12}. This does not seem to have been done as of yet, and it is important for some laboratory to invest some efforts along this line.

Javan.   Methods for such frequency or wavelength reproducibility have already been demonstrated with the recent extension of absolute frequency measuring techniques into the near infrared. For instance, if the frequency of a laser is compared to that of a cesium clock, which already offers one part in 10^{12} accuracy, one would automatically have at hand a length standard of the same accuracy. In that case, one would wish to define the speed of light as a dimensionless parameter and define the length standard by means of frequency measurements, utilizing an existing atomic clock.

Jacquinot. I agree with this point of view. The trouble is that the technique of absolute frequency measurements in the infrared and optical regions is not, as of yet, universally available in laboratories where standards are needed.

Javan. Yes, that is true for now.

Hall. In considering the length as a standard one would probably want to use length metrology, and for us that means observations by means of interferometric techniques. The limitations in that case are due to physical imperfections of the optical components, which seem to limit the accuracy to one part in 10^8 or 10^9. In principle, at least, one can perhaps improve this limit by one or two orders of magnitude using laser tricks and interferometers with long arms, such as a 30-m interferometer.

Schawlow. Since Javan is giving me the signal that we had better bring the session to a close, let us have just one more comment.

Mallozzi. Another application of high-power lasers which has not been mentioned here, is the use of laser-generated shocks to enhance the material properties of solids. Very recently, using this technique, we have increased the strength of a 50-mil steel plate by 30%. It is also very likely that improvements in hardness and resistance to corrosion can be obtained similarly.

Schawlow. How is this done?

Mallozzi. It was done by using a laser beam to generate a shock wave at the surface of the material, thereby increasing the strength by shock hardening. Shock hardening has been around for a long time but we will probably be able to do this more cheaply with lasers.

Townes. Many of the ideas that have been mentioned this afternoon have a long road to travel. The question is whether what we see in the future is real or merely a mirage. I think that is what we must try to resolve.

PICOSECOND NONLINEAR OPTICS

N. Bloembergen
Division of Engineering and Applied Physics, Harvard University
Cambridge, Massachusetts

ABSTRACT

The first part of this paper is an introductory review of the experimental techniques for generation and detection of picosecond light pulses and of their application to the measurement of very short characteristic times. A brief survey of nonlinear processes induced by picosecond pulses, including third harmonic generation, is given. In the second part, transient Raman scattering and transient self-focusing and frequency broadening are described with special emphasis on recent experimental results.

I. INTRODUCTION

The earliest form of light pulses has been produced in ancient history, for example, when tribesmen on the American continent would interrupt the smoke signal from a camp fire by holding a blanket in front of it. The same technique may well have been used even earlier in our ancient host country, Iran. The duration of these pulses was on the order of 1 sec, determined by the speed of muscular reaction and well matched to the resolving time of the human eye of about 0.1 sec.

Faster mechanical shutters and electro-optic shutters subsequently reduced the pulse time to a fraction of a microsecond or about 10^{-7} or perhaps 10^{-8} sec, before the advent of lasers in 1960. In the early pulsed solid

21

state lasers, ruby or Nd^{3+}:glass, the duration of the output pulse was determined by the duration of the flash discharge which provided the pump light.
This duration was typically 10^{-3} sec. The output energy was typically
between 0.1 and several joules depending on the size of the laser material
in the oscillator or subsequent amplifiers. The introduction of Q switching,
by means of a Kerr cell shutter or by a rotating mirror in the laser cavity,
reduced the pulse duration to about 10^{-8} sec. Finally, the introduction of
mode locking by DeMaria and co-workers in 1966 was successful in achieving
light pulses lasting only a few picoseconds or less. An excellent review [1]
containing references to essentially all pertinent early work on picosecond
pulses appeared in January 1969.

It is important that the shortening of the pulses is accompanied by a
concomitant rise in the power level. The total energy output of the mode-
locked laser is roughly constant and is determined largely by storage in the
active ions in the laser host material. Therefore, a small laser may put
out 0.01 J per picosecond pulse. This energy may be amplified up to 1 J by
subsequent laser amplifier rods.

One thus has available power fluxes in the light pulse on the order of
10^{12} W, or 1 terawatt. This power is approximately equal to the total output
of all electric generation stations on earth [2]. Even in the pulse from a
small mode-locked laser with a pulse energy of 10^{-2} J in, say, 4×10^{-12} sec,
one may, by focusing the output beam, easily exceed power flux densities of
10^{12} W/cm^2 in the focal region. Such a flux density corresponds to amplitudes of the electric field in the light wave which exceeds 3×10^7 V/cm.
Any dense material subjected to such a field for a time longer than about
10^{-10} sec will break down electrically and will be transformed into a high-
density high-temperature plasma. In fact, the production of neutrons in
thermonuclear reactions in such a light-initiated plasma has been reported
[3]. During the picosecond pulse the material does not have time to break
down.

Thus these ultrashort light pulses can serve a dual function. They may
be used to extend the range over which measurements of characteristic
times of physical and chemical processes can be performed by several
orders of magnitudes. Direct time measurements in the domain between
10^{-11} and 10^{-13} sec were previously not possible. Picosecond pulses may
also be used to study material properties at extremely high electromagnetic
energy densities. Very strong optical nonlinearities become evident, and
these often, but not always, display transient characteristics which are
different from those excited by light signals of longer duration. After the
general methods for generation, detection, measurement of ultrashort times,
and picosecond nonlinear processes have been briefly reviewed in Secs. II
and III, some recent results on the transient stimulated Raman effect and the
transient self-focusing will be discussed in Secs. IV and V.

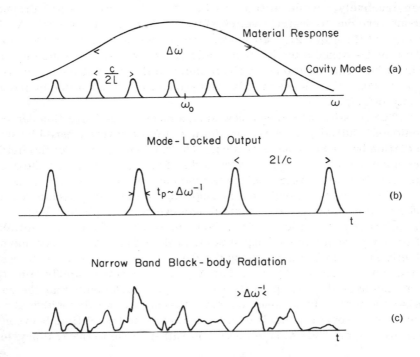

Figure 1. (a) The frequency response curves of the active medium and longitudinal modes of the optical resonator. (b) The time response of the output intensity (envelope function) for perfect mode locking. (c) The time of the intensity (envelope) from a large number of modes with random relative phases.

II. GENERATION AND DETECTION OF PICOSECOND PULSES

In Fig. 1(a) the frequency response curve of an active laser medium is sketched. For Nd^{3+} ions in a glass matrix the width $\Delta\nu$ of this response is about 100 cm^{-1}. For Cr^{3+} ions in ruby at room temperature it is about 6 cm^{-1}. For a typical laser cavity of length ℓ many longitudinal modes with a frequency separation $\delta\nu = c/2\ell$ lie under this broad response curve. Consider a steady-state regime where a very large number of these modes are excited simultaneously with about equal amplitudes and phase differences. The field may be expressed as

$$E = \sum_{n} A_0 \exp(-in\,\delta\omega\,t + in\omega)\,\exp(i\omega_0 t)$$

where n runs over about $\Delta\omega/\delta\omega$ integers. At $t_0 = \varphi/\delta\omega$ all modes interfere constructively,   while at time $t = t_0 + \pi/\Delta\omega$ the envelope has dropped to about zero due to destructive interference between the modes. At time $t = t_0 + 2\pi/\delta\omega = t_0 + 2\ell/c$ all modes are, of course, again exactly in phase. This mode–locked situation thus leads to a time response which is shown in Fig. 1(b). The duration of a single pulse in the train of the pulses is about $t_p \sim \pi/2\Delta\omega$. This is essentially the optimum value permitted by the uncertainty principle.

The question now arises how such a mode–locked situation can be achieved. Initially the active medium will emit "narrow band blackbody" radiation by spontaneous emission. The random behavior of the field envelope in a time domain is sketched in Fig. 1(c). Note that the average distance between zero crossings is again on the order of $\Delta\omega^{-1}$.

The spontaneous emission is amplified in the laser medium and subsequently attenuated by a cell containing a dye solution placed close to one of the mirrors in the laser cavity, shown in Fig. 2. The dye is saturable and after many passes the strongest noise peak of Fig. 1(c) has become sufficiently strong to begin bleaching the dye. The dye is assumed to respond to a frequency spectrum broader than $\Delta\omega$, and to recover rapidly after the large peak has passed. This peak therefore grows much faster than the rest of the noise and it quickly becomes the amplified signal, a pulse with a duration of about $\Delta\omega^{-1}$, repeating itself with the turn–around travel time between the mirrors $2\ell/c$. This simple picture is rather close to the real physical sit-

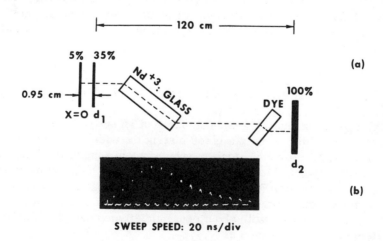

SWEEP SPEED: 20 ns/div

Figure 2.   Schematic of a mode–locked laser and its pulse train output (after DeMaria et al., Ref. 1).

uation [4], and it shows how the nonlinear characteristics of the saturable absorber provide the mode locking necessary to get the output with an extreme amplitude modulation.

The intensity in the pulse is sufficiently high that other nonlinear distortions may occur simultaneously. One may expect a phase modulation caused by the real part of the nonlinear susceptibility describing the quadratic Kerr effect. Whereas with a ruby mode-locked laser the ideal situation is often approached, the pulses from a mode-locked Nd^{3+}:glass laser often show a frequency chirp, i.e., the phase varies in a nonlinear fashion during the duration of the pulse, which in turn is longer than the limit Δw^{-1} set by the uncertainty principle.

It should be clear from the foregoing remarks that the observations of the frequency spectrum does not provide a measure of the pulse duration. In fact, a single pulse of the train in Fig. 1(b) shows essentially the same spectral intensity distribution as the continuous signal in Fig. 1(c). Both are the Fourier transform of the lowest order auto correlation function $<E(t)E(t - \tau)>$. The cw narrow band blackbody signal may be considered as a random sequence of picosecond pulses with random duration, phase, and amplitude.

The measurement of the shape of the pulse cannot be made directly because the response of detectors with associated circuitry is not sufficiently fast to follow picosecond variations. The experimental techniques to determine the pulse duration are based on the determination of higher-order correlation functions by means of nonlinear effects, as first realized by Armstrong et al., who used second harmonic generation [5]. Similar schemes were independently used by Giordmaine and Kaiser and by Weber. Giordmaine [6] et al. developed a convenient, widely applied method of two photon absorption in a fluorescent dye. More recently higher-order nonlinear effects have been utilized, including three-photon absorption by fluorescent molecules [7] and frequency mixing of three light waves [8]. An essential point in all of these and other picosecond measurement techniques is the splitting of the pulse into two parts, or comparison of the pulse with a second pulse which is given a variable delay with respect to the first one. Note that a delay of 10^{-12} sec corresponds to a change in light path in air of 0.03 cm, which is an easily measured quantity. The two-photon absorption technique which has the advantage of simplicity is schematically represented in the lower right-hand corner of the diagram in Fig. 3. A pulse is split into two parts which then traverse in opposite directions through a dye which does not absorb at the fundamental frequency, but can be excited by a two-photon absorption process. The subsequent fluorescent signal scattered sideways and recorded by the camera is a measure of the strength of the double photon transitions. Where the two pulses overlap in space-time the signal is stronger than elsewhere along the track, because the signal is a nonlinear

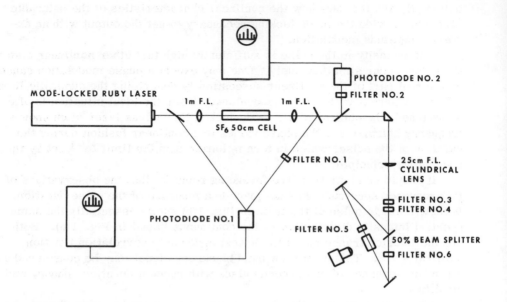

Figure 3.   Schematic of experimental arrangement to measure the transient stimulated Raman scattering in SF_6 gas. The duration of the laser and Stokes pulses is measured by the two-photon absorption fluorescent technique, in which two picosecond pulses traverse the cell in opposite directions. By means of filters 3-6 both pulses may be selected to be at the laser frequency, both at the Stokes frequency, or one at the laser and the other at the Stokes frequency (after Carman and Mack, Ref. 22).

function of the intensity. The width of the brighter spot, examples of which are shown in Fig. 4, is a measure for the duration of the pulse t_p.

The detailed shape of the contrast ratio is important for a more precise evaluation of the nature of the pulse. If the pulse does not have the minimum duration, i.e., for $t_p \, \Delta w > 1$, it may be said that a single pulse exhibits multimode characteristics. Numerous papers on the correlation properties of the field in a pulse have been written and the errors in some early papers are now well understood. A good comprehensive discussion of this question with an extensive bibliography may be found in Ref. 9. In a recent application one part of the pulse is converted to the second harmonic $w_2 = 2w_1$, the other half is the fundamental frequency w_1. The two pulses are then recombined to produce a combination signal at $2w_2 - w_1 = 3w_1$ in a fluid, with the wave vectors so chosen that phase matching is achieved. This signal is a measure for the correlation function $<E^4(t)E^{*4}(t)E(t - \tau)E^*(t - \tau)>$. It is

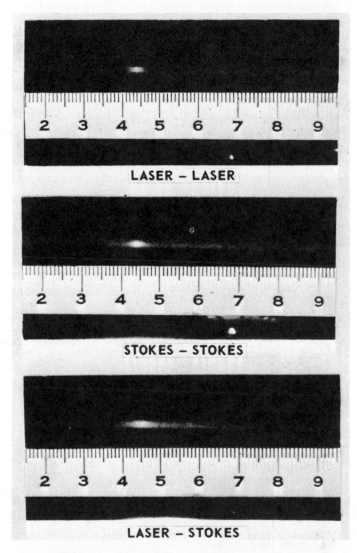

Figure 4. Raman Stokes pulsewidth and delay--SF$_6$. Two-photon fluorescence signals, recorded with the arrangement shown in Fig. 3 (after Carman and Mack, Ref. 22). Note the lateral displacement in the bright spot when the laser pulse meets the Stokes pulse, indicating the delay of the stimulated Stokes pulse with respect to the laser pulse.

thus seen that nonlinear optical processes play an essential role in both the generation and detection of picosecond pulses. It should be mentioned that the stimulated Raman effect is an exponential nonlinear process and also gives detailed information about the pulse shape, as discussed further in Sec. IV and the references given there.

While the method of generation leads, in general, to a train of pulses, it is possible to isolate one individual pulse by means of a fast Kerr cell shutter [1]. The experimental arrangement to accomplish this is shown in Fig. 5. The light beam is polarized in such a manner that the initial pulses in the train are deflected by the Glan prism towards a spark gap. This triggers the discharge of the transmission line capacitor at a predetermined pulse height in the train. The Kerr cell voltage suddenly rises in a time short compared to the time between pulses, and the next pulse in the train

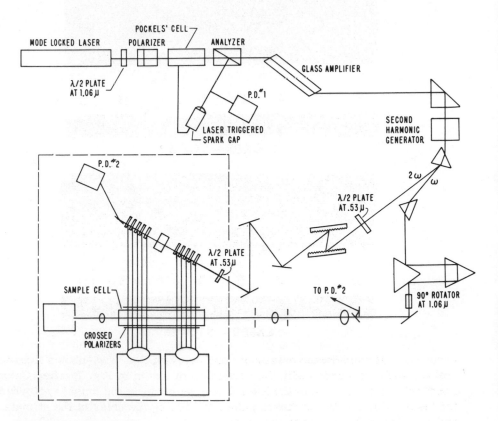

Figure 5.   Experimental arrangement to measure the birefringence and self-focusing by a single picosecond pulse (after Reintjes, Ref. 44).

is transmitted because its polarization is turned by $90°$.   Before the follow-ing pulse comes along the capacitor is discharged and the Kerr cell voltage has returned to a low value.

The presence of a frequency chirp in pulses from a Nd:glass laser makes it possible to compress these pulses in time by passing the pulse between two parallel diffraction gratings [10], as is also shown in Fig. 5. The transit time of different Fourier components increases approximately linearly with the wave length.   This permits the low frequency at the end of the pulse to catch up with the high-frequency components at the beginning. The pulse may be sharpened by a factor of 3 or more down to about 0.5×10^{-12} sec.

The availability of such short light pulses clearly affords an opportunity to observe the short characteristic time of other fast processes.   Shelton and Armstrong [11] measured the decay rate of the excited state of the saturable dye used in the mode locking by probing it with a second weak pulse with a variable delay.   The attenuation of the weak probing pulse was measured as a function of the delay time.   Thus the recovery rate of the absorption after bleaching by the strong pulse was observed.

More refined is the operation of a picosecond Kerr cell light shutter, first demonstrated by Duguay and Hansen [12].   It is well established that anisotropic molecules in a liquid traversed by a linearly polarized light beam will orient themselves preferentially with their axis of easy polariza-bility parallel to the electric field.   This minimizes the dielectric free energy.   The fluid thus becomes birefringent.   In CS_2 the molecular reorien-tation time is so fast that this birefringence is established with a character-istic time $\tau = 2 \times 10^{-12}$ sec.   Now consider a weak light signal polarized by a polarizer at $45°$ with respect to the polarization of the strong laser pulse. A crossed analyzer will not pass this light, unless the birefringence induced by the strong laser pulse is synchronous with this signal.   Thus the experi-mental arrangement shown in Fig. 6 acts with a resolution time of a few picoseconds, and has been used by Duguay et al. to measure the short relaxation time of several fast fluorescent materials [12].   The signal is the fluorescence induced by the green light obtained by splitting off part of the picosecond pulse train and doubling its frequency in a KDP crystal.   Such methods have considerable promise to probe intermediate products in fast chemical reactions, and have been used to analyze the relaxation of excited levels of several organic molecules [13].

III.   BRIEF SURVEY OF PICOSECOND NONLINEAR EFFECTS

Since the instantaneous intensity is high in a picosecond pulse, most nonlin-ear phenomena observable with pulses of longer duration are also observable

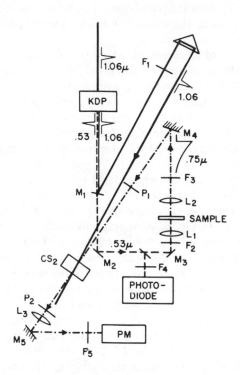

Figure. 6. Experimental arrangement to measure fast fluorescent decay times by means of a picosecond Kerr cell shutter (after Duguay and Hansen, Ref. 12).

in the picosecond regime. There are, however, important exceptions especially among those phenomena having characteristic time constants which are much longer than the pulse duration. They may not be sufficiently developed in the transient regime.

It is well known that the lowest-order optical nonlinearity corresponds to an electric polarization which is a quadratic function of the electric field amplitude. It occurs in media lacking inversion symmetry and is responsible for second harmonic generation, rectification of light, and parametric sum and difference frequency generation. These same phenomena have been observed with picosecond pulses and are reviewed in Ref. 1. In second harmonic generation it is important to have phase matching between the fundamental and second harmonic. One wishes this matching condition to be fulfilled not only for the central longitudinal mode, but as well for all the other modes equally spaced in frequency. This imposes the additional condition that $\partial \omega / \partial k$ should be equal for the fundamental and second harmonic. This condition developed from a consideration of the modes in the frequency domain is also immediately evident from a consideration of the physical situation in the time domain. One wishes the generated SH pulse to remain in synchronism with the fundamental pulse. This requires equal group velocities at the two frequencies. In general, the phase matching and group velocity matching cannot be simultaneously satisfied. Depending on the mismatch, different shapes for the harmonic or parametrically generated pulses can be expected. The most extensive theoretical analyses of these transient pulse shapes phenomena have been made by Akhmanov and co-workers [14].

It is well known that stimulated Raman and Brillouin scattering may be considered as the parametric down conversion of a laser wave into a Stokes-shifted light wave and an optical or acoustical phonon, respectively. The transient response of these phenomena has been analyzed theoretically in several papers. The theoretical and experimental results for the transient stimulated Raman effect will be discussed in more detail in Sec. IV.

Light waves can also interact with periodic thermal variations. This phenomenon may be called thermal Rayleigh scattering. Since thermal diffusion is a relatively slow process, a steady state cannot be developed in a single pulse. There is, however, some instantaneous thermal response. Consider the spatial interference pattern of the two picosecond pulses propagating in slightly different directions in an absorbing medium [15]. In the maxima, energy will be absorbed and for short enough relaxation times for energy dissipation the temperature of those volume elements in the material will rise, while in the minima a much smaller rise occurs. An index diffraction grating is developed because of the intrinsic temperature dependence $(\partial n / \partial T)_\rho$. Even if the molecules remained in the excited state during the picosecond pulse, a change in index of refraction would occur because the molecular polarizability is different in the excited state. Although there is no time for thermal expansion and diffusion during a picosecond pulse, such processes will take place during the passage of the whole pulse train. The integrated effect of the whole train is more difficult to analyze because of the superposition of fast and slow processes [16]. Observations with a single pulse are preferable.

Among the nonlinear properties described by a polarization which is a cubic function of the field amplitudes, one may mention third harmonic generation, two-photon absorption, intensity-dependent index of refraction, and frequency mixing leading to the combination frequency $2\omega_2 - \omega_1$. All these effects have been observed with picosecond pulses and some applications were already described in Sec. II. The self-focusing and self-phase modulation of picosecond pulses caused by the intensity-dependent index of refraction will be discussed in Sec. V.

The picosecond pulses are particularly useful to observe the third harmonic generation and other high-order nonlinear effects in absorbing media. They require a high peak intensity for their observation, but the total energy

content of the laser signal must be limited lest the heat dissipation evapor-
ates the material. The electronic nonlinearity responsible for third harmonic generation (THG) has been observed in a variety of semiconductors and
metals in this manner [17]. The intriguing question of conservation of angular momentum in the generation of left circularly polarized TH by a right
circularly polarized laser beam propagating along the cubic axis of an absorbing silicon crystal has been resolved experimentally and theoretically.

Picosecond pulses may also be used to extend the range of investigations
of the phenomena of self-induced transparency and light echos. The criteria
for a $n\pi$-pulse, that $\hbar^{-1}\mu \int E(t')dt' = n\pi$ and $t_p \ll T_2$, could presumably be
satisfied even in media with a homogeneous line width of 10 cm^{-1} or more.

The full exploitation of these and other picosecond phenomena depends
on the availability of well-controlled reproducible single picosecond pulses.
During the past few years rapid progress has been made in reliable generation of reproducible pulses of variable frequency. Tunability is achieved by
utilizing harmonic generation and Stokes-shifted frequencies from stimulated
scattering effects, but more important is the continuously variable tuning
range from picosecond pumped dye lasers.

As examples of recent progress the stimulated transient Raman scattering and self-focusing will be discussed in Secs. IV and V.

IV.   TRANSIENT STIMULATED RAMAN SCATTERING

Since the duration of a picosecond pulse is often shorter than the phase-mem-
ory time of a molecular vibration, the Raman scattering occurs on a transient basis. The standard time proportional transition probability as expressed by Fermi's Golden Rule cannot be used, as the light scattering process is over before the molecular quantum-mechanical state has reached a
steady-state value.

A review of the steady-state stimulated Raman effect [18] showed that a
quantitative confirmation of the theory by experiment is only possible if the
influence of other nonlinear effects is carefully eliminated. In particular
the self-focusing of the light should be avoided if a quantitative interpretation
of the experimental results is desired. The same requirements must also
be imposed on experiments of the transient stimulated Raman effect.

Figure 7.   Transient stimulated vibrational Raman scattering in gases
(a) interleafed incident and transmitted laser pulse train showing depletion
due to Raman scattering in SF_6, (b) generated Stokes pulse train, (c) Stokes
emission in CO_2 from self-trapped filaments, induced by picosecond pulses
(after Mack, Carman, Reintjes, and Bloembergen, Ref. 22).

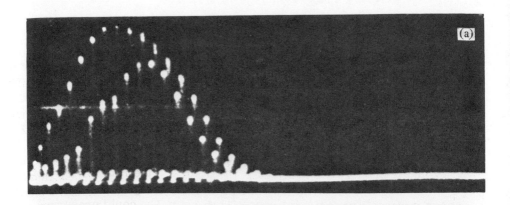

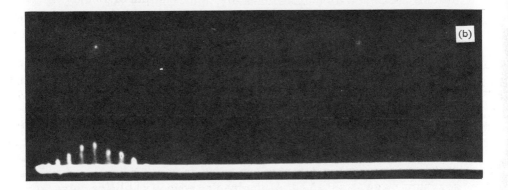

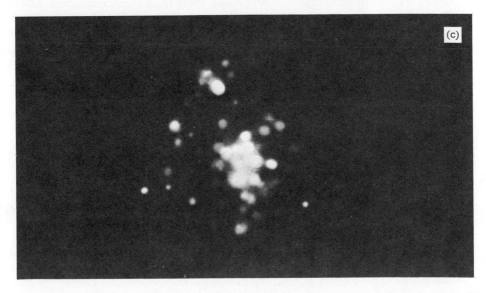

The theory of the transient effect has been investigated in detail by many authors [14, 19-21]. In general, one has a set of five coupled partial differential equations between the complex amplitudes of the laser field, Stokes field, and anti-Stokes field, respectively, and the diagonal and off-diagonal density matrix elements of molecular system. The latter is simplified to a 2×2 matrix, taking account of the lower and upper energy level of the Raman transition involved. This system of equations may be reduced to more manageable proportions if depletion of the laser power and production of anti-Stokes power may be ignored and the upper level of the Raman transition never receives a significant fraction of the molecular population. These conditions can be satisfied experimentally, although they may often be violated if self-focusing occurs. Figure 7 shows significant depletion in the early part of a laser pulse train and self-focusing even though the medium in this case was a gas.

When the system of equations is reduced to two parametrically coupled waves, one for the complex amplitude of the Stokes wave and one for that of the vibrational wave corresponding to the off-diagonal matrix element of a long-wavelength optical phonon, closed solutions for the temporal and spatial behavior have been obtained which may be evaluated numerically [20]. The form of the equations is identical to that of the voltage and current spikes propagating along a lossy transmission line. The essential physical features of the transient Stokes behavior are shown in Fig. 8 where a rectangular laser pump pulse is assumed. At the leading edge of the laser pulse no molecular vibration is excited. The Stokes input is assumed constant as given, for example, by the spontaneous scattering process or by a fixed input into a Stokes pulse amplifier. During the pulse the Stokes wave and molecular vibration grow quasiexponentially. The rigorous solutions are expressed in terms of a modified Bessel function with an imaginary argument. At the trailing edge of the laser pulse, the Stokes field stops abruptly with the pump pulse, but the molecular vibration decays with its characteristic damping τ, which has been assumed larger than the pulse duration $\tau > t_p$. More careful analysis assuming more realistic shapes of the laser pulse shape shows the following features:

1.   The transient gain regime is determined by the condition $(t_p/\tau)G_{ss} < 1$, where $G_{ss} = g_s \ell$ is the steady-state Raman gain coefficient and g_s is the steady-state gain per unit length proportional to the peak spontaneous Raman differential cross section and the laser intensity $|E_L(t)|^2$.

2.   The transient gain is smaller than the steady-state gain and is asymptotically given by $\exp(2 \, G_{ss}t_p/\tau)^{1/2}$. It is determined by the integrated power or total energy in the laser pulse and by the total integrated (rather than the peak) spontaneous scattering cross section.

3.   The maximum gain of the Stokes pulse occurs with some delay with respect to the laser pulse due to the "flywheel" effect of the molecular

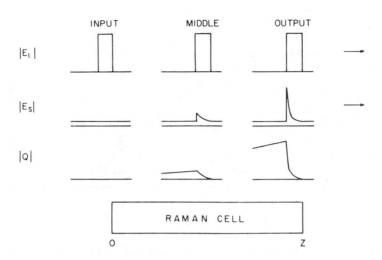

Figure 8.   Schematic diagram indicating the spatial dependence of the laser, Stokes, and optical phonon amplitudes in various parts of the Raman cell. The temporal behavior at a fixed position in the cell is similar (after Carman, Shimizu, Wang, and Bloembergen, Ref. 20).

vibration.
    4.   The Stokes pulse has a duration which is shorter than the laser pulse. For realistic laser pulse shapes (Gaussian), the shortening is typically 20-50%. The results are quite sensitive to the detailed shape of the laser pulse.

    These features have been confirmed in detail in a series of experiments by Carman, Mack, and Reintjes [22]. The best controlled results have been obtained in SF_6 gas. Whereas in gases excited with longer pulses the stimulated Brillouin effect is usually dominant; this effect is so transient for a picosecond pulse that it cannot build up to significant levels. As a consequence picosecond pulses have stimulated several Raman transitions, including both vibrational and rotational transitions, in many gases. In SF_6 one dominant Raman transition occurs and self-focusing and anti-Stokes production can be completely avoided. The experimental arrangement was shown in Fig. 2. The duration and delay of the Stokes pulses is detected by means of the two-photon absorption-fluorescence technique as shown in Fig. 3. The bright spot is narrower when two Stokes pulses traverse the cell in opposite direction (ss trace) than when two laser pulses are present ($\ell\ell$ trace). When a laser pulse meets a Stokes pulse (ℓs trace), obtained by

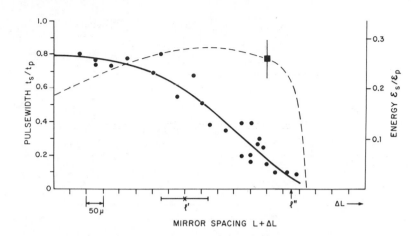

Figure 9.   Normalized Stokes pulse width (full curve) and energy (dashed curve) as a function of mirror spacing of the Stokes resonator (after Colles, Ref. 23).

suitable filtering, the bright spot is shifted by a fraction of a millimeter. For a laser pulse duration $t_p \sim 2.5$ psec the Stokes pulse was observed to have a duration of 1.5 psec.   For $t_p = 15$ psec the Stokes pulse was only 9 psec wide and its maximum was shifted by about 4 psec.   These results are consistent with a Gaussian pulse shape and rule out more slowly decaying wings.

More dramatic Stokes pulse shortening has recently been obtained by Colles [23], who amplified a Stokes pulse repetitively in a short Raman cell, pumped by consecutive pump pulses from a laser train.   Mirrors transparent for the laser pump are used to reflect the Stokes pulse.   The mirror spacing of the Stokes resonator may now be adjusted to obtain optimum action by consecutive laser pulses.   The results in Fig. 9 show that the total energy in a single Stokes pulse is usually about 30% of that in a single laser pulse, but that the Stokes pulse duration, measured by the two-photon technique, is a sensitive function of this spacing.   The shortest Stokes pulses are obtained when the Stokes pulse round-trip time in the cavity is slightly longer than the time between two consecutive laser pulses.   A Stokes pulse ten times shorter than the laser pulse has been observed.   The upper limit of its duration of 3×10^{-13} sec equals the resolving time of the experimental technique.   The peak power density in this Stokes pulse is consequently three times larger than that in a laser pulse.   In this arrangement significant depletion of laser pump power can take place.   The molecular vibration decays completely between two consecutive pulses and each passage acts as an independent pulse amplifier. Since the group velocity of the Stokes pulse in the Raman liquid is slightly larger than that of the laser pulse, best results are obtained if the Stokes pulse begins with some delay and starts "eating its way" forward in the laser pulse. This effect is similar to that observed by Maier, Kaiser, and Giordmaine [24] for a backward moving Stokes pulse pumped by a long laser pulse. A quantitative theory for this situation has recently been developed by Akhmanov et al [25].

Kaiser and co-workers [26] and also Alfano and Shapiro [27] have demonstrated directly the existence of the vibrational excitation by probing the Raman medium with a second weaker picosecond pulse, following the first pulse with a variable delay. The second pulse is at the second harmonic to facilitate experimental discrimination, and both Stokes and anti-Stokes sidebands are induced by the decaying optical phonon excitation. Kaiser and co-workers obtain a phase memory time for the optical phonons in diamond, which is the inverse of the frequency width of the spontaneously scattered Raman light. This might be expected as the measurement in the time domain is the Fourier transform of the observation in the frequency domain. Shapiro and Alfano report, however, some discrepancy in the case of calcite between the measurements in the time and frequency domain. It is possible that the measurement with strictly coherent light pulses could make a distinction between homogeneous and inhomogeneous dephasing mechanisms. The characteristic times observed in these experiments are of the T_2 type related to the off-diagonal element of the density matrix. The change in level populations is characterized by a time T_1, which does not have a one-to-one correspondence with the observed spectral width. In such a case new information about the material system is obtained. An example is the fluorescence decay, determined by the lifetime of excited states of dye molecules.

The initial buildup of transient Stokes pulses from spontaneous emission noise requires some additional comment. The spontaneously scattered Stokes light is an exact replica in time of the laser pulse and its spectrum is a convolution of the spectral density of the laser pulse with the Raman molecular spectral response function. For $t_p < \tau$, the spontaneous noise input into a particular Stokes mode is reduced. Thus the effective Stokes gain to reach a prescribed output level must be larger. For a Nd: glass laser the spectral width of the pump may well be a factor of e^3 larger than the width of the molecular response. This could explain, in part, why the apparent threshold for stimulated Raman build-up is higher for Nd : glass than for ruby laser radiation. It also explains why the computed curve for the Stokes gain with a random laser pump of about 100 cm^{-1} bandwidth lies about a factor of e^4 below that for a monochromatic pump [20]. Note that in the absence of color dispersion, a random pump should be just as effective in producing gain as a coherent pump. This feature might make it possible to achieve par-

parametric pumping with an incoherent pump in the x-ray region of the spectrum. A small amount of color dispersion would thwart this effect, as the phase relation between pump, signal, and idler initially necessary to produce gain would not be maintained as the three pulses travel through the dispersive medium. In the absence of color dispersion the phase φ_s of the Stokes radiation adjusts itself so that everywhere in the laser and Stokes pulses the phase difference $\varphi_L - \varphi_s$ equals the phase of the driven coherent vibration which carries a phase memory through the pulse duration. Although the gain in a dispersionless medium is not changed by a broad random pump spectrum, the spontaneous noise level or Stokes input into a particular mode is reduced by a factor of $\Delta\omega_L\tau$, compared to that for a coherent pump.

The foregoing theoretical results hold also for the transient stimulated Brillouin scattering. Since the characteristic damping times for acoustic phonons in that case are typically several orders of magnitude larger, transient effects can be tested with much longer pulses of nanosecond duration. Whereas for picosecond pulses the backward gain is essentially zero, since the backward pulse does not experience much of the pump field for a cell length $\ell \gg \tau_p c$, for a nanosecond pulse in a short cell with $\ell \ll \tau_p c$, the characteristic transient response has recently been observed for backward Brillouin scattering. The delay and sharpening of the pulse are again characteristic features [28].

Finally it should be pointed out that Stokes pulse sharpening also occurs for long pulses with $t_p \gg \tau b_{ss}$, even when the gain has an essentially steady-state response. For such slow time variations we find that the time dependence of Stokes signal $|E_s(t)|^2$ is proportional to $\exp[C\,|E_L(t)|^2]$. Since the laser intensity occurs in the exponent the time variation of the Stokes signal is much steeper. This feature has recently been verified experimentally [29] and the general solution of time-dependent coupled parametric equations describes the phenomena well, provided the occurrence of other non-linear processes can indeed be excluded. This section may be summarized with the conclusion that the transient stimulated Raman, Rayleigh, and Brillouin scattering are well understood.

V.   TRANSIENT SELF-FOCUSING AND FREQUENCY BROADENING

A good review [30] of self-focusing contains references to the most pertinent work through 1967. A great deal of work has been published since then. The starting point for the interpretation of the phenomena has always been the nonlinear wave equation with a polarization which is a cubic function of the field amplitude and which describes a quadratic Kerr effect. If we denote the slowly varying complex amplitude of the light wave by

$$E(t, z, r) = A(t, z, r)\, \exp[iks(t, z, r)]$$

where s is the eikonal or phase function, and if we express the nonlinear polarization in terms of an intensity-dependent index of refraction δn and the group velocity of the light is denoted by v, the equation can be written in the form

$$2\left(\frac{1}{v}\frac{\partial s}{\partial t} + \frac{\partial s}{\partial z}\right) + \left(\frac{\partial s}{\partial r}\right)^2 = \frac{2\delta n}{n_0} + \frac{1}{k^2 A}\left(\frac{\partial^2 A}{\partial r^2} + \frac{1}{r}\frac{\partial A}{\partial r}\right) \tag{1}$$

$$\frac{1}{v}\frac{\partial A}{\partial t} + \frac{\partial A}{\partial z} + \frac{\partial s}{\partial r}\frac{\partial A}{\partial r} + \frac{A}{2}\left(\frac{\partial^2 s}{\partial r^2} + \frac{1}{r}\frac{\partial s}{\partial r}\right) = 0 \tag{2}$$

The intensity-dependent index has a dynamic time response, which is characteristic for different physical mechanisms in the material. One should write

$$\delta n = \Sigma\,\delta n_i(t) \tag{3}$$

The following mechanisms over which the summation i must be carried out have been identified:

 1. Reorientation of anisotropic molecules:

$$\frac{\partial(\delta n_{an})}{\partial t} + \frac{1}{\tau_{an}}\,\delta n_{an} = \frac{1}{\tau_{an}}\,n_{2,an}A^2 \tag{3a}$$

 2. Pure electronic hyperpolarizability:

$$\delta n_{el} = n_{2,el}A^2 \tag{3b}$$

 3. Libration of molecular groups:

$$\frac{\partial^2(\delta n_\ell)}{\partial t^2} + \frac{1}{\tau_\ell}\frac{\partial(\delta n_\ell)}{\partial t} + \omega_\ell^2(\delta n_\ell) = \omega_\ell^2\,n_{2,\ell}A^2 \tag{3c}$$

 4. Electrostriction.

 5. Thermal index changes, induced by absorption.

Although mechanisms 4 and 5 can be quite strong and are usually dominant effects for (de)focusing in very long laser signals, they play an insignificant role during an isolated picosecond pulse. Their influence during an entire pulse train cannot be ruled out, however.

The dominant effect during Q-switched laser pulses for both self-focusing and frequency broadening in most molecular liquids was identified very early [31] to be the Kerr effect associated with molecular reorientation. The value of $n_{2,\text{an}}$ lies typically between 10^{-11} and 10^{-12} e.s.u. This should be compared with the electronic quadratic Kerr effect, for which typically $n_{2,\text{el}} \sim 10^{-14}$ e.s.u. Even in fluids with isotropic molecules, such as CCl_4 or liquid argon, the effect of molecular association or collisional encounters is thought to be more important than the pure electronic δn_{el} associated with isolated atoms or molecules. Thus even in these media the first effect appears to be dominant.

It was believed originally that the self-focusing in a picosecond pulse should be ascribed to the electronic effect [32], which has an instantaneous response as indicated by Eq. (3b). The response time for this nonresonant mechanism is essentially equal to the period of a light cycle, $\sim 10^{-15}$ sec. It is clear from the magnitudes quoted that even for a picosecond pulse with $t_p/\tau_{\text{an}} \sim 0.1$ the anisotropic Kerr effect will be dominant. The response will have a transient character and the features of self-focusing will be different from that in the steady state.

The set of coupled nonlinear partial differential equations (1-3) admit an almost infinite variety of solutions. The difficulty lies in choosing meaningful approximations that conform reasonably closely to the initial boundary conditions set by the experiments. Since the solutions are very sensitive to initial conditions and points of instability exist, it is a difficult matter to obtain satisfactory agreement between theory and experiment. The control of experimental conditions is very important in a meaningful test of the validity of the equations (1 - 3).

This set of equations could be extended further in several ways. Solutions have been discussed with the addition of terms in A^4 and A^6 to the right-hand side of Eq. (3). These terms would account for saturation of the molecular reorientation and two- or three-photon absorption in the fluid. They would tend to stabilize certain solutions. There is, however, firm experimental evidence that δn does not exceed 4×10^{-3} in the filamentary track, while saturation would require $\delta n > 0.1$. More consequential is the omission of terms representing frequency dispersion of the linear index of refraction. This effect could formally be taken into account by adding terms proportional to

$$\frac{1}{2} \frac{\partial^2 k}{\partial \omega^2} \left(\frac{\partial A}{\partial t} + v \frac{\partial A}{\partial z} \right) \text{ and } \frac{1}{2} \frac{\partial^2 k}{\partial \omega^2} \left(\frac{\partial s}{\partial t} + v \frac{\partial s}{\partial z} \right)$$

in Eqs. (1) and (2), respectively. This would so increase the mathematical complexity that no solutions in closed form have yet been obtained, but the effect of linear dispersion may certainly not be ignored in the discussion of frequency broadening under experimental conditions, as will be explained below.

It is perhaps instructive to enumerate the approximations made in various theoretical models which have been used as framework for the discussion of the experimental observations.

A.   Steady-State cw Self-Trapping

In this original case of Chiao, Garmire, and Townes [33] one takes s = constant, $\partial/\partial z = 0$ in Eqs. (1)-(3). The solution so obtained is not stable against small perturbations.

B.   Steady-State Self-Focusing

The self-focusing distance may be estimated from the eikonal equation. If diffraction effects are ignored and a parabolic dependence of δn as a function of r is assumed, one finds that $\partial s/\partial r$ is proportional to r. The self-focusing distance as a function of incident power and effective beam radius a was given by Kelley [34] in the form

$$z_{sf}(t) = \frac{n_0 a^2}{4(c/n_2)^{1/2}} \; \frac{1}{P^{1/2}(t - z/c) - P_{crit}^{1/2}} \tag{4}$$

where the critical power for self-trapping [33] is determined by

$$P_{crit} = \frac{\lambda^2 c}{32\pi^2 n_2} \tag{5}$$

and P is assumed constant.

C.   Moving Self-Focus

If the power P is not constant, but a slowly varying function of time, the same equation may be used, but now the self-focusing distance becomes time dependent. This leads to the moving self-focus model of Lugovoi and Prokhorov [35]. The self-consistent construction of the focal point as a function of z and t leads to a U-shaped curve [36] with a part where the focal point moves backward, $\partial z/\partial t < 0$.

It is now generally accepted that a carefully controlled single-mode Q-switched ruby laser indeed leads to this situation. In the apparent filamentary track forward and backward moving focal spots have been identified [35, 37] and there is no appreciable phase modulation or frequency broadening.

These results are in sharp contrast to those obtained with multimode Q-switched lasers. It is now believed that such laser shots act as a random succession of picosecond pulses. This is in agreement with the considerations in Sec. I. It should be noted that in this case only a small fraction of the total light energy (about 1%) takes part in the filament formation.

D.   Transient Self-Trapping

The duration of a single picosecond pulse is about three or four orders of magnitude shorter than a Q-switched pulse. If one were to replace the mechanism of the intensity-dependent index of refraction by reorientation by that for the electronic Kerr effect, the index response would again be instantaneous and again a U-shaped curve for moving foci could be expected, although its width would be much contracted.

As discussed above, one may still expect the reorientation quadratic Kerr effect to be dominant, but for $t_p \gtrsim \tau_{an}$ the index response will have a transient character. Akhmanov et al. [30] have shown that the solution of Eqs. (1)-(3) is now possible with $A(t - z/v, r)$ and $s = $ constant. In the limit $t_p \ll \tau_{an}$ the input radius will be undeformed, as the material has had no time to respond at the leading edge. Towards the end of the pulse the index change has accumulated to such an extent that it balances the appreciable diffraction accompanying the contracted radius. This solution is schematically represented in Fig. 10. In the frame co-moving with the pulse, the radial dependence is stationary. So is the corresponding intensity distribution. The intensity on the axis acquires a skewed and narrowed shape, as only the trailing part of the pulse is self-focused. Again only a fraction of the total energy is within the radial contour sketched in the figure and takes part in the focusing. The question of stability in this quasistationary solution is not so important, as there is no time for an instability to develop. If the pulse duration were lengthened, the radius would eventually increase again. The solution may also be regarded as the first half of a focal spot which remains in the tail of the pulse. Recent computer calculations by Shimizu [38] and by Fleck and Kelley [39] have reproduced the features sketched qualitatively in Fig. 10. The solutions are quite sensitive to choice of input radius and intensity of the short pulse.

E.   Self-Phase Modulation

If one assumes a steady-state filament of constant radius one may find a solution with $A = A_1(t - z/v)A_2(r)$ and $\partial s/\partial r = 0$. This solution leads to a relation first used by Shimizu [40], $\partial s/\partial z \approx 2 \, \delta n/n_0$, or a phase which increases linearly with z. Since $\delta n \propto A_1^2 (t - z/v)$, the resulting phase modulation of the pulse yields a modulated spectral density. For an instantaneous response

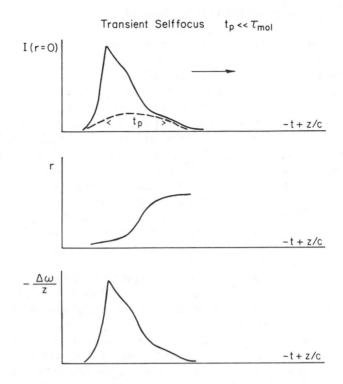

Figure 10.   Qualitative features of transient self-focusing in the limit that the pulse duration is very short compared with the response time of the physical mechanism responsible for the intensity-dependent index of refraction. Top: Intensity dependence of pulse on axis. The intensity in the trailing edge is increased by self-focusing. The incident pulse contour is given by the dotted line for comparison. Middle: A radial contour of the pulse containing a fixed amount of energy. This contour is quasistationary and leads to a behavior reminiscent of steady-state self-trapping for the trailing edge. Bottom: The frequency shift is entirely on the Stokes side and is proportional to the instantaneous intensity in the extreme transient case.

of the index of refraction, i. e., for a long pulse duration, the resulting spectrum is symmetric and the modulation freuqency $\Delta\omega \sim t_p^{-1}$. The pulse duration so measured is usually a factor of 3 to 5 shorter than the independently measured duration of the picosecond pulse at the input. The transient self-trapping would provide a natural mechanism for an effective shortening of the pulse.

F.   Transient Self-Phase Modulation

If $t_p \gtrsim \tau_{an}$ the spectral broadening should become asymmetric with a higher intensity on the Stokes side. Detailed calculations [41] have been made and the computed spectra can be fitted well to the observations provided one adjusts the pulse duration and effective intensity and radius appropriately. In the limit of extreme transient response $t_p \ll \tau_{an}$ there would only be broadening on the Stokes side. The frequency, rather than the phase, would now be proportional to the instantaneous intensity, as sketched in Fig. 10. The extreme transient phase modulation would lead to a phenomenon, which might be called "self-chirping." Computer calculations show that most of the phase modulation takes place in the self-focused trailing edge of the pulse [38]. The effective pulse duration here is considerably shortened.

It is usually stated that amplitude modulation of the pulse makes a negligible contribution compared to phase modulation. Once the frequency broadening has proceeded to say 500 or 1000 cm^{-1}, the linear dispersion of the medium cannot be ignored. For $\Delta\omega = 10^3$ cm^{-1}, the frequency difference $(\partial n/\partial \omega)\Delta\omega$ in media with normal dispersion will lie between 10^{-4} and 10^{-3}. This implies that after a distance of about 10^3 or 10^4 wavelengths, i.e., between 0.1 and 1 cm path length, the phase relation between the frequency components is destroyed. In other words, a sizeable conversion from phase modulation to amplitude modulation has taken place. It is then useful to describe further frequency broadening as the beating of different Fourier components in the laser beam. For example, ω_L and $\omega_L - \Delta\omega$ combine to create the anti-Stokes component $2\omega_L - (\omega_L - \Delta\omega) = \omega_L + \Delta\omega$. It must be kept in mind that the same physical nonlinearities, associated with the quadratic Kerr effect are involved in this amplitude modulated broadening. The discussion of frequency modulation can, in principle, be carried out either in the frequency [31] or in the time domain [40]. There is complete equivalence between these two approaches, although sometimes physical initial conditions will favor one method or the other [41].

In general, both amplitude and phase will have complicated dependences on r, z, and t, which are in turn sensitive functions of the intensity, radius, and spectral propoerties of the incident pulse. This is the reason why, in spite of a large experimental effort by many different research groups [36, 42-44], the combined questions of self-focusing and frequency broadening are not completely resolved. Each group utilizes the same store of physical effects mentioned above, but by placing the emphasis on different terms in the equations and making different approximations, the semantics often appear irreconcilable.

Experimental evidence, recently obtained by Carman and Reintjes [44] with the arrangement shown in Fig. 5, will now be presented to show that the Kerr effect associated with molecular reorientation is indeed the dominant mechanism for the self-focusing in a single picosecond pulse. A single pulse is switched out of a train from a mode-locked Nd : glass laser.   The single pulse is amplified and partly converted to second harmonic radiation in a KDP crystal.   The frequency chirped green pulse is separated from the infrared pulse by a prism and is compressed to an effective duration of 0.5-1 psec.   It is split by 10 glass beam splitters into 10 probing rays which pass through crossed polarizer and analyzer sheets, not shown in Fig. 5. They probe the birefringence induced by the powerful infrared pulse which passes through the cell at right angles to the green beams.   There is a variable delay in the "infrared arm" and the angle of the "green arm" is adjusted so that each green probing just crosses the infrared pulse at the appropriate point in the cell.

The strong birefringence in the filamentary self-focused regions can thus be observed for a single pulse.   It was found both from the velocity of propagation required for synchronism and from the amount of the green light transmitted by the induced birefringence in the 5-μ-thick filament that δn in the filament was less than 0.2×10^{-2}.   This rules out saturation effects and is in agreement with other recent findings.   The first filaments were generally formed within 1-2 cm from the entrance window of the collimated laser pulse.   They could be followed and identified for about 5 cm.   No new filaments were formed after 10 cm and they all disappeared between 16 and 20 cm from the entrance window.   If the intensity of the incident pulse was critically adjusted, only one or two filaments were formed.   The length of the streak is a convolution of the probing pulse duration and the duration of the birefringence.   The picture shown in Fig. 11 was taken with CS_2.   The self-focusing track is identified over about 5 cm.   Note the two spots corresponding to the

Figure 11.   Photograph of birefringence induced in CS_2 by a picosecond pulse, obtained with the experimental arrangement shown in Fig. 5 (after Reintjes, Ref. 44).

light reflected from the front and the back of each beam splitter. The probing light also covers the area between the two dark traces, but the birefringence there is not coincident with the arrival time of the probing light. It could be concluded that the birefringence in CS_2 decays with a characteristic time $\tau \lesssim 2.2$ psec. If other fluids, such as toluene and nitrobenzene, are used, the birefringence persists much longer, with $\tau_{an} \sim 10 - 30$ psec. This was determined from the exponential decay of the birefringence as a function of the time delay between the infrared pulse and the green probing pulse at a given point in the cell. There is no observable increase in birefringence at the beginning, when the pulses overlap and the instantaneous electronic Kerr effect is measured. It may be concluded that in toluene and nitrobenzene the dominant contribution to the self-focusing comes from molecular reorientation. This is a fortiori true for CS_2 with a shorter τ_{an} and a larger value of $n_{2, an}$. One recognizes in the method described here to determine the characteristic time of the physical process responsible for self-focusing the same general features of experimental techniques discussed in Sec. II. It is concluded that the transient self-trapping described in Sec. V D and schematically represented in Fig. 10 (top and middle) is probably correct.

The frequency broadening obtained with the Nd:glass laser pulses by imaging filaments at the exit window on the slit of a spectrograph was surprisingly symmetric and not at all in agreement with the picture at the bottom of Fig. 10. It appears as if a faster mechanism takes over in the self-focused trailing part of the pulse. One might reasonably expect that the same mechanism which is responsible for the self-focusing would also make a dominant contribution to the self-phase modulation. It is conceivable, however, that the effective pulse duration in the self-focused trailing edge is sufficiently shortened, that another faster mechanism starts to contribute significantly to the intensity dependent index, and thus determines the phase modulation. Svelto and co-workers [43] have measured the r and z dependence of the frequency broadening. They show that their data are consistent with broadening caused by the librational contribution to the intensity-dependent index, if an effective pulse duration of about 1 psec is assumed. Our observations on frequency broadening are in qualitative agreement with those by Svelto et al. The combined interpretation of self-focusing and subsequent frequency broadening by picosecond pulses requires that the former proceeds by transient molecular reorientation, while the latter is then mainly determined by a faster librational process, which takes over after the focused intensity pulse has been sufficiently shortened. Ultimately these results should be obtainable from the complete numerical solution of Eqs. (1)-(3), which should also incorporate the linear dispersion.

Detailed explanations must also be given why the minimum diameter of the filament should be about 5 μ, why the frequency broadening does not continue beyond a maximum characteristic for each fluid, and why the filament eventually expires in the fluid. It is customary to invoke losses due to stimulated scattering and other nonlinear processes. Such suggestions have, however, an ad hoc character and emphasize both the theoretical and experimental difficulties presented by these phenomena which are very localized both in space and time.

VI.   CONCLUSION

The transient nature of stimulated scattering processes, which can be studied separately and, in particular, can be isolated from the phenomena of self-focusing, is well understood. Good agreement between theory and experiment exists for the stimulated Raman scattering from picosecond pulses. The Kerr effect associated with molecular reorientation of optically anisotropic molecules has been identified as a dominant mechanism for self-focusing of picosecond pulse in molecular fluids. A comprehensive understanding of all observations, which includes frequency broadening, length and diameter of filament, in addition to the transient self-trapping, is still lacking.

ACKNOWLEDGMENTS

The author is indebted to R. L. Carman, M. J. Colles, R. Polloni, J. Reintjes, and F. Shimizu for several discussions and communication of some of their results before publication.

ADDENDUM (AUGUST 1972)

At the 7th International Conference on Quantum Electronics, Montreal, May 1972, it was proposed by Bloembergen [45], that a filament cannot contract further than the size at which the field strength in the center of the filament approaches the threshold for dielectric breakdown. For picosecond pulses this threshold breakdown strength has been estimated by Yablonovitch and Bloembergen [46] to lie near 1 or 2 $\times 10^7$ V/cm for many transparent dielectrics. This estimate is in good agreement with the experiment of Brewer and Lee [32]. Their data show that the filament diameter and the self-focusing power vary with the ability of the material to support reorientation of molecular groups, but the electric field strength inside the filament is not correlated with this mechanism. The filament diameter may be estimated from $d \approx \lambda/4n_2^{1/2} |E^{br}|$, where E^{br} is the threshold for breakdown by a picosecond pulse. Taking $|E^{br}| \approx 10^7$ V/cm and an appropriate value for n_2 in Eqs. (3)-(5), one finds $d \approx 3$-10 μ for most materials.

The filament diameter is stabilized by the negative real part of the index of refraction of the electron density in the incipient plasma. When this density does not exceed 10^{17} electrons/cm^3 there is no catastrophic breakdown. The absorption in the plasma causes a temperature rise of less than $50°C$ during a few picoseconds. This absorption will limit the length of the filament to several centimeters. Under certain other conditions it is possible for the plasma density to rise catastrophically above 10^{18} electrons/cm^3, in which case a hot spark appears and a fossile damage track is left behind in solids.

The exponential nonlinearity associated with picosecond avalanche ionization appears to provide a solution for the questions about the filament dimensions posed at the end of this paper.

REFERENCES

1. A. J. DeMaria, W. H. Glenn, M. J. Brienza, and M. E. Mack, Proc. IEEE 57, 2 (1969).
2. It should be noted that the prefix "tera-" derives from the Greek word "τερασ" or monstrosity, and not from the Latin word for earth "terra."
3. N. G. Basov, P. G. Kriukov, S. D. Zakharov, Yu V. Senatski, and S. V. Tchekalin, IEEE J. Quantum Electron. QE-4, 864 (1968); G. W. Gobeli, J. C. Bushnell, P. S. Peercy, and E. D. Jones, Phys. Rev. 188, 300 (1969).
4. N. G. Basov, P. G. Kriukov, and V. S. Letokhov, Optics and Laser Technol. 2, 126 (1970); V. S. Letokhov, Zh. Eksp. Teor. Fiz. 55, 1077 (1968) [Sov. Phys. JETP 28, 562 (1969)].
5. J. A. Armstrong, Appl. Phys. Letters 10, 16 (1967).
6. J. A. Giordmaine, P. M. Rentzepis, S. L. Shapiro, and K. W. Wecht, Appl. Phys. Letters 11, 218 (1967).
7. P. M. Rentzepis, C. J. Mitschele, and A. C. Saxman, Appl. Phys. Letters 17, 122 (1970).
8. D. H. Auston, Appl. Phys. Letters 18, 249 (1971); Optics Commun. 3, 272 (1971).
9. H. E. Rowe and T. Li, IEEE, J. Quantum Electron. QE-6, 49 (1970).
10. E. B. Treacy, Physics Letters 28A, 34 (1968).
11. J. W. Shelton and J. A. Armstrong, IEEE J. Quantum Electron. QE-3, 696 (1967).
12. M. A. Duguay and J. W. Hansen, Appl. Phys. Letters 15, 192 (1969); Optics Commun. 1, 254 (1969).
13. P. M. Rentzepis, M. R. Topp, R. P. Jones, and J. Jortner, Phys. Rev. Letters 25, 1742 (1970).
14. S. A. Akhmanov, D. P. Krindach, A. V. Migulin, A. P. Sukhorukov, and R. V. Khokhlov, IEEE J. Quantum Electron. QE-4, 598 (1968);

happens to the light beam beyond the focal point.

In 1967 Dyshko, Lugovoi, and Prokhorov (DLP) [11] made calculations without limiting the z value. It was shown that at a given power $P >> P_{cr}$ the beam broke into annular zones with subsequent focusing into separate foci on the beam axis. The general picture will be the same if the initial distribution of the laser beam is not Gaussian but it must have only one maximum. The position and the number of foci on the beam axis are determined by laser power (Fig. 1).

In experiments the power is time dependent, so the observed picture must be quite different [12]. With pulse power increasing in time, the foci will increase in number and move toward the laser. When the pulse power reaches its maximum, the foci stop moving. When the pulse power decreases the foci will move in the opposite direction. Hence, the tracks of these moving foci will be filaments directed along the beam axis. The process of self-focusing throughout the pulse will proceed as follows. At first, when the power just exceeds the critical value, one moving focus appears. With further power increase, a second focus appears, and so on. The number of foci is maximum at a maximum pulse power.

One can find the characteristic time for the focus to pass through the output end (plane) of the medium. For typical experimental conditions this time is about 0.5×10^{-10} sec [12], which is consistent with the observations

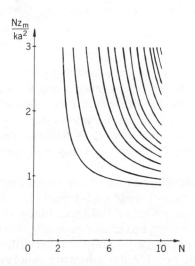

Figure 1.   Foci position along the beam axis versus initial relative field strength N. The case of a Gaussian initial intensity distribution with a fixed radius a and an initial plane phase front. The positions z_m of various foci for each value are given in units of $\kappa a^2/N$, where κa^2 is the diffraction divergence length of the initial beam in a nonlinear medium.

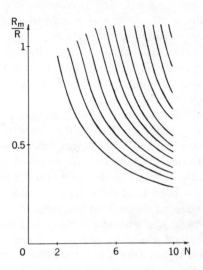

Figure 2.   Foci position on the beam axis versus initial relative field strength N. The case of a Gaussian initial intensity distribution with a fixed radius a and a spherically converging initial phase front with a linear focal distance $R = \kappa a^2/4$. Foci positions z_m for each N value are given in units of R.

by Brewer and Lifsitz [3] on self-focusing processes in liquids. In many experiments self-focusing is observed using a focusing lens. Therefore, in 1969 DLP [13] developed the theory for the case when the initial phase front is spherical. Here the general picture is the same, but the foci are located in the focal area of the lens as well as between the lens and its focus (Fig. 2). Their position depends on power and they move toward the lens when power increases.

In connection with the advent of picosecond pulses it is also interesting to consider the question of their self-focusing. This problem was discussed in Ref. 14; nonlinearity versus field was taken in the form of Eq. (1). With ultrashort pulses the light beam can be much shorter than the cell length. Therefore, the picture of self-focusing must differ from that given in Refs. 11 and 12. Solutions provide the following conclusions. Self-focusing takes place even with ultrashort pulses when the light beam is much shorter than the length of self-focusing. The points of the medium where foci occur coincide with those for a stationary case. So they are formed at certain instants of time and split on formation (Fig. 3). They move in the direction of the pulse propagation and in the opposite direction.

Figure 3.   A qualitative picture of the time dependence of foci positions z_m for a pulse propagating in a Kerr-type nonlinear medium.

We have to note that in the work by Loy and Shen [15] the effect of the pulse width on the picture of self-focusing is also considered.

For picosecond pulses the transient processes can be important when the pulse duration is shorter than, or comparable with, the relaxation time of the nonlinear refractive index.   In this case the process of foci formation will be associated with the losses due to transient processes.   But this effect is not important if the nonlinear refractive index has an electronic origin as in this case the relaxation time is 10^{-15} sec.   There is some evidence that for picosecond pulses the transient processes are not important, because for picosecond pulses [16] the radial dimension of filaments (foci) is equal to 5 μm as for the case of nanosecond pulses.   Of course, other experiments must be performed to check this assertion.

The dimension of focal regions is of great interest.   The longitudinal dimension is of special concern when we study picosecond pulses.   The problem is that a picosecond pulse size is about 1 mm along the axis of light propagation, which can be comparable with the longitudinal dimension of the foci.   Thus, when a powerful picosecond pulse propagates in a nonlinear medium there is a "bright" core in it which is formed by a series of foci. Energy is put into this core from the beam and dissipates due to nonlinear absorption.   The beam leaves the focal region with a considerable divergence.

The following two points should be emphasized here:

1.   The core carries only a small part of the total pulse energy.

2.   The core is not a waveguide since there is no wave propagation in it and it cannot exist without the main part of the pulse.

The multifoci theory predicted in Refs. 9-12 did not give foci dimensions, because the simple refractive index dependence in form (1) cannot provide an answer to this question. In order to obtain a finite focal dimension, it is necessary that either a saturable nonlinear refractive index or losses be introduced.

In several papers [17-19] numerical calculations were made with the saturable nonlinear refractive index in the form

$$n = n_0 + \frac{\frac{1}{2} n_0 n_2 |E|^2}{(1 + |E|^2 / |E_s|^2)} \tag{3}$$

where $|E_s|^2 \approx 1/n_2$.

This kind of field dependence of the refractive index imposes limitations on the field magnitude in the focal region. But in this case foci come out to be small compared to the wavelength. Indeed, in the case of saturation the field in the focal regions $|E_F|^2 > |E_s|^2$, and the value $E_F{}^2 = 4/n_2 (\kappa a)^2$. Hence, $a < \lambda$. So in this case the parabolic equation cannot be used. It was mentioned [12] that nonlinear absorption can be the factor limiting the energy density in the focal region. The dimension of the focal regions will depend upon nonlinear absorption and can greatly exceed the wavelength. Experiments show that the foci are large compared with the wavelength. In this case we can use the parabolic equation and also neglect Kerr nonlinearity saturation. Experimental proof of this situation is the slight variation of the refractive index in the focal region (10^{-3}). In this connection in 1971 DLP [20] took the field dependence of the refractive index in form (1). To take nonlinear absorption into account, an imaginary part of the refractive index must be introduced. In our work three types of nonlinear absorption were considered:

1.   three-photon absorption,
2.   absorption due to the energy pumping into the first Stokes component, and
3.   two-photon absorption.

In each of these three types of nonlinear absorption a multifoci structure takes place. But the position and the number of foci depend on the type and magnitude of nonlinear absorption. The multifoci structure is well resolved when nonlinear absorption occurs only in the focal regions and does not prevent their formation. It is interesting to point out that there are critical values for absorption parameters for the second and the third types of nonlinear absorption. If the value of such a parameter is less than the critical value, the field in the focal region is infinitely large. A finite value of

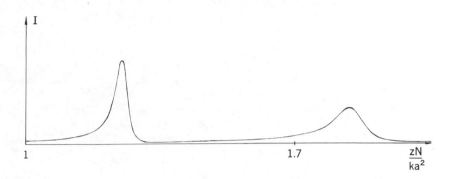

Figure 4.   Axial intensity of the beam stationary in time versus the distance between a given cross section and the initial plane for the case of two-photon absorption in a medium and an initial plane phase front of the beam.   The z values are given in units of $\kappa a^2/N$.

the field in the focal region is obtained if this parameter equals or exceeds the critical value.   But if the parameter appreciably exceeds the critical value the self-focusing is strongly diminished by absorption (Fig. 4).   Therefore, a multiphoton absorption for $k \geq 3$ is of special interest because in this case there is no critical value for the absorption parameter.

Consider the case of a three-photon absorption.   The field dependence of the refractive index is taken in the form

$$n = n_0 \left[1 + \frac{1}{2} \left(n_2 |E|^2 + im_4 |E|^4 \right) \right] \qquad (4)$$

In the numerical solution of the parabolic equation we introduced the following designation:

$$r_1 = \frac{r}{a}; \quad z_1 = \frac{z}{l_x}; \quad l_x = \frac{\kappa a^2}{N}; \quad N = \frac{E_0}{E_{cr}};$$

$$E_{cr} = \frac{1}{[n_2(\kappa a)^2]^{1/2}}; \quad \mu_4 = \frac{m_4 E_0^2}{n_2} N^2$$

The z dependence of intensity (r = 0) is shown in Fig. 5 for N = 6, μ_4 = 0.05. In this case the longitudinal dimension of the first focal region (along z axis) is found to be 6×10^{-2} cm.

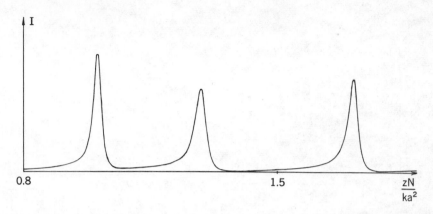

Figure 5.   Axial intensity of the beam stationary in time versus distance z between a given cross section and the initial plane for the case of three-photon absorption in a medium with N = 6, μ_4 = 0.05 and an initial plane phase front of the beam.   The z values are given in units of $\kappa a^2/N$.

The calculations show that if $N \geq 4$ and $\mu_4 \leq 10^{-2}$ the maximum field magnitude of the first focal region and its diameter can be determined by the following formulas [20]:

$$|E_F|^2 \approx 0.15 \, \frac{n_2}{m_4} \qquad (5)$$

$$d_F \approx 1.1 \, \lambda \, \frac{\sqrt{m_4}}{n_2} \qquad (6)$$

In the general case of k-photon absorption ($k \geq 3$) formulas (5) and (6) can be generalized in the following form [20]:

$$|E_F|^2 \approx \left[\frac{0.15 n_2}{m_{2k-2}}\right]^{1/(k-2)} \qquad (5a)$$

$$d_F \approx 0.15 \, \lambda \left[\frac{m_{2k-2}}{(0.15 n_2)^{k-1}}\right]^{1/(2k-4)} \qquad (6a)$$

From (5a) and (6a) the following relation can be obtained:

$$|E_F|^2 \, d_F^2 \approx 0.15 \, \frac{\lambda^2}{n_2} \qquad (7)$$

This formula allows one to calculate the field strength in the focal region if the diameter d_F and n_2 are known experimentally. As an illustration we shall take the results obtained for CS_2 liquid assuming a three-photon nonlinear absorption for this case. For CS_2 the value of $n_2 = 10^{-11}$ cgse, the diameter of the focal region is 5 μm, $\lambda = 0.7$ μm. Hence, according to (6) and (7), m_4 is estimated to be 4.3×10^{-21} cgse and $E_F = 5.6 \times 10^6$ V/cm, which corresponds to the peak intensity of 90 GW/cm^2.

III.   COMPARISON OF THE MULTIFOCI THEORY WITH THE EXPERIMENTAL DATA

A multiphoton absorption plays an important role in focusing processes, and therefore it is necessary to go into more detail about this problem.

Multiphoton processes in atoms have been studied in greater detail than those in solids and liquids. This is due, first of all, to the fact that in the case of atoms (or molecules) experiments can be run at low pressures when other processes are not important. On the other hand, the results of investigations of multiphoton ionization of atoms can be used in studying self-focusing in solids, liquids, and gases.

The probability of multiphoton ionization [21, 22] of atoms was shown to be $W = A\rho^k$, where $k \le m$. Here m is the number of quanta necessary for atomic ionization and ρ is the energy density. The k value is greatly changed when there is resonance with intermediate levels. Thus, for example, in case of a multiphoton ionization in Xe a change [22] of neodymium laser frequency by 9 cm^{-1} will reduce the k value from 10.8 to 5.8. So these experiments are indicative of the important role of intermediate levels, particularly in solids and liquids exhibiting wide absorption bands. The probability of multiphoton ionization [22] of Xe, Ar, and Kr, ranges from 10^5 sec to 10^8 sec^{-1} at a light intensity of 2.5×10^{11} W/cm^2. In the case of solids and liquids this probability, due to wide absorption bands, must be of the same order or higher than that for inert gases. Assume, for example, this value to be 10^5 sec^{-1} at an intensity of 10^{11} W/cm^2. For solids and liquids the light intensity in the focal regions is of the order of 10^{11} W/cm^2. If the focal region changes its position during 10^{-10} sec, then within this time electrons are produced with a density of 10^{18} cm^{-3} due to multiphoton absorption. Multiphoton ionization in the focal region gives rise to an avalanche ionization.

Now it is known [23, 24] that self-focusing precedes gas breakdown. Moreover, in 1971 Richardson and Alcock [25] when studying gas breakdown in inert gases observed that filament formation preceded the gas breakdown. The filaments were 10 μm in radius and 120 μm in length and the electron concentration was $10^{18}/cm^3$. These experimental results can be easily interpreted. The filament is a focal region where high field intensity produces electrons by multiphoton absorption. So these experiments support the idea of the importance of multiphoton absorption in self-focusing phenomena.

The experimental results and the multifoci theory should be compared within those assumptions that have been made in the theory. First, the experiments should be carried out with single-mode lasers, since according to the Bespalov-Talanov theory [26], amplitude-phase perturbation of a plane electromagnetic wave causes the beam to break into several filaments.

In 1969 Loy and Shen [15] using a single-mode Q-switched ruby laser observed self-focusing in liquids. These authors have found that many filaments of self-focusing are formed when the laser is operated in a multimode regime. In case of single-mode operation, the picture is simplified. From their experiments with a single-mode laser they have concluded that many effects inherently related to self-focusing are consistent with the moving foci picture. The observations were made from the cell end. In 1970 Korobkin et al. [27] investigated self-focusing phenomena in liquids using a single-mode ruby laser and a high-speed streak camera. The picture taken from the end of the cell is given in Fig. 6. The foci are located within 5 μm of the laser beam axis and each focus appears at a certain value of the laser power. Another picture was taken from the side of the cell, that is, at a right angle to the propagation axis of the laser beam (Fig. 7). In this case tracks of moving foci were observed. It is noteworthy that when the laser power achieves its maximum, self-focusing is destroyed due to disturbance of the liquid.

Self-focusing in glasses has been observed by several authors [28-29]. In this case optical damage due to self-focusing was observed in the form of long, thin filaments. In 1970 Lipatov et al. [30] investigated this phenomenon in glasses. It was supposed that filament structure was formed when the laser power was time dependent. If we produce a pulse with power constant in time the picture must be generally simplified because instead of the filament structure we must have damage in certain spots where the foci are located. Therefore, in the experiment [30] a rectangular-pulse single-mode ruby laser was used with pulse duration of 10^{-8} sec. For a Kerr-type mechanism the foci should be fixed in space and damage should take place at a certain number of fixed spots. This picture was realized experimentally. At a certain power only one damage spot was observed. At a higher power two and then three damage spots were observed (Fig. 8). It is of interest to note that when we used an ordinary single-mode ruby laser we observed damage in the glass in the form of one long, thin filament.

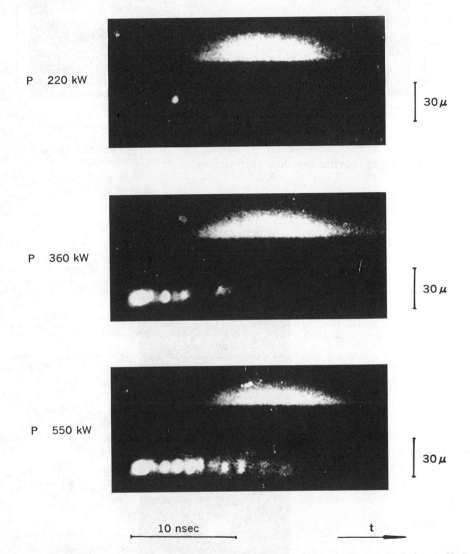

Figure 6. Streak photograph of foci in CS_2 taken from the end of the cell for different peak powers of a giant-pulse ruby laser. Foci appear successively as the laser power increases in time.

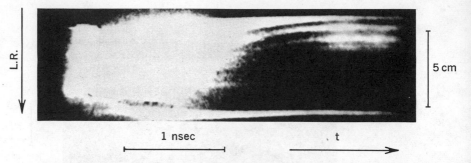

Figure 7.   Streak photograph of moving foci in CS_2 from the side of the cell. The tracks of moving foci are filaments.

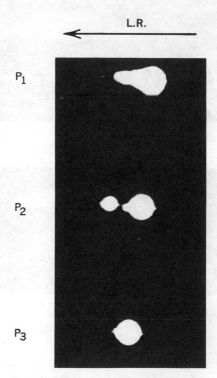

Figure 8.   Photograph of optical damage in glass due to self-focusing for different powers ($P_3 < P_2 < P_1$) of a rectangular-pulse ruby laser. In this case damage takes place in certain spots instead of long, thin filaments as is the case with time-dependent pulse power.

62

To check the multifoci theory, in 1971 Korobkin et al. investigated propagation of a picosecond pulse train in liquid CS_2 using a high-speed streak
camera [31].   From the multifoci theory one would expect strong scattering
of laser light at each moment when the focus velocity is zero.   Therefore,
the experimental picture at a right angle to the propagation axis must consist
of discrete bright spots for each picosecond pulse.   The position and the
number of these spots must be different for each pulse because each of the
successive pulses has a higher peak power.   This was observed experimentally (Fig. 9).   It is interesting to point out that there are additional scattering
centers which correspond to those created by a previous picosecond pulse.

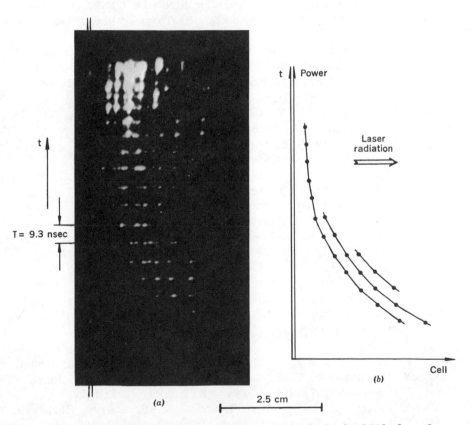

Figure 9.   Streak photograph of propagation of mode-locked Nd-glass laser
radiation in CS_2.   (a) shows laser light scattered at $90°$ to the incident power.
The experimental points of the focal region positions and the theoretical
curves are plotted in (b).

There is a good agreement between the experimental data and the theoretical curve for the foci position versus $N = E_0/E_{cr}$.

REFERENCES

1. V. N. Lugovoi, Dokl. Akad. Nauk SSSR 176, 58 (1967) [Sov. Phys. Dokl. 12, 866 (1968)].
2. N. F. Piliptetskii and A. R. Rustamov, ZhETF Pis. Red. 2, 88 (1965) [JETP Letters 2, 55 (1965)].
3. R. G. Brewer and J. R. Lifsitz, Phys. Letters 23, 79 (1966).
4. R. Y. Chiao, M. A. Johnson, S. Krinsky, H. A. Smith, C. H. Townes, and E. Garmire, IEEE J. Quantum Electron. QE-2, 467 (1966).
5. R. Y. Chiao, E. Garmire, and C. H. Townes, Phys. Rev. Letters 13, 479 (1964).
6. T. F. Volkov, in Plasma Physics and the Problem of Controlled Thermonuclear Reactions (Pergamon Press Ltd., 1960), Vol. IV, p.114.
7. G. A. Askar'yan, Zh. Eksp. Teor. Fiz. 42, 1567 (1962) [Sov. Phys. JETP 15, 1088 (1962)].
8. V. I. Talanov, Izv. Vuzov, Radiofizika 7, 564 (1964).
9. S. A. Akhmanov, A. P. Sukhorukov, and R. V. Khokhlov, Usp. Fiz. Nauk 93, 19 (1967) [Sov. Phys. Usp. 10, 609 (1968)].
10. P. L. Kelley, Phys. Rev. Letters 15, 1005 (1965).
11. A. P. Dyshko, V. N. Lugovoi, and A. M. Prokhorov, ZhETF Pis. Red. 6, 665 (1967) [JETP Letters 6, 146 (1967)].
12. V. N. Lugovoi and A. M. Prokhorov, ZhETF Pis. Red. 7, 153 (1968) [JETP Letters 7, 117 (1968)].
13. A. P. Dyshko, V. N. Lugovoi, and A. M. Prokhorov, Dokl. Akad. Nauk SSSR 188, 792 (1969) [Sov. Phys. Dokl. 14, 976 (1970)].
14. A. A. Abramov, V. N. Lugovoi, and A. M. Prokhorov, ZhETF Pis. Red. 9, 675 (1969) [JETP Letters 9, 419 (1969)].
15. M. T. Loy and Y. R. Shen, Phys. Rev. Letters 22, 994 (1969).
16. R. Cubeddu, R. Polloni, C. A. Sacchi, O. Svelto, and F. Zaraga, Phys. Rev. Letters 26, 1009 (1971).
17. J. H. Marburger and E. L. Dawes, Phys. Rev. Letters 21, 556 (1968).
18. E. L. Dawes and J. H. Marburger, Phys. Rev. 179, 862 (1969).
19. N. E. Zakharov, V. V. Sobolev, and N. S. Synakh, Zh. Eksp. Teor. Fiz. 60, 136 (1971) [Sov. Phys. JETP 33, 77 (1971)].
20. A. P. Dyshko, V. N. Lugovoi, and A. M. Prokhorov, Zh. Eksp. Teor. Fiz. 61, 2305 (1971) [Sov. Phys. JETP 34, 1235 (1972)].
21. N. B. Delone and L. V. Keldysh, Preprint #11, P. N. Lebedev FIAN (1970).

22. G. A. Delone, N. B. Delone, V. A. Kovarsky, and N. F. Perelman, Short Report #8, P. N. Lebedev FIAN (1971).
23. V. V. Korobkin and A. J. Alcock, Phys. Rev. Letters $\underline{21}$, 1433 (1968).
24. F. V. Bunkin, I. K. Krasyuk, V. M. Marchenko, P. P. Pashinin, and A. M. Prokhorov, Zh. Eksp. Teor. Fiz. $\underline{60}$, 1326 (1971) [Sov. Phys. JETP $\underline{33}$, 717 (1971)].
25. M. C. Richardson and A. J. Alcock, Appl. Phys. Letters $\underline{18}$, 357 (1971).
26. V. I. Bespalov and V. I. Talanov, ZhETF Pis. Red. $\underline{3}$, 471 (1966) [JETP Letters $\underline{3}$, 307 (1966)].
27. V. V. Korobkin, A. M. Prokhorov, R. V. Serov, and M. Ya. Shchelev, ZhETF Pis. Red. $\underline{11}$, 153 (1970) [JETP Letters $\underline{11}$, 94 (1970)].
28. M. Hercher, J. Opt. Soc. Am. $\underline{54}$, 563 (1964).
29. Yu. M. Zverev, E. K. Maldutis, and V. A. Pashkov, ZhETF Pis. Red. $\underline{9}$, 108 (1969) [JETP Letters $\underline{9}$, 61 (1969)].
30. M. I. Lipatov, A. A. Manenkov, and A. M. Prokhorov, ZhETF Pis. Red. $\underline{11}$, 444 (1970) [JETP Letters $\underline{11}$, 300 (1970)].
31. N. N. Ilichev, V. V. Korobkin, V. A. Korshunov, A. A. Malutin, T. G. Okroashvili, and P. P. Pashinin, ZhETF Pis Red. $\underline{15}$, (1972) [JETP Letters $\underline{15}$, 133 (1972)].

RECENT RESULTS ON SELF-FOCUSING AND TRAPPING

Fujio Shimizu* and Eric Courtens
IBM Zurich Research Laboratory
8803 Rüschlikon, Switzerland

ABSTRACT

Both theoretical and experimental results on the self-focusing and trapping of optical beams in nonlinear media are presented. For relatively long single-mode pulses the steady-state theory describes well the experimental results. For short pulse excitation, it is necessary to account for the relaxation time of the medium. The theory then seems adequate for the description of observed filaments, without having to resort to saturation of the nonlinearity.

I. INTRODUCTION

This paper considers the propagation of intense optical beams in media exhibiting a nonlinear dielectric response. Since the invention of giant pulse lasers the field has had a lively history. As early as 1962, Askar'yan [1] mentioned the possibility that electromagnetic radiation could produce its own waveguide in nonlinear media. Detailed calculations of self-trapped filaments were given by Chiao, Garmire, and Townes [2], and by Talanov [3] in 1964. Soon after, several authors reported observations which were

---

*Present address: Department of Applied Physics, University of Tokyo

ascribed to filament formation [4]. Propagating intense ruby laser beams through Kerr liquids, they could see both narrow streaks of light upon side observation of the medium, and minute spots at the exit window.

In most experiments being conducted with large diameter beams, a focusing action must precede the occurrence of trapping. Kelley [5], and independently Keldysh [6], calculated the lenslike, or self-focusing, effect of these media. However it was found that in the presence of a cubic nonlinearity only, and at powers above the critical focusing power, the beam came to a catastrophic collapse at focus. To prevent this unphysical situation, saturation of the nonlinearity was introduced in the formalism, but satisfactory agreement with current trapping experiments could not be reached [7]. The power required to saturate the nonlinearity was an order of magnitude higher than experimental trapping powers, and moreover the calculations did not lead to filaments, but to solutions which diverged past the focal region. This suggested that another explanation could be given for the experimental findings.

Since experiments had been conducted with fairly short laser pulses and tracks had not been time resolved, it was thought that they could be due to moving foci [8]. Experimentalists started looking for this behavior and, under carefully controlled conditions, in particular using single-mode lasers, the evidence for moving foci was found. The sharp spot observed at the exit window could also be seen far inside the cell [9]. Side pictures taken with streak cameras exhibited focal-spot motion [10], and time-resolved photodetection at two points in the cell gave similar results [11].

In spite of this, many questions still remain. Considerable evidence for filaments has been found through other observations performed under different experimental conditions. For instance, the sharp angular distribution of anti-Stokes rings suggests that these originate from filamentary structures [12], and the emission angle of these rings does not correspond to phase matching with bulk refractive indices [13]. Self-phase modulation observations [14] are in agreement with other trapping data, in particular with the measured power and size of filaments. In view of the controversy, some experiments were performed with the purpose of showing unambiguously the existence of filaments. One of these used fine metal meshes or glass plates immersed at various places along the cell [15]. It gave quite conclusive evidence for filaments. Also, using picosecond pulses, induced birefringence tracks were shown to fly through the cell with the speed of light and over distances of several centimeters [16]. Thus there exists a large body of observations which can hardly be described by the moving foci theory.

Therefore the following questions should be answered. Can the self-focusing of a beam lead to self-trapping? In particular, what is the exact behavior of the focusing itself? And what is the mechanism which could stabilize filaments to the observed size? We shall attempt to answer some of these questions here.

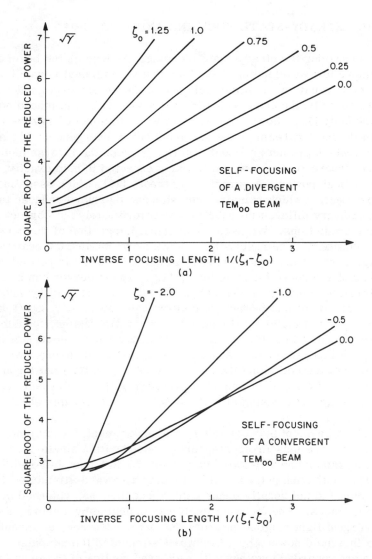

Figure 1. The reduced distance $\zeta - \zeta_0$ equals $(Z-Z_0)/np$, where Z is the axial distance measured from beam waist, and p is the confocal parameter, n being the refractive index of the medium; $\zeta_0 = Z_0/p$ is the entrance face position. This was selected as the point where the peak intensity was equal to 400 times the peak input intensity. The reduced power is $\gamma = 2P/P_0$, where P_0 is given in Ref. 19.

II.   STEADY-STATE FOCUSING AND TRAPPING

Let us consider the steady-state focusing. As long as the beam size is larger than a few wavelengths it is clear that the simplest possible parabolic
diffraction equation, with the inclusion of the lowest-order index nonlinearity,
should describe the phenomenon rather well. This was indeed confirmed experimentally [17]. However it is worthwhile finding out how the focusing
length depends on beam power and input beam divergence. The approximate
theory, which predicts a linear relation between the inverse focusing length
and the square root of power [5], is not expected to hold exactly, especially
near critical power. A paraxial ray theory [18], which essentially assumes
a cross section which remains Gaussian and therefore reduces the problem
to an ordinary differential equation, is unfortunately not correct in its quantitative predictions. We used a numerical integration of the parabolic differential equation. The results for a lowest-order Gaussian beam are shown
in Fig. 1 where the square root of a reduced power γ is plotted versus the
inverse of a reduced focusing length [19]. These curves can be fitted to an
empirical formula which assimilates them to branches of hyperbolas. For
sufficiently divergent beams the approximate law $\sqrt{\gamma} - \sqrt{\gamma_0} \approx 1/L$ holds well,
though γ_0 is not the critical power. For parallel beams, focusing at critical
power occurs at infinity, and for convergent beams it occurs at finite distance. With a given cell measuring the threshold focusing power for various
positions of the cell along the beam, we take a vertical cut through this set
of curves. If the cell is long enough, there will be a flat bottom to this cut,
corresponding to focusing at critical power and at lengths shorter than the
cell length.

The agreement between theory and experiment is excellent, as shown
in Fig. 2. The measurements were obtained with a single-mode ruby laser
emitting rather short Q-switched pulses of 7 to 8 nsec full-width at half-intensity. With such pulses, the disturbing backward-stimulated Brillouin
scattering did not usually occur before focusing, as evidenced by our oscilloscope traces. The focusing itself was indicated by a short pulse of strongly divergent light originating from the exit window and, at essentially the
same threshold power, by a backward-stimulated Raman pulse. These signals were recorded together with a delayed replica of the input laser pulse,
and from the known delays it could be checked that they occurred simultaneously with the peak of the input pulse. For long cells (longer than 2 np
where p is the confocal parameter in free space) [Fig. 2(b)], the curve has
a flat bottom corresponding to critical power. In this flat bottom, only a
backward Raman pulse can be seen. This not only indicates that indeed no
moving focus did cross the exit window, as predicted by theory, but also
that no filament originating from the focus region was able to reach the window. It is worthwhile investigating in more detail what the numerical integration predicts for focusing at critical power.

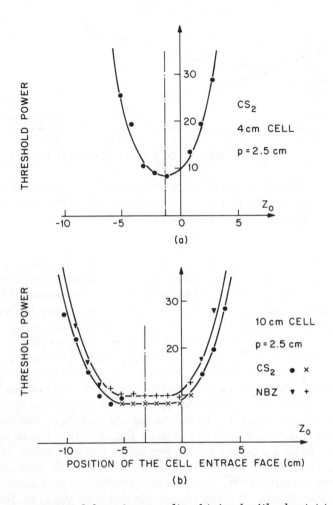

Figure 2.   Experimental focusing results obtained with short (a) or long (b) cells. The full points are threshold powers for the observation of forward divergent light. The crosses correspond to backward Raman scattering only, with no forward divergent light.

Figure 3 shows the evolution of the beam radius as a function of distance for various powers around critical. The input face is at $\zeta_0 = -0.5$, and the beam radius is defined as the $1/e$ radius of the field amplitude (the cross section does not remain Gaussian). The critical power γ_{cr} is between 7.48 and 7.49. One sees that, as the power approaches critical power from below, the minimum radius ρ_m decreases and the beam takes a filamentary aspect.

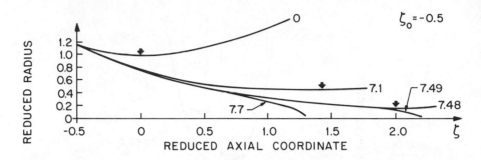

Figure 3.   Evolution of the beam radius for a convergent beam ($\zeta_0 = -0.5$) at powers near critical power. The parameter of the curves is γ. The linear propagation ($\gamma = 0$) is shown for comparison.

At critical power, catastrophic focusing occurs, and for practical purposes we can say that the collapse takes place at a finite distance given by $\zeta_f = -1/\zeta_0$. To get an idea of the degree of filamentation of the beam, we can form the ratio $(\zeta_2 - \zeta_1)/2\rho_m^2$, where ζ_2 and ζ_1 are the abscissa where the radius has increased to $\sqrt{2}$ times its minimum value ρ_m. This ratio, which is unity for linear propagation, is greater than 20 for $\gamma = 7.48$. The field distribution at the waist is, to better than 1% accuracy, that of the Chiao, Garmire, and Townes [2] filament. This shows that these quasi-steady-state filaments could, in principle, be formed by the focusing of convergent beams. As ζ_0 is made less negative, the filament region is rejected towards infinity according to the relation $\zeta_f = -1/\zeta_0$. The detailed region of the filament for $\zeta_0 = -0.5$ is shown in Fig. 4. This figure shows the equiphase surfaces for power values just below and just above catastrophic focusing. In order to make their curvature visible, the equiphase surfaces have been magnified fifty times in the direction of propagation. The $1/e$ radius is also shown. We can appreciate that with $\gamma = 7.48$, the surfaces are practically plane. From this figure, and from Fig. 3, we also see that the power range over which a filament can be formed is rather narrow, at most a few percent. This should not have prevented us from seeing these filaments in the experiments, since our laser was rather stable in its power output. Most probably steady-state filaments are rapidly destroyed by stimulated scattering.

III.   TRANSIENT THEORY

Let us now try to answer the second question. What stabilizes filaments

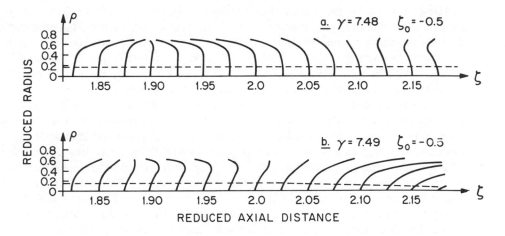

Figure 4.   Equiphase surfaces in the filament region. Note that the scale of the surfaces is enlarged about 50 times in the ζ direction, so that the optical rays cannot be drawn perpendicular to the indicated surfaces.

excited by multimode or mode-locked lasers? Or, in other words, is there a mechanism other than saturation which is able to prevent beam collapse under short pulse excitation? It will be shown that the relaxation time τ associated to the index nonlinearity has a major stabilizing effect. The idea was already proposed by several authors [20]. Fleck and Kelley performed numerical calculations but failed to obtain the expected behavior. The qualitative idea is the following. The index at some point along the pulse is not determined by the instantaneous intensity but by the integrated intensity of the preceding section of pulse, the integral being taken over a length of the order of τ. The leading edge of the pulse keeps expanding by diffraction. As the rest of the pulse contracts, it assumes a horn shape with decreasing radius toward the tail. This variation of radius along the pulse effectively diminishes the focusing effect and eventually the diffraction force and the focusing action balance along this horn.

Quantitatively, the phenomenon is described by the following two equations:

$$4i\left(\frac{\partial \mathcal{E}}{\partial \zeta} - \frac{1}{v}\frac{\partial \mathcal{E}}{\partial t}\right) + \frac{1}{\rho}\frac{\partial}{\partial \rho}\left(\rho \frac{\partial \mathcal{E}}{\partial \rho}\right) + \chi \mathcal{E} = 0 \tag{1a}$$

$$\frac{\partial \chi}{\partial t} + \chi = \gamma |\mathcal{E}|^{2} \tag{1b}$$

where t is normalized to the relaxation time.

We immediately note that the inclusion of relaxation does not change the scaling property of the diffraction equation. Any solution can still be scaled by the beam parameter. Before describing numerical results, let us mention that there exists a steady-state solution of these equations. The form of the equations is such that there exists a solution whose cross section is constant in time and whose shape is preserved all along the pulse, but whose radius increases exponentially with distance in the propagation direction.

The cross section has finite area for powers above the steady-state critical power, and the exponential rate of change of radius is an increasing function of power. For a long pulse (i.e., one for which the intensity changes slowly), this solution can be used to estimate the radius change for which diffraction and focusing balance. For shorter pulses this estimated radius change is an upper bound to the actual change. This intuitively proves the existence of a limiting radius, but does not necessarily imply that the beam becomes trapped. It is indeed known that in the case of instantaneous saturation the beam actually shoots back almost to the original radius once the focal point is past [7]. It will however become apparent from our numerical integration results that the introduction of a relaxation time produces the required irreversibility to prevent the shooting back. The beam actually tends to stabilize to its minimum radius, and its behavior thereafter can be estimated from the scaling properties of the equations. As the front edge expands by diffraction, the radius of every part of the pulse expands proportionally. Since the front edge is much bigger than the minimum radius, the contracted part of the pulse travels without changing its radius over distances which are much longer than the corresponding diffraction distance. In this sense one may say that the beam is trapped. We thus see that the general behavior of the trapped solution can be understood by a mere inspection of Eqs. (1). Let us discuss the numerical results.

The equations were integrated by a finite difference scheme for various Gaussian input pulses of the form $\mathcal{E} = \exp[-(t/T_0) - \rho^2]$. Three different pulse lengths were tried: $\sqrt{2}T_0 = 0.5$, 3, and 20. The input power was selected so that the minimum radius did not become too small for numerical integration. For the largest input power used, the minimum radius was approximately 1/20th of the input radius. This ratio, which is limited by computational feasibility, is also a realistic one. For a filament size of $5\,\mu$ it corresponds to an input radius of 100μ. For this case, the expected propagation length of the filament, which is of the order of the diffraction length of the input beam, is about 27 cm for a ruby laser pulse in CS_2.

We show in Figs. 5-7 the results for a 3τ pulse with peak input $\gamma = 40.5$, which is 5.4 times the steady-state critical power. Figure 5 shows the evolution of the power density at beam center. Figure 6 shows the 1/e power radius of the same pulse. The curves are drawn for equal intervals of $\zeta = 0.133n$ with $n = 1, 2, \ldots, 8$ from top to bottom. The base line has been shifted vertically for each curve in order to improve the visibility. It is

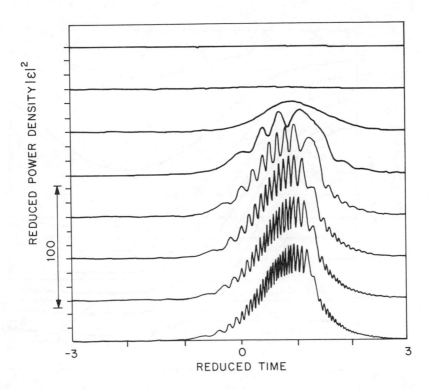

Figure 5.   Evolutions of the power density at the center of a 3τ pulse. Curves are from top to bottom at points $\zeta = 0.133n$, $n = 1, 2, \ldots 8$, respectively. Vertical scales are normalized to the input values. Zero in the horizontal scale corresponds to the center of the input pulse.

clearly seen that both the radius and the power density stabilize, except for some oscillations. Although its radius is not strictly constant, the latter part of the pulse, extending from 0 to 2, can be interpreted as a filament. Over this section the radius reaches its minimum value almost simultaneously around $\zeta = 0.4$, and stays essentially constant up to $\zeta = 1.33$, where the integration was terminated. This remarkable stabilization is due to the relaxation time whose averaging effect prevents rapid temporal change over such a length of pulse.

In addition to a gradual radius change in the filament, we find a faster oscillation which is coupled to the longitudinal phase modulation. The period of this oscillation is correlated to each 2π change in phase, and the pulse gradually develops a complicated fine structure. The oscillation amplitude remains approximately constant. In the early stage of filament formation

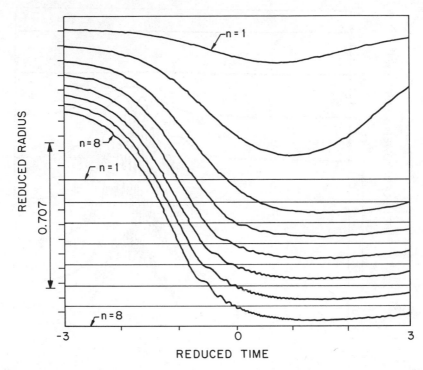

Figure 6.   Evolution of the beam radius for the pulse of Fig. 5.

this oscillation is accompanied by a small energy loss towards the wing, particularly in the tail where it amounts to 20% in the first cycle. However this loss decreases rapidly and even at $\zeta = 1.33$ there is no evidence for filament destruction or deformation by energy losses.

In spite of the rather complicated filament shape, the frequency spectrum is quite similar to the result of plane-wave calculations, and reproduces well experimental observations in CS_2 [14]. Figure 7 shows the spectra at various radial points, observed at position $\zeta = 0.6$. The spectra are shown at equal intervals from center to wing, the top curve corresponding to the center of the pulse.

The results on 0.5τ pulses are essentially the same. A stable filament forms which is about 0.15 long. The spectrum is more asymmetric. For 20τ pulses the trapping is less stable. Results obtained with input power $\gamma = 12$ show that a small portion of the pulse tail diverges soon after the focal point. A filamentary structure, whose length is of the order of 4, is still seen in the preceding part of the pulse. It propagates rather stably.

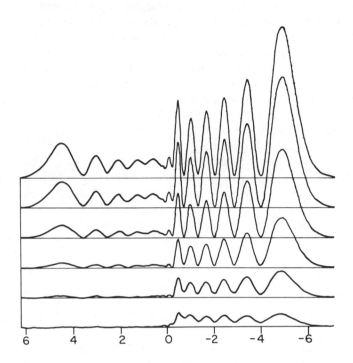

Figure 7.   Frequency spectrum (in reduced units) at various radial positions of the pulse. The curves are drawn at equal intervals in $\rho = 0.0125$. The top curve corresponds to the center. The observation point is $\zeta = 0.6$. The input laser frequency is indicated by zero.

The theoretical analysis of Eqs. (1) predicts that the limiting radius rapidly decreases with increasing input power. The actual numerical integration shows that a 30% change in input power changes the filament radius by a factor of 2 for the 3τ pulse shown previously. However experiments always report filament radii which are rather constant and of the order of a few microns. We believe this can be reconciled with our calculation without even introducing saturation. The most plausible explanation is that the pulse is destroyed by other nonlinear effects once its radius becomes smaller than a certain size which would then be the approximately observed radius. In the preceding part of the pulse there is always a portion which has a limiting radius of this size. This part can be considered as a filament and will remain trapped as long as the pulse is relatively short. The following part does not play an essential role, although some filament characteristics may be affected by the nature of the disturbance. Two pieces of evidence support this interpretation. In most cases observed filaments are accompanied by strong stimulated Raman scattering, and a clean spectrum, as shown in Fig. 7, is dominant only when the laser is operated at a power very close to the threshold of filament observation.

It should be noted once more that the leading edge of the pulse is essential to stable filament propagation. The leading edge governs the filament radius, and the filament keeps its size only because the leading edge expands slowly in view of its larger radius. The entire pulse, from the leading edge to the filament, has to be treated as a single object, though the experiment will only pick up the filament.

In the above discussion we have seen how successful the simple nonlinear diffraction equation is. So far the results are in good agreement with observations, both for steady-state and short pulse focusing, provided the relaxation time is properly accounted for in the latter case. Looking toward the future, we still see the possibility for much experimental and theoretical development in this field.

REFERENCES

1. G. A. Askar'yan, Zh. Eksp. Teor. Fiz. 42, 1567 (1962) [Sov. Phys. JETP 15, 1088 (1962)].
2. R. Y. Chiao, E. Garmire, and C. H. Townes, Phys. Rev. Letters 13, 479 (1964); Phys. Rev. Letters 14, 1056 (1956).
3. V. I. Talanov, Izv. Vuzov, Radiofizika 7, 564 (1964).
4. P. Lallemand and N. Bloembergen, Phys. Rev. Letters 15, 1010 (1965); Y. R. Shen and Y. Shaham, Phys. Rev. Letters 15, 1008 (1965); N. F. Pilipetskii and A. R. Rustamov, ZhETF Pis. Red. 2, 88 (1965) [JETP Letters 2, 55 (1965)]; E. Garmire, R. Chiao, and C. Townes, Phys. Rev. Letters 16, 347 (1966); G. Hauchecorne and G. Mayer, Compt. Rend. 261, 4014 (1965).
5. P. L. Kelley, Phys. Rev. Letters 15, 1005 (1965).
6. L. V. Keldysh, Paper at Session of Division of General and Applied Physics, USSR Academy of Sciences, 1964.
7. T. K. Gustafson, P. L. Kelley, R. Y. Chiao, and R. G. Brewer, Appl. Phys. Letters 12, 165 (1968); W. G. Wagner, H. A. Haus, and J. H. Marburger, Phys. Rev. 175, 256 (1968); E. L. Dawes and J. H. Marburger, Phys. Rev. 179, 862 (1969).
8. A. L. Dyshko, V. N. Lugovoi, and A. M. Prokhorov, ZhETF Pis. Red. 6, 655 (1967) [JETP Letters 6, 146 (1967)].
9. M. M. T. Loy and Y. R. Shen, Phys. Rev. Letters 22, 994 (1969).
10. V. V. Korobkin, A. M. Prokhorov, R. V. Serov, and M. Ya. Shchelev, ZhETF Pis. Red. 11, 153 (1970) [JETP Letters 11, 94 (1970)].

11. M. M. T. Loy and Y. R. Shen, Phys. Rev. Letters 25, 1333 (1970).
12. E. Garmire, Phys. Letters 17, 251 (1965).
13. K. Shimoda, Japan. J. Appl. Phys. 5, 86 (1966).
14. F. Shimizu, Phys. Rev. Letters 19, 1097 (1967); T. K. Gustafson, J. P. Taran, H. A. Haus, J. R. Lifsitz, and P. L. Kelley, Phys. Rev. 177, 306 (1969).
15. M. M. Denariez-Roberge and J. P. Taran, Appl. Phys. Letters 14, 205 (1969).
16. J. Reintjes and R. L. Carman (private communication).
17. M. Maier, G. Wendl and W. Kaiser, Phys. Rev. Letters 24, 352 (1970).
18. W. G. Wagner, H. A. Haus, and J. H. Marburger, Phys. Rev. 175, 256 (1968).
19. E. Courtens, Phys. Letters 33A, 423 (1970).
20. J. A. Fleck, Jr. and P. L. Kelley, Appl. Phys. Letters 15, 313 (1969); V. A. Aleshkevich, S. A. Akhmanov, A. P. Sukhorukov, and A. M. Khachatryan ZhETF Pis. Red. 13, 55 (1971). [JETP Letters 13, 36 (1971)]; M. M. T. Loy and Y. R. Shen, Phys. Rev. A 3, 2099 (1971).

SHORT PULSE NONLINEARITIES

W. Kaiser
Physik - Department der Technischen Universität München
8 München 2, Arcisstr. 21, Germany

ABSTRACT

Several measuring techniques for picosecond pulses and the state of the art of their generation is reviewed. Single bandwidth-limited pulses of several picoseconds duration are now readily available in the laboratory. Recent investigations with these pulses are discussed. The relaxation times of molecular and lattice vibrations are measured directly using probe pulses of known pulse shape.

It has been pointed out in the earlier part of this book that the quality and the understanding of picosecond pulses is in a somewhat unsatisfactory state at the present time. Although the author of this article is well aware of the many difficulties, he wishes to show that under well-controlled experimental conditions single picosecond pulses of reproducible duration and peak power have been generated and studied in detail. The first part of this paper is of an applied nature to introduce the reader of other fields to short light pulses. It reviews briefly various experimental techniques for testing picosecond pulses and it gives an account of the present state of the art in our laboratories. The second part reports on some recent investigations with picosecond light pulses: molecular and lattice relaxation times are directly observed in liquids and solids, respectively.

81

An important tool for monitoring a train of picosecond pulses is a fast photodiode in conjunction with a fast oscilloscope. A time resolution of 0.3 to 0.5 nsec is readily available with commercial equipment. A detection system of this sort gives some valuable information:

1.   The energy of the pulse is obtained from a calibrated detector if the pulse train consists of single pulses (see below) separated by the pulse round-trip time T;

2.   Multiple pulses within the time T may be detected;

3.   The investigation of the background between the intense pulses is possible if one pulse is isolated from the pulse train with the help of a fast switch (see below).

The oscilloscope trace gives no information of the pulse structure on a time scale below 300 psec. In this latter case the two-photon fluorescence technique (TPF) is most valuable [1]. When two picosecond light pulses-- coming from opposite directions --traverse an appropriate liquid (transparent to the incident frequency and absorbent at twice the frequency) an increase in fluorescence is observed at the place where the two pulses meet compared to the range where the two pulses travel individually. The width of the brighter area is a measure of the duration of the light pulse. If a track of approximately 10 cm is photographed, the existence of multiple pulses within a time interval of 1/2 nsec is readily tested, complementing the oscilloscope data. Additional information on the picosecond pulses are obtained when the two-photon fluorescence is observed quantitatively, i.e., with a calibrated photographic plate in conjunction with a camera of high resolution. In Fig. 1 the intensity profile of the TPF track is shown schematically for two special cases. In Fig. 1(a) a smooth maximum with peak intensity 3 is superimposed on a broad background of intensity unity. This contrast ratio of 3:1 is predicted for perfect mode locking [2]. In fact, microdensitometer traces similar to Fig. 1(a) [3] suggest: (a) the major part of the energy is contained in the pulse and not in the background; (b) the pulses have no subpicosecond structure; and (c) the pulse duration t_p is obtained from the width of the fluorescence maximum. Quite different is the situation when the TPF traces are observed of the type depicted in Fig. 1(b) (narrow spike superimposed on broad maximum) [4]. In this case the picosecond pulse is of complicated substructure and should be disregarded in many applications.

The frequency spectrum of picosecond pulses gives additional information on the mode-locked pulses. Calibrated plates should be used to obtain the true intensity spectrum. If the frequency width, $\Delta\nu$, times the pulse duration, t_p, is close to unity, then the bandwidth-limited pulses are free of subpicosecond substructure and frequency chirp. Bandwidth-limited pulses have been found experimentally [3] and are used in the experiments discussed

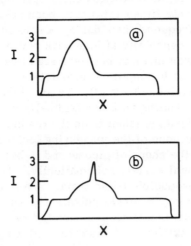

Figure 1.   Two photon fluorescence pattern, intensity versus distance (schematic).   For a discussion see text.

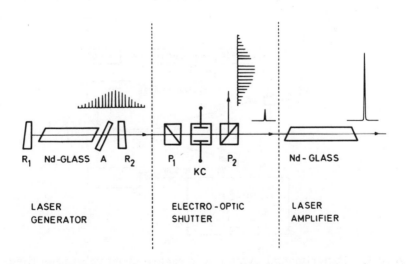

Figure 2.   Experimental system to generate intense single–picosecond light pulses (schematic).

84   W. Kaiser

below. In most existing mode-locked glass systems, however, large frequency chirps [5] (twenty times $1/t_p$) have been noticed. According to our
results, substantial frequency broadening occurs in the middle and at the end
of the pulse train and especially at high peak power of the pulses, where the
nonlinearity of the laser medium becomes significant [6,7]. Transformation
of frequency modulation to amplitude modulation may result in the subpicosecond structure mentioned above [8]. To avoid these difficulties and to
ascertain that we are dealing with bandwidth-limited pulses, we worked with
individual picosecond pulses taken from the leading part of the pulse train [7].
It should be noted that none of the measuring techniques discussed so far
gives information on the shape of picosecond light pulses. Figure 2 shows
part of our experimental system schematically. The passively switched laser
(left) generates a reproducible pulse train. A following electro-optic shutter
allows the transmission of a single pulse. The shutter is operated in such a
way that pulses of equal power values pass the system in each operation [9].
The single pulse is amplified in a laser amplifier by a factor of approximately
20. We have made a detailed study of our single picosecond pulses [3]. In
brief, the experimental data are as follows: Pulse duration at half-maximum
intensity t_p = 8 psec, pulse energy (before amplification) 10^{-4} J, frequency
width $\Delta\nu = 10^{11}$ Hz giving $\Delta\nu t_p \simeq 1$. A contrast ratio of 3.0 ± 0.2 of the two-
photon fluorescence technique and quantitative measurements of a three-pho-
ton fluorescence system indicate that we are working with a single picosecond
pulse with a background power more than three orders of magnitude below
the peak intensity of our pulses [7].

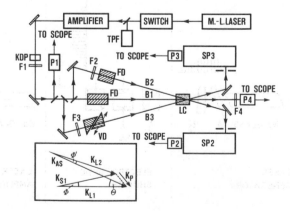

Figure 3.   Experimental system to measure short relaxation times. Three
light beams, B1, B2, and B3 (a pump and two probe beams) traverse the
same region of the sample LC. Inset: Schematic of phase-matched probe
scattering at the anti-Stokes frequency.

Having picosecond pulses of high quality at our disposal we felt confident that quantitative measurements of short relaxation times were now possible. In the following part of this article investigations of the molecular relaxation time in liquids and of the decay time of TO lattice phonons in diamond will be discussed.

Our total experimental system [10] is depicted schematically in Fig. 3. The mode-locked glass laser is followed by a switch (to cut a single pulse out of the pulse train), a two-photon fluorescence system, an amplifier rod, and a KDP crystal for upconversion to $0.53~\mu$.

The main (pumping) pulse traverses the medium, a liquid cell (LC), or a solid crystal generating stimulated Raman scattering. The conversion efficiency of laser into Stokes light is monitored by the two photodiodes P1 and P4. Two beam splitters preceding the Raman medium provide two additional light beams of small intensity (10^{-3} of the incident pulse). The optical path of B3 contains a variable delay (VD) of high precision, while the light beam B2 with fixed delay (FD) serves as a reference beam. The light scattered coherently by the Raman vibrations is detected with the help of spectrometers and photmultipliers (P2 and P3). In the lower part of Fig. 3, the phase-matching diagrams of the scattering process is shown schematically. The laser and the Stokes wave, with wave vectors K_{L1} and K_{S1}, respectively, generate the molecular or lattice vibrations with wave vector K_p. The probing beam with wave vector K_{L2} is scattered by the material excitation K_p to give an anti-Stokes wave K_{AS}. In our experiments, the angles φ and θ are very small (several degrees) and $\varphi' = 0$.

Before discussing our experimental results several theoretical calculations should be presented. We have made detailed computer calculations of the transient stimulated Raman process appropriate to our experimental situation [10]. In Fig. 4 the calculated molecular amplitude Q is plotted as a function of time (normalized to the pulse duration t_p). Zero time corresponds to the peak of the pump pulse. The solid lines are calculated for pump pulses of Gaussian time dependence. The important parameter in Fig. 4 is t_p/τ, the ratio of the pulse duration t_p to the relaxation time τ of the material excitation. In liquid (two-level) systems the dephasing time τ corresponds to T_2 in the Bloch equation. In solids where fundamental TO lattice vibrations are excited, the value of τ corresponds to the energy relaxation of the lattice modes. It is readily apparent from Fig. 4 that the value of t_p/τ drastically influences the time dependence of the material excitation. For large values of τ (small values of t_p/τ) the material excitation rises rapidly, reaches a maximum after the peak of the pumping pulse, and decays exponentially with a time constant equal to τ. Probing the exponential decay of Q provides a direct measure of the value τ (see below). For small relaxation times τ (e.g., $t_p/\tau = 32$) the material excitation rises and decays very rapidly. In fact, the Q pulse is substantially narrower than the electromag-

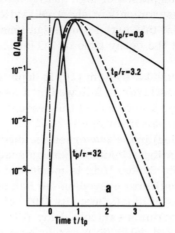

Figure 4.   Calculated vibrational amplitude versus time. t_p is the duration of the exciting and probe pulse; τ is the relaxation time.   Solid curves are calculated for Gaussian-shaped light pulses.

netic pump pulse; it will be used to obtain information on the shape of the probe pulse (see below).

Experimentally we observe the signal scattered by the material excitations.   This signal is calculated by convolution of the material excitation with

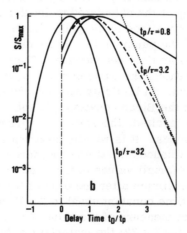

Figure 5.   Calculated anti-Stokes scattering signal versus delay time of the probe pulse.

the probe pulse. In Fig. 5, calculated anti-Stokes scattering signals are presented as a function of delay time of the probe pulse. The solid curves are calculated for a Gaussian shape of the probe and pumping pulse. Two limiting cases should be discussed here. For long relaxation times (e. g., $t_p/\tau = 0.8$) the anti-Stokes signal reaches a maximum at a delay time $t_D \simeq t_p$ and decays exponentially with the characteristic relaxation time τ of the medium. Measurements of this type, providing a direct observation of τ, will be presented below. For short relaxation times (e. g., $t_p/\tau = 32$) the scattered signal shows a parabolic time dependence on the semi-log scale of Fig. 5, i.e., gives a Gaussian time dependence. This finding indicates that scattering experiments using media with small values of τ allow a determination of the shape of the electromagnetic probe pulse. It should be emphasized that the rapidly rising and decaying material excitation is well suited to analyze the low intensity wings of the electromagnetic probe pulse [11].

Experimental results [10] obtained on two liquids with widely differing values of τ are presented in Fig. 6. The scattered anti-Stokes signal is plotted as a function of delay time t_D for liquid carbon tetrachloride and ethyl alcohol. First, the relatively slow relaxation of the centrosymmetric CCl_4 vibration ($\omega_p/c = 459$ cm^{-1}) should be discussed. As expected from our calculations (see Fig. 5) the anti-Stokes signal rises to a maximum value at $t_D \simeq t_p \simeq 8$ psec and decays exponentially. The value of the relaxation time τ obtained in this way comes out to be $\tau = 4.0 \pm 0.5$ psec. This number

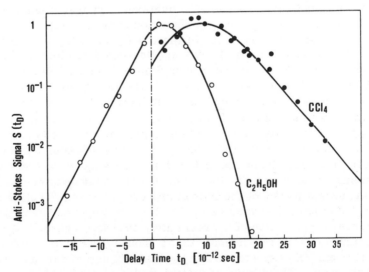

Figure 6.   Measured anti-Stokes scattering signal versus delay time for ethyl alcohol and carbon tetrachloride.

must be compared with spontaneous Raman data of the same molecular vibration. While earlier measurements of the Raman linewidth suggested $\Delta\omega/c \simeq 10$ cm^{-1} ($\tau \simeq 0.5$ psec), more recent investigations revealed a distinct isotope structure with a linewidth of the isotope components of $\Delta\omega/c \simeq 1.5$ cm^{-1} ($\tau \simeq 3.6$ psec). Obviously, our measured relaxation time τ corresponds to the individual isotope component of largest abundance. We turn now to the discussion of the scattering data obtained from C_2H_5OH. Spontaneous Raman data of the 2928-cm^{-1} vibration (investigated here) indicate a linewidth $\Delta\omega/c \simeq 20$ cm^{-1} suggesting a relaxation time of $\tau \simeq 0.26$ psec. The parameter t_p/τ has now the value 32. According to our theoretical discussions (see Fig. 5), the scattered signal gives us information on the shape of the probe pulse. In fact, the experimental curve shown in Fig. 6 is a direct picture of the probe pulse. The pulse rises like a Gaussian pulse and decays exponentially. Note that for a positive delay the beginning of the probe pulse is measured. The asymmetry of the pulse reported here has been expected because of the finite relaxation time of the switching dye in our mode-locked laser system. It should be emphasized that we are able to measure the shape of the probe pulse three orders of ten below the maximum. Measurements of the shape of picosecond pulses (total pulse train) have been made using compressed [12] pulses and a fifth-order nonlinear interaction process [13].

Stimulated Raman scattering provides strong material excitations in a narrow frequency interval and in a small angular range. In solids where the fundamental TO lattice modes can be excited, a small region in k-space is involved in the Raman process. Calculations show that a small number of modes are excited to very large occupation numbers [14]. It was thought of interest to investigate the decay of these hot phonons by the technique outlined above and compare the results with spontaneous data close to thermal equilibrium. In Fig. 7 the optical phonon density is plotted as a function of the steady-state gain $G_{SS} = gI_Lz$, where g is the gain factor of the medium, I_L the laser intensity, and z the interaction length. In our specific experiment on diamond we know g from the literature and the value of I_L for our pulses. The upper scale gives the interaction length z in mm. The curves of Fig. 7 are calculated according to the theory of transient stimulated Raman scattering [14]. It is seen from Fig. 7 that for increasing relaxation times (i.e., decreasing parameter t_p/τ) the generated phonon density decreases because of the transient nature of the Raman process. For the TO phonon in diamond (at 77 $^\circ$K) we have a parameter $t_p/\tau = 2.4$. In our experiment we worked with a power conversion of approximately 10^{-1} (see broken lines in Fig. 7). The calculations clearly indicate that a phonon density of approximately $N_0 = 10^{17}$ cm^{-3} is expected within the last 1/2 mm of our diamond crystal 7 mm in length. The density of optical modes participating in our experiment is estimated from the equation: $N_m = k^2\Delta k\Delta\Omega/(2\pi)^3$. A value of $N_m = 10^7$

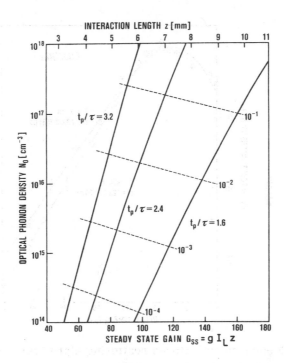

Figure 7.   Calculated density of TO-phonons in diamond versus steady-state gain; dotted lines represent power conversion efficiencies.

modes per cm^3 is calculated using our experimental data $k = 2 \times 10^4$ cm^{-1}, $\Delta k = 20$ cm^{-1} and $\Delta \Omega = 0.2$ sr.   From the numbers obtained so far we are now in the position to calculate the occupation number of optical phonon modes $n_0 = N_0/N_m = 10^{10}$, i.e, 10^{10} quanta are generated per mode by the stimulated scattering process.   This value has to be compared with the thermal occupation number of the same modes of 10^{-3} and 10^{-11} at $300°K$ and $77°K$, respectively.   We see from these estimates that for a duration of several picoseconds, TO modes are excited in diamond to an occupation number exceeding the equilibrium value ($300°K$) by 13 orders of 10.   The decay of these very hot phonons was measured directly using the experimental system discussed in connection with Fig. 3.

Our experimental results [14] of the scattered anti-Stokes intensity versus delay time t_D of the probe pulse are presented in Fig. 8 for crystal temperatures of $295°K$ and $77°K$.   The scattered signal rises to a maximum at $t_D \simeq t_p \simeq 8$ psec on account of the transient excitation process and decays

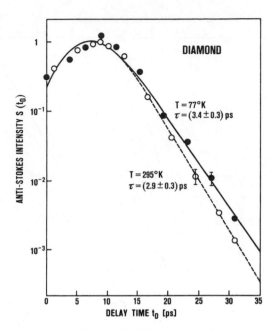

Figure 8.   Measured anti-Stokes scattering signal versus delay time for crystal temperatures of 295°K and 77°K.

exponentially with a characteristic relaxation time of 2.9 ± 0.3 psec and 3.4 ± 0.3 psec for 295°K and 77°K, respectively. The corresponding linewidths are calculated to be 1.83 ± 0.2 cm^{-1} (300°K) and 1.56 ± 0.15 cm^{-1} (77°K). Comparing these numbers with literature data of the spontaneous Raman linewidth of 1.65 to 2.2 cm^{-1} (300°K) and 1.48 cm^{-1} (77°K), we find good agreement between these different experimental investigations. Our data give strong evidence that our hot phonons decay with the same time constant as the phonon close to thermal equilibrium.

To understand this result we compare the vibrational amplitude of our hot phonons with the thermal vibrations of the lattice. In addition we discuss the possible effects of parametric phonon breakdown in our experiments.

For our hot phonons with 10^{10} quanta of $h\nu_p$ per mode and 10^7 modes/cm^3 we calculate a vibrational amplitude of $(\overline{\Delta r^2})^{1/2}$ of 10^{-4} Å. On the other hand zero point fluctuations of $1/2\, h\nu_p$ in the optical branch with 8.8×10^{22} modes/cm^3 give rise to an average vibrational amplitude of 5×10^{-2} Å. Together with the modes of the acoustic branches, the vibrational amplitude is approximately 10^{-1} Å. These numbers indicate that the thermal amplitudes are considerably larger than the amplitudes generated in our experiment. It is expected, therefore, that our hot phonons decay through the same anharmonic terms as the thermally excited TO phonons.

The possibility of phonon breakdown in crystals has been discussed in recent publications. It was suggested that the decay of the optical phonons might be accelerated through a parametric process involving acoustic phonons produced by the decay of the optical phonons. The decay rate of TO phonons of occupation number n_0 can be written in the form

$$dn_0/dt \simeq (n_0/\tau)(1 + n_{-K} + n_{+K}) \tag{1}$$

where n_K and n_{-K} are the nonequilibrium occupation numbers of the acoustic modes interacting with the TO phonons. Equation (1) is appropriate for diamond at $77°K$ where the thermal equilibrium values and reverse reactions are negligible. Using published curves for the density of states of diamond we estimate $N_A = 10^{20}$ acoustic modes per cm^3 at a frequency $\nu_p/2$ and for a frequency interval of 1.5 cm^{-1}. Since 10^{17} TO phonons per cm^3 decay in 2×10^{17} acoustic phonons per cm^3, we estimate n_K, $n_{-K} \simeq 10^{-3}$. The small values of the acoustic occupation numbers suggest that the decay of our TO phonons is not altered significantly by the parametric breakdown process.

In conclusion we wish to say that well-defined picosecond pulses are now available. Various linear and nonlinear experimental techniques exist which allow a continuous control of the quality of our short light pulses. The direct quantitative observation of molecular and lattice relaxation times is an example for the application of picosecond light pulses.

The author gladly acknowledges the important contributions made by A. Laubereau, A. Penzkofer, and D. von der Linde.

ADDENDUM (SEPTEMBER, 1972)

During the past year considerable progress has been made in the analysis and the applications of picosecond light pulses. A brief summary should be given here. The speed of image converter cameras has been pushed to a time resolution of approximately 1 psec [15]. While these cameras will not be available to most laboratories, it is comforting to see directly individual picosecond light pulses with small background intensity. Recently it was shown (in a very simple experiment) that the peak intensity of single picosecond light pulses can be readily determined from measurements of the energy transmitted through a saturable absorber [16]. An extension of this technique [17] allows the determination of the ratio of pulse-to-noise energy.

Two techniques have been reported for the generation of subpicosecond light pulses. The first method uses the whole mode-locked pulse train and generates a Stokes shifted pulse train through the transient stimulated Raman effect [18]. The second system is based on the pulse shortening of saturable absorbers [19]. Using a multiple-absorber-amplifier system, light pulses were shortened from 8 to 0.7 psec in five transits.

It was discussed above that the dephasing time of molecular vibrations in liquids can be determined by measuring the coherent light scattering of a delayed probe pulse. Now, it was demonstrated that a different time constant, the vibrational life time of molecules in liquids, can be obtained from incoherent light scattering of a probe pulse [20]. A normal mode of a molecule is excited above the thermal equilibrium value using the stimulated Raman effect. The anti-Stokes signal of the probe pulse is a direct measure of the (decaying) excess population. Experiments of this sort are of special interest since there is no other experimental technique which provides the vibrational life time of well-defined normal modes of molecules in liquids.

REFERENCES

1. J. A. Giordmaine, P. M. Rentzepis, S. L. Shapiro, and K. W. Wecht, Appl. Phys. Letters 11, 216 (1967).
2. H. P. Weber, Physics Letters 27A, 321 (1968); J. R. Klauder, M. A. Duguay, J. A. Giordmaine, and S. L. Shapiro, Appl. Phys. Letters 13, 174 (1968).
3. D. von der Linde, O. Bernecker, and W. Kaiser, Opt. Commun. 2, 149 (1970).
4. D. J. Bradley, G. M. C. New, and S. J. Caughey, Physics Letters 30A, 78 (1969); S. L. Shapiro and M. A. Duguay, Physics Letters 28A, 698 (1969).
5. E. B. Treacy, Appl. Phys. Letters 17, 14 (1970); D. H. Auston, Opt. Commun. 3, 272 (1971).
6. M. A. Duguay, J. W. Hansen, and S. L. Shapiro, IEEE J. Quantum Electron. QE-6, 725 (1970).
7. D. von der Linde, IEEE J. Quantum Electron., to be published.
8. J. A. Giordmaine, M. A. Duguay, and W. Hansen, IEEE J. Quantum Electron. QE-4, 252 (1968); E. B. Treacy, IEEE J. Quantum Electron. QE-5, 454 (1969); R. A. Fisher, P. L. Kelley, T. K. Gustafson, Appl. Phys. Letters 14, 140 (1969); A. Laubereau und D. von der Linde, Z. Naturforschung 25a, 1626 (1970); O. Svelto, Appl. Phys. Letters 17, 83 (1970).
9. D. von der Linde, O. Bernecker, and A. Laubereau, Opt. Commun. 2, 215 (1970).
10. D. von der Linde, A. Laubereau, and W. Kaiser, Phys. Rev. Letters 26, 954 (1971).
11. D. von der Linde and A. Laubereau, Opt. Commun. 3, 279 (1971).
12. E. B. Treacy, Appl. Phys. Letters 14, 112 (1969).

13. D. H. Auston, Appl. Phys. Letters 18, 249 (1971).
14. A. Laubereau, D. von der Linde, and W. Kaiser, Phys. Rev. Letters 27, 802 (1971).
15. N. G. Basov, M. M. Butslov, P. G. Kriukov, Yu. A. Matveets, E. A. Smirnova, S. D. Fanchenko, S. V. Chekalin, and R. V. Chikin, Conference on Nonlinear Optics, Minsk, June 1972; E. G. Arthurs, D. J. Bradley, B. Liddy, F. O'Neill, A. G. Roddie, W. Sibbett, and W. E. Sleat, 10th International Conference on High Speed Photography, Nice, September 1972.
16. A. Penzkofer, D. von der Linde, and A. Laubereau, Opt. Commun. 4, 377 (1972).
17. R. J. Harrach, T. D. Mac Vicar, G. I. Kachen, and L. L. Steinmetz, Opt. Commun. 5, 175 (1972).
18. M. J. Colles, Appl. Phys. Letters 19, 23 (1971).
19. A. Penzkofer, D. von der Linde, A. Laubereau, and W. Kaiser, Appl. Phys. Letters 20, 351 (1972).
20. A. Laubereau, D. von der Linde, and W. Kaiser, Phys. Rev. Letters 28, 1162 (1972).

ULTRASHORT LASER PULSES AND APPLICATIONS

Michel A. Duguay
Bell Telephone Laboratories
Murray Hill, New Jersey

This discussion will cover a few topics in the field of ultrashort laser pulses and their applications.

Perhaps the first thing that should be said is that there does not exist at the present time a really good generator of picosecond laser pulses. Most experiments in this field have been performed with the help of the famous mode-locked Nd:glass laser. But this laser is still pretty much in the state it was in back in 1966 when it was first developed by DeMaria, Stetser, and Heynau [1], and there is a pressing need to improve it or to find a better laser. What we want is a laser producing the following "textbook"-type pulse: 1-GW peak power, smooth Gaussian-shape 1 psec at half-height, TEM_{00} spatial mode structure, perfect reproducibility from shot to shot. By contrast, a typical mode-locked Nd:glass oscillator generates pulses with the following characteristics: 1/2-GW peak power, complicated noiselike pulse shape about 10 psec at half-height, highly nonuniform spatial power distribution (i.e., many high-order modes), utterly irreproducible from shot to shot.

In collaboration with Hansen and Mattick, we have used the Nd:glass laser at Bell Labs to do some new experiments in ultrahigh-speed photography [2-4]. One of these has been the direct photography of green light pulses in flight through a scattering medium. The setup is shown in Fig. 1. The ultrafast shutter is essentially an electrodeless Kerr cell; birefringence is induced in a CS_2 cell by a powerful infrared laser pulse via the optical Kerr effect. The Kerr shutter time-dependent transmission is

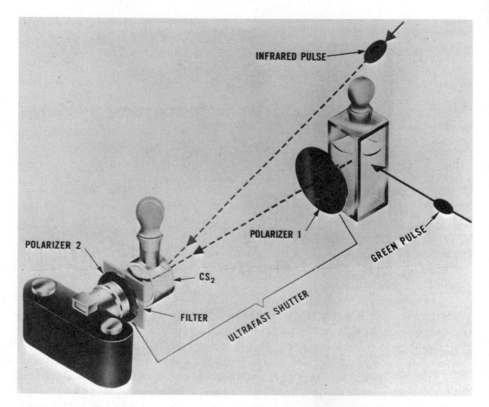

Figure 1.   Setup used to photograph a green laser pulse in flight through a water cell.   A drop of milk was added to the water to enhance light scattering and thus render the pulse brightly visible from the side.   The ultrafast shutter is essentially an electrodeless Kerr cell.   A powerful ultrashort infrared laser pulse is used to induce Kerr birefringence in the liquid, thus opening the shutter for about 10 psec.   The filter greatly attenuates the infrared pulse and prevents it from damaging the camera.   Both pulses are generated simultaneously by a Nd:glass laser.

proportional to the squared time profile of the infrared pulse.   Since our laser produces pulses about 10 psec in duration, our shutter opening time is 7 psec (see Ref. 3 for more details).   The photograph obtained with this set-up is shown in Fig. 2.   There a green pulse has been photographed as it was flying through a water cell.   The green pulse is derived from the infrared one by second harmonic generation.   A drop of milk was added to the water

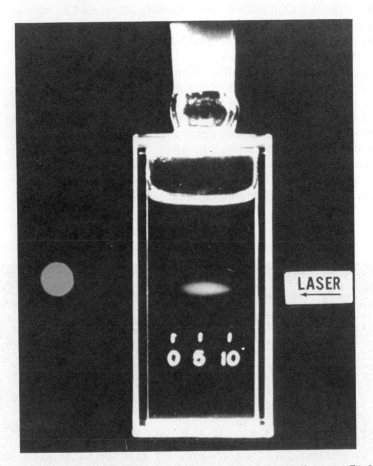

Figure 2. Stop-motion photograph of ultrashort green pulse "in flight" through a water cell. The green pulse was moving from right to left. The scale is in millimeters. The camera shutter opening time was about 10 psec. The round spot is the impression left on the high-speed Ektachrome film by an infrared laser pulse used to activate the ultrafast shutter. A Nd:glass laser generates the infrared pulse, which in turn gives rise to the green pulse by passing through a nonlinear optic crystal (not shown). This photograph is thus simultaneously an illustration of second harmonic generation from the infrared (red spot) to the green.

to increase light scattering and render the green pulse more brightly visible from the side.

The red round spot to the left of the cell in Fig. 2 is the impression made by the infrared pulses (incompletely attenuated by the filter shown) on the high-speed Ektachrome film used. Thus Fig. 2 constitutes in addition a pictorial representation of second harmonic generation from the infrared (red round spot) to the green.

Fig. 2 is only a first step. The picture of the green pulse is blurred due to a linear motion of about 2 mm during the 7-psec opening time of the shutter. From this picture it is not therefore possible to deduce the precise shape and substructure of the green pulse. The green spot represents a type of convolution of the infrared pulse (which drives the shutter) and the green pulse derived from the former by SHG. A time smearing factor of 3 psec also has to be thrown in [3]. As a result one can only deduce the overall duration of the pulse from Fig. 2. It turns out, in fact, that the length of the green spot is approximately equal to the spatial length of the infrared pulse.

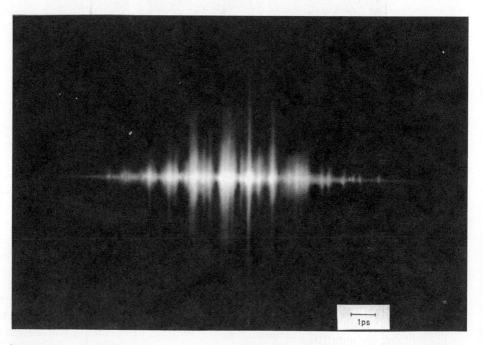

Figure 3.   Artist's view of what a photograph of the green pulse in Fig. 2 might look like if taken with an infinitely fast camera shutter. The precise nature of this substructure remains a mystery.

The infrared and green pulses are known to possess a substructure [5] from studies using two-photon fluorescence [6]. Fig. 3 shows an artist's conception of what the green pulses might look like if photographed by an infinitely fast shutter. The pulses would appear white and might display the noiselike substructure shown, with peaks ≈0.3 psec in duration. If a laser capable of producing 0.2-psec pulses existed, a shutter with an equally short opening time could be built, possibly using glass as the active Kerr medium [5]. In the meantime Fig. 3 remains a dream, and the precise shape of these pulses remains a mystery.

One advantage of this photography as a method of displaying picosecond pulses is the ability to detect small subsidiary pulses that might accompany the main pulse. In Fig. 4, we have photographed the main pulse, as before, but by changing the delay between infrared and green pulses we have also photographed ahead of and behind it. Large satellite pulses are clearly visible there. Their distance from the main pulse is twice the mirror to dye cell distance inside the dye mode-locked laser. Efforts to chase away the large satellites were unsuccessful. Other work has shown, however, that they may be eliminated by placing the mode-locking dye directly in contact with one of the laser mirrors [7,8]. Fig. 5 shows the main pulse and its ever-present large satellites, plus some smaller pulses which come and go erratically from shot to shot. The positions of these small pulses was not related to any laser internal dimension. The overall picture which we

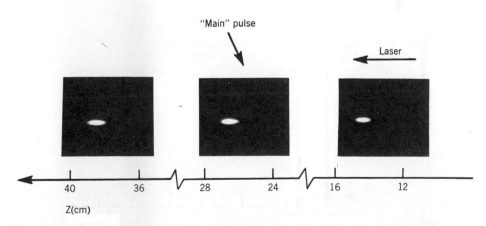

Figure 4. Three pictures were taken at different delays to show the "main pulse" and the satellites at 2d/c = 12 cm on either side. The satellites are due to the spacing d between the mode-locking dye cell and one laser mirror and could not be eliminated by changing various laser parameters. The scale is in equivalent centimeters of air.

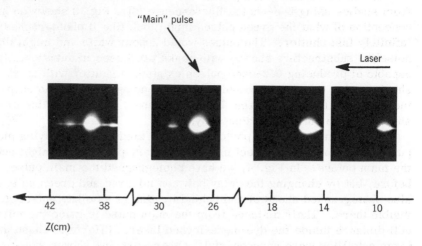

Figure 5. This mosaic was taken in water with exposures brighter than in Fig. 4, but equally short. The main pulse and the dye satellite pulses are grossly overexposed giving an idea of their size relative to the smaller subsidiary pulses which are clearly visible on the original photographs and usually amount to ~1% of the main pulse height. The subsidiary pulse on the rightmost photograph was unusually large (10% main pulse) and appeared only once out of ~100 shots.

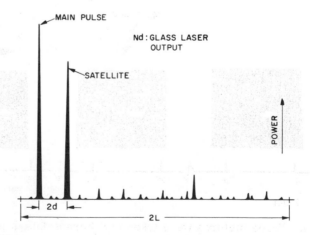

Figure 6. Schematic model of the infrared laser beam generated by a dye mode-locked Nd:glass laser. The larger pulses are known to possess a substructure not shown here.

reconstruct from these photographs is shown in Fig. 6. This represents the
laser output in a given $2L/c$ time interval, where L is the optical length of
the laser cavity. Quantitative measurements of the background on each side
of the main pulse have been made by Auston [8]. His measurement of a
pulse autocorrelation function of fifth order is shown in Fig. 7. The central
peak gives approximately the overall shape of the infrared pulses produced
by the mode-locked Nd:glass laser. In Auston's laser the average background level is $\sim 10^{-4}$, relative to the main pulse, with peak excursions
$\times 10^{-2}$-10^{-3}. In our laser the peak excursions of the small subsidiary pulses
(Fig. 6) are 3-7%.

   Such differences in the quality of mode locking between similar lasers
have been frequently reported. Clearly, some ill-defined parameter is
having an important influence on mode locking. One of these is probably the
lasing spectral width $\Delta\theta$ at the time just prior to dye Q switching in the laser.
According to recent theoretical calculations [9,10], reliable mode locking
can only be achieved if that width is less than $1/\tau$, where τ is the relaxation
time of the mode-locking dye. For the Kodak #9860 dye, $\tau \approx 10$ psec, and
this would require that $\Delta\nu$ should not exceed 3 cm^{-1}. The fact that in
practice $\Delta\nu$ varies between 5 and 30 cm^{-1} could explain the unreliable and
imperfect mode locking depicted in Fig. 6. Forcing $\Delta\nu$ to be $<$3 cm^{-1} might
considerably improve the mode locking [9,10].

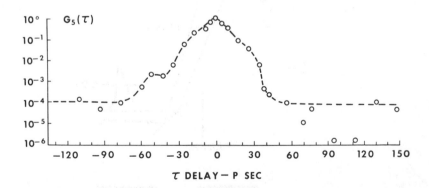

Figure 7. The fifth-order correlation function measured by Auston (Ref. 8)
gives approximately the average intensity versus time profile of the infrared
pulses generated by the dye mode-locked Nd:glass laser. The leading pulse
edge corresponds to $\tau > 0$. The mean pulse width turns out to be ~ 12 psec.
Noise level of the detection system is 10^{-6}, so that the 10^{-4} background seen
is real. Each point is the average of 8 shots. On a given shot the intensity
for a given τ may reach 10^{-3}, on a rare occasion 10^{-2}.

Turning now to applications, let us look at the ranging experiment shown in Fig. 8. Two slides, with the words "front" and "back" engraved on them, are illuminated by 7-psec green pulses. By opening the ultrafast shutter at the appropriate time, the echo from one or the other can be selectively recorded. This suggests a new way of reading a book without turning the pages! A subnanosecond access time computer memory could be built along this principle.

Another interesting possibility is to see through partially transmitting obstacles, as shown in Fig. 9. There, by opening the camera shutter at the right time, one selectively records the echo from the target, effectively "seeing through" the tissue. In many parts of the human body, veins can be seen through the skin, suggesting that the skin has sufficient transmission to allow gate-ranging through it. This could become a valuable medical application. At present, the 10-psec pulses from the Nd:glass laser are too long to allow resolution through millimeter thick skin. For this purpose subpicosecond pulses are needed. The Raman laser built by Colles [11] or some future mode-locked dye laser might answer this need. Ultrafast gating with subpicosecond pulses would give 10 - 100 μm accuracy in ranging, making it interesting for industrial applications.

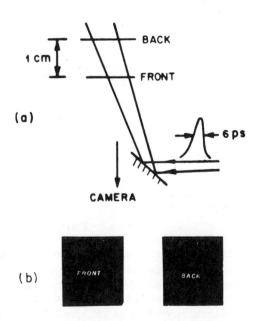

Figure 8. (a) Setup used to picture-range two glass slides on which the words "front" and "back" had been engraved. The camera setup is as in Fig. 1. (b) Each slide is selectively photographed by properly adjusting the delay.

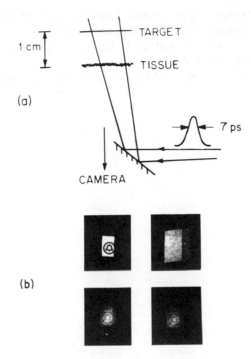

Figure 9.   (a) A piece of thin paper tissue masks a target carrying the drawing of a bell plainly visible below when the tissue is removed.   (b) The upper right photograph is taken in room light illumination; the target is not visible.   By picture-ranging the bell can be seen in the original photographs (bottom) although with greatly reduced contrast and sharpness.

REFERENCES

1.   A. J. DeMaria, D. A. Stetser, and H. Heynau, Appl. Phys. Letters 8, 174 (1966).
2.   M. A. Duguay and J. W. Hansen, IEEE J. Quantum Electron. QE-7, 37 (1971).
3.   M. A. Duguay and A. T. Mattick, Appl. Optics 10, 2162 (1971).
4.   M. A. Duguay, American Scientist 59, 550 (1971).
5.   M. A. Duguay, J. W. Hansen, and S. L. Shapiro, IEEE J. Quantum Electron. QE-6, 725 (1970).
6.   J. A. Giordmaine, P. M. Rentzepis, S. L. Shapiro, and K. W. Wecht, Appl. Phys. Letters 11, 216 (1967).

7.   D. J. Bradley, B. Liddy, and W. E. Sleat, Optics Commun. 2, 391 (1971).
8.   D. H. Auston, Appl. Phys. Letters 18, 249 (1971).
9.   V. S. Letokhov (private communication).
10. B. Ya. Zel'Dovich and T. I. Kuznetsova, Preprint #67, P. N. Lebedev FIAN (1971).
11. M. J. Colles, Appl. Phys. Letters 19, 23 (1971).

PLASMA HEATING AND CONTROLLED FUSION

SOME ASPECTS OF CONTROLLED FUSION BY USE OF LASERS*

Ray E. Kidder
Lawrence Livermore Laboratory
University of California
Livermore, California

ABSTRACT

The application of lasers to controlled fusion is considered, particular attention being given to fusion reactors based on laser-heated DT pellets. The requirements a laser must meet to be useful for this purpose are discussed. A decoupling effect resulting from the fact that long-wavelength light can only penetrate and heat plasma electrons at relatively low density is described, which suggests that light of wavelength as great as $10\,\mu$ may not be suitable for pellet heating.

I. INTRODUCTION

The basic task of controlled fusion is to heat a fusionable plasma to high temperature, and to confine it for a long period of time. Sufficient thermo-nuclear energy will then be released to pay back the investment made in heating the plasma, and provide enough additional energy for the production

---

*Work performed under the auspices of the U.S. Atomic Energy Commission.

of power.   The laser can in principle be applied to either the heating or confinement role.

We shall comment only briefly on the application of lasers to plasma confinement. We will then consider the plasma heating application in somewhat more detail, with a view toward determining the characteristics a laser should have to be useful for this purpose.

II.   PLASMA CONFINEMENT

The application of lasers to plasma confinement has received very little attention to date. It would appear at first glance to be an attractive possibility, because the directionality of a laser beam allows the electromagnetic confining stress to be precisely controlled in space, and to be derived from a source that is relatively remote from the intense radiation source represented by the hot thermonuclear plasma. In addition, the laser beam can be precisely controlled in time, and might serve to heat as well as to confine the plasma.

However, in order that the electromagnetic stress be sufficient to confine the plasma, the energy density of the laser beam must be comparable with the plasma pressure, a situation in which a variety of light beam-plasma instabilities can be strongly excited. The confinement question is therefore likely to be complicated, but deserves further consideration.

III.   PLASMA HEATING

The application of lasers to plasma heating depends on the ability of the hot plasma to absorb laser light. The absorption mechanism is free-free inverse bremsstrahlung absorption, possibly strongly enhanced by ion-density fluctuations resulting from the above-mentioned beam-plasma instabilities [1]. We shall not dwell further on the details of the beam-plasma interaction at this point [2], but will begin with the relationship of the laser to a fusion power plant.

A.   Fusion Power Plant Energy Schematic

The energy schematic of a fusion power plant employing laser heating of the thermonuclear fuel is shown in Fig. 1. The numbers given in Fig. 1 for laser power, electric power, and waste heat are based on assuming 40% conversion of electricity (for both reactor heat and laser heat), and 30% con-

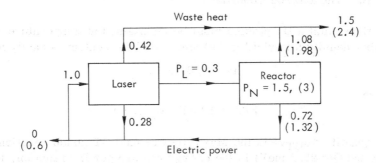

Figure 1.   Laser-reactor power flow diagram.   Numbers denote power in arbitrary units.   Those not in parentheses correspond to zero net electric power output. P_L and P_N denote laser output power and thermonuclear power, respectively.

version of electricity into laser light.   The corresponding efficiency of conversion of heat into light is therefore 17%.   The 40% efficiency represents modern steam-electric plant performance.   The 30% laser efficiency has been achieved with the electric discharge CO_2 laser.   These numbers therefore represent the current state-of-the-art, and not what may ultimately be achievable.   (Laser efficiencies approaching 50% have recently been reported with an electric discharge CO laser at liquid-nitrogen temperatures, for example.)

The conclusion that emerges from an inspection of the power flow diagram of Fig. 1 is that the thermonuclear power release should be large compared with the laser output power required to heat the fuel, i.e.,

$$P_N/P_L = \alpha \gg 1 \tag{1}$$

For example, with the multiplication factor α equal to 10 the laser represents more than 60% of the power plant load, and the overall plant efficiency (important in coolant cost and thermal pollution considerations) is only 20%. We will therefore assume that an α of at least 10 is a requirement of a laser-heated fusion plant, and that further improvements in laser and plant efficiencies are utilized to improve plant performance and reduce capital cost rather than to permit a reduction in α.

The laser that is required must be an efficient (>30%) and reliable unit providing very high average power output (an appreciable fraction of the electrical output of the fusion power plant.

B.   The Lawson Condition

The time τ that a DT plasma must be maintained at temperature T and (ion) number density n to yield $(\alpha/\beta\epsilon)$ times its heat content is easily found to be

$$n\tau = (\alpha/\beta\epsilon)F(T) \tag{2}$$

where

$$F(T) = 12\ kT/<\sigma_{DT}v>Q \tag{3}$$

The quantity $<\sigma_{DT}v>$ is the Maxwell-averaged DT fusion reaction cross section; and $Q(= 22.2\ meV)$ is the energy release per DT reaction, including the exothermic neutron capture reaction in a tritium-breeding lithium blanket.

$$D + T = \alpha + n + 17.6\ MeV \tag{4}$$

$$\underline{Li^6 + n = \alpha + T + 4.6\ MeV} \qquad\qquad$$
$$22.2\ MeV \tag{5}$$

The factor F is a function of temperature alone, as shown in Fig. 2, having the value 5×10^{13} sec/cm^3 at a temperature of 10 keV.

If we now identify ϵ as that fraction of the laser light that is converted into heat content of the plasma characterized by (n, T), and allow for the possibility that the heat content may be multiplied β-fold by thermonuclear self-heating of the plasma, we may identify the quantity α in Eq. (2) with the ratio P_N/P_L of the thermonuclear output power to laser input power as defined by Eq. (1).

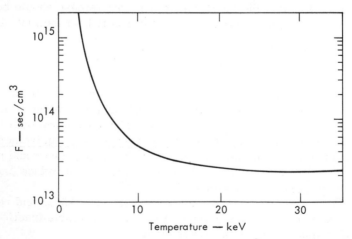

Figure 2.   Graph of the Lawson factor F(sec/cm^3) versus ion temperature T(keV) for equimolar DT fuel.

A conventional (relatively low density, non-laser-heated) fusion reactor is characterized by $\beta \sim 1$, $\epsilon \sim 1$. To compensate for a 40% steam-electric plant efficiency, α must exceed 2.5 in order to provide a net power gain. If the fuel temperature is 10 keV, we find from Eq. (2) that $n\tau \sim 10^{14}$ sec/cm^3. This is the often-quoted "Lawson Condition" for the production of fusion power.

Assuming that the same $\beta\epsilon$ product (~ 1) applies to the laser-heated fusion reactor, but using the value $\alpha (= 10)$ arrived at in the preceding section, we require that

$$n\tau = 5 \times 10^{14} \text{ sec/cm}^3 \tag{6}$$

for a practical laser-heated fusion reactor.

C. The Lawson Condition Applied to a Freely Expanding Plasma

We will now specialize our discussion to the case of laser-heated solid DT pellets, an attractive approach to controlled fusion because no magnetic fields are required for plasma confinement. The use of lasers to heat lower density ($n \sim 10^{17}$ cm^{-3}) magnetically confined plasmas has been considered elsewhere [3].

The confinement time τ of a laser-heated (pellet) plasma freely expanding in vacuum is proportional to the time required for a rarefaction wave, traveling with (isothermal) sound speed c_s, to traverse the plasma radius R. Taking the proportionality factor to be $1/4$, the relation between R and τ becomes

$$R = 4c_s\tau \tag{7}$$

Since the ion density n is equal to $(2.4 \times 10^{23})\rho$, ρ being the plasma density, we obtain that

$$n\tau = (6 \times 10^{22}/c_s)\rho R \tag{8}$$

Assuming a temperature of 10 keV, then $c_s = 9 \times 10^7$ cm/sec, $F = 5 \times 10^{13}$ sec/cm^3, and we obtain from Eqs. (2) and (8)

$$n\tau = (5 \times 10^{13})\alpha/\beta\epsilon \tag{9}$$

or

$$\rho R = (0.075)\alpha/\beta\epsilon \tag{10}$$

Using the values $\alpha = 10$, $\beta_\epsilon = 1$, we obtain the equivalent results:

$$n\tau = 5 \times 10^{14} \text{ sec/cm}^3, \quad \rho R = 0.75 \text{ g/cm}^2 \tag{11}$$

We can now write approximate relations for the fuel mass M, the confinement time τ, and the required laser pulse energy W in terms of the dimensionless parameters α, β_ϵ, and η, where η is defined as the ratio of the plasma density to that of solid DT (0.2 g/cm^3).

$$M = (4\alpha/\beta_\epsilon)^3/\eta^2 \qquad \text{mg} \tag{12}$$

$$\tau = (\alpha/\beta_\epsilon)/\eta \qquad \text{nsec} \tag{13}$$

$$W = (4\alpha)^3/(\beta_\epsilon)^4\eta^2 \qquad \text{MJ} \tag{14}$$

It should be noted that the required pulse energy W is a sensitive function of the parameters α, β_ϵ, and η, which will vary widely depending on the particular manner in which the pellet is heated. (This may account for the widely varying estimates of required laser pulse energy that have been obtained by different investigators.) With this word of caution, let us illustrate these results with two examples in Table I: "Breakeven" ($\alpha = 1$, $\beta_\epsilon = 0.5$, $\eta = 10$), and "Power Plant" ($\alpha = 10$, $\beta_\epsilon = 1$, $\eta = 250$ [4]).

TABLE I.

	R (μ)	τ (psec)	W (MJ)
Breakeven	700	200	10
Power Plant	140	40	1

We conclude that a pellet-heating fusion-plant laser is required to provide very energetic ($\sim$ megajoule) subnanosecond pulses on a repetitive basis. High f-number focusing optics are desirable to maximize the distance to the intensely radiating pellet, so that good beam quality will be required to allow the beam to be focused on a small distant pellet.

D. Coupling of Pellet Corona With Pellet Core

When a DT pellet in vacuum is illuminated from many sides by a powerful focused laser pulse, the surface of the pellet is strongly heated and the pellet is quickly engulfed in a hot low-density plasma that flows outward from the heated surface. The density of this expanding plasma decreases with increasing distance (r) from the center of the pellet, and the laser pulse cannot penetrate inside the surface ($r = r_L$) at which the density (n) of free electrons of the plasma attains a critical value (n_L), which depends on the (vacuum) wavelength (λ_L) of the laser light. For example,

$$n(r_L) = n_L(\lambda_L) = \begin{cases} 10^{19}, & \lambda_L = 10.6 \ \mu \\ 10^{21}, & \lambda_L = 1.06 \ \mu \end{cases} \tag{15}$$

It is therefore, natural to describe the plasma electrons as being divided intwo two spatially overlapping groups: The outer or "corona" electrons, with an effective temperature (T_h), that are or were being directly heated by the absorption of laser light; and the inner or "core" electrons, with temperature (T_c), that are not. The region of overlap between the corona and core electrons will be of thickness (λ), approximately the range of a radially inward-directed laser-heated electron of average energy ($3kT_h/2$). A sketch of the situation is shown in Fig. 3.

The heating of the core electrons by the laser pulse is thus an indirect process consisting of two steps:

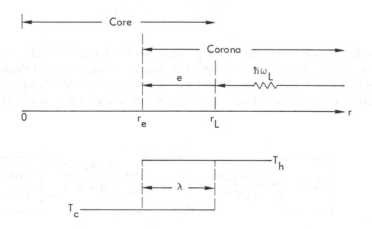

Figure 3.   Corona and core regions.

1. Corona electrons are heated by absorbing laser light via inverse bremsstrahlung (possibly strongly enhanced by instability-produced ion-density fluctuations [1]).

2. Core electrons are then heated by Coulomb collisions with corona electrons in the region where they overlap each other, as illustrated in the energy flow diagram, Fig. 4.

It should now be clear that if the corona electrons are heated by the laser light at a significantly faster rate than they can be cooled by the inner core electrons, then their temperature will increase rapidly and decouple or "run away" from that of the core. This is, of course, a most undesired effect, since the objective is to heat the dense plasma of the core to thermonuclear temperatures, and not merely to produce a tenuous corona of very hot electrons.

It can be shown [5] that the coupling of the corona temperature T_h with the core temperature T_c can be described by the simple relation

$$dT_c/dt = (T_h - T_c)/\tau_c \tag{16}$$

where

$$\tau_c = \frac{2}{3} \frac{1 + \ln(\rho_0/\rho_L)}{\ln(1 + 2\lambda_e/r_L)} \tau_{eq} \tag{17}$$

$$n_L \lambda_e = \frac{(kT)^2}{\pi e^4 (Z+1) \ln\Lambda} \tag{18}$$

$$\tau_{eq} = \frac{3}{4} \sqrt{\frac{m}{\pi}} \frac{(kT)^{3/2}}{e^4 n_L \ln\Lambda} \tag{19}$$

The time τ_{eq} denotes the energy equipartition time [6] of the corona electrons, ρ_0/ρ_L is the ratio of the core density to the critical density, λ_e is the effective mean free path of the laser-heated corona electrons, T is the mean electron temperature $(T_h + T_c)/2$, and $\ln\Lambda$ is the "Coulomb logarithm" [6].

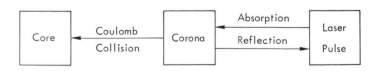

Figure 4.   Heating of core electrons.

The dependence of the coupling $(1/\tau_c)$ on the mean electron temperature (T) is sketched in Fig. 5.

We note that the coupling between the corona and core electrons decreases rapidly, $\propto (T_d/T)^{3/2}$, if the temperature (T) much exceeds the value (T_d) at which maximum coupling occurs. If the corona continues to be strongly heated by light absorption, this decreased coupling results in an accelerating rate of temperature increase of the corona electrons, and a "runaway" situation arises where the core temperature increases very little while the corona temperature rises rapidly to a high value.

To illustrate these results with an example, let $Z = 1$, $n_L = 10^{19}$ cm^{-3}, $r_L = 0.3$ cm, $\ln\Lambda = 10$, and $\rho_o/\rho_L = 10^4$. We then find that:

$$\left.\begin{array}{c} T_d = 1.3 \text{ keV} \\[2mm] \tau_{min} = 2.4 \text{ nsec} \end{array}\right\} 10.6\ \mu$$

The dependence of the maximum-coupling temperature and the minimum equilibration time on the laser frequency ω_L is:

$$T_d \propto \omega_L^{2/3}, \quad \tau_{min}^{-1} \propto \omega_L \tag{20}$$

The heating rate of the core therefore scales as

$$\dot{T}_c \propto T_d/\tau_{min} \propto \omega_L^{5/3} \tag{21}$$

Using these relations to scale the results of the previous example to a tenfold greater laser frequency, we obtain

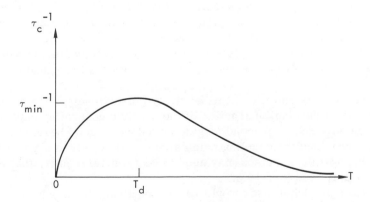

Figure 5.   Core-corona coupling versus temperature.

$$T_d = 6 \text{ keV}$$
$$\tau_{min} = 0.24 \text{ nsec}$$
$$\left. \right\} (1.06 \ \mu)$$

We thus see that 1.06-μ radiation can heat a pellet to five times the temperature in 1/10th the time, as compared with 10.6-μ radiation heating the same pellet. That is, the heating rate achievable, before severe decoupling of the corona from the core occurs, is 50-fold greater in the case of 1.06-μ radiation. Moreover, the maximum coupling of 1.06-μ radiation occurs at a convenient (high) temperature. Although the heating rate achievable with 1.06-μ light appears sufficient to raise small dense pellets of DT to thermonuclear temperatures, the same cannot be said of 10.6-μ radiation in view of this 50-fold penalty.

The decoupling effect is based on the following: (i) long-wavelength light can only penetrate and heat electrons at low density, and (ii) these low density electrons couple poorly with the dense core that it is desired to heat, especially when they are hot. If long-wavelength light can penetrate into denser plasma than has been assumed (via the relativistic decrease in the plasma frequency at very high intensity, for example), or if the coupling between the corona and core electrons is greater than has been assumed (via plasma instabilities set up in the overlap region by counterstreaming corona and core electrons), then the decoupling effect may be considerably reduced. Thermoelectric effects, which we have ignored, may also influence the degree of coupling.

IV.   SUMMARY

The approach to controlled fusion via laser-heated pellets is attractive because no magnetic confinement of the plasma is required. The walls of the fusion reactor can consist of liquid lithium, situated an appropriately large distance from the reacting thermonuclear pellet. No coils, superconducting or otherwise, need be employed. The problems of heating and confinement are effectively solved. However, as is usually the case, one has to pay a price.

The price is an efficient repetitively pulsed laser whose average power represents a substantial fraction of the electric generating capacity of the power plant. Subnanosecond megajoule pulses of good beam quality seem to be required, and the coupling calculations we have given suggest that the wavelength of these pulses may need to be relatively short, that is, considerably less than 10 μ.

Although such a laser would represent a very large advance beyond the present state of the art, there does not appear to be any obstacle that would, in principle, rule it out. However as Rose has pointed out [7], today's short-pulse lasers are fragile and expensive devices, and probably the most difficult task will be that of making a laser (that produces peak powers of 10^{15} W) both cheap enough and reliable enough for practical power plant use.

ADDENDUM (SEPTEMBER, 1972)

If a resonable approximation to isentropic compression of a solid DT pellet can be achieved by the method described by Nuckolls [8], it becomes possible to consider compressions η as large as 10^4. The work w per electron of the pellet (regarded as a fully degenerate electron gas) that must be expended to achieve (isentropically) a compression η is given by

$$w = 3\eta^{2/3} \text{ eV/electron} \tag{22}$$

which amounts to only 1.4 keV per electron at a compression of 10^4, an entirely acceptable cost in energy invested.

This higher value of compression (10^4, as opposed to the value 250 used in arriving at the "Power Plant" numbers listed in Table I) allows us to consider the following more favorable example, which we shall designate "Power Plant (SHC)" (Super-High Compression). The parameters describing this example are taken to be: $\alpha = 50$, $\beta\epsilon = 1$, $\eta = 10^4$. The corresponding values of R, τ, and W are given in Table II.

TABLE II.

	R (μ)	τ (psec)	W (MJ)
Power Plant (SHC)	20	5	0.1

The Power Plant cited earlier would require a laser producing megajoule pulses with 50% efficiency (according to a rough criterion that α times laser efficiency must exceed 500), whereas Power Plant (SHC) would require a laser producing 100-kJ pulses with 10% efficiency, a much more modest requirement. The necessity for extremely high fuel compression is well-illustrated in this comparison. The conclusions contained in Sec. IV remain basically unchanged, however.

REFERENCES

1. W. L. Kruer, P. K. Kaw, J. M. Dawson, and C. Oberman, Phys. Rev. Letters 24, 987 (1970).
2. R. E. Kidder, Proceedings of the International School of Physics. Enrico Fermi, Course XLVIII, edited by P. Caldirola and H. Knoepfel, (Academic, New York, 1971).
3. J. M. Dawson, A. Hertzberg, R. E. Kidder, G. C. Vlases, H. G. Ahlstrom, and L. C. Steinhauer, Proceedings of IAEA Conference on Controlled Fusion, Madison, Wisconsin, 1971 (International Atomic Energy Agency, Vienna, 1971), Vol. 1, p. 673.
4. N. G. Basov, for example, has estimated that values of η between 10^2 and 10^3 can be achieved by means of spherical hydrodynamic plasma implosions. (Paper presented at the European Physical Society Meeting on "Laser-Plasma Interaction", September, 1971, Hull, England.)
5. R. E. Kidder and J. W. Zink, Nuclear Fusion 12, 325 (1972).
6. L. Spitzer, Jr., Physics of Ionized Gases (Interscience, New York, 1962).
7. D. J. Rose, Science 172, 797 (1971).
8. J. Nuckolls, L. Wood, A. Thiessen, and G. Zimmerman, Nature 239, 139 (1972).

CONTROLLED FUSION USING LONG-WAVELENGTH LASER HEATING

WITH MAGNETIC CONFINEMENT

John M. Dawson and W. L. Kruer
Plasma Physics Laboratory, Princeton University
Princeton, New Jersey

Abraham Hertzberg and George C. Vlases
Aerospace Research Laboratory, University of Washington
Seattle, Washington

Harlow G. Ahlstrom and Loren C. Steinhauer[*]
Department of Aeronautics and Astronautics, University of Washington
Seattle, Washington

Ray E. Kidder
Lawrence Radiation Laboratory
Livermore, California

ABSTRACT

The recent development of efficient high-pulse-power and high-energy
10.6-μ N_2-CO_2 lasers has created new and attractive possibilities for
controlled fusion processes. In particular, by combining magnetic confine-

*Present address: Department of Mathematics, Massachusetts Institute of
Technology, Cambridge, Massachusetts.

ment techniques with this type of laser, new reactor configurations become feasible and are discussed in this paper. Plasmas at densities of 10^{17} to 10^{19} particles/cm^3 can be heated to thermonuclear temperatures and confined utilizing available magnetic field technology. The basic physics of these configurations is discussed and scaling parameters developed. In addition, some of the engineering problems of reactors are explored for the various proposed configurations. For example, with field strengths of 500 kG, a Lawson condition straight reactor of approximately 1000 m is possible if end losses are not inhibited. Techniques for reducing end losses are suggested which could result in a Lawson length of 200-300 m.

I. INTRODUCTION

The use of lasers to create high-density high-temperature plasmas was first suggested by Basov [1] and independently by Dawson [2] who pointed out that it was possible for high-pulse-power lasers to create kilovolt plasmas useful for thermonuclear research. These papers stimulated interest in the possibility of creating a new class of thermonuclear reactors utilizing pulsed laser devices. For example, Hertzberg and his colleagues [3] suggested the use of an array of lasers to create an implosion to increase density and aid in confining the plasma. Other investigators suggested the use of the laser in connection with the filling of various thermonuclear machines [4, 5]. With the demonstration by Basov [6] and his colleagues of neutron production by laser radiation and verification and extension of these results [7-9], an increased interest has developed. Most recent approaches take advantage of subnanosecond laser pulses produced by mode locking of the 1-μ neodymium glass laser to confine the plasmas inertially [10]. However, with the development of the high-pulse-power long-wavelength (10. 6-μ) N_2-CO_2 system operating on the microsecond time scale, a new approach has become feasible [11-14]. This relatively long wavelength can effectively heat plasmas in the density range of 10^{17} to 10^{19} particles/cm^3. These plasmas can be confined by available magnetic field technology. For the lower densities, fields of 200 to 300 kG are required; for the higher densities, fields of 2 to 3 MG are needed. This combination is particularly attractive and some of the new reactor possibilities will be discussed in this paper.

Efficient production of high pulse energy with electrical discharge CO_2-N_2 laser systems has been demonstrated. Also, extensions of the gas dynamic laser principle allow an alternate approach to high pulse energy [15]. The physics and engineering characteristics of the N_2-CO_2 system are relatively well understood, making it possible to scale lasing systems to large pulse energies. Electrical excitation involving electron beam preionization has further increased the pulse energy and energy density [16]. For example, pulse energy in the kilojoule range at energy densities of 50 J/liter at 1 atm has been reported by AVCO [17]. In general, the pulse times can be tailored to fit the needs of the devices considered in this paper. By mode locking, even shorter pulse times may be available, though at lower efficiencies, for application to inertial confinement approaches.

The present electrical conversion efficiency of about 30% for electrical discharge N_2-CO_2 lasers (quantum efficiency 40%) compares very favorably with that of the 1-μ neodymium glass lasers, which is less than 1%. It is possible that by using regenerative schemes some of the waste energy in the N_2-CO_2 lasing system can be recovered to further increase the efficiency. Finally, it has been recently shown for the N_2-CO_2 laser that coherent light can be converted from heat with Carnot cycle efficiency indicating even higher efficiencies are possible in the future [18].

A number of reactor possibilities which utilize this laser technology can be considered. The simplest is an axial device operated in a pulsed mode. To summarize, for a temperature of 10 keV, plasma densities of 10^{17}, 10^{18}, and 10^{19} particles/cm^3 require confining fields of 2.8, 9, and 28 x 10^5 G, respectively. The first of these is within the capability of steady magnetic field technology and the second within the technology of pulsed fields. The confinement times required to meet the Lawson criteria ($n\tau \approx 10^{14}$) are in the range of 10 to 1000 μsec which match very will with the operating ranges of the pulsed high-power CO_2 lasers described above. The size of such a device, therefore, depends on the strength of the magnetic field that can be used and on the ability to inhibit end losses. For example, for a field of 500 kG, it appears that a working reactor of less than 1000 m is possible without inhibition of end losses. The energy required is 1-10 MJ. The length and energy required can be reduced significantly by reducing end losses, as discussed in Sec. IV. The simplicity of geometry is an attractive feature and lends itself well to a combined fission-fusion scheme. In the following sections features of the overall configuration, physics of the laser-plasma interaction, and scaling laws for a working reactor are discussed.

The authors feel that the overall capabilities of this approach, even with operation at conservative limits, leads to a class of machines that compares favorably from a number of viewpoints with the concepts now currently being studied for pure magnetic confinement or laser inertial confinement. The relative simplicity of this geometry further enhances our capability of carrying out experiments on a modest scale which are readily extrapolable to full-scale reactor machinery.

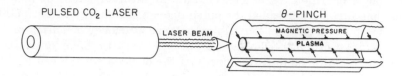

Figure 1.   Schematic diagrams of laser augmented pinch.

II. GENERAL CONFIGURATIONS

The configurations most suitable for laser heating are open-ended linear
devices analogous to that shown schematically in Fig. 1. The confining
magnetic field may either be pulsed, as shown in the figure, or steady as
provided, for example, by a superconducting magnet. The field lines may
be straight, or magnetic mirrors may be incorporated at the ends or periodically along the length of the device. "Stabilized Z pinch" configurations are
also possible. In the following discussion the pulsed field configurations will
be referred to as "laser-augmented pinches" and the steady field cases as
"laser-heated solenoids."

A. Laser-Augmented Pinches

This configuration is similar to that produced by conventional θ or stabilized
Z pinches, but here an arbitrary portion of the plasma energy is supplied by
an axially directed pulsed laser. This leads to two major advantages over a
conventional pinch. First, it provides independent control over the final
values of electron density (n_e), temperature (T), and plasma column radius
(r_1). Thus the coil "filling factor," or fraction of the coil volume occupied
by the plasma, can be made arbitrarily large without the use of a high-voltage shock-wave capacitor bank as has been recently proposed [19]. The
confining field annulus needs only be large enough to reduce cross-field
diffusion losses to a value below end losses. Therefore, the ratio of magnetic energy to plasma energy, an important parameter in energy balance calculations, can be made on the order of unity rather than 0.01-0.05 as is the
case in conventional θ pinches.

The second advantage gained with laser heat addition is the ability to
simultaneously create thermonuclear temperatures and very high densities
without the use of very-high-voltage low-inductance capacitor banks. This
reduces the required confinement time and hence the minimum reactor length.
The limiting factor becomes the maximum B field that can be imposed without

damaging the coil.

While the augmented θ pinch appears to be somewhat more straightforward than an augmented stabilized Z pinch, it can be shown that if the magnetic field strength at the coil surface is taken to be the limiting factor, a Lawson-condition Z pinch is about three times shorter than a θ pinch. This results from the higher B at the plasma surface, and hence higher density, at which the device may operate. The calculation includes the effect of the axial field which must be provided to make the Z pinch MHD stable.

The addition of laser energy and of magnetic field energy to the plasma-field system within the coil can be programmed in a number of ways. For example, after a slight initial pinch to get the plasma off the wall, the laser energy and field energy can be added simultaneously at a rate such that the plasma pressure $p = (n_e + n_i)kT$ at all times equals the magnetic pressure $B^2/8\pi$. In this case the filling density is approximately equal to the final density and all the plasma energy is supplied by the laser while the capacitor bank provides only confining field energy. Another case is obtained by strongly pinching the plasma to a thin column in quasiequilibrium in the conventional manner. Laser energy is then added to the column causing it to expand and increase in temperature, until it again occupies most of the coil volume. A thermodynamic analysis [20] shows that for a θ pinch about 70% of the laser energy goes into plasma heating and the balance into magnetic field energy. The first of these schemes would probably be preferable for a long device in order to match absorption and column length; in the second the laser energy absorption occurs at higher densities which is advantageous for small-scale devices built for preliminary experiments.

B. Laser-Heated Solenoid

While the laser-augmented pinch concept described above improves on conventional pinches in the two respects cited, it still suffers from the disadvantage, from a reactor point of view, of requiring that the pulsed coil be located inside the neutron-moderating blanket. This problem is circumvented in the "laser-heated solenoid" concept by providing a steady B field with a superconducting magnet placed outside the blanket. Thus neither a pulsed main confining field bank nor a high-voltage shock-wave capacitor bank feeding a normalconducting coil is needed. The plasma vessel is filled uniformly with a cold or weakly preionized gas. The laser beam is responsible for heating the plasma to thermonuclear temperatures. The penalty incurred by having a cool plasma blanket surrounding the thermonuclear core is that of greater radial energy loss. Classical heat conduction across the B field, however, does not impose an unduly severe energy loss, and scaling calculations given below indicate that the column diameter required to reduce the losses to a value lower than the end losses can be made no larger than the minimum diameter required by laser beam considerations.

Scaling calculations (Sec. IV) indicate that field strengths on the order of 400–500 kG are desirable for reasonable reactor lengths. On the other hand, while it appears unlikely that superconducting coils will be built in this range in the near future, superconducting compounds with critical fields above 400 kG have been reported. However, it is nevertheless possible to operate the "laser-heated solenoid" at fields of 400–500 kG, with supercon- ducting magnets on the order of 200–250 kG, by addition of a "flux conserver" in the form of a simple metallic tube such as Be-Cu placed between the plasma and the blanket. The expanding laser heated plasma will compress the field between its boundary and the flux conserver. The magnetic forces on this inner coil amount only to the difference in magnetic pressure between the inner and outer field regions, and will exist only on millisecond time scales. The inner coil, having no feed slot, is inherently stronger than a θ-pinch coil and would be relatively inexpensive to replace after prolonged exposure to neutrons. The inner coil serves the additional purpose of preventing the superconductor from going normal as the plasma expands and perturbs the field.

The minimum length of a reactor from both concepts described above can be reduced by inhibiting end losses or by the use of a fission blanket. Techniques for accomplishing this are discussed in Sec. IV.

III.   PHYSICS OF LASER-PLASMA INTERACTION

A.   Absorption Mechanisms

Inverse Bremsstrahlung. Central to the problem of heating plasmas by means of lasers is the absorption of radiation. The simplest absorption process is that of inverse bremsstrahlung or resistive absorption due to electron-ion collisions. The absorption length given by this process is

$$\ell_{ab} = \frac{5 \times 10^{27} T_e^{3/2}}{n_e^2 Z \lambda^2} \left(1 - \frac{\lambda^2}{\lambda_p^2}\right)^{1/2} \tag{1}$$

where λ in the wavelength of the laser light in cm, λ_p is the wavelength of radiation at the plasma frequency

$$\lambda_p = \frac{2\pi c}{\omega_p}, \quad \omega_p = \left(\frac{4\pi n_e e^2}{m_e}\right)^{1/2}$$

T_e is the electron temperature in eV, n_e is the electron density in cm^{-3}, and Z is the ionic charge (Z should be replaced by the mean ionic charge $\Sigma_i(Z_i^2 n_i)/\Sigma_i(Z_i n_i)$ if there is more than one species of ion present). The laser radiation penetrates the plasma only if its frequency exceeds the plasma frequency, i.e., only if

$$n_e \lambda^2 < 10^{13} \tag{2}$$

For $\lambda/\lambda_p < 1$, $(1-\lambda^2/\lambda_p^2)^{1/2} \approx 1$, this factor is ignored in the following paragraphs. Thus, ℓ_{ab} varies as n_e^{-2} and $T_e^{3/2}$ for fixed λ. Figure 2 shows a plot of absorption length for various densities and temperatures for 10-μ radiation and Z = 1. Also shown is the magnetic field at which the magnetic pressure equals the plasma pressure.

For optimum absorption the maximum plasma frequency should be near

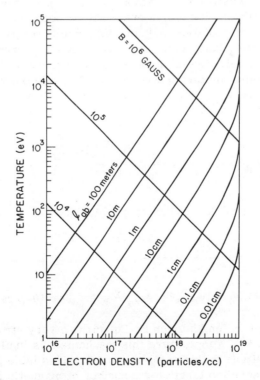

Figure 2. Absorption length and required confining field for $\lambda = 10.6\ \mu$ and Z = 1 as a function of T_e and n_e.

the laser frequency. However, the plasma density which may be used is determined by the plasma pressure which can be confined. Therefore, it is advantageous to use the longest possible wavelength which can be obtained at the required powers and efficiencies.

An important quantity for long straight reactors is the product of the density times the length $(n_e\ell)$. Thus it is of interest to calculate $n_e\ell_{ab}$. From Eq. (1)

$$n_e\ell_{ab} \approx \frac{5 \times 10^{27}T_e^{3/2}}{n_e\lambda^2 Z} \tag{3}$$

For the case of density near critical, as given by Eq. (2) and Z = 1

$$n_e\ell_{ab} \approx 5 \times 10^{14}T_e^{3/2} \tag{4}$$

Inverse Synchrotron Absorption. There are absorption processes other than inverse bremsstrahlung. Among these is inverse synchrotron absorption [21]. The absorption length due to this process depends on the polarization of the incident radiation and upon its angle of propagation with respect to the magnetic field. It is stronger for propagation perpendicular to B and becomes rather inefficient if the angle to the perpendicular is greater than v_t/c (v_t is the electron thermal velocity). For a Maxwellian distribution and propagation perpendicular to B (or within the angle v_t/c to $\perp$) this absorption length is roughly given by

$$\ell_{ab} = 7.6 \times 10^8 \frac{B}{n_e} \left(\frac{m_o c^2}{kT_e}\right)^{j-3/2} \left(\frac{1}{j}\right)^{j-3} \tag{5}$$

where j is the harmonic number, $j = \lambda_c/\lambda$, where

$$\lambda_c = \frac{2\pi c m_o}{eB} = \frac{10^4}{B}$$

Equation (5) applies only if $j^2 kT/m_o c^2 \leq 1$. For 10-μ radiation $\lambda = \lambda_c$ for $B = 10^7$ G.

The absorption length given by Eq. (5) is a very strong function of j and $kT/m_o c^2$. For keV temperature this mechanism is ineffective for harmonics above 5 or 6. Table I illustrates this point. This table gives the inverse synchrotron absorption length for a number of magnetic fields, the electron densities, electron temperatures, and wavelengths. We see that for $\lambda = 10 \mu$ it is difficult to get appreciable inverse synchrotron absorption for fields

TABLE I. INVERSE SYNCHROTRON ABSORPTION LENGTHS

$\lambda(\mu)$	$B(G)$	$n_e(cm^{-3})$	$kT_e(eV)$	$\ell_{ab}(cm)$
10	5×10^5	10^{18}	10^3	2.5×10^{24}
25	5×10^5	10^{18}	10^3	4×10^9
10	5×10^5	3×10^{17}	10^4	2.5×10^6
25	5×10^5	3×10^{17}	10^4	4×10^3
10	10^6	4×10^{18}	10^3	1.8×10^{12}
25	10^6	4×10^{18}	10^3	33
10	10^6	10^{18}	10^4	2.3×10^4
25	10^6	10^{18}	10^4	0.42
10	2×10^6	4×10^{18}	10^4	13
25	2×10^6	4×10^{18}	10^4	1.7×10^{-3}

below 10^6 G and temperatures of 10^4 eV. However, for higher fields and
temperatures the absorption is quite strong. If an efficient high-powered
laser existed with wavelength of 25 μ or larger then quite large absorption
could be achieved by this mechanism.

Heterodyning. Radiation pulses at longer wavelengths would be desirable
for heating low-density devices. One possibility is that efficient, powerful
lasers with wavelengths from 100-1000 μ may be developed. A second possibility is that of heterodyning two CO_2 lasers operating on different vibrational-rotational transitions. Difference frequencies corresponding to
wavelengths ranging from about 70 to 1000 μ are, in principle, attainable.

Anomalous Absorption. If radiation of sufficient intensity falls on a
plasma it can create instabilities which lead to enhanced absorption. Such
effects are particularly efficient if the frequency of the radiation is near
either the plasma frequency or twice the plasma frequency [22, 23] and can
lead to absorption lengths of only a few wavelengths. Unfortunately, for
10-μ radiation this requires densities in the range 2.5×10^{18} to 10^{19}. The
magnetic fields required to confine such plasmas at thermonuclear temperatures are greater than 10^6 G and appear impractical. Again, if high-efficiency lower-frequency lasers were developed, this type of absorption could be
important. For sufficiently intense radiation there is a weaker instability
which occurs even when the frequency is large compared to the plasma
frequency. This instability has been investigated by Silin [24, 25] and roughly
requires that the oscillating velocity of the electrons due to the electric field
of the wave, E_0, be equal to their thermal velocity. Thus it requires

$$\frac{eE_o}{m_o \omega v_{T_e}} \approx 1$$

The threshold E field and hence laser power is independent of the plasma density. For keV plasmas and 10-μ radiation the critical power is 10^{14} W/cm^2. Numerical simulation of this effect at frequencies somewhat higher than the plasma frequency has shown its existence and that it effectively heats electrons.

Related to this is the anomalous absorption of radiation in a plasma containing large ion-density fluctuations (ion acoustic turbulence). This effect is included in the theory of Dawson and Oberman [26] for the ac resistivity of a plasma. For frequencies well above the plasma frequency and assuming isotropic ion turbulence, this theory gives an absorption length of

$$\ell_{ab} = \frac{c\omega^3}{(2\pi)^3 \omega_p^6} \left(\int \frac{f_o'(\omega/k)\,(|n(k)|^2)d^3k}{k^2 n^2} \right)^{-1}$$

where f_o' is the derivative of the one-dimensional velocity distribution. If the ion-density fluctuations are large in a relatively narrow spectral range about k_o this gives an absorption length

$$\ell_{ab} \approx \frac{c\omega^3 k_o^2 n_o^2}{(2\pi)^3 \omega_p^6 f_o'(\omega/k_o) <|\delta n(k_o)|^2>}$$

where $<|\delta n(k_o)|^2>$ is the mean square density fluctuation due to the ion turbulence. For ω/k_o of the order of the electron thermal velocity we obtain an absorption length of roughly

$$\ell_{ab} \approx \frac{c}{\omega_p} \frac{\omega^5}{\omega_p^5} \frac{n^2}{<|\delta n(k_o)|^2>} \tag{6}$$

For a plasma of density 2×10^{17} and turbulent density fluctuations of the order of 0.2 this formula gives an absorption length of 30 m. This should be compared with an absorption length of 1.25×10^3 m for classical inverse bremsstrahlung for such a plasma at 10 keV. From this it appears that nonthermal levels of ion turbulence can materially reduce the absorption length for plasma frequencies appreciably below the laser frequency.

However, it will be difficult to make this work if $\omega/\omega_p > 10$ because of the fifth-power dependence on ω/ω_p.

B. Details of the Absorption Process

One of the properties of inverse-bremsstrahlung absorption is a significant increase in the absorption length as the plasma is heated. The result is the propagation of an absorption or bleaching "wave" into the plasma at a hole-burning velocity as successive opaque layers are heated and rendered more transparent. This occurs for both static heating and when radial expansion is allowed and has been studied in Refs. 27 and 28.

With rapid addition of laser energy to the plasma the electron tempera-ture exceeds that of the ions. However, the equilibration time based on classical calculations [29] is roughly 0.1 to 0.3 of the required confinement time so that sufficient equilibration should occur. Thermal conduction will easily give a uniform temperature in the long plasma column within the confinement time.

C. Refraction of the Beam and Self-Focusing

Laser-Augmented Pinch. It has been assumed in previous discussions that the axially directed laser beam is not refracted out of the plasma before being absorbed. Light rays traversing a highly ionized plasma are deflected toward the region of lower density. It is thus clear that the electron density must have a minimum on axis and increase in the radial direction before decreasing at the plasma-field boundary in order to prevent refraction losses. The required depth of the density minimum is very small; it needs only be sufficient to overcome the diffraction spreading of the laser beam.

To illustrate some salient features, consider a TEM_{00} beam propagating along a plasma column with a parabolic (unfavorable) density distribution. Let ζ equal the ratio of density to critical density. A ray optics calculation shows that the distance the beam will travel before one-half its energy has been refracted out of the column is given by

$$\ell_{refraction} = 1.14\, a \left(\frac{1 - (1 - \gamma/2)\,\zeta_{axis}}{\gamma \zeta_{axis}} \right)^{1/2} \tag{7}$$

where a is the plasma radius and γ is the ratio of the laser beam radius to the plasma radius. For $\zeta_{axis} = 3 \times 10^{-2}$ and $\gamma = 1$, $\ell_{refraction} \approx 10a \approx 10\,cm$. However, the inverse-bremsstrahlung absorption length at 10 keV is given by

$$\ell_{ab} = 70 \frac{(1-\zeta)^{1/2}}{\zeta^2} \approx 8 \times 10^4 \text{ cm}$$

Even for $\gamma \ll 1$, say $\gamma = 10^{-2}$, $\ell_{\text{refraction}} = 100a$ which is still much less than the absorption length.

On the other hand, it is shown in Ref. [28] that a parallel beam is always trapped in a region of the plasma with a density minimum. Solutions for the phase propagation have been worked out for a typical density distribution and from this case it was shown that the effective absorption length can also easily be reduced by a factor of 2.

There are several approaches to operating with a favorable density distribution for the confinement and absorption of the laser beam. If the density distribution has a minimum initially, the laser energy addition will be straighforward and refraction effects will be favorable. Such a distribution exists early in the dynamic phase of a θ pinch; trapping of 337-μ radiation in this case has been observed experimentally [30]. If heating is begun when the density minimum exists, the minimum on axis will persist.

Laser-Heated Solenoid. The problem of beam trapping in this case is simpler than for the laser-augmented pinch. If the gas is initially nonconducting, a breakdown wave [31] will propagate along the column towards the source; self-focussing in such a wave is well documented. If the plasma has been preionized, heating in the center with an initially uniform density will lead rapidly to a density profile with a minimum on axis.

Small-scale experiments on beam refraction for both the laser-augmented pinch and the laser-heated solenoid are in progress at the University of Washington.

IV.   SCALING FOR A REACTOR

A.   The Lawson Condition

In order to have a successful pulsed reactor the reaction must produce enough energy to initiate a new pulse. If we need α times as much energy as is required to heat the plasma ($\alpha \geq [1/\eta] - 1$, where η is the overall cycle efficiency) then we must confine the plasma for a time τ such that

$$n_D n_T <\sigma v> Q\tau = \alpha \left(\frac{3}{2}kT\right)(n_D + n_T + n_e)$$

where σ is the cross section for reactions between deuterium and tritium,

$<\sigma v>$ is the average of σ and v over a Maxwellian, Q is the energy released in the reaction. This equation neglects bremsstrahlung emission, which is permissible for conditions considered. We take Q to be 22. 2 MeV, including the 17. 6 MeV obtained directly from the DT reaction plus 4. 6 MeV from breeding tritium from lithium 6. If we assume we have the optimum mixture with

$$ n_D = n_T = \frac{n_e}{2} \equiv \frac{n}{2}, \text{ then } n\tau = \frac{12\alpha kT}{<\sigma v>Q} \tag{8} $$

A plot of $n\tau/\alpha$ versus T is shown in Fig. 3; $n\tau/\alpha$ has a shallow minimum at about 20–30 keV. We see from this plot that for α's of about 3, $n\tau$ must be about 10^{14}.

For a steady or quasi-steady-state reactor where the reaction supplies the energy to heat incoming fuel and maintain the plasma temperature, the energy confinement must satisfy Eq. (8) with α at least equal 6 since one-sixth of the reaction energy (3. 5 MeV) is carried by charged (He) reaction products which can deposit their energy in the plasma. However, in order for these products to maintain the temperature they must be stopped in the plasma. The stopping is mainly by collisions with electrons, and the stopping time is given by

$$ \tau_s = \frac{4.5 \times 10^7 \, T_e^{3/2}}{n_e} \text{ sec} \tag{9} $$

with the velocity given by $v = v_o \exp(-t/\tau_s)$. The stopping distance is

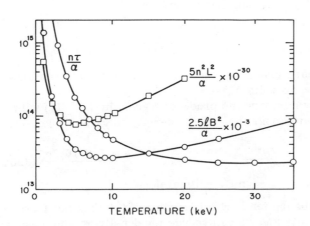

Figure 3.   Scaling parameters for a Lawson condition reactor.

$$\ell_s = v_o \tau_s = \frac{4.5 \times 10^{16} \, T_e^{3/2}}{n_e}$$

The fraction of the energy which the particle deposits in traveling the distance x is given by

$$f = 1 - \frac{v(x)^2}{v_o^2} = \frac{x}{\ell_s} \left(2 - \frac{x}{\ell_s} \right)$$

Thus, in a distance of 0.3 ℓ_s, 50% of the reaction products' energy is deposited.

B.   Length Scaling--Unstopped Ends

The field required to confine a $\beta = 1$ plasma is

$$B = 6.3 \times 10^{-6} \, [(n_e + n_i) \, T]^{1/2} \text{ Gauss} \tag{10}$$

For uninhibited flow out the end the confinement time is roughly given by

$$\tau = \frac{\ell}{2v_T}$$

where ℓ is the coil length and v_T the ion thermal velocity.   The Lawson condition, Eq. (8), can thus be written

$$n\ell = \frac{24 \times 10^6 \, \alpha \, (kT)^{3/2}}{<\sigma v> Q} \tag{11}$$

(Use of a more accurate similarity solution for plasma loss out the ends gives similar results.)

We may compare the product $n\ell$ given by Eq. (11) with that associated with absorption near the critical density as given by Eq. (4):

$$\left(\frac{n\ell}{n\ell_{ab}} \right)_{\text{near critical density}} = \frac{5 \times 10^{-8} \, \alpha}{<\sigma v> Q} \approx 25\alpha \tag{12}$$

Thus, the absorption length is much shorter than the Lawson length near the critical density and we can reduce the density at which we heat by the factor 25α (if N_P passes through the plasma are provided for then we can reduce the density by the factor $25\alpha N_P$). This means that we can effectively heat plasmas of densities in the 10^{17} range to thermonuclear temperatures if these plasmas satisfy the Lawson condition.

Since equilibrium requires a balance between magnetic and plasma pressure, Eq. (11) can be written as

$$\ell = \frac{(1.9 \times 10^{-3}) (kT)^{5/2}}{<\sigma v> Q} \frac{\alpha}{B^2} = f(T) \frac{\alpha}{B^2} \tag{13}$$

A plot of $f(T) = \ell B^2/\alpha$ versus T is shown in Fig. 3; $f(T)$ has a shallow minimum at T = 10 keV. The length for this temperature is thus

$$\ell_c = 0.8 \times 10^{16} \frac{\alpha}{B^2} \text{ cm} \tag{14}$$

which gives the minimum reactor length as a function of α and B.

As discussed above, the absorption length must be of the order of the reactor length at the final temperature for efficient absorption of laser radiation. For a $\beta = 1$ plasma and a density not too near the critical density, Eq. (1) becomes

$$\ell_{ab} = \frac{3.5 \times 10^7 (kT_e)^{7/2}}{B^4 \lambda^2} \tag{15}$$

At 10 keV, therefore,

$$\frac{\ell_{ab}}{\ell_c} = \frac{7.07 \times 10^5}{\alpha (B\lambda)^2} \tag{16}$$

For $\alpha = 3$ ($\eta \geq 1/4$), $\lambda = 10.6\ \mu$ and B = 500 kG (which corresponds to a density of 3×10^{17}), the reactor length become $\ell_{ab} = 0.71\ell_c$.

C. Length Scaling with Inhibited End Losses

Even with optimistic assumptions concerning the size of the magnetic field and the efficiency factor α, it appears that a simple straight device of this type would have to be on the order of 1 km long to work. There are, however, a number of relatively simple possibilities for reducing this length.

First, the plasma might be confined in the axial direction by means of material end plugs. For such a situation, the heat conduction to the cold gas at the ends must be overcome. The time it takes for heat to diffuse a distance ℓ is given by

$$t = \frac{(Z + 1)n_e \ell^2}{2 \times 10^{19} (kT_e)^{5/2}}$$

If the plasma is confined, the reaction will maintain the temperatures provided the energy confinement time satisfies Eq. (8) with $\alpha = 6$. Thus, we find for the length (taking $L = 2\ell$)

$$L^2 = \frac{9.60 \times 10^{20} (kT_e)^{7/2}}{<\sigma v>(1 + Z)n_e^2 Q} \tag{17}$$

A plot of $n^2 \ell^2/\alpha$ versus T is shown in Fig. 3 for $Z = 1$. This curve has a minimum of $16 \times 10^{42} \alpha$ at a temperature of about 7 keV. At this minimum the product $n\ell$ is

$$n\ell_{conduction} = 4 \times 10^{21} \sqrt{\alpha} \tag{18}$$

This product is smaller than that given by inertial confinement (acoustic flow out the ends) by the factor

$$\frac{n\ell_{conduction}}{n\ell_{inertia}} = \frac{0.4}{\sqrt{\alpha}} \tag{19}$$

For realistic α's of about 4 the critical length is about 1/5 of that for inertial confinement or about 200 m for a field of 500 kG and a density of 4×10^{17} and a temperature of 7 keV.

Getting the laser beam into a plasma which is confined by heavy cold material might present some problems. However, since the time required for the plasma to leave the tube is of the order of 10^{-4} sec, it appears that it will be possible to first heat the plasma with the laser and then to close off the ends. The inertia of the plug can easily prevent the expansion of the plasma for it need only have 10^2-10^3 times the mass per cm^2 contained in the plasma ($10^{-2} g/cm^2$) to effectively inhibit the plasma expansion.

A second possibility for reducing loss out the ends of a straight system is to apply mirrors along the length of the system. If the mirrors are spaced about a mean free path apart, then the particles will perform a random walk from one mirror to another, and the loss out the end will be diffusion-type process. A second way to look at it is the following: As the plasma flows along the field, it sees a modulated field, and so it appears to be magnetically pumped. This dissipates energy which must come from the flow. The flow is rather dissipative and is of the nature of diffusion. If the mean free path of a particle is less than or about equal to the distance between mirrors, then the magnetic pumping arguments give for the diffusion coefficient [31]

$$D \approx \frac{15\nu}{k^2 M^2}$$

where k is the wave number associated with the mirrors, ν is the ion–ion collision frequency, and M is the modulation in the magnetic field, $M = \Delta B/B$; B is the mean magnetic field, and ΔB is the maximum deviation from the mean.

As a specific example we will consider the case of B = 300 kG with mirror fields of 500 kG (M = 0.3) one mean free path apart. We take the plasma to have a density of 2×10^{17} and a temperature of 5 keV. The mean free path is 120 cm which we set equal to $2\pi k^{-1}$. The diffusion coefficient is then found to be

$$D = 3 \times 10^{10} \text{ cm}^2/\text{sec}$$

In order to have an $n\tau$ of 10^{15}, the half-length of the system must be 1.3×10^4 cm.

This length is shorter than the stopping distance for the reaction products. However, here the mirrors offer a large advantage since they trap most of these (70% for this case) so that their energy is delivered to the plasma. An appreciable fraction of the energy of the untrapped reaction products (20%) will also be deposited in the plasma.

It is well known that simple mirror configurations are unstable. The system could be made stable in the minimum average B sense through the use of the periodic multipole configuration of Furth and Rosenbluth [33]. However, even for the case of simple mirrors estimates of the minimum disruption time appear to be within a factor of 10 of that required to meet the Lawson criterion. (Instability growth rates $\gamma \approx kV_{Ti}$ where k is the wave number for the mirrors and V_{Ti} is the ion thermal velocity; we assume 15e foldings are required for disruption of the column.) Stabilization or reduction of instability growth rates by a number of mechanisms such as by the inertia and viscous dissipation provided by a surrounding cold dense plasma might suffice.

One final interesting possibility for reducing the length is to surround the plasma column with a blanket of U^{238}. The 14.1-MeV neutrons from the DT reaction can fission U^{238} directly, releasing 200 MeV of energy per fission. Thus, a significant reduction of the Lawson length can be achieved.

One of the secondary neutrons given off by the fission must be used to breed tritium from Li^6. The others can be used to produce plutonium although efficient breeding of this is not necessary since U^{238} directly becomes a fuel by this method. Such a reactor should be absolutely safe since the U^{238} will not sustain a reaction without the supply of 14.1-MeV neutrons from the fusion reaction.

D. Plasma Radius Scaling

The plasma radius required for these long straight plasmas is determined by the requirement that the radial loss of plasma and energy not exceed the axial loss, as this has already been chosen to be just tolerable. The following values are then required for the radii and energies under the assumptions given; B = 500 kG, a plasma $\beta \approx 1$, and the assumptions given below.

(a) r = 0.03 cm   classical diffusion, no heat loss perpendicu-
     E = 3 J/cm   lar to B (plasma surrounded by vacuum)

(b) r = 0.2 cm   classical ionic heat conduction perpendicular
     E = 180 J/cm   to B to a cold plasma surrounding the
                          thermonuclear plasma

(c) r = 12 cm
     E = 7 × 10⁵ J/cm   Bohm diffusion of either plasma or heat

This leads to overall plasma energies for a Lawson condition reactor of 6×10^4 to 3×10^5 J for the laser-augmented pinch with classical diffusion for the case of stoppered ends ($\ell = 200$ m) and unstoppered ends ($\ell = 1000$ m) respectively. For the laser-heated solenoid with classical ionic heat conduction transverse to B, the corresponding figures are 3.6×10^6 and 18×10^6 J. The figures given for the laser-augmented pinch are somewhat unrealistic because of the difficulties associated with focusing the beam into a spot less than 0.6 mm in diameter with a sufficiently small ray entrance angle to ensure beam trapping by the density minimum on axis. However, it should be quite possible from an optics standpoint to focus the radiation on a 2-mm radius column and have it trapped. Thus, plasma energies of a few megajoules for both the laser-augmented pinch and laser-heated solenoid configurations appear to be sufficient.

V. CONCLUSIONS

The advent of the high-efficiency high-pulse energy 10.6-μ N_2-CO_2 laser opens up the possibility of producing thermonuclear plasmas in the density range 10^{17}-10^{19}. It is possible to confine these plasmas with available magnetic fields in the range 300-500 kG. If such lasers were used to heat a plasma in a long straight solenoid with free flow of the plasma out the ends, it appears that a working reactor would be of the order of 10^5 cm long (using realistic B fields and laser efficiencies). If some steps are taken to inhibit flow out the ends of the device, reactors of 200 to 300 m length with a plasma energy investment of several megajoules should be possible.

The authors feel that even in the conservative limits this approach compares favorably in magnetic energy requirements and plasma energy requirements with many of the other approaches now being considered. Therefore, it is worthy of serious consideration involving further detailed analysis and experimental studies.

ACKNOWLEDGMENTS

One of the authors, J. M. Dawson, would like to acknowledge many stimulating discussions with Dr. W. M. Hooke. He is also indebted to Dr. H. Dreicer for the suggestion of using inverse synchrotron absorption.

This work was supported jointly by the U. S. Atomic Energy Commission [Contract No. AT (30-1)-1238], the National Science Foundation (Grant GK No. 28562) and the National Aeronautics and Space Administration (Grant No. NGL 48-002-044).

ADDENDUM (SEPTEMBER 1972)

In this addendum we report further work on reduction of the machine length, and experimental confirmation of beam trapping predictions.

In Sec. IV, the average confinement time for a fluid element in the machine is estimated to be $\tau = \ell/2v_T$. Recent experimental evidence obtained in the 5-m linear Scyllac section [34] and in the 8-m Culham linear θ pinch [35] indicates a lifetime roughly two to three times longer than $\ell/2v_T$. This reduction in end loss results from the natural magnetic mirroring which occurs when the plasma pressure near the ends begins to drop, and is in good agreement with an earlier flow model by Wesson [36]. Thus, our Eq. (11) should read

$$n\ell \approx 12 \times 10^6 \frac{\alpha (kT)^{3/2}}{<\sigma v>Q}$$

and the length of a 500-kG machine at 10 keV becomes 500 m. Further length reduction by means of magnetic mirrors is proposed in the paper (Sec. IV). Recent Scyllac experiments [19] with simple end mirrors of mirror ratio up to 3 indicate an enhanced neutron yield of about 50%, although some tendency towards instability was observed.

Further work on end stoppering by use of cold gas has been carried out [37] using a detailed flow model. For any cold gas pressure above 1 atm, the plasma escape velocity becomes subsonic. At pressures of a few hundred psi the plasma is effectively confined within the coil for long times and cools by conduction to the ends. In one sample calculation, the neutron energy yield from a 400-m device operating at 5 keV and $n_e = 6 \times 10^{17}$ was increased by an order of magnitude over that obtained with unstoppered ends, and was well above breakeven with realistic efficiencies.

Convincing evidence of favorable refraction or beam trapping, as discussed in Sec. III, has been obtained in experiments at the University of Washington [38]. A long pulse TEA laser was used to axially illuminate a small (4 cm × 19 cm) θ pinch with quasi steady radiation intensity over several half-cycles. The transmitted beam was monitored at various distances from the axis to determine absorption and refraction. Simultaneous measurements of n_e were made. The results indicate that beam trapping and absorption occur during periods of favorable density gradients, and strong refraction out of the column occurs when unfavorable gradients develop within the beam radius.

Further work on anomalous absorption and heterodyning is in progress with encouraging results.

In conclusion, advances made in the last year in our understanding of the physical phenomena involved continue to indicate that long-wavelength laser heating of magnetically confined plasma columns is a viable approach for fusion feasibility and merits further investigation.

REFERENCES

1. N. Basov and O. Krokhin, Proceedings of the 3rd International Congress on Quantum Electronics, edited by N. Bloembergen and P. Grivet (Columbia University Press, New York 1963).
2. J. M. Dawson, Phys. Fluids 7, 981 (1964).
3. J. W. Daiber, A. Hertzberg, and C. E. Witliff, Phys. Fluids 9, 617 (1966).
4. A. F. Haught and D. H. Polk, Phys. Fluids 13, 2825 (1970).
5. M. Lubin, H. S. Dunn, and W. Friedman, Proceedings IAEA Conference on Plasma Physics and Controlled Nuclear Fusion, Novosibirsk, 1968 (International Atomic Energy Agency, Vienna, 1968), Vol. I, p. 945.
6. N. G. Basov, S. D. Zakharov, P. G. Kryukov, V. Senatskii, and S. V. Chekalin, JETP Letters 8, 14 (1968).
7. G. W. Gobeli, J. C. Bushnell, P. S. Peercy, and E. D. Jones, Phys. Rev. 188, 300 (1969).
8. F. Floux, D. Cognard, L-G. Denoeud, G. Piar, D. Parisot, J. Bobin, F. Delobeau, and C. Fauquinon, Phys. Rev. A 1, 821 (1970).
9. S. W. Mead, R. E. Kidder, J. E. Swain, F. Ranier, and J. Petruzzi, Appl. Opt. 11, 345 (1972).
10. L. Wood and J. Nuckolls, Paper presented at AAAS Meeting, Philadelphia, Dec. 1971.

11. J. M. Dawson, A. Hertzberg, R. E. Kidder, G. C. Vlases, H. G. Ahlstrom, and L. C. Steinhauer, Proceedings of IAEA Conference on Plasma Physics and Controlled Nuclear Fusion, Madison, Wisconsin, (International Atomic Energy Agency, Vienna, 1971), Vol. 1, p. 673.
12. Alan Hill, Appl. Phys. Letters $\underline{12}$, 324 (1968).
13. G. J. Dezenberg, E. L. Roy, and W. B. McKnight, IEEE J-Quantum Electron. $\underline{QE-8}$, 58 (1972).
14. A. J. Beaulieu, Proc. IEEE $\underline{59}$, 667 (1971).
15. W. H. Christiansen, AIAA Bull., Paper 71-572 (1971).
16. C. A. Fenstermacher, M. J. Nutter, J. P. Rink, and K. Boyer, Bull. Amer. Phys. Soc. $\underline{16}$, 42 (1971).
17. Physics Today $\underline{25}$ (No. 1), 17 (1972).
18. A. Hertzberg, E. W. Johnston, H. G. Ahlstrom, AIAA Bull., Paper 71-106 (1971).
19. F. Ribe, Bull. Amer. Phys. Soc. $\underline{16}$, 1290 (1971).
20. G. C. Vlases, Phys. Fluids $\underline{14}$, 1287 (1971).
21. B. A. Trubnikov and A. E. Bazhanova, Plasma Physics and the Problem of Controlled Thermonuclear Reactors (Pergamon, London, 1959), Vol. III, p. 141.
22. P. K. Kaw and J. M. Dawson, Phys. Fluids $\underline{12}$, 2586 (1969).
23. W. L. Kruer, P. K. Kaw, J. M Dawson, and C. Oberman, Phys. Rev. Letters $\underline{24}$, 987 (1970).
24. V. P. Silin, Zh. Eksp. Teor. Fiz. $\underline{51}$, 1842 (1966) [Sov. Phys. JETP $\underline{24}$, 1242 (1967)].
25. V. P. Silin, Zh. Eksp. Teor. Fiz. $\underline{57}$, 183 (1969) [Sov. Phys. JETP $\underline{30}$, 105 (1970)].
26. J. M. Dawson and C. Oberman, Phys. Fluids $\underline{6}$, 394 (1963).
27. L. C. Steinhauer and H. G. Ahlstrom, Phys. Fluids $\underline{14}$, 81 (1971).
28. L. C. Steinhauer, Ph. D. thesis (University of Washington, 1970) (unpublished).
29. L. Spitzer, Physics of Fully Ionized Gases, 2nd ed. (Interscience, New York, 1962), p. 135.
30. R. Turner and T. Poehler, Phys. Fluids $\underline{13}$, 1072 (1970).
31. Ya. B. Zeldovich and Yu. P. Raizer, Physics of Shock Waves and High Temperature Hydrodynamic Phenomena (Academic, New York, 1966), Vol. I, p. 338.
32. J. M. Dawson, N. K. Winsor, E. C. Bawers, and J. L. Johnson, Proceedings of IV European Conference on Controlled Fusion and Plasma Physics, Rome, 1971 (CNEN), p. 9.
33. H. P. Furth, and M. N. Rosenbluth, Phys. Fluids $\underline{7}$, 764 (1964).
34. K. S. Thomas, H. W. Harris, F. C. Jahoda, G. A. Sawyer, and R. E. Siemon, Proceedings of the A. P. S. Topical Conference on Pulsed High β Plasmas, Garching, July 1972.

35. H. A. Bodin, J. McCartan, I. K. Pasco, and W. H. Schneider, Phys. Fluids 15, 1341 (1972).
36. J. A. Wesson, Proceedings of IAEA Conference on Plasma Physics and Controlled Nuclear Fusion Research Culham, 1965 (IAEA, Vienna, 1966), Vol. 1, p. 223.
37. H. Willenberg, M. S. thesis, (U. of Washington, June 1973) (unpublished).
38. N. Amherd and G. C. Vlases, Proceedings of the A. P. S. Topical Conference on Pulsed High β Plasmas, Garching, July 1972.

PHOTON MACHINES

A. Hertzberg, W. H. Christiansen, and E. W. Johnston*
Aerospace Research Laboratory
University of Washington
Seattle, Washington

H. G. Ahlstrom
Department of Aeronautics and Astronautics
University of Washington
Seattle, Washington

ABSTRACT

The basic thermodynamics of thermal lasers of the gas-dynamic type are reviewed, and it is shown that an efficient coherent photon generator can be developed on a closed-cycle principle. The efficiency limits of such a device are explored, and the results of the analysis indicate that the production efficiency of coherent radiation from heat can, in the limit of high component efficiency, be equal to that of the production of work. An indispensable element of any power transmission system also involves an engine capable of transforming the transmitted energy into useful shaft power. It is shown that a closed-cycle system may also be developed in principle which can transform the transmitted laser radiation into shaft power with an efficiency approaching one.

---

*Currently with the Bendix Corporation, Ann Arbor, Michigan.

141

I. INTRODUCTION

One of the dreams of engineering is the wireless transmission of power.
The concept has been a constantly recurring theme since the original contribution of Hertz. Significantly, the ideas of Hertz led to the development of
wireless telegraphy and all that followed. This development was indeed the
wireless transmission of sufficiently powerful signals so that information
could be transmitted great distances. The development of efficient very high
frequency oscillators has led to the techniques of beaming energy so that, in
a limited sense, wireless power transmission has been achieved. However,
with the development of the laser, much higher frequencies can be generated,
and with relatively small optical systems energy may be beamed with small
losses for many thousands of kilometers. It thus becomes a practical engineering possibility to consider the remote operation of machinery by radiant
energy power transmission. A host of applications immediately suggest
themselves. These range from supplying power to satellites from power
plants on earth, or the reverse, to the remote operation of vehicles of various types such as aircraft supplied by a power plant at a distant point or even
in space with possible ecological advantages.

The authors are fully aware of the apparent radical nature of these
statements and of the many attendant problems of costs, atmospheric transmission, limitations on pointing, and beam divergence. Nonetheless, it
would have been facetious prior to the existence of the laser to even reach
this level of discussion. Without the laser, a practical technology did not
exist by which large amounts of coherent optical energy could be generated
with any degree of efficiency. Recently, this has changed dramatically and
we can confidently look forward to future developments of equal significance.

To explore the possibility of power transmission using lasers, this
paper will examine the theoretical efficiency limits by which coherent radiation can be generated, transmitted, or reused as available power. Specifically, thermal lasers of the gas-dynamic type using N_2 and CO_2 gas mixtures
will be examined using a conceptual coherent photon generator involving
rapid expansion processes leading to laser action. It will be shown that the
efficiency of such a photon generator can be, in principle, very high; and in
the limiting case, the entire amount of shaft power provided can be converted
directly to laser radiation. Although practical restraints will limit such
high efficiencies, the photon generator does allow us to examine the factors
that limit efficiency and to set the goals that we should try to achieve in
practice.

An indispensible part of any power transmission system involves not
only a generator, but also an engine or device which can receive the light
energy and transform it to a usable form. One of the principal barriers to
radiant power transmission is the limited thermal efficiency that results if
the radiant energy is only used as thermal energy. Here the normal limita-

tions of materials will limit the efficiency of reconversion to between 30 and 40%. What is needed, therefore, is an engine which can be operated directly by radiation at reasonable energy densities and with the possibility of higher efficiency.

In order to achieve this capability, a device called the photon engine will be described. It can provide a useful function in controlling and transforming coherent radiation, and it also forms the basis of a device called the photon capacitor (see Appendix). As stated previously, all of these devices are developed around the operating principle of fluid mechanical thermal lasers, often called gas-dynamic lasers, utilizing N_2 and CO_2 gas mixtures as the working fluid.

In the following sections thermal lasers of the N_2-CO_2 type will be described in sufficient detail to develop the concept of a generator which can, in the most efficient manner possible, convert heat to coherent radiation. Utilizing this as a basis, a description will be given of the photon engine which can reconvert laser radiation back to useful work with an efficiency approaching one. This will thus provide the two indispensable elements of any power transmission system.

II. THERMAL LASERS

The suggestion that an electronic population inversion could be obtained in a gas was first put forward by Javan [1] in 1959 following the historic paper of Schawlow and Townes [2]. This work led to the development of the first successful gas laser in which electrically excited helium metastables were used to preferentially pump a specific level in neon so that a population inversion was achieved in the gas [3]. This gas laser has now become highly developed as one of the most convenient sources of cw laser radiation available and has laid the groundwork for all of the developments in gas lasers which have followed.

In 1962, while examining the processes leading to noneqilibrium in various rapid expanding flows, Hurle, Hertzberg, and Buckmaster [4] pointed out that is should be possible to create a population inversion in electronic states by the rapid adiabatic expansion of a gas. If a sufficient equilibrium population of the necessary excited states was achieved, the rapid expansion could create conditions whereby the population of an upper state would remain essentially frozen, provided that the rate of deexcitation of the upper state was slow compared to the cooling rate. If, at the same time, the lower state could be depopulated either by collisions or by radiative transfer as a result of the rapid cooling, a population inversion could be achieved. The upper and lower levels would of necessity have to have different relaxation rates. This work was later expanded and submitted for publication in 1964 [5]. In that paper, the basic geometry of such fluid mechanical lasers was described and has proved consistent with present developments. At about the same time, Basov et al. [6] independently proposed the use of rapid heating or cooling to produce a population inversion. However, gas lasers were still in their infancy, and many of the systems which have proved so valuable had not yet been created. However, with the development of the N_2-CO_2 laser by Patel [7] in 1964, a new generation of very-high-power gas lasers became possible. This laser system, which emits in the infrared, typically attains a population inversion in a glow discharge with a gas mixture of N_2 and CO_2 and a suitable catalyst to help depopulate the lower lasing level. In fact, the basic processes bear many similarities to those in the helium-neon laser. As in that laser, a metastable level of a working gas (in this case N_2) is populated by electron impact excitation. The first excited level of N_2 is in near resonance with the upper level of the 10.6-μ transition in CO_2, and transfers its vibrational energy to unexcited CO_2 molecules (see Fig. 1), thereby pumping that level much in the same way as electrically excited helium metastables pump the neon upper levels. Since the cross section for excitation of the metastable first vibrational level of N_2 by electron impact is very high, electrical energy is easily and efficiently transferred to the N_2 so that a population inversion is produced in the CO_2.

This particular lasing system has proved to be most suitable for application of the rapid expansion scheme suggested by Hurle and Hertzberg since the lasing levels exhibit different relaxation rates, as required for thermal pumping. The identification and experimental verification of this important possibility was carried out at the AVCO Everett Research Laboratory and

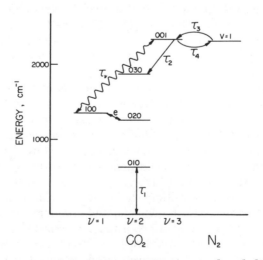

Figure 1.   Simplified N_2 and CO_2 energy level diagram.

has recently been reported by Gerry [8]. The AVCO group was the first to identify the possibility of a high energy density system, and to then develop the basic technology whereby large amounts of radiation could be extracted from a flowing gas-dynamic system. Independently, Konyukhov and Prokhorov [9], as well as Basov et al. [10], identified the possibility of obtaining a population inversion utilizing thermal pumping of an N_2-CO_2 mixture.

An inversion is achieved in the gas-dynamic system by the following process. A mixture of N_2-CO_2 containing a small amount of water vapor or helium is heated in the plenum of a conventional convergent-divergent nozzle to a temperature of approximately $2000^\circ K$. The heating can conveniently be accomplished in a number of ways, such as with a shock tube, or in an arc jet, or in a rocket motor where the products of combustion include appropriate amounts of CO_2 and N_2. When the N_2-CO_2 gas mixture is so heated, an equilibrium distribution of excited states is produced in both the N_2 and the CO_2. The first vibrational level of N_2 is thermally pumped; and in the case of pure N_2, the vibrational energy in that state represents about 9% of the internal energy (at $2000^\circ K$) or about 7% of the flow energy (see Fig. 2). If the gas is expanded rapidly enough through the convergent-divergent nozzle, the cooling rate in the nozzle can be very fast compared to the relaxation rate of N_2. Hence, it is possible to freeze the populations of the excited vibrational levels of the N_2 in the supersonic expansion part of the nozzle and to maintain a vibrational temperature which is closer to the plenum temperature than to the translational temperature of the expanding gas.

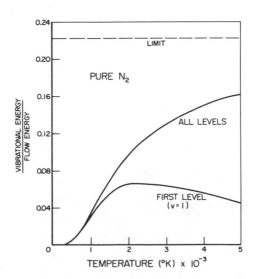

Figure 2. Temperature dependence of N_2 vibrational energy divided by flow energy.

Thus, in the supersonic flow region of the nozzle, the fundamental condition necessary to produce a population inversion exists; that is, there exist two very dissimilar temperatures in the gas. The lower laser level is thermally associated with the very-low translational temperature ($\sim 300^\circ$K) which results from the expansion. All of the lower energy levels of CO_2 would also normally have a population distribution corresponding to this low translational temperature. However, the mechanism that produces the inversion is the near-resonant energy transfer which occurs from the vibrationally excited N_2 to the asymmetric vibrational stretch mode of CO_2. A population inversion is then achieved since the upper lasing level of the CO_2 tries to achieve a number density associated with a much higher temperature than the lower laser level, as can be seen in Fig. 3.

It should be pointed out that very short nozzles are required to rapidly cool the gas in a time that is short compared to the upper-level relaxation time. Hurle and Hertzberg [5] pointed out that one of the most rapid possible expansions which could be attained in a practical sense would be via a Prandtl-Meyer expansion fan obtained by a free jet expansion. In a more practical sense and in a system capable of higher mass handling capacity, they also pointed out that approximately the same results can be achieved by utilizing a grid nozzle of the type first suggested by Ludwieg [11].

The gas kinetic processes involved in this system have been described in detail by Gerry [8]. In addition, detailed studies of the kinetic processes in these machines have been carried out and verified experimentally by Christiansen and Tsongas [12], and Konyukhov et al [13]. Basov et al. [14] and Anderson [15] have also numerically solved the coupled flow-kinetics problem. In fact, our knowledge of these types of machines has grown quite

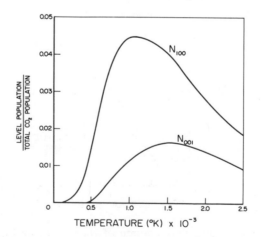

Figure 3.   Temperature dependence of CO_2 level population.

rapidly, and it is possible to predict with surprising accuracy the amount of energy available for lasing in a machine of any given geometry or size. This fact is of the utmost importance since it provides the basis for calculating the limiting thermodynamic performance of these machines.

Unfortunately, the apparent efficiency of the thermally excited laser described by Gerry is relatively low in that only about 1 to 2% of the enthalpy of the gas can be extracted as laser radiation. Therefore, these efficiencies do not compare well with the conversion efficiency of conventional electrically excited N_2-CO_2 lasers where efficiencies approaching 30% have been reported [16]. However, it should be pointed out that electrical energy, like shaft energy, must be purchased with an efficiency of approximately 30 to 40%, and hence the overall thermal efficiency of an electrical system operating at its maximum potential is at most about 12%. These efficiencies are not particularly encouraging if one wishes to consider the electrically excited lasing systems as a proper tool for the development of power transmission systems.

It was realized that the amount of energy extracted by the laser barely affects the stagnation temperature of the gas; therefore, an energy recovery system based on a diffuser would be able to repump the gas thermally so that any following expansion could reexcite the system. The system described by Gerry utilized a diffuser for recovering a sufficient amount of the total pressure of the cavity to make it convenient to discharge the laser to the surrounding atmosphere. The group at the University of Washington suggested that a logical extension of this laser system would be to close the cycle.

III. A CLOSED-CYCLE PHOTON GENERATOR

In a photon generator (closed-cycle gas-dynamic laser), the gas would be expanded via a supersonic grid nozzle into the lasing cavity, lasing energy would be extracted, the gas returned to near stagnation temperature with an efficient diffuser, and then recirculated through a heat exchanger and adiabatic compressor to its original lasing configuration. The basic closed-cycle system would look very similar to a conventional supersonic closed-cycle wind tunnel and is shown in Fig. 4.

One of the chief drawbacks to the practical achievement of such a system is that to secure effective lasing actions, the temperature of the gas has to exceed the normally available limitation imposed by materials for high-temperature compressors. For example, it has been established that approximately $1000°K$ is the limit for an uncooled turbine to run continuously for long periods of time. As seen from Fig. 2, below $1200°K$ very little energy is available in the N_2 vibrational system and hence a high-temperature compressor must be inserted into this laser system. High-temperature

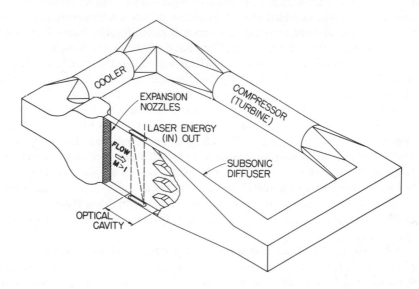

Figure 4.   Schematic diagram of photon generator (photon engine).

compressors of the Comprex type have been developed to supply hypersonic
wind tunnels for materials testing purposes [17].

　　For calculations on the closed-cycle gas-dynamic laser system, certain
assumptions can be made about the nature of the components and the gas-dynamic processes which facilitate the analysis. In the plenum, where all the
gases involved are in thermodynamic equilibrium, the number distribution
in the various energy states is characterized by a single temperature. As
the gas flows through the nozzle, full thermal equilibrium is initially maintained; however, as the flow velocity increases, the vibrational temperature
tends to remain high as the translational temperature drops. The vibrational temperature of the N_2 will assume some characteristic value appropriate to the dimensions of the nozzle and the gas mixtures employed. For
ease of calculation the sudden freezing approximation can be used. In
this approximation thermal contact between the vibrational system and the
translational system is instantaneously decoupled at some point in the expansion nozzle. In an actual case there is always some energy transfer between
the two systems and hence some entropy production, but this loss is small
compared to other losses in the cycle. With the sudden freezing approximation the total entropy of the system remains constant.

　　The heat transfer to and from the walls, with the exception of the heat
exchanger, can be neglected. This assumption is quite reasonable for a
relatively large machine in which the walls are modestly insulated and at

equilibrium with the gas temperature. The flow losses in the system (e. g. , turning vane losses and viscous effects in the nozzle and the duct) can be combined together into an overall pressure drop associated with the diffuser. This assumption is quite conventional in calculating a closed-cycle wind tunnel since this loss far outweighs all the others. The compressor is assumed to be adiabatic and the heat exchanger isobaric.

In addition to the diffuser loss, another loss is present in the system which is not normally present in a supersonic wind tunnel. As was pointed out previously, the N_2 vibrational temperature can be maintained at a significantly higher level than the translational temperature in the test section. In the process of lasing, only part of the N_2 vibrational energy can be extracted as radiation; and, because of the processes maintaining the lower laser level at the translational temperature, a significant amount of this energy (approximately 60% of the energy frozen in the N_2) is given up as heat added to the supersonic stream. Assuming that the laser is coupled to an efficient cavity so that most of the radiation energy is removed from the cavity and not absorbed by the mirrors, a very straightforward calculation can be made of the total pressure loss associated with the lasing action. This loss may be large in view of the fact that heat is being added to a relatively high Mach number stream.

The remaining processes in the cycle are assumed to be in thermodynamic equilibrium. The total pressure deficiencies are made up by the action of the adiabatic compressor and heat is withdrawn at constant pressure in the cooler. Figure 5 shows a cycle temperature distribution for a perfect

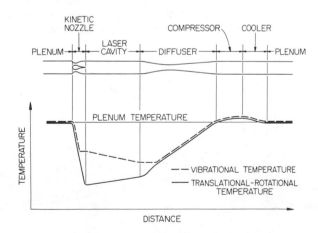

Figure 5. Temperature distribution for a closed-cycle laser with a constant-area lasing cavity, a perfect diffuser, an isentropic compressor, and an isobaric cooler.

diffuser, an isentropic compressor, an isobaric cooler, and a constant-area lasing cavity.

Preliminary numerical calculations were made for a closed-cycle operating with a lasing medium composed essentially of N_2 in a constant-area laser cavity. The cycle was assumed to be efficient enough to extract all the available laser energy from the flowing gas. The calculations for Fig. 6 were based on a plenum temperature of $1700°K$, a vibrational freezing temperature of $1417°K$ and a Mach number of 4 at the entrance to the laser cavity. Figure 6 shows the efficiency for converting thermal energy to laser energy as a function of the overall pressure drop associated with the diffuser. It has been assumed that thermal energy can be converted to work with an efficiency of 40%. This is the efficiency with which the heat removed by the cooler can be converted into work going into the compressor, thereby providing regeneration.

From Fig. 6 it is seen that even with a perfect diffuser the conversion efficiency is around 19%. The reason for this low efficiency is the large loss in total pressure during lasing (about 25%). Efforts to increase the output of laser energy by varying the operating temperatures in a constant-area lasing cavity resulted in a lower production efficiency. The results shown in Fig. 6 are typical of the performance of closed-cycle gas-dynamic lasers using N_2-CO_2 mixtures. While small variations are possible by adjusting the parameters of the system (such as varying the mixture ratio or the Mach number of the flow), these results may be regarded as typical. They reveal that our chief concern must lie in reducing the total pressure loss due to lasing and developing high-efficiency diffusers to work in laser configurations. Indeed, these results are typical of any closed system in which the energy extracted is only a small percentage of the circulating energy. As can be demonstrated by elementary calculations, such systems are particularly

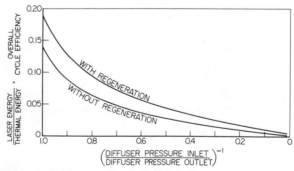

Figure 6. The effect of diffuser recovery on the overall cycle efficiency of the photon generator.

sensitive to component efficiency. For example, this system is sensitive to the diffuser pressure recovery ratio. As pointed out earlier, this loss tends to dominate in such flow systems. The inclusion of the other minor component losses with the diffuser loss yields approximately the same result. For example, with a pressure recovery of one-half in the diffuser, the compressor pressure ratio required to make up this loss is only 2. Compressor efficiencies in this region can be high (> 90%). However, unless the total pressure drop due to lasing can be significantly reduced, a high-efficiency system is not obtainable. These results therefore motivated a deeper study of the thermodynamics of closed-cycle lasers in the hope of improving their efficiency.

It is shown in Sec. V that by properly adjusting the ratio of the vibrational temperature to the translational temperature of the gas, the loss due to lasing can be minimized, a condition can be achieved in which no heat has to be rejected by the system as a consequence of lasing, and with a perfect diffuser the compressor supplies work equal to that of the laser radiation exhausted. This complete conversion of shaft work into laser radiation is a limiting condition and is not easily achieved in practice. Since the shaft work must be obtained from thermal energy, it is clear that it is possible to generate coherent radiation from a very idealized system with approximately the same efficiency with which electricity can be generated from thermal energy.

Of course the significant technical problem lies in the inefficiencies associated with the diffuser, especially with the high Mach numbers normally associated with gas-dynamic lasers. In the case of a nonisentropic diffuser, the shaft work that must be supplied to the compressor is greater than the energy of the laser radiation extracted and the difference is rejected as heat through the cooler.

Many practical difficulties will have to be overcome in building a machine of this type, but it is felt that the problems are within the limits of technical feasibility. In addition, the potential for very high efficiency makes devices of this type worth serious further consideration.

IV. THE PHOTON ENGINE

As stated previously, an effective power transmission system must consist of both a generator and an engine. The actual transmission will be accomplished by a suitable system of optics to collimate the radiation and a corresponding system of optics at the point of application to direct the energy flux into the engine. The photon generator and the photon engine are similar to the electric generator and the electric motor in that in both cases the motor and the generator can interchange roles. It will be shown that coherent radiation will be absorbed and shaft work efficiently produced.

The modification is explained as follows. With reference to Fig. 4, flow is expanded from a plenum through a supersonic nozzle. However, a conventional supersonic nozzle rather than a grid nozzle is used since thermal equilibrium is required in the absorbing section which replaces the laser cavity. The stagnation temperature of the flow is selected so that at the supersonic Mach number in the absorbing section, the equilibrium static temperature of the mixture is about 600° K. The laser radiation is coupled to this section and if the intensity is sufficiently high the following processes will take place. As the mixture of N_2-CO_2-He enters the absorption region, the lower level is thermally populated to the degree shown in Fig. 7, while the upper laser level has a much smaller population density. In this figure the energy states of a volume moving with the fluid velocity in the absorbing section are followed at constant input intensity for a specific case. Therefore, the variations of energy states are shown as a function of time rather than distance. This distribution of states is the reverse of population inversion and the laser radiation is absorbed. Both the lower laser level and the upper laser level now attempt to find a new equilibrium in the presence of the radiation field. For a sufficiently high radiation intensity the population of the lower laser level n_L will decrease and the population of the upper laser level n_U will increase until the two are equal. The gas is said to be bleached when $n_U = n_L$. However, since the gas contains N_2 there is an efficient transfer of the energy of the upper laser level to the first vibrational level of N_2. Since this is a resonant V-V transition it is quite rapid and the final bleaching condition occurs when the populations of the two laser levels are equal and the vibrational temperature of N_2 is equal to the temperature of the upper laser level of CO_2. Now, since the energy is stored in the $N_2(v = 1)$ state, translational-rotational energy is removed from the gas

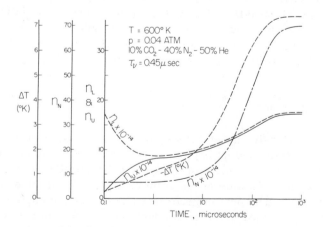

Figure 7.   Temperature and population variations of the important levels in a N_2-CO_2-He mixture during bleaching.

producing a cooling of the supersonic flow. For each N_2 molecule excited to the $v = 1$ state only 40% of the energy came from the laser. The remainder came from T - V transitions populating the lower laser level and reducing the translational temperature. Depending on the mixture ratio and temperature of these gases, the cooling can be significant (varying between 1 and 5% of the translational temperature).

In reverse of the case of heating of a supersonic gas, the cooling of a supersonic gas will result in a total pressure increase. After the bleaching process, the gas is coupled to a grid diffuser and the gas is returned to a new equilibrium by the process of sudden unfreezing. After the diffuser the gas mixture would be at a final stagnation pressure somewhat higher than the corresponding pressure upstream. As the gas returns around the cycle, heat is again extracted from the cooler; but since the total pressure is now increased, the compressor now becomes a turbine and shaft work can be extracted from the system. This device is therefore called a photon engine.

In carrying out calculations, it was determined that a specific balance between the translational-rotational temperature and the vibrational temperature is required. Indeed, with the balance of temperatures suggested above, the efficiency of the photon engine will be zero or negative. The radiation will be absorbed but the entropy rise will be so dramatic that work cannot be extracted; in fact, a compressor would be required to run the cycle. However, if temperatures are correctly selected the efficiency will be positive and can be very high. Therefore, the concept must be modified so that the initial expansion is nonequilibrium and creates a special relationship between the translational-rotational temperature and the vibrational temperature. Also, during the process of absorption the area of the channel is varied to keep this temperature balance at the optimum ratio. As can be seen in Sec. V, this leads to the possibility of a highly efficient photon engine by which the entire amount of radiation absorbed can be transformed into shaft energy. In an actual photon engine the efficiencies of its components would not allow the realization of the indicated potential efficiency of 100%.

It should also be pointed out that the idea of supersonic cooling to obtain a closed-cycle system in which a gas is pumped around a loop is not new. However, Shapiro has remarked that the heat-transfer mechanism of a conventional heat exchanger placed in a supersonic flow involved frictional losses which in general would exceed any increase in total head due to cooling. In the case of the photon engine, the cooling takes place in the volume without friction and this limitation does not necessarily apply.

V. THERMODYNAMIC LIMITS OF PHOTON GENERATORS AND ENGINES

In this section the thermodynamics of the photon generator (the gas-dynamic laser) and the photon engine are examined to determine the fundamental limits on their operation. Now the closed-cycle gas-dynamic laser is very similar to a closed-circuit supersonic wind tunnel. In a wind tunnel there are losses due to friction, heat transfer, the supersonic diffuser, and the compressor. All losses that are not associated with the fundamental processes of generating or utilizing the photons to do work are ignored. These dissipative processes in flow machinery are understood in relation to photon generation (Sec. III), but in this section are ideally reduced to zero. The purpose of this section is to examine those factors affecting the efficiency that are intrinsic to the lasing process itself.

In view of the significant developments in the N_2-CO_2 laser, the analysis will be further restricted to N_2-CO_2 system in which energy is principally stored in N_2 and is resonantly transferred to CO_2. The analysis can be extended to include other lasers such as the N_2-CO system, but in this development it will be clearer to retain one model. The laser transition is to an intermediate lower laser level, and the energy from the lower laser level to the ground state is rejected as waste heat. The gas mixture is assumed to be predominantly N_2 with just enough CO_2 and catalyst to provide the laser action, but not so much as to contribute appreciably to the total enthalpy of the gas. A one-dimensional flow process is considered and the thermodynamic state of the gas is assumed to be characterized by the pressure and two temperatures, T_V and T. The temperature T_V characterizes the population distribution of all vibrational states of N_2 as well as the upper laser level of CO_2. The temperature, T, describes the distribution of the translational and rotational states and the lower laser level. The temperatures T and T_V are defined by a Boltzmann population of the upper level at T_V, $n_U = A \exp(-\varepsilon_U/kT_V)$, and a lower level at T, $n_L = A \exp(-\varepsilon_L/kT)$, where ε_L and ε_U are the energy levels of the lower and the upper laser levels, respectively. The nonequilibrium represented by the difference of the two temperatures is produced by a supersonic expansion. It is also assumed that the freezing process is perfect so that once T_V is greater than T, T_V is changed only by the removal of laser radiation. Indeed, the losses associated with collisions and fluorescence in CO_2-rich mixtures can be neglected, because the percentage of CO_2 is assumed to be small. Reducing the CO_2 concentration has been shown by Christiansen and Tsongas [12] to be realizable in practice and supports the utilization of this simplification to study the ideal thermodynamic limits. As the flow is recompressed, equilibrium is reestablished when $T = T_V$.

The efficiency of the closed-cycle laser will be calculated by following a unit mass of gas through the cycle. The analysis starts at the lasing stage where a small amount of laser energy per unit mass, δq_L, is radiated by the system in a constant-area duct. Conservation of energy requires that

$$- \delta q_L = \delta \varepsilon_V + \delta h_{T,R} + \delta \left(\frac{U^2}{2}\right) \tag{1}$$

where $\delta\epsilon_V$ is the change in the vibrational energy per unit mass flow, $\delta h_{T,R}$ is the change in the translational-rotational enthalpy per unit mass flow characterized by the translational temperature T and gas constant R, and U is the directed velocity. From the thermodynamic relation

$$T \, \delta s_{T,R} = \delta h_{T,R} - \rho^{-1} \delta p \qquad (2)$$

where ρ, p, and $s_{T,R}$ are the mass density, static pressure, and the change in the entropy of the translational-rotational system, respectively. Solving for $\delta h_{T,R}$ and substituting into Eq. (1)

$$-\delta q_L = \delta \epsilon_V + T \, \delta s_{T,R} + \rho^{-1} \delta p + \delta (\frac{U^2}{2}) \qquad (3)$$

Conservation of mass and momentum in one-dimensional motion gives

$$\rho^{-1} \delta p + \delta (\frac{U^2}{2}) = 0 \qquad (4)$$

Equation (3) then reduces to

$$-\delta q_L = \delta \epsilon_V + T \, \delta s_{T,R}$$

or

$$-\delta q_L = T_V \delta s_V + T \, \delta s_{T,R} \qquad (5)$$

where δs_V is equal to the change in the entropy of the vibrational system.

It is assumed that the collisions with N_2 molecules which excite the CO_2 molecules to the upper laser level from the ground state are exactly resonant to a good approximation. As a result of lasing, an energy $\delta \epsilon_V$ is taken out of the vibrational subsystem at temperature T_V, a part of it, $\epsilon_L \epsilon_U^{-1} \delta \epsilon_V$, goes to the translational-rotational subsystem at temperature T, and $(\epsilon_U - \epsilon_L) \epsilon_U^{-1} \delta \epsilon_V$ becomes laser radiation $-\delta q_L$.

From the above it is seen that

$$-\delta q_L = (\epsilon_U - \epsilon_L) \epsilon_U^{-1} \delta \epsilon_V$$

or

$$\delta s_V = - (\alpha T_V)^{-1} \delta q_L \qquad (6)$$

where $\alpha = (\epsilon_U - \epsilon_L)\epsilon_U^{-1}$, the quantum efficiency of the laser transition. For CO_2, $\epsilon_L = 1388$ cm^{-1} and $\epsilon_U = 2349$ cm^{-1} so that $\alpha = 0.40$. The assumptions made in this development are appropriate to the N_2-CO_2 laser. At present there are no other gas-dynamic lasers to which these assumptions are appropriate, but α is left as a parameter. (Actually, even in the CO_2 gas-dynamic laser, one could lase at 9.6 μ where the lower laser level is the $02^0 0$ state with $\epsilon_L = 1286$ cm^{-1} and $\alpha = 0.44$.) Substituting Eq. (6) into Eq. (5)

and solving for $\delta s_{T,R}$

$$\delta s_{T,R} = -(\frac{T_V}{T})\,(1-\alpha)\,\delta s_V \tag{7}$$

The total entropy change $(\delta s_V + \delta s_{T,R})$ for the flowing gas due to lasing is therefore

$$\delta s_{LG} = \delta s_V - (\frac{T_V}{T})\,(1-\alpha)\,\delta s_V$$

or $\hspace{8cm}$ (8)

$$\delta s_{LG} = [(1-\alpha)(\frac{T_V}{T}) - 1](\alpha T_V)^{-1}\delta q_L$$

The requirement that the gas lase is guaranteed by the existence of a popula-tion inversion, i.e., $n_U/n_L > 1$ which for the definitions of T and T_V reduces to $T_V/T > \epsilon_U/\epsilon_L$ or $T_V/T > (1-\alpha)^{-1}$. Thus Eq. (8) shows that a change in energy of δq_L due to lasing is accompanied by an increase in entropy propor-tional to δq_L.

The other operations of the cycle restore the energy and entropy to their prelasing values and must be calculated to complete the analysis. After lasing, the flow is assumed isentropic to stagnation conditions at the exit of the diffuser where it enters the isentropic compressor whose function is to return the stagnation pressure to its original value. The amount of work the compressor has to do to bring about this recovery is now calculated.

Suppose that the stagnation temperature and pressure are T_0 and p_0 before entering the expansion nozzle and the latter is changed by δp_0 as a result of lasing. Since the change in energy is $-\delta q_L$ the change in entropy δs_{LG} can be expressed in terms of stagnation conditions as

$$\delta s_{LG} = -T_0^{-1}\delta q_L - Rp_0^{-1}\delta p_0 \tag{9}$$

where R is the gas constant per unit mass. Combining Eqs. (8) and (9) gives

$$Rp_0^{-1}\delta p_0 = -\{[(1-\alpha)(\frac{T_V}{T})-1](\alpha T_V)^{-1} + T_0^{-1}\}\delta q_L \tag{10}$$

from which the change in stagnation pressure δp_0 due to the removal of laser radiation δq_L can be determined. With δh_w equal to the change in enthalpy and $-\delta p_0$ the change in pressure in the isentropic compressor, the following thermodynamic relation must be satisfied,

$$T_0^{-1}\delta h_w + Rp_0^{-1}\delta p_0 = 0 \tag{11}$$

From Eq. (10), δh_w, the work done by the compressor to restore the total pressure to p_0, is found to be

$$\delta h_w = \{(\frac{T_o}{\alpha T_V})[(1-\alpha)(\frac{T_V}{T})-1] + 1\}\delta q_L \qquad (12)$$

With the stagnation pressure back to its preexpansion value, the final step in the cycle is to return the stagnation temperature to T_0 and thereby return the unit mass of gas to the reservoir ahead of the expansion nozzle. This final processing is accomplished by an isobaric cooler. The amount of heat δh_Q given to the cooler can be easily determined from the relation

$$-\delta s_{LG} = \frac{\delta h_Q}{T_o} \qquad (13)$$

with Eq. (8) it becomes

$$\delta h_Q = -(\frac{T_o}{\alpha T_V})[(1-\alpha)(\frac{T_V}{T})-1]\delta q_L \qquad (14)$$

The thermodynamic consequences of taking a unit mass around this very idealized closed-cycle gas-dynamic laser are (a) laser energy δq_L is radiated, (b) work δh_w is done on the gas by the compressor, and (c) heat is rejected by the cooler. For a fixed value of δq_L the value of δh_w and δh_Q depend on α, T, T_V, and T_0. The quotient T_V/T is not arbitrary however, but must be large enough to achieve a population inversion if the gas is to lase. The necessary condition for a population inversion in Eq. (13) shows that $\delta h_Q < 0$, which means that the designation of the heat exchanger as a cooler is correct.

Define the efficiency of the conversion of work δh_w into laser radiation δq_L as η_{GDL}, then

$$\eta_{GDL} \equiv \delta q_L / \delta h_w$$

$$\eta_{GDL} = \{(\frac{T_o}{\alpha T_V})[(1-\alpha)(\frac{T_V}{T})-1] + 1\}^{-1} \qquad (15)$$

Note that this efficiency is different from the overall cycle efficiency reported in Fig. 6 where the efficiency of converting heat into work was also included. Because $\eta_U > \eta_L$, not only is $0 < \eta_{GDL} < 1$, but η_{GDL} can be made arbitrarily close to unity by appropriate choice of conditions, giving virtually complete conversion of work into laser radiation. Thermodynamically speaking, this near complete conversion is possible because the change in the entropy of the gas during lasing can be made arbitrarily small, thereby requiring only an arbitrarily small transference of heat to the cooler.

The system discussed above is a device for converting work into laser radiation. A device is now discussed for getting work from laser radiation which is designated here as a photon engine. What is envisaged is similar to a closed-cycle gas-dynamic laser run in reverse; laser energy is absorbed in a constant-area duct with frozen flow, work is extracted by an isentropic turbine section, and heat is removed in a cooler.

The earlier analysis prior to the introduction of the lasing condition also applies to the photon engine. Now, however, δq_L is negative and there is an absorption condition defined by $\eta_U < \eta_L$. The efficiency of the photon engine is defined as

$$\eta_{PE} \equiv \frac{\delta h_W}{\delta q_L}$$

or   (16)

$$\eta_{PE} = 1 + (\frac{T_o}{\alpha T_V})[(1-\alpha)(\frac{T_V}{T})-1]$$

In contrast to the laser, η_{PE} can be negative. For example, if $T_o > T_V = T$, then $\eta_{PE} = 1 - T_o/T < 0$. A negative value for the efficiency of the photon engine means that the entropy of the gas is increased so much during the laser radiation absorption process that additional energy in the form of work done on the gas (making $\delta h_W < 0$) must be supplied to help transfer entropy to the cooler.

Now consider the case when the laser energy per unit mass flow either given off by the laser or absorbed by the flow is not infinitesimal but is a finite quantity. Laser radiation can be removed in the laser cavity as long as the lasing condition, $T_V/T > (1-\alpha)^{-1}$ is satisfied. As laser energy is removed from the flowing gas in a constant-area duct, however, the quotient T_V/T is reduced until it reaches $(1-\alpha)^{-1}$ at which point lasing ceases. Figure 5 shows a cycle temperature distribution for this case. Similarly for the photon engine, absorption in a constant-area duct continues until T_V/T increases to $(1-\alpha)^{-1}$ and absorption stops. For a constant-area duct, the amount of laser energy obtainable per cycle depends on the value of T_V/T at the entrance to the lasing duct, this energy being greater for larger values of T_V/T. Similarly for the photon engine, the laser energy absorbed per cycle depends on the value of T_V/T at the entrance to the constant-area absorption duct. However, for this case the capacity increases with decreasing T_V/T. From Eqs. (15) and (16), it is seen that efficiencies decrease for values of T_V/T significantly different from $(1-\alpha)^{-1}$. For a constant-area lasing and absorption duct, greater energy transference results in lower average conversion efficiencies.

This situation is altered if a variable-area duct is used in place of the constant-area duct. Suppose that the area in the radiation extraction section of the laser increases in the flow direction such that T_V/T is maintained constant over its entire length. Then T_V/T can be maintained at a value slightly greater than $(1-\alpha)^{-1}$, resulting in a high efficiency. At the same time a finite amount of laser energy per cycle can be extracted. A finite amount of work can also be obtained at a high efficiency from the photon engine by using an absorption duct whose area decreases in the flow direction.

This development has shown that the thermodynamic limit for the efficiency of producing finite amounts of laser radiation from shaft work or for generating shaft power from laser radiation is unity. This limit is achieved only if the machine is properly loaded and certain idealized assumptions of component efficiencies can be realized.

VI. CONCLUDING REMARKS

In the above section, the thermodynamics of thermal lasers of the gas-dynamic type have been examined. These studies have led to the design of a conceptual device in which the limitation of transforming heat into coherent radiation can be examined. By exploring the basic thermodynamic relationships controlling the operation of this device, it is concluded that a closed-cycle gas-dynamic laser is possible in which all of the shaft energy supplied can be turned into laser radiation. Hence, it is possible in principle to convert heat into coherent radiation with approximately the same efficiency with which heat may be converted into work.

By modifying the closed-cycle gas-dynamic laser system, it is shown that this system can be operated in reverse and the incoming radiation may be used to pump the gas in the loop so that shaft power can be extracted. By carefully controlling the temperature distribution in this machine, laser energy can be converted into useful shaft energy with an efficiency approaching unity.

The authors are aware that these machines are conceptual in nature and that practical and useful devices would require component efficiencies pushing the state of the art. However, these machines have demonstrated that the limits of the efficiency are high and, therefore, it is hoped that they will stimulate thinking leading to the development of practical devices which will make the concept of radiant energy power transmission a reality.

ACKNOWLEDGMENTS

The authors are honored to acknowledge the many contributions of their colleagues. R. Leon Leonard contributed much of the discussion of the photon capacitor and completed the calculations presented here. George Mullaney contributed to the development of the coherent radiation generator. We are also indebted to our many colleagues who have contributed valuable discussions that helped guide our thinking. This work was performed under the auspices of NASA Grant No. NGL48-002-044 monitored by A. Gessow.

Editors' Note. The authors have recently proposed a positive displacement engine cycle, which would bear many similarities to a reciprocating engine cycle, to replace the high velocity flow cycle described above. By postulating a closely tailored piston motion, coherent radiation can be added during compression with a vanishingly small entropy rise. Work could be withdrawn from the cycle reversibly with a theoretical efficiency approaching one. The advantages of such an approach are that flow velocities are subsonic and one does not require a supersonic diffuser beyond present technological capability. Difficulties arise from unfavorable kinetic rates in the N_2-CO_2 system, but other systems such as N_2-CO are under investigation.

APPENDIX A

The Photon Capacitor. The photon capacitor is a new device based on the kinetic processes in a N_2-CO_2 laser. The photon capacitor absorbs laser energy, stores it, and emits laser radiation after suitably processing the storage medium. The energy is stored in the vibrational levels of a N_2-CO_2 gas mixture which is flowing at near sonic velocities. Since the laser energy is stored, the device is called a photon capacitor.

In order for the concept of the photon capacitor to be realized, the absorption of resonant radiation and the saturation of that absorption at high incident laser intensities is very important. Hence, the photon capacitor has many similarities to the gas-dynamic CO_2 laser. One significant application of the photon capacitor would be to store radiation from a multimode CO_2 laser and then reemit the radiation in a low-order diffraction-limited beam. The photon capacitor appears to have the potential for accomplishing this beam improvement at somewhat higher overall efficiency than present mode-selection techniques in high-power lasers.

The photon capacitor is shown in Fig. 8. Multimode laser light from a large CO_2 laser is incident upon a heated mixture of CO_2, N_2, and a catalyst, usually helium. The temperature of the absorbing mixture is selected to maximize the number of molecules in the first excited vibrational energy

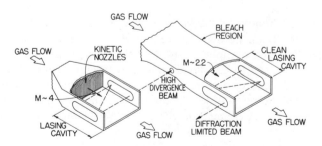

Figure 8. Photon capacitor configuration.

state of CO_2 (the 100 symmetric stretching state). Since the laser radiation is produced by 001-100 transition, resonant absorption occurs; and provided the transition rate due to the absorption is large compared to collisional deactivation times of the 001 level, the number of molecules in the 001 level will increase until it becomes equal to the number in the 100 level. The gas is then said to be bleached because resonant absorption no longer occurs and the absorbing CO_2 becomes transparent to 10.6-μ radiation. This phenomenon will occur in pure CO_2. But when N_2 is present, an additional collisional deactivation mechanism occurs in the resonant exchange with the $v = 1$ level of N_2, and the resonant absorption is then a mechanism for populating the $N_2(v = 1)$ level and storing significant amounts of vibrational energy. In an N_2-CO_2 mixture total bleaching occurs when $n_U = n_L$ and when the upper level is in collisional equilibrium with the first vibrational level in N_2.

Thus it is clear that significant amounts of vibrational energy can be stored in N_2-CO_2 at relatively low translational temperatures. In order to extract the stored vibrational energy, the flow is simply expanded supersonically to a higher Mach number so that the lower laser level is depopulated and an inversion is produced. The stored energy can then be extracted. However, the starting condition for the expansion is not equilibrium at some relatively high temperature for which

$$n_U/n_L = \exp(-1380/T)$$

as in a gas-dynamic laser, but the bleached condition where $n_U = n_L$. Thus, the requirement that the gas kinetics be frozen is not nearly so difficult to achieve. As a result, the final Mach number is only about one-half of the final Mach number necessary for a gas-dynamic laser. The consequence of this lower Mach number and the less stringent freezing condition is a much more uniform flow than is possible in the gas-dynamic laser and, therefore, a lasing medium capable of correspondingly greater mode purity at high power levels.

Kinetics of the Photon Capacitor.   The kinetic processes in a photon capacitor can be separated into two regions, the absorption region where energy is invested in the 001 level of CO_2 and the $v = 1$ level of N_2, and the expansion region where the population inversion is generated.

For the absorption region, the important vibrational levels are indicated in Fig. 1. The assumption is made that 100, 020, and 010 levels of CO_2 are in equilibrium at the translational temperature [18] for all time scales important to the problem. The latter assumption is violated during the resonant absorption process, but again becomes valid to within 5% as the absorption process nears the bleached condition.

The rate equations can be written in the following form:

$$dn_U/dt = - \tau_\nu^{-1}(n_U - n_L) + \tau_2^{-1}(n_{U_e} - n_U) + \tau_3^{-1} n_N - \tau_4^{-1} n_U$$

$$dn_N/dt = \tau_4^{-1} n_U - \tau_3^{-1} n_N$$

$$[1 + 2 \exp(721/kT) + 3 \exp(102/kT)] (dn_L/dt) = \tau_\nu^{-1}(n_U - n_L) - \tau_2^{-1}$$

$$- \tau_2^{-1} (n_{U_e} - n_U) + 2 \exp(721/kT) \tau_1^{-1} (n_{L_e} - n_L)$$

where kT is expressed in amounts of cm^{-1}, τ_{1-4} are the collisional decay times for the transitions as indicated on Fig. 1, n_N is the number density of N_2 molecules in the $v = 1$ level, and the subscript e refers to the population at equilibrium at the translational temperature T. The complexity of the third equation results from the assumption of local equilibrium among the 100, 020, and 010 levels. τ_ν is given by

$$\tau_\nu = (\pi h \nu \, \tau \Delta \nu)/(\lambda^2 I)$$

where ν and λ are the frequency and wavelength of the resonant (laser) radiation, τ is the linewidth, which is determined by pressure broadening in this paper, and I is the intensity (power per unit area) of the incident laser radiation. The definition of τ_ν has been restricted here to transitions for which the upper and lower levels have a degeneracy of unity.

With the assumption of constant pressure, temperature, and τ_ν, the rate equations are a set of three linear first-order differential equations with constant coefficients, and require only the solution of a cubic in order to determine the relaxation of the system. A solution for one set of conditions is shown in Fig. 7 and is typical of the nature of the solution. At time $t = 0$, the gas mixture is in thermodynamic equilibrium at the translational temperature. At that time, the laser is switched on at full intensity. In a time scale comparable to τ_ν, the upper and lower laser levels come to nearly the same population, essentially independent of the other gases and behaving as a two-level system. This process reduced n_L below its equilibrium value and n_U is increased above its equilibrium value.

Consequently, on a time scale comparable to τ_1, the 001 level refills and cools the ground state; and on a time scale comparable to τ_4, the 001 level transfers vibrational energy to the $N_2(v = 1)$ level in which it is stored. The absolute filling time for the $N_2(v = 1)$ level is longer than τ_4, because filling is dependent upon transfer both from the CO_2 (000) to CO_2 (100) and upon CO_2 (001) to $N_2(v = 1)$ which are collisional transitions. Varying the mixture ratio of CO_2, N_2, and catalyst affords considerable opportunity to affect the process, but its physical nature does not change.

In a steady-state radiation field as $t \to \infty$ the final state of the gas is

$$n_{U_\infty} = (\tau_\nu n_{U_e} + \tau_2 n_{L_e})/(\tau_\nu + \tau_2)$$

$$n_{N_\infty} = (P_{N_2}/P_{CO_2}) n_{U_\infty}$$

where P_{N_2} and P_{CO_2} are the partial pressures of N_2 and CO_2, respectively, and for $\tau_\nu \gg \tau_2$, this approaches the limit

$$n_{L_\infty} = n_{U_\infty} = n_{L_e}$$

In the above it was assumed that the bleaching process took place at constant temperature. The temperature change of the gas due to bleaching will be that due to an energy extraction.

$$Q_{out} = \Delta E_{100-000}(n_U + n_N - n_{N_e}) = n_{total}\, C_V \Delta T$$

The functional form of the temperature change is

$$\Delta T/t \sim (n_U + n_N)/n_{total}$$

So for mixtures with typical amounts of He as a catalyst, the temperature changes are of the order of 1 to 5%. Consequently, the assumption of constant temperature for determination of collisional decay times is valid. However, the assumption $\tau_\nu = $ const depends not only upon the constant temperature and pressure but also upon constant laser intensity. Since the laser is contributing the energy to excite the photon capacitor, this assumption is invalid. The variation of I, however, is spatial rather than temporal so that the above results may be considered valid for local regions in the gas. The problem is more difficult for $\tau_\nu \neq$ const.

As discussed earlier, the expansion and energy extraction from the photon capacitor are kinetically similar to the gas-dynamic laser. The kinetic model for that process has been discussed elsewhere, hence is not repeated here. One question which should be considered is: how long can the photon capacitor store "light" or, more correctly, vibrational energy, and with what efficiency will it perform?

For the systems considered, wall effects are assumed negligible because of surface-to-volume considerations. The principal loss in the photon capacitor is due to the collisional leakage from the $\nu = 3$ to the $\nu = 2$ mode indicated schematically in Fig. 1 by τ_2. For high efficiencies and long storage times, τ_2 must be long compared to the storage time and this is accomplished by "storing" after the gas-dynamic expansion at low pressure. One must still accept energy loss due to τ_2 during the bleaching process, and this loss was $\sim 50\%$ for the operating conditions of Fig. 7. More appropriate selection of mixtures, however, will markedly decrease that loss. Since the photon capacitor is similar to the photon engine, the high-efficiency operating conditions should be similar. It has previously been shown that for the photon engine to even run as an engine the energy must be added when the flow is supersonic.

REFERENCES

1. A. Javan, Phys. Rev. Letters 3, 87 (1959).
2. A. L. Schawlow and C. H. Townes, Phys. Rev. 112, 1940 (1958).
3. A. Javan, W. R. Bennett, Jr., and D. R. Herriott, Phys. Rev. Letters 6, 106 (1961).
4. I. R. Hurle, A. Hertzberg, and J. D. Buckmaster, CAL Report No. RH-1670-A-1 (Cornell Aeronautical Laboratory, Buffalo, N.Y., 1962).
5. I. R. Hurle and A. Hertzberg, Phys. Fluids 8, 1601 (1965).
6. N. G. Basov and A. N. Oraevskii, Zh. Eksp. Teor. Fiz. 44, 1742 (1963) [Sov. Phys. JETP 17, 1171 (1963)].
7. C. K. N. Patel, Phys. Rev. Letters 13, 617 (1964).
8. E. T. Gerry, IEEE Spectrum 7, 51 (1970).
9. V. K. Konyukhov and A. M. Prokhorov, ZhETF. Pis. Red. 3, 436 (1966) [JETP Letters 3, 286 (1966)].
10. N. G. Basov, A. N. Oraevskii, and V. A. Shcheglow, Zh. Tekh. Fiz. 37, 339 (1967) [Sov. Phys. Tech. Phys. 12, 243 (1967)].
11. J. K. Royle, A. G. Bowling, and J. Lukasiewicz, Report No. AERO 2221, S. D. 23 (Royal Aircraft Establishment, Farnborough, Hants, 1947).
12. W. H. Christiansen and G. A. Tsongas, Phys. Fluids, 14, 2611 (1971).
13. V. K. Konyukhov, I. V. Matrosov, A. M. Prokhorov, D. T. Shalunov, and N. N. Shirokov, ZhETF. Pis. Red. 10, 84(1969) [JETP Letters 10, 53 (1969)].
14. N. G. Basov, V. G. Mikhailov, A. N. Oraevskii, and V. A. Shcheglov, Zh. Tekh. Fiz. 38, 2031 (1968) [Sov. Phys. Tech. Phys. 13, 1630 (1969)].
15. J. D. Anderson, Jr., Phys. Fluids 13, 1983 (1970).
16. C. K. N. Patel, Sci. Am. 219, 22 (1968).
17. R. C. Weatherston and A. Hertzberg, J. of Eng. for Power, Trans. ASME, Series A, 89, 217 (1967).
18. R. L. Taylor and S. Bitterman, Rev. Mod. Phys. 41, 26 (1969).

X-RAY EMISSION FROM LASER GENERATED PLASMAS*

P. J. Mallozzi, H. M. Epstein, R. G. Jung, D. C. Applebaum,
B. P. Fairand, and W. J. Gallagher
Battelle Memorial Institute
Columbus Laboratories
505 King Avenue, Columbus, Ohio

ABSTRACT

Theoretical and experimental studies of laser-produced high-Z radiative
plasmas were conducted. A theoretical model and computer code have been
developed which include the effect of index of refraction, bremsstrahlung,
recombination radiation, line radiation, single fluid shock hydrodynamics,
separate kinetic temperatures for electrons and ions, and electron thermal
conductivity. Its validity extends to highly nonequilibrium regimes such as
the "time-dependent corona" regime. Computer calculations have been performed which predict that large fluxes of x rays can be generated by placing
high-Z targets in the laser focus, with efficiencies in excess of 20% for
converting laser light into x rays being feasible at 10^{14} W/cm^2 of laser light
incident over 1 nsec. The experiments were performing using 10-100 J,
1-2 nsec, 1.06 μ pulses generated with Battelle's HADRON/CGE VD-640

---

*This research was supported by the Advanced Research Projects Agency.
Some results from an ongoing research project for Battelle Memorial Institute under its Physical Sciences Program have been incorporated into this
article.

neodymium-doped glass laser, isolated from backscattered light by a special exploding mirror surface. The x-ray yield was measured with several techniques, including scintillator detectors, material damage, microcalorimeter, scintillator camera, and thermoluminescence detectors. Conversion efficiencies of laser light into x rays of at least 10%, probably 15-20%, and possibly higher were obtained with iron slab targets oriented at 45° with respect to the incident beam. The x-ray yield above 1 keV is at least 5% of the incident laser light.

I.   INTRODUCTION

Progress in the development of high power lasers during the past several years has opened the door to several new areas of research. The best known, of course, is the remarkable possibility of achieving controlled thermonuclear fusion by means of laser heated plasmas.

There is a second and newer application of high power lasers, however, which from the point of view of basic laser and plasma physics is equally exciting. The new problem is to generate an intense x-ray source using lasers. The simplest approach is to place a small particle in the focus of a high power laser or array of high power lasers, and thereby heat it to such a high temperature that the energy it radiates lies in the x-ray regime.

There are many points of similarity between the x-ray problem and the fusion problem. Both, for example, involve the laser heating of plasmas to the multikilovolt regime. But there is an important difference. The fusion work is mostly directed to the heating of fusionable materials, which by nature have a low atomic number, Z. The x-ray work, on the other hand, involves the heating of high-Z materials, which are far more effective in converting laser light into x rays.

The research reported here has conclusively demonstrated that x rays can be generated in sufficient quantities and in an appropriate spectral range to satisfy a number of important requirements for laboratory x rays.

In the early phase of the study, the Battelle laser-generated plasma computer code FLASH was generalized to account for time dependent ionization, and conversion efficiencies greater than 20 percent of laser light into x rays were calculated assuming certain initial plasma conditions. At the same time, a HADRON/CGE VD-640 laser was installed by Battelle-Columbus, and a special isolator device was developed to eliminate the troublesome problem of glass damage caused by amplification of laser light backscattered from targets [1].

In the next phase, the total x-ray yield and effective temperature were studied with 1-nsec laser pulses preceded by a variety of "feet" (i.e., prepulses) ranging from approximately 1 to 15 nsec. The purpose here was to identify the prepulses that would establish optimum initial plasma profiles when the laser beam was focused on particular solid targets. In addition to the foot study, tests were conducted to establish the feasibility of preventing debris from the laser generated plasma from striking x-ray irradiation samples.

The results were extremely successful. X rays with an energy above ~ 1 keV were routinely delivered, behind a beryllium debris shield, at sufficiently high fluxes to <u>vaporize lead</u> over an area of several cm^2. Efficiencies of at least 5% for converting laser light to x rays above 1 keV were obtained. The overall conversion efficiency of laser light to x rays that was obtained was at least 10%, probably 15-20%, and possibly 30-40%. Most remarkably, these results were all obtained using 1-2-nsec laser pulses with less than 100 J total energy. There is no question that these conversion efficiencies, which are already satisfactory, could be increased with the use of larger laser pulses.

In the third phase of the study, attention was focused on incorporating this information into a study of the feasibility and cost of building a 10,000 J neodymium-doped glass laser suitable for x-ray production. As part of the evaluation, the Sovirel rod in the last stage was replaced with a rod of Owens-Illinois ED-2.1 glass. The higher damage threshold of the ED-2.1 glass immediately enabled 240 J to be delivered in a single 1-2-nsec pulse. When two 1-2-nsec pulses separated by ~ 10 nsec were passed through the system, 350 J were obtained in two ~ 175 J pulses. This increased the short pulse capability of the Battelle laser by a factor of 3.5, since the pulses can in principle be emitted along separate optical paths and be simultaneously and routinely delivered to targets.

The plasmas most effective for x-ray production consist of high-Z materials raised to the multikilovolt regime. The dynamics of a high-Z plasma and its radiative output involve a variety of collisional and radiative processes in addition to the usual phenomena that occur in low-Z plasmas. The new processes make possible higher conversion efficiencies, but also add considerable complexity to the problem. In view of this complexity it will be valuable, before discussing the theoretical and experimental work in detail, to establish the basic story of what happens when a high-intensity laser pulse strikes a high-Z target: A discussion of the basic phenomenology is provided in the next two sections, where for definiteness, 1000 J of pulsed neodymium laser light are assumed.

II. MACROSCOPIC CHARACTERIZATION

The first problem to consider is to ensure that laser light is <u>admitted</u> into the plasma and not reflected away at the surface. Whenever the laser frequency ω_L is less than the plasma frequency $\omega_p=(4\pi n_e e^2/m)^{1/2}$ of the target plasma, the incident light is reflected. If the reverse is true, the light will

enter the plasma. The critical electron density at which the plasma
frequency equals the frequency of neodymium laser light is 10^{21} cm^{-3}, so
that to avoid reflection

$$n_e \leq 10^{21} \text{ cm}^{-3} \tag{1}$$

This description of light reflection is naive and not always true, but it is a
useful introductory statement of a problem that will be dealt with in more
detail in Sec. IV.

The second problem to consider is whether the laser light is <u>absorbed</u>
by the plasma. In order to ensure a high conversion efficiency of laser
light into x rays, most of the light must be absorbed rather than transmitted
through the plasma. This condition imposes a lower bound on the electron
density of about 10^{19} or 10^{20}. Thus, there is a working range of one or two
factors of ten of electron density in which the laser light enters the plasma
without being reflected and is also mostly absorbed before passing through.

Now where within this working range is it most desirable to operate?
The answer is provided by the objective of maximizing the x-ray yield.
Since all of the processes known to contribute x radiation, namely, bremsstrahlung, recombination radiation, and line radiation, increase with the
square of electron density, it is desirable to work as close as possible to
the upper end of the working range. This establishes 10^{21} as a sort of
"favorite" electron density for x-ray production.

Now what should the plasma size be? This is established by heat
capacity considerations based on complete absorption of the 1000 J of
incident laser light. The characteristic volume and side length of a plasma,
of electron density 10^{21}, that lies in the multikilvolt regime with 1000 J of
absorbed energy are

$$L^3 \sim 10^{-3} \text{ cm}^3 \tag{2}$$

$$L \sim 10^{-1} \text{ cm} \tag{3}$$

This is a reasonable approximation for plasmas of any atomic number, Z.

It is now possible to determine what the pulse length of the laser should
be. This parameter is established by the characteristic hydrodynamic
expansion time of a 0.1-cm-thick multikilovolt plasma, which is $\sim 10^{-9}$ sec.
The approximation is valid for plasmas of any n_e and Z within a factor of 3
or so. If a pulse substantially longer than 1 nsec is used, the plasma will
expand and go transparent during the irradiation, and will convert the
absorbed energy into x rays with a reduced efficiency. A shorter pulse, on
the other hand, is probably acceptable, but there is usually no reason to
make the problem more difficult by employing a pulse which is unnecessarily
short. The appropriate pulse width of the laser is thus

$$\tau_L \sim 10^{-9} \text{ sec} \qquad\qquad (4)$$

What about heat conduction and temperature uniformity? Heat is conducted in the laser plasma mainly by electron thermal conductivity. This process is similar to the conductivity of an ordinary gas except that the carrier particles are electrons rather than atoms. Calculations show that electron thermal conductivity will generally render the electron temperature uniform within 1 nsec if the final electron temperature is greater than ~5 keV, but will have difficulty in doing so if the final electron temperature is only 1 or 2 keV. When the conductivity fails, other processes, such as shock heating and the tendency of the incident light to be transmitted through hot regions and absorbed in cold regions, will usually step in and establish a reasonably uniform temperature. Reradiated energy, on the other hand, is not an effective agent for temperature equalization. Most of it escapes from the plasma, except at low temperatures.

We have thus far arrived at a self-consistent set of laser and macroscopic plasma parameters for maximizing x-ray production for the case of 1000 J of neodymium laser light. But neither the numbers nor their stated limitations are absolute. For example, it appears possible to heat plasma with electron densities greater than 10^{21} by means of a two-step process, the first step of which was pointed out by Dawson, Kaw, and Green [2]. The idea is that no plasma is uniform, so that reflection of an external laser beam at the critical density point entails a two-way passage of the light through an outer skin with an electron density ranging from zero to 10^{21}. This "underdense" portion of the plasma is often capable of absorbing most of the incident energy, which is then partly transported to the "overdense" portion of the plasma by electron thermal conductivity. Thus, the numbers obtained above are mainly logical guidelines for an initial study.

III. MICROSCOPIC CHARACTERIZATION

In any case, we are now in a position to characterize the microscopic nature of the plasma. The overriding feature in this realm is that the plasma particles are not in local thermodynamic equilibrium (LTE). One nanosecond is long enough for the electrons to equilibrate among themselves by Coulomb collisions and be characterized by an electron kinetic temperature, T_e. Similarly, there is enough time for the ions to equilibrate among themselves at an ion kinetic temperature, T_i. But there is generally insufficient time for T_e and T_i to become equal. What usually happens is that the incident laser light is absorbed by the electrons, which rise in temperature relatively rapidly while slowly heating the ions by electron-ion collisions.

Now this in itself constitutes a breakdown of LTE. But it is a trivial aspect of its breakdown, both from an analytic point of view and from the point of view of x-ray production. This is not to say that it is unimportant. Indeed, in a fusion plasma it is of paramount importance, since the desired product--fusion energy--would be sharply suppressed by a lagging ion temperature. But in the x-ray problem the desired product depends mainly on electron temperature, which is only slightly affected (in high-Z plasmas) if the ion temperature lags.

There is another aspect of the breakdown of LTE which has a far more serious effect on x-ray production. This is the question of whether or not the relative ion populations (i.e., the relative number of, say, 15, 16, and 17 times ionized tungsten) are governed by the Saha Equations, as they should be for a plasma in LTE. The importance of this question is due to the strong dependence of x-ray emission on ionization potential.

The Saha Equations may be written in the approximate form [3]

$$\frac{f_{s+1}}{f_s} = 3.0 \times 10^{39} \frac{(kT_e)^{3/2}}{n_e} \exp\left(-\frac{\chi_s}{kT_e}\right) \tag{5}$$

where f_s and f_{s+1} are the ion fractions that are s and s+1 times ionized, n_e is the density of free electrons, and kT_e is the electron temperature; χ_s is the ionization potential of the s times ionized ion. In this report, cgs units are used unless otherwise stated. In the case of a plasma which is optically thin, the ionization state described by the Saha Equations results from the competitive balance between collisional ionization by two-body electron-ion collisions, and recombination by so-called three-body recombination. The latter process is a radiationless three-body "collision" involving one ion and two electrons, in which one electron combines with the ion, while the other is scattered away with a direction and velocity which guarantees energy and momentum conservation. These processes are statistical inverses of one another (i.e., one transforms into the other under time reversal); this is to be expected, since the Saha Equations pertain to an ion distribution which is in LTE.

The Saha Equations definitely break down. This may be seen by applying the following (very approximate) rule of thumb given by McWhirter [4]:

$$n_e \lesssim 10^{16} \left[(T_e)_{eV}\right]^{7/2} \text{ cm}^{-3} \tag{6}$$

The quantity $(T_e)_{eV}$ is the electron temperature expressed in electron volts. Whenever Eq. (6) is satisfied, the Saha Equations break down and the ion distribution is out of equilibrium. This is generally the case in the present problem, since use of $T_e \gtrsim 1000$ eV in Eq. (6) gives $n_e \lesssim 3 \times 10^{26}$, which is easily satisfied by the key electron density 10^{21}. On a microscopic level, the state of affairs that prevails when Eq. (6) is satisfied is that the relative ion populations are determined by the competitive balance between collisional ionization--the same ionization process as in the Saha regime--and two-body radiative recombination, in which a free electron recombines with an ion and the excess energy and momentum are carried off by a photon. These processes are not statistical inverses of one another: the ionization process is purely collisional, whereas the recombination process involves a photon.

Now when Eq. (6) is satisfied, the ion distribution is often still in a quasisteady, though nonequilibrium, state. The ionization equations in this case are as tractable as those for equilibrium, the main difference being that the Saha Equations are replaced by another set of quasisteady equations, generally known as "Coronal Equilibrium" equations. The exact form of these equations depends on the formulas selected for the fundamental cross-sections. The Coronal Equilibrium equations implied by the formulas given in Sec. IV are

$$\frac{f_{s+1}}{f_s} = 3.3 \times 10^{-16} \left(\frac{1}{1+d}\right) \frac{\Delta_o(s)(kT_e)^{3/4}}{\chi_s^{11/4}} \exp\left(-\frac{\chi_s}{kT_e}\right) \tag{7}$$

where $\Delta_o(s)$ is the number of electrons in the outer Bohr shell ($\Delta_o=1$ for hydrogenlike ions, $\Delta_o=6$ for oxygenlike ions, etc.), d is a correction term, and the other quantities are defined as in Eq. (5). The term "quasisteady", as applied to Eqs. (5) and (7), refers to the fact that the f_s's at any instance are specified by the instantaneous values of T_e and n_e.

Unfortunately, in a laser-heated plasma, quasisteady conditions do not always prevail. This may be established by a second rule of thumb, also given by McWhirter [4], which applies whenever Eq. (6) is satisfied; this second rule of thumb states that quasisteady conditions break down if the characteristic time for hydrodynamic expansion or contraction, τ_h, satisfies the relation

$$\tau_h \lesssim \frac{10^{12}}{n_e} \tag{8}$$

If Eq. (8) is satisfied, the plasma is in an extremely nonequilibrium state. To specify the instantaneous ionization state, it is necessary to know not only the local free electron conditions, but their entire past history as well! The only recourse in such a situation is to solve rate equations which include the principal microscopic processes that contribute to the ionization rate, and calculate the time development of the plasma from an initial state.

The ionization of a multikilovolt laser-generated plasma requires the time-dependent treatment. This may be seen by inserting the critical electron density 10^{21} into Eq. (8). In the underdense plasma "skin", where $n_e < 10^{21}$, the full time-dependent treatment is required, although in the overdense "core" (which can be heated by shocks and thermal conduction), a quasisteady treatment is generally valid.

IV.   THEORETICAL MODEL

A theoretical model and computer code have been developed to predict the time development of a high-Z radiative plasma interacting with a laser beam. The computer code is designated by the name FLASH, and is continually being extended and improved. The version to be discussed in this report is FLASH 3, which is one dimensional but accounts for the time dependence of ionization.

A.   Absorption and Reflection

In the model for absorption and reflection of laser light, the light enters the surface of the plasma and penetrates, via the laws of geometrical optics, to the critical density point, where 100% reflection is assumed to occur: It then retraces its path and exits from the plasma. On the way in and on the way out it is assumed to be absorbed at the rate predicted by the Dawson and Oberman [5] microwave conductivity formula, suitably modified by the approximate index of refraction [6]

$$n = \left(1 - \frac{\omega_p^2}{\omega^2}\right)^{1/2}$$

(9)

where ω_p is the plasma frequency. Thus, the beam intensity attenuates as $dI/dx = -\alpha I$, with the absorption coefficient given by

$$\alpha = 6.2 \times 10^{-25} \frac{n_i^2 \, \overline{Z^*} \, \overline{Z^{*2}} \ln \Lambda}{\omega^2 (1 - \omega_p^2/\omega^2)^{1/2} (kT_e)^{3/2}} \ cm^{-1}$$

(10)

where

$$\omega_p^2 = 3.182 \times 10^9 n_i \overline{Z^*} \ sec^{-2}$$

(11)

The light which survives absorption during the two-way passage constitutes reflection. In these formulas, n_i denotes ion density, T_e denotes electron temperature, and ω is the frequency of the laser light. The quantity $\overline{Z^*}$ is the mean ion effective charge, $\overline{Z^{*2}}$ is the mean square ion effective charge, and $\ln\Lambda$ is the so-called Coulomb logarithm, which is generally set equal to 5 or 10 throughout a given problem. The justification for this extension of microwave absorption theory to the optical regime was provided by Mallozzi and Margenau [7], who showed that the primary physical process underlying the absorption of laser light in plasmas, namely, free-free absorption (inverse bremsstrahlung), is the quantum analogue of the process which underlies classical microwave absorption: The formulas for free-free absorption and microwave absorption were shown to be identical when $h\nu \ll kT$.

B.   Constitutive Equations

The laser energy which is absorbed in each zone of the plasma is converted into electron kinetic energy. Now for a plasma in LTE, the energy which the electrons absorb is immediately partitioned, in a fixed and rate-independent way, among the internal degrees of freedom in each zone. However, in the present plasma, the conditions in each zone must be established by solving an appropriate set of rate equations. These equations are given next, and provide a time-dependent link between the total energy density ϵ, the total pressure p, the kinetic electron and ion temperatures T_e and T_i, and the relative ion populations f_s.

The relative ion populations are described by the equations

$$\frac{df_0}{dt} = \overline{Z^*}n_i(-f_0S_0 + f_1\alpha_1 + \overline{Z^*}n_if_1\beta_1)$$

$$\vdots$$

$$\frac{df_s}{dt} = \overline{Z^*}n_i(-f_sS_s - f_s\alpha_s - \overline{Z^*}n_if_s\beta_s$$
$$+ f_{s-1}S_{s-1} + f_{s+1}\alpha_{s+1} + \overline{Z^*}n_if_{s+1}\beta_{s+1})$$

$$\vdots$$

$$\frac{df_Z}{dt} = \overline{Z^*}n_i(-f_Z\alpha_Z - \overline{Z^*}n_if_Z\beta_Z + f_{Z-1}S_{Z-1}) \tag{12}$$

together with the normalization condition

$$\sum_{s=0}^{Z} f_s = 1 \tag{13}$$

and the definition

$$\overline{Z^*} = \sum_{s=1}^{Z} f_s Z_s \tag{14}$$

The quantity

$$Z_s = 0, 1, \ldots Z \tag{15}$$

and denotes effective ion charge. The coefficients S, α, and β appearing in Eq. (12) account respectively for collisional ionization, radiative recombination, and three-body recombination, and are given by

$$S(Z_s) = 4.9 \times 10^{-24} \frac{\Delta_o(Z_s)(kT_s)^{1/4}}{\chi^{7/4}(Z_s, g)} \exp\left(-\frac{\chi(Z_s, g)}{kT_e}\right) \tag{16}$$

$$\alpha(Z_s) = 1.5 \times 10^{-8} \ (1+d) \frac{\chi(Z_s-1, g)}{(kT_e)^{1/2}} \tag{17}$$

$$\beta(Z_s) = 1.6 \times 10^{-63} \frac{\Delta_o(Z_s-1)}{\chi^{7/4}(Z_s-1, g)(kT_e)^{5/4}} \tag{18}$$

In these formulas, $\chi(Z_s, g)$ denotes the ground-state ionization potential of an ion of effective charge Z_s, and $\Delta_o(Z_s)$ is the number of electrons in the outermost occupied Bohr shell. The quantity d is a correction for dielectronic recombination [8]. The formulas for S and α are essentially those given in Ref. 4, except for d, and their use leads Eq. (12) to reduce to the Coronal Equilibrium Equations (7) in the limit of low particle density and slow time variation. The formula for β was derived from S by requiring Eq. (12) to reduce to the Saha Equations (5) in the limit of high particle density and slow time variation. This procedure leads to a low estimate for β: The reason is that formula (16) underestimates S at high densities, since it ignores the existence of excited states.

The ionization equations given above must be solved simultaneously with the following equations for energy and pressure:

$$\epsilon = \epsilon_e + \epsilon_i + \epsilon_I \tag{19}$$

$$p = \frac{2}{3}\left(\epsilon_e + \epsilon_i\right) \tag{20}$$

$$\epsilon_e = \frac{3}{2}\,\overline{Z^*}\,n_i k T_e \tag{21}$$

$$\epsilon_i = \frac{3}{2}\,n_i k T_i \tag{22}$$

$$\epsilon_I = n_i \sum_{s=0}^{Z} f_s U_s = n_i \sum_{j=1}^{Z} f_s \sum_{j=1}^{s} \chi(Z_j - 1, g) + const \tag{23}$$

$$\frac{d(\epsilon_e + \epsilon_I)}{dt} = -(\epsilon_e - \overline{Z^*}\,\epsilon_i)\frac{1}{\tau_{ei}} + \left[\frac{\partial(\epsilon_e + \epsilon_I)}{\partial t}\right]_{rerad} \tag{24}$$

$$\frac{d\epsilon_i}{dt} = (\epsilon_e - \overline{Z^*}\,\epsilon_i)\frac{1}{\tau_{ei}} \tag{25}$$

Eq. (19) expresses the total energy density ϵ (erg/cm^3) as a sum of the electron and ion kinetic energy densities ϵ_e and ϵ_i, defined by the ideal-gas formulas (21) and (22), plus the total ionization energy density ϵ_I. The latter quantity is defined in Eq. (23) in terms of the ground-state binding energies U_s per ion, and alternately, the ground-state ionization potentials. The binding energies and ionization potentials can be supplied to FLASH as inputs, or calculated internally using a fundamental atomic model. Equation (20) expresses the pressure in terms of the total kinetic energy density, in accord with the ideal-gas picture for pressure. Eq. (24) gives the rate of change of $\epsilon_e + \epsilon_I$. The first term on the right-hand side accounts for the change due to elastic electron-ion collisions: This term affects only ϵ_e. The second term accounts for the change in $\epsilon_e + \epsilon_I$ due to reradiation, and equals the negative of the total radiated power per cm^3: this term in general affects both ϵ_e and ϵ_I. Eq. (25) describes the change in ϵ_i, which is attributed to elastic electron-ion collisions.

These equations all ignore the potential energy between free electrons, the effect of inelastic collisions on ϵ_i, and the reabsorption of emitted radiation; they also assume that the ions are mostly found in the ground state. The quantity τ_{ei} which appears in Eqs. (24) and (25) is the electron-ion equipartition time, and is given by [9]

$$\tau_{ei} = 1.5 \times 10^{26} \frac{(kT_e)^{3/2} A}{n_i \overline{Z*2} \ln\Lambda} \tag{26}$$

where A is the atomic mass number (A=1 for hydrogen). This formula, as well as most others in FLASH 3 pertaining to collisional processes, is restricted to situations where T_i is not more than a few hundred times higher than T_e. Such conditions almost always apply.

C.   Conduction and Hydrodynamics

Heat conduction in FLASH 3 is accounted for by the following expression for the heat flow vector:

$$F = 1.3 \times 10^{-5} \frac{T_e^{5/2}}{\ln\Lambda} \frac{\overline{Z*}}{\overline{Z*2}} \nabla T_e \ erg/cm^2 sec \tag{27}$$

This expression arises from electron thermal conductivity. Ion thermal conductivity is ignored.

Hydrodynamic motion is predicted by the hydrodynamic equations for single fluid one-dimensional motion. The Lagrangian method is employed, with shock computations stabilized by an "artificial viscosity" term. Energy density and pressure are given by the time-dependent model described earlier, modified to allow for a changing n_i. Two terms are added to the right-hand side of Eq. (24) to allow for heat conduction and the absorption of laser light, which affect ϵ_e, but not ϵ_i or ϵ_I. Heating due to compression and artificial viscosity is assigned to ϵ_e and ϵ_i [i.e., added to the right-hand sides of Eqs. (24) and (25)] in proportion to the instanteous ratio of ϵ_e to ϵ_i.

D.   Reradiation

The energy radiated by the plasma arises from three mechanisms: free-free radiation (bremsstrahlung), free-bound radiation (recombination radiation), and line radiation. FLASH 3 calculates the contribution from each process and assumes that the emitted radiation escapes from the plasma.

The free-free and free-bound radiation emitted into the spectral band $\Delta(h\nu) = h\nu_2 - h\nu_1$ is given by

$$P_{ff}^{1,2} = 1.2 \times 10^{-19} n_i^2 (kT_e)^{1/2} \left[\exp\left(-\frac{h\nu_1}{kT_e}\right) - \exp\left(-\frac{h\nu_2}{kT_e}\right) \right] \overline{Z^*} \cdot \overline{Z^{*2}} \ \text{erg/cm}^3 \text{sec}$$

$$(28)$$

$$P_{fb}^{1,2} = 5.5 \times 10^{-9} n_i^2 \frac{\overline{Z^*}}{(kT_e)^{1/2}} \sum_{s=1}^{Z} \sum_{n=1}^{6} f_s \chi_{sn}^2 (Z_s-1) \frac{2n^2 - \Delta_n(Z_s)}{n} \exp\left(\frac{\chi_n(Z_s-1)}{kT_e}\right)$$

$$\times \left[\exp \frac{-h\nu_1 H[h\nu_1 - \chi_n(Z_s-1)] - \chi_n(Z_s-1)H[\chi_n(Z_s-1) - h\nu_1]}{kT_e} \right.$$

$$\left. - \exp \frac{-h\nu_1 H[h\nu_2 - \chi_n(Z_s-1)] - \chi_n(Z_s-1)H[\chi_n(Z_s-1) - h\nu_2]}{kT_e} \right] \text{erg/cm}^3\text{sec}$$

$$(29)$$

where $\Delta_n(Z_s)$ is the number of nth shell electrons, $\chi_n(Z_s-1)$ is the ionization potential of an electron captured into that shell (the capture decreases Z_s by 1), and H is the "Heaviside" step function, defined by

$$H(x) = 0, \quad x < 0$$

$$H(x) = 1, \quad x \geq 0 \qquad (30)$$

Eqs. (28) and (29) are the usual formulas for free–free and free–bound radiation found in the literature [9] except that integrations over $h\nu$ have been performed with Gaunt factors set equal to unity. The summation over n in Eq. (29) extends over the first six shells, which are the ones most likely to influence two–body recombination.

Line radiation is given by

$$P_L = 6.4 \times 10^{-24} n_i^2 \frac{\overline{Z^*}}{(kT_e)^{1/2}} \sum_{s=0}^{Z} f_s F_s \exp\left(-\frac{\chi_r(Z_s)}{kT_e}\right) \ \text{erg/cm}^3\text{sec} \qquad (31)$$

where $\chi_r(Z_s)$ is a characteristic excitation energy for the sth ion species and F_s is a characteristic absorption oscillator strength. Eq. (31) is based on a model where each ion species emits a single spectral line of photon energy $h\nu = X_r(Z_s)$, which is produced when ground-state ions suffer inelastic two-body collisions with electrons and the excited states subsequently decay; the plasma is assumed optically thin to the emitted radiation. This picture of line radiation is admittedly approximate and lacking in fine structure. However, when combined with the FLASH time-dependent ionization model, the frequency-averaged line emission is a substantial improvement over what might be calculated with either the Saha or Coronal Equilibrium ionization models. Work is underway to develop a more detailed model for line radiation, including self-absorption effects.

V.   SAMPLE CALCULATIONS

The purpose of this section is to present the results of two selected FLASH 3 calculations (Figs. 1 - 4). The input parameters for these computer runs are given in Table I, together with the input parameters of nine additional runs, which are available on request. The computer runs cover two materials, three laser power densities, two laser wavelengths, and three initial plasma profiles. The problems are all one dimensional, with the laser beams normally incident on infinite plasma slabs at uniform initial temperatures of 8.6 eV and with initial ion density profiles which are linear. These profiles are given in Table II.

The main purpose in choosing these problems as the first to be analyzed was to illustrate the effect on x-ray production of basic interaction phenomena such as conduction and time-dependent ionization: Thus, simple slabs were studied rather than a spread of different target geometries. Another intention was to vary crucial parameters corresponding to tunable laboratory quantities such as laser power density and the initial plasma profile. The latter quantity is related to the ionizing prepulse which strikes a solid target before arrival of the main pulse. One might wish that more materials were run. However, iron (Z=26) and calcium (Z=20) both yield high conversion efficiencies, and the basic plasma equations promise similar behavior for higher-Z targets if slight changes in initial plasma profiles are made: The main difference is that harder x-ray spectra are generally attainable with higher-Z targets.

An analysis of the computer runs and a comparison with earlier work [10 - 12] will be presented in Sec. VII. The runs will also be employed in interpreting the experimental results which are presented there.

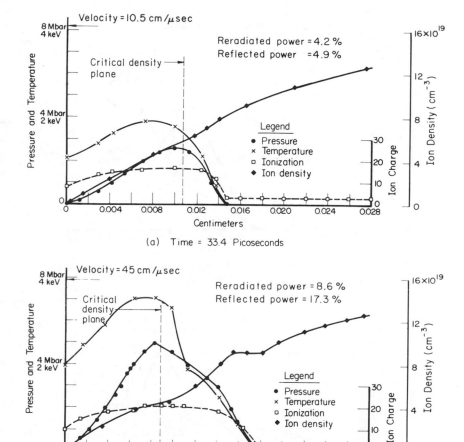

Figure 1.   Nonequilibrium iron calculation: time-dependent ionization. Incident laser flux = 1. 05 × 10^{14} W/cm^2; wavelength = 1. 06 μ; initial conditions: plasma profile A.   (See Tables I, II, pages 184, 185).

179

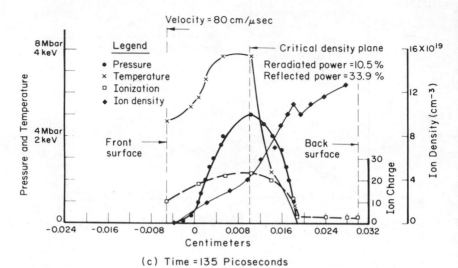

(c) Time = 135 Picoseconds

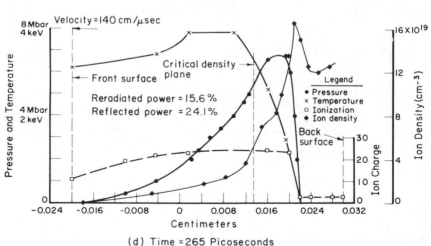

(d) Time = 265 Picoseconds

Figure 1 (continued).

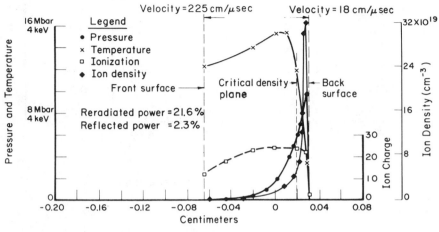

(e) Time = 500 Picoseconds

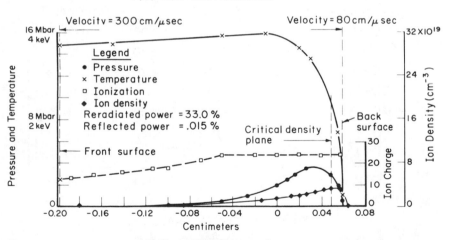

(f) Time = 1000 Picoseconds

Figure 1 (continued).

Spectral Region, keV	Free-Free, %	Free-Bound, %	Line, %	Sum Over Processes, %
0.0 – 0.5	0.04	0.01	3.71	3.76
0.5 – 1.0	0.02	0.01	1.89	1.92
1.0 – 2.0	0.03	0.03	1.28	1.34
2.0 – 4.0	0.03	0.04	0	0.07
4.0 – 8.0	0.02	0.02	0.01	0.05
8.0 – 16.0	0.00	0.01	0	0.01
16.0 – 1024	0.00	0.00	0	0.00
Total	0.14	0.12	6.89	7.15

(a) Time = 135 Picoseconds

Spectral Region, keV	Free-Free, %	Free-Bound, %	Line, %	Sum Over Processes, %
0.0 – 0.5	0.07	0.01	3.96	4.04
0.5 – 1.0	0.05	0.03	2.30	2.38
1.0 – 2.0	0.06	0.06	2.76	2.88
2.0 – 4.0	0.06	0.13	0	0.19
4.0 – 8.0	0.03	0.06	0.12	0.21
8.0 – 16.0	0.01	0.01	0	0.02
16.0 – 1024	0.00	0.00	0	0.00
Total	0.28	0.30	9.14	9.72

(b) Time = 265 Picoseconds

Figure 2. Nonequilibrium iron calculation: time-dependent ionization. Time-integrated reradiated energies as percent of time-integrated input. Incident laser flux = 1.05 x 10^{14} W/cm^2; wavelength = 1.06 μ; initial conditions: plasma profile A. (See Tables I, II, pages 184, 185).

Spectral Region, keV	Free-Free, %	Free-Bound, %	Line, %	Sum Over Processes, %
0.0 – 0.5	0.13	0.02	4.70	4.85
0.5 – 1.0	0.09	0.06	2.71	2.86
1.0 – 2.0	0.11	0.13	4.39	4.63
2.0 – 4.0	0.10	0.29	0	0.39
4.0 – 8.0	0.05	0.13	0.32	0.50
8.0 – 16.0	0.01	0.03	0	0.04
16.0 – 1024	0.00	0.00	0	0.00
Total	0.49	0.66	12.12	13.27

(c) Time = 412 Picoseconds

Spectral Region, keV	Free-Free, %	Free-Bound, %	Line, %	Sum Over Processes, %
0.0 – 0.5	0.19	0.02	11.00	11.21
0.5 – 1.0	0.17	0.06	2.14	2.37
1.0 – 2.0	0.15	0.22	4.33	4.70
2.0 – 4.0	0.14	0.50	0	0.64
4.0 – 8.0	0.09	0.20	0.81	1.10
8.0 – 16.0	0.03	0.04	0	0.07
16.0 – 1024	0.00	0.00	0	0.00
Total	0.77	1.04	18.28	20.09

(d) Time = 1000 Picoseconds

Figure 2 (continued).

TABLE I. PARAMETERS FOR COMPUTER RUNS

	1	2	3	4	5	6	7	8	9	10	11
Material	Iron	Iron	Calcium	Calcium	Calcium	Calcium	Calcium	Calcium	Calcium	Calcium	Calcium
Wavelength (μ)	1.06	1.06	1.06	1.06	1.06	1.06	1.06	1.06	1.06	1.06	10.6
Laser Pulse Width (nsec)	1	1	1	0.5	0.5	0.75	1	1	1	1	0.1
Laser Power Density (W/cm^2)	1.05×10^{14}	1.05×10^{14}	1.05×10^{14}	1.05×10^{14}	1.05×10^{14}	1.05×10^{13}	1.05×10^{15}	1.05×10^{14}	1.05×10^{13}	1.05×10^{14}	1.05×10^{14}
Initial Plasma Profile	A	A	A	A	A	A	B	B	B	C	A
Ionization Dynamics	Time Dep.	Equil.	Time Dep.	Equil.	Equil.	Time Dep.	Time Dep.	Time Dep.	Time Dep.	Time Dep.	Time Dep.
Conduction	Yes	Yes	Yes	Yes	No	Yes	Yes	Yes	Yes	Yes	Yes
Figures	1,2	3,4	------------------------------------Available on request------------------------------------								
Pages	179–183	186–190									

184

TABLE II.   INITIAL PLASMA PROFILES
FOR COMPUTER RUNS

Plasma Profile	Slope (ions/cm^3 per cm)	Thickness (cm)
A	5.1×10^{21}	0.030
B	5.1×10^{20}	0.30
C	1.02×10^{20}	1.50

VI.   DESCRIPTION OF EXPERIMENTS

There is more to be said concerning the theoretical models and calculations presented in Secs. IV and V. But it is best to delay that discussion until the experimental work is presented.

The purpose of the present section is to describe the general experimental approach and the principal experimental techniques employed in the study. The discussion includes a description of the laser and the modifications made on it, and of the special pulse and target problems associated with efficient x-ray production. It also describes the diagnostic methods which were employed and the type of information which was obtained with each method. The discussion of the laser is cursory, since the characteristics of the VD-640 are readily available from the manufacturer. But the laser modifications are discussed in some detail and form an integral part of the study.

A.   The Laser

The experiments were performed with 1.06 μ-wavelength laser pulses generated with Battelle's HADRON/CGE VD-640 neodymium-doped glass laser. A photograph of the laser is shown in Fig. 5. This instrument consists of a Q-switched oscillator (located at the far end of the room), followed by six amplifying stages. The final stage contains a 64-mm-diam rod. The capability of the laser ranges from 100 J in ~1 nsec, up to 450 J in 30 nsec. This range of outputs is achieved by generating a standard 4-J 30-nsec pulse with the oscillator and passing it through a "pulse clipper", which is located between the oscillator and the first amplifying stage. The pulse clipper consists of a laser-triggered spark gap which switches a Pockels cell on for 1-2

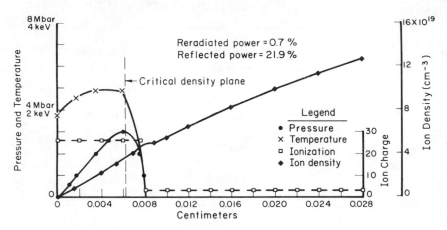

Time = 33.4 Picoseconds

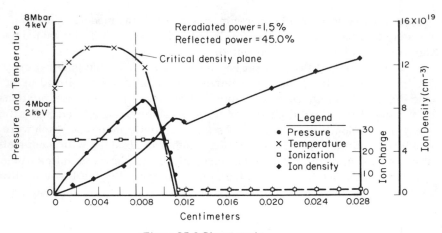

Time = 83.9 Picoseconds

Figure 3:   Equilibrium iron calculation.   Incident laser flux = 1. 05 × 10^14 W/cm^2; wavelength = 1. 06 μ; initial conditions: plasma profile A.

nsec near the peak of the 30-nsec pulse.   The width of the clipped pulse is essentially maintained as it passes through the amplifiers.

B.   Laser Modifications

However splendid the VD-640 laser might be, the first experiments were

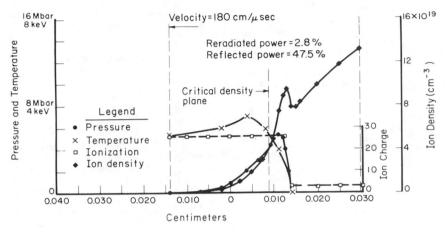

Time = 143 Picoseconds

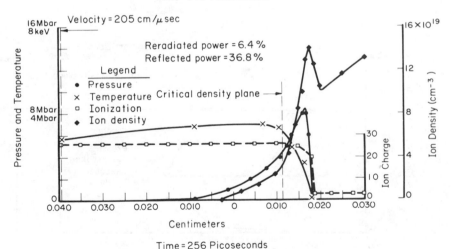

Time = 256 Picoseconds

Figure 3 (continued).

frustrated by the troublesome problem of damage to laser rods caused by
laser light backscattered from the target. This problem was well known from
laser-generated thermonuclear fusion work, and had forced many experiments to be performed at power levels far below the capability of the available laser systems [13]. In the VD-640 laser operating in the 100 J/1 nsec
mode, glass damage will occur if as little as 0.01 J is reflected back along
the optical path by the target. The 0.01-J pulse, although distributed

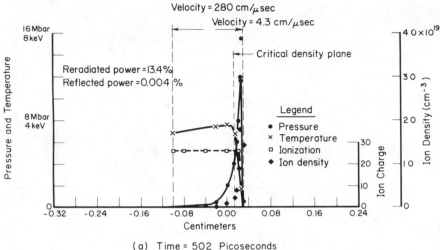

(a) Time = 502 Picoseconds

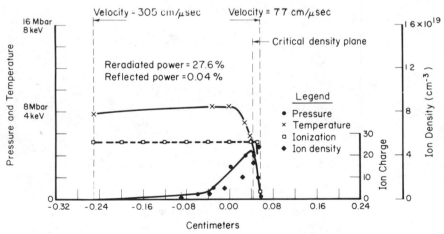

(b) Time = 1000 Picoseconds

Figure 3 (continued).

over approximately 30 cm^2 of glass when it reenters the last amplifier, will converge and be amplified as it retraces the optical path, and at some point will exceed the 10-J/cm^2 damage threshold as it funnels through smaller and smaller rods. There are certain combinations of pulses and targets for which the backscattered light is acceptably low, but this does not solve the problem. A malfunction in equipment might still deliver a dangerous pulse,

Spectral Region, keV	Free- Free, %	Free- Bound, %	Line, %	Sum Over Processes, %
0.0 - 0.5	0.04	0.01	0.37	0.42
0.5 - 1.0	0.02	0.02	0.01	0.05
1.0 - 2.0	0.03	0.05	0.00	0.08
2.0 - 4.0	0.03	0.16	0	0.19
4.0 - 8.0	0.02	0.05	0.00	0.07
8.0 - 16.0	0.00	0.59	0	0.59
16.0 - 1024	0.00	0.02	0	0.02
Total	0.14	0.90	0.38	1.42

(a) Time = 137 Picoseconds

Spectral Region, keV	Free- Free, %	Free- Bound, %	Line, %	Sum Over Processes, %
0.0 - 0.5	0.09	0.02	0.77	0.88
0.5 - 1.0	0.06	0.05	0.01	0.12
1.0 - 2.0	0.07	0.12	0.01	0.20
2.0 - 4.0	0.06	0.41	0	0.47
4.0 - 8.0	0.03	0.15	0.00	0.18
8.0 - 16.0	0.00	1.43	0	1.43
16.0 - 1024	0.00	0.04	0	0.04
Total	0.31	2.22	0.79	3.32

(b) Time = 277 Picoseconds

Figure 4.   Equilibrium iron calculation.   Time-integrated energies as percent of time-integrated input.   Incident laser flux = 1.05 x 10^{14} W/cm^2; wavelength = 1.06 μ; initial conditions: plasma profile A.

and in any case it is undesirable to be limited in the experiments that can be performed.

It would appear that the laser could be protected most simply by introducing an electro-optical shutter after the last stage.   The shutter would transmit the forward pulse but be switched off before the return of the backscattered pulse.   Unfortunately, the construction of Kerr or Pockels cells large enough to handle large diameter beams is very difficult and expensive. In a somewhat more attractive scheme a large multipass Faraday rotator and polarizers are used [14].   But even Faraday rotators have drawbacks for large lasers, owing to the difficulty of establishing the large volume, strong, uniform magnetic fields which are required.

To avoid these difficulties, a new method for isolating Q-switched

Spectral Region, keV	Free-Free, %	Free-Bound, %	Line, %	Sum Over Processes, %
0.0 - 0.5	0.18	0.04	1.23	1.45
0.5 - 1.0	0.11	0.12	0.03	0.26
1.0 - 2.0	0.13	0.25	0.03	0.41
2.0 - 4.0	0.11	0.80	0	0.91
4.0 - 8.0	0.05	0.26	0.01	0.32
8.0 - 16.0	0.01	2.51	0	2.52
16.0 - 1024	0.00	0.07	0	0.07
Total	0.59	4.05	1.30	5.94

(c) Time = 428 Picoseconds

Spectral Region, keV	Free-Free, %	Free-Bound, %	Line, %	Sum Over Processes, %
0.0 - 0.5	0.31	0.07	1.00	1.38
0.5 - 1.0	0.17	0.26	0.11	0.54
1.0 - 2.0	0.20	0.48	0.15	0.83
2.0 - 4.0	0.17	1.44	0	1.61
4.0 - 8.0	0.10	0.39	0.00	0.49
8.0 - 16.0	0.03	3.44	0	3.47
16.0 - 1024	0.00	0.17	0	0.17
Total	0.98	6.25	1.26	8.49

(d) Time = 1000 Picoseconds

Figure 4 (continued).

lasers was developed [1]. The technique is readily applicable to Q-switched lasers of any aperture, power, and pulse width.

The conceptual basis of the new isolator is outlined in Figs. 6 and 7. A key element is a special mirror designed to be highly reflecting to light incident from the front and to be destroyed by light incident from the back. The mirror coating consists of high-reflectivity copper deposited over high-absorptivity germanium.

In Fig. 6, an incident pulse with energy E_0 strikes a beamsplitter, which transmits most of the pulse to the copper mirror and reflects the portion RE_0 through the long leg to the mirror backside. The path length of the back leg is sufficiently long that the main pulse completely reflects off the front of the special mirror before the small rerouted portion strikes it from the back. The small pulse blows the mirror film off the glass plate, as

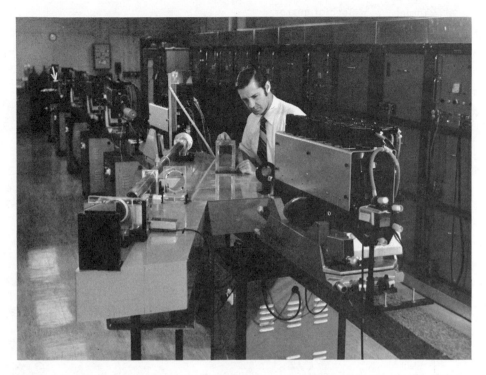

Figure 5.   The VD-640 laser facility.   The Battelle isolator is in the foreground.   The arrow indicates the position of the nanosecond pulse shaper.

shown in Fig. 7(a).   Fig. 7(b) shows the light scattered back along the optical path striking the cloud of mirror-coating debris.

A laser isolator based on these principles has been designed and installed between the last two stages of the Battelle laser, and is visible in the
foreground in Fig. 5.   The device includes a precautionary feature, not
shown in Fig. 6, in which two lenses are inserted between the beamsplitter
and the special mirror for the purpose of focusing the beam through a small
iris and recollimating it.   The iris is enclosed in a vacuum to suppress
attenuation from air breakdown, and serves the function of limiting the solid
angle of acceptance of light which may be diffusely scattered into the return
path by the debris cloud.   In all tests conducted thus far, light backscattered
by the target was attenuated by a minimum of 99% in passing through the
isolator.   The 2- or 3-dB attenuation in the forward direction is easily
recovered by increasing the amplification of the final stage.   One drawback

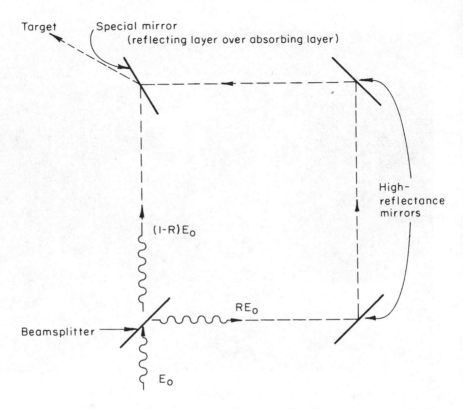

Figure 6.   Conceptual design of isolator (splitting of incident pulse.)

of this system is that a laser pulse entering the isolator at a level which is
unexpectedly low by a factor of 4 or more will yield insufficient energy in
the back leg to blow off the mirror.   But the laser is not likely to suffer
reflection damage at these lower pulse levels.   In any case, additional
mirror development is expected to substantially improve the low pulse limit
for mirror blowoff, and the problem can be eliminated altogether by inserting a supplemental nonlinear absorber to protect against the weaker backscattered pulses, or else, at a sacrifice of damage to the mirror blank, by
vaporizing the mirror using an electrical discharge.   In the present system
the mirror blanks are undamaged and can be recoated at reasonable cost.
    The device described above is actually a highly versatile instrument
which is capable of simultaneously serving many functions besides isolation.

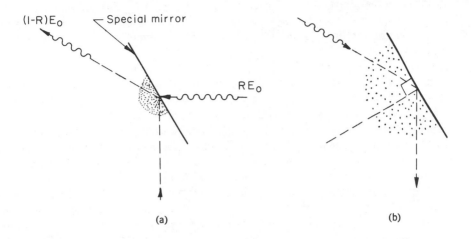

$(I-R)E_0$ Special mirror

RE_0

(a)   (b)

Figure 7.   (a) Special mirror is removed by light pulse routed to mirror backside. (b) Light backscattered along optical path by plasma strikes mirror debris cloud.

These include beam homogenization, superradiance supression, and divergence sharpening.

The special mirror, for example, serves a secondary function of suppressing hot spots in the laser beam. This is accomplished by placing the front surface of the mirror at a position where the average flux approaches the damage threshold of the reflective coating. The hot spots will exceed the damage threshold and be reduced in energy, which promotes a more homogeneous beam and helps protect downstream laser rods from hot-spot damage. The same effect affords protection from accidentally large pulses.

The lens-iris-lens combination can be used to limit the beam divergence, and do it in such a way that it actually provides some isolation by itself and suppresses amplified spontaneous emission to boot. Light backscattered from the main pulse by the plasma will be relatively diffuse and be stopped in large part by an iris designed to barely accommodate the normal divergence. Amplified spontaneous emission also has a relatively high divergence and will be sharply reduced in passing through the iris. Use of the iris in the indicated way can replace procedures based on separating the laser amplifiers by long distances from one another and from the target.

The varied uses for the isolator components suggests that its placement at a number of strategic points in an amplifier chain can alleviate many of the problems of multistage high-gain systems. As a byproduct, the lens-iris-lens combination can serve as a beam expander and collimator to step up the beam diameter between stages. Use of collimated beams in the laser rods can reduce the degradation of beam divergence by pump-induced inhomogeneities.

One of the objectives of the present study was to assess the feasibility and cost of building a large parallel rod multistage laser for x-ray production. The study showed that problems such as isolation and amplified spontaneous emission can be solved at a reasonable cost, and that the critical factor in the overall cost is the output energy that can be delivered from each final stage in 1-2 nsec pulses. As part of the evaluation of the output per stage, two 100-J 1-2 nsec pulses were delivered through the last stage of the Battelle laser, with the pulses spaced sufficiently in time to prevent glass damage. This doubled the usual short pulse capability of the Batelle laser since the pulses were emitted along separate optical paths. This permits them to be simultaneously delivered to targets after appropriate time delays.

Even better performance was obtained when the Sovirel rod in the last stage was replaced with a rod of Owens-Illinois ED-2.1 glass. The higher damage threshold of the ED-2.1 glass immediately enabled 240 J to be delivered in a single 1 - 2 -nsec pulse. The two-pulse output reached 350 J in two ~175-J pulses. Up to 500 J were produced in the 30-nsec mode. These data are included in a plot of the gain characteristics given in Fig. 8, where the single-pulse and double-pulse data are seen to overlap. The overlap indicates that the "three-level laser effect", where populations build up in the lower laser level and restrict the output, is not important for 1 - 2-nsec pulses. It is thus likely that the 500 J seen in the 30-nsec mode can be delivered in two 1 - 2-nsec pulses, and possibly in one.

C. X-Ray Production

Once the isolation problem was solved, laser pulses with energies ranging from ~10 to 100 J and a pulse width of ~1.5 nsec were focused down on a variety of targets, including spheres, slabs, and wires, and the emitted x rays were observed. The basic experimental configuration is outlined in Fig. 9, which shows the case of a thick slab irradiated at a 45° angle of incidence. The focal length of the lens is 7 cm, and it focuses the beam down to a spot size of ~100 to 200 μ diameter. Allowing for the ~15% energy loss suffered in passing through a lens and chamber window, this corresponds to a range of power density centering on 10^{14} W/cm^2. The same range was employed in the computer calculations.

This program has concentrated on slab targets, in spite of the fact that the best results can probably be achieved with special target designs [10]. This approach was useful in gaining basic data. It also proved that even

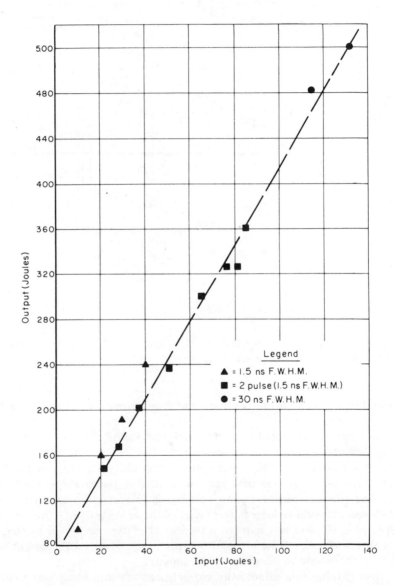

Figure 8. Amplifier gain characteristic of 64-mm stage of the VD-640 laser with ED-2.1 glass. Flash tube voltage is 11 keV.

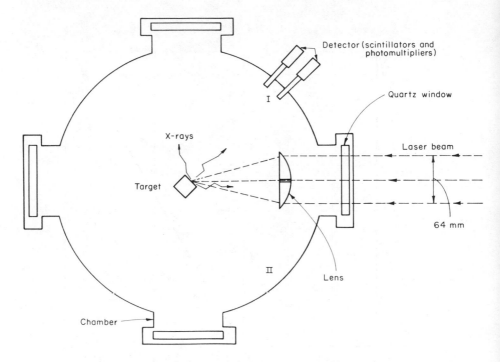

Figure 9.   Basic experimental configuration used for x-ray production.
Detectors are shown at position I.

simple targets can yield high conversion efficiencies, at least for the case
of soft x rays.

The first experiments, performed with infinitely thick slabs, produced
a tiny flux of x rays on the second shot. But, when slabs thought to be optimally thick were used, no x rays were detected. The reason for this
mysterious behavior was that several tenths or hundredths of a joule of
amplified spontaneous emission was being delivered to the target over an
$\sim 10^{-4}$-sec interval before the arrival of the main pulse. This amount of
energy is sufficient to vaporize a parcel of matter designed to reach multikilovolt temperatures with the 100-J main pulse, and 10^{-4} sec is sufficient
time for the vapor to essentially disappear.

Recognizing this effect, the experiments in this study were designed for
thick slabs and many of the optimizing effects associated with using a variety
of thin targets were simulated with prepulses. The prepulses or "feet" that
were tried ranged in pulse width from 1 to 20 nsec, and struck the surface
at power densities of about 10^{11} W/cm^2. The function of the foot is to

produce a plasma so that the main pulse strikes a plasma rather than the solid. The thickness and density profile of the initial plasma can to a large extent be controlled by tuning the foot parameters. A typical combination of foot and main pulse is shown in Fig. 10. Further discussion of the foot is provided in Sec. VII and in the Appendix.

D. X-Ray Diagnostics

The characteristics of the emitted x rays were measured by several diagnostic methods, including differential absorption foil techniques, thin-film and thinister resistance calorimeters, material damage threshold calorimeters, a bent crystal x-ray spectrometer, calibrated scintillation detectors, thermoluminescence detectors, and x-ray pinhole cameras. The purpose was to assess the yield and time-integrated spectrum and to provide information for comparisons with the models.

1. Differential Foils In the studies made with differential foils, the x-ray power was usually monitored by two scintillation detectors placed at angles nearly perpendicular to the target face (cf. Fig. 9). In addition, the laser power and foot power were simultaneously recorded for each shot. Various thicknesses of beryllium, carbon, and aluminum foils were placed

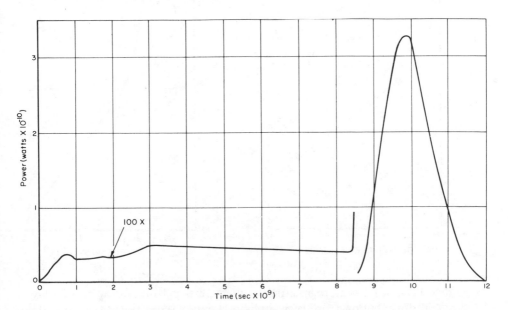

Figure 10. Laser pulse with foot.

in front of the scintillators to block the reflected laser light and to serve as differential x-ray absorption foils. The ratio of x-ray transmission through foils of two different thicknesses was reduced to a "bremsstrahlung temperature" by the technique developed by Jahoda et al. [15], and was identified with the plasma electron temperature.

Now this technique is strictly applicable only to a spectrum consisting of bremsstrahlung and the high-frequency tail of the free-bound continuum. But when proper care is exercised in interpreting the results, the technique is also valid in situations such as the present one, where line emission is important.

The relative transmission through several thicknesses of beryllium of the x rays emitted from an iron slab target with ~40 J incident on 10^{-4} cm^2 is shown in Fig. 11(a). A low-resolution spectrum derived from the transmission curve has the shape of a plasma bremsstrahlung distribution curve with kT ~ 0.8 keV, except in the strong line regime below 1.25 keV [Fig. 11 (b)]. This indicates that consistent results can be obtained with two foil temperature measurements when care is exercised in selecting foils of the

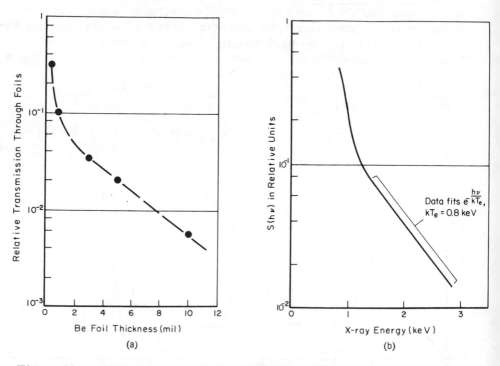

Figure 11.   (a) X-ray transmission through beryllium foils.   (b) X-ray spectral distribution from iron slabs.   Derived from x-ray transmission through beryllium foils.

proper thickness. A derivation of the spectral distribution function $S(h\nu)$ and the range of thicknesses over which the Jahoda temperature will be relatively independent of the choice of foils is given next.

The absorption cross section $\rho\mu$ of beryllium may be approximated by

$$\rho\mu(h\nu) = 980(h\nu)^{-3} \text{ cm}^{-1} \tag{32}$$

where $h\nu$ is the photon energy in keV. The energy detected behind a thickness x of beryllium is thus

$$E(x) = \int_0^\infty S(h\nu) \exp[-\rho\mu(h\nu)x] \, d(h\nu)$$

$$= \int_0^\infty S(h\nu) \exp[-980(h\nu)^{-3}x] \, d(h\nu) \tag{33}$$

from which we derive

$$\frac{dE(x)}{dx} = -\int_0^\infty \frac{S(h\nu)}{x} \{980(h\nu)^{-3}x \exp[-980(h\nu)^{-3}x]\} \, d(h\nu) \tag{34}$$

Now for each foil of thickness x there corresponds a photon energy $h\nu_x$ determined by

$$980(h\nu_x)^{-3}x = 1 \tag{35}$$

The quantity in the curly brackets has a rather sharp maximum at that energy, so that

$$\frac{dE(x)}{dx} \approx -\frac{a}{x} S(h\nu_x) \tag{36}$$

which may be written

$$S(h\nu_x) \approx -\frac{1}{a} \frac{dE(x)}{dx} x \tag{37}$$

This equation provides the basis for deriving Fig. 11(b) from Fig. 11(a) and establishes that the range of foil thicknesses corresponding to the plasma bremsstrahlung-shaped portion of Fig. 9(b) ($1.3 < h\nu < 3$ keV) is ~1 to 10 mil. Consistent two foil temperatures can be obtained with foils in this range. However, extrapolations of the breamsstrahlung curve to determine total conversion efficiencies are unreliable and were not used here.

2. Bent Crystal X-Ray Spectra [16]. The spectral lines were measured with a bent crystal x-ray diffraction spectrometer. A bent KAP crystal with

a 1-in. radius was positioned at 36 and 39 cm from the x-ray source. The
x rays were diffracted from the convex face of the crystal onto a strip of
Kodak NS 2T film with a 3-inch radius of curvature. The spectrometer
chamber was optically isolated from the target chamber by a 3000-Å pary-
lene window coated with 100 μg/cm^2 of aluminum.

3. Calorimetry. Three calorimetric techniques were employed in this
study: thermoluminescence detectors (TLD's), material damage threshold
calorimeters, and a thinister (thin thermister on a glass backing).

The TLD's were tried first, but were completely saturated at
reasonable distances by the high intensity of soft x rays. This problem was
eliminated by placing large thicknesses of absorber over the TLD's, and
with 10 mil of beryllium the TLD's agreed within a factor of 2 with results
obtained from calibrated scintillators. This corroborated the hard x-ray
results. But it rendered the method useless for soft x rays, which were the
main component of the output.

A technique which proved more useful was the material damage threshold calorimeter shown in Fig. 12. This new technique, which relates x-ray
fluence to vaporization of a thin lead film, was developed for the special
purpose of providing lower bound conversion efficiencies for nanosecond

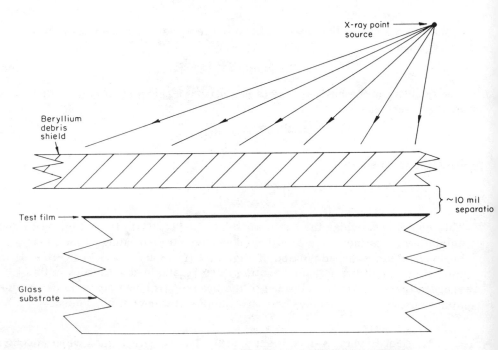

Figure 12.   Material damage threshold calorimeter.

pulses of soft x rays. A frequently employed configuration consists of a 2000-Å lead film deposited on a glass substrate and placed behind a 1-mil beryllium debris shield. The effect of thermal radiation from the beryllium shield has been shown to be negligible. A 2000-Å lead thickness was preferred because it absorbed a reasonable fraction of the x rays and was sufficiently thin to thermally equilibrate in 1 nsec. Laser irradiations of this film with 1 - 2-nsec pulses showed that complete film removal occurs only when the absorbed laser fluence is sufficient to vaporize the entire thickness of lead. The threshold amount of absorbed x-ray energy required to vaporize the lead will essentially be the same, since the thermal equilibration time is less than the pulse width.

VII. ANALYSIS OF RESULTS

The purpose of this section is to analyze and compare the results of the computer calculations and the experiments.

The experiments were performed for a range of incident power densities from 10^{13} to $\sim 2 \times 10^{14}$ W/cm^2 and for a variety of elements, including iron and calcium. The computer calculations were performed for the same range of power densities, and also covered iron and calcium.

The input conditions in the computer calculations differed in a few respects from those actually employed in the experiments. The experiments were mostly performed with a peaked pulse focused at 45° incidence to a finite spot on a slab, whereas the calculations dealt with a square wave pulse covering an infinite plane at normal incidence. But these differences turn out to be immaterial for most of the cases studied.

The key to relating theory with experiment lies in the foot. When the role of the foot is included, the two methods of study fall into basic agreement, both in the overall prediction of conversion efficiencies and in the point-by-point description of how these efficiencies come about. An approximate model for the plasma created by the foot is given in the Appendix. The model assumes that the plasma is optically thick to laser radiation and to its own radiation, in contrast to its behavior during the main pulse. It is intended for high-Z slabs which are irradiated by neodymium laser light at an intensity of $I_f = I_{foot}$ of $\sim 10^9$ to 10^{13} W/cm^2, with pulse widths ranging from approximately 1 to 20 nsec or more.

The model given in the Appendix for the foot plasma has been used to compute the average ion density $(n_i)_{av}$, the temperature T, and the thickness d of the initial plasma profile for iron as a function of the intensity I_f and pulse width τ_f of the foot. The results are given in Figs. 13 and 14. Now the computed thicknesses of most of the initial plasmas generated in the experiments are somewhat greater than the ~ 0.01- to 0.02-cm focal spot diameter. This and other geometrical aspects of the experiments can be

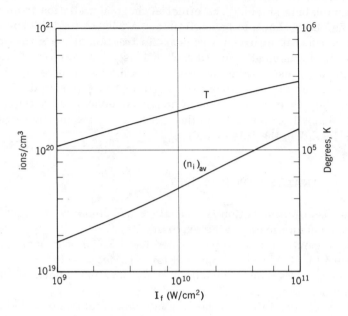

Figure 13.   Temperature and average ion density of initial iron plasma as predicted by Eqs. (A-27) and (A-29).

dealt with in an approximate way by assigning a linear density gradient, or slope, to each of the initial plasmas. A reasonable first estimate is to assign an ion density slope of $(n_i)_{av}/d$ to the plasma with $(n_i)_{av}$ taken from Fig. 13 and d taken from Fig. 14. Slopes calculated on this basis are given in Fig. 15.

A large number of experiments were performed with iron at an incident power density of $\sim 10^{14}$ W/cm^2 for the main pulse. The foot employed in most of these experiments was $\sim 10^{11}$ W/cm^2 over ~ 10 nsec. The corresponding parameters for the initial plasma profile are found from Figs. 13-15.

$$N_i \sim 4 \times 10^{18} \text{ ions/cm}^2 \tag{38a}$$

$$(n_i)_{av} \sim 1.5 \times 10^{20} \text{ ions/cm}^3 \tag{38b}$$

$$\text{Slope} \sim 6 \times 10^{21} \text{ ions/cm}^3 \text{ per cm} \tag{38c}$$

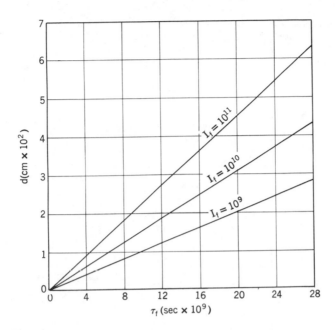

Figure 14.   Thickness of initial iron plasma as predicted by Eq. (A-28).

$$d \sim 0.025 \text{ cm} \tag{38d}$$

The calculated value of N_i implies that $\sim 10^{15}$ ions are located in the focal volume and heated by the main pulse.   This quantity of matter will be raised to an electron temperature of ~ 5 keV upon absorbing a 100-J main pulse, so that the foot is well "tuned" with respect to "selecting" an initial plasma with an optimum heat capacity.   The calculated value of $(n_i)_{av}$ implies that the average electron density will start at $\sim 10^{21}$, and then approach several times 10^{21} as heating by the main pulse causes additional ionization. Although the laser light is absorbed in the underdense region, higher elec- tron densities are achieved by conduction past the critical density point. Finally, the calculated slope and plasma thickness are essentially the same as in the computer calculations presented in Sec. V and designated by "Plasma Profile A." This is all in accordance with the macroscopic design principles discussed in Sec. II.

With regard to the calculations for the main pulse, the first computer run in Sec. V predicts that most of the x rays are emitted in the form of spectral lines, and that the conversion efficiency of laser light into x rays is

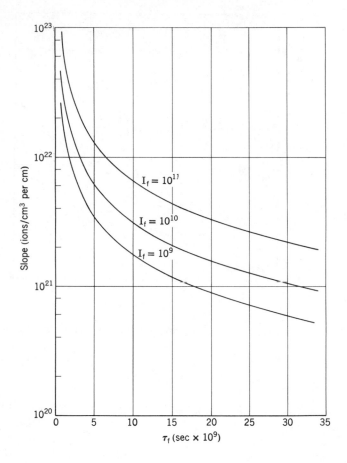

Figure 15. Slope of ion density profile of initial iron plasma calculated by procedure described in text.

at least 20%. This computation uses plasma profile A and an incident power density of 10^{14} W/cm^2, chosen to correspond to the foot and main pulse of the principle experiments. The prediction that reradiation is predominantly spectral lines is attributed to the time-dependent model for ionization. In this model the average degree of ionization, and hence the average energy $h\nu = \chi_r$ of the available line transitions, substantially lags the electron temperature kT_e during the heating process. The average value of χ_r/kT_e is therefore much lower than the one predicted by the Saha Equation. This raises the power of line radiation, but lowers its average frequency. The second computer run presented in Sec. V is a "control" run which is performed with the same input conditions as the first, except for the unphysical assumption that the ion populations are predicted by the Saha Equations. Calculations of x-ray conversion efficiencies based on this type of model have recently appeared elsewhere, but these results can be highly misleading, as seen from Figs. 3 and 4. In spite of the fact that relatively high conversion efficiencies are predicted, most of the radiation is erroneously calculated to be free-bound radiation, and furthermore, the average photon energy is too high by a factor of 10. It is noteworthy that all of the computer runs, except the unphysical ones designated "Equilibrium Calculation," predict that most of the radiation is in the form of spectral lines. It should be mentioned at this point that in the bent crystal spectrometer measurements made with the Lockheed Group, the radiation appears to be mostly lines in the spectral interval studied ($\sim$ 0.7 to 1.3 keV). A typical spectrum is shown in Fig. 16, which was taken at a crystal angle of $20°$ and a viewing angle normal to the target.

The high conversion efficiencies predicted in the computer calculations at 10^{14} W/cm^2 were largely due to energy conducted past the critical density point. The underdense region becomes heated to several kilovolts, but the bulk of the radiation is produced by spectral lines emanating from the relatively cool ($\lesssim$1 keV) leading edge of the thermal diffusion front. It is gratifying that this description was supported by the experiments. The high-energy x rays emitted from the hot less-dense outer region of the plasma were observed through thick layers of absorber. This high-temperature component is consistent with a conversion of $\sim$1% of the incident laser energy to free-free and free-bound radiation with an electron temperature of $\sim$5 keV. It is unnecessary to invoke plasma instabilities to account for it. The cold regime was analyzed by transmission measurements through thin foils of carbon and beryllium. A typical spectral envelope of this regime, which was obtained by unfolding the beryllium transmission curve [Fig. 11(a)], is shown in Fig. 11(b). A key feature of this spectrum is the sharp increase in x-ray intensity as the frequency decreases to the regime where L-lines are predicted. The slope of the upper part of the curve indicates an electron temperature less than 1 keV in the region where the lines are predicted.

Another significant feature of the experiment is the change in reflection which occurred when the foot was varied. Fig. 17 shows that most of the light is reflected for feet which are 1 to 2 nsec long, and that the reflection drops to less than 50% for foot lengths much greater than 10 nsec. The decrease in reflectivity was accompanied by an increase in the observed temperature. Now it is tempting to attribute the temperature increase to the simple fact that a decrease in reflectivity implies a greater power delivered to the plasma. This is undoubtedly part of the explanation, but the computer calculations refute this as the major cause. The computer runs were performed with initial conditions corresponding to a large range of foot lengths,

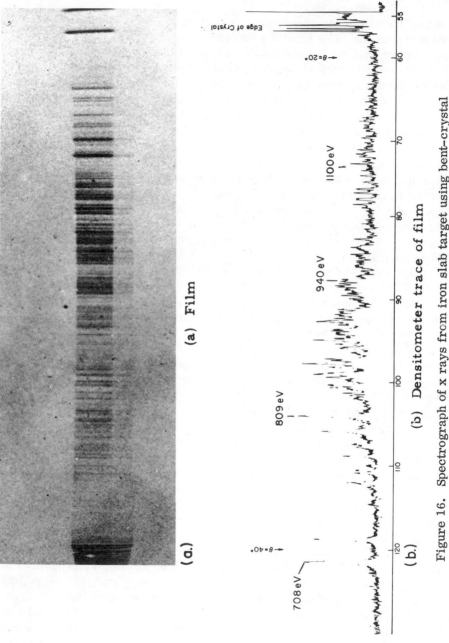

(a.)

(a) Film

(b) Densitometer trace of film

Figure 16. Spectrograph of x rays from iron slab target using bent–crystal spectrometer. Laser flux is incident on iron slab at $\sim 10^{14}$ W/cm^2.

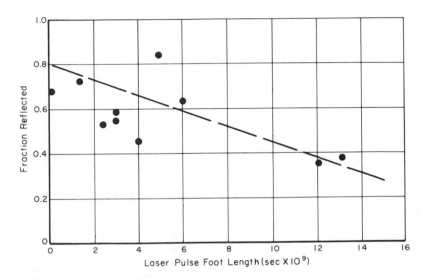

Figure 17.   Laser light reflection versus foot length.   Laser flux is incident on iron slab at $\sim 10^{14}$ W/cm^2.

and predict the same reflection and temperature trends as were observed. But they also imply the existence of a critically short foot. The plasma in that case behaves much like the foot plasma described in the Appendix (essentially blackbody). It now becomes clear how the often invoked blackbody temperature limitation for laser-generated plasmas (~ 180 eV at 10^{14} W/cm^2) is circumvented by a sufficiently long foot. This explanation is also valuable in interpreting the decrease in temperature with atomic number shown in Fig. 18, where all of the shots were made with a foot which gave good conversion efficiencies for iron. Scaling considerations based on the FLASH formulas and the foot formulas given in the Appendix predict that a foot which leads to a 20% conversion efficiency for iron will approach a blackbody response for the main pulse at higher Z. This insight provides a clue for obtaining high conversion efficiencies above 1 keV with the heavier elements. A satisfactory procedure might be to use a longer and less powerful foot. The end result would probably be a higher-energy spectrum than obtained from iron, but with comparable conversion efficiencies.

We now turn to the question of optical thickness. The FLASH radiation model assumes that the plasma is optically thin to its own radiation. Transmission measurements through 1 mil of beryllium show that the x rays above ~ 1 keV are most intense in the direction normal to the target (90°), and drop

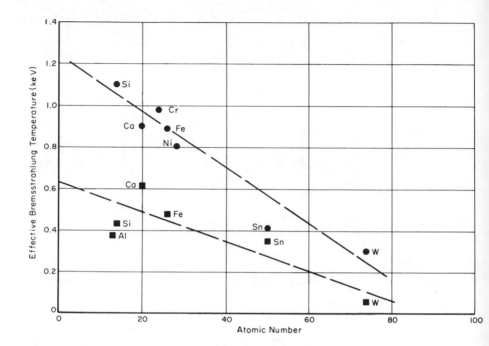

Figure 18.   Effective bremsstrahlung temperature versus atomic number.
Upper curve corresponds to ~ 10^{14} W/cm^2. Lower curve corresponds to
~ 0.4 x 10^{14} W/cm^2.

off by up to a factor of 2 in intensity at $0°$. This demonstrates the existence
of a noticeable but not troublesome self-absorption effect above ~1 keV. The
plasma probably cannot be considered optically thin much below 1 keV, so
that the model should probably be modified there to include self-absorption.

The key test of the overall understanding is to verify the high conversion
efficiencies which are predicted. This has been done in several ways. One
method was the material damage threshold calorimeter technique illustrated
in Fig. 12, which established lower bounds for the x-ray yield and provided a
permanent record of the x-ray damage. In a typical verification shot for conversion efficiency, a 22-J laser pulse struck an iron slab in the usual experimental configuration (Fig. 19). A material damage threshold calorimeter
with a 1-mil beryllium foil and a 2000-Å lead film was placed 0.43 cm from
the plasma at a viewing angle of $0°$. The lead was vaporized down to the
substrate. Since the density and total energy of vaporization of lead are 11
g/cm^3 and 1200 J/g, the x-ray fluence absorbed by the lead at that point was

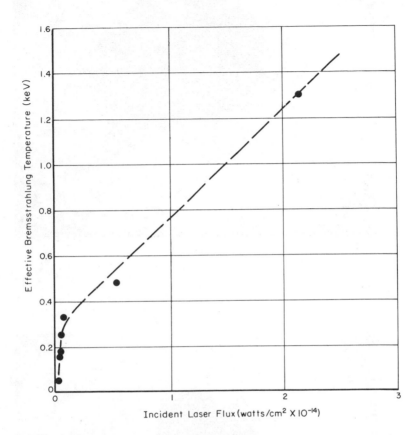

Figure 19.   Effective bremsstrahlung temperature versus laser flux incident on iron slab.

0.26 J/cm^2. A photograph of the x-ray damage to the lead is shown in Fig. 20(a). Figure 20(b) shows the damage to the same film caused by ~1/4 J/cm^2 of laser light absorbed from the laser pulse.

Now the lead absorbs only a minor fraction of the incident fluence. The rest is stopped by the beryllium or passes through both the beryllium and the lead. A plot of the fraction absorbed in lead versus photon energy is given in Fig. 21, and shows that any spectrum of x rays in the several keV range or lower will at most be 20.2% absorbed in the lead. The lower bound fluence emitted at 0° is thus 0.26/0.202 = 1.29 J/cm^2. Since 0° is the angle of minimum flux, a lower bound estimate for the x-ray fluence emitted into the experimentally observable 2π sr solid angle is 2π(0.43)^{2}1.29=1.50 J, which corresponds to a conversion efficiency of 6.8%. The sharp dropoff

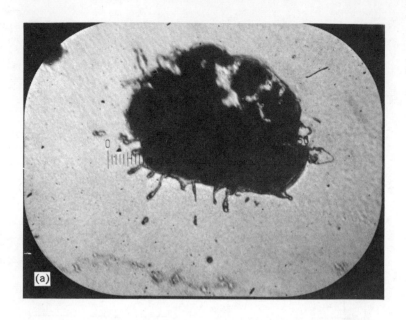

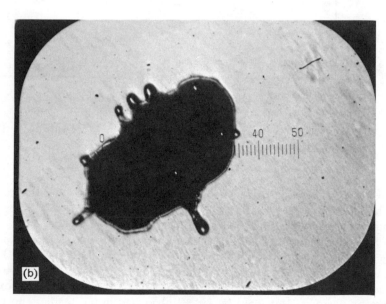

Figure 20. (a) X-ray damage to 2000-Å lead film behind 1-mil beryllium debris shield (20x). (b) Laser damage to 2000-Å lead film (20x).

210

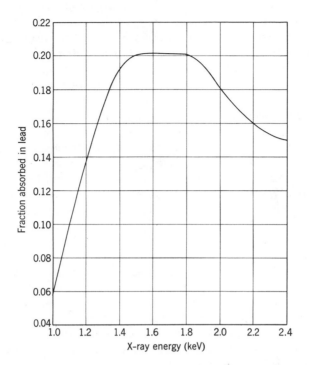

Figure 21.   Fraction of x rays absorbed in 2000-Å lead film after passing through 1-mil beryllium foil versus energy.

of the curve in Fig. 21 near 1 keV implies that most of the 6.8% is above 1 keV.   The appropriate number to compare with theory is the lower bound conversion efficiency of 13.6% for x rays emitted in 4π [17].

In another method, a thermister calorimeter at $90°$ yielded an upper bound conversion efficiency of $\sim 20\%$ above ~ 1 keV based on transmission through 1 mil of beryllium.   This conversion efficiency is based on isotropic emission over 4π sr, and is consistent with the 13% lower bound determined at $0°$.

VIII.   CONCLUSIONS

The main conclusions drawn from this study are the following:

1. Conversion efficiencies of 6–10% into 2π over ~1 keV have been demonstrated with nanosecond laser pulses on slabs. Slabs of proper thickness will permit use of the full 4π.

2. Slab targets have several advantages for large laser systems. Focusing is simplified, the output is reproducible, and the initial plasma profile is conveniently established with a foot. Amplified spontaneous emission does not disrupt the target. Large outputs can be obtained in a desired pulse shape with a multiple-pulse technique, with the pulses sequentially targeted at separate points on the slab.

3. An understanding of the basic phenomenology is essential for optimizing the output. The bulk of the x rays are produced by spectral lines emanating from the relatively cool ($\lesssim$ 1 keV) leading edge of a thermal diffusion front. A hard component is emitted from a hot low-density region with ~1% conversion efficiency, and corresponds to free-free plus free-bound radiation at ~5 keV electron temperature. The spectrum and conversion efficiencies are strongly modified by the lag in ionization due to nonequilibrium stripping. Nonequilibrium stripping raises line output, lowers free-bound output, and lowers average photon energy. Conversion efficiencies are generally higher than with the Saha Equilibrium ionization model. Strong laser absorption can be predicted without invoking instabilities. The output can be varied by tuning the initial plasma conditions with a foot. Higher-Z elements promise to yield higher average photon energies.

4. A diminished debris problem due to high conversion efficiency has been demonstrated. Over 1 J/cm^2 of soft xrays were delivered over a usable area, free of debris.

5. The laser must be protected from glass damage caused by light backscattered from targets. An isolation technique has been developed which is applicable to lasers of any aperture.

6. The nanosecond output of the final stage of the VD–640 can be increased by a factor of 5 without glass damage. The single pulse output was increased from 100 to 240 J, and 350 J was produced in a double-pulse mode. At least 500 J is available in the double-pulse mode without serious modification.

Further development promises to provide a unique soft x-ray tool for medical applications such as cancer therapy, and may also ultimately lead to a brute-force x-ray laser. An x-ray laser which produces 1 J or so of coherent x rays would greatly aid the life sciences by providing phase information in x-ray diffraction studies of genetic materials. The interaction of a laser beam with high-Z matter is also useful for controlled thermonuclear fusion. Although most of the emphasis in laser-generated fusion to date has been on low-Z materials, use of a high-Z material in conjunction with deuterium and tritium may lower the threshold laser energies required for controlled thermonuclear fusion.

APPENDIX: MODEL FOR PLASMA GENERATED BY FOOT

The hydrodynamic equations for compressible inviscid fluid flow are

$$\frac{d\rho}{dt} + \rho \vec{\nabla} \cdot \vec{v} = 0 \qquad \text{(A-1a)}$$

$$\rho \frac{d\vec{v}}{dt} + \vec{\nabla}p = 0 \qquad \text{(A-1b)}$$

$$\rho \frac{d}{dt} (U + \frac{1}{2} v^2) + \vec{\nabla} \cdot (p\vec{v} + \vec{Q}) = 0 \qquad \text{(A-1c)}$$

Here $d/dt = \partial/\partial t + \vec{v} \cdot \vec{\nabla}$ is the hydrodynamic time derivative, $\vec{v}$ is the fluid velocity, and ρ, U, and p are the mass density, internal energy density per gram, and scalar pressure, respectively. The quantity $\vec{Q} = -K\nabla T$ accounts for energy transport by heat conduction. To account for the effects of radiation one must include the radiation pressure p_r in the momentum equation, the radiation energy per gram U_r in the energy equation, and add the radiation energy transport vector $\vec{J}$ to the heat conduction vector $\vec{Q}$ in the energy equation. In other words, one makes the transformation

$$p \rightarrow p + p_r \qquad \text{(A-2a)}$$

$$U \rightarrow U + U_r \qquad \text{(A-2b)}$$

$$\vec{Q} \rightarrow \vec{Q} + \vec{J} \qquad \text{(A-2c)}$$

To obtain the particular form of the hydrodynamic equations for the foot problem, we ignore Eqs. (A-2a) and (A-2b) in view of the relations

$$p_r \ll p \qquad \text{(A-3a)}$$

$$U_r \ll U \qquad \text{(A-3b)}$$

which hold for any temperatures of interest. We also assume that the flow is one dimensional and quasisteady in some region in the vicinity of the solid surface, which allows the y, z, and t variations to be dropped. The resulting equations are

$$v \frac{d\rho}{dx} + \rho \frac{dv}{dx} = 0 \qquad \text{(A-4a)}$$

$$\rho v \, \frac{dv}{dx} + \frac{dp}{dx} = 0 \tag{A-4b}$$

$$\rho v \, \frac{d}{dx} \, (U + \frac{1}{2} \, v^2) + \frac{d}{dx} \, (pv + Q + J) = 0 \tag{A-4c}$$

These equations will be solved for the case where the fluid velocity is steeply raised from its value v_s at the solid surface to a value near the asymptotic velocity for continuous isothermal fluid flow, which is given by

$$v_\infty = \sqrt{\frac{p_\infty}{\rho_\infty}} = \sqrt{\frac{(\overline{Z^*_\infty} + 1)kT_\infty}{M}} \quad . \tag{A-5}$$

where M is the ion mass. The isothermal temperature is limited by blackbody radiation into space and is given by

$$I_f \approx \sigma T_\infty^4 \tag{A-6}$$

where I_f is the foot intensity and σ is the Stefan–Boltzmann constant. It is to be understood that quantities denoted by the subscript ∞ are applicable only to a given distance from the surface, after which quasisteady conditions break down and Eqs. (A-4) no longer apply. This point will be discussed later.

Eq. (A-4a) immediately integrates to

$$\rho v = \dot{m} = constant \tag{A-7}$$

which when substituted into Eq. (A-4b) gives

$$\dot{m} \, \frac{dv}{dx} + \frac{dp}{dx} = 0 \tag{A-8}$$

This integrates to

$$\dot{m}v + p = K_1 \tag{A-9}$$

where the constant K_1 may be evaluated with use of Eq. (A-7) and the boundary condition (A-5).

$$K_1 = \dot{m}v_\infty + p_\infty = \dot{m}v_\infty + \rho_\infty v_\infty^2 = 2\dot{m}v_\infty \qquad \text{(A-10)}$$

Eq. (A-9) may thus be written in the form

$$p = \dot{m}(2v_\infty - v) \qquad \text{(A-11)}$$

The ratio of vapor pressure at the solid surface to the asymptotic pressure p_∞ is

$$\frac{p_s}{p_\infty} = 2 - \frac{v_s}{v_\infty} \qquad \text{(A-12)}$$

In view of the fact that

$$\frac{v_s}{v_\infty} \ll 1 \qquad \text{(A-13)}$$

the pressure varies across the region of interest by less than a factor of two. Its maximum value occurs adjacent to the surface and is given approximately by

$$p_s \approx 2\dot{m}v_\infty \qquad \text{(A-14)}$$

whereas it approaches the minimum value

$$p_\infty = \dot{m}v_\infty \qquad \text{(A-15)}$$

with increasing distance.

Eq. (A-4c) is also readily integrated, and with use of Eq. (A-7) it takes the form

$$\dot{m}(U + \tfrac{1}{2}v^2) + \dot{m}\frac{p}{\rho} + Q + J = K_2 = \text{constant} \qquad \text{(A-16)}$$

or equivalently

$$\dot{m}(h + \tfrac{1}{2}v^2) + (Q + J) = K_2 \qquad \text{(A-17)}$$

where h is the enthalpy per unit mass.

$$h = U + \frac{p}{\rho} \qquad (A\text{-}18)$$

The rate of mass removal $\dot{m}$ may be estimated with use of the approximate boundary conditions

$$Q_\infty + J_\infty \approx -I_f \qquad (A\text{-}19)$$

$$Q_s + J_s \approx -\dot{m}H \qquad (A\text{-}20)$$

where H is the energy per gram required to vaporize the solid. Eq. (A-19) neglects the energy reradiated into space (which appears in practice to never exceed one-half the incident energy). Eq. (A-20) assumes that most of the energy absorbed by the solid ends up in vapor rather than a liquid or a heated solid. Use of the boundary condition (A-19) in Eq. (A-17) leads to an expression for the constant K , which when inserted back into Eq. (A-17) together with the boundary condition (A-20) gives

$$\dot{m}(h_s + \frac{1}{2} v_s^2) - \dot{m}H = \dot{m}(h_\infty + \frac{1}{2} v_\infty^2) - I_f \qquad (A\text{-}21)$$

This equation may be solved for $\dot{m}$.

$$\dot{m} \approx \frac{I_f}{H + (h_\infty - h_s) + \frac{1}{2}(v_\infty^2 - v_s^2)} \qquad (A\text{-}22)$$

In the regimes of interest, $h_\infty \gg h_s$, $h_\infty \gg H$, and $v_\infty \gg v_s$, so that Eq. (A-22) reduces to

$$\dot{m} \approx \frac{I_f}{h_\infty + \frac{1}{2} v_\infty^2} \qquad (A\text{-}23)$$

The denominator may be written in a more useful form with the help of Eqs. (A-18), (A-5), and (A-6), and the approximation that the ionization energy per gram, U_I, is double the kinetic energy per gram, U_K. This approximation is reasonable for high-Z plasmas which are not fully ionized. Thus,

$$h_\infty + \frac{1}{2} v_\infty^2 = U_\infty + \frac{p_\infty}{\rho_\infty} + \frac{1}{2} v_\infty^2 = U_\infty + \frac{3}{2} v_\infty^2$$

$$= U_K + U_I + \frac{3}{2} v_\infty^2 \approx 3U_K + \frac{3}{2} v_\infty^2$$

$$= \frac{9}{2} (\overline{Z_\infty^*} + 1) \frac{kT_\infty}{M} + \frac{3}{2} (\overline{Z_\infty^*} + 1) \frac{kT_\infty}{M}$$

$$= 6 (\overline{Z^*} + 1) \frac{kT_\infty}{M}$$

$$= 6 (\overline{Z_\infty^*} + 1) \frac{k}{M} \sigma^{-1/4} I_f^{1/4} \tag{A-24}$$

which leads to our key result

$$\dot{m} = \frac{M\sigma^{1/4}}{6(\overline{Z_\infty^*} + 1)k} I_f^{3/4} \tag{A-25}$$

The total number of ions/cm^2 removed by the foot during the pulsewidth τ_f is

$$N_i = \frac{\dot{m}}{M} \tau_f \approx \frac{\sigma^{1/4}}{6(\overline{Z_\infty^*} + 1)k} I_f^{3/4} \tau_f \tag{A-26}$$

Ready formulas for the plasma temperature, plasma thickness, and average ion density are obtained from Eqs. (A-5), (A-6), and (A-26).

$$T_\infty \approx \sigma^{-1/4} I_f^{1/4} \tag{A-27}$$

$$d \approx v_\infty \tau_f = \left(\frac{(\overline{Z_\infty^*} + 1)k}{M} \right)^{1/2} \sigma^{-1/8} I_f^{1/8} \tau_f \tag{A-28}$$

$$(n_i)_{av} \approx \frac{N_i}{d} = \frac{M^{1/2} \sigma^{3/8}}{6(\overline{Z_\infty^*} + 1)^{3/2} k^{3/2}} I_f^{5/8} \tag{A-29}$$

As mentioned earlier, the fluid flow is quasisteady only out to a given distance from the surface. However, formulas (A-25) and (A-26) are independent of that fact, and only require that the quasisteady equations and the associated asymptotic description be valid out to that point. Formula (A-27) is also generally applicable if the sharply varying region near the solid surface is ignored. The only formulas that might be changed due to nonsteady flow are (A-28) and (A-29).   But formula (A-27) essentially eliminates that possibility by requiring that most of the absorbed energy be tied up in internal energy, which prevents the average fluid velocity from increasing much beyond v_∞.

ADDENDUM (OCTOBER, 1972)

The following comments and reviewing remarks will be useful in relating these results to the overall goals and objectives of x-ray laser research.

An intense point source of x rays with a high conversion efficiency of laser light into x rays has been developed.   The x rays emanate from a spot of approximately 100 μ diameter, and a large fraction of the x rays (sometimes 50% or more) are emitted in the form of spectral lines.

An important implication of these properties is the fact that many of the applications that are generally cited in justification of an x-ray laser can be accomplished with a point source (i.e., "spatially coherent" source) of x-ray lines, or even with a point source of broad-band x rays.   For example, a significant fraction of the x rays can be focused to another point or into a parallel beam by means of critical-angle or Bragg-angle reflectors.   It is anticipated that further development will result in a spectrum which is controllable and tunable throughout most of the 0.1- to 100-keV x-ray range. This could provide hospitals with x-ray machines for taking ultrasharp, ultrafast radiographs, and performing swift and precise irradiations of cancer tissue.   It would also serve molecular biology by greatly speeding x-ray diffraction studies of genetic materials.   These things could all be accomplished without laser action taking place, and are merely a few of the many possibilities.

It is calculated that the x-ray emission in a single spectral line can exceed 1% of the incident laser light.   This could also be true when a substantial amount of amplification through stimulated emission occurs.   In the latter case the device would constitute an actual x-ray laser, which could provide a narrow beam if the plasma shape were a suitably long cylinder or an x-ray cavity were constructed.   Indeed, in addition to the fact that it can serve many of the functions of an x-ray laser, the present device is only one step away from a true x-ray laser.   Large outputs of x-ray lines are being produced in a highly nonequilibrium plasma, and theory says that a substantial amount of stimulated emission can occur with some adjustments of plasma temperature and target configuration.

Statements are often made that picosecond or terawatt pumping pulses are required for an x-ray laser.   This appears to be untrue.   Calculations based on the theory presented in this paper indicate that a nanosecond laser pulse with 100 to 1000 J of energy should be more than adequate; a list of wavelengths at which laser action might be expected to occur will be presented in a forthcoming report. There is some possibility that an x-ray laser based on these ideas may have already been constructed at the University of Utah using 30-J 20-nsec laser pulses [18].

Another route to an x-ray laser is to use the x rays generated by the present technique to knock out inner-shell electrons from atoms in a separate laser medium and thereby create a population inversion [19].

As a final point, it should be mentioned that the present device is additionally unique in that it provides copious quantities of soft x rays. These can be used to irradiate surface tumors without damage to underlying tissue, and can also be injected into the body by means of an x-ray "hypodermic needle" (i.e., a thin hollow x-ray "light pipe"), thereby avoiding damage to intervening tissue. Preliminary experiments at Battelle have demonstrated the feasibility of the x-ray light pipe.

The medical applications of this research are being conducted in cooperation with M. Muckerheide and R. Uecker of the Wausau Clinic, Wausau, Wisconsin.

REFERENCES

1. P. J. Mallozzi, H. M. Epstein, C. T. Walters, D. C. Applebaum, W. J. Gallagher, and J. E. Dennis, J. Appl. Phys. **42**, 4531 (1971).
2. J. Dawson, P. Kaw, and B. Green, Phys. Fluids **12**, 875 (1969).
3. J. W. Bond, Jr., K. M. Watson, and J. A. Welch, Jr., Atomic Theory of Gas Dynamics (Addison-Wesley, Reading, Mass., 1965).
4. R. W. P. McWhirter, in Plasma Diagnostic Techniques, edited by R. H. Huddlestone and S. L. Leonard (Academic, New York, 1965), Chap. 5.
5. J. Dawson and C. Oberman, Phys. Fluids **5**, 517 (1962).
6. P. Mallozzi and H. Margenau, Ann. Phys. **38**, 117 (1966).
7. P. Mallozzi and H. Margenau, Astrophys. J. **137**, 851 (1963).
8. A. Burgess, Astrophys. J. **139**, 776 (1964).
9. T. F. Stratton, in Plasma Diagnostic Techniques, edited by R. H. Huddlestone, and S. L. Leonard (Academic, New York, 1965), Chap. 8.
10. P. Mallozzi, Proposed Research Program to Space and Missile Systems Organization (Battelle Columbus Laboratories, Columbus, Ohio, 1967).
11. M. J. Bernstein and G. C. Comisar, J. Appl. Phys. **41**, 729 (1969).
12. J. W. Shearer and W. S. Barnes, Report No. UCRL-71733, Lawrence Radiation Laboratory, University of California, Livermore (1969).
13. Laser Focus **6** (No. 11), 41 (1970).
14. P. K. Fong, Rev. Sci. Instrum. **41**, 1434 (1970).
15. F. C. Jahoda, E. M. Little, W. E. Quinn, G. A. Sawyer, and T. F. Stratton, Phys. Rev. **119**, 843 (1960).

16. These studies were performed in conjunction with L. F. Chase, W. Jordan, and T. Miller of Lockheed Palo Alto.

17. The x rays radiated into the slab will be absorbed by iron atoms and should produce an L-fluorescence yield of order ~0. 01% with respect to the main pulse. Preliminary analysis of the bent crystal spectrometer results appears to confirm the presence of atomic L-lines in this quantity.

18. J. G. Kepros, E. M. Eyring, and F. W. Cagle, Proc. Nat. Acad. Sci. 69, 1744 (1972).

19. M. A. Duguay and P. M. Rentzepis, Appl. Phys. Letters 10, 350 (1967).

LASERS IN CHEMISTRY

PHOTOCHEMISTRY OF SINGLE VIBRONIC STATES:

AN APPLICATION OF NONLINEAR OPTICS

Edward S. Yeung and C. Bradley Moore
Department of Chemistry, University of California
Berkeley, California

ABSTRACT

Lifetimes, energy transfer, and quenching are observed for single vibronic states of gaseous formaldehyde (D_2CO). Excitation is accomplished by either single photons of a tunable uv laser or by simultaneous absorption of two photons from a Nd:YAG laser.

I. INTRODUCTION

Intense tunable laser sources promise to open many new areas of research in photochemistry. Dramatic advances have already occurred in flash photolysis. The domain of time resolution has been pushed from micro-seconds down to picoseconds; simultaneously the spectral width of the photolysis flash has been reduced by orders of magnitude [1]. The laser has made possible a new sort of photodissociation spectroscopy which allows direct study of dissociative molecular states [2]. It has been possible to study many sorts of vibrational and electronic relaxation processes using the laser to excite single molecular quantum states [3]. High concentrations of

223

excited species are often produced such that collisions between excited species may be studied. For example, the vibrational relaxation of HCl by photolyticly produced Cl atoms has been observed [4]. In these experiments advantage has been taken of the fortuitous coincidences between fixed-frequency laser lines and molecular absorption frequencies. Tunable lasers will permit the excitation and study of almost any desired molecular quantum state. Thus the tunable laser will make many new sorts of experiments possible and allow a much greater wealth of information to be obtained from experiments developed with fixed-frequency lasers.

The application of tunable lasers in chemistry, or more particularly photochemistry, is just beginning. In the visible region Sackett and Yardley have used a tunable dye laser to resolve spectral structure in the radiative lifetimes of NO_2 molecules [5]. In this article the application of tunable lasers in photochemistry is illustrated by a study of fluorescence decay times of single electronic-vibration states of formaldehyde. A tunable uv laser pumps single bands in the vibrational progression of the absorption spectrum of D_2CO. Levels whose one-photon absorption is very weak due to symmetry are excited by simultaneous absorption of two visible photons. With these techniques radiationless decay--predissociation, energy transfer, and quenching--may be studied for each individual quantum state.

Fluorescence has long been a useful tool in the study of the behavior of molecules in excited states [6]. In order to obtain the most information in such studies, it is essential to be able to excite the molecules to known eigenstates and to observe the decay from these states. Past monochromatic excitation methods have used fixed-frequency sources, e.g., atomic lines and single-frequency lasers, which limit the number of molecular states that can be studied. Spectrally separating out a portion of a broadband light source has been tried [7], but the spectral radiance as well as the spectral resolution of the excitation are greatly reduced. So, there has always been a difficult compromise between monochromaticity and tunability. The development of dye lasers [8] and nonlinear optics [9], makes it possible to produce an intense tunable uv pulse of short duration and small spectral bandwidth, suitable for selectively exciting molecular states. Such light pulses are produced in a nonlinear crystal by summing the frequencies of a high-power ruby laser and a laser-pumped dye laser. The spectral bandwidth is much smaller than the spacing between vibronic levels in gaseous formaldehyde so that known single levels are excited.

The process through which a molecule simultaneously absorbs two photons to reach an excited state without requiring the presence of an intermediate state is predicted by second-order time-dependent perturbation theory. As early as 1931, calculations of two-photon absorption cross sections were made [10]. However, it was not until the development of lasers that high intensities of light were available to make such processes observable [11]. The selection rules for two-photon absorption allow one to reach states of the same symmetry as the ground state, thus complementing the available states in one-photon absorption. A number of vibronic levels in formaldehyde are greatly weakened by symmetry selection rules in one-photon absorption, but are strongly allowed in two-photon absorption. Fluorescence from these levels to odd symmetry vibrational levels of the ground electronic state is allowed. The observation of time-resolved fluorescence from two-photon excitation with a laser of modest power demonstrates the feasibility of high-resolution two-photon spectroscopy of dilute gases with a tunable laser.

II. EXPERIMENTAL

A. One-Photon Excitation

A continuously tunable monochromatic uv source is used to excite known vibronic levels of the first singlet excited state in D_2CO. The optical arrangement used to produce the uv radiation has been described [12] and is shown in Fig. 1. Ultraviolet light with 100–200 kW power in 7 nsec and a spectral bandwidth of 1 Å is produced by summing a ruby laser with a tunable dye laser [8]. The giant pulse ruby laser generates second harmonic radiation in a KDP (potassium dihydrogen phosphate) crystal (KDP 1). The ruby second harmonic (2ω) is separated from the ruby fundamental (ω) by M1 and proceeds via M2 and M3 into the dye cell to pump the dye laser. The dye laser cavity is made up of a grating (Littrow operation) and a 16% front reflector. Tunable dye laser light (ω_D) is reflected by M2 onto the summing crystal (KDP 2). Meanwhile, the ruby fundamental is reflected by M1 and M4 and made collinear with the dye laser by M5. The summation frequency of the two lasers is produced by the summing crystal, is separated from the

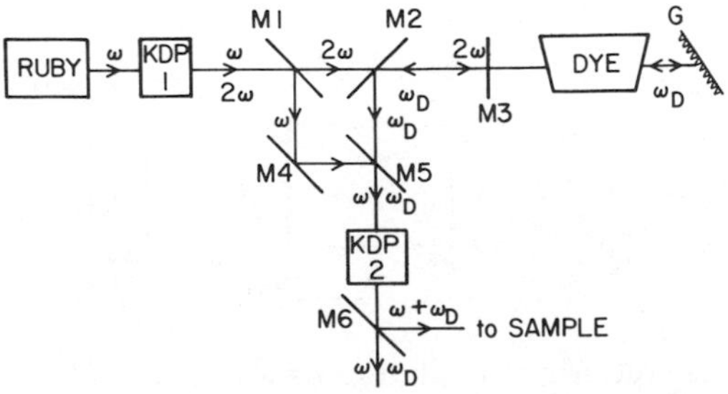

Figure 1. Apparatus for producing tunable uv light. M1: T at 2ω, R at ω; M2: T at 2ω, R at ω_D; M3: T at 2ω, 16% R at ω_D; M4: R at ω; M5: T at ω_D, R at ω; M6: T at ω and ω_D, R at $\omega + \omega_D$; G: grating.

visible beams by M6, and passes into a 1-in. diam stainless-steel gas cell
with 1/8-in. thick quartz windows.

Fluorescence perpendicular to the excitation light passes through a 200-
Å bandpass optical interference filter, is detected by an Amperex 56AVP
photomultiplier, and displayed on a Tektronix 7704 oscilloscope. The great
increase in sensitivity due to using an interference filter instead of a monochromator [12] has made it possible to study the fluorescence intensity at
much longer times after excitation and at much lower gas pressures. A
long-lived component of the fluorescence has been observed for the excitation wavelengths tried. In view of this, all fluorescence decay curves have
been analyzed as double exponential decays:

$$I(t) = A_0 \exp\left(\frac{-t}{\tau_A}\right) \;+\; B_0 \exp\left(\frac{-t}{\tau_B}\right) \tag{1}$$

B.   Two-Photon Excitation

The experimental arrangement for two-photon excitation is shown in Fig. 2.
A Chromatix Model 1000 Nd:YAG laser is chosen because of its high available repetition rate and its convenient output wavelengths. The two
frequency-doubled outputs at 669 and 679 mμ are Q-switched pulses about
180 nsec wide with peak powers of 4 kW. The 1.25-mm-diam TEM_{00} laser
beam is focused by a 105-mm-focal length lens (L) into the center of the gas

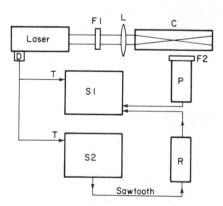

Figure 2.   Apparatus for two-photon experiment. C: gas cell; F1: Corning
2-58 filter; F2: two Corning 5-57 filters; L: lens; S1: Tektronix 7704
scope; S2: Tektronix 585A scope; R: voltage divider; T: trigger pulse;
D: diode monitor.

cell described above. A Corning CS-2-58 filter (F1) is placed between the laser and the lens to cut out stray light from the laser flashlamp. Calculations show that the intensity at the focal point is of the order of 120 MW/cm^2. The fluorescence is detected by a photomultiplier as in the previous section. However, instead of the interference filter, two Corning CS-5-57 filters (F2) are used so that about 20% of the total fluorescence is detected.

The internal diode monitor of the laser is used both to trace the shape of the excitation pulse and to trigger the scopes. Figure 3 shows a typical excitation pulse (lower trace) together with the fluorescence signal (upper trace) represented by individual photons. The ratio of the number of observed photons to the number of dark photons is 25:1 for the lowest pressures and can easily be increased by cooling the photomultiplier.

In actual experiments, the laser is operated at 15 pps. In order to accomodate 10 signal traces on a single polaroid photograph, a Tektronix 585A scope (SCOPE 2) is triggered at the start of the camera to provide a ramp bias voltage of 0-150 V in 2 sec. This bias voltage passes into a voltage divider (R) and is combined by the "ADD" mode in a Tektronix 7704 scope (SCOPE 1) with the signal pulse. This serves to step each successive signal trace up across the oscilloscope screen with the spacing between them adjustable by the voltage divider. At each wavelength and each pressure of D_2CO, at least 1000 photons are needed to provide a good estimate of the

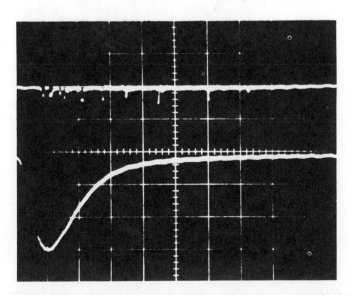

Figure 3. Typical two-photon experiment. Top trace--signal photons as vertical spikes. Bottom trace--laser pulse shape. Time scale left to right 100 nsec/cm.

lifetime, so that typically 100–200 signal traces are needed for analysis.

The laser pulse is stable enough so that an average of 10 traces of the diode monitor before and after the experiment can be used for the laser pulse shape throughout the experiment. The fluctuation in peak intensity is less than ±5%. The pulse shape is conveniently fitted to two exponential terms

$$I = -a_1 \exp\left(\frac{-t}{c_1}\right) + a_2 \exp\left(\frac{-t}{c_2}\right) \tag{2}$$

where the first term contributes to the rising part of the laser pulse and the second term the falling part. The fit is better than 3% accurate over the whole pulse. This shape is convoluted with an exponential decay for comparison to the observed fluorescence.

The experimentally recorded signal photons are time ordered with the aid of a computer. Channel width chosen varies from 10 to 20 nsec. The number of counts per channel is plotted against the channel number and is compared with the theoretically calculated curves for different exponential decay lifetimes. The lifetime is determined by best fit. A typical theoretical convolution curve is shown as the solid line in Fig. 4 with the

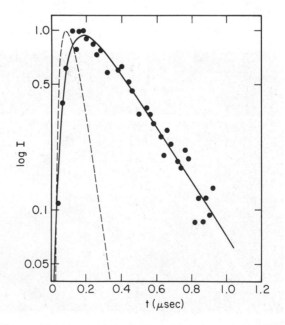

Figure 4. Analysis of two-photon data at 2.5 Torr and 6790 Å excitation. Dashed line: effective excitation curve; solid line: theoretical convolution curve counts per channel, normalized.

experimental counts per channel plotted on top. The dashed curve gives the excitation pulse intensity squared, because this is a two-photon process. The uncertainty in the lifetime so determined is 15%. A description of the convolution analysis is given in Appendix I.

III. KINETIC SCHEME

The observation of two different components of the fluorescence decay on excitation into a single vibronic level indicates the presence of energy transfer. We propose the following kinetic scheme which fits the experimental observables. A two-level system is considered where A is the initially excited vibronic state and B is the state (or group of states having approximately similar properties) which is populated by energy transfer from A. From the detailed analysis of the decay curves, it is unlikely that a third state with an intermediate lifetime or with a longer lifetime contributes to the fluorescence.

$$D_2CO \xrightarrow{h\nu} A \qquad\qquad \text{absorption}$$

$$A \xrightarrow{k_a} A' \qquad\qquad \text{predissociation, or intersystem crossing}$$

$$M + A \xrightarrow{k_{qa}} A'' + M \qquad\qquad \text{quenching}$$

$$M + A \xrightarrow{k_e} B + M \qquad\qquad \text{energy transfer}$$

$$B \xrightarrow{k_b} B' \qquad\qquad \text{predissociation, or intersystem crossing}$$

$$M + B \xrightarrow{k_{qb}} B'' + M \qquad\qquad \text{quenching}$$

In the above scheme, M is any collisional partner of D_2CO in these experiments. The values k_a and k_b include the natural radiative lifetimes of the two states, and can be taken as the nonradiative lifetimes in the limit where the radiative quantum yield is much less than unity. A', A'', B', and B'' are species which do not fluoresce. The rate equations for such a system are

$$-\frac{d(A)}{dt} = [k_a + (k_e + k_{qa}) (M)](A) \qquad\qquad (3)$$

$$\frac{d(B)}{dt} = k_e(M)\,(A) - [k_b + k_{qb}(M)\,]\,(B) \tag{4}$$

Solving these:

$$A(t) = A_0 \exp\{-[k_a + (k_e + k_{qa})\,(M)\,]t\} \tag{5}$$

$$B(t) = \frac{k_e\,(M)\,(A_0)}{k_b + k_{qb}(M) - k_a - (k_e + k_{qa})\,(M)} \tag{6}$$

$$\times \left[\exp\{-[k_a + (k_e + k_{qa})\,(M)]t\} - \exp\{-[k_b + k_{qb}(M)]t\} \right]$$

Comparing these with Eq. (1), one finds that A_0 is identical with the experimentally determined constant, and that

$$1/\tau_A = k_a + (k_e + k_{qa})\,(M) \tag{7}$$

$$1/\tau_B = k_b + k_{qb}\,(M) \tag{8}$$

The experimental value for τ_A is much smaller than that for τ_B, so that the fluorescence at long times is completely dominated by τ_B.

On integration of Eqs. (5) and (6) from $t = 0$ to $t = \infty$,

$$A_{total} = \frac{A_0}{k_a + (k_e + k_{qa})\,(M)} \tag{9}$$

$$B_{total} = \frac{k_e\,(M)\,(A_0)}{[k_a + (k_e + k_{qa})\,(M)]\,[k_b + k_{qb}\,(M)]} \tag{10}$$

and

$$\frac{B_{total}}{A_{total}} = \frac{k_e\,(M)}{k_b + k_{qb}\,(M)} \tag{11}$$

Stern–Volmer plots of the inverse lifetimes for the two decaying components are shown in Figs. 5 and 6. From Eqs. (5) and (6), the extrapolated zero pressure lifetimes give k_a and k_b, respectively, and the slopes give $(k_e + k_{qa})$ and k_{qb}, respectively.

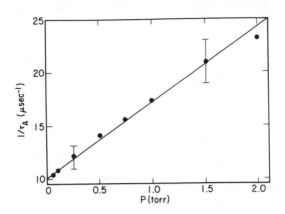

Figure 5. Stern-Volmer plot of lifetime data for excitation at 3139 Å--initial state, A.

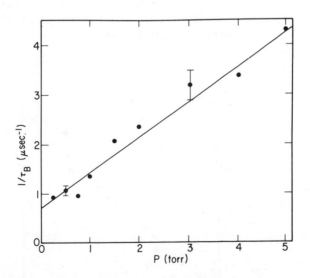

Figure 6. Stern-Volmer plot of lifetime data for excitation at 3139 Å--energy-transfer populated state, B.

From the experimental curves, one finds that

$$A_{total} = A_0 \tau_A \tag{12}$$

$$B_{total} = B_0 \tau_B \tag{13}$$

and in Fig. 7, B_{total}/A_{total} is plotted against

$$F_2(P) = \frac{P}{k_b + k_{qb} (M)} \tag{14}$$

It should be noted that Fig. 7 is a plot to determine k_e. The closeness of the data points to a straight-line fit provides evidence that the kinetic scheme proposed is indeed correct.

The interpretation of the value k_e found here must be made with caution. In the case when the radiative transition probabilities of the two states are the same, the value k_e represents the true energy transfer rate from A to B. Otherwise, k_e must be multiplied by the ratio of the Einstein A coefficients of state A to state B to give the true rate. Note that energy transfer can also occur from the lower state B and from states excited by two-photon absorption. However, the low fluorescence intensities in these experiments do not allow such processes to be observed.

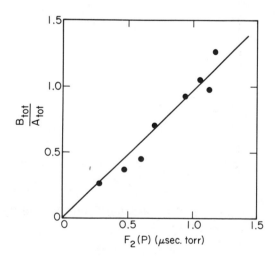

Figure 7.   Kinetic analysis of band at 3139 Å. B_{tot}/A_{tot} versus F_2 (P).

IV. RESULTS AND CONCLUSIONS

Properties of some of the vibronic states of the first singlet excited state of D_2CO excited by either the one-photon or the two-photon technique are listed in Table I. A general trend is that the larger the amount of vibrational energy the state has, the shorter the zero-pressure lifetime and the larger the quenching constant (which includes any energy transfer processes). The shortening of the zero-pressure lifetimes as one increases the vibrational energy is too rapid to be explained by a shortening in the radiative lifetimes of the states alone. Further, the fluorescence quantum yield of these highly excited levels is known to be much less than unity [13]. We conclude that the main contribution to the lifetimes at zero pressure is from nonradiative processes, presumably predissociation of formaldehyde. By such a tuned-laser excitation of the various vibrational levels, one can study predissociation or intersystem crossing as a function of both the total vibrational energy and the particular normal mode. This kind of information should provide considerable insight into the dynamics of molecular photochemical processes.

The rate of energy transfer from the 2^34^1 state in D_2CO is 0.95 μsec^{-1} $Torr^{-1}$, assuming equal radiative rates for the two states. This means that quenching is six times as fast as energy transfer in depopulating the state. Judging from the quenching constants and zero-pressure lifetimes, the states populated by energy transfer from the 2^34^1 state and the $2^14^25^1$ state are low-lying vibrational levels of the excited singlet. It might be possible eventually to identify these secondary states once the lifetimes and the quenching constants of most of the vibrational levels are measured and spectrally resolved fluorescence data are taken.

A rough estimate of the matrix elements for two-photon absorption of the 2^1 state of D_2CO gives $\delta = 10^{-52}$ cm^4 sec $molecule^{-1}$ $photon^{-1}$, and a slightly smaller value for the 2^14^2 state. A determination of the absolute absorption cross section is not possible in these experiments without a knowledge of the fluorescence quantum yield. However, the fluorescence technique has the advantage of increasing the sensitivity of detection of two-photon absorptions by a few orders of magnitude, while at the same time giving direct lifetime measurements. The excitation of a single vibronic state of a molecule in the gas phase by a two-photon technique and the measurement of its properties is a significant advance in two-photon techniques. The same two-photon experiment can be done using two photons from a tunable dye laser, one photon each from a dye laser and a ruby laser (a simple adaptation of the arrangement used in the one-photon experiment here), or the tunable output of a parametric oscillator [15]. This allows one to reach molecular states not accessible by the usual single-photon absorption methods. Much weaker signals than the ones here can be analyzed using a time-amplitude converter coupled with a multichannel height analyzer.

TABLE I. PROPERTIES OF SOME VIBRONIC STATES OF THE
FIRST SINGLET EXCITED STATE OF D_2CO

D_2CO Band[a]	Excitation[b] Wavelength (Å)	Excess Energy (cm^{-1})	$\tau_A^{o\ c}$ (μ sec)	$k_{qa}+k_e^{\ c}$ (μsec^{-1}torr^{-1})	$\tau_B^{o\ c}$ (μsec)	$k_{qb}^{\ c}$ (μsec^{-1}torr^{-1})
4^3	3471/I	501	1.50	2.0	5.0	0.74
2^1	6790/II	1154	0.54	1.0		
$2^1 4^2$	6690/II	1594	0.42	1.2		
$1^1 2^1 4^1$	3164/I	3304	0.105	5.4	0.79	0.51
$2^3 4^1$	3139/I	3544	0.099	7.1	1.41	0.71
$2^1 4^2 5^1$	3115/I	3802	0.076	9.4	0.52	0.47

[a]Assignments from Ref. 18. Superscripts represent the number of quanta in the particular vibrational mode: v_2, C=O stretch; v_4, inversion; v_5, asymmetric C–H stretch; v_1, symmetric C–H stretch.
[b]Process I is one–photon excitation; process II is two–photon excitation.
[c]All constants accurate to ±10%. τ_A^o and τ_B^o are zero pressure lifetimes from Stern–Volmer plots.

With the tuned-laser excitation technique, single vibronic levels of not only formaldehyde and its isotopic species, but also of other molecules, can be studied. This should give insights into the mechanisms of both energy transfer and nonradiative processes in molecules. The two-photon method also allows one to assign the proper symmetries to various vibronic states when the polarizations of the exciting photons are controlled [16].

APPENDIX I

The two-photon transition probability is proportional to the product of the intensities of the two interacting light beams [17]. In this experiment, both of the photons are derived from the same light beam; the amount of absorption is then proportional to the square of the intensity of the laser beam. Given a laser pulse shape as in Eq. (2), one obtains the effective excitation pulse shape

$$I(\text{eff}) = A_1 \exp(\frac{-t}{E_1}) + A_2 \exp(\frac{-t}{E_2}) + A_3 \exp(\frac{-t}{E_3}) \tag{A1}$$

where

$$A_1 = a_1^2, \quad A_2 = a_2^2, \quad A_3 = -2a_1 a_2,$$

$$E_1 = \frac{c_1}{2}, \quad E_2 = \frac{c_2}{2}, \quad E_3 = \frac{1}{1/c_1 + 1/c_2}$$

A generalized convolution integral [18] gives the relation between the observed emission, $R(t)$, as a function of the exciting light pulse, $S(t')$, and the true time course of fluorescence, $F(t-t')$.

$$R(t) = \int_0^t S(t')F(t-t') \, dt' \tag{A2}$$

The integral can be solved in closed form if one assumes that the fluorescence follows a single exponential decay:

$$F(t-t') = (\frac{1}{\tau}) \exp(\frac{-(t-t')}{\tau}) \tag{A3}$$

For the excitation pulse $S(t) = I(\text{eff})$ in Eq. (A1),

$$R(t) = \sum_{n=1}^{3} \frac{A_n E_n}{E_n - \tau} \exp(\frac{-t}{E_n}) - \left(\sum_{n=1}^{3} \frac{A_n E_n}{E_n - \tau}\right) e^{-\frac{t}{\tau}} \tag{A4}$$

The time-ordered photon counting technique has been widely used in obtaining fluorescence lifetimes from very weak signals [7]. Here, the experimental curve is compared with various theoretical curves given by Eq. (A4) for various values of τ. The fluorescence lifetime is determined by the value of τ that gives the best fit, as in Fig. 4. It is evident that the more photons included in the analysis, the more accurate the determined lifetime is.

ADDENDUM (SEPTEMBER, 1972)

The above-described work has lead to the demonstration of a particularly simple and efficient method of isotopic separation using lasers [19]. The formaldehyde spectrum is sharp and well resolved since the lifetimes of individual states are in the range of 10^{-5} to 10^{-9} sec. In most regions the quantum yield for dissociation is near unity and near the band origin the products are predominantly the stable molecules hydrogen and carbon monoxide. Thus a laser tuned to the absorption frequency of a single isotopic molecule will yield stable isotopic molecular products. Isotopic separation of carbon, oxygen, and hydrogen is possible in this way. Spectroscopic and photochemical studies of other molecules can be expected to reveal molecules suitable for the separation of other elements.

REFERENCES

1. P. M. Rentzepis, M. R. Topp, R. P. Jones, and J. Jortner, Phys. Rev. Letters <u>25</u>, 1742 (1970); G. Porter and M. R. Topp, Proc. Roy. Soc. (London) <u>A315,</u> 163 (1970).
2. K. R. Wilson, Excited State Symposium edited by J. N. Pitts, Jr. (Environmental Resources, Inc. Riverside, Calif., 1970).
3. C. B. Moore, Am. Rev. Phys. Chem. <u>22</u>, 387 (1971).
4. N. C. Craig and C. B. Moore, J. Phys. Chem. <u>75</u>, 1622 (1971).
5. P. B. Sackett and J. T. Yardley, Chem. Phys. Letters <u>6</u>, 323 (1970).
6. J. G. Calvert and J. N. Pitts, Photochemistry (Wiley, New York, 1966
7. B. K. Selinger and W. R. Ware, J. Chem. Phys. <u>53</u>, 3160 (1970).
8. B. H. Soffer and B. B. McFarland, Appl. Phys. Letters <u>10</u>, 266 (1967) B. B. Snavely, Proc. IEEE <u>57</u>, 1374 (1969).
9. G. C. Baldwin, An Introduction to Nonlinear Optics (Plenum, New York 1969), p. 155.
10. M. Goeppert-Mayer, Ann. Physik <u>9</u>, 273 (1931).
11. W. Kaiser and C. G. B. Garrett, Phys. Rev. Letters <u>7</u>, 229 (1961).
12. E. S. Yeung and C. B. Moore, J. Am. Chem. Soc. <u>93</u>, 2059 (1971).

13. R. D. McQuigg and J. G. Calvert, J. Am. Chem. Soc. 91, 1590 (1969).
14. V. Sethuraman, V. A. Job, and K. K. Innes, J. Mol. Spectry. 33, 189 (1970).
15. S. E. Harris, Proc. IEEE 57, 2096 (1969).
16. P. R. Monson and W. M. McClain, J. Chem. Phys. 53, 29 (1970).
17. W. L. Peticolas, Ann. Rev. Phys. Chem. 18, 233 (1967).
18. L. Hundley, T. Coburn, E. Garwin, and L. Stryer, Rev. Sci. Instrum. 38, 488 (1967).
19. E. S. Yeung and C. B. Moore, Appl. Phys. Letters 21, 109 (1972).

STIMULATION OF CHEMICAL REACTIONS BY LASER RADIATION

N. G. Basov, E. M. Belenov, E. P. Markin, A. N. Oraevskii, and
A. V. Pankratov
P. N. Lebedev Physical Institute of the Academy of Sciences
Moscow, USSR

I. INTRODUCTION

The application of lasers opens wide possibilities for the selective stimulation of chemical reactions that proceed in the presence of vibrationally excited molecules [1-5].

Chemical reactions are known to take place only when the energy of reactants exceeds $\mathcal{E}^*$--the activation energy of a process. Considering the energetic effect, chemical conversions can be divided into two classes: endothermic reactions that require a certain amount of energy, and exothermic ones that proceed with heat output. Obviously, endothermic reactions always require a certain amount of activation energy. But a majority of exothermic conversions do not occur without activation. This is associated with the fact that in the course of a chemical reaction (independent of its energetic profit), the reactants must be brought very close together. At short distances, forces of repulsion appear between particles. To overcome them a certain energy is needed corresponding to the energy of activation. In a Maxwell velocity distribution of reacting molecules (at temperature T), the number of collisions initiating the reaction does not exceed the fraction $(\mathcal{E}^*/kT)\exp(-\mathcal{E}^*/kT)$ of the total number of collisions. It is natural to assume that to break a bond or to initiate a reaction, not only the energy of the translational motion can be spent, but also the energy of internal degrees of freedom: vibrational and rotational energies. Moreover, according to contempo-

239

rary concepts, in chemical conversions the vibrational molecular energy plays the main role. For instance, dissociation of an unexcited molecule is extremely small even if the translational energy exceeds the energy of the molecular bond. Dissociation occurs in molecules when the vibrational energy exceeds the energy of the bond [1, 6, 7]. Chemical conversions usually proceed by the following scheme. The interacting molecules "rise" through the vibrational levels up to the activation energy $\mathcal{E}^*$. At energy $\mathcal{E} = \mathcal{E}^*$ the molecules may effectively react. Hence, the typical time of a reaction process is defined by the time it takes a molecule to reach the threshold energy $\mathcal{E}^*$.

Existing physical methods of influencing chemical processes lead, mainly, to the excitation of all degrees of freedom of the molecule. Translational, vibrational, and rotational degrees of freedom are usually in thermodynamic equilibrium. Besides there being an unproductive waste of energy, reactions with equilibrium excited molecules characteristically proceed in the direction of breaking the weakest bond.

A new approach to the problem of chemical conversions of substances would be to consider the possibility of influencing not a molecule as a whole, but its individual bonds. Such a method of selective excitation can be realized by means of lasers. Molecules that are excited by laser radiation to excited mode temperatures of tens of thousands of degrees may come into direct chemical reactions at rates significantly greater than the rates of vibrational–translational relaxation. Thus, besides proceeding in a specific direction, the laser–chemical reactions are characterized by a rapid break between vibrational and translational temperatures [1, 2, 4, 7, 8, 9, 10]. Under certain conditions, thermodynamic equilibrium is broken between internal degrees of freedom as well; a molecule thus enters the chemical process with an excited mode that is resonating with laser radiation.

As an example, let us consider the reactive ability of a molecule of the A–B–C type. If bond A–B is the weakest one, the molecule, as a rule, reacts with the dissociation of this bond. With radiative excitation of the B–C bond in an A–B–C molecule, bond B–C with reacting fragment C becomes chemically active, not bond A–B with reactional center A. The chemical properties of an A–B–C molecule and a selectively excited (A–B–C) molecule are different. Thus, laser–chemical reactions stand apart from the reactions with "equilibrium" molecules, and can be designated as a new class of reactions of selectively excited molecules. It is here that we find the path toward the control of chemical processes by the knowledge of how to excite the required bonds in a molecule and how to maintain a local excitation up to the moment of a chemical reaction.

A characteristic feature of laser radiation is its high spectral monochromacity. This provides the possibility of selective excitation and subsequent separation of molecules with nearly equal vibrational frequencies, for instance, molecules of a different isotopic content. The process is reduced to the initiation of a chemical reaction (or dissociation) of a molecule of a given isotope by laser radiation with a subsequent chemical separation of the mixture products [3].

We have reason to believe that the method of laser initiation will substantially influence a number of the main chemical productions, will open the possibility of the direct reduction and synthesis of new chemical compounds, and will be of interest in molecular biology, i.e., will find application wherever a chemical process proceeds in the presence of vibrationally excited molecules.

This work deals with the conditions of selective excitation of high vibrational levels of molecules by laser radiation, and a number of experimental results in the field of stimulated chemical reactions are given.

II. KINETICS OF LASER HEATING OF MOLECULAR BONDS AND CHEMICAL REACTION RATES

From the moment of laser photon resonant absorption to the beginning of a chemical reaction the molecule undergoes a number of transformations such as the population of higher vibrational levels, the establishment of equilibrium between intermolecular degrees of freedom, dissociations, and vibrational-translational relaxation.

The equations describing the above processes are of the form

$$\frac{\partial X_n^k}{\partial t} = \left(\frac{\partial X_n^k}{\partial t}\right)_{VT} + \sum_j \left(\frac{\partial X_n^j}{\partial t}\right)_{VV} + \left(\frac{\partial X_n^k}{\partial t}\right)_d + \left(\frac{\partial X_n^k}{\partial t}\right)_r + \left(\frac{\partial X_n^k}{\partial t}\right)_q \tag{1}$$

where

$$\left(\frac{\partial X_n^k}{\partial t}\right)_{VT} = Z\left[{}^kP_{n-1}X_{n-1}^k - {}^kP_{n,n-1}X_n^k + {}^kP_{n+1,n}X_{n+1}^k - {}^kP_{n,n+1}X_n^k\right] \tag{2}$$

$$\left(\frac{\partial X_n^k}{\partial t}\right)_{VV}^{kj} = Z\left[\left(\sum_\ell {}^{kj}Q_{n-1,n}^{\ell+1,\ell} X_{\ell+1}^j\right)X_{n-1}^k - \left(\sum_m {}^{kj}Q_{n,n-1}^{m,m+1} X_m^j\right)X_n^k\right.$$
$$\left. + \left(\sum_\ell {}^{kj}Q_{n+1,n}^{\ell,\ell+1} X_\ell^j\right)X_{n+1}^k - \left(\sum_m {}^{kj}Q_{n,n+1}^{m+1,m} X_{m+1}^j\right)X_n^k\right] \tag{3}$$

$$\left(\frac{\partial X_n^k}{\partial t}\right)_d = -{}^kW_{nd}X_n^k$$

$$\left(\frac{\partial X_n^k}{\partial t}\right)_r = -{}^kW_{nr}X_n^k$$

(4)

$$\left(\frac{\partial X_n^k}{\partial t}\right)_q = \left[{}^kS_{n-1,n}X_{n-1}^k - {}^kS_{n,n-1}X_n^k + {}^kS_{n+1,n}X_{n+1}^k - {}^kS_{n,n+1}X_n^k\right]$$

(5)

$X_n^k(t)$ is the concentration of molecules in the n level of k-type vibrations, ${}^kP_{nm}$ is the probability of vibrational-translational molecular transition of the n vibrational state of the k mode in a collision (V-T process), Z is the number of molecular collisions per sec, ${}^{kj}Q_{n,n+1}^{m+1,m}$ is the probability of transformation of a vibrational quantum from an oscillator of the j mode in the m+1 state to an oscillator of the k mode in the n state (V-V process), ${}^kW_{nd}$, ${}^kW_{nr}$ are the rates of molecular dissociation and chemical reaction in the (n, k) state, respectively, ${}^kS_{mn}$ is the transition rate of the k oscillator from the state m to n under the action of a laser field.

The system of kinetic equations (1) is nonlinear, and in the general case the solution cannot be shown. Nevertheless, the vibrational character of processes described by Eqs. (1) seems clear. Let us introduce the times of the relaxation processes: τ_{VT}^k--the time of vibrational-translational relaxation of k type vibrations; τ_{VV}^{kj}--the time of establishment of equilibrium via vibrational-vibrational exchange of excitation between the k- and j-modes of a molecule; τ_d^k, τ_r^k--times of molecular decay in the processes of dissociation and chemical reaction; τ_q^k-- time required to excite the k vibrational branch of a molecule by a laser field.

Under typical conditions the following inequality is fulfilled:

$$\tau_{VV}^{kk} \ll \tau_{VV}^{kj} \ll \tau_{VT}^k$$

(6)

Because of Eq. (6), in the system of molecules the equilibrium is established most quickly in each vibrational mode; then intermolecular equilibrium is established, and finally equilibrium between vibrational and translational degrees of freedom. With the times τ_d^k and τ_r^k everything is more complex. A dissociation or a chemical reaction proceeds effectively only in the case when a molecule stores a rather large amount of energy: the probability of such a process is of the order of $\exp(-\mathcal{E}^*/\theta)$, where $\mathcal{E}^*$ and θ are the energy of activation and the average molecular energy, respectively. With thermal

heating of a system, the number of active $(\theta \simeq \mathcal{E}^*)$ molecules is comparatively small, and hence, such processes are very slow. By contrast, with
selective excitation of molecules by a laser field, θ may be comparable to
$\mathcal{E}^*$, and the processes of dissociation and chemical reaction become among
the fastest processes of the system. Thus, molecular heating by resonant
laser radiation, described by kinetic equation (1), proceeds in the following
manner. At small intensities q of a laser field (τ_q^k is larger than all characteristic times) the system is heated preserving the total thermodynamic
equilibrium. Such heating does not differ from thermal heating. With an
increase of the field intensity ($\tau_{VV}^{kj} < \tau_q^k < \tau_{VT}^{kj}$) the vibrational temperature
of a molecule deviates from the translational one but the energy distribution
inside the molecule remains at equilibrium. Finally, in a rather strong
field the equilibrium is broken between intramolecular degrees of freedom.
Only a molecular bond resonant with the field is selectively heated. The
latter case seems to be optimal for laser chemistry. Below we shall discuss
the conditions for such heating and shall estimate the necessary parameters
of the laser radiation.

III. EXCITATION OF HARMONIC VIBRATIONS OF A MOLECULE
IN A LASER FIELD

Assume that a chemical reaction proceeds in a system with quite a small
activation energy $\mathcal{E}^*$, so that the molecular vibrations with energy $\mathcal{E} \leq \mathcal{E}^*$
may be considered harmonic. In the case of rather high intensity of radia-
tion q all the terms of the kinetic equation can be neglected, except the field
term. Omitting the index k, for simplicity, from Eq. (1) we obtain the
following system of equations:

$$\frac{\partial X_n}{\partial t} = \left(S_{n-1} X_{n-1} - S_{n,n-1} X_n + S_{n+1,n} X_{n+1} - S_{n,n+1} X_n \right) - W_{nr} X_n \qquad (7)$$

For the harmonic oscillator,

$$S_{n-1,n} = S_{n,n-1} = \frac{q}{\hbar w} \sigma_{n-1,n}, \quad \sigma_{n-1,n} = n\sigma_{10} \qquad (8)$$

where q is the laser radiation flux density, σ_{mn} is the cross section of the
resonant transition m → n, and $\hbar w$ is the energy of the vibrational quantum
which we assume to be equal to the quantum energy of the radiation.
    System (7) can be solved using the following assumptions: (a) The energy
of the vibrational quantum $\hbar w$ is small compared to the activation energy $\mathcal{E}^*$.
In this case, from the discrete populations X_n of the levels one can pass to a
continuous distribution function $x(t, \mathcal{E})$. Formally the transition is realized

244   N. G. Basov et al.

by letting $\hbar \omega$ approach zero. (b) The molecules excited to an energy $\mathcal{E} = \mathcal{E}^*$ react effectively. In this case the levels with energies $\mathcal{E} > \mathcal{E}^*$ are practically unpopulated.

With these assumptions the system (7) is reduced to the following diffusion equation:

$$\frac{\partial X}{\partial t} = -\frac{\partial J}{\partial \mathcal{E}} \tag{9}$$

where $J = -q\sigma_{10}\mathcal{E}(\partial X/\partial \mathcal{E})$ is the flux of particles in energy space and with the boundary conditions

$$X(t, \mathcal{E}^*) = 0, \quad J(t, 0) = 0 \tag{10}$$

Let us express $X(t, \mathcal{E})$ in the form

$$X(t, \mathcal{E}) = \sum_{\mu} T_{\mu}(t) X_{\mu}(\mathcal{E})$$

From (9) and (10) we find

$$X(t, \mathcal{E}) = \sum_{\mu} C_{\mu} \exp(-\gamma_{\mu} t) \, J_0[2(\gamma_{\mu} \mathcal{E}/q\sigma_{10})^{1/2} \tag{11}$$

where

$$\gamma_{\mu} = \frac{Z_{\mu}^2 q \sigma_{10}}{4\mathcal{E}^*} \tag{12}$$

and Z_{μ} is the zero of the Bessel function $J_0(Z)$.

In the expansion (11) we shall neglect all the terms except the first one; in this case, due to (12) the rate constant of the chemical reaction is given by

$$\gamma = \frac{Z_1^2 q \sigma_{10}}{4\mathcal{E}^*}, \quad Z_1 \simeq 2.4 \tag{13}$$

Expression (13) for γ has a simple physical meaning. γ^{-1} is the time during which a molecule interacting with the field should absorb energy $\mathcal{E} \simeq \mathcal{E}^*$. Taking into account that on the average $d\mathcal{E}/dt \sim q<\sigma_{mn}>$ and assuming that $<\sigma_{mn}>$ can be identified with σ_{10} from $\int_0^{1/\gamma} q<\sigma_{mn}> \simeq \mathcal{E}^*$ we obtain a relationship similar to (13).

The time $\tau = 1/\gamma$ for the chemical reaction to take place does not depend, according to (13), on pressure, and is inversely proportional to the laser radiation flux q. Therefore, under the conditions for which the harmonic approximation is valid one can always find sufficiently large fields (or sufficiently small pressures) for which a molecule may enter a chemical process that is excited only in an absorbing bond. Let us indicate a numerical example. For $q \simeq 10^2$ W/cm^2, $\sigma_{10} \simeq 10^{-16}$ cm^2, and $\mathcal{E}^* \simeq 1.5$ eV, we have [from (13)], $\gamma \simeq 10^5$ sec^{-1} [11]. Thus, at pressures $\lesssim 100$ Torr, for time $\tau_{VV}^{kj} \simeq 10^{-4}$-$10^{-5}$ sec intermolecular relaxation has no time to be established, and the reaction proceeds with the participation of the excited bond only.

IV.   EXCITATION OF MOLECULAR VIBRATIONS IN A LASER FIELD TAKING INTO CONSIDERATION V-V RELAXATION EFFECTS

Let us assume that the excited vibrational bond of a molecule is almost harmonic for a vibrational exchange up to the levels where the chemical reaction proceeds with a probability close to unity. Actually, anharmonicity is quite significant for the absorption of laser radiation, and absorption is possible only between the ground and the first molecular levels. Such a situation, in view of the high monochromaticity of laser radiation, is expected under actual conditions of laser initiation of a chemical reaction.

Let us write a system of equations that fulfills the above approximation:

$$\left(\frac{\partial X_n^k}{\partial t}\right) = \left(\frac{\partial X_n^k}{\partial t}\right)_{VV}^{kk} + \left(\frac{\partial X_n^k}{\partial t}\right)_q + \left(\frac{\partial X_n^k}{\partial t}\right)_r \tag{14}$$

The expression for the field integral $(\partial X_n^k/\partial t)_q$ is given in Eqs. (7)-(9). The expression

$$\left(\frac{\partial X_n^k}{\partial t}\right)_{VV}^{kk} = \frac{1}{\tau_{VV}^{kk}} \left\{(n+1)\left(1+\frac{\theta^k}{\hbar\omega}\right)X_{n+1}^k - \left[(n+1)\frac{\theta^k}{\hbar\omega} + n\left(1+\frac{\theta^k}{\hbar\omega}\right)\right]X_n^k + n\frac{\theta^k}{\hbar\omega}X_{n-1}\right\} \tag{15}$$

where $\theta^k = \sum_n n X_n^k$ is the total number of vibrational quanta at time t of the collision integral, describing a process of V-V relaxation of the k-branch of the molecule can easily be obtained from (3), if one assumes that for a harmonic oscillator

$$Q_{n+1,n}^{kk\,\ell-1,\ell} = (n+1)\ell/Z \;\tau_{VV}^{kk} \tag{16}$$

Considering the diffusion approximation of Eqs. (14) and (15) for a distribution function $X(t, \mathcal{E})$ (omitting index k) we obtain the following kinetic equation

$$\frac{\partial X}{\partial t} = \frac{1}{\tau_{VV}} \frac{\partial}{\partial \mathcal{E}}\left(X + \theta\frac{\partial X}{\partial \mathcal{E}}\right)\mathcal{E} + q\sigma_{10}\,u\left(\mathcal{E}-\mathcal{E}_*\right)\frac{\partial}{\partial \mathcal{E}}\left(\mathcal{E}\,\frac{\partial X}{\partial \mathcal{E}}\right) \tag{17}$$

where $u(\mathcal{E}) = 1$ for $\mathcal{E} < 0$, $u(\mathcal{E}) = 0$ for $\mathcal{E} > 0$, θ is the "vibrational temperature" of the excited bond, and $\mathcal{E}_* \simeq \hbar\omega$ is the boundary energy, above which the molecules are populated only by resonant excitation transfer.

Expressing the solution of (17) in the form $X(t, \mathcal{E}) = \exp(-\gamma t) X_0(\mathcal{E})$ where γ is the rate of the chemical reaction, we rewrite (17) in the form

$$\kappa X_0 = \frac{\partial J}{\partial \mathcal{E}}, \quad J = -\mathcal{E}\left[X_0 + \left(\theta + X\ u(\mathcal{E}-\mathcal{E}_*)\frac{\partial X_0}{\partial \mathcal{E}}\right)\right] \tag{18}$$

$$X_0(\mathcal{E}^*) = 0, \quad J(0) = 0$$

where

$$\kappa = \gamma \tau_{VV}, \quad X = q \sigma_{10} \tau_{VV}$$

For $X = 0$ (laser field absent) Eq. (18) has the following solution:

$$X(\mathcal{E}) = C_0 \exp\left(-\frac{\mathcal{E}}{\theta}\right) {}_1F_1\left(-\kappa, 1, \frac{\mathcal{E}}{\theta}\right) \tag{19}$$

where C_0 is a constant of integration, ${}_1F_1(-\kappa, 1, \mathcal{E}/\theta)$ is a confluent hypergeometric function. For $X \neq 0$, the solution of (18) differs from (19), but physically it is clear that if the activation energy $\mathcal{E}^*$ is much more than the quantum energy $\mathcal{E}_* = \hbar\omega$, the difference is insignificant. Therefore, let us assume that for $\mathcal{E} > \mathcal{E}_*$ the solution of Eq. (18) is given by formula (19). For $\mathcal{E} < \mathcal{E}_*$ the solution of (18) is also expressed through the hypergeometric function

$$X_0(\mathcal{E}) = C_1 \exp\left(-\frac{\mathcal{E}}{\theta + X}\right) {}_1F_1\left(-\kappa, 1, \frac{\mathcal{E}}{\theta + X}\right) \tag{20}$$

The chemical reaction constant $\gamma = \kappa/\tau_{VV}$ is obtained from the condition ${}_1F_1(-\kappa, 1, \mathcal{E}^*/\theta) = 0$. Assuming the asymptotic $(\mathcal{E}^*/\theta \gg 1)$ behavior of the hypergeometric function

$$ {}_1F_1\left(-\kappa, 1, \frac{\mathcal{E}^*}{\theta}\right) \sim 1 - \kappa\frac{\theta}{\mathcal{E}^*}\ \exp\left(\frac{\mathcal{E}^*}{\theta}\right) \tag{21}$$

we obtain the following dependence of γ on the "vibrational temperature" θ:

$$\gamma = \frac{1}{\tau_{VV}}\ \frac{\mathcal{E}^*}{\theta}\ \exp\left(-\frac{\mathcal{E}^*}{\theta}\right) \tag{22}$$

The vibrational temperature in turn is determined through the reaction constant γ and the flux density q. To find this dependence one can do the following: Let us equate the energy absorbed from the field

$$q\sigma_{10}\hbar\omega\,[\,X_0(0) - X_0(\hbar\omega)\,]$$

and the energy

$$\mathcal{E}^*\gamma \int_0^{\mathcal{E}^*} X_0(\mathcal{E})\;d\mathcal{E}$$

that is carried away by molecules during the chemical reaction:

$$q\sigma_{10}(\hbar\omega)^2 \frac{\partial X(0)}{\partial\mathcal{E}} \simeq \mathcal{E}^*\gamma \int_0^{\mathcal{E}^*} X(\mathcal{E})\;d\mathcal{E} \tag{23}$$

[a stricter condition for the energy balance can be directly obtained from the kinetic equation (18). Constants C_0 and C_1 are included in Eq. (23). For $\hbar\omega < \theta < \mathcal{E}^*$ the constant C_0 is determined by the normalization condition

$$C_0 \int_0^{\mathcal{E}^*} \exp\left(-\frac{\mathcal{E}}{\theta}\right)\,{}_1F_1\left(-\kappa, 1, \frac{\mathcal{E}}{\theta}\right) d\mathcal{E} \simeq 1$$

and is approximately equal to $1/\theta$. The constant $C_1 \simeq C_0$ is found from the condition of continuity of the molecular flux $J(\mathcal{E})$ at the point $\mathcal{E} = \mathcal{E}_*$. Then Eq. (23) has the form

$$q\sigma_{10}(\hbar\omega)^2 \frac{1}{\theta(\theta + X)} \simeq \gamma\mathcal{E}^* \tag{24}$$

By combining (23) and (24) we finally obtain

$$\gamma \simeq \frac{1}{\tau_{VV}} \frac{X}{\mathcal{E}^*} \left(\frac{\hbar\omega}{\mathcal{E}^*}\right)^2 \frac{(\mathcal{E}^*/\theta)^2}{1 + X/\theta}$$

$$\tag{25}$$

$$\exp\left(-\frac{\mathcal{E}^*}{\theta}\right) \simeq \frac{X}{\mathcal{E}^*} \left(\frac{\hbar\omega}{\mathcal{E}}\right)^2 \frac{\mathcal{E}^*/\theta}{1 + X/\theta}$$

It follows from (24) and (25) that for small fields ($X \to 0$) the chemical reaction constant is actually proportional to the radiation flux density. On the other hand, for large fields ($X \to \infty$) the vibrational temperature and the reaction constant tend to the limits

$$\theta_\infty = \frac{\mathcal{E}^*}{2 \, \ln(\mathcal{E}^*/\hbar\omega)} \tag{26}$$

$$\gamma_\infty = \frac{2}{\tau_{VV}} \left(\frac{\hbar\omega}{\mathcal{E}^*}\right)^2 \, \ln(\mathcal{E}^*/\hbar\omega) \tag{27}$$

which are determined by the saturation of the $0 \to 1$ transition by the laser field. For $\mathcal{E}^* \simeq 2$ eV, $\hbar\omega \simeq 0.1$ eV from (26) and (27) we obtain $\theta_\infty \simeq 0.33$ eV, $\gamma \simeq 1.5 \times 10^{-2}/\tau_{VV}$. Thus, before the onset of vibrational-translational relaxation processes and before the system approaches inter-molecular equilibrium, most of the molecules will enter into chemical reaction, (assuming the defect of quantum energy for the various modes is large compared to the translational temperature). The required fluxes of laser radiation are comparatively low. In fact, the saturation of θ and γ corresponds to the condition $\chi/\theta \sim 1$. Hence, for $\tau_{VV} \simeq 10^{-6}$ sec, $\sigma_{10} \simeq 10^{-16}$ cm^2 we obtain $q \simeq 10^2$ W/cm^2.

V. EXCITATION OF MOLECULAR VIBRATIONS IN A LASER FIELD TAKING INTO ACCOUNT V-V AND V-T PROCESSES

Let us assume that a laser pulse duration τ is comparable to or exceeds the times of vibrational-vibrational and vibrational-translational relaxations. In this case the excited k type of vibrations are cooled by interaction with all degrees of molecular freedom. Such a situation is not, however, an analog of thermal heating, since the selectively excited mode is not in thermal equilibrium with the nonresonant degrees of freedom, and may have a temperature which significantly exceeds the average temperature T_0 of the gas.

Let us consider the collision integrals of the kinetic equation. To the V-T relaxation corresponds a collisional integral $(\partial X_n^k/\partial t)_{VT}$, which has the form for a harmonic oscillator

$$\left(\frac{\partial X_n^k}{\partial t}\right)_{VT} = \frac{1}{\tau_{VT}^k} \left\{ (n+1)X_{n+1}^k - \left[(n+1)\,\exp\left(-\frac{\hbar\omega^k}{T}\right) + n\right]X_n^k + nX_n^k\,\exp\left(-\frac{\hbar\omega^k}{T}\right) \right\} \tag{28}$$

where T is the translational temperature. V-V exchange between the k mode and the j mode is described by the integral of collisions

$$\left(\frac{\partial X_n^k}{\partial t}\right)_{VV}^{kj} = -\frac{1}{\tau_{VV}^{kj}}\left\{(n+1)\left(1+\frac{\theta^j}{\hbar\omega}\right)X_{n+1}^k - \left[(n+1)\frac{\theta^j}{\hbar\omega} + n\left(1+\frac{\theta^j}{\hbar\omega}\right)\right]X_n^k + n\theta^j_{\hbar\omega}\,X_{n-1}^k\right\} \tag{29}$$

where θ^j is the temperature of the j mode. In order for the problem to be solved, it is also necessary to take into account the interaction of the nonresonant degrees of molecular freedom. However, if the duration τ of the laser pulse is of the order of (or greater than) the cross-relaxation time, such an effect can be considered simply by ascribing the same temperature to the nonresonant degrees of freedom. Taking this temperature to be T, and passing from the discrete equations (28) and (29) to the diffusion approximation we obtain

$$\left(\frac{\partial X^k}{\partial t}\right)_{VT} = \frac{1}{X_{VT}^k}\frac{\partial}{\partial \mathcal{E}}\left[\mathcal{E}\left(X^k + T\frac{\partial X^k}{\partial \mathcal{E}}\right)\right] \tag{30}$$

$$\left(\frac{\partial X^k}{\partial t}\right)_{VV}^{kj} = \frac{1}{\tau_{VV}^{nj}}\frac{\partial}{\partial \mathcal{E}}\left[\mathcal{E}\left(X^k + T\frac{\partial X^k}{\partial \mathcal{E}}\right)\right] \tag{31}$$

With respect to (30) and (31) the kinetic equation for $X^k(t,\mathcal{E}) = X(t,\mathcal{E})$ is written in the following manner:

$$\frac{\partial X}{\partial t} = \frac{1}{\tau_{VV}}\frac{\partial}{\partial \mathcal{E}}\left[\mathcal{E}\left(X + \theta\frac{\partial X}{\partial \mathcal{E}}\right)\right] + \frac{1}{\tau^0}\frac{\partial}{\partial \mathcal{E}}\left(X + T\frac{\partial X}{\partial \mathcal{E}}\right) + q\sigma_{10}\,u(\mathcal{E}-\mathcal{E}_*)\frac{\partial}{\partial \mathcal{E}}\left[\mathcal{E}\frac{\partial X}{\partial \mathcal{E}}\right] \tag{32}$$

where the introduced time τ^0 of cross-relaxation is determined by the ratio

$$\frac{1}{\tau^0} = \frac{1}{\tau_{VT}} + \sum_{j\neq k}\frac{1}{\tau_{VV}^{kj}} \tag{33}$$

For $X = \exp(-\gamma t)X_0(\mathcal{E})$ the solution of Eq. (32) has the form

$$X_0(\mathcal{E}) = \begin{cases} C_0 \exp\left(-\frac{\mathcal{E}}{\theta}\frac{1+\beta}{1+\beta T/\theta + X/\theta}\right)\,_1F_1\left[\frac{\kappa}{1+\beta}, 1, \frac{\mathcal{E}}{\theta}\left(\frac{1+\beta}{1+\beta T/\theta + X/\theta}\right)\right], & \mathcal{E}<\hbar\omega \\[3ex] C_1 \exp\left(-\frac{\mathcal{E}}{\theta}\frac{1+\beta}{1+\beta T/\theta}\right)\,_1F_1\left[\frac{\kappa}{1+\beta}, 1, \frac{\mathcal{E}}{\theta}\left(\frac{1+\beta}{1+\beta T/\theta}\right)\right], & \mathcal{E}>\hbar\omega \end{cases} \tag{34}$$

where $\beta = \tau_{VV}^{kk}/\tau^0$.

Hence we can easily obtain the following relationships for the "vibrational temperature" θ and rate γ of reactions:

$$\gamma \simeq \frac{1}{\tau^{kk}_{VV}} \frac{\mathcal{E}^*}{\theta} \frac{(1+\beta)^2}{(1+\beta T/\theta)} \exp\left(-\frac{\mathcal{E}^*}{\theta} \frac{1+\beta}{1+\beta T/\theta}\right) \tag{35}$$

$$\frac{\chi}{\theta}\left(\frac{\hbar\omega}{\theta}\right)^2 \frac{(1+\beta)^2}{(1+\beta T/\theta)(1+\beta T/\theta + \chi/\theta)} \simeq \beta\left[\frac{1+\beta T/\theta}{1+\beta} - \frac{T}{\theta}\right] + \gamma\, \tau_{VV} \frac{\mathcal{E}^*}{\theta} \tag{36}$$

Let us consider particular cases of expressions (35) and (36). For $\chi = 0$ (no laser field) neglecting the energy lost to the chemical reaction in (36), we find $\theta = T$. Now (35) gives a typical expression for the chemical reaction constant with thermal heating of the system:

$$\gamma = \frac{1}{\tau} \frac{\mathcal{E}^*}{T} \exp\left(-\frac{\mathcal{E}^*}{T}\right), \quad \frac{1}{\tau} = \frac{1}{\tau^0} + \frac{1}{\tau_{VV}} \tag{37}$$

For $\chi \neq 0$, but in the absence of j-k vibrational exchange and V–T relaxation ($\beta = 0$), from (35) and (36) we obtain formulas (25) and (24).

Finally, taking into account V–V and V–T processes when the main energy losses of the resonant mode are associated with the thermal gas heating, from (35) and (36) it follows that

$$\gamma \simeq \frac{1}{\tau_{VV}} \frac{\mathcal{E}^*}{\theta} \exp\left(-\frac{\mathcal{E}^*}{\theta}\right) \tag{38}$$

$$\frac{\chi}{\theta}\left(\frac{\hbar\omega}{\theta}\right)^2 \frac{1}{1+\chi/\theta} = \frac{\tau_{VV}}{\tau^0} \tag{39}$$

From (39) it follows that for low fluxes q of laser radiation the vibrational temperature varies as $q^{1/3}$. Because of saturation of the resonant transition for large values of q the temperature tends to the finite limit

$$\theta_\infty = \hbar\omega\left(\frac{\tau^0}{\tau_{VV}}\right)^{1/2} \tag{40}$$

which is equal, for instance, for $\tau^0/\tau_{VV} \simeq 10^2$ and $\hbar\omega = 0.1$ eV to a value of about $10,000°K$. To reach the maximum values of temperature θ_∞ and rate γ_∞ comparatively low ($\lesssim 10^2$ W/cm^2) radiative intensities are required.

The process of cross-relaxation, which is connected with the energy transfer to the nonresonant degrees of freedom, seems to be detrimental. However, for pulse duration τ of the order of the cross-relaxation time τ^0 the energetic losses are still small, and the energy absorbed from the field is distributed approximately equally between the excited mode and other degrees of freedom. In connection with this the change of the average temperature T (particularly in the case of a polyatomic molecule) should be negligible in comparison with θ.

VI. EXCITATION OF MOLECULAR VIBRATIONS BY RAMAN SCATTERING

A number of important molecules (for instance, N_2 and O_2) have no intrinsic dipole moment, and therefore their direct excitation by a light field may turn out to be ineffective. At the same time these molecules are active during Raman scattering. The probability of Raman scattering is known to be appreciable only for the condition $\Omega = \omega_s + \omega$, where Ω, ω_s, and ω are the frequencies of the absorbed photons, scattered photons, and the molecular vibrational frequency, respectively. This equality expresses the conservation of energy in an elementary scattering process: a photon with energy $\hbar\Omega$ is absorbed from the external field; at the same time a photon of Stokes frequency ω_s is radiated, and the energy of molecular vibrations is increased by $\hbar\omega$. By means of this process the vibrational branch of the molecule with frequency $\omega = \Omega - \omega_s$ is selectively heated in the process of Raman scattering.

The transition rate $S_{k,k+1}$ of a molecule between the levels k and k+1 in the field

$$E = E_\Omega \cos\Omega t + E_s \cos\omega_s t \qquad (41)$$

is given by

$$S_{k,k+1} = kS_{10}$$

$$S_{10} = \frac{6\pi T^* c^2}{\hbar^2 \omega_s^2} q_\Omega q_s Q_0 \qquad (42)$$

Here E_Ω, E_s, q_Ω, and q_s are intensities of the field and flux densities of the main and Stokes radiation, respectively, c is the velocity of light, Q_0 is the total cross section for Raman scattering, and T^* is the characteristic time of relaxation of the excited mode.

With respect to (42) the field term of the kinetic equation has the form

TABLE I. LASER-CHEMICAL REACTIONS PRODUCED BY INFRARED
RADIATION. (Time of Irradiation is less than 0.5 sec [2]).

No.	Reac-tants	Partial Pressure (Torr)	Inten-sity[a] (Watts)	Lumin-escence[a]	Pressure of Reaction Products (Torr)	Reaction Products, Volume %
1.	N_2F_4 NO	100 100	20	No	212	FNO - 34% NF_3 - 21 N_2 - 24 F_2 - 20
	N_2F_4 NO	100 100	40	Yes	262	FNO - 28 NF_3 - 3 N_2 - 24 F_2 - 46
2.	N_2F_4 NO N_2	100 200 460	70	Yes	--	NO_2, N_2O N_2
3.	N_2F_4 NO CF_4	100 100 200	20	--	--	NF_3 - 10 NO - 10 CO_2 - 10 N_2
4.	N_2F_4 N_2O	300 150	40	Weak (n = 3)	512	NO_2 - 49 NF_3 - 38 N_2O - 5 FNO_3, N_2
5.	N_2F_4 H_2	50 100	50	Yes	--	N_2, HF
6.	N_2F_4 CH_4	228 114	50	Yes	534	CF_4 - 22 HF, N_2
	N_2F_4 CH_4	150 150	40		393	CF_4 - 23 HF, N_2, C

(continued)

TABLE I. (continued).

No.	Reactants	Partial Pressure (Torr)	Intensity[a] (Watts)	Luminescence[a]	Pressure of Reaction Products (Torr)	Reaction Products, Volume %
7.	N_2F_4 BCl_3	114 114	50 (n = 3)	Yes (n = 1)	289	BF_3, Fluorinated chlorides
8.	SiH_4	228	50 (n = 3)	Yes (n = 1, 2)	284	H_2 - 34% Si
9.	SiH_4 BCl_3	112 112	40 (n = 3)	Yes (n = 1, 2, 3)	234	$BHCl_2$, SiH_3Cl
	SiH_4 BCl_3	300 100	40 (n = 5)	--	431	B_2H_6, $BHCl_2$, $SiCl_4$, $SiHCl_3$, $SiCl_3H$, SiH_2Cl_2
10.	SiH_4 SF_6	300 150	40	Yes	698	SiF_4 - 20 H_2S - 70
11.	HNF_2	200	40	Yes	--	HF, N_2, F_2
12.	HNF_2 CH_4	250 100	40	Yes	250	CF_4, C_2F_6, HF, N_2
13.	HNF_2 SiH_4	250 100	40	Yes	460	SiF_4 - 17 H_2 - 47 N_2, HF
14.	HNF_2 H_2	100 30	40	Yes	110	N_2, HF

[a]n = number of irradiations.

$$\left(\frac{\partial X}{\partial t}\right)_q = (S_{10} \hbar w) \frac{\partial}{\partial \mathcal{E}} \left(\mathcal{E} \frac{\partial X}{\partial \mathcal{E}}\right) \tag{43}$$

and it does not differ in form from the field term of the dipole molecule [see Eq. (9)]. Therefore, substituting $S_{10} \hbar w$ for $q \sigma_{10}$ one can go from the formulas describing dipole molecular heating to formulas describing molecular heating for Stokes scattering of a laser field.

Note that Stokes component generation in an excited gas is apparently a small effect. We think the case of gas heating in the given E_Ω and E_S fields is of more interest at relatively low pressure ($p \sim 10^{-3}$ Torr) when the relaxation rate $1/T^*$ is small. For instance, fields E_S and E_Ω can be obtained by transmitting the radiation of the main frequency through dense gases or liquids composed of molecules of the same type as the excited ones.

Let us now make some numerical estimates. Consider the case in which the field (41) excites transitions only between the ground state and the first molecular level, and the population of the higher levels is determined by the processes of vibrational-vibrational exchange. According to (25) and (24) with $\theta/S_{10} \hbar w \tau_{VV} \sim 1$, the reaction rate constant γ and the vibrational temperature are close to their maximum values of $\gamma = \gamma_\infty$ and $\theta = \theta_\infty$. Hence, for $\theta_\infty \simeq 10 \hbar w$ and $\tau_{VV} \sim 10^{-7}$ sec, it follows that $S_{10} \simeq 10^8$ sec^{-1}. Assuming $\hbar w_s \simeq 1$ eV, $q_\Omega \simeq q_s = q$, $T^* \simeq 10^{-9}$ sec we find that the flux required for saturation of γ and θ is $q \lesssim 10^2$ MW/cm^2. For a laser pulse duration of $\sim 10^{-6}$ sec a significant fraction of the molecules can enter the chemical reaction.

VII. EXPERIMENTAL DETERMINATION OF THE PHOTOCHEMICAL ACTION OF LASER RADIATION

We have studied CO_2 laser infrared radiation ($\lambda = 10.6$ μ) interacting with the systems N_2F_4-NO and SF_6-NO in which tetrafluorohydrazene (frequencies of valence vibrations 934 cm^{-1} and 946-959 cm^{-1}) and hexafluorosulphur (deformational molecular vibration is of 943 cm^{-1}) show resonant absorption. In the absence of radiation in the mixture of N_2F_4 and NO the conversions begin at a temperature higher than $600°$K (during 5 min heating); N_2F_4 does not react with NO but decomposes with the formation of nitrogen trifluoride. In the mixture of SF_6 and NO heated quickly to $1000°$K, no reaction takes place.

The behavior of both systems upon irradiation depends on the radiation intensity. Reactions do not take place for short exposures and below intensities of about 20 W. With intensities higher than 20 W and exposures less than 0.1 sec the instantaneous reactions occur accompanied by luminescence. In the system N_2F_4-NO, the products of the reaction are: FNO, F_2, N_2,

NF_3, and NO_2, the percentage of each depending upon the experimental conditions [4]. Conversion of N_2F_4 and NO is 100%. In the SF_6 + NO system, thionyl fluoride (SF_2O) is formed.

Calculation shows that if the absorbed laser energy supplies the heating of the reacting systems, for the mixture SF_6 + NO the temperature would increase to $1000^{\circ}K$ while for the system SF_4 + NO, $500^{\circ}K$. At such temperatures and with heating times that are 2-3 orders of magnitude longer than the light exposures, thermal reactions do not occur in either system.

For the system N_2F_4 + NO it has also been shown [1, 4, 10] that tetrafluorohydrazene does not react when nitrogen oxide is not present, or is present only in small concentrations in the irradiated system. The process begins only at comparable concentrations of N_2F_4 and NO. This fact indicates the absence of preliminary dissociation of tetrafluorohydrazene under these conditions. This does not mean that tetrafluorohydrazene cannot dissociate in the ir radiation; dissociation can be expected at power levels above 100W.

It is important to note that in reactions such as N_2F_4 with NO, N_2F_4 with N_2O, and many others, products are obtained which differ from those obtained by thermal activation of the process. Thus, in both systems in the absence of radiation, the decomposition of N_2F_4 dominates: $3N_2F_4 = 4NF_3 + N_2$. Nitric oxide and nitrous oxide do not react with N_2F_4. In a laser-chemical reaction the properties of N_2F_4 are changed. The formation of FNO in the reaction $N_2F_4 + 4NO \rightarrow 4FNO + N_2$ indicates that N_2F_4 acquires a fluorinating ability, i.e., the activation of the N-F bond, the valence vibration of which is resonant with the ir radiation frequency and which was chemically inactive in the unexcited molecule. A list of some observed laser induced reactions is given in Table I.

One can conclude that under the action of the laser ir radiation, vibrationally excited molecules initiate chemical reactions. The reactions are normally exothermic, and can be classified as superfast processes.

Thus, resonant interaction of coherent radiation permits one to heat molecular vibrational degrees of freedom to high temperatures, to direct a chemical reaction, to reduce energy losses and to obtain products which cannot be produced by thermal heating of reactants.

REFERENCES

1. N. G. Basov, E. P. Markin, A. N. Oraevskii, and A. V. Pankratov, Dokl. Akad. Nauk SSSR 198, 1043 (1971) [Sov. Phys. Dokl. 16, 445 (1971)].
2. N. G. Basov, E. P. Markin, A. N. Oraevskii, A. V. Pankratov, and A. N. Skachkov (to be published).

3. S. W. Mayer, M. A. Kwok, R. W. F. Gross, and D. J. Spencer, Appl. Phys. Letters 17, 516 (1970).
4. N. G. Basov, V. I. Igoshin, E. P. Markin, and A. N. Oraevskii, Kvant. Elektron. 1 (No. 2), 3 (1971) [Sov. J. Quant. Electron, 1, 119 (1971)].
5. A. N. Oraevskii and V. A. Savva, Brief Communications in Physics, FIAN 7, 50 (1970).
6. E. V. Stupochenko and A. I. Osipov, Zh. Fiz. Khim. 32, 1673 (1958).
7. N. V. Karlov, Yu. N. Petrov, A. M. Prokhorov, and O. N. Stel'makh, ZhETF Pis. Red. 11, 220 (1970) [JETP Letters 11, 135 (1970)].
8. N. D. Artamonov, V. T. Platonenko, and R. V. Khokhlov, Zh. Eksp. Teor. Fiz. 58, 2195 (1970) [Sov. Phys. JETP 31, 1185 (1970)].
9. Yu. V. Afanas'ev, E. M. Belenov, E. P. Markin, and I. A. Poluektov, ZhETF Pis. Red. 13, 462 (1971) [JETP Letters 13, 331 (1971)].
10. N. G. Basov, E. P. Markin, A. N. Oraevskii, A. V. Pankratov, and A. N. Skachkov, ZhETF Pis. Red. 14, 251 (1971) [JETP Letters 14, 165 (1971)].
11. Ya. B. Zel'dovich and Yu. P. Raizer, Elements of Gasdynamics and the Classical Theory of Shock Waves [Academic Press; New York (1968)].

PHOTOFRAGMENT SPECTROSCOPY: LASERS IN, LASERS OUT*

Graham Hancock and Kent R. Wilson†
Department of Chemistry, University of California, San Diego
La Jolla, California

ABSTRACT

Lasers are in two ways an integral part of photofragment spectroscopy, a new branch of molecular spectroscopy. First, they form part of the actual experimental apparatus, in which a molecular beam is crossed with pulses of polarized laser light, and the recoiling photodissociation products are detected with a mass spectrometer. We discuss the use of various solid-state and dye lasers, with particular emphasis on the description of a frequency quadrupled neodymium laser system, which produces 10-nsec pulses of 1 J at $37,550$ cm^{-1} (2662 Å). Second, photofragment spectroscopy can directly probe the processes involved in photodissociation lasers. We

*Supported by the National Science Foundation, by the Office of Naval Research (under contract number N00014-69-A-0200-6020), and by the Environmental Protection Agency, Office of Air Programs. Use of computer facilities supported by the National Institutes of Health and the National Science Foundation are gratefully acknowledged.

†Alfred P. Sloan Research Fellow.

describe the results of experiments which have particular interest to the understanding of such lasers, and to the production of new laser systems. Thus, photofragment spectroscopy shows the need for a reinterpretation of the molecular states involved in the I_2 main visible continuum, and therefore removes the mystery as to the failure of various attempts to make a photodissociation laser based on this absorption. The feasibility of a Br_2 photodissociation laser is discussed, and the population inversions responsible for the first photodissociation lasers are probed for a series of alkyl iodides. Finally, an interpretation of the photofragment spectrum of ICN is given, revealing that lasing action from either the I or CN photodissociation fragment is to be expected.

I. INTRODUCTION

Spectroscopists have for years probed the discrete band spectra of a large number of molecules, providing a wealth of detail about the bound molecular states involved in these transitions. In contrast, there exists comparatively little information about the dissociative states of molecules (i.e., states from which the molecule breaks apart into separating fragments) in spite of the fact that in theory the number of dissociative states is far greater. Photofragment spectroscopy is a technique which we have developed to study dissociative transitions in detail. It provides direct information about unbound states in a conceptually simple way; instead of observing the absorption of photons in a molecular system, it provides a means of detecting the recoiling fragments from a dissociative event, of characterizing them in mass, in angular distribution, and in kinetic and internal energy, and provides details of the molecular dynamics of the dissociation process by which they are formed.

Photofragment spectroscopy involves lasers in two ways. In an applied sense, lasers are used to produce the intense beam of photons necessary to dissociate a large enough number of molecules to ensure a sensible signal-to-noise ratio in the detection of the photofragments. In this respect, the success of the experiments depends heavily upon advances in laser technology. We have used a variety of high-power laser systems at different photon energies, the wavelength of the available radiation usually being the controlling factor in the choice of molecules for which study is feasible. The strength of chemical bonds is generally such that they break on absorption of uv light, and for some time the lack of powerful laser sources covering this region has limited progress. This situation has been recently improved by the use of a new system, described below, producing 1-J pulses at $37,550$ cm^{-1} (2662 Å).

The second connection of the technique with laser systems concerns the prediction and explanation of photodissociation lasers. In the early 1960's, several groups [1-3] suggested the possibility of using molecular photodissociation to produce an inverted population of electronically excited fragments. Molecules such as TlBr [3] and some alkali halides [2] were potential candidates for laser systems, the inversion being formed by the production of an electronically excited thallium or alkali atom. It was not until 1964 that the first photochemical laser was discovered, operating on a transition different from those initially suggested. Kasper and Pimentel [4] dissociated CF_3I and CH_3I with a broad-band flashlamp, and observed lasing action from the $^2P_{1/2}$ excited state of iodine atoms.

Sec. II will briefly describe the apparatus of photofragment spectroscopy, and indicate the types of information available from its use. Further sections will explore the application of the technique to the understanding and development of photochemical laser systems. As will be shown, photofragment spectroscopy can measure the distribution of internal energy in photodissociation fragments, and thus can be used to predict whether or not an inverted population will be produced. In addition, as shown by Zare and Herschbach [5], the fragment angular and translational energy distributions, directly observable in the photofragment spectrum, influence laser gain through Doppler broadening.

II. APPARATUS AND METHOD

A. General

Figure 1 shows a cutaway drawing of the important interaction region of the photofragment spectrometer; a full description of the apparatus is published elsewhere [6]. A molecular beam of the molecules to be studied is formed in a stainless-steel high-vacuum chamber. A short ($\sim$ 10-30 nsec) pulse of polarized light from a laser perpendicularly crosses the molecular beam, and photofragments recoiling orthogonally to the two beams are detected as a function both of mass and of flight time, by a quadrupole mass spectrometer placed a few centimeters from the interaction region. The orientation of the electric vector $\vec{E}$ of the incoming radiation can be varied with respect to the detection direction by rotation of a half-wave plate, placed in the laser path. This is almost equivalent to a rotation of the detector about the laser beam axis; only fragments recoiling in a fixed direction with respect to $\vec{E}$ pass into the mass spectrometer and are detected. The photofragment spectrum can thus be recorded as a function of several variables; photon energy and flux, fragment mass, fragment recoil kinetic energy (determined from the time fragments take to reach the mass spectrometer), and recoil angle with respect to the electric vector $\vec{E}$. An on-line computer controls the experiment, stores and averages the data, and communicates the results back to the experimenter via a storage oscilloscope.

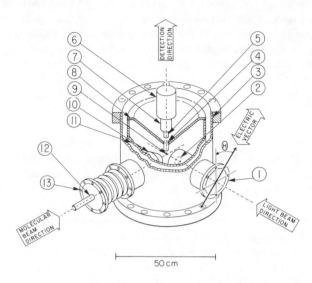

50 cm

Figure 1. Cutaway drawing of photofragment spectrometer. The beam of
molecules to be photodissociated enters from the left and is crossed perpendicularly by pulses of polarized light from a laser. The photodissociation
fragments which recoil upward are detected by a mass spectrometer as a
function of mass, of photon energy, of photon flux, of time t after the laser
pulse, and of angle of recoil Θ measured from the electric vector of the light.
(The Θ shown in the drawing represents a negative angle of recoil.) The
interaction region and the mass spectrometer are in separately pumped
chambers connected by a small liquid-nitrogen-cooled (LN_2) tube, which
collimates the fragments. The numbered components are (1) port for laser
beam (various solid state and laser-pumped dye lasers are used), (2) lens to
match diameter of laser beam to that of molecular beam, (3) outer wall of
bakeable ultrahigh vacuum chamber, (4) LN_2-cooled fragment collimating
tube, (5) mass spectrometer electron bombardment ionizer, (6) mass spectrometer electron multiplier, (7) quadrupole section of mass spectrometer,
(8) LN_2-cooled partition between interaction and detection chambers,
(9) interaction region, (10) LN_2-cooled molecular beam collimator and oven
shield, (11) LN_2-cooled inner wall of interaction chamber, (12) molecular
beam oven with capillary slits, and (13) molecular beam port.

    From a knowledge of the dissociation energy of the molecule being
studied, the measured kinetic energy distribution of the fragments, and
energy conservation, the distribution of total internal energy in the products
of photodissociation can be determined. For diatomic molecules, this

distribution directly reveals the electronic states of the atomic fragments. For polyatomics, although the distribution of <u>total</u> fragment internal energy is still determined, the situation is complicated by the existence of vibrational and rotational degrees of freedom in the fragments, with the possibility of energy also being partitioned into these modes.

The angular distributions of the recoiling fragments provide important information on the symmetries of the ground and dissociative states of the parent molecules [5, 7, 8]. For diatomic molecules the upper state can often be unambiguously characterized in this way, for the angular distribution is linked to the molecular state symmetries by the change in the quantum number Ω, the component of the total angular momentum vector along the internuclear axis. The flux of recoiling fragments has a $\cos^2\theta$ distribution for $\Delta\Omega = 0$, a $\sin^2\theta$ distribution for $\Delta\Omega = \pm 1$, θ being the angle between the electric vector $\vec{E}$ of the light and the fragment recoil direction in the center-of-mass coordinate system. With polyatomic molecules, one finds mixtures of $\cos^2\theta$ and $\sin^2\theta$, depending on the angle between the transition dipole moment and fragment recoil directions within the molecule [8]. Information on dissociative state lifetimes and the molecular dynamics of the dissociation process can also be derived from their effects on the angular distribution [8]. Measurements obtained in the laboratory coordinate system can be converted to center-of-mass distributions by the appropriate transformation. Angular distributions in the laboratory recoil angle Θ (see Fig. 1) differ from those in the center-of-mass system mainly by a simple rotation of several degrees, otherwise essentially retaining their shape.

B. Use of Lasers

We have extensively used both neodymium-glass and ruby Q-switched lasers in photofragment spectroscopy. Fundamental radiation from both these systems is of rather low energy to break chemical bonds, neodymium-glass being at 9415 cm^{-1} (10, 620 Å), and ruby at 14, 405 cm^{-1} (6943 Å), although dissociation of I_2 in the weak continuum reached by the ruby laser has been observed [6, 9]. Frequency doubling these systems produces more useful photon energies. Second harmonic neodymium-glass, using a KDP doubling crystal, gives us ~ 0.1 J per pulse in the green, at 18, 830 cm^{-1} (5310 Å), while second harmonic ruby, using an ADP crystal, gives ~ 0.15 J pulses in the near uv, at 28, 810 cm^{-1} (3471 Å). The second harmonic ruby radiation is also used to pump a diffraction grating tunable dye laser, covering the visible spectrum at a somewhat lower output energy. Particularly useful dyes include acridone (lasing at $\sim 22, 990$ cm^{-1}) and 7-diethylamino 4-methyl coumarin ($\sim 21, 690$ cm^{-1}).

Recent experiments utilize a new laser system for the photochemically more rewarding uv region. The laser, designed and constructed for us

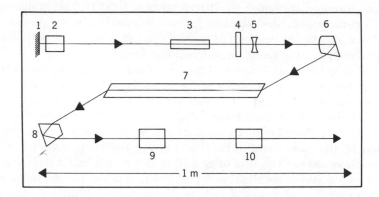

Figure 2.   Schematic diagram of the 37,550-cm^{-1} (2662-Å) laser system, designed and built under contract by W. F. Hagen of American Optical Company.   A 5.1-cm-long 0.63-cm-diam neodymium-doped YAG rod (3) is flashlamp pumped in a cavity formed by a rear reflector (1) and sapphire flat front reflector (4).   A Pockels cell and polarizer comprise the Q-switching unit (2);   the resultant oscillator produces 150 mJ at 9388 cm^{-1} in a 15-nsec pulse.   The emergent beam, after being expanded to a diameter of 1.5 cm by the combination of a negative lens (5) and telescopic folding prism (6), passes into a flashlamp-pumped neodymium-glass amplifier rod (7) through a Brewster angle face.   Amplification yields a 10-J pulse in the near infrared.   The folding prism (8) diverts the beam into a KDP frequency doubler (9) which produces an output of 3 J at 18,775 cm^{-1} (5324 Å). Finally, an ADP crystal (10) frequency doubles the green radiation to a 1-J pulse at 37,550 cm^{-1} (2662 Å).

under contract by W. F. Hagen at American Optical Corporation, relies upon frequency quadrupling the near-infrared output of neodymium to produce over 1 J per pulse at 37,550 cm^{-1} (2662 Å).   Figure 2 shows a diagram of the laser.   A small Q-switched neodymium-YAG oscillator produces a pulse at 1.0648 μm of narrow bandwidth and low divergence, and this is amplified by a larger neodymium-glass rod, yielding a 10-J pulse of ~15 nsec duration. The high radiance (power per square centimeter per unit solid angle) and narrow bandwidth produce good phase matching through the KDP second harmonic generator crystal, resulting in a 3-J pulse at 18,755 cm^{-1} (5324 Å). Fourth harmonic generation is accomplished with an ADP crystal.   Here, the narrow bandwidth from neodymium-YAG is critical, as the high dispersion of ADP in the uv permits good phase matching only over a narrow spectral range.   A pulse of 1.4 J at 37,550 cm^{-1} (2662 Å) has been produced;

however, normal operation is at 1 J or less per pulse at a repetition rate of 1 pulse/min. Thus, a wide range of previously unattainable photochemical processes can now be studied. This laser is also being used as a pumping source for a powerful dye laser which we expect will ultimately be tunable throughout the near uv, visible, and near-ir regions.

Photofragment spectroscopy performed prior to 1970 in this laboratory on the molecules Cl_2, Br_2, I_2, IBr, NO_2, NOCl, and ethyl nitrite is discussed in two recent articles [6,9], and references therein. Since then, further results on I_2 [10,11], IBr [12], several alkyl iodides [13], and ICN [14] have been obtained. Similar experiments on Cl_2 and NO_2 have been performed by Diesen and co-workers at Dow Chemical, Midland [15]. Photolysis mapping, a technique which measures the angular distribution of photodissociation products (but not their translational energies), has been used at Columbia by Bersohn and co-workers to study I_2 and Br_2, cadmium dimethyl, and carbonyl compounds [16].

III. APPLICATIONS TO THE STUDY OF PHOTODISSOCIATION LASERS

A. I_2

For many years the accepted interpretation of the main visible continuum in I_2 at energies above $\sim 20,000$ cm^{-1} has involved a transition from the ground state to the B $^3\Pi_{0u}^+$ upper state, with subsequent production of equal quantities of ground state $(^2P_{3/2})$ and excited state $(^2P_{1/2})$ iodine atoms [17,18]. With this assignment, it was recognized [19] that a photodissociation laser operating on the 1.315 μm atomic transition in iodine $(^2P_{1/2} \rightarrow {}^2P_{3/2})$ should be possible using broad-band flashlamp pumping of the molecule in the visible region, a negative absorption coefficient being assured due to the favorable statistical weights of the two equally populated atomic states. In practice, however, the prediction was never realized--no lasing action was observed [19]. Photofragment spectroscopy has now provided the explanation for this, producing results which have led to a reinterpretation of the states involved in the main visible continuum [10].

A beam of I_2 is crossed with pulses of polarized 21,510-cm^{-1} light from a laser-pumped tunable dye laser. Two peaks in the distribution are seen, differing both in recoil velocities and in angular distributions of the fragments. The upper panel of Fig. 3 shows the faster of the two peaks, found to be a maximum at recoil angles perpendicular to the electric vector of the light. Thus $\Delta\Omega = \pm 1$, and since $\Omega = 0$ for the ground state of I_2, $\Omega = 1$ in the upper state. The solid curve represents the predicted flight time distribution for a transition producing both iodine fragments in the $^2P_{3/2}$ ground

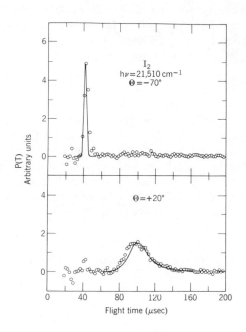

Figure 3.   Photofragment spectrum of I_2 at $21,510$ cm^{-1}.   The upper panel is the $^1\Pi_{1u}$ state, dissociating to ground-state atoms.   The lower panel is the B $^3\Pi_{0u}^+$ state, dissociating to one ground- and one excited-state atom. The horizontal scale is fragment flight time over a 3.9-cm flight path.   The vertical scale is proportional to counts per unit time at the mass spectrometer detector divided by flight time in order to correct for the velocity dependence of the ionization probability.   The open circles are from the measured signals; the solid lines are theoretically calculated line shapes and positions normalized to the heights of the measured peaks.   Normalization of the areas shows that the $^1\Pi_{1u}$ absorption is at least as strong as that of the B state.

state.   The only $\Omega = 1$ states correlating [18, 20] with ground-state atoms are the $^3\Pi_{1u}$ and $^1\Pi_{1u}$.   The $^3\Pi_{1u}$ state should lie at lower energies [21], and is almost certainly that state associated with the long-wavelength bands and continuum [18]: photofragment spectroscopy has, in fact, confirmed the existence of a dissociative 1u state in this region $(14,405$ cm$^{-1})$ [6, 9].   The faster peak thus corresponds to absorption to the $^1\Pi_{1u}$ state.

The slower peak, showing a maximum parallel to the electric vector, and thus corresponding to an $\Omega = 0$ upper state, is illustrated in the lower panel of Fig. 3.   Here the solid curve represents the predicted distribution

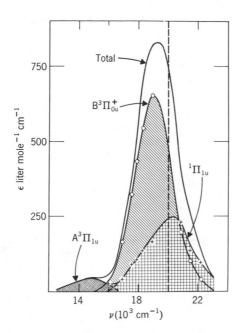

Figure 4. The absorption spectrum of I_2 in the visible region, showing the total absorption curve, and the contributions to this from transitions to the A $^3\Pi_{1u}$, $^1\Pi_{1u}$ and B $^3\Pi_{0u}^+$ states (Ref. 22 gives the sources from which these curves were calculated). Excited iodine atoms can only be produced following absorption to the B state, and then only at photon energies greater than the dissociation limit of I_2 into $I(^2P_{1/2})$ + $I(^2P_{3/2})$ atoms, the limit being indicated by the dashed line in the figure. Several attempts have been made to use the B state continuum as the basis for a photodissociation laser. Photofragment spectroscopy, as illustrated in Fig. 3, demonstrates that the failure of these attempts is due to the $^1\Pi_{1u}$ state, which overlaps the B state. The $^1\Pi_{1u}$ absorption is much stronger than had been previously believed and, as it leads to dissociation into ground–state atoms, it prevents an iodine atom population inversion.

for a transition producing one ground-state $(^2P_{3/2})$ and one excited-state $(^2P_{1/2})$ atom. Only one possible upper state can correlate with these results, the well-known B $^3\Pi_{0u}^+$. The properly scaled ratio of the areas of the two peaks gives a measure of the importance of the 1u state compared to that of the B state; at $21,510$ cm^{-1} over 50% of the I_2 molecules dissociate via the 1u continuum to produce ground-state atoms. Figure 4 (based on the work of several investigators [22]) gives the total absorption curve for I_2 in the visible region and shows approximately the relative intensities of the various transitions. Note that the only source for excited I atoms is the B state, to the right of its dissociation limit indicated by the dashed line. It is clearly seen from the results that a population inversion in atomic iodine is not to be expected. The lack of stimulated emission is thus due to the unexpected strength of the $^1\Pi_{1u}$ transition, producing a preponderance of ground-state I atoms.

Examples are known of lasing mechanisms in which the population inversion is formed following the absorption of two photons [23]. Absorption in molecular iodine involving two photons has been observed in photofragment spectroscopy experiments using a second harmonic neodymium-glass system at $18,830$ cm^{-1}, with both recoiling I atoms being produced in the $^2P_{1/2}$ excited state [24]. While it may well be possible to pump hard enough to overcome one-photon processes giving ground-state atoms and thus to produce a population inversion, this particular transition seems an unlikely route to a practical laser. Photofragment spectroscopy, however, may provide the tool to discover other two-photon processes leading to practical systems.

B. Br_2

We believe that it is probably possible to make a Br_2 photodissociation laser, and that previous attempts have failed because of the use of too much pumping light. Figure 5 shows the molecular states involved in the photodissociation spectrum of bromine and summarizes the results obtained by photofragment spectroscopy [9, 25, 26]. The two most important states appear to be the $^1\Pi_{1u}$ and $^3\Pi_{0u}^+$, distinguished in the approximate absorption spectrum of Fig. 5 by the dashed lines, the $^1\Pi_{1u}$ continuum peaking at higher energies. A third state, the $^3\Pi_{1u}$, is observed underlying the banded region. The relative positions of these states implies that there is an intriguing possibility for the construction of a Br_2 photodissociation laser [25]. Broad-band pumping of Br_2, which has been attempted [27], is not successful, since the larger absorption peak is the $^1\Pi_{1u}$ continuum, which produces only ground-state atoms. However, with pumping light between the threshold for excited Br atom production at $19,600$ cm^{-1} (5100 Å) and the onset of strong $^1\Pi_{1u}$ absorption at $\sim21,300$ cm^{-1} (4700 Å), the predominant process should be excitation to the $^3\Pi_{0u}^+$ state, with production of one ground-state and one excited-state Br atom, fulfilling, with the degeneracies of the atomic states involved, the

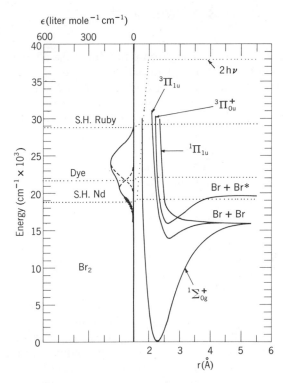

Figure 5. Br_2 absorption spectrum and potential curves. The absorption shown in the visible and near uv consists mainly of three continua, with the two principal peaks outlined by dotted lines. The higher-energy one can be assigned to $^1\Pi_{1u}$, dissociating to two ground-state Br $(^2P_{3/2})$ atoms, and the lower-energy one to $^3\Pi_{0u}^+$, dissociating to one ground-state atom and one excited Br* $(^2P_{1/2})$ atom. The photofragment spectrum of Br_2 has been measured at the three photon energies shown. Second harmonic ruby at 28,810 cm^{-1} picks up the short-wavelength tail of the $^1\Pi_{1u}$ continuum, producing two ground-state atoms, with an angular distribution peaking at $\Theta \approx 90°$. Illumination with a laser-pumped dye laser at 21,690 cm^{-1} gives two peaks, one corresponding to $^1\Pi_{1u}$, and the other to $^3\Pi_{0u}^+$ (translational energy corresponding to Br + Br*, angular distribution peaking at $\Theta \approx 0°$). Illumination with the second harmonic of a neodymium-glass laser at 18,830 cm^{-1} in the discrete region of the spectrum gives ground-state atoms with $\sim \sin^2 \Theta$ distribution, probably arising from a $^3\Pi_{1u}$ continuum hidden under the discrete region, as well as atoms from processes involving two photons. The photofragment spectrum makes it clear that in order to produce a Br_2 photodissociation laser, one should not illuminate with a full spectrum flash, as has been attempted, but instead one should use less light, pumping preferably only within the 19,600 cm^{-1} (5100 Å) to 21,300 cm^{-1} (4700 Å) $^3\Pi_{0u}^+$ state window.

267

necessary condition for stimulated emission. At 20,000 cm^{-1}, for example, the ratio of the absorption coefficient [28] of the $^3\Pi_{0u}^+$ state to that of the $^1\Pi_{1u}$ is ~10:1. The attraction of such a laser system, like the similar IBr system [27], is that the active medium would suffer no net photodecomposition, and would thus not have to be constantly replenished inside the cavity.

The experiment to test this proposal would be relatively simple. Br_2 could be pumped using either a laser at ~20,000 cm^{-1}, or a broad-band flashlamp source with an appropriate filter to pass light only in the Br_2 "window" (~19,600-21,300 cm^{-1}).

C. Alkyl Iodides

The first photodissociation laser [4], based on the flash photolysis of CF_3I and CH_3I, led to the discovery that some other alkyl and fluorinated alkyl iodides could also produce inverted populations of iodine atoms upon absorption of light in the quartz uv region, but that other closely related molecules do not lase at all [19,29]. We have recently measured [13] the photofragment spectra of several of these compounds at 37,550 cm^{-1} (2662 Å). Figure 6 illustrates the results for CH_3I, with the distribution presented on the bottom scale as a function of E_t, the center-of-mass recoil translational energy of the photofragments, obtained by an approximate transformation from the laboratory flight time distribution. From the known dissociation energy of CH_3I into alkyl and iodine fragments, the distribution of energy, E_{int}, partitioned into internal excitation of the products, can be found, and is also presented in the figure. Two peaks are clearly seen, separated by ~6500 cm^{-1}, close to the 7603-cm^{-1} spin-orbit splitting of the iodine atom ^2P states. This suggests that the peaks are due to production of ground- and excited-state iodine atoms, with the larger peak at higher internal energy corresponding to the formation of the excited $^2P_{1/2}$ atom, and the smaller peak having only enough internal energy to be from ground-state atoms, with the peaks in the ratio of ~5:1.

The photofragment spectra of three other members of the alkyl iodide series have been measured at 37,550 cm^{-1}: C_2H_5I, n-C_3H_7I and iso-C_3H_7I. It is interesting to contrast the behavior of the two propyl iodides in their production of excited iodine atoms, for the n-propyl species has been shown to produce the I-atom 7603-cm^{-1} lasing line when flash photolyzed, whereas the iso-propyl iodide does not [19,29]. Figure 7 compares the results obtained for the dissociation of the two different propyl iodide molecules at 37,550 cm^{-1}. Separation into two peaks, readily distinguishable in the CH_3I spectrum (and just resolved in the spectrum of C_2H_5I) is now no longer apparent. The significant difference between the n-propyl and iso-propyl spectra is that the fragment internal energy distribution is much broader in

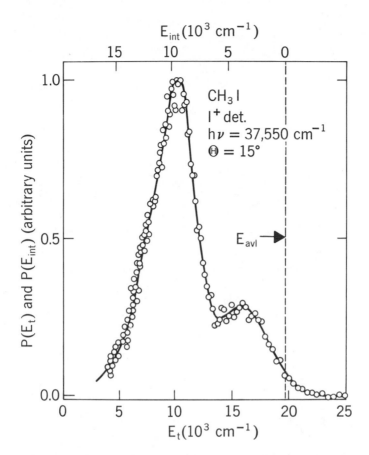

Figure 6. The photofragment spectrum of CH_3I taken at $37,550$ cm^{-1} (2662 Å) with the lab recoil angle Θ set at $15°$, the maximum in the angular distribution. The flux of particles arriving at the mass spectrometer is given as a function of E_t, the total center-of-mass translational energy of the dissociating fragments, and of E_{int}, their total internal energy. E_{avl} is the energy available for partitioning into translation and internal degrees of freedom, and is given by

$$E_{avl} = E_{int} + E_t = E_{par} + h\nu - D_0^0$$

in which E_{par} is the most probable initial thermal internal energy of the parent molecule, $h\nu$ is the photon energy, and D_0^0 is the dissociation energy of ground-state parent molecules into ground-state fragments. The left hand peak is almost certainly due to the production of excited I atoms, the source of the iodine photodissociation laser.

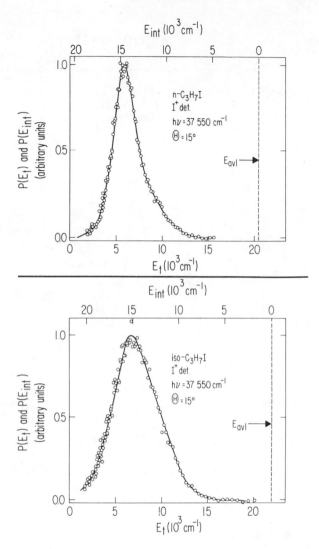

Figure 7. The photofragment spectrum of n-propyl iodide (upper panel) and iso-propyl iodide (lower panel), taken at $37,550$ cm^{-1} (2662 Å) with the laboratory recoil angle set at $15°$, the maximum of the angular distribution. The flux of particles arriving at the mass spectrometer is shown as a function of the total center-of-mass translational energy of the fragments, E_t, and of their total internal energy, E_{int}. Values of E_{avl} for the two molecules are unequal, due to the difference in their dissociation energies. Why n-propyl iodide lases and has a narrow internal energy distribution while iso-propyl iodide does not lase and has a wider energy distribution is an intriguing and as yet unanswered question.

the latter case.   Given the lack of stimulated emission in iso-propyl iodide photolysis, this broad distribution might be interpreted as a composite of two closely separated peaks in the spectrum, corresponding to formation of ground- and excited-state iodine atoms in a ratio unfavorable to the production of a population inversion, while the narrower n-propyl peak might correspond to production mainly of excited atoms.

D. ICN

The photofragment spectrum of ICN has recently been measured [14] at $37,550$ cm^{-1}, and the energy distribution of the fragments is presented in Fig. 8. It is immediately apparent that a large fraction of the available energy appears as internal excitation, and that ICN is a good candidate for a new photodissociation laser system.

Several explanations of these results can be put forward, and experiments proposed to probe the alternative explanations in more detail.   The first possible interpretation offers the promise of a laser system closely related to that of the alkyl iodides.   The peak at E_{int} = 9100 cm^{-1} could be due to the production of excited $(^2P_{1/2})$ iodine atoms with only $\sim$1500 cm^{-1} of internal vibrational and rotational energy appearing in the $CN(X^2\Sigma^+)$ electronic ground state.   This result would be close to that predicted by a quasidiatomic model developed in our group [30].   In any interpretation, the smaller broad peak at lower E_{int} must, by energy conservation, correspond to the production of I and CN in their ground electronic states, but with fairly high CN vibrational + rotational excitation.   The relative size of the two peaks indicates that an atomic iodine population inversion could be produced, with lasing at 7603 cm^{-1}.   The construction of a photodissociation laser operationg on the $^2P_{1/2} \rightarrow {}^2P_{3/2}$ iodine transition does not appear to have been attempted for ICN, perhaps because of the tendency of $(CN)_x$ polymer to form on the walls of reaction vessels used in ICN photolysis [31].

The second possible interpretation is that the sharp peak at E_{int} = 9100 cm^{-1} corresponds to the process

$$ICN \xrightarrow{h\nu} I(^2P_{3/2}) + CN(A^2\Pi, \ v = 0)$$

The energy match in this case is almost exact.   If this interpretation is correct, a new type of infrared laser, based on the CN (A $\rightarrow$ X) transition, should result; one arising from the <u>electronic</u> excitation of a <u>molecular</u> photofragment.

The third possibility is that both I and CN are formed in their ground electronic states, but with the CN containing two distinct distributions of vibrational + rotational energy.

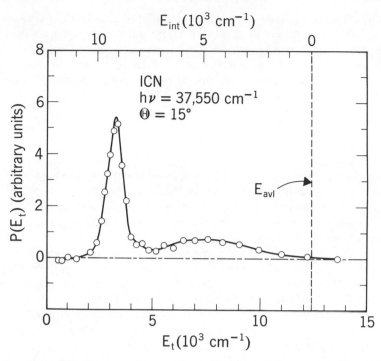

Figure 8. The photofragment spectrum of ICN, taken at 37,550 cm^{-1}
(2662 Å). On the basis of this photofragment spectrum, one would expect
ICN photodissociation to lead to infrared lasing, although, to our knowledge,
this has not yet been attempted.

Previous studies on ICN do not offer a full explanation of the details of
the photofragment spectrum. The early fluorescence experiments of
Jakovleva [31] were not designed to observe emission from the A $^2\Pi$ state.
Basco et al [32], in a flash photolysis study, concluded that CN was initially
produced vibrationally cold by one of two processes:

$$ICN \xrightarrow{h\nu} CN(X^2\Sigma^+, \ v = 0) + I(^2P_i)$$

or, in the opinion of the authors, a less likely step

$$ICN \xrightarrow{h\nu} CN(A^2\Pi, \ v = 0) + I(^2P_i)$$

followed by fluorescence or collisional quenching of CN to the ground state:

$$CN(A^2\Pi, \; v = 0) \longrightarrow CN(X^2\Sigma^+, \; v = 0) + h\nu$$

Again experimental conditions precluded the observation of any CN(A → X) fluorescence. The former mechanism is supported by the spectral assignment of King and Richardson [33] although recently an alternative assignment has been proposed [34].

We plan to clarify the mechanism of the photodissociation of the ICN molecule, first, by attempting to observe any CN(A → X) or $I(^2P_{1/2} \to {}^2P_{3/2})$ fluorescence from the system after illumination with 37,550 cm^{-1} radiation, and perhaps second by trying to produce an ICN photodissociation laser.

IV. CONCLUSION

The early prediction that lasers could revolutionize photochemistry [35] is proving to be true in the study of the photodissociation of simple molecules. Photofragment spectroscopy has utilized the development of laser technology to probe the energetics and dynamics of several of these processes, and is now starting to produce meaningful results in the interpretation and prediction of photodissociation laser systems. The technique has shown that a population inversion is not produced by the dissociation of I_2 following single-photon absorption in the visible region. A Br_2 photodissociation laser has been proposed, pumped by narrow bandwidth radiation at ~20,000 cm^{-1}. The photofragment spectra of some simple alkyl iodides have provided detailed information on the energy partitioning in the dissociating fragments, information useful in the understanding of their role as laser systems. Finally, the photofragment spectrum of ICN points to the possibility of producing stimulated emission in the near infrared from either the I or the CN fragment.

REFERENCES

1. S. G. Rautian and I. I. Sobel'man, Zh. Eksp. Teor. Fiz. 41, 2018 (1961) [Sov. Phys. JETP 14, 1433 (1962)].
2. J. R. Singer and I. Gorog, Bull. Am. Phys. Soc. 7, 14 (1962).
3. G. Gould in Quantum Electronics, Proceedings of the Third International Congress, edited by P. Grivet and N. Bloembergen (Columbia University Press, New York, 1964), p. 459.
4. J. V. V. Kasper and G. C. Pimentel, Appl. Phys. Letters 5, 231 (1964).

5. R. N. Zare, Ph. D. Thesis (Harvard University, Cambridge, Mass., 1964); R. N. Zare and D. R. Herschbach, Appl. Opt. Suppl. 2, 193 (1965).

6. G. E. Busch, J. F. Cornelius, R. T. Mahoney, R. I. Morse, D. W. Schlosser, and K. R. Wilson, Rev. Sci. Instrum. 41, 1066 (1970).

7. R. N. Zare, Mol. Photochem. 4, 1 (1972).

8. G. E. Busch and K. R. Wilson, J. Chem. Phys. 56, 3638 (1972).

9. K. R. Wilson in Excited State Chemistry, edited by J. N. Pitts, Jr. (Gordon and Breach, Inc., New York, 1970), p. 33.

10. R. J. Oldman, R. K. Sander, and K. R. Wilson, J. Chem. Phys. 54, 4127 (1971).

11. R. T. Mahoney, R. K. Sander, and K. R. Wilson (to be published).

12. S. Okuda, M. Sc. thesis [University of California (San Diego), San Diego, Calif., 1970]; R. K. Sander and K. R. Wilson, J. Chem. Phys. (to be published); J. H. Ling and K. R. Wilson (to be published).

13. S. J. Riley and K. R. Wilson (to be published).

14. J. H. Ling and K. R. Wilson (to be published).

15. R. W. Diesen, J. C. Wahr, and S. E. Adler, J. Chem. Phys. 50, 3635 (1969); R. W. Diesen, J. C. Wahr, and S. E. Adler, J. Chem. Phys. 55, 2812 (1971).

16. J. Solomon, J. Chem. Phys. 47, 889 (1967); C. Jonah, P. Chandra, and R. Bersohn, J. Chem. Phys. 55, 1903 (1971); J. Solomon, C. Jonah, P. Chandra, and R. Bersohn, J. Chem. Phys. 55, 1908 (1971).

17. R. S. Mulliken, J. Chem. Phys. 4, 620 (1936); G. W. King, Spectroscopy and Molecular Structure (Holt, Rinehart and Winston, Inc., New York, 1964), p. 163.

18. L. Mathieson and A. L. G. Rees, J. Chem. Phys. 25, 753 (1956).

19. J. V. V. Kasper, J. H. Parker, and G. C. Pimentel, J. Chem. Phys. 43, 1827 (1965).

20. R. S. Mulliken, Phys. Rev. 36, 1440 (1930).

21. R. S. Mulliken, Phys. Rev. 57, 500 (1940).

22. The total extinction coefficients shown in Fig. 4 are from the data of E. Rabinowitch and W. C. Wood [Trans. Faraday Soc. 32, 540 (1936)] for transitions above 16,000 cm^{-1}, and from the measurements by J. Ham [J. Am. Chem. Soc. 76, 3886 (1954)] for absorption below 16,000 cm^{-1}. Relative strengths of the $^1\Pi_{1u}$ and B $^3\Pi_{0u}^+$ transitions below the B state dissociation limit are from the data of A. Chutjian [J. Chem. Phys. 51, 5414 (1969)], A. Chutjian and T. C. James [J. Chem. Phys. 51, 1242 (1969)], and J. B. Tellinghuisen {Ph. D. thesis [University of California (Berkeley), Berkeley, Calif., 1969], available as UCRL Report No. UCRL-19112 (1969)}. Above this limit, in the main visible continuum, the points are from the photofragment spectrum of I_2, Ref. 10. A more precise version of this delineation of the I_2 states can be found in a

recent study by J. B. Tellinghuisen and L. Brewer [J. Chem. Phys. 56, 3929 (1972)].

23. P. P. Sorokin and J. R. Lankard, J. Chem. Phys. 51, 2929 (1969).
24. G. E. Busch, R. T. Mahoney, R. I. Morse, and K. R. Wilson, J. Chem. Phys. 51, 837 (1969).
25. G. E. Busch, R. T. Mahoney, and K. R. Wilson, IEEE J. Quantum Electron. QE-6, 171 (1970).
26. R. J. Oldman, R. K. Sander, and K. R. Wilson (to be published).
27. C. R. Guiliano and L. D. Hess, J. Appl. Phys. 40, 2428 (1969).
28. A. A. Passchier, J. D. Christian, and N. W. Gregory, J. Phys. Chem. 71, 937 (1967).
29. M. A. Pollack, Appl. Phys. Letters 8, 36 (1966).
30. K. E. Holdy, L. C. Klotz, and K. R. Wilson, J. Chem. Phys. 52, 4588 (1970).
31. A. Jakovleva, Acta Physicochim. URSS 9, 665 (1938).
32. N. Basco, J. E. Nicholas, R. G. W. Norrish, and W. H. J. Vickers, Proc. Roy. Soc. (London) 272 A, 147 (1963).
33. G. W. King and A. W. Richardson, J. Mol. Spectry, 21, 339 (1966).
34. J. W. Rabalais, J. M. McDonald, V. Scherr, and S. P. McGlynn, Chem. Rev. 71, 73 (1971).
35. J. G. Calvert and J. N. Pitts, Jr., Photochemistry (John Wiley & Sons, Inc. New York, 1966), p. 722.

LASER-INDUCED FLUORESCENCE AND ABSORPTION STUDIES OF

MOLECULAR ENERGY TRANSFER IN GASES*

George W. Flynn†
Department of Chemistry, Columbia University
New York, New York

ABSTRACT

Laser-excited infrared fluorescence has been observed in methyl fluoride (CH_3F). Following excitation by a Q-switched CO_2 laser operating on the P(20) line of the 9.6-μ band, the major fluorescence emission in CH_3F occurs from vibrational states with energies near 3000 cm^{-1}. The relaxation rate for these high vibrational states has been found to be similar to that expected for collisional deactivation of low-lying vibrational states of CH_3F with energies near 1000 cm^{-1}. Equilibrium between the 3000 and 1000 cm^{-1} states appears to be established in times of the order of a few microseconds at a pressure of one Torr of CH_3F. Vibrational relaxation has been studied in ethylene (C_2H_4) using a laser double-resonance technique. Following excitation by a Q-switched CO_2 laser, the ν_7, v=1 vibrational level of ethylene returns to thermal equilibrium in two steps. The first step is a rapid

---

*Work supported by the National Science Foundation and the Army Research Office (Durham).

†Alfred P. Sloan Fellow.

equilibration of excited vibrational states with each other, and the second step is a much slower equilibration of these excited states with the translational degrees of freedom. A mode-locked He-Ne laser operating at 6328 Å has been used as the excitation source for a photon-counting apparatus capable of measuring fluorescence decay times in the 1.5-20-nsec range. The technique employs signal averaging for enhanced sensitivity. Fluorescence decay times have been measured for the B $^1\Pi_u$ state of diatomic potassium (K_2) with this apparatus.

I. INTRODUCTION

The study of molecular energy transfer in gases has been aided greatly by the availability of high repetition rate, short-pulse lasers having moderate to large output power [1-11]. Laser fluorescence [1-7, 9-11] and double-resonance techniques [12, 13] have been used successfully in the past few years to study vibrational relaxation processes in many molecular systems. An understanding of such relaxation processes is important for the development and improvement of laser devices. Progress in the field of molecular energy transfer is also a necessary prerequisite to an improved understanding of the mechanisms and processes which determine the reaction pathways for chemical reactions in polyatomic molecules. In the present work three different methods for studying molecular energy transfer with pulsed lasers are described. The first is an infrared fluorescence experiment in methyl fluoride gas; the second is a laser double-resonance study of coupled vibrational relaxation processes in gaseous ethylene; and the third is a visible fluorescence investigation of diatomic potassium gas in equilibrium with potassium atoms.

II. INFRARED FLUORESCENCE IN CH_3F

The apparatus used to study laser-excited infrared fluorescence is shown in Fig. 1. A Q-switched CO_2 laser, which can be operated on a single vibration-rotation line of either the 9.6- or 10.6-µ band, is used to excite gaseous molecules in an aluminum sample cell with NaCl windows. Fluorescence signals are detected with either a Au:Ge or InSb liquid-nitrogen-cooled infrared element, amplified, and averaged over many laser pulses. The overall response time of the detection apparatus is 1 µsec. Laser pulses are typically 0.5 µsec wide and contain about 1-3mJ of energy. Various interference filters are employed to isolate fluorescence signals originating from a particular infrared band of the sample. A calibrated capacitance manometer is used to measure gas pressures.

The normal C^{12} isotopic species of methyl fluoride strongly absorbs

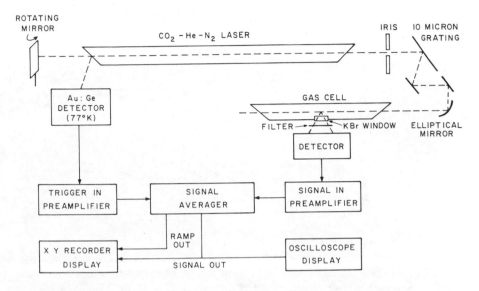

Figure 1. Apparatus diagram for laser-induced infrared fluorescence experiments. The detector is typically an InSb $(77^\circ K)$ or a Cu:Ge $(4^\circ K)$ device.

the P(20) 9.6-μ CO_2 laser transition [14–18]. The molecular vibrational
transition in CH_3F which corresponds most closely to the CO_2 laser line is
the ν_3, v=0 → v=1 transition [14–19]. Thus the laser pulse promotes a large
number of molecules to the lowest-lying vibrational state (ν_3, v=1) of CH_3F
which lies about 1050 cm^{-1} above the ground state. A typical CH_3F fluorescence decay curve following excitation by the P(20) 9.6-μ CO_2 laser line is
shown in Fig. 2. The wavelength of this fluorescence radiation is shorter
than 5 μ and thus emanates from states with energy greater than 2000 cm^{-1}.
In a separate experiment employing a chopped laser, a monochromator, and
a phase-sensitive detector, the majority of the fluorescence was found to
originate from the ν_1, v=1 and ν_4, v=1 states at 2964 and 2982 cm^{-1},
respectively. Thus a collision or radiative process must be responsible for
promoting molecules from ν_3, v = 1 at 1050 cm^{-1} to the ν_1 and ν_4 states.

The total fluorescence decay rate $(1/\tau)_{tot}$ may be written as the sum of
three terms

$$(1/\tau)_{tot} = (1/\tau)_{col} + (1/\tau)_{dif} + (1/\tau)_{rad} \tag{1}$$

where $(1/\tau)_{col}$ is the contribution to the decay rate due to collisional deactivation of a given vibrational state, $(1/\tau)_{dif}$ is the contribution to the decay

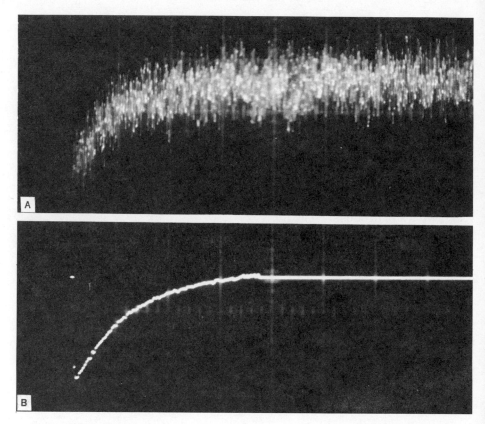

Figure 2.   CH₃F fluorescence decay curve (a) before averaging (b) after averaging.   The vertical scale is 2 V/cm for both traces and the horizontal scale is 0.1 msec/cm.   The CH₃F pressure is 11.01 Torr.

rate arising from diffusion of excited molecules to the wall of the fluorescence cell followed by deactivation at the wall, and $(1/\tau)_{rad}$ is the contribution to the decay rate arising from spontaneous emission processes.   For pressures up to about 1 atm, $(1/\tau)_{col}$ increases linearly with pressure (binary collision region) and $(1/\tau)_{dif}$ increases approximately linearly with (1/pressure).   The radiative rate will be pressure independent in the absence of radiative trapping.   For molecules radiative trapping effects in the infrared are usually small due to the high density of rotational states and the weak vibrational transition moments [20]. Thus a study of the pressure dependence of $(1/\tau)_{tot}$ allows $(1/\tau)_{col}$ and $(1/\tau)_{dif}$ to be extracted from the fluorescence decay rate. These two quantities give respectively the binary collisional deactivation rate and the diffusion coefficient for the excited vibrational state. For the methyl fluoride fluorescence shown in Fig. 2, the collisional deactivation rate is 0.83 msec^{-1} Torr^{-1}. Experiments are presently underway to measure the decay rate in the region 0.1-5 Torr so that an accurate value of the excited-state diffusion coefficient can be obtained.

The collisional deactivation rate observed for the CH_3F fluorescence is similar to the vibrational deactivation rate measured for CH_3F using ultrasonic techniques [21]. The ultrasound experiments were performed at $373°K$ and are not directly comparable to the present room-temperature fluorescence studies. Generally, however, the temperature coefficient for vibration-translation energy transfer rates measured by ultrasound techniques is not large. The similarity between the ultrasound and fluorescence rates strongly suggests that both experiments are measuring the same vibrational deactivation process. The ultrasonic technique is usually sensitive to only the relaxation of the lowest vibrational states (ν_3, $v = 1$, 1050 cm^{-1} and ν_6, $v = 1$, 1200 cm^{-1} for CH_3F). This implies that the fluorescing vibrational states near 3000 cm^{-1} and the low-lying vibrational states near 1000 cm^{-1} remain in equilibrium during the fluorescence decay process. Since the fluorescence rise time following laser excitation is very rapid ($\lesssim 10^{-6}$ sec at 1 Torr CH_3F), this equilibrium must be established by collision processes having very large cross sections.

In the pressure range 5-50 Torr CH_3F the fluorescence intensity has been found to increase linearly as the square of the laser input power. In these experiments the laser was operated cw and chopped mechanically while the fluorescence was monitored with phase-sensitive detection at the laser chopping frequency. Since the vibrational energy of a fluorescing molecule is nearly three times that of a laser photon, it might be expected that the fluorescence intensity would increase as the cube of the laser power. This would be true if the fluorescing molecules were excited from the ground state by a series of one-photon absorption processes such as $v=0 \rightarrow v=1$, $v=1 \rightarrow v=2$, $v=2 \rightarrow v=3$ (ν_3 mode) followed by rapid collisional equilibration of ν_3, $v=3$ with ν_1 and ν_4. The observed two-photon dependence of the fluorescence intensity could arise if the molecules were promoted from $v=1$ to $v=3$ of ν_3 by successive absorptions such as $v=1 \rightarrow v=2 \rightarrow v=3$ (ν_3 mode). Rapid equilibration of ν_3, $v=3$ with ν_1 and ν_4 would then lead to the observed fluorescence. The data which have been obtained so far cannot completely eliminate this possibility; however, it is a very unlikely mechanism, since the measured absorption coefficient [14-18] of CH_3F is too large to be attributed purely to "hot-band" absorption. The majority of laser photons must be absorbed by the lowest ν_3 transition $v=0 \rightarrow v=1$.

A variety of absorption and collisional energy transfer processes can be used to explain the dependence of the fluorescence intensity on the square of the laser power and also to account for the rapid equilibrium which is established between states near 3000 cm^{-1} and those near 1000 cm^{-1}. One of these is the simultaneous occurrence of ground-state and "hot-band" absorptions which can lead to promotion of molecules to combination states such as $\nu_3 + \nu_6$, $\nu_3 + \nu_5$, $\nu_3 + \nu_2$ as well as to ν_3. Collisions of the type

$$CH_3F(\nu_3) + CH_3F(\nu_3 + \nu_6) \rightarrow CH_3F(0) + CH_3F(2\nu_3 + \nu_6) + \Delta E \qquad (2)$$

($\Delta E \sim 0$) would then lead to the population of very highly excited vibrational states. The cross sections for these nearly resonant energy transfer processes are expected to be very large [22-25] leading to very rapid and efficient production of $2\nu_3 + \nu_j$ molecules following laser excitation. Since only the $2\nu_3$ energy is supplied by the laser pulse while the ν_j energy is supplied thermally, the population of such states will depend on the square of the laser power and the temperature. Rapid equilibrium between 3000 and 1000 cm^{-1} states is probably maintained by ladder collisions of the type

$$CH_3F(n\nu_3) + CH_3F(0) \rightarrow CH_3F[(n-1)\nu_3] + CH_3F(\nu_3) + \Delta E \qquad (3)$$

($\Delta E \sim 0$). Again, the cross section for processes such as (3) is expected to be large [22-25].

Other possible collision mechanisms which do not require hot-band absorption exist for excitation of CH_3F molecules to high vibrational states. For example, a simple calculation shows that the collision path

$$2CH_3F(\nu_3) \rightarrow CH_3F(2\nu_3) + CH_3F(0) + \Delta E \qquad (4)$$

($\Delta E \sim 0$) can be an important one if the cross section is near gas kinetic. The slow decay of $CH_3F(\nu_3)$ molecules following laser excitation (0.83 msec^{-1} Torr^{-1}) allows ample time for collisions of the type (4) to occur. Energy for further excitation of molecules from $2\nu_3$ to the fluorescing states would have to be supplied thermally to account for the square dependence of the fluorescence intensity on laser power.

For example, a process such as

$$CH_3F(2\nu_3) + CH_3F(\nu_6) \rightarrow CH_3F(2\nu_3 + \nu_6) + CH_3F(0) + \Delta E \qquad (5)$$

would lead to excitation of vibrational states similar to those described by (2). Finally, collision paths such as

$$CH_3F(2\nu_3) + CH_3F(0) \rightarrow CH_3F(0) + CH_3F(\nu_3 + \nu_6) - 147 \text{ cm}^{-1} \quad (6)$$

may be important. In SF_6 where the energies supplied or absorbed thermal-ly are very much less than 147 cm^{-1}, fluorescence experiments indicate that collision processes of the type (6) are extremely efficient [26].

Although much work remains to be done, the preliminary results of these experiments in CH_3F suggest that substantial progress can be made toward the understanding of mechanisms for vibrational energy transfer in polyatomic molecules by the use of laser fluorescence techniques.

III.   INFRARED DOUBLE RESONANCE IN C_2H_4

Laser infrared double resonance was first used by Rhodes [12] to study vibra-tional relaxation in CO_2. An experimental investigation of relaxation pro-cesses in SF_6 has also been carried out using this method [13]. Figure 3 shows the double-resonance apparatus used in the present studies of ethylene

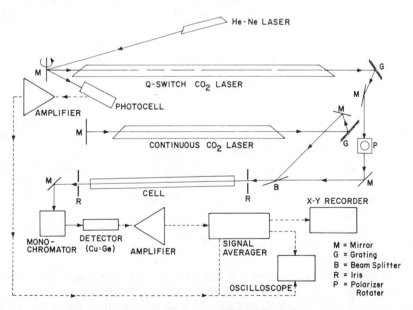

Figure 3. Laser-laser double-resonance apparatus. The polarizer rotator P is employed so that both cw and Q-switched laser beams can be joined at the beam splitter B and still allow 80% of both beams to be transmitted to the cell. The cell is typically 2 m long and its temperature can be varied from 50 to 300°C. The He-Ne laser is used to derive a trigger signal for the signal averager.

(C_2H_4). A low-power continuous CO_2 laser operating on the 10.6-μ P(14) transition monitors the absorption of a Q-branch band of the ν_7 (0 → 1) transition in C_2H_4 [27]. No saturation effects are caused by this laser which is used simply to probe the vibrational state populations of ethylene. A second high-power Q-switched CO_2 laser operating on the 10.6-μ P(26) line is used to irradiate the same ν_7 (0 → 1) vibrational transition but in a different part of the rotational manifold than the cw laser. Following excitation of the gas by the Q-switch pulse, rapid equilibration of C_2H_4 rotational states causes a sudden change in the absorption of the cw laser. With the present apparatus the rotational equilibration rate cannot be measured due largely to detector response time and sensitivity limitations. The relatively slow relaxation of the vibrational states, however, can be readily monitored by observing the changes in absorption of the cw laser. In ethylene a substantial amount of translational heating occurs after the vibrational states have come into equilibrium with the translational degrees of freedom. Since this heating is not uniform in the sample cell, a very slow decay of absorption can usually be observed due to thermal transport processes [13,26,28]. To eliminate this effect, the heat capacity of the sample is increased by adding large amounts of rare gas to small amounts of ethylene. Sample pressures are typically 0.1-5 Torr of C_2H_4 and 15-100 Torr of Ar, Kr, or Xe.

Following excitation of the ν_7 band of C_2H_4 by the Q-switch laser, the absorption of the cw laser returns to its steady-state value in two steps. The first step is a rapid exponential decay lasting about 10^{-5} sec at a pressure of 1 Torr C_2H_4 and 30 Torr Kr. The second step is a slower exponential process with a decay time of approximately 10^{-4} sec at similar C_2H_4 and Kr pressures. Figure 4 shows a plot of the fast decay rate as a function of total krypton gas pressure for a sample with constant C_2H_4 pressure. The intercept of this plot gives the rate for fast vibrational deexcitation arising from C_2H_4-C_2H_4 collisions while the slope is a measure of the deactivation rate due to C_2H_4-Kr collisions. Spontaneous emission decay of vibrational states is not expected to contribute significantly to the intercept rate due to the weakness of infrared transitions. A plot similar to Fig. 4 can also be used to determine the deactivation rate for the slow vibrational relaxation process due to C_2H_4-C_2H_4 collisions and also due to C_2H_4-rare-gas collisions.

The slow rate observed in the double-resonance experiments almost certainly corresponds to the vibrational deactivation of the lowest state in C_2H_4 (ν_{10}, v=1) which is 826 cm^{-1} above the ground state. Relaxation of this state requires that all of the 826 cm^{-1} of vibrational energy be converted into translational and/or rotational degrees of freedom (V-T/R transfer). The rate measured by double resonance agrees with that determined by ultrasonic techniques [29] which are usually sensitive to only relaxation of the lowest vibrational state [30,31]. It should be mentioned that direct V-T/R deactivation of states such as ν_7 (949.2 cm^{-1}), ν_4 (1027 cm^{-1}), and ν_8

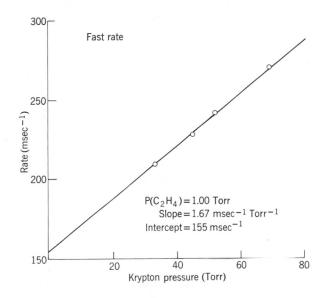

Figure 4. Two decay rates are observed in the double-resonance study of ethylene. The faster rate is plotted here as a function of added rare-gas pressure for a fixed ethylene pressure (1.00 Torr). The intercept gives the C_2H_4-C_2H_4 deactivation rate for the fast process.

(943 cm^{-1}) may also contribute to or even dominate in the V–T/R transfer process. For example, although the direct deactivation of v_7 requires that about 125 cm^{-1} more vibrational energy be taken up into translational and/or rotational degrees of freedom than the corresponding deactivation of v_{10}, the vibration amplitude for the v_7 fundamental is larger than that of the v_{10} fundamental [32,33]. Since both the molecular vibration amplitude and the energy deficit are important in determining the efficiency of vibration-translation energy transfer processes [34], direct deactivation of v = 1, v_7 may well contribute to the slow rate measured by double resonance and also to the ultrasonic rate.

The fast decay process observed with double resonance has not been measured before. One report of double dispersion in an ultrasonic study of C_2H_4 has been made, but the lifetime values reported do not correspond to those measured in the present work [29]. The process which is probably giving rise to this fast decay rate for the v=1 molecules is an equilibration of the v_7 and v_{10} modes in ethylene by a mechanism such as

$$C_2H_4(\nu_7) + C_2H_4(0) \rightleftharpoons C_2H_4(0) + C_2H_4(\nu_{10}) + 125 \text{ cm}^{-1} \qquad (7)$$

Analysis of the pressure dependence of the fast decay rates shows that this equilibration process requires approximately 80 kinetic theory collisions between C_2H_4 molecules. Since both the ν_7 and ν_{10} modes are infrared active, the temperature dependence of the fast rate can be used to test the validity of the mechanism (7) [22, 35]. Contributions to the fast equilibration rate from processes such as

$$C_2H_4(\nu_7) + C_2H_4(0) \rightleftharpoons C_2H_4(0) + C_2H_4(\nu_8) + 6 \text{ cm}^{-1} \qquad (8)$$

cannot be ruled out but are probably too fast to be measured with the present apparatus.

A detailed analysis of the effects of rare gases on the slow and fast rates is being prepared and will be presented shortly [36].

IV. MODE-LOCKED LASER-EXCITED FLUORESCENCE IN K_2

The apparatus shown in Fig. 5 has been used to study the fluorescence lifetime of the $B\,^1\Pi_u$ state of diatomic potassium (K_2) [37]. Except for the mode-locked He–Ne laser, which replaces the usual pulsed nitrogen lamp, the experimental arrangement is standard for a photon-counting system capable of nanosecond resolution. The laser delivers one pulse of width 0.8 nsec (determined with a subnanosecond response photodiode and sampling oscilloscope) every 23 nsec. The peak pulse power is approximately 0.1 W. The laser radiation is absorbed by K_2 molecules which fluoresce from the $B\,^1\Pi_u$ state at wavelengths above the 6328-Å laser line. Scatter from the laser is usually eliminated with an interference filter although equivalent results have been obtained with a 1/4 m Jarrell-Ash monochromator. The density of K_2 molecules in equilibrium with K atoms is controlled by varying the temperature of the oven containing the potassium cell from approximately 200 to 450°C.

The photon-counting detection electronics used to monitor the fluorescence decay act like a relatively inefficient nanosecond signal averaging device. A beam splitter is used to feed part of a given laser pulse to a trigger circuit which goes to the start channel of a time-to-amplitude converter (TAC). The remainder of the laser pulse enters the potassium cell where it excites fluorescence in K_2. The first photon emitted by the sample which enters the 56TVP red-sensitive phototube (capable of counting single-photon events) sends a pulse to the stop channel of the TAC. The TAC then sends a pulse through a very linear (interface) amplifier to a pulse-height

analyzer.   The amplitude of the output TAC pulse is directly proportional to
the time between the start and stop pulses which enter the TAC.   Thus the
height of the TAC output pulse is directly proportional to the time between
laser and fluorescence events.   The pulse-height analyzer stores one count
for each pulse supplied by the TAC.   The channel number where a given
count is stored increases with increasing pulse height so that the channel
numbers of the pulse-height analyzer correspond to the times between laser
and fluorescence photons.   If the laser intensity is kept sufficiently low, only
one fluorescence photon will be emitted in a time interval corresponding to
several fluorescence lifetimes.   In such a situation the arrival of late
photons from the tail of the fluorescence decay curve is not affected by the
arrival of early photons from the more intense part of the fluorescence
decay.   When this is true, the number of pulses accumulated in each channel
of the pulse-height analyzer gives an accurate measure of the fluorescence
intensity at a given point in time following the laser pulse.

For fluorescence decay times of order 5 to 10 nsec it would be possible
in principle to collect one photon from every laser pulse (every 23 nsec).
Unfortunately, the minimum recycle time for the TAC is about 1 μsec while

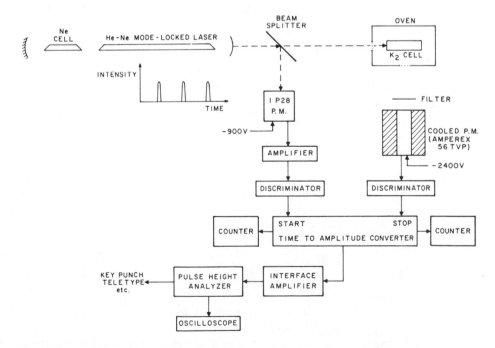

Figure 5.   Apparatus diagram for photon counting experiments.   The 1P28
phototube and associated electronics are used to trigger the TAC start
channel.   The minimum TAC recycle time is 1 μsec.

the pulse-height analyzer is an ancient one which requires 50-100 μsec to process an input pulse. Thus the limiting data collection rate for the present apparatus is 10-20 kHz. Modern analyzers are capable of processing data almost at the same rate at the TAC. The practical limitation on the data collection rate is, therefore, about 1 MHz.

Although the time resolution of the TAC and pulse-height analyzer is a fraction of a nanosecond, the fastest decay rate which can be measured with this apparatus is limited to about 1.5 nsec as will be seen below. This limit is probably set by the spread in transit times between photon arrival at the photocathode surface and pulse output from the phototube at the end of the dynode chain. The finite laser pulse width (0.8 nsec) could also be a contributing factor to this limit.

The operation of the photon counting system was tested by replacing the K_2 fluorescence cell with a piece of tissue paper. The scattered laser pulses were then monitored by the phototube and associated electronics. The result of this experiment is shown in the upper half of Fig. 6. The laser pulses were found to be about 1.5 nsec wide after processing by the apparatus. This particular trace is useful for setting a lower limit on the fluorescence lifetimes which can be measured, but it is also very valuable for deconvolution of lifetime data which approaches the limiting apparatus resolution time.

A typical fluorescence decay for the B $^1\Pi_u$ emission is shown in the lower half of Fig. 6. The time required to collect this data was approximately 10-20 min, and typical partial pressures of K_2 were 50 mTorr. Excellent signal to noise ratio is apparent in these traces for lifetimes ranging from 4-9 nsec depending on the oven temperature.

It has been suggested [37] that the total fluorescence decay rate for the B $^1\Pi_u$ state can be written

$$(1/\tau)_{tot} = (1/\tau)_0 + (1/\tau)_{col} \tag{9}$$

where $(1/\tau)_0$ is the spontaneous emission rate and $(1/\tau)_{col}$ is the rate of deactivation of the B $^1\Pi_u$ state due to collisions with K atoms. The K atoms can deactivate excited K_2 molecules by the near-resonant process

$$K_2^* \ (v{\sim}7, \ B) + K \rightarrow K_2 \ (v', \ X) + K^* + \Delta E \tag{10}$$

The major states populated by the laser are the v=7, 8 vibrational levels of B $^1\Pi_u$ and K^* is the ^2P excited electronic state of the atom. For ΔE to be near zero, v' must be of order 50.

For a deactivation process such as (10) the collisional deactivation rate can be written

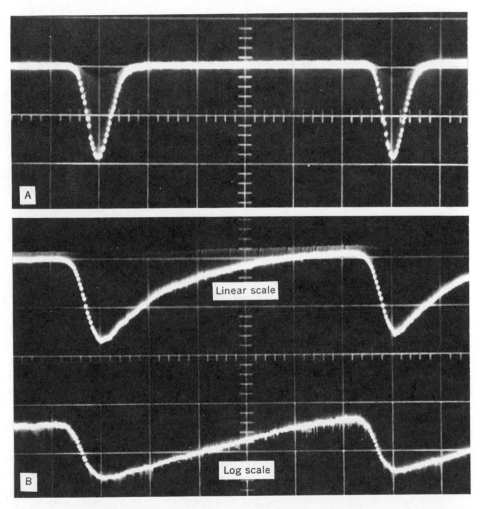

Figure 6. Typical results obtained with photon counting apparatus of Fig. 5. (a) was obtained by replacing the K_2 cell and oven with an inert scattering block. (b) is an actual fluorescence decay (repeated twice) of the B $^1\Pi_u$ state of K_2. The horizontal scale is 3.9 nsec/cm.

$$(1/\tau)_{col} = (12R/\pi M_K)^{1/2} \sigma \rho_K T^{1/2} \tag{11}$$

where R is the gas constant, M_K the atomic weight of potassium, σ the cross section for collisional deactivation, ρ_K the potassium atom density, and T the absolute temperature. A plot of $(1/\tau)_{tot}$ versus $\rho_K T^{1/2}$ will thus have as its intercept the spontaneous emission rate $(1/\tau)_0$ and a slope proportional to the cross section for collisional deactivation. The preliminary data obtained with the photon-counting apparatus agrees within 25% with two previous determinations of the spontaneous lifetime measured with a phase shift technique [37,38]. The general pressure dependence observed in the present experiments is in agreement with that of Zare [37] although the collision cross section is somewhat larger. The photon-counting results have not been corrected for finite response time of the electronics. This will cause about a 20% change in the fastest rates measured. Also the K_2 sample which was used in these experiments may have some H_2 impurity present which could lead to an increase in the collisional cross section. Further experiments are underway to obtain an accurate spontaneous emission and collisional deactivation rate with a very pure potassium sample.

The technique described here can be exceptionally useful in studies of very rapid molecular energy transfer. It will be interesting to measure K_2 deactivation rates in the presence of added rare gases, H_2, and N_2. Of course, these experiments are not limited to K_2. The same apparatus has been used to excite fluorescence in C-Phycocyanin, a photosynthetically active algal protein with tetrapyrrolic chromophores, phycocyanobilin [39, 41]. Energy transfer studies in this system are particularly interesting since C-Phycocyanin is involved in photosynthesis by accepting the excitation energy of phycoerythrin and then transferring to chlorophyll [42]. The lifetime measured for this system ($\sim$1.7 nsec) was very close to the limit of resolution of the photon-counting apparatus.

Further experiments are planned with a mode-locked argon laser as the exciting source. It should be possible eventually to couple a mode-locked argon laser with a closely matched dye laser cavity as an exciting source for this type of fluorescence monitoring system. Such an apparatus would offer relatively monochromatic wavelength-tunable subnanosecond excitation along with nanosecond resolution of fluorescence and high repetition rate. The variety of energy transfer experiments in atomic, molecular, and biophysical systems which could be performed with such an apparatus is almost limitless.

ACKNOWLEDGMENTS

The experiments described in this paper are the work of the author's students, Eric Weitz (CH_3F), Raymond Yuan and Jack Preses (C_2H_4), and

Rina Giniger and J. Thomas Knudtson (K_2). The infrared fluorescence and double resonance work was done in collaboration with Professor A. M. Ronn of the Department of Chemistry, Polytechnic Institute of Brooklyn. I am indebted to all of these people not only for their fine work, but also for the stimulation and excitement of their thinking. I would like to thank Professor Richard Zare for many informative discussions concerning K_2 fluorescence and also for the loan of a K_2 cell and several timing modules.

REFERENCES

1. L. O. Hocker, M. A. Kovacs, C. K. Rhodes, G. W. Flynn, and A. Javan, Phys. Rev. Letters 17, 233 (1966).
2. J. T. Yardley and C. B. Moore, J. Chem. Phys. 45, 1066 (1966).
3. J. T. Yardley and C. B. Moore, J. Chem. Phys. 46, 4491 (1967).
4. C. B. Moore, R. E. Wood, Bei-Lok Hu, and J. T. Yardley, J. Chem. Phys. 46, 4222 (1967).
5. R. D. Bates, Jr., G. W. Flynn, and A. M. Ronn, J. Chem. Phys. 49, 1432 (1968).
6. J. T. Knudtson, R. D. Bates, Jr., T. N. Rescigno, G. W. Flynn, and A. M. Ronn, Bull. Am. Phys. Soc. 14, 65 (1969).
7. J. T. Yardley, J. Chem. Phys. 49, 2816 (1968).
8. K. B. Eisenthal, Chem. Phys. Letters 6, 155 (1970).
9. R. Velasco, Ch. Ottinger, and R. N. Zare, J. Chem. Phys. 51, 5522 (1969).
10. S. E. Johnson, K. Sakurai, and H. P. Broida, J. Chem. Phys. 52, 6441 (1970).
11. Richard D. Bates, Jr., George W. Flynn, J. Thomas Knudtson, and A. M. Ronn, J. Chem. Phys. 53, 3621 (1970).
12. C. K. Rhodes, M. J. Kelly, and A. Javan, J. Chem. Phys. 48, 5730 (1968).
13. J. I. Steinfeld, I. Burak, D. G. Sutton, and A. V. Nowak, J. Chem. Phys. 52, 5421 (1970); 51, 2275 (1969).
14. T. Y. Chang, C. H. Wang, and P. K. Cheo, Appl. Phys. Letters 15, 157 (1969).
15. E. Weitz, G. W. Flynn, and A. M. Ronn, J. Appl. Phys. 42, 5187 (1971).
16. E. Weitz, G. W. Flynn, and A. M. Ronn, Bull. Am. Phys. Soc. 16, 42 (1971).
17. W. L. Smith and I. M. Mills, J. Mol. Spectry. 11, 11 (1963).
18. R. G. Brewer and R. L. Shoemaker, Phys. Rev. Letters 27, 631 (1971).
19. M. Kovacs, D. Ramachandra Rao, and A. Javan, J. Chem. Phys. 48, 3339 (1968).
20. M. A. Kovacs and A. Javan, J. Chem. Phys. 50, 4111 (1969).

21. P. G. T. Fogg, P. A. Hanks, and J. D. Lambert Proc. Roy. Soc. (London) A219, 490 (1953).
22. R. D. Sharma and C. A. Brau, Phys. Rev. Letters 19, 1273 (1967).
23. B. H. Mahan, J. Chem. Phys. 46, 98 (1967).
24. J. C. Stephenson, R. E. Wood, and C. B. Moore, J. Chem. Phys. 48, 4790 (1968).
25. R. D. Sharma and C. A. Brau, J. Chem. Phys. 50, 924 (1969).
26. R. D. Bates, Jr., J. T. Knudtson, G. W. Flynn, and A. M. Ronn, Chem. Phys. Letters 8, 103 (1971).
27. W. L. Smith and I. M. Mills, J. Chem. Phys. 40, 2095 (1964).
28. R. D. Bates, Jr., George W. Flynn, J. T. Knudtson, and A. M. Ronn, J. Chem. Phys. 57, 4174 (1972).
29. R. Holmes and W. Tempest, Proc. Phys. Soc. 78 1502 (1961); G. H. Hudson, J. C. McCoubrey, and A. R. Ubbelohde, Proc. Roy. Soc. A 264, 289 (1961); P. G. Couran, J. D. Lambert, R. Salter, and B. Warburton, Proc. Roy. Soc. 244A 212 (1958).
30. T. L. Cottrell and J. C. McCoubrey, Molecular Energy Transfer in Gases (Butterworths, London, 1961).
31. G. M. Burnett and A. M. North, Transfer and Storage of Energy by Molecules, Vol. 2 (Wiley, New York, 1969).
32. G. C. Golike, I. M. Mills, W. B. Person, and B. Crawford, J. Chem. Phys. 25, 1266 (1956).
33. W. L. Smith and I. M. Mills, J. Chem. Phys. 40, 2095 (1964).
34. R. N. Schwartz, Z. I. Slawsky, and K. F. Herzfeld, J. Chem. Phys. 20, 1591 (1952).
35. W. A. Rosser, Jr., A. D. Wood, and E. T. Gerry, J. Chem. Phys. 50, 4996 (1969).
36. Raymond Yuan, Jack Preses, George Flynn, and A. M. Ronn, to be published.
37. W. J. Tango and R. N. Zare, J. Chem. Phys. 53, 3094 (1970).
38. G. Baumgartner, W. Demtroder, and M. Stock, Z. Physik 232, 462 (1970).
39. C. Ó hEocha, Biochemistry 2, 375 (1963).
40. H. L. Crespi, L. J. Boucher, G. D. Norman, J. J. Katz, and R. C. Dougherty, J. Am. Chem. Soc. 89, 3642 (1967).
41. W. J. Cole, D. J. Chapman, and H. W. Siegelman, J. Am. Chem. Soc. 89, 3643 (1967).
42. C. S. French and V. K. Young, J. Gen. Physiol. 35, 873 (1952).

SPECTROSCOPY

A. Narrow Resonances and Precision Applications

MODERN METHODS IN PRECISION SPECTROSCOPY:

A DECADE OF DEVELOPMENTS

Ali Javan
Department of Physics, Massachusetts Institute of Technology
Cambridge, Massachusetts

ABSTRACT

The past decade of developments in precision laser spectroscopy has made
available a wide range of new approaches applicable to the study of atomic
and molecular processes. This article reviews the various developments
and discusses the application to highly accurate observations.

I. INTRODUCTION

Early in the 1960's, when a variety of potential laser applications were
being considered, it became evident that the application to highly precise
spectroscopic studies with an accuracy surpassing the previous measurements had to await new discoveries and the development of a whole new set
of technologies. A decade of efforts in that direction have now come to
fruition and the necessary tools are at hand to reexamine with an unsurpassed accuracy a variety of heretofore unobservable features of atomic and
molecular resonances.

295

There are two major prerequisites to the application of monochromatic radiation in high-resolution spectroscopy. The first relates to our ability to resolve individual spectral lines and accurately reset the frequency of the monochromatic radiation to the exact center frequency of each line. In this case, the observed breadth of the spectral line profile becomes the limiting factor in the resolution and frequency resetability. The second prerequisite, however, relates to our ability to measure precisely the frequency (or the wavelength) of the monochromatic radiation, once its frequency is reset to the center of the spectral line profile. This requires the availability of a method to allow accurate comparison of the frequency (or the wavelength) of the monochromatic radiation with that of the primary, or a secondary, standard.

Conventional optical and infrared spectroscopy is based on precision wavelength measurements utilizing grating instruments or other dispersive means which enable comparison of one wavelength with another. On the other hand, precision measurements in the microwave region have relied on comparison of microwave frequencies by means of frequency-mixing methods. The ultimate accuracy of the microwave frequency measurements is limited only by the frequency purity and the stability of the microwave radiation. In contrast, the optical wavelength measurement accuracy is, in practice, determined by the instrumental resolving width and not the ultimate purity and stability of the monochromatic light sources, which can be considerably beyond the resolving powers of even the best optical instruments. Moreover, precise comparison of two widely different wavelengths is generally subject to cumbersome systematic errors arising from diffraction fringe shifts and other causes. Despite their limitations, the existing wavelength measuring methods are more than adequate in conventional types of optical spectroscopy where the widths of the observed spectral lines are broad and exceed the resolving widths of the high-precision wavelength measuring devices.

With the advent of the gas laser, it became possible to explore a variety of novel effects arising from the nonlinear interaction of monochromatic radiation fields and Doppler broadened atomic transitions. These explorations revealed several types of nonlinear processes which can be used to produce highly pronounced and well defined resonant features, superimposed on the broad Doppler profile of an optical transition and capable of defining the center of the transition with great accuracy. The characteristic linebreadths of these resonances originate from natural and collision broadening which, at a low gas pressure where the collision broadening is small, can be orders of magnitude below the Doppler linewidth. Accordingly, the nonlinear processes made it possible to achieve sizable reductions in the breadths of the optical spectral profiles in low pressure gaseous systems. And correspondingly, the application of these processes enabled a dramatic improvement in high-resolution optical spectroscopy capable of an absolute accuracy much beyond the earlier limits of the classical approaches.

The nonlinear effects, together with their consequent spectroscopic implications, were initially demonstrated at MIT. In the demonstrations, several approaches were adopted and applied in separate experiments, to resolve the small spectral splittings due to hyperfine interactions and the isotope shifts, in a number of optical transitions in atomic Ne and in atomic Xe [1-6]. In all these examples, the splittings of the optical transitions were considerably below their classical Doppler line widths. The nonlinear optical resonances, on the other hand, appeared well resolved and non-overlapping due to their extremely narrow widths. In some of the experiments, the level splittings were accurately determined by measuring small frequency separations between the centers of the nonlinear optical resonances. The measurements were done by mixing in a photomultiplier the frequencies of several oscillating lasers tuned individually to the center frequencies of each of the closely-spaced nonlinear resonances; their frequency spacings, which lay in the radio frequency region, were correspondingly measured by a radio receiver.

In addition to the nonlinear effects, there are other line-narrowing processes which can now be applied in high resolution studies of spectral lines in the optical and the infrared regions. An important example is the traveling wave line-narrowing process which is observed when a high gain amplifying transition results in many fold amplification of its own spontaneous emission. This effect can be applied to resolve and determine precisely [7] the structure of a closely-spaced set of transitions which would normally be overlapping and unresolved.

More important, these early demonstrations made us aware that in order to obtain spectroscopic information with an absolute accuracy surpassing the previous optical measurements, in addition to the high resolution we still required a radically different approach to the art of measurement in optics--one that would have an ultimate accuracy much beyond the limits offered by the classical methods of wavelength measurement. This had to wait until the extension of microwave frequency measuring methods into the infrared region became a reality.

The art of frequency measurement in the microwave region is rooted in the technology of microwave-rectifier frequency-mixer diodes. In addition to the frequency measurements, this technology forms the basis on which all types of microwave spectrometers and, more generally, microwave receivers, are founded. The extension of this art and all of its consequent applications into the infrared region posed a major challenge ever since the early days when the monochromatic laser light sources became available.

In this discussion, I will review our past decade of activities, first in the area of extending the microwave detection and measurement technology into the optical region, and second, in the area of exploring the nonlinear effects and other processes which result in considerable reduction in the linewidths of the optical and infrared transitions. These activities, which progressed along separate courses of development, were pursued concurrently and in parallel. This discussion will emphasize the relationship of these two areas in accurate spectroscopic observation, as well as their separate applications in other fields of studies.

II. EXTENSION OF MICROWAVE FREQUENCY MEASURING METHODS INTO THE OPTICAL REGION

The early attempts to extend the microwave frequency-mixing technologies into the optical regions started late in 1964 at MIT. A long-term goal was established at that time to develop a high-speed point-contact diode with a nonlinear current-voltage characteristic responding to the flow of alternating currents at the infrared frequencies. This required development of a high-speed diode with a microscopically small shunt capacitance, C, across the point contact and a sufficiently low series resistance, R, corresponding to an RC time constant smaller than the period of an infrared cycle, i.e. $RC \ll 10^{-14}$ or 10^{-15} sec.

Prior to the early experiments, it was known that the high-frequency response limit of an ordinary silicon point-contact diode extended generally into the submillimeter region. Accordingly, in our initial experiments, an exhaustive search was made to extend the high frequency limit of the semiconductor point-contact diodes, including the silicon point-contact diode, beyond the submillimeter region [8-12]. While these early attempts paved the way towards eventually extending the microwave frequency mixing and rectification methods into the near-infrared region, the high frequency limit of performance of the microwave types of semiconductor point-contact diodes was found to lie in the 50 to 100-μm region of the far infrared and hence these diodes were not useful at shorter wavelengths. A major accomplishment of these early attempts was, however, to extend successfully the absolute frequency measuring methods from the microwave region into the far-infrared range of the electromagnetic spectrum.

Further extension of the frequency mixing and the rectification method into the infrared and the near-infrared region became possible with the advent of a new type of optical element consisting of a point-contact diode which operates on the principle of quantum mechanical electron tunnelling across a potential barrier formed by a thin dielectric layer, on the order of several Å thick. The RC time constant of the diode is sufficiently small to allow its use as a lumped circuit element at the infrared frequencies. The frequency-mixing capabilities of the diode were first established in the far-infrared region [12], and later it was used to mix widely different frequencies in the infrared [13]. While the status of this new optical element is still in its infancy, its application in absolute frequency measurements in

the infrared has already been firmly established.   The discussion in this
section is divided into five subsections as follows:
A.   The early experiments; absolute frequency measurements in the
far infrared.
B.   Infrared metal-dielectric-metal tunneling diode; absolute frequency
measurements in the near infrared.
C.   Spectroscopic applications.
D.   Precision wavelength measurements and the speed of light.
E.   Parametric generation of infrared radiation and other applications
of the infrared diode.

II-A.   The Early Experiments: Absolute Frequency Measurements in
the Far Infrared

In the initial frequency-mixing experiment, a commercial model of a millimeter wave cross-guide mixer was used to mix V-band klystron radiation
(in the 70 GHz region) with the output of an HCN laser oscillating on either
of the 337-μm or 311-μm far infrared laser transitions [8].   The frequency-
mixing element in the cross-guide mixer consisted of a point-contact
silicon diode.   In the experiment, the laser radiation and the klystron output
were simultaneously coupled to the diode.   A radio frequency beat signal (in
the 30 MHz region) was observed across the diode as the klystron was tuned
to a frequency whose 13th or 14th harmonic differed respectively from the
frequency of the 337-μm or 311-μm laser radiation by the beat frequency.
Here, the incident laser radiation played the role of a strong local oscillator, enabling superheterodyne detection of a weak current flowing through
the diode at a far-infrared frequency corresponding to the high-order harmonic of the V-band klystron.   The next experiment attempted to obtain a
similar frequency-mixing signal using the DCN laser lines in the 190-μm
region together with the V-band klystron; here the 22nd and 23rd harmonics
of the klystron frequency were tuned to coincide, respectively, with the
frequencies of two of the DCN laser transitions at 194 and 190 μm [9].   The
beat signals obtained in that experiment were sufficiently above the noise to
enable accurate frequency measurements; however, the observed signal
intensities were considerably less than those obtained in the previous HCN
laser frequency measurements in the 300-μm region.
At that stage of the experimentation, a large signal-to-noise ratio was
of considerable importance because of its further implications about the
feasibility of an extension of the technique toward yet higher frequencies.
Therefore, the next step was to find a way to improve the observed frequency-mixing signal-to-noise ratio.   This was achieved by adopting a different
approach in which the frequency-mixing was done in two steps, each of which
corresponded to a low-order harmonic mixing (instead of the 22nd or 23rd

order mixing). An inspection of the HCN and DCN laser frequencies indicated that the difference between the frequency of the 2nd harmonic of the 337-μm HCN laser line and the fundamental frequency of 190-μm DCN laser line fell in the microwave region; this difference frequency was in the region of the 3rd harmonic of the V-band klystron already available in the laboratory. Thus, in the next experiment, the outputs of the 337-μm and the 190-μm lasers together with the V-band klystron radiation were coupled onto an improved model of the silicon diode mixer. An enormously large beat signal was observed when the klystron was tuned to the appropriate frequency [10]. In this case, twice the HCN laser frequency minus three times the klystron frequency equalled the DCN laser frequency.

This experiment constituted the first frequency multiplier chain with a laser link, in which a laser frequency that could be compared with a microwave standard frequency was used as a link for absolute frequency measurement of another laser lying at an appreciably higher frequency. This technique was then applied to establishing a multiplier chain up to the 118-μm transition of the H_2O laser [10], and subsequently up to the 84-μm transition of the D_2O laser [11]. Needless to say, considerable efforts were invested in improving the frequency mixing element and the coupling of the laser radiation to it. The later models of the crystal mixer in no way resembled the model used in the initial experiments. Instead of a millimeter wave-guide, the mixer diode was mounted in an open structure and was coupled to the laser radiation by means of a lens or a parabolic mirror. The making of the diode itself was, at this stage, an art. Figure 1 gives an oscilloscope

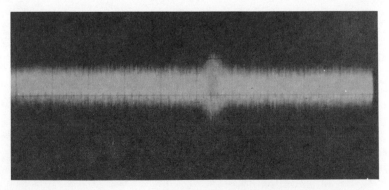

Figure 1.   Zero beat signal arising from frequency mixing of a 118 μ H_2O laser, a 337 μ HCN laser, and a microwave intermediate frequency. This tracing is obtained by displaying the signal across the diode directly on the oscilloscope. The scale is 2mv/cm.

tracing of the zero beat signal obtained by mixing the 337-μm and the 118-μm laser radiation together with the output of the V-band klystron. In this case, the klystron frequency is swept periodically in the region of the zero beat in synchronism with the oscilloscope sweep; the signal across the diode is directly displayed on a wide-band oscilloscope with a bandwidth of several Hz. The HCN laser and the klystron powers were each about 30 mW, and the H_2O laser power was 2 mW.

Despite the large signal-to-noise ratio obtained in the laser harmonic frequency-mixing experiments up to the 84-μm region, there was no way to tell whether the silicon point-contact response could in fact extend to much higher frequencies. Among the various factors, the response time of a semiconductor diode is determined by its carrier lifetime as well as the electron transit time across the depletion layer. However, even at a high impurity concentration in which the carrier lifetime is short, and under favorable circumstances with regard to the electron transit time, it is not a priori evident that a sufficient amount of infrared frequency voltage may be developed across the point-contact junction to allow the flow of alternating current through it at an infrared frequency. Achievement of a diode usable as a lumped circuit element at infrared frequencies required the development of an element with a microscopic configuration dictated by the short wavelength of the infrared radiation and the required small RC time constant. The feasibility of a practical experimental model was not evident in the early infrared experiments; in fact, all attempts to construct an infrared semiconductor point-contact diode capable of laser harmonic frequency mixing in the region much beyond 84 μm failed completely. The attempts included experiments with highly doped semiconductors such as different types of ion-implanted GaAs with very high impurity concentrations.

It is of interest to point out that when the output of a 10.6-μm CO_2 laser is coupled to a semiconductor point-contact element, a DC voltage is generally found to develop across it. This observation is not indicative of the element's response to an actual flow of an infrared frequency current through its point-contact and a subsequent rectification due to a nonlinear voltage-current characteristic: a slow process, such as the one arising from the thermoelectric effect, can also produce a DC voltage across the element. Similarly, with the CO_2 laser oscillating on two adjacent P-branch transitions differing in frequency by an amount δ lying in the 54-GHz region, a frequency-mixing beat signal can be observed across the diode as microwave klystron radiation in the 54-GHz region is also coupled to the diode and its frequency is tuned to δ. Again, this experiment is not indicative of the element's high-speed capabilities at the infrared frequencies; it only proves that its response may be as fast as the microwave difference frequency δ.

II-B. An Infrared Metal-Dielectric-Metal Tunneling Diode; Absolute
Frequency Measurements in the Near Infrared

The event which subsequently made possible the extension of microwave
frequency mixing and rectification into the infrared and the near infrared
was the successful demonstration that a special element consisting of a
metal-to-metal point-contact diode, operating on the principle of electron
tunneling across a thin dielectric layer, can be made with a speed of
response at least as high as about 10^{12} Hz [12]. The demonstration was
done by means of a frequency-mixing experiment in the far infrared in which
the output of a 337-μm HCN laser and that of a V-band klystron were coupled
to the new element and a 13th-order harmonic frequency-mixing signal was
observed [12], as in the initial experiment with the silicon diode. After this
experiment, which took place in the summer of 1968, it was only a matter of
time until in the next experiment [13], which was completed by the summer
of 1969, we showed that the frequency mixing capabilities of our diode
extended up to the 9.3-μm region of the infrared. And by the following
summer, we were able to show that its response extended all the way up to
the 5-μm region of the CO laser lines [14]. The success in these experiments strongly indicated that the element's harmonic frequency mixing
potentials may yet extend to the shorter wavelengths, possibly the visible or
the near-UV range of the spectrum.
    The configuration of the metal-to-metal point-contact infrared diode is
as follows: A thin tungsten wire antenna--about a few μm in diameter and
several mm long--is mounted at the end of a coaxial cable, its tip pointed by
means of a standard etching technique and contacted mechanically with a
polished metal post mounted on a differential screw to allow fine adjustment
of the point-contact [12], (see Fig. 2). The metal post can be made of
nickel, cold-rolled steel or other metals. The pointed tip of the wire antenna is less than 1000 Å in diameter, (see Fig. 3) and its contact with the
metal post occurs through a thin oxide layer, on the order of several Å
thick. In a series of as yet unpublished experiments [15], we have shown in
great detail that the quantum mechanical electron tunneling across the thin
oxide layer is responsible for the nonlinear current-voltage characteristic
of the diode which is capable of responding to the infrared frequencies. As
infrared radiation is coupled to the diode's antenna, an alternating voltage at
the corresponding frequency is developed across the oxide layer. This gives
rise to an alternating current flowing through the diode at the corresponding
frequency, but with a distorted waveform due to the nonlinear characteristic
The distorted wave can have an average rectified value as well as current
components at high-order harmonics of the frequency of the applied infrared
field. When two or several radiation fields are coupled to the diode, the
total current will contain additional components at frequencies corresponding
to the mixtures (i.e. sums and differences) of the frequencies of the applied

Figure 2.   A magnified photograph of the metal-to-metal point-contact infrared diode.

fields, as well as the mixtures of their various harmonics; this can occur for frequency mixtures below the diode's high-frequency cutoff limit, which is determined by the electron tunneling time across the oxide layer and the RC time constant of the element.

The rectified current component flowing through the diode gives rise to a DC voltage across it. Also, as will be discussed in more detail in Sec. II-E, the current components at the various harmonics and frequency mixtures which are generated in the point-contact can excite propagating current

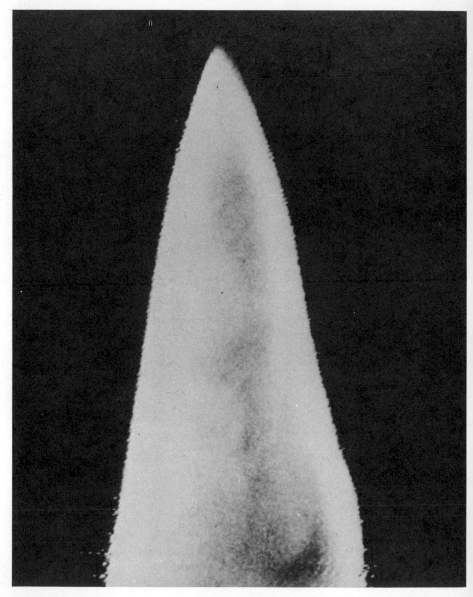

Figure 3.   Electron micrograph of tip of the etched tungsten wire antenna (magnification 4.4 x 10^3).

waves in the thin wire antenna which can subsequently radiate electromagnetic waves at the corresponding frequencies, which may lie in the infrared. In the harmonic frequency-mixing experiments, however, only the frequency mixtures corresponding to a zero beat note (or a beat note at a convenient intermediate frequency in the tens of MHz) is detected directly across the diode, in the form of an alternating voltage at the beat frequency.

In the next experiment mentioned above [13], where the diode's high-frequency response was shown to extend at least up to the 9.3-μm region of the infrared, the output of a CO_2 laser oscillating on the R(12) line of the CO_2 9.3-μm band was coupled to the diode together with the output of a water-vapor laser oscillating on the 28.0-μm transition and also the radiation from a K-band klystron; here, the third harmonic of the H_2O transition differs from the CO_2 laser line by an amount δ lying in the K-band region. The frequency-mixing signal was observed across the diode as the klystron's frequency was tuned to δ. This observation conclusively showed that the alternating currents at frequencies at least as high as the third harmonic of the H_2O laser can flow through the diode's point-contact, demonstrating its high speed frequency-mixing capabilities anywhere below and up to the 9.3-μm range of frequency [13].

In the next experiment [14], which showed the harmonic frequency mixing capability of the diode up to the 5.3-μm region, the radiation from a CO laser oscillating on the P(13), $v = 7 \rightarrow 6$ line was mixed in the diode with 10.6-μm CO_2 P(20) laser radiation and microwave radiation at the difference frequency of the CO laser and the second harmonic of the CO_2 laser. The observed signal and its dependence on the intensities of the incident fields, compared with those observed previously at lower frequencies, indicated that for all practical purposes the diode's frequency response could be assumed flat from the radio frequency and the far infrared up to at least the 5.3-μm region of the near infrared.

Meanwhile, in a series of experiments at NBS, Evenson and his collaborators observed, in a similar metal-to-metal diode, frequency mixing between the 28.0-μm H_2O laser and the 337-μm HCN laser together with an intermediate microwave radiation [16]. Their later experiments included frequency mixing of three lasers and an intermediate microwave frequency in the same diode [17]. Their recent efforts have been directed towards frequency mixing of the methane stabilized He-Ne 3.39-μm laser line and a CO_2 laser oscillating on a 10.2-μm transition. The frequency-mixing signal in this experiment is expected to be sizable and of the same magnitude as the similar signals obtained at lower frequencies. Activities along these lines have also begun at the National Physical Laboratories in Great Britain and at other laboratories.

With the extension of the microwave frequency mixing methods into the infrared and the near infrared, it is now possible to perform frequency measurements in these regions with an absolute accuracy limited by that of the basic microwave frequency standard. The accuracy of the existing Cs clock is somewhat better than one part in 10^{12}. Accordingly, with this clock as the basic standard, precise frequency measurements can now be performed throughout the infrared with an absolute accuracy as high as about one part in 10^{12}. It is likely that before too long, with the aid of the nonlinear spectroscopy reviewed in the following section, optical clocks of considerably higher accuracies will become a reality. Our frequency measuring method would then allow application of such improved optical clocks in precise measurements, not only in the optical and the infrared regions, but also in the microwave range of the spectrum.

In summary, a frequency multiplier chain can now be constructed in which a reference frequency marker is produced in the form of an alternating frequency current flowing through the diode at any desired frequency in the far-infrared, infrared and near-infrared regions of the electromagnetic spectrum. In the multiplier chain, the frequency of the reference marker is synthesized following several successive frequency-mixing steps along the lines described above. In each step, the fundamental or harmonic of a microwave klystron frequency is added to or subtracted from the fundamental or harmonic of a laser frequency; the resulting frequency is then added to or subtracted from another laser frequency used in the next step of the synthesis. In a highly accurate frequency multiplier chain, all of the successive frequency mixing stages are operated simultaneously. In such a chain, it is possible to apply a phase-locking scheme, already demonstrated at MIT [18], to stabilize and maintain the infrared or the far infrared frequency of the reference marker at a specified frequency, within the same degree of short and long term stability as that of the basic microwave clock used in the chain. Similar phase-locking schemes have been commonly used in the microwave frequency multiplier chains and are necessary in highly accurate measurements. In routine measurements, however, starting from a crystal controlled radio frequency oscillator calibrated against a frequency standard, a less elaborate frequency multiplier chain can be assembled with ease, capable of measuring frequencies throughout the infrared with an absolute accuracy ranging anywhere from one part in 10^6 to about one part in 10^{10}.

II-C.   Spectroscopic Applications

Despite the short history of frequency mixing in the far infrared and the infrared, a large number of transition frequencies have already been determined to within an accuracy ranging from a few parts in 10^6 to a few parts in 10^7. As of the time of this conference, the measured transitions range from 300 μm to 5 μm. With the application of the line-narrowing effects reviewed in Sec. III of this manuscript, a host of these transitions can now be measured with several orders-of-magnitude improved accuracy. Efforts along these lines are currently on the way at a number of centers, internationally. It is also certain that in the near future, frequency measurements will be performed in the regions beyond 5 μm, extending into the visible and the near ultraviolet portion of the spectrum. (This can already be achieved if desired, via direct generation of harmonic radiation in a crystal with nonlinear refractive index, starting from near-infrared radiation and multiplying it into the shorter wavelength region.)

The initial spectroscopic applications have been mostly incidental and preliminary, but the examples clearly indicate the potentials of the method. The first application followed the initial experiment in which the 337-μm HCN laser transition was mixed with a high-order harmonic of a microwave radiation frequency. At that time, the 337-μm laser oscillation was thought to originate from an electronic transition in the CN free radical. The frequency-mixing experiment, which immediately gave an accurate measure of the laser line's frequency, indicated that the assignment of the laser transition was incorrect [8]. Subsequent measurements of other oscillating far-infrared lines obtained from the same laser showed that they all belonged to specific rotation-vibration transitions of the HCN molecule in its ground electronic state [19]. Because of the Coriolis coupling and other perturbations, there are in total eleven important spectroscopic parameters which define the corresponding HCN rotation-vibration spectrum; with some additional measurements, the values of all these parameters can now be determined with much improved precision. Similarly, the frequency measurements in the 190-μm region provided complete identification of the DCN laser lines and some preliminary high-precision spectroscopic information [20].

In a different area, the precise frequency measurements of the oscillating P- or R-branch transitions in both the 10.6-μm and the 9.3-μm bands of the CO_2 molecules have made available accurate values of their corresponding band centers and rotational constants [21]; the measured frequencies are obtained from the frequency-mixing results at MIT [13], NBS [17], and BTL [22]. Recently, the absolute frequencies of some 30 oscillating laser lines belonging to the 10.5-μm band in the N_2O molecule have been measured at MIT [23]. The results have provided precision band parameters in N_2O, as well as making available a host of additional transitions of known frequencies useful as secondary frequency standards. As for the atomic systems, in an as yet unpublished experiment at MIT [24], the 93.5-μm $3^1D \rightarrow 3^1P$ transition frequency in atomic He has been recently measured and found to be $3.1297915 \times 10^{12} \pm 3 \times 10^6$ Hz. Other measurements in He are currently being planned at MIT; the aim is to reexamine, with much improved accuracy, the fundamental processes underlying the interaction between the two electrons in He; these processes include relativistic as well as electrodynamic effects. We are also actively performing an experiment to observe and measure the precise center frequencies of extremely narrow nonlinear resonances in several electronic transitions of singly-ionized He and the neutral hydrogen atom lying in the 5-μm and the 10-μm regions. Among general spectroscopic information, these measurements are expected to provide an accurate value for the Rydberg constant. These types of experiments are also in progress in other laboratories.

II-D.   Precision Wavelength Measurements and the Speed of Light

In the past, precise wavelength measurements in the visible region of the spectrum have been possible with absolute accuracies as high as one part in 10^7 to one part in 10^8. In fact, the krypton standard line in the visible region, obtained in special gaseous discharges operating at liquid N_2 temperature, is reproducible with an accuracy slightly better than one part in 10^8.

In the far-infrared and the infrared regions of the spectrum, however, the application of the previous wavelength measuring methods always suffered from a great variety of practical problems, making precise wavelength measurements with an accuracy approaching anywhere close to the limits of the instrumental resolving power most difficult and generally impractical. With conventional light sources, for instance, the previous wavelength measurements in the far infrared were limited at the best to an accuracy of a few parts in thousand.

With the availability of the coherent gas laser light sources in the far-infrared and the infrared regions, it is now possible to improve upon our ability to perform wavelength measurements in these regions of the spectrum. This is possible mainly because of the high intensity of the monochromatic gas laser light sources and their high degree of spatial coherence which enable observation of interference fringes over a fairly long pathlength. This allows determination of a laser wavelength in the far infrared and the infrared with an accuracy considerably better than the limits obtainable previously. Despite the fact that the absolute frequency measurements in the far infrared and the infrared regions provide accuracies much beyond the capabilities of the wavelength measuring methods, there are still a variety of applications where adequately accurate information can be obtained with improved wavelength measurements in these regions of the spectrum.

Over the past several years, an experimental program at MIT has explored the potential application of a long-arm scanning Michelson interferometer to the accurate comparison of two widely different laser wavelengths, one of them lying in the far infrared or the infrared and the other consisting of a He–Ne 6328-Å laser having an accurately calibrated wavelength with respect to a Kr standard. The measurements are done by simultaneous fringe counting at the two wavelengths, while the scanning arm of the interferometer is varied over a long path-length [25]. In the interferometer, highly accurate optical components, having large apertures to minimize the diffraction effect, are employed. The measurements are done with an on-line computer used for control and data processing. The method has already enabled routine wavelength measurements to within one part in 10^6 in the 100-μm to 80-μm region of the far infrared, and about one order of magnitude better in the 10-μm region. Our efforts are currently aimed at establishing this method's ultimate limit of accuracy in the 10-μm region, which is expected to be about one part in 10^8. The details are discussed by Kurnit in his lecture at this symposium [26].

Baird and his collaborators at the National Research Council of Canada, Ottawa, are currently exploring a different method for precision laser wavelength measurements in the infrared. In their approach the infrared wavelength is up-converted parametrically into the visible region. This is done by mixing the infrared laser radiation with visible laser radiation in a crystal with a nonlinear refractive index. In this way, the up-converted radiation component at the sum of the two laser frequencies falls in the visible region. The final measurements are done by determining the wavelength of the up-converted light and that of the visible laser. This method is particularly advantageous, since the measurements rely on wavelength comparison in the visible region where fringe shifts due to diffraction effect which introduce systematic errors in the measurements are somewhat less cumbersome.

In another current series of activities at the National Bureau of Standards, the methane stabilized 3.39-μm He-Ne laser wavelength is being compared in a Fabry-Perot interferometer with the Kr standard line. In all these measurements, the ultimate accuracy is limited to about one part in 10^8, or possibly a factor of two better. This limitation is imposed by the wavelength reproducibility of the Kr standard, as well as the inherent fringe shift errors which, to varying degrees, are commonly present in all of the wavelength measuring methods. The methane-stabilized 3.39-μm He-Ne laser has been discussed by Hall at this conference [27].

Extension of the absolute frequency measurements in the far infrared and the infrared region has made possible the precise measurement of the speed of light by simultaneous measurements of the wavelength and the frequency of monochromatic laser radiation. The first measurement of this type was demonstrated at MIT in 1969 using the 84-μm gas laser transition in the D_2O molecules [11, 24]. In that experiment the measured value for the speed of light was found to be 299 792.7 ± 2.0 km/sec, in which the accuracy was dominantly due to wavelength measuring errors of the scanning Michelson interferometer mentioned above. This early value agrees with the best previous determinations of this quantity with nearly comparable accuracy. A more precise value of c can now be obtained via frequency and wavelength

measurements in the near infrared, where wavelength measurements of
higher accuracy can be performed with less difficulty.

We have already noted that the frequency measuring methods are
inherently more accurate than the methods of determining wavelengths.
From this, an inescapable conclusion follows that the speed of light must be
ultimately defined as a dimensionless parameter with an arbitrarily assigned
value. In this way, a precise measure of the frequency of a monochromatic
wave would uniquely fix to the same accuracy, the number corresponding to
its vacuum wavelength obtainable from $\lambda \nu = c$. This would allow wavelength
measurements to be performed in terms of the standard of the frequency,
without the necessity of a separate length standard. A number most
naturally suitable for the speed of light definition can be obtained by measuring the wavelength λ of a stable laser radiation to within the accuracy of the
existing krypton standard, and at the same time determining the frequency ν
to a higher accuracy, readily obtainable with an existing frequency measuring method. The subsequent number obtained from $c = \lambda \nu$ can then be adopted for the speed of the light, once its last digits falling within the errors of
the best measurements have been agreed upon and rounded off to within the
accepted error. Accordingly, the numerology for the speed of light definition now lies before the standards laboratories internationally, requiring
independent measurements at different centers. The major problem is in
the determination of the wavelength to the required accuracy, and not in the
frequency measurements which can now be performed routinely in the
frequency regions as high as the near infrared.

Ultimately in the future, in highly accurate observations, we can expect
precise frequency measurements to replace direct measurements of wave-
lenghts. We must also note that in astronomical observations, distances
have always been determined in terms of the elapsed times using accurate
frequency standards as the time keeping clocks. Accordingly, the number
for the speed of light would ultimately lose much of its significance in highly
precise measurements, since it would merely serve to transform one set of
numbers into another set. This is an unnecessary step which may well be
dispensed with.

II-E. Parametric Generation of Infrared Radiation and Other
Applications of the Diode

In the decades following World War II, the state of the art in the microwave
detection and frequency-mixing technology has reached a high degree of
perfection; this technology can now be extended and applied directly in the
far-infrared and infrared regions of the spectrum.

It was noted that when the metal-to-metal point-contact diode is coupled
simultaneously to several radiation fields, the current components at the

synthesized frequencies of the incident fields generated at the point contact can excite propagating current waves in the diode's antenna and hence emit electromagnetic radiation at the corresponding synthesized frequencies. In particular, the synthesis of an infrared laser frequency, ν, and the fundamental or the harmonics of a tunable klystron frequency, ν_m, can provide tunable frequency radiation at the infrared sideband frequencies, $\nu_s = \nu \pm n\nu_m$, with n as an integer. This is an attractive application because of the broad-band characteristics of the diode which would allow generation of radiation not only in the infrared, but also in the longer wavelength region. Such a novel tunable frequency light source, which is capable of operating at room temperature, can be of importance in a variety of spectroscopic applications.

In a recent experiment at MIT this effect has been observed for the first time [28]. In the experiment, the diode was subjected simultaneously to the output of a 10.6-μm laser at a frequency ν and microwave klystron radiation at a frequency ν_m. The infrared side-band radiations emitted from the diode's antenna were collected by means of a lens and focused on a He-cooled Cu:Ge infrared detector. A superheterodyne detection system was used to observe the first and the second infrared side bands, corresponding to $\nu_s = \nu \pm n\nu_m$ with n = 1 and 2, incident on the detector. An important feature of the experiment was the application of a dc bias voltage on the diode, which enabled optimization of the emitted infrared radiation at a given side band. This was achieved by adjusting the bias voltage, in order to shift the operating points on the diode's current-voltage, I-V, characteristics into a region most favorable for the frequency synthesis. It may be of interest to note that in the diode, the current component at a given side band defined by the integer n, is proportional to the (n+1)th order derivative of the I-V characteristics. The nature of these characteristics, which is dictated by the quantum mechanical tunneling of the electrons through the thin oxide layer, is such that the behavior of its various derivatives versus the voltage differ sizably from one another. From details of this consideration, it follows, for instance, that the radiation from the diode at the first side-band frequency is enhanced appreciably as the value of the bias voltage is increased from zero; the second side-band, on the other hand, decreases as the bias voltage is increased [28]. Prior to the experiment, the diode's I-V characteristics and its high order derivatives were studied extensively. These studies, in addition to the information useful in the generation of the side-band radiation, have provided a crucial test of the quantum mechanical electron tunneling process in the diode. The results agree with a detailed theory describing the effect. (This work is currently in preparation for publication [15].)

With the application of the phase-locking method [18] mentioned in Sec. II-B, it is possible to stabilize the infrared frequency of the component of the diode current at a given side band against a stable reference marker phase-locked on a frequency standard. In this way, it is now possible to obtain infrared radiation emitted from the diode at a specified frequency, to within the same degrees of long and short term frequency stability as those of the best available standard clocks. Such a light source, slaved on an accurate standard, can be useful in fundamental experiments.

As a room temperature fast-response detector element, the diode can be used in sensitive superheterodyne receivers throughout the infrared and the far-infrared regions. In microwave terminology, our present method of coupling the infrared radiation to the diode suffers from a sizable insertion loss; (i.e., the coupling is far from optimum). Despite this, with a 10.6-μm laser local oscillator, superheterodyne detection sensitivity as low as about 5×10^{-14} watts can be obtained at one Hz bandwidth.

As a fast-response detector element, other applications of the diode include its use in detection of very short duration infrared laser pulses. This is a new area which has recently become of interest. For this application, the diode can be gated and used as a fast switch in the measurement of the shapes of very short duration infrared pulses.

The microscopic configuration of our diode, together with the simplicity of its structure, would allow in principle, the application of microelectronic technology in its manufacture. With microelectronics it should be possible to deposit on a substrate an array of diodes whose antennae are, for instance, distributed in parallel, forming a grating capable of receiving the incident radiation over a large area. In such a diode array, it should be possible to obtain orders-of-magnitude improvements in coupling the incident radiation to each diode, and thus obtain a considerable reduction in the overall insertion-loss factor. These types of configuration can in principle be made under highly controlled conditions with the existing photolithographic methods or other approaches, such as the scanning electron beams. This possibility points toward a promise that a great deal more can evolve by introducing the concept of our diode into the art of integrated optics.

III.   LASER-INDUCED LINE-NARROWING EFFECTS

We now turn attention to a different but related area of laser application in spectroscopy where a variety of effects, including the nonlinearities of atomic or molecular resonances, are used to obtain sizable reductions in the widths of the optical or infrared spectral lines. The past decade of developments in this area, and our recent success in the art of precision frequency measurements, can now be applied together in ultrahigh-resolution spectroscopy of a type never before possible.

In the early experiments, the line-narrowing effects were applied at MIT in a number of examples demonstrating several approaches to resolve and determine precisely the structures of a number of selected optical transitions [1–7]; in these examples, the line structures consisted of closely spaced components which would normally appear overlapping and unresolved. Since these early demonstrations, the exploitation of the line-narrowing effects in spectroscopic studies, in particular line-narrowing arising from the nonlinear processes, has become a particularly active field.

The spontaneous emission originating from a Doppler-broadened transition has the well-known Gaussian intensity distribution versus frequency ω given by $\exp[-(\omega-\omega_0)^2/\Delta\omega^2]$ where $\Delta\omega = (\bar{v}/c)\omega_0$ with $\bar{v}$ the average thermal velocity and c the speed of light. Near room temperature, the Q of a Doppler-broadened resonance, defined by $Q = (\omega_0/\Delta\omega) = c/\bar{v}$, ranges from about 10^5 to about 10^6, depending on the atomic mass. These were the highest-Q optical resonances obtainable prior to the advent of gas laser spectroscopy.

Laser-induced line-narrowing effects can be, in general, subdivided roughly into three categories:

A. Traveling wave line-narrowing in a high gain medium.
B. Nonlinear interaction of a standing-wave monochromatic radiation field with a Doppler-broadened resonance. This category can in turn be subdivided into two areas.
  1. Nonlinear standing-wave resonances manifested on the induced polarization: Lamb-dip observed in stimulated emission or in absorption.
  2. Nonlinear standing-wave resonances manifested on the change induced in level population.
C. Laser-induced line-narrowing effects in coupled Doppler-broadened transitions.

III-A. Traveling-Wave Line Narrowing in a High-Gain Medium

In a high-gain medium with a cylindrical geometry of length L, the spontaneous emission observed end-on along the axis of the cylinder will be proportional to the exponential gain: $\exp[LG(\omega)]$, where $G(\omega)$ is gain per unit length at frequency ω. The frequency distribution of the exponential gain is narrower than that of $G(\omega)$. The effect is sizable for $LG(\omega_0) \gg 1$, where ω_0 is the center frequency corresponding to maximum G. For a Doppler-broadened transition responsible for the gain, $G(\omega) = G_0\exp[-(\omega-\omega_0)^2/\Delta\omega^2]$. In this case, the narrowed profile of the amplified spontaneous emission given by the exponential gain has a (1/e) width given by $\Delta\omega/\sqrt{G_0L}$. Accordingly, in a high gain medium where, e.g. G_0 can be, say, 1/cm and for L of 1 meter, the width of the amplified spontaneous emission is narrowed by one order of magnitude.

An important example of this effect is observed in a high-current pulsed nitrogen discharge. In this system, the transitions belonging to the 2nd positive band in N_2, $C^3\Pi_u \rightarrow B^3\Pi_g(0,0)$, can be obtained in the amplifying phase with extremely high gains [29]. These transitions, which occur in the 3370-Å region, consist of a large number of fine-structure components, many of which are closely spaced with a separation on the order of or less than their 9-mÅ room temperature Doppler widths. Figure 4 gives a photographic plate of a portion of the amplified spontaneous emission spectrum of this band, obtained [7] by means of a high-resolution 20-ft cross-grating echelle spectrograph with an instrumental resolving width of 3 mÅ. From analysis of this plate and others obtained similarly, it is seen that the traveling-wave line-narrowing effect due to high gain has narrowed the width of each amplified spontaneous emission line to a value below the 3-mÅ instrumental width. This experiment, which was performed several years ago, provided the first dramatic application [7] to obtain a near complete resolution of a rather complex fine-structure spectrum having a number of closely spaced components which could barely be resolved previously. (The earlier attempts included observations in which the N_2 light source, consisting of a low pressure cw gaseous discharge, was cooled to the liquid-N_2 temperature to reduce the Doppler width).

The traveling-wave line narrowing is in essence the inverse of the same process which gives rise to self-reversal of an emission line occurring whenever the spontaneous emission is reabsorbed by atoms in the lower level of the same transition. There exist a host of recent atomic and molecular gas lasers in which very large gains have been observed. The

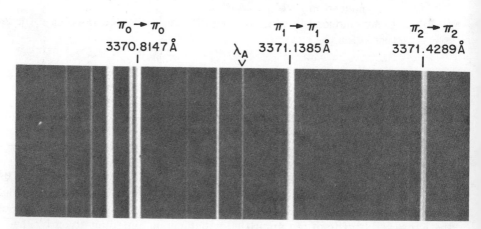

Figure 4. N_2 high-gain stimulated emission spectrum $C^3\Pi_u \rightarrow B^3\Pi_g(0,0)$. This is obtained with a 20-ft cross-grating echelle spectrograph which achieves a resolution of ~5 × 10^5 and provides a dispersion of 0.1496 Å/mm in the 3700 Å region.

examples include the high-gain rotation-vibration transitions in the chemical HF laser oscillating in the 3-μm region, in which sizable line-narrowing has been already observed [30]. Another example is the pure rotational transitions in an excited vibrational state of HF molecules. Extremely high gains have been observed in these transitions by means of an optical pumping method in a low pressure pure HF gas [31]. These systems are all excellent candidates for highly precise spectroscopic studies. The only system exploited for this purpose is, so far, the N_2 system mentioned above.

There exist a number of additional important processes which can play major roles in the process of traveling-wave line-narrowing in a high-gain medium. These include the effect of non-resonant feedback [32], and the nonlinear effects which become of importance in the limit where the intensity of the amplified spontaneous emission is built up to a value sufficient to produce saturation (see the discussion in this proceedings by Parks [33]).

III-B.   Nonlinear Interaction of a Monochromatic Standing-Wave Field with a Doppler-Broadened Resonance

Consider a Doppler-broadened atomic transition centered at a frequency ω_0, interacting with a monochromatic field at a frequency ω lying within the Doppler width of the transition. The natural width of the transition, γ, which is assumed to be considerably less than the Doppler width, is given by $\gamma = (\gamma_a + \gamma_b)/2$, where γ_a and γ_b are the radiative decay rates of the two levels of the transition. When the incident monochromatic field is in the form of a running wave, its interaction is most strong with atoms whose velocity component v along the propagation direction can Doppler shift ω to the proximity of ω_0, to within the natural width, γ (i.e. $|\omega(1-v/c) - \omega_0| \approx \gamma$). Accordingly, a monochromatic radiation field at a frequency ω can selectively excite atoms whose velocity components lie within a narrow range given by $\delta v = (\gamma/\omega)c$, and centered at a prescribed value v_c given by $v_c = [(\omega-\omega_0)/\omega]c$.

When the incident field at the frequency ω is in the form of a standing wave consisting of two traveling waves propagating in the opposite directions, two different groups of atoms selectively interact with each of the two traveling waves, respectively. The velocity components along the axis of the standing wave for the two groups are opposite to one another. The magnitudes of the velocity components for both groups are, however, the same and lie within the narrow range, $\delta v = (\gamma/\omega)c$, centered at $(|\omega-\omega_0|c/\omega)$. As the frequency of the incident field, ω, is tuned within the Doppler profile, the two velocity groups overlap when ω approaches ω_0 to within γ, for which $(|\omega-\omega_0|c/\omega) \approx \delta v$. For $\omega = \omega_0$, the two groups identically merge into one consisting of atoms whose velocities are essentially perpendicular to the axis of the standing-wave field.

In the linear approximation valid for weak fields, the two velocity group contribute additively to the expression describing the interaction between the standing wave field and the Doppler-broadened resonance. Accordingly, in this limit, the interaction remains unaffected as ω is tuned across the Doppler profile in the region encompassing ω_0. However, the nonlinear terms in the interaction, which become of importance in the limit of intense field, strongly depend on the overlap of the two velocity groups, and hence suffer a sizable change as ω is tuned across ω_0. The change occurs in a frequency region of width γ centered at ω_0.

Two different experimental approaches have enabled application of the above behavior of the nonlinear terms in the interaction to the observation of extremely narrow optical resonances. These are described below in Secs. III-B. 1 and III-B. 2.

III-B. 1   Nonlinear Standing-Wave Resonances Manifested on the Induced Polarization: The Lamb Dip Observed in Absorption or Stimulated Emission

The influence of the overlap of the two velocity groups discussed above on the polarization induced by the applied standing-wave field was first predicted by Lamb [34]. He showed that the lowest order nonlinear term in the polarization, which appears as a term proportional to the third power of the standing wave's electric field amplitude, has a Lorentzian dependence on the frequency ω. The Lorentzian is centered at ω_0 and has a half-width $\gamma = (\gamma_a + \gamma_b)/2$, which can be considerably below the Doppler width, $\Delta\omega$). The effect was observed for the first time on the frequency dependence of the output power of a single-mode He-Ne gas laser [35]: due to the nonlinear terms, the laser intensity versus its frequency shows a narrow power dip centered at the peak of the Doppler line profile. In the initial spectroscopic applications, the effect was applied to a precise determination of small isotope shifts in several laser transitions in Ne [36], and subsequently in the study of a variety of collision-broadening effects which contribute to the width of the power dip [37]. Examples of later spectroscopic applications of the Lamb dip observed on the laser power output include an elegant measurement of the isotope shifts in Xe [38].

The analysis of Lamb [34] is applicable to the interaction of a standing-wave field with a Doppler-broadened absorbing or amplifying transition. The effect manifested on the nonlinearities of an absorbing transition [39], has been applied in a variety of ways to obtain very high-Q nonlinear resonances The Lamb dip in methane obtained with the 3.39-μm He-Ne laser [40, 27], and the similar resonances in I_2 obtained with the 6328-Å He-Ne laser [41], have been particularly attractive in wavelength or frequency standard applications. As a typical illustrative example, the Lamb dip observed in a low pressure absorber gas consisting of NH_2D molecules [42], will be described here. Several of the NH_2D transitions belonging to the ν_2 rotation-vibration band can be Stark tuned by application of a dc electric field into resonance with several of the CO_2 10.6-μm laser lines. In the experiment [42], the output of a single-mode CO_2 laser oscillating on a transition lying close in frequency to that of a Stark tunable NH_2D absorption line, is transmitted through a polarizer, $\lambda/4$ isolator plate, and a beam splitter, and then passed through an absorption cell containing the NH_2D molecules at a low pressure (see Fig. 5). The standing-wave field is produced by reflecting the laser beam transmitted through the absorption cell back upon itself. The absorption of the standing-wave field versus the frequency tuning of the absorption line is observed by detecting the laser beam reflected from the beam splitter, and recording it versus dc Stark-electric field. Figure 6 shows a recorder tracing obtained by modulating the dc Stark field at an audio frequency by a small amplitude and using a phase-sensitive amplifier synchronized to the modulation frequency. Accordingly, the recorder tracing is proportional to the first derivative of the saturated absorption versus frequency tuning of the resonance line. This experiment has enabled observation of very fine details of the NH_2D spectrum in the infrared [42]. These details are several orders of magnitude beyond the resolving powers of conventional infrared spectrometers.

In cases where the atomic or molecular resonances consist of several closely spaced transitions whose Doppler widths are overlapping, the narrow nonlinear standing-wave resonances observed in absorption (or amplification) appear as a set of completely resolved lines, each of them located at the center of its corresponding Doppler-broadened transition. The widths of the narrow nonlinear resonances can be several orders of magnitude

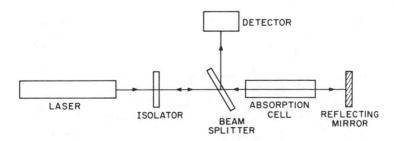

Figure 5.   Block diagram of a typical system for observation of the Lamb dip in absorption. For instance, in the NH_2D experiment described in the text, the absorption cell contains low-pressure NH_2D gas and the laser is a CO_2 laser.

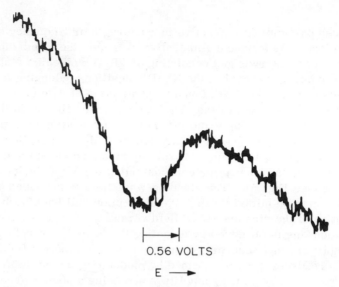

0.56 VOLTS

E ⟶

Figure 6.   The NH$_2$D Lamb dip.   The recorder tracing shows the derivative of the saturated absorption versus the applied dc Stark field.   The half-width at half maximum corresponds to 1.4 MHz and is primarily due to electric field inhomogeneity.

below the Doppler width, enabling complete resolution of otherwise over-lapping transitions not resolvable previously.

Consider now two closely spaced Doppler-broadened transitions which share a common energy level [e.g. Fig. 7(a)] interacting with a standing-wave monochromatic field.   As the frequency of the standing-wave field is tuned across the two overlapping Doppler profiles, two narrow nonlinear

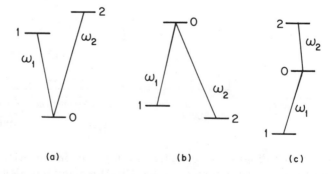

(a)   (b)   (c)

Figure 7.   Three different configurations for two coupled transitions sharing a common energy level.

Lamb-dip resonances appear at the center frequencies of the two transitions, as discussed above. Because of the common energy level shared by the two transitions, however, an additional nonlinear resonance appears at a frequency midway between the two Lamb-dip resonances. The new resonance, initially predicted by Schlossberg and Javan [43], has an intensity which is proportional to the arithmetic mean of the intensities of the two Lamb dips, as long as the Doppler profiles of the two transitions are closely overlapping. This effect has recently been observed in the Na D-lines using a tunable dye laser [44]. To illustrate the origin of the effect, let us first assume the standing-wave frequency ω to be tuned exactly to $(\omega_1 + \omega_2)/2$, with ω_1 and ω_2 representing the center frequencies of the two coupled transitions. We note that the group of atoms whose velocities Doppler shift one of the two traveling-wave components into resonance with, say, ω_1, happen to have the required velocity to Doppler shift the other traveling wave component into resonance with ω_2. In this case both transitions are simultaneously coupled to the applied field via the same group of atoms. However, for $|\omega - (\omega_1 + \omega_2)/2| \gg \gamma$, two different groups of atoms in the velocity distribution interact respectively with each of the two traveling-wave components, so that each atom is most strongly coupled to the applied field only via one of the two coupled transitions. Because of this, the nonlinear terms in the expression describing the interaction between the standing-wave monochromatic field and the coupled transition contain a narrow resonance of width γ centered at the mid-frequency, $(\omega_1 + \omega_2)/2$ [43].

At the limit of large standing-wave field intensities, a variety of interesting power-broadening effects influence the line shape of the nonlinear resonances. (For a review of these and other line-shape effects see the discussion in these proceedings by Feld [45].)

With the recent advent of tunable lasers, nonlinear laser spectroscopy of the type described above can in principle be applied to absorbing or amplifying transitions which lie within the frequency ranges covered by well-behaved tunable lasers [46].

III-B.2 Nonlinear Standing-Wave Resonances Manifested on the Change Induced in Level Population

An essential requirement for application of the methods described above is that the absorption coefficient (or the amplification factor) of the transition interacting with the monochromatic standing-wave field shall be sufficiently large to produce a detectable absorption (or amplification) of the incident standing-wave field. It is most desirable, however, to devise a method which would enable observation of the line-narrowing effect in a medium at a low gas pressure where the atomic (or molecular) density may be below that necessary to yield a detectable change in the incident laser power interacting with the absorbing (or amplifying) transition. Among various applications, such a method would enable ultrahigh-resolution spectroscopic studies in dilute media at very low gas pressures, where the level shifts due to pressure effects, caused by the interatomic or intermolecular collisions, can be negligibly small. The following approach, which has recently been demonstrated in a preliminary but important example [47], provides the way to explore these types of applications.

To describe the method, let us again consider the interaction of a monochromatic standing-wave field and a Doppler-broadened transition. Assume, for example, that the transition is in the absorbing phase with a lower-level population appreciably larger than that of the upper level. As the intensity of the standing-wave field is increased, the populations in the two levels change appreciably due to the saturation effect. In this case, the upper-level population increases and the lower-level population decreases. Because of the velocity selective characteristic of the interaction of monochromatic radiation with a Doppler-broadened resonance, and the fact that the standing wave consists of two oppositely propagating running waves, the population change in each level occurs over two narrow regions within the distribution of the atomic velocity components along the axis of the standing-wave field. For $|\omega-\omega_0| \gg \gamma$, the velocity components in the two regions have opposite directions, but their magnitudes are both the same and centered at a value ($|\omega-\omega_0|c/\omega$), lying within a narrow width given by $\delta v = (\gamma/\omega)c$. (Note that for the Doppler-broadened line considered here, where $\gamma \ll \Delta\omega$, $\delta v \ll \bar{v}$, where $\bar{v}$ is the average velocity.)

The lower curve in Fig. 8 indicates the distribution of the velocity component along the axis of the standing-wave field for atoms in the upper level of the absorbing transition and for the case $|\omega-\omega_0| \gg \gamma$. The two narrow-width Lorentzian curves superimposed on the broad background curve represent the change in the velocity distribution induced by the laser field as discussed above. Accordingly, the area under the two Lorentzian curves give the total population change of the upper level. As ω is tuned closed to the center frequency of the Doppler profile, ω_0, the Lorentzian curves, which are symmetrically placed on either side of $v = 0$, move toward each other and they overlap for $|\omega-\omega_0|$ approaching γ. Due to the nonlinearities of the interaction, the area under the two Lorentzians no longer add linearly as the overlap begins to set in. The deviation is in the direction of enhancing the effect of saturation, which generally causes the level population change to occur at a value below that described by the terms proportional to the laser intensity. The upper curve in Fig. 8 shows the velocity distribution curve in the presence of the standing wave for $\omega = \omega_0$ where the overlap is complete. In this case, the saturation effect is enhanced, i.e. the area under the Lorentzian curve, which gives the corresponding population change induced by the laser field, is less than the sum of the areas under the nonoverlapping Lorentzian curves for the detuned case $|\omega-\omega_0| \gg \gamma$.

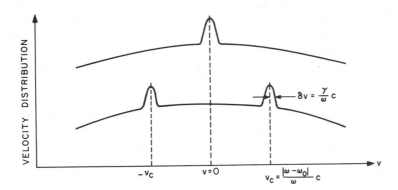

Figure 8.   Velocity distribution of the upper (or the lower) level of an absorbing (or amplifying) transition in the presence of a saturating standing-wave field.   The lower curve corresponds to detuned case $|\omega-\omega_0| \gg \gamma$ and the upper curve corresponds to exact resonance, $\omega = \omega_0$.

(Here, for simplicity, we have assumed implicitly that $|\omega-\omega_0| \ll \Delta\omega$ so that interaction occurs in the region of $v \ll \bar{v}$ where the thermal velocity distribution is flat.)   The difference, to the lowest order in the laser electric field amplitude, E, appears in the expression describing the population change, as a term proportional to E^4.   However, in the limit of large field intensity where an expansion in the power series is no longer valid, the total change induced in the level population will differ sizably for $\omega = \omega_0$ compared to $|\omega-\omega_0| \gg \gamma$.   It can be shown that this difference, i.e. the difference between the value for $|\omega-\omega_0| \gg \gamma$ minus the value for $\omega = \omega_0$, has an asymptotic value which is 30% of the total population change induced by the standing wave for $|\omega-\omega_0| \gg \gamma$.   This limit can be approached at laser intensities corresponding to $\mu E/\hbar \gtrsim \gamma$, where μ is the transition matrix element.   This condition can readily be satisfied for most cases of interest at a moderate laser field intensity.

A sensitive method to probe the change in level population is to observe, if possible, photons emitted spontaneously from the level of interest.   In particular, the detection sensitivity is large if the spontaneous emission can be detected in the absence of other spurious radiation reaching the photon detector.   (The presence of spurious radiation, such as scattered laser radiation, etc., can produce added noise in the detector.)   In our case, one convenient method is to observe the spontaneous emission from the upper (or the lower) level to a third level corresponding to a transition at a frequency appreciably different from the laser frequency.   This would allow simple use of an optical filter to prevent the laser radiation from reaching the detector.

Since the spontaneous emission intensity is proportional to the total level population integrated over its entire velocity distribution, it can be used to detect the population change induced by the laser field. Accordingly, when the laser radiation is applied, the spontaneous emission from the upper (or the lower) level changes by an amount ΔI. As the laser frequency is tuned across the Doppler profile, ΔI suffers a resonant change over a narrow region centered at $\omega = \omega_0$. The width of the resonance is essentially given by γ, with some contributions from power broadening (which can become large only for $\mu E/\hbar \gg \gamma$). Accordingly, in cases of interest where $\gamma \ll \Delta\omega$, this method enables observation of line narrowing in systems where the absorption (or the amplification) can be extremely weak.

There exist a number of variations of the above method with interesting important features not discussed above. In particular, if the transition belongs, e.g., to a rotation-vibration band consisting of closely spaced rotational structures, collisional coupling between adjacent levels plays an important role. The following experiment, demonstrating the first application of the method, clarifies these features.

In the CO_2 10.6-μm absorption band, narrow nonlinear resonances can only be expected at reduced pressures where collision broadening is small. However, at a low pressure, the 10.6-μm absorption coefficient in CO_2 is much too small to allow applications of the Lamb-dip method. The laser-induced change in the level population, on the other hand, is applicable in this case and enables the observation of extremely narrow resonances in each of the individual P or R branch transitions of the 10.6-μm band. This is done by subjecting the low-pressure CO_2 absorber gas to the standing-wave radiation of a CO_2 laser oscillating in a single mode of a preselected P or R branch transition of its 10.6-μm (or 9.3-μm) band. As a reminder we note that the upper levels of both the 10.6-μm and 9.3-μm bands belong to the (001) vibrational mode lying at about 2300 cm^{-1} above the ground vibrational state of CO_2, the (000) state. The lower levels of these two bands belong to the (100) and (020) vibrational modes which are in Fermi resonance with one another and lie at about 1400 cm^{-1} above the ground vibrational state.

The level population change induced by the laser field in the (100) vibrational state is detected by observing the change in the intensity of the 4.3-μm spontaneous emission over the entire (001) $\rightarrow$ (000) band. The observation is made as the laser frequency is tuned across the Doppler profile of the corresponding 10.6-μm (or the 9.3-μm) absorption line. In the presence of the laser field, the intensity of the 4.3-μm emission from the absorber gas increases over the background thermal radiation by an amount ΔI, which is a nonlinear function of the laser field intensity, as described above. For a sufficiently intense laser field, ΔI diminishes resonantly by a sizable amount as the laser frequency ω is tuned across the center frequency ω_0 of the absorbing line. From the above discussions, it follows that the center of this resonance occurs at $\omega = \omega_0$ and its linewidth is determined in this case by the characteristic collision broadening, the power broadening of the absorbing transition, and the broadening effect due to the molecular transit time across the diameter of the incident laser beam.

For a molecular rotation–vibration transition, a change in the population of an individual rotational level is accompanied by a change in the total population of the corresponding vibrational state. This is due to the collisional coupling between the rotational levels, which tends to maintain a thermal population distribution among them. Because of this, the population of all the rotational levels of a given vibrational state vary in proportion to the change induced in the population of an individual level belonging to the specific transition which interacts with the laser field. Accordingly, as the laser frequency is tuned across the line profile of the absorbing transition, the frequency-dependent change in level population, including the resonant behavior due to the standing-wave effect, occurs in all of the collisionally coupled rotational levels. This results in the intensity change, ΔI, for the entire 4.3-μm (001) $\rightarrow$ (000) spontaneous emission band to show the nonlinear standing wave resonance versus the frequency detuning $(\omega-\omega_0)$.

In the preliminary experiment, the observations were made in the 0.005 to 0.8 Torr pressure region. Half-widths as low as 200 kHz have been observed. The limiting widths observed at the low pressure was mostly due to the power broadening which can be considerably reduced by further experimental refinements. In a preliminary attempt, with the aid of an automatic feedback control, the laser frequency has been stabilized at the line-profile centers of the various absorbing transitions to within an accuracy exceeding one part in 10^{10}. Figure 9 gives a typical tracing of the resonance for the P(18) CO_2 10.6-μ line.

The narrow resonance obtained on the individual P- or R-branch transitions of the 10.6-μm or the 9.3-μm absorption bands make available excellent secondary frequency standards useful in providing a link between the far infrared and the near infrared or the visible regions. Much further work needs to be done to exploit precision "clock" application of these resonances. In this respect, it is of interest to note that the pressure shifts in these transitions are found to be extremely small. This is of particular importance in frequency resetability experiments.

It is important to note that in a system in which the spontaneous emission photons fall in the visible or the ultraviolet regions, the above method can be applied at extremely low gas pressures. In the visible or the uv range, excellent photon counters are available. In fact, the method would be applicable even if the level populations are so low that only a few photons per second emitted from the levels of interest reach the detector.

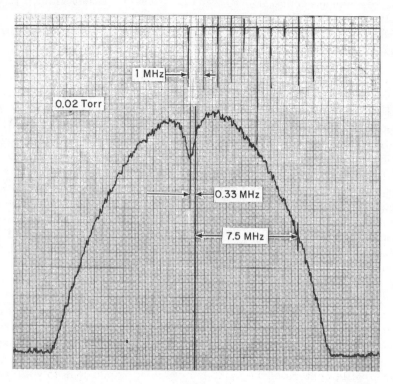

Figure 9:   A typical recorder tracing of the CO_2 nonlinear resonance versus frequency.   It shows the change induced in the (001) vibrational band population by an incident CO_2 P(18) 10.6-μm line.   The gas pressure was 0.02 Torr.

III-C.   Laser-Induced Line Narrowing in Coupled Doppler-Broadened Transitions

We shall now briefly review a different feature of the velocity selective property of a monochromatic radiation field interacting with a Doppler-broadened resonance consisting of a pair of coupled transitions sharing a common energy level.   The resultant line narrowing is applicable both in (1) precise measurements of the closely spaced structure of an atomic resonance and (2) accurate determination of the center frequency of an optical transition.

Let us consider an illustrative example consisting of a pair of nearly overlapping Doppler-broadened transitions arising from the coupled level configuration shown in Fig. 7(a) (i.e., a V-type three-level system). We assume arbitrarily that the two upper levels, 1 and 2, are more heavily populated than the lower level, 0, so that both transitions show optical gains. The center frequencies of the (1,0) and (2,0) transitions will be denoted by ω_1 and ω_2, respectively, and their homogeneous widths will be assumed to be dominated by radiative decay processes; the rates of the radiative decays of the three levels will be designated by γ_0, γ_1, and γ_2 respectively. Consider this system subjected to an intense monochromatic laser radiation in the form of a running-wave at a frequency Ω_L close to ω_2 but differing from it by an amount larger than the homogeneous width of the (2,0) transition, $\gamma = (\gamma_0 + \gamma_2)/2$. The propagation direction of the laser field will be designated by (+) and the reverse direction by (−). Let us further assume a weak running-wave field at a frequency Ω tunable in the region close to ω_1, to be applied for the purpose of probing the gain profile of the (1,0) transition in the presence of the intense laser field interacting with the (2,0) transition. As the intensity of the intense field is increased to a value where saturation of the (2,0) transition becomes appreciable, the gain profile of the coupled transition appears modified. The observed change, however, is dependent on the relative propagation directions of the probe and the intense fields. For the probe field co-directional with the intense saturating field (i.e., propagating in the (+) direction), the modified (1,0) gain profile shows a resonant decrease over a narrow interval which is centered at Ω_1^+, given closely by:

$$\Omega_1^+ = \omega_1 + (\Omega_L - \omega_2) \tag{1}$$

For the probe field propagating in the reverse direction, the change in the gain profile appears over a narrow range centered on the opposite side of ω_1, given similarly by:

$$\Omega_1^- = \omega_1 - (\Omega_L - \omega_2) \tag{2}$$

These equations hold for the case $\omega_1 \simeq \omega_2$ considered here. More general expressions valid for arbitrary ω_1 and ω_2, are derived below. This effect was initially discussed in detail by Schlossberg and Javan, who showed that the widths of the narrow resonant changes in the (1,0) gain profile differed in the two cases [43]. For the probe field propagating in the same direction as the saturating field, the width, Γ^+, was shown to be independent of γ_0 and given by $\Gamma^+ = (\gamma_1 + \gamma_2)/2$. For oppositely propagating fields, the width, Γ^-, was shown to be broader than Γ^+ and dependent on γ_0. Explicitly, Γ^- is given by $\Gamma^- = (\gamma_1 + \gamma_2 + 2\gamma_0)/2$. As explained below, the explicit expression for Γ^- agrees with that obtainable from simple inspection of the problem;

but the exact expression for the narrower width Γ^+, corresponding to the co-directional propagation, has a somewhat anomalous appearance requiring detailed algebraic derivation. It should be noted that there are a number of important cases in which Γ^- is considerably smaller than Γ^+.

Let us for the moment ignore the intricacies of the exact linewidth and instead, focus attention on the spectroscopic application of the narrow resonant change induced in the $(1, 0)$ gain profile for the probe and the saturating fields propagating in the same direction. In this case, the frequency corresponding to the peak of the induced resonance given by Eq. (1) satisfies the relation $\Omega_1^+ - \Omega_L = w_1 - w_2$. Accordingly, by tuning the probe field to the frequency of the peak of the narrow resonant change, Ω_1^+, and measuring its separation from Ω_L, the level spacing $w_1 - w_2$ can be measured to within a small fraction of Γ^-, which can be orders of magnitude below the Doppler width. We note, incidentally, that in this expression, the exact laser frequency, Ω_L, is unimportant, as long as it lies within the Doppler width of the transition.

In the first spectroscopic application [2], the above method was applied at MIT to accurate determination of the closely spaced hyperfine splitting of the upper level of the 3.37-μ laser transition in the odd isotope of xenon with atomic number 129. In the experiment, rather than tuning the frequency of the probe field, the frequency separation of the closely spaced hyperfine components were Zeeman tuned by application of a dc magnetic field. This was done in the presence of two laser fields separated in frequency by a fixed amount, Δ, lying in the radio-frequency region. In the transition, the Zeeman effect produced a number of pairs of coupled components arising from the various hyperfine Zeeman sublevels whose splittings depended on the values of the magnetic field. Well-resolved resonances of narrow width $\Gamma^+ = 0.6$ MHz were observed as the splittings of the appropriate pairs of coupled transitions were tuned to Δ. (The Doppler width of the transition is about two hundred times larger than the observed Γ^+.) These resonances were used to obtain an accurate value of the zero field hyperfine splitting [2]. Figure 10 gives the observed resonances due to several of these pairs of coupled transitions. The 0.6-MHz width of the resonances is in full agreement with the expression $\Gamma^+ = (\gamma_1 + \gamma_2)/2$. (In the transition, $\gamma_1 = \gamma_2 = 0.6$ MHz, while γ_0 is at least one order of magnitude larger. Accordingly, in this case, the value of Γ^- is at least one order of magnitude larger than the observed Γ^+ width.)

In an important series of experiments, the above method has been recently applied by Brewer [48], to molecular resonances in the infrared, utilizing the Zeeman or the Stark effects to split the resonance lines. In his experiments, he has been able to observe fine details of small Zeeman splittings arising from weak rotational magnetic moments in molecules which are primarily in a $^1\Sigma$ electronic state.

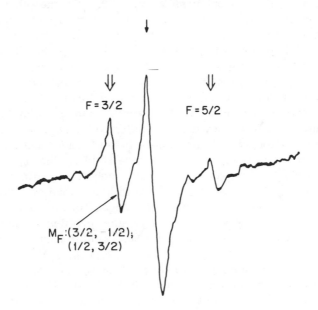

Figure 10.   Nonlinear resonances belonging to the 3.37-μm transition in Xe obtained via Zeeman tuning.   The widths of the resonances correspond to 0.6 MHz.   The Xe laser was made with an isotopically enriched sample of Xe gas consisting of a mixture of the even Xe isotopes and the odd isotope ^{129}Xe.   The laser was oscillating in two modes spaced by 57.1 MHz.   The resonances indicated by double arrows arise from Zeeman splittings of the upper level of the 3.37-μ transitions in ^{129}Xe for hyperfine components F = 5/2 and F = 3/2.   The resonance with single arrow is a similar resonance due to even isotopes which have no hyperfine structure (J = 2).

        Before reviewing further applications, we shall now discuss some aspects of the physical processes which contribute to the line-narrowing effect considered here.   Due to the saturation of the (2, 0) transition by the intense laser field at frequency Ω_L, the population of level (0) increases. This occurs over a narrow region of the velocity distribution corresponding to atoms whose velocities Doppler shift Ω_L to the proximity of w_2, to within the natural width of the (2, 0) transition, $\gamma = (\gamma_2 + \gamma_0)/2$.   The narrow region in the velocity distribution is centered at $v_c = [(\Omega_L - w_2)/w_2]c$ and has a width of $\delta v = (\gamma/w_2)c$; here v_c represents the algebraic value of the component along an axis collinear with the intense laser field with the positive direction pointing toward the propagation direction.   For $\Omega_L > w_2$, which requires a negative Doppler shift, $v_c = [(\Omega_L - w_2)/w_2]c > 0$.   For $\Omega_L < w_2$, requiring a positive Doppler shift, $v_c < 0$.   On the other hand, the probe field at the frequency Ω interacts most strongly with the (1, 0) transition via atoms whose velocity components along the propagation direction can Doppler shift Ω to the proximity of w_1--to within the (1, 0) natural width which is given by $\gamma' = (\gamma_1 + \gamma_0)/2$.   For the probe field propagating in the (+) direction, the algebraic value of the corresponding velocity components, (referred to the same axis as before), are centered at $v_c' = [(\Omega - w_1)/w_1]c$, and lie within the narrow width given by $\delta v' = (\gamma'/w_1)c$. As Ω is tuned to a frequency corresponding to $v_c = v_c'$, the atoms which interact most strongly with the intense laser field via the (2, 0) transition also interact most strongly with the probe field via the (1, 0) transition. However, when Ω is tuned to a frequency such that v_c' differs from v_c by an amount appreciably larger than the two widths δv and $\delta v'$, the atoms coupled most strongly to the probe field are no longer coupled also to the intense laser field.   Accordingly, we expect the gain profiles of the (1, 0) transition to be modified over a narrow region whose center frequency can be obtained from $v_c = v_c'$, which is $\Omega = w_1 + (\Omega_L - w_2)(w_1/w_2)$.

Since we are assuming w_1 to lie close to w_2, the expression for the center frequency in this limit reduces to the expression for Ω_1^+ given by Eq. (1).   As for the width of the narrow resonances, we note that the two velocity groups centered at v_c and v_c' have an appreciable overlap for Ω detuned from Ω_1^+ by an amount for which $|v_c - v_c'| = (\delta v + \delta v')/2$.   This corresponds identically to $|\Omega - \Omega_1^+| = [\gamma' + \gamma(w_1/w_2)]/2$.   From this we might expect the width of the corresponding resonant change to be about $[\gamma' + \gamma(w_1/w_2)]/2$, which in our case where $(w_1/w_2) \approx 1$ reduces to $(\gamma_1 + \gamma_2 + 2\gamma_0)/2$.   We note that our actual width given by $\Gamma^+ = (\gamma_1 + \gamma_2)/2$, fundamentally differs from this roughly estimated width in that Γ^+ is actually independent of γ_0.   This is a rather pleasant feature since in practice there are important cases where γ_1 and γ_2 are considerably less than γ_0, resulting in very narrow resonances with widths $\Gamma^+ = (\gamma_1 + \gamma_2)/2$; the Xe example cited above is an excellent example [2].

On the other hand, the situation is entirely different for the case where the probe field propagates along a direction opposite to that of the intense field: in this case, the argument for the center of the corresponding resonance and the expected width follows identically as before, with the exception that for the same positive axis defined previously, v_c' in this case is given by $v_c' = [(w_1 - \Omega)/w_1]c$.   From this, the center of the corresponding resonance is found to be:

$$\Omega_1^- = [w_1 - (\Omega_L - w_2)(w_1/w_2)]$$

This expression, which holds for arbitrary w_1 and w_2, will reduce to

Eq. (2) for $\omega_1 \approx \omega_2$. The expected width obtained, as before, from $|v_c - v_c'| = (\delta v + \delta v')/2$, will be:

$$\Gamma^- = |\Omega - \Omega_1^-| = [\gamma + \gamma'(\omega_1/\omega_2)]/2 \approx (\gamma_1 + \gamma_2 + 2\gamma_0)/2, \text{ for } \omega_1 \approx \omega_2$$

Unlike Γ^+, the actual expression for Γ^-, obtained via detailed calculation, is dependent on γ_0 [43]. In fact, the exact expression for Γ^- is identically the same as our rough estimate of $\Gamma^- = [\gamma + \gamma'(\omega_1/\omega_2)]/2$, obtained above. The origin of the seemingly anomalous width behavior lies in the details of the radiative processes occurring in a three level system coupled to two monochromatic radiation fields. The importance of these details were, in fact, recognized some time ago in a theoretical treatment which dealt with homogeneously-broadened coupled resonances [49]. From Ref. 49 we note that the change in the (1, 0) gain profile is determined not only by a change in the level populations emphasized above, but also by additional radiative processes which include two-quantum Raman type transitions between levels 2 and 1, with level 0 as a resonant intermediate state. These processes have different functional dependences on the frequencies of the two applied fields, and hence, contribute differently to the overall line profile. Accordingly, for an atom of a given velocity, the relative importance of these processes will depend on the exact values of the Doppler-shifted frequencies of the probe and pump fields. We further note that the signs of the two Doppler shifts are either the same or differ from one another, depending on the relative propagation directions of the pump and the probe fields. As a result, the velocity average of the radiative processes contributing to the gain at the (1, 0) transition is weighted differently in the case in which the probe and the pump fields propagate in the same direction, as opposed to the case in which they propagate in opposite directions. This results in a change in the (1, 0) gain profile which differs in the two cases: For co-directional fields, the width of the change in the absorption profile happens to be less than the corresponding width for oppositely propagating fields; this difference turns out to be γ_0. However, the areas (i.e., magnitude × frequency) of the two resonances are always equal in the two cases.

In the above discussions, we have assumed ω_1 to be close to ω_2. For the case where ω_1 differs appreciably from ω_2, the two resonant changes are centered at $\Omega_1^\pm = \omega_1 \pm (\Omega_L - \omega_2)(\omega_1/\omega_2)$ as noted above. The expressions for the exact widths are also modified in this case. The corresponding expression for Γ^- agrees with the expression obtained from simple inspection, which predicts $\Gamma^- = [\gamma + \gamma'(\omega_1/\omega_2)]/2$, but as before, Γ^+ differs from Γ^- by γ_0: $\Gamma^+ = \Gamma^- - \gamma_0$. Accordingly, we note that Γ^- becomes independent of γ_0 only for ω_1 close to ω_2.

Figure 7(b) and 7(c) give two other types of level configurations for a pair of coupled transitions. In these cases, and for ω_1 appreciably different from ω_2, the line-narrowing effect can be conveniently observed on the line profile of the spontaneous emission from level (0), obtained in the presence of the intense monochromatic field interacting with the (2, 0) transition. Note that the resulting spontaneous emission spectrum directly follows the spectrum of emission stimulated by a weak probe field tuned through the coupled transition when the lower level population of the coupled transition is ignored.   Accordingly, the spontaneous emission line profile, observed along the same direction as the laser field or opposite to it, appears modified over narrow regions of differing widths centered at Ω_1^+ and Ω_1^-, identically as discussed above.   If the intense laser field is in the form of a standing wave which consists of two running waves propagating along opposite directions, the spontaneous emission profile will appear modified over both narrow regions (centered at Ω_1^+ and Ω_1^-) simultaneously.

An excellent example for observing the effect in spontaneous emission has been the 1.15-μm transition of Ne arising from the $2s_2 \rightarrow 2p_4$ transition coupled to a cascade transition originating from its lower level, say $2p_4 \rightarrow 1s_4$ which lies at 6096 Å.   Unlike the example of the Xe transition given above, the difference in the broad and the narrow widths is small in this case, hence the effect is not very dramatic.   In fact, in the early observation of the effect [3], the difference between the two widths was not detected, and hence overlooked.   The effect was later observed by Holt [50]. A detailed theoretical analysis of the effect, which includes the limit of highly intense saturating field, has been given by Feld and Javan [51].   The discussions in Ref. 51 clarify the relationships between the various similar but seemingly different treatments of the problem in spontaneous emission, given independently by several authors [52-55].

In a recent application [56], the laser-induced line narrowing in spontaneous emission has been applied to resolve the hyperfine structure in the ^{21}Ne isotope which has a nuclear spin of 3/2.   This experiment gives detailed information on the structure of all the three levels of the coupled transitions $2s_2 \rightarrow 2p_4$ and $2p_4 \rightarrow 1s_4$, from which an accurate measure of the ^{21}Ne nuclear quadrupole moment is obtained.   Figure 11(b) gives the classical Doppler profile of the $2p_4 \rightarrow 1s_4$ transition in ^{21}Ne in the absence of the laser-induced line narrowing.   This tracing is obtained by means of a scanning Fabry-Perot interferometer over a region of 4090 MHz.   In this tracing, the hyperfine components are completely overlapping and unresolved.   Figure 11(a) gives the laser-induced line-narrowing results obtained over the same identical region using the same Fabry-Perot interferometer.   This is obtained in the presence of a 1.15-μ laser radiation saturating the $2s_2 \rightarrow 2p_4$ transition.   The tracing shows clearly the resolved hyperfine components achieved by the line-narrowing effect.

In the above discussions, for the sake of illustration, the emphasis is placed on the application of the line narrowing to the determination of the structure of an optical transition.   The effect can also be applied to finding

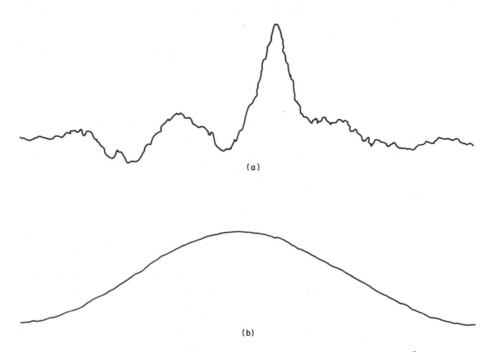

Figure 11.   Spontaneous emission profile of the $2p_4 \rightarrow 1s_4$ (6096 Å) line in ^{21}Ne.   (a) Resolved hyperfine components obtained by laser–induced line narrowing in which the $2s_2 \rightarrow 2p_4$ coupled transition is satured by radiation from the 1.15-μ He–Ne laser.   (b) The unresolved ordinary spontaneous emission spectrum.   Both (a) and (b) were taken with the same scanning Fabry–Perot.   The total frequency ranges in both (a) and (b) are 4090 MHz.

the center frequency of the coupled transition to within an accuracy deter-mined by the size of Γ^+ and Γ^- which can be orders of magnitude below the Doppler width.   This may be seen by noting from Eqs. (1) and (2) that the (1, 0) line center is exactly midway between the two narrow resonances induced by the intense laser field: $\omega_1 = (\Omega_1^+ + \Omega_1^-)/2$.   In fact, in the first spectroscopic application of the effect in spontaneous emission, this feature was used to determine accurately the isotope shifts in Ne giving rise to a displacement of the center of the transitions corresponding to the two iso-topes, ^{20}Ne and ^{22}Ne.

In the high–intensity limit where power–broadening effects contribute to the line shape, a host of interesting radiative behaviors are encountered. These and other important line-shape phenomena have been analyzed in great detail by Feldman and Feld [57]. A review of the various effects and other line-shape details, including the application to a novel unidirectional amplifier [58], is given in these proceedings by Feld [45].

The above presentation gives only the highlights of the various developments; it is not intended to be a complete review.

REFERENCES

1.   A. Szőke and A. Javan, Phys. Rev. Letters 10, 521 (1963).
2.   H. R. Schlossberg and A. Javan, Phys. Rev. Letters 17, 267 (1966).
3.   R. H. Cordover, P. A. Bonczyk, and A. Javan, Phys. Rev. Letters 18, 730 (1967).
4.   R. H. Cordover, T. Jaseja, and A. Javan, Appl. Phys. Letters 7, 322 (1965).
5.   J. S. Levine, P. A. Bonczyk, and A. Javan, Phys. Rev. Letters 177, 540 (1969).
6.   M. S. Feld, J. H. Parks, H. R. Schlossberg, and A. Javan, in Physics of Quantum Electronics, edited by P. L. Kelley, B. Lax, and P. E. Tannenwald (McGraw-Hill, New York, 1965), p. 567.
7.   J. H. Parks, D. Ramachandra Rao, and A. Javan, Appl. Phys. Letters 13, 142 (1968).
8.   L. O. Hocker, A. Javan, D. Ramachandra Rao, L. Frenkel, and T. Sullivan, Appl. Phys. Letters 10, 147 (1967).
9.   L. O. Hocker, D. Ramachandra Rao, and A. Javan, Phys. Letters 24A, 690 (1967).
10.   L. O. Hocker, and A. Javan, Phys. Letters 26A, 255 (1968).
11.   L. O. Hocker, J. G. Small, and A. Javan, Phys. Letters 29A, 321 (1969).
12.   L. O. Hocker, D. R. Sokoloff, V. Daneu, A. Szőke, and A. Javan, Appl. Phys. Letters 12, 401 (1968).
13.   V. Daneu, D. R. Sokoloff, A. Sanchez, and A. Javan, Appl. Phys. Letters 15, 398 (1969).
14.   D. R. Sokoloff, A. Sanchez, R. M. Osgood, and A. Javan, Appl. Phys. Letters 17, 257 (1970)
15.   A. Sanchez and A. Javan, to be published.
16.   K. M. Evenson, J. S. Wells, L. M. Matarrese, and L. B. Elwell, Appl. Phys. Letters 16, 159 (1970)
17.   K. M. Evenson, J. S. Wells, and L. M. Matarrese, Appl. Phys. Letters 16, 251 (1970).
18.   J. G. Small and A. Javan, to be published. For a summary description of the phase-locking procedure and the experimental results, see: J. G. Small, J.-P. Monchalin, M. J. Kelly, F. Keilmann, A. Sanchez,

S. K. Singh, N. A. Kurnit, and A. Javan, Proceedings of 26th Annual Symposium on Frequency Control, June 1972, p. 248 (available from: Electronic Industries Assoc., 2001 Eye St. N. W., Washington, D. C.)

19. L. O. Hocker and A. Javan, Phys. Letters 25A, 499 (1967). See also D. R. Lide and A. G. Maki, Appl. Phys. Letters 11, 2 (1967).

20. L. O. Hocker and A. Javan, Appl. Phys. Letters 12, 124 (1968).

21. D. R. Sokoloff and A. Javan, Bull. Am. Phys. Soc. 15, 505 (1970).

22. T. J. Bridges and T. Y. Chang, Phys. Rev. Letters 22, 811 (1969).

23. D. R. Sokoloff and A. Javan, J. Chem. Phys. 56, 4028 (1972).

24. J. S. Levine, A. Sanchez, and A. Javan, Physical Review (to be published).

25. V. L. Daneu, L. O. Hocker, A. Javan, D. Ramachandra Rao, A. Szőke, and F. Zernike, Phys. Letters 29A, 319 (1969).

26. N. A. Kurnit, in this volume.

27. J. L. Hall, in this volume.

28. A. Sanchez, S. K. Singh, and A. Javan, Appl. Phys. Letters 21, 240 (1972). (This work was completed after the Esfahan Symposium.)

29. D. A. Leonard, Appl. Phys. Letters 7, 4 (1965); E. T. Gerry, Appl. Phys. Letters 7, 6 (1965).

30. J. Goldhar, R. M. Osgood, Jr., and A. Javan, Appl. Phys. Letters 18, 167 (1971).

31. N. Skribanowitz, I. P. Herman, R. M. Osgood, Jr., M. S. Feld, and A. Javan, Appl. Phys. Letters 20, 428 (1972).

32. R. V. Ambartsumyan, N. G. Basov, P. G. Kryukov, and V. S. Letokhov, Zh. Eksp. Teor. Fiz. 51, 724 (1966) [Sov. Phys. JETP 24, 481 (1967)]; R. V. Ambartsumyan, P. G. Kryukov, and V. S. Letokhov, Zh. Eksp. Teor. Fiz. 51, 1669 (1966) [Sov. Phys. JETP 24, 1129 (1967)].

33. J. H. Parks, in this volume.

34. W. E. Lamb, Jr., Phys. Rev. 134A, 1429 (1964).

35. A. Szőke and A. Javan, Ref. 1. See also R. A. McFarlane, W. R. Bennett, Jr., W. E. Lamb, Jr., Appl. Phys. Letters 2, 189 (1963).

36. A. Szőke and A. Javan, Ref. 1. See also R. H. Cordover, T. Jaseja, and A. Javan, Ref. 4.

37. A. Szőke and A. Javan, Phys. Rev. 145, 137 (1966). R. H. Cordover and P. A. Bonczyk, Phys. Rev. 188, 696 (1969). For a theoretical analysis see: B. L. Gyorffy, M. Borenstein and W. E. Lamb, Jr., Phys. Rev. 169, 340 (1968).

38. J. Brochard and R. Vetter, J. Physique 1, 79 (1969); J. Brochard and R. Vetter, J. Physique 3, 250 (1967).

39. P. H. Lee and M. L. Skolnick, Appl. Phys. Letters 10, 303 (1967); V. S. Letokhov, ZhETF Pis. Red. 6, 597 (1967) [JETP Letters 6, 101 (1967)].

40. R. L. Barger and J. L. Hall, Phys. Rev. Letters 22, 4 (1969).

41.   G. R. Hanes and K. M. Baird, Metrologia 5, 32 (1969).
42.   R. G. Brewer, M. J. Kelly, and A. Javan, Phys. Rev. Letters 23, 559 (1969); M. J. Kelly, R. E. Francke, and M. S. Feld, Chem. Phys. 53, 2979 (1970).
43.   H. R. Schlossberg and A. Javan, Phys. Rev. 150, 267 (1966). For an errantum correcting misprints and minor algebriac errors, see Phys. Rev. A 5, 1974 (1972).
44.   T. W. Hänsch, M. D. Levenson, and A. L. Schawlow, Phys. Rev. Letters 26, 949 (1971).
45.   M. S. Feld, in this volume.
46.   A. L. Schawlow, in this volume.
47.   C. Freed and A. Javan, Appl. Phys. Letters 17, 53 (1970).
48.   R. G. Brewer, in this volume.
49.   A. Javan, Phys. Rev. 107, 1579 (1957).
50.   H. K. Holt, Phys. Rev. Letters 20, 410 (1968).
51.   M. S. Feld and A. Javan, Phys. Rev. 177, 540 (1969).
52.   G. E. Notkin, S. G. Rautian, and A. A. Feoktistov, Zh. Eksp. Teor. Fiz. 52, 1673 (1967) [Sov. Phys. JETP 25, 1112 (1967)].
53.   H. K. Holt, Phys. Rev. Letters 19, 1275 (1967).
54.   M. S. Feld and A. Javan, Bull. Am. Phys. Soc. 12, 1053 (1967).
55.   M. S. Feld and A. Javan, Phys. Rev. Letters 20, 578 (1968). See also; A. Javan, in Quantum Optics and Electronics, Les Houches, 1964, edited by C. DeWitt, A. Blandin, and C. Cohen-Tannoudji (Gordon and Breach, New York, 1965), p. 383; R. Rose and J. A. White, Bull. Am. Phys. Soc. 13, 172 (1968).
56.   T. W. Ducas, M. S. Feld, L. W. Ryan, Jr., N. Skribanowitz and A. Javan, Phys. Rev. A 5, 1036 (1972).
57.   B. J. Feldman and M. S. Feld, Phys. Rev. A 1, 1375 (1970); B. J. Feldman and M. S. Feld, Phys. Rev. A 5, 899 (1972).
58.   N. Skribanowitz, M. S. Feld, R. E. Francke, M. J. Kelly and A. Javan, Appl. Phys. Letters 19, 161 (1971). See also N. Skribanowitz I. P. Herman, R. M. Osgood, Jr., M. S. Feld, and A. Javan, Ref. 31.

NARROW RESONANCES INDUCED IN AN ABSORBING GAS

BY AN INTENSE LIGHT FIELD

V. S. Letokhov
Institute of Spectroscopy of the Academy of Sciences
Moscow, USSR

I. INTERACTION OF RADIATION WITH AN INHOMOGENEOUSLY-BROADENED LINE

The width of the absorption spectral lines of the atoms or molecules in a gas is determined by the joint action of three main effects:

1. radiative broadening: $\Delta\nu_{rad}$
2. collision broadening: $\Delta\nu_{coll}$
3. Doppler broadening: $\Delta\nu_{Dop}$

For atomic and molecular transitions in the visible and infrared range at low gas pressures ($\lesssim$ several Torr) the following inequality usually holds:

$$\Delta\nu_{Dop} >> \Delta\nu_{rad}, \ \Delta\nu_{coll} \tag{1}$$

i.e., the Doppler width significantly exceeds the radiative and collisional widths. The Doppler broadening is inhomogeneous, i.e., the spectral line location of each particle depends on its velocity $\vec{v}$ ($\omega = \omega_0 + \vec{k} \cdot \vec{v}$), and the spectral line of the whole ensemble of moving particles consists of the spectral lines of individual particles according to the velocity distribution

335

$W(\vec{v})$. The width of the absorption line for an individual particle is determined by the broadening due to radiative decay and to collisions, which limit the duration of the particles' interaction with the field, that is, the homogeneous broadening $2\Gamma = \Delta\nu_{coll} + \Delta\nu_{rad}$.

For inhomogeneous broadening, the light field interacts only with those particles in resonance with it. The fraction of these particles interacting with the field depends upon the ratio of the homogeneous and Doppler widths and, strictly speaking, upon the spatial configuration of the light field. If the monochromatic field is <u>isotropic</u> (for instance, if it is a set of waves traveling in different directions inside the scattering cavity), then all the particles can interact with the field [Fig. 1(a)]. On the other hand, for a <u>plane</u> traveling wave $\mathcal{E} \cos(\nu t - \vec{k} \cdot \vec{r})$ only those particles within the homogeneous width 2Γ with resonant frequency $\nu = \omega_0 + \vec{k} \cdot \vec{v}$ interact with the field [Fig. 1(b)]. In other words, only those particles having a definite velocity projection in the direction of the wave vector of the traveling wave interact with the field:

$$|\omega_0 - \nu + \vec{k} \cdot \vec{v}| \lesssim \Gamma \qquad (2)$$

Note that the resonance width depends upon the extent to which the field is a plane wave. The moving particle feels the light field in a certain macroscopic region. For example, for the case of a spherical wave of radius R, an additional broadening occurs [Fig. 1(c)] due to the change in the Doppler shift for each particle by an amount [1]:

$$\Delta\nu_{sp} \simeq \frac{ku}{\sqrt{kR}} \qquad (3)$$

where u is the average velocity of particles, $ku \simeq \Delta\nu_{Dop}$. Therefore the movement of the particle yields, in addition to the inhomogeneous Doppler broadening, a spatial or gemetrical broadening of the resonance.

A. Spatial (Geometrical) Broadening

This type of broadening is explained physically by the fact that because of the movement of the particle, the interaction with the field depends not only on its velocity $\vec{v}$, but also on its position $\vec{r} = \vec{r}_0 + \vec{v}(t - t_0)$. In the limiting case of a plane wave of beam diameter a, the curvature is determined only by diffraction $(R \sim a^2 k)$, and the spatial broadening according to (3) is reduced to transit broadening due to the definite time of transit through the beam, $\Delta\nu_{sp} \simeq u/a$. In the limiting case of an isotropic field with average radius of curvature $R \sim 1/k$, the spatial broadening according to (3) exceeds the Doppler width.

So, a monochromatic light wave with a small divergence, i.e., a light

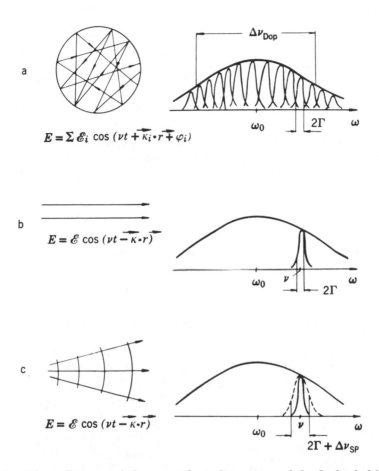

Figure 1.   The influence of the spatial configuration of the light field on resonant interaction width for a Doppler spectral line: (a) the isotropic monochromatic field; (b) the plane coherent wave; (c) the spherical coherent wave.

field with spatial and temporal coherence, can interact with a small fraction of the atoms or molecules within a Doppler-broadened transition. Consequently, this field can change the state and sharply discriminate this small fraction of particles with respect to the rest whose velocity does not satisfy the resonance condition (2).   Let the intensity of the light field be sufficient to transfer a certain fraction of particles into the excited state.   The probability of excitation into the upper level W_{12} of a particle with velocity $\vec{v}$ by a traveling wave $\mathscr{E} \cos(\nu t - \vec{k} \cdot \vec{r})$ in the simplest case of equal quenching of the upper and lower levels is

$$W_{12}(\vec{v}) = \frac{G^2}{2} \frac{\Gamma^2}{(\Omega - \vec{k} \cdot \vec{v})^2 + \Gamma^2(1 + G^2)} \tag{4}$$

where $\Omega = \nu - w_0$ is the detuning of the traveling wave frequency with respect to the transition frequency of a stationary particle, $G^2 = (\mu \mathcal{E}/\hbar\Gamma)^2$ is the saturation parameter of the transition, μ is the matrix element of the transition dipole moment.

The resonance width of the particles also depends on the degree of saturation. Therefore, there is another broadening mechanism, which will be described in the next section.

B. Power Broadening

The extent of power broadening according to (4) is determined by the expression

$$\Delta\nu = 2\,\Gamma\sqrt{1 + G^2} \tag{5}$$

This mechanism of broadening is well known in spectroscopy in the radio frequency range [2].

The excitation of particles with a definite velocity changes the equilibrium velocity distribution of particles in each level. In the velocity distribution of particles in the lower level, $N_1(\vec{v})$, there is a lack of particles with velocities satisfying the resonance condition:

$$N_1(\vec{v}) = N_1^0(\vec{v})\left(1 - \frac{G^2}{2} \frac{\Gamma^2}{(\Omega - \vec{k} \cdot \vec{v})^2 + \Gamma^2(1 + G^2)}\right) \tag{6}$$

where N_1^0 is the initial distribution of particles. However, in the upper level there is an excess of these particles. As a result the shape of the Doppler line is distorted. There appears a "hole" in it at the field frequency ν [3], the width of which is given by Eq. (5). This width is equal to the homogeneous width, corrected for power broadening (Fig. 2). The depth of the dip is equal to the difference between the initial (linear) and saturated (nonlinear) absorption coefficients and is determined by the degree of saturation. The saturated absorption coefficient is expressed by averaging over the Maxwell velocity distribution probability of excitation $W_{12}(\vec{v})$ given by Eq. (4) [4, 5]:

$$\kappa(\Omega) = \kappa_0(\Omega)/\sqrt{1 + G^2} \tag{7}$$

Thus, the light wave can create a deformation in the Doppler contour

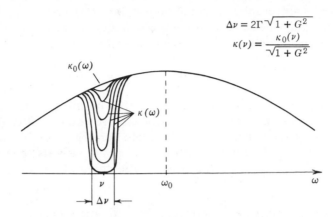

$$\Delta \nu = 2\Gamma \sqrt{1 + G^2}$$

$$\kappa(\nu) = \frac{\kappa_0(\nu)}{\sqrt{1 + G^2}}$$

Figure 2. The shape of a Doppler contour saturated by a strong traveling wave at frequency ν.

which contains information about the narrower homogeneous width.   To produce this condition the light wave should satisfy the following three conditions:

1. monochromaticity or temporal coherence;
2. directionality or spatial coherence;
3. intensity sufficient to saturate the transition.

The radiation of a continuous-wave gas laser [6] has all these properties, and this is why progress in the field of quantum electronics has resulted in the development of nonlinear superhigh-resolution spectroscopy within the Doppler line.

The possibility of "hole burning" within the amplification contour was mentioned in the first papers on lasers. Schawlow [7] noted this possibility for luminescent crystals whose spectral lines are inhomogeneously broadened at low temperatures. Bennett considered this effect in some detail for the Doppler-broadened emission profile of the gas laser [3].

Note that in the case of luminescent crystals, where the particles are stationary, a "hole" may be burned into the inhomogeneous distribution without the use of spatially coherent light (conditions 1 and 3 are sufficient). However, this "simplification" lessens the possibilities of using the narrow "holes" in luminescent crystals for high-resolution spectroscopy and other applications. The fact is that the formation of a narrow hole in the inhomogeneous distribution at the frequency of the external field is not sufficient for spectroscopy with resolution inside the inhomogeneous line.   In the

Doppler case one can achieve such resolution by finding the Doppler line center with an accuracy up to the hole width or by resolving, for instance, two overlapping spectral lines centered at ω_{01} and ω_{02} and with a separation which is considerably smaller than the Doppler width $\Delta\nu_{Dop}$:

$$\Gamma \lesssim |\omega_{01} - \omega_{02}| \ll \Delta\nu_{Dop} \tag{8}$$

In the case of a Doppler-broadened line it is possible to take advantage of the directionality with respect to molecular velocity by utilizing two suitably oriented plane traveling waves. The hole burning in the inhomogeneously-broadened line of the luminescent crystal is not sensitive to the spatial shape of the light field, and therefore all the effects considered below are absent.

II. NONLINEAR ABSORPTION OF TWO OPPOSITELY DIRECTED TRAVELING WAVES

Let the light field be a standing wave, i.e., the superposition of two plane traveling waves with the same frequency and amplitude:

$$E = \mathcal{E} \cos(\nu t - \vec{k}\cdot\vec{r}) + \mathcal{E} \cos(\nu t + \vec{k}\cdot\vec{r}) \tag{9}$$

In general, when such a field resonantly interacts with a Doppler-broadened absorbing (amplifying) transition two groups of particles whose velocity satisfies the following conditions are selectively saturated:

$$|\nu - \omega_0 \pm \vec{k}\cdot\vec{v}| \lesssim \Gamma \tag{10}$$

On the Doppler contour these two velocity groups occupy two spectral ranges located symmetrically with respect to the center of the line. When saturation occurs each traveling wave burns out its "hole" [Fig. 3(a)]. However, if the field frequency is tuned to the center of the line, then both traveling waves interact with the same group of particles. In this case the degree of saturation increases, and one deep hole is burnt out [Fig. 3(b)]. If the field frequency ν is scanned through the Doppler profile, one will observe a dip in the saturated absorption (amplification) coefficient as the field is tuned through line center. Note that in the case of an inhomogeneously-broadened transition of stationary atoms no such dip in the absorption profile would occur. This dip is very important, as it determines the center of the Doppler contour where the nonlinear interaction of the gas with the standing wave occurs.

In the case of weak saturation the dependence of the standing wave saturated absorption (amplification) coefficient as a function of frequency has the form

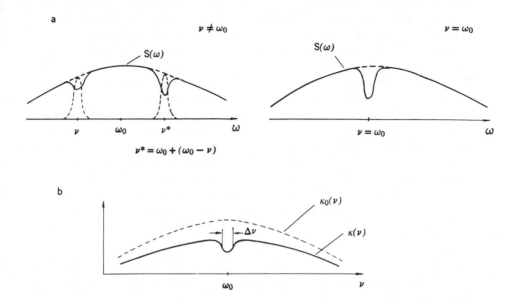

Figure 3.   The shape of a Doppler contour $S(\omega)$ saturated by a strong stand-ing wave (a) at frequency ν off line center ($\nu \neq \omega_0$, two separate holes) and (b) at line center ($\nu = \omega_0$, single joint hole);   (c) the coefficient of saturated absorption $\kappa(\nu)$ of standing wave as a function of standing wave frequency ν.

$$\kappa(\Omega) = \kappa_0(\Omega)\left[1 - \frac{G^2}{2}\left(1 + \frac{\Gamma^2}{\Gamma^2 + \Omega^2}\right)\right] \tag{11}$$

where $\Omega = \nu - \omega_0$ is the detuning of the field frequency with respect to the center of the line, 2Γ is the homogeneous linewidth determined by the total relaxation of particles in the upper and lower levels [Fig. 3(c)].   This expression can be obtained by calculating the resultant hole as a function of frequency as the two holes overlap.   Mathematically this is described by the convolution operation, and at first glance it seems that the resonance width at the center of the Doppler contour (for instance, the overlapping of two Lorentzian holes) will be equal to twice the width of one of the holes $2(2\Gamma)$, and not 2Γ.   In fact, this is not so, as the frequency separation between the two holes is twice the distance between the hole and the center of the line. As a result, in the general case the width of the dip $\Delta\nu_{dip}$ is described by a simple expression:

$$\Delta\nu_{dip} = \frac{1}{2}\left(\Delta\nu_{left\ hole} + \Delta\nu_{right\ hole}\right) \tag{12}$$

where $\Delta \nu_{\text{left hole}}$ and $\Delta \nu_{\text{right hole}}$ are the widths of the left and right holes.

The dip formation was considered for the first time by Lamb [8] in his analysis of the interaction of a standing wave in a laser cavity with a Doppler-broadened amplification line of an active gas medium. In the gas laser this effect leads to the production of an intensity minimum in the center of the amplification line when the laser frequency is tuned to line center. The existence of the resonant intensity minimum was shown experimentally [9, 10] with a classic device of quantum electronics -- the He-Ne laser [6]. Since that time the Lamb dip has been studied in detail in many papers. The theorists tried to extend Lamb's calculation first carried out for a weak degree of saturation [8] to the case of an intense field. The experimentalists by studying the Lamb dip at various lasing transitions obtained very important information concerning the parameters of the amplifying media. However, the most important and interesting results were obtained in the investigations aimed at achieving very narrow, frequency stable dips.

To achieve the extremely narrow and frequency stable resonance dips it was proposed to use the saturation of <u>absorption</u> (but not of amplification) of <u>molecular</u> transitions at <u>low gas pressures</u> [11]. This idea is very simple, but rather effective. The dip in the amplification line results in a resonance decrease of intensity, but a similar dip in the absorption line yields a resonant peak in intensity. This peak of transmission is often called the inverted Lamb dip. To use it, two facts are important. First, the dip in the absorption line can be made hundreds of times narrower than that in the amplification line. In fact, the absorption in contrast to the amplification can occur on transitions from the ground state, or near to it, to an excited, long-lifetime state. As a result, the radiative width can be negligibly small. Since the population of the ground state is relatively large without any excitation, the absorption can be sizable at very low pressure. Because of this the collisional width also becomes insignificant. Second, because of the low pressure and absence of excitation the frequency of the resonance peak can be made rather stable.

If the nonlinear absorbing cell is placed inside the laser cavity, then the total amplification of the two media has a resonance peak in the center of the absorption line ω_b with a width $2\Gamma_b$ (Fig. 4). This amplification peak results in a peak in the laser output. The first observations of the inverted Lamb dip were performed using a He-Ne laser with an absorbing cell in the cavity [12, 13]. As was mentioned above, it is most profitable to use a molecular absorbing cell [11, 13], since in this case one can obtain the most narrow and stable resonances. The first experiment of this type was carried out by Barger and Hall [14] using a He-Ne laser oscillating at 3.39 µm and a methane absorbing cell as proposed in Ref. 11 and 13 (the P(7) transition of the ν_3 band in CH_4 coincides with the 3.39 µm line of the He-Ne laser [15]). The width of the inverted dip achieved with this system was 0.3 MHz, i.e., the

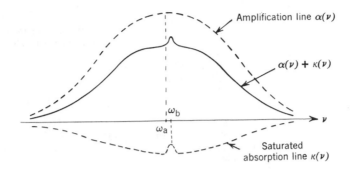

Figure 4. The shape of the total amplification of a two-component medium inside a cavity (amplifying medium and nonlinear absorbing gas cell).

width was 10^3 times less than the Doppler width of the methane absorption line.

To obtain the Lamb dip it is not necessary to use a standing wave. Moreover, it is not an optimum method for obtaining large amplitude peaks. The fact is that at strong saturation the inverted dip amplitude decreases and its maximum value is about 26%. This is due to the fact that at strong saturation the nonlinear absorption is very small and the overlapping of already deep holes can decrease the saturated absorption only slightly.   This disadvantage is eliminated by quasi-traveling wave techniques [16] using two waves of the same frequency but different amplitudes traveling in opposite directions:

$$E(t) = \mathcal{E}_1 \cos{(\nu t + \vec{k} \cdot \vec{r})} + \mathcal{E} \cos{(\nu t - \vec{k} \cdot \vec{r})} \qquad (13)$$

The traveling wave with amplitude $\mathcal{E}$ is strong, and it saturates the absorbing transition. The reverse wave is weak ($\mu\mathcal{E}_1/\hbar\Gamma \ll 1$) and does not cause saturation. The weak probe field linearly interacts with molecules within the homogeneous width at the mirror-image frequency of $\omega_0 + (\omega_0 - \nu)$. Only the strong field burns out a hole in the Doppler contour (Fig. 5). However, when the field frequency ν coincides with the center of the line, the weak field (reverse wave) interacts with the molecules saturated by the strong field (direct wave). As a result the weak field absorption in the center of the line sharply decreases, and a very sharp peak of transmission arises.

The dependence of the absorption coefficient of the reverse wave on frequency is described by a simple expression [17]:

$$\kappa(\Omega) = \kappa_0(\Omega)\left[1 - S\,\mathcal{L}\left(2\frac{\nu - \omega_0}{\Delta\nu_{dip}}\right)\right] \qquad (14)$$

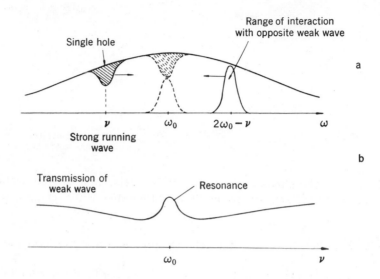

Figure 5. The formation of a narrow transmission resonance by a quasi-traveling wave: (a) the shape of the Doppler contour and (b) the transmission of the weak wave.

where $\Omega = \nu - \omega_0$, $\mathcal{L}(x) = 1/(1 + x^2)$ is the normalized Lorentzian form, $\Delta\nu_{dip}$ is the dip width at half maximum given by

$$\Delta\nu_{dip} = \Gamma_b(1 + \sqrt{1 + b\mathcal{E}^2}), \quad b\mathcal{E}^2 = G^2 \qquad (15)$$

S is the dip amplitude found using the relation

$$S = 1 - \frac{1}{\sqrt{1 + b\mathcal{E}^2}} \qquad (16)$$

Note that the dip width according to (12) is equal to the half-sum of the hole width $2\Gamma_b\sqrt{1 + b\mathcal{E}^2}$ burnt out by the strong field and the homogeneous width $2\Gamma_b$ corresponding to the interval of frequencies over which the weak field interacts.

The experimental data agree with the calculation for this simple model. Figure 6 illustrates the width dependence of the narrow molecular resonance within the Doppler line of the SF_6 molecule as a function of the degree of saturation. The experiment used the P(18) line of the CO_2 laser at 10.6 μm. This line coincides in frequency with a rotational-vibrational transition of

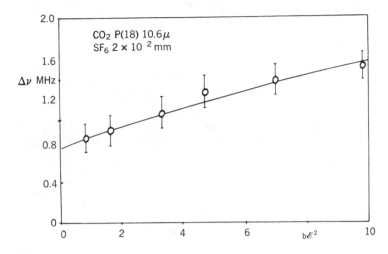

Figure 6. The power broadening observed by the quasi-traveling wave technique in SF_6 using the P(18) transition of the CO_2 laser at 10.6 μm (solid curve -- theoretical).

the ν_3 band of the SF_6 molecule [18, 19]. The solid curve is the theoretical dependence calculated according to (15).

The amplitude of the peak S according to (16) is determined by the difference between the absorption coefficients of the weak traveling wave off resonance and on resonance [$1/\sqrt{1+b\mathcal{E}^2}$ in units of κ_0, according to (7)]. The essential point is that the amplitude of the transmission peak grows monotonically as the strong wave intensity increases in contrast to the case of saturation of absorption by a standing wave. The peak amplitude has also been studied experimentally for the more general case of opposite traveling waves of arbitrary relative intensity [20]. The results of such an experiment carried out on the SF_6 molecule using the P(18) line of the CO_2 laser at 10.6 μm are given in Fig. 7. Here the amplitude of resonance A is denoted in units of the weak traveling wave intensity, i.e., $A = S\mathcal{E}_1^2$ where S is the amplitude of the peak in the absorption coefficient of the weak traveling wave. The solid curves are calculated using the relations describing the absorption difference off resonance and on resonance:

$$A = \mathcal{E}_1^2 \left[\frac{\kappa_0}{\sqrt{1+b\mathcal{E}^2}} - \frac{\kappa_0}{\sqrt{1+b(\mathcal{E}^2+\mathcal{E}_1^2)}} \right] \qquad (17)$$

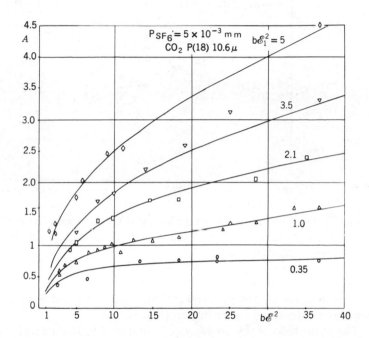

Figure 7.   Dependence of resonance amplitude on the degree of saturation $b\mathscr{E}^2$ by a strong wave for different degrees of saturation by an opposite wave (solid curves--theoretical)[20].

It is seen that this rather simple model is in good agreement with experiment.

In these experiments a resolution of 0.30–0.50 MHz is achieved. As an illustration of the possibilities of this method, Fig. 8 shows the structure of an anomalously broadened resonance observed with the P(16) line of the CO_2 laser at 10.6 μ in the ν_3 band of the SF_6 molecule. At a low pressure of SF_6 (5 mTorr) in a long cell (L = 120 cm) it is possible to resolve the structure of this resonance [17, 20]. It consists of three components of different intensities which are separated by 1.0 and 1.25 MHz. The width of the outer resonances at half maximum is 0.3 MHz, while the width of the central

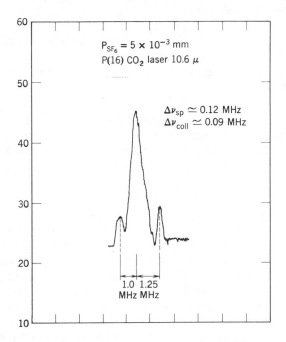

Figure 8.   The structure of an anomalously broadened resonance in SF_6 observed using the P(16) line of the CO_2 laser (10. 6 μm) with a resolution of 0.2 MHz.

resonance is 1.3 MHz.   The central resonance probably consists of several unresolved components.   In this experiment the resolution is limited by geometrical (spatial) broadening and collisional broadening.   The geometrical broadening is due to the finite diameter a of each beam and the small angle φ between them:

$$\Delta\nu_{sp} \simeq \frac{u}{a} + \varphi\frac{u}{\lambda} \tag{18}$$

where u is the average velocity of molecules and λ is the wavelength.   In this experiment u = 2 × 10^4 cm/sec (SF_6 at 300°K), a = 0.5 cm, φ = 4 × 10^{-3} rad yielding a total geometrical broadening of $\Delta\nu_{sp} \simeq$ 0. 12 MHz. The collisional broadening was $\Delta\nu_{coll} \simeq$ 0. 09 MHz at 5 × 10^{-3} Torr pressure SF_6 (the collisional broadening is equal to 17 kHz/mTorr [17, 21]).   The total width of the resonance is 0. 2 MHz which is close to the experimental value, taking into account some power broadening.

III.   NONLINEAR ABSORPTION OF TWO SPATIALLY SEPARATED
       TRAVELING WAVES

At a very low gas pressure the collisional mean free path of the molecules
$\ell_{f.p.}$ can become larger than the diameter a of the light beam.   In this case
the hole burnt in the Doppler contour by the traveling wave is spread in
space by a distance of the order of the mean free path of the molecules.   The
result is that two spatially separted traveling waves can interact resonantly
with each other because of molecules crossing both beams without collisions.
In other words, the absorption of one beam depends on the field in the other
beam.   This effect occurs if the distance between the beams d and their
diameters a_1 and a_2 is less than the mean free path of the molecules:

$$d + a_1 + a_2 \lesssim \ell_{f.p.} \tag{19}$$

This situation resembles the problem concerning the interaction of a
beam of atoms or molecules with two separated oscillating fields considered
by Ramsey [22].   However, in the optical range there exist essential differences.   First, due to Doppler broadening only a small fraction of the molecules interact with the beam, and second, the size of the spatial region of
interaction is several orders of magnitude larger than the wavelength.
Therefore in the optical case it is not possible to make use of the known
results of Ramsey, and it was necessary to make the calculation again [23].
       The polarization of the molecules in the second beam crossing from the
first beam can be represented in three parts:

$$P_2(\mathcal{E}_1, \mathcal{E}_2) = P_2^{self}(\mathcal{E}_2) + P_2^{conv}(\mathcal{E}_1) + P_2^{mix}(\mathcal{E}_1, \mathcal{E}_2) \tag{20}$$

The self polarization P_2^{self} depends only on the field in the second beam.   It
describes the interaction (the hole burning and polarization) with molecules
whose velocity satisfies the condition

$$\left| w_0 - \nu + \vec{k}_1 \cdot \vec{v} \right| \lesssim \Gamma \tag{21}$$

where 2Γ is the linewidth determined by the transit time of the molecules
through the beam.   The convective polarization P_2^{conv} is taken into the
second beam from the first one by those molecules whose velocity satisfies
the resonance condition

$$\left| w_0 - \nu + \vec{k}_2 \cdot \vec{v} \right| \lesssim \Gamma \tag{22}$$

The mixed polarization P_2^{mix} depends on both fields.   The power absorbed
by the molecules in the second beam, which have crossed both beams is
described by the relation

$$W = \langle E_2 \frac{dP_2}{dt} \rangle_{t,\vec{r},\vec{v}} \tag{23}$$

where it is necessary to perform the averaging over time, over the region of interaction $\vec{r}$ (over the diameter and the length of beams), and over the molecular velocity $\vec{v}$ (over the transit time through the beam and the Doppler contour).

The interaction of the beams is essentially different for the case of parallel and antiparallel traveling waves. In the case of parallel traveling waves $(\vec{k}_1 = \vec{k}_2)$ the cross absorption is sensitive to the phase difference between the beams Ψ [Fig. 9(a)] and is described by the following expression [23]:

$$\kappa_{12} = \kappa_0 \left[1 - (b_1 \mathcal{E}_1^2 + b_2 \mathcal{E}_2^2) + f\left(\frac{d}{a}\right) \frac{\mathcal{E}_1}{\mathcal{E}_2} \left(\mathcal{E}_1^2 + \mathcal{E}_2^2 \right) \cos \Psi \right] \tag{24}$$

The value of the interference term depends on the distance between the beams d and the ratio of the amplitudes of the fields. The dependence on the distance between the beams is rather sharp. The function $f(d/a)$ is almost zero at $d \simeq a$. This is explained by the fact that the molecules belonging to the hole of the first beam expand in a transverse direction by an amount $\Delta z \simeq d\lambda/a$ during transfer into the second beam, even if they started from the same point in the first beam. For $d > a$ the amount of expansion is $\Delta z > \lambda$ and a spatial averaging of the interference effect takes place. The

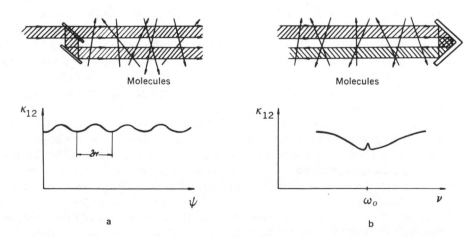

Figure 9. The cross absorption of two beams as a function of (a) phase difference for parallel waves and (b) frequency for antiparallel waves.

enhancement of the effect for $\mathcal{E}_1 \gg \mathcal{E}_2$ is explained by the fact that in the strong field of the first beam molecules are being significantly polarized, and this polarization becomes predominant in comparison with that induced by the weak second beam. At the same time the cross absorption of the parallel waves does not depend upon the detuning of the field frequency with respect to the center of the Doppler line. This is quite natural as both beams interact with the same group of molecules on the Doppler contour.

In the case of antiparallel traveling waves ($\vec{k}_1 = -\vec{k}_2$) the interference effect vanishes due to averaging over the region of interaction of length $\gg \lambda$, but in this case the cross absorption depends on the detuning $\Omega = \nu - \omega_0$ of the field frequency with respect to the center of the line (Fig. 9(b))[23]:

$$\kappa_{12} = \kappa_0 \left\{ 1 - b_2 \mathcal{E}_2^2 - b_1 \mathcal{E}_1^2 \left[1 - g\left(\Omega \frac{2a}{u} \right) \right] \right\}$$ (25)

where $g(x)$ is an even function having its maximum ($g = 1$) at the point $x = 0$ and decreases as x increases with a half-width $x \simeq 1$. This function describes the resonant decrease of the cross absorption at the center of the line, which can be considered as the Lamb dip in the spatially separated opposite traveling waves. The resonance width is two times less than the dip width in a standing wave of the same diameter. Note that for the interaction of the separated antiparallel traveling waves, in contrast to the parallel waves, there is no need to have coherence of interaction. For such interaction the hole burning in the first waves is sufficient, i.e., it is similar to the case of the quasi-traveling wave method considered above.

The experimental observation of the nonlinear resonance and interference effects in the separated beams is very difficult since very low pressures are necessary. Under these conditions, the amount of absorption is rather small and the apparatus should be highly sensitive and stable. Nevertheless, an experiment was recently carried out with separated antiparallel beams of a CO_2 laser in a low-pressure SF_6 cell [24]. The scheme of the experiment is shown in Fig. 10. The CO_2 laser operated on the P(16) line of the 10.6 μm band. For this line the SF_6 molecule has a large absorption coefficient ($\kappa_0 = 1.3$ cm^{-1} Torr^{-1}), which allows one to work at low pressures of SF_6, where the mean free path of the molecules is greatest. The laser radiation was sent into an external SF_6 cell 120 cm long, in the form of two spatially separated opposite traveling waves. The intensity modulated wave $I_1(\Omega)$ saturated the absorption. The opposite wave I_2 was a weak beam probing the state of molecules in the neighborhood of the strong beam. The frequency was slowly scanned across the Doppler absorption contour of SF_6. A narrow transmission peak appeared at the center of the line. However, this effect can occur without "spatial transfer" of a hole, since due to diffraction there is always an overlapping region of two waves, i.e., the region of the usual standing wave. To discriminate the effect connected

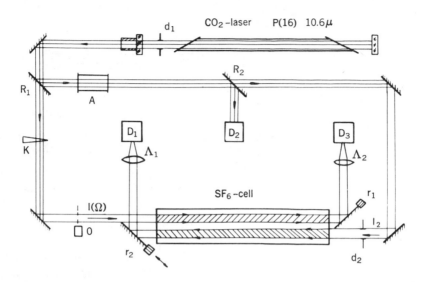

Figure 10.   The experimental apparatus for the observation of a narrow
resonance induced by two spatially-separated opposite beams (d_i--dia-
phragms, D_i--photodetectors, R_i--semitransparent mirrors, r_i--movable
mirrors, Λ_i--lens, K--optical wedge, A--attenuator in form of SF_6 high-
pressure absorbing cell).

with the hole transfer, the dependence of the resonance peak amplitude on
the distance between the beams d at <u>various pressures</u> of SF_6 was measured.
The experimental curves normalized per unit maximum intensity are given
in Fig. 11.   In the region of relatively higher pressures the peak arises due
to the inevitable diffraction overlapping of the beams.   However, with
decreasing pressure the dependence of the resonance amplitude shifts into
the region of large distances between the beams.   This shift is equal to
approximately 0.8 mm at a pressure of 4.5 mTorr, and it is inversely
proportional to pressure.   This is explained naturally by the "hole" transfer
from one wave to another in a distance equal to the mean free path of the
molecules ($\ell_{f.p.}$ = 4.5 ± 1.0 mm mTorr).   If the pressure is further
decreased to $2-3 \times 10^{-4}$Torr, one can probably observe the nonlinear interference effect for the interaction of parallel traveling waves.
     Note that for hole transfer, broadening should occur due to collisions
with small scattering angles.   For hole broadening even weak collisions
with scattering at an angle $\alpha = \Gamma/ku$ are essential, i.e., $\alpha \simeq \lambda/a$ at very
low pressures.   In our opinion, the investigation of the separated beams is

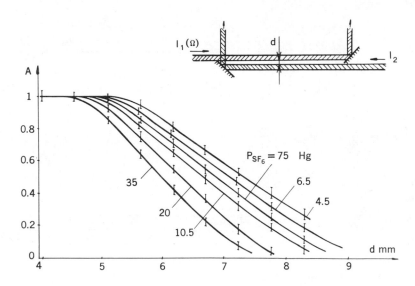

Figure 11. The amplitude of the narrow dip induced by spatially separated opposite beams, saturating an absorbing cell of SF_6 as a function of distance d between the beams for different SF_6 pressures.

very promising for the study of these effects by the nonlinear absorption method.

IV.   NARROW AMPLIFICATION PEAKS WITHIN THE DOPPLER
       ABSORPTION LINE

In the interaction of a coherent light wave with an absorption line one can obtain not only a narrow hole, but also can invert the level populations within a narrow frequency region of the Doppler contour to produce a narrow amplifying resonance. This possibility was considered in 1965 [25], before the work on narrow Lamb dips in saturated absorption. The idea was as follows.

    Let the molecules traverse a coherent light beam, the frequency of which is within the Doppler absorption line. If the mean free path of the molecules $\ell_{f.p.}$ is much larger than the diameter of the beam a, and if the absorption line corresponds to the transition of molecules to an excited long lived state, then the molecules interact with the field coherently during the whole transit time $\tau \simeq a/u$. When the field of the light wave satisfies the condition

$$\frac{\mu\mathcal{E}}{\hbar}\,\tau \simeq \pi \tag{26}$$

where μ is the transition dipole moment, a population inversion of the
molecules in resonance with the field takes place. This inversion is similar
to the inversion of the spin population by a $180°$ pulse, well known in nuclear
magnetic resonance. In the present case, the pulse character of inversion
is achieved because of a definite transit time of the molecules crossing the
optical beam, and not because of the switching on of the field. The inhomogeneous character of the absorption line broadening causes the deformation
to occur only over a narrow portion of the gain profile (Fig. 12).

In this way molecules satisfying the resonant condition and the inversion
condition (26) are pumped into the amplifying phase. Note that in such an
experiment the amplified wave occurs after the molecules have traversed
the pump beam. It is not so obvious that this scheme can be applied to gas
molecules at thermal equilibrium, since the inverted molecules which
traverse the light beam are always accompanied by noninverted molecules.
However, even in this case amplification can be achieved by means of a
"tubular" light beam [26] (Fig. 13). In such a geometry all the molecules
entering the internal hollow region of the beam first transit the light field,
inverting the population of the molecular levels if condition (26) is satisfied.

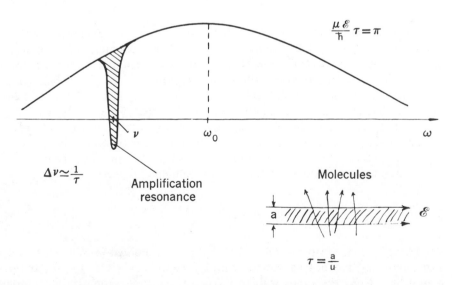

Figure 12.   The formation of a narrow amplification resonance in a Doppler
absorption line by a "$180°$ pump pulse".

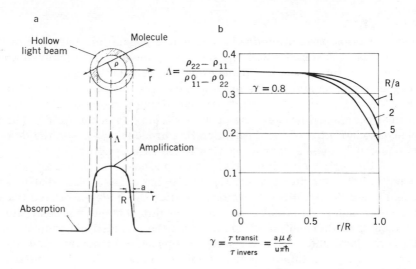

Figure 13. The formation of a narrow amplification resonance for the interaction of low-pressure gas molecules with a hollow light beam (R--inner
radius, R + a--external radius): (a) geometry of amplification region;
(b) radial distribution of inversion degree Λ.

It is worth noticing that the "180° inversion condition" cannot be satisfied
simultaneously for all molecules inside the beam because of frequency
detuning, thermal distribution of the molecular velocities, and the various
transit times for molecules with ρ > 0 (see Fig. 13).

The spatial inversion distribution in the beam cross section is illus-
trated in Fig. 13. The maximum inversion $\Lambda = (\rho_{22} - \rho_{11})/(\rho_{11}^{0} - \rho_{22}^{0})$
(ρ_{ii} and ρ_{ii}^{0} are the populations of the i-th level in and out of the field,
respectively) is achieved in the center of the beam when the inversion time
$\tau_{inv} = \pi \hbar/\mu \mathscr{E}$ is chosen approximately equal to the time necessary for
molecules to penetrate the tubular shell (shaded region, Fig. 13(a); i.e.,

$$\gamma = \frac{\tau_{tr}}{\tau_{inv}} \simeq 1 \qquad (27)$$

where $\tau_{tr} = a/u$, with $u = (2kT/m)^{1/2}$ the most probable molecular velocity.
For particles in exact resonance with the field ($\omega_{0} - \nu + \vec{k} \cdot \vec{v} = 0$) the value of
Λ is 0.36. Near the inner wall of the beam the inversion diminishes due to
the increase in the fraction of molecules entering into the beam with a large

shift ρ with respect to the center of the beam. The extent of this decrease depends upon the ratio of the internal radius R of the beam to the thickness a of its wall. Inversion is not achieved outside the beam because of the presence of molecules which do not transit the light beam.

The degree of inversion Λ depends essentially upon the ratio of the inversion time τ_{inv} to the transit time τ_{tr}, i.e., upon the parameter γ. Due to the difference in the thermal velocities of the molecules, the maximum possible inversion $\Lambda = 1$ is not achieved. For a gas in thermal equilibrium interacting with a tubular light beam the degree of inversion depends on the parameter γ as shown in Fig. 14 (dotted curve) for the case of a two-level non-degenerate system. In fact, molecular rotational-vibrational transitions are usually highly degenerate. Because of this level degeneracy several assemblies of two-level particles with various dipole moments μ_m interact with the field. This degeneracy also makes it difficult to produce simultaneous population inversion of all molecules satisfying the resonance condition. This problem was considered in detail in Ref. 27. The dependence of inversion on the parameter γ for P, Q, and R-branch transitions with angular momentum J = 10 is shown in Fig. 14. It is evident that the inversion conditions noticeably differ for the various branches. The maximum inversion Λ_{max} also depends on the angular momentum J of the transition, as shown in Fig. 15. For J = 1 a significant inversion is achieved only for the

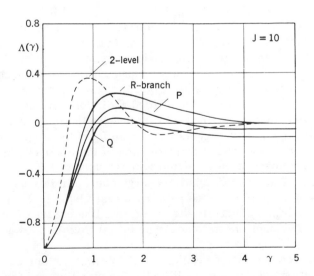

Figure 14. The degree of population inversion of a molecular gas as a function of the saturation parameter $\gamma = \tau_{tr}/\tau_{inv}$ for a nondegenerate two-level system (dotted curve) and for $J = 10 \rightarrow J' = 9, 10, 11$ P, Q, R-branch transitions.

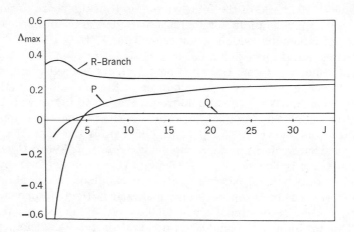

Figure 15.   The maximum degree of population inversion in a molecular gas as a function of angular momentum J for P, Q and R branch transitions.

R branch.   As J increases the values of Λ_{max} for P and R branches approach each other and become smooth.   Furthermore, Λ_{max} for the Q branch is considerably less than that for the P and R branches.   This is explained by the different distributions of μ_m for the P, Q, and R branches.   For P and R branches the value of μ_m is concentrated near the definite value μ_o, and for the Q branch μ_m is distributed more uniformly.

The narrow amplifying peaks induced by the coherent beam in the gas absorption line allow one to create an unusual type of cw laser [25, 26] operating in a two-level scheme (Fig. 16).   This is a gas laser which is coherently pumped by its own amplified radiation.   In the low-pressure gas absorption cell the tubular pump beam inverts the molecules.   The inverted molecules in the hollow region of the beam form the amplifying medium of the laser.   After significant amplification of the laser output by an optical amplifier, a telescope and a diaphragm are used to form the hollow tubular pump beam.   The amplifier is the energy source in this oscillator.   The properties of such a laser are rather unusual [28]: the frequency of oscillation is automatically tuned to the absorption line center of the gas with high accuracy.   For oscillation to occur an initial field is necessary with an amplitude $\mathcal{E} > 0$ and so on.

The narrow amplifying peaks are of interest, in our opinion, both for high-resolution spectroscopy within the Doppler line and for the development of gas lasers with high frequency stability.

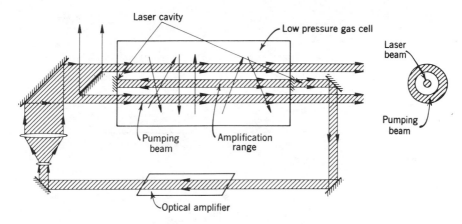

Figure 16.   The principal scheme for a two–level cw gas laser which is pumped by its own amplified radiation.

V.   NARROW RESONANCES IN THE TOTAL LEVEL POPULATION

In Sec. IV we considered the formation of a narrow dip in the absorption line due to the interaction of gas molecules with a light field in several different configurations (standing wave, quasi–traveling wave, spatially separated waves, and hollow beam).   Simply speaking, the formation of this dip is a result of the fact that the depth A_0 of the joint hole at line center is greater than the average depth A_{av} of the two holes in the off–resonance case:

$$A_0 > \frac{1}{2} (A_{left} + A_{right}) = A_{av} \tag{28}$$

For example, in the standing wave case for weak saturation ($G^2 \ll 1$)

$$A_0 = 2G^2 > A_{av} = \frac{1}{2} (G^2 + G^2) \tag{29}$$

In the quasi–traveling wave case considered in Sec. III:

$$A_0 = \left(1 - \frac{1}{\sqrt{1+G^2}}\right) > A_{av} = \frac{1}{2}\left[\left(1 - \frac{1}{\sqrt{1+G^2}}\right) + 0\right] \tag{30}$$

Can resonant changes occur for the total population in the upper or lower levels as the laser frequency is tuned through the Doppler line center? Strictly speaking, this effect does not depend on the Lamb dip because it is determined not by the shape of the Doppler contour distorted by the strong

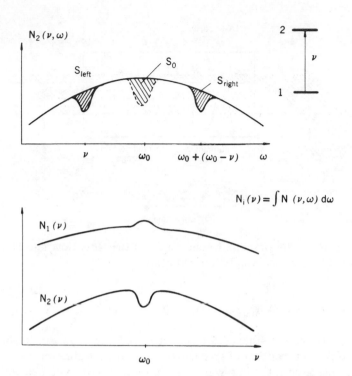

Figure 17. The formation of a narrow resonance of total population at lower $N_1(\nu)$ and upper $N_2(\nu)$ levels. (ν--frequency of standing wave, ω_0--Doppler line center)

field but by the integral over the whole Doppler line. For example, the total population of the upper level is determined by the net <u>area</u> of the holes. The total level population will undergo a resonant change at Doppler line center if the area of the joint hole, S_0 on resonance ($\nu = \omega_0$) differs from the total area S_{tot} of left and right holes of the off-resonance case ($\nu \neq \omega_0$) [Fig. 17 (a)]:

$$S_0 < S_{tot} = S_{left} + S_{right} \qquad (31)$$

In the case of a weakly saturating standing wave or a quasi-traveling wave at any level of saturation, $S_0 = S_{tot}$ and as a result the effect vanishes. However for a strong standing wave such an effect occurs at saturation due to $(b\mathcal{E}^2)^2 = (G^2)^2$ and higher-order terms, where $b\mathcal{E}^2$ is the saturation parameter. This result can be easily proven by noting that the extent of saturation by a standing wave at line center is twice as large as that off line center.   Then the ratio of hole areas on resonance and out of resonance is given by

$$\frac{S_0}{S_{tot}} \sim \frac{(2\Gamma\sqrt{1+2G^2})(1-1/\sqrt{1+2G^2})}{2(2\Gamma\sqrt{1+G^2})(1-1/\sqrt{1+G^2})} = \frac{1}{2}\frac{\sqrt{1+2G^2}-1}{\sqrt{1+G^2}-1} \tag{32}$$

The depth of the dip at low saturation is proportional to $(G^2)^2$, that is, it appears at higher saturation than the Lamb dip. As the degree of saturation increases the depth of the dip approaches the value $1 - 1/\sqrt{2} \simeq 0.3$ asymptotically. Such behavior distinguishes this effect from the Lamb dip.

The total population of the lower level, in contrast, has a resonant peak at the Doppler line center [Fig. 17(b)]. This time, its relative magnitude is less by a factor of $2\Gamma/\Delta\nu_{Dop}$ because the lower level is ordinarily highly populated.

The resonant change in the total level populations can be detected by monitoring the intensity change of the total spontaneous emission from either level. This possibility was noted first in Ref. 29, and the application of this technique for detecting narrow resonances was named the "nonlinear-fluorescent cell method." The observation of the narrow resonances by means of the spontaneous emission intensity is a very useful method for gases with a low coefficient of absorption, since it is always advantageous to use fluorescence techniques for detection of small absorption variations. In particular, in Ref. 29 it was noted that this method is applicable for detection of a narrow resonance in the Doppler line of an atomic or molecular beam, when the absorption is extremely small. The proposal of the nonlinear-fluorescent cell method was made independently in a successful experiment performed by Freed and Javan [30]. In this experiment a CO_2 absorbing cell was saturated at low pressure by CO_2 laser radiation. The absorption coefficient of CO_2 at room temperature is approximately 10^{-6} cm^{-1}. This method is applicable for any of the P- and R-branch lines of CO_2, N_2O, and other molecular lasers. A useful feature in the molecular case is the fact that molecules in all the collisionally-coupled rotational levels of the excited vibrational state contribute to the spontaneous emission [30].

Let us note here that the change in the total number of molecules in the upper or lower level can be detected by other methods. In particular, it is possible to measure the absorption coefficient for coupled transitions between the upper level of the saturated transition and a higher state, including transitions to the continuum (Fig. 18). These transitions can have larger cross sections and shorter wavelengths and therefore the nonlinear effects can be observed with increased sensitivity. This method allows one to obtain additional information about the upper state even if the transition overlaps those of other states because of Doppler broadening. The resonant

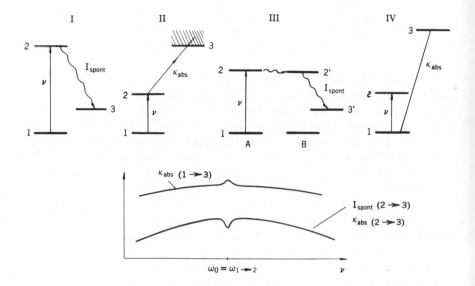

Figure 18. Different methods for the observation of resonant change of total population at upper and lower levels: I--the intensity of total spontaneous emission; II--the absorption to upper states, including continuum; III--the intensity of spontaneous emission for a two component molecular gas; IV-- the absorption of molecules in the lower level.

change of total level population can also be used to enhance the selectivity of the two-step photoionization and photodissociation processes [31]. This effect allows one to eliminate a limitation of the selectivity caused by the Doppler broadening. It is also possible to add another gas which has a fast luminiscent level close to the excited level of the saturated transition. In this case the narrow resonance in the total number of upper-level molecules of the absorbing gas can be observed through changes in the intensity of luminescence of the added molecules. This modification is useful for absorbing molecules having low probability of luminescence.

VI. NARROW OPTO-ACOUSTIC RESONANCES

Narrow resonances induced by the saturated absorption of a low-pressure gas can be detected without the use of ordinary techniques of infrared

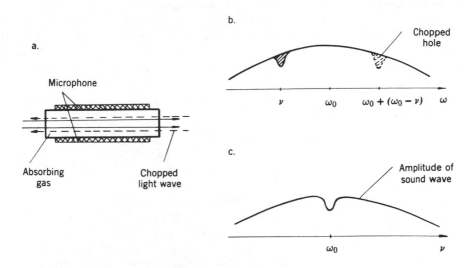

Figure 19. The formation of a narrow opto-acoustic resonance: (a) the experimental scheme; (b) the Doppler contour; (c) the dependence of acoustic wave intensity versus frequency of light waves.

radiation detection. For this one can use the opto-acoustic effect, i.e., the generation of acoustic waves in the absorbing gas when the saturation is produced by an intensity-modulated light beam [32]. For this purpose the absorbing cell is illuminated by two opposite waves of the same frequency, one of which is intensity modulated at an audio frequency Ω [Fig. 19(a)]:

$$E(t) = \mathcal{E} \cos(\nu t - \vec{k} \cdot \vec{r}) + \mathcal{E} \cos(\Omega t) \cos(\nu t + \vec{k} \cdot \vec{r}) \qquad (33)$$

Because of absorption there is bulk heating of the gas and a subsequent increase in the gas pressure. The intensity of the acoustic wave is determined by the modulated part of the absorbed radiation power, i.e., by the depth of the modulated hole in the Doppler contour [Fig. 19(b)]. As the laser is tuned through the line center this value decreases resonantly because of the absorption saturation by the nonmodulated direct wave. The result of the narrow opto-acoustic resonance is shown in Fig. 19(c).

The advantage of the opto-acoustic technique is its high sensitivity. The sensitivity is determined by the detection threshold of the acoustic wave, that is by the Brownian motion of a microphone diaphragm ($I_{thr} \simeq 10^{-9}$ W [33]). On the other hand, the dip formation requires a radiation intensity approximately equal to the absorption saturation intensity I_{sat}.

Hence the modulated part of the intensity equals about $0.1\ I_{sat}$. The absorption saturation intensity for a low-pressure molecular gas equals $I_{sat} \simeq 10^{-2} - 10^{-1}\ W/cm^2$ and therefore it is possible to detect the narrow resonance with an absorption coefficient of $10^{-6}-10^{-7}$ for a single transit of the gas cell.

The serious disadvantage of the opto-acoustic method in the case of low-pressure molecular gases is the long detection time. In order to generate an acoustic wave it is necessary that the vibrational-translation energy transfer rate be satisfied by

$$W_{VT} \gg \Omega \tag{34}$$

For low-pressure molecular gases W_{VT} is very small. For example, in $SF_6\ W_{VT} > 5 \times 10^3\ sec^{-1}$ at 1 Torr pressure [34, 35]. Therefore for the gas cell at 10^{-2} Torr the modulation frequency must be less than 10Hz. On the other hand, at a low pressure the cooling rate is large because of a significant value for the thermal diffusivity coefficient. The result of this process is that a significant part of the energy is expended in heating the walls and not in the bulk heating of the gas. In principle this difficulty can be eliminated by using the heating of the cell walls for acoustic effect.

VII.   APPLICATIONS OF NARROW RESONANCES

Narrow resonances in the Doppler absorption line have at least two important applications. (1) The practical application is the absolute high-precision stabilization of laser frequencies. (2) The fundamental application is nonlinear high-resolution spectroscopy and precise measurements. Let us consider these very briefly.

The narrow, frequency-stable resonance can be used for the high-precision stabilization of laser frequencies. It was for this practical purpose that the first proposals for lasers with nonlinear low-pressure gas absorbing cell were made [11-13]. The inverted Lamb dip, which is produced by saturating a gas inside an absorption cell in the laser cavity, produces a peak in the laser output at line center. This peak can be used to stabilize the laser frequency by means of a servo system. The first such experiment was performed [36] with the He-Ne laser at $\lambda = 6328$ Å and a Ne absorbing cell [36]. The long-term frequency stability achieved in such a system was 10^{-9}. As mentioned above, molecular rotational-vibrational transitions are much more attractive for achieving narrow, frequency-stable resonances. The first such experiment was performed succesfully in Ref. 14 with the He-Ne laser at $\lambda = 3.39$ μm with an internal low-pressure CH_4 cell. The width of the inverted Lamb dip was 0.3 MHz, i.e., 10^3 times less than the Doppler width. It was found that the pressure shift of the P(7) transition of the ν_3 band of CH_4 is abnormally small (17 kHz $Torr^{-1}$). The methane system has a long-term stability of 10^{-13} and a reproducibility better than 10^{-11} [37]. This is the best result achieved so far in the optical range.

The narrow resonance of the effective amplification in a gas laser with an internal nonlinear absorbing cell can significantly stabilize the laser frequency without using an automatic system. In Refs. 11, 38, and 39, it was shown that such a laser operating at the proper conditions (narrow width and large amplitude of the inverted Lamb dip) can have strong nonlinear frequency pulling because of the narrow effective amplification resonance. This effect of "self-stabilization of frequency" is important for achieving high laser frequency stability for the following reason. The linewidth of a maser amplifier is much less than the bandwidth of the cavity. This feature guarantees a relatively small maser frequency shift for significant instability of the cavity's resonance frequency Ω_{cav} [Fig. 20(a)]. In lasers the reverse situation ordinarily occurs and the laser frequency varies significantly due to instability in the optical length of the Fabry-Perot cavity [Fig. 20(b)]. However in a laser with an internal nonlinear absorbing cell a situation similar to that of the maser case can be achieved [Fig. 20(c)]. It is important also that the time constant of self-stabilization is very small (10^{-5}-10^{-6} sec). Therefore this effect is useful in suppressing fast frequency fluctuations which cannot be eliminated by a servosystem because of the relatively large time constants of the feedback loop.

Combining the internal and external absorbing cell methods, that is, the output power peak and the frequency self-stabilization, and the narrow resonances of absorption and spontaneous emission, the frequency stabilization of all important cw lasers in the visible, infrared, and submillimeter regions with a reproducibility better than 10^{-11} will undoubtedly be performed in the near future. With these methods one can build a set of good secondary frequency and wavelength laser standards whose frequency can be compared with a primary laser standard using nonlinear techniques of frequency conversion. It is quite conceivable to look forward to the creation of a primary laser frequency standard with a frequency reproducibility of 10^{-13}-10^{-14}. However the discovery of such a system requires a great deal of routine work and, of course, a little good luck.

The discovery of the narrow resonance in the Doppler contour also provided possibilities for the development of nonlinear high-resolution molecular spectroscopy. At present this field is developing using lasers which have near coincidences with molecular resonances, occasionally using the Stark and Zeeman effects to bring the molecular resonance into coincidence with the laser frequency [40-43]. Therefore the molecules studied are random. Significant progress in frequency-tunable infrared lasers has occurred recently. The spin-flip Raman laser [44, 45] is particularly promising. By means of such tunable lasers the systematic investigation of

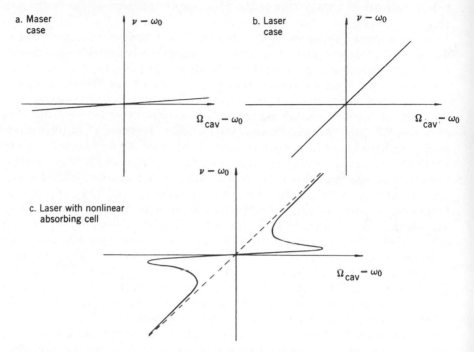

Figure 20. Dependence of oscillation frequency ν on cavity frequency Ω_{cav}: (a) maser case, $\Delta\nu_{ampl} \ll \Delta\nu_{cav}$; (b) ordinary laser case, $\Delta\nu_{ampl} \gg \Delta\nu_{cav}$; (c) laser with nonlinear absorbing cell in the self-stabilization frequency regime [ω_0--central frequency of amplification (absorption in case b) line; $\Delta\nu_{ampl}$--width of amplification line, $\Delta\nu_{cav}$--bandwidth of cavity].

rotational structure of molecular vibrational states with resolution of 10^8-10^9 can begin.

Using a set of laser secondary frequency standards and nonlinear techniques of absolute frequency measurement [46] it is possible to measure molecular transitions with an accuracy better than 10^8-10^9. Let us note that traditional techniques of linear absorption spectroscopy by means of unique instruments have an accuracy of 10^6 [47]. Of course, the measurements of molecular transition frequencies with an accuracy of 10^8-10^9 allows an increase in the precision of speed of light measurements by the spectrocopic method [48].

The possibility of precision measurement of the transition frequencies

of several simple molecules (for example, H_2, HD, D_2) is very attractive.
If theorists can calculate the transition frequencies of these molecules ν_i
with the same accuracy, i. e. , to express them through the fundamental
constants $\nu_i = f_i(e, \hbar, c,$ m, M), it will be possible to obtain more precise
values for the fundamental constants. To enhance the precision it is possible
to use a number of independent transitions, greater than the number of fundamental constants. The absorption saturation techniques are applicable to
the HD molecule. As to homonuclear molecules (H_2 and D_2) their rotational
and vibrational transitions can be measured by Raman scattering techniques
with relatively low precision because of the high pressures necessary in such
experiments. The progress of cw high-power visible lasers will allow one
to use lower pressures and therefore increase the accuracy. Moreover, by
means of strong fields it is possible in principle to observe induced infrared
absorption [49].

REFERENCES

1. V. S. Letokhov, Zh. Eksp. Teor. Fiz. 56, 1748 (1969) [Sov. Phys.
   JETP, 29 937 (1969)].
2. R. Karplus and J. Schwinger, Phys. Rev. 73, 1020 (1948).
3. W. R. Bennett, Jr., Phys. Rev. 126, 580 (1962).
4. S. G. Rautian, Doctoral dissertation (Fiz. Inst. Akad. Nauk, 1966);
   Pro. FIAN 43, XXX (1968).
5. D. H. Close, Phys. Rev. 153, 360 (1967).
6. A. Javan, W. R. Bennett, Jr., and D. R. Herriott, Phys. Rev. Letters
   6, 106 (1961).
7. A. L. Schawlow, Advances in Quantum Electronics, edited by
   J. R. Singer (Columbia University Press, New York, 1961), p.50.
8. W. E. Lamb, Jr., Phys. Rev. 134A, 1429 (1964).
9. R. A. McFarlane, W. R. Bennett, Jr., and W. E. Lamb, Jr., Appl.
   Phys. Letters 2, 189 (1963).
10. A. Szöke and A. Javan, Phys. Rev. Letters 10, 521 (1963).
11. V. S. Letokhov, ZhETF Pis. Red. 6, 597 (1967) [JETP Letters 6, 101
    (1967)].
12. P. H. Lee and M. L. Skolnick, Appl. Phys. Letters 10, 303 (1967).
13. V. N. Lisitsyn and V. P. Chebotayev, Zh. Eksp. Teor. Fiz. 54, 419
    (1968)[Sov. Phys. JETP 27, 227 (1968)].
14. R. L. Barger and J. L. Hall, Phys. Rev. Letters 22, 4 (1969).
15. M. S. Feld, J. H. Parks, H. R. Schlossberg and A. Javan, Physics
    of Quantum Electronics, edited by P. L. Kelley, B. Lax, and
    P. E. Tannenwald (McGraw-Hill, New York, 1966).
16. V. S. Letokhov and V. P. Chebotayev, ZhETF Pis. Red. 9, 364 (1969)

[JETP Letters 9, 215 (1969)].

17. N. G. Basov, O. N. Kompanets, V. S. Letokhov and V. V. Nikitin, ZhETF Pis. Red. 9, 568 (1969) [JETP Letters 9, 345 (1969)]; Zh. Eksp Teor. Fiz. 59, 394 (1970) [Sov. Phys. JETP 32, 214 (1971)]; Preprint #37, P. N. Lebedev FIAN (1970).

18. O. R. Wood and S. E. Schwartz, Appl. Phys. Letters 11, 88 (1967).

19. H. Brunet and M. Perez, Compt. Rend. 267, 1084 (1968).

20. O. N. Kompanets and V. S. Letokhov (to be published).

21. P. Rabinowitz, K. Keller and J. T. LaTourrette, Appl. Phys. Letters 14, 376 (1969).

22. N. F. Ramsey, Molecular Beams (Clarendon Press, Oxford, 1956).

23. V. S. Letokhov and B. D. Pavlik. Abstracts of Allunion Symposium on Gas Laser Physics, Novosibirsk, June, 1969; Preprint #140, P. N. Lebedev FIAN (1969); Optics and Spectroscopy 32, 455 (1972); Optics and Spectroscopy 32, 573 (1972).

24. O. N. Kompanets and V. S. Letokhov, ZhETF Pis. Red. 14, 20 (1971) [JETP Letters 14, 12 (1971)].

25. N. G. Basov and V. S. Letokhov, ZhETF Pis. Red. 2, 6 (1965) [JETP Letters 2, 3 (1965)

26. N. G. Basov and V. S. Letokhov, ZhETF Pis. Red. 9, 660 (1969) [JETP Letters 9, 409 (1969)]; Preprint #80, P. N. Lebedev FIAN (1969).

27. V. S. Letokhov, B. D. Pavlik and S. P. Fedoseev, Preprint #86, Institute of Spectroscopy, Moscow (1971); Optics and Spectroscopy (to be published).

28. V. S. Letokhov and B. D. Pavlik, Zh. Eksp. Teor. Fiz. 53, 1107 (1967)[Sov. Phys. JETP 26, 656 (1968)]; Zh. Tekh. Fiz. (to be published).

29. N. G. Basov and V. S. Letokhov, Proceedings of URSI Conference on Laser Measurements, Warsaw, September, 1968; Electron Technology 2, 15 (1969).

30. C. Freed and A. Javan, Appl. Phys. Letters 17, 53 (1970).

31. R. V. Ambartsumian and V. S. Letokhov, Reports on OSA/IEEE III International Conference on Laser Engineering and Applications, Washington, June, 1970; IEEE J. Quantum Electron (to be published).

32. M. L. Veingerov, Proc. Acad. Sci. USSR 19, 687 (1938); Zavodskai Laboratory 13, 426 (1947).

33. L. B. Kreuzer, J. Appl. Phys. 42, 2934 (1971).

34. I. Burak, A. V. Nowak, J. I. Steinfeld and D. G. Sutton, J. Chem. Phys. 51, 2275 (1969).

35. J. D. Lambert, D. G. Parks-Smith and J. L. Stretton. Proc. Roy. Soc. (London) A282, 380 (1964).

36. V. P. Chebotayev, I. M. Beterov and V. N. Lisitsyn, IEEE J. Quantum Electron. QE-4, 788 (1968).

37. R. L. Barger and J. L. Hall, Proceedings of the 23rd Annual Symposium on Frequency Control, Fort Monmouth, New Jersey, May 1969, p. 306.
38. V. S. Letokhov, Zh. Eksp. Teor. Fiz. 54, 1244 (1968) [Sov. Phys. JETP 27, 665 (1968)].
39. V. S. Letokhov and B. D. Pavlik, Kvantovaya Elektronika 1, 53 (1971) [Sov. J. Quant. Electron. 1, 36 (1971)]; Preprint #96, P. N. Lebedev FIAN (1970).
40. R. G. Brewer, M. J. Kelly and A. Javan, Phys. Rev. Letters 23, 559 (1969).
41. R. G. Brewer, Phys. Rev. Letters 25, 1639 (1970).
42. E. E. Uzgiris, J. L. Hall and R. L. Barger, Phys. Rev. Letters 26, 289 (1971).
43. T. W. Hänsch, M. D. Levenson and A. L. Schawlow, Phys. Rev. Letters 26, 946 (1971).
44. C. K. N. Patel, E. D. Shaw and R. J. Kerl, Phys. Rev. Letters 25, 8 (1970).
45. A. Mooradian, S. R. J. Brueck and F. A. Blum, Appl. Phys. Letters 17, 481 (1970).
46. A. Javan, Ann. N. Y. Acad. Sci. 168, 715 (1970).
47. D. H. Rank, D. P. Eastman, B. S. Rao and T. A. Wiggins, J. Opt. Soc. Am. 52, 1 (1962).
48. D. H. Rank, J. Mol. Spectr. 17, 50 (1965).
49. E. U. Condon, Phys. Rev. 41, 759 (1932).

LASER SATURATION SPECTROSCOPY IN COUPLED DOPPLER-
BROADENED SYSTEMS: HOW TO FIND A NEEDLE IN A HAYSTACK

M. S. Feld
Department of Physics, Massachusetts Institute of Technology
Cambridge, Massachusetts

ABSTRACT

The theory and applications of laser-induced line narrowing in coupled
Doppler-broadened systems are reviewed, and several new features are
presented. The theoretical discussions include effects due to intense fields,
standing-wave fields, and systems with degenerate levels. The applications
discussed include spontaneous emission line narrowing, mode crossing, and
optical pumping leading to unidirectional gain. In the latter application it is
shown that amplification can be achieved even in the absence of population
inversion, and this is verified experimentally. Future applications are also
discussed.

I. INTRODUCTION

The advent of lasers has opened a new chapter in the spectroscopy of atomic
and molecular gases. Previously, spectral resolution was limited by the
thermal motion of the atoms themselves (Doppler broadening), even with the
best instruments available. It is now possible to produce spectral lines
hundreds of times narrower by means of laser saturation techniques, allowing unsurpassed resolution. In the following discussion the principles of

369

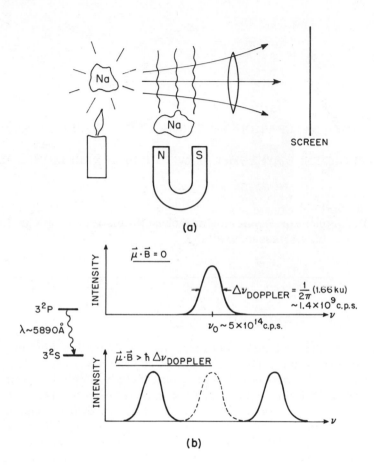

Figure 1. Professor Zeeman's experiment.

these techniques and several examples of their application will be presented

To illustrate the limitations of conventional spectroscopy, consider the
way Professor Zeeman demonstrated the Zeeman effect to his class [1],
using only a magnet, a Bunsen burner and some sodium. One piece of sodium
is used as a source of yellow "D-line" light. The second piece, placed in a
strong magnetic field, serves as an absorber. A lens is used to focus the
source image onto a screen [Fig. 1(a)]. When the magnetic field is off, the
intense D-line radiation, produced by transitions between the 3^2P level and
the ground level, is strongly reabsorbed. Therefore, little light is projected
on the screen. However, if the magnetic field is strong enough to split the

Figure 2. Zeeman experiment. The insert shows the projection of the
magnet poles and sodium vapor on the screen. The author is grateful to
Harry Anderson of MIT for providing this photograph.

Zeeman components by the order of a Doppler width, the absorption lines
split apart, so that the second piece of sodium no longer acts as an absorber
[Fig. 1(b)]. Accordingly, the source radiation now illuminates the screen.

A modified version of the experiment is shown in Fig. 2. Here, the
source is a sodium lamp (left-hand side). The absorber, placed between the
poles of an electromagnet, is sodium vapor produced by gently heating a drop
of sodium acetate solution (placed at the tip of a glass rod) with a bunsen
burner. In the projected image (inset, Fig. 2) absorbing sodium vapor
extends above and below the shadow of the magnet poles. Between the poles,
where the field is intense, the yellow light is transmitted; above and below
it is strongly absorbed. This is indeed a dramatic way to demonstrate the
Zeeman effect without a spectrometer.

The point of this discussion is that ordinary spectroscopic techniques
are inherently limited in resolution by Doppler broadening. In the case of
the sodium D-lines the Doppler width is about 1400 MHz, requiring a magnetic field of at least 1000 G. Laser saturation techniques are much more

powerful: In a similar experiment using lasers which is discussed later, it will be seen that the Zeeman components can be resolved with a magnetic field of only a few gauss.

Linear Spectroscopy

In the spirit of review, it may be instructive to begin with some early laser spectroscopy experiments (1962-64) in which Doppler widths did limit the resolution. In these experiments [2] a 3.39-μ He–Ne laser, which can be magnetically tuned over a few thousand MHz, was used to probe the absorption spectra of simple organic molecules in the 3-μ region. These absorption lines all correspond to the C–H stretching vibration at 2960 cm^{-1}. To increase the sensitivity, and also to obtain additional information on molecular parameters, molecules were chosen which exhibited strong Stark effect. The sample gas was subjected to an applied electric field containing a small audio–frequency component (Fig. 3). The detected signals were fed into a phase–sensitive amplifier tuned to the Stark modulation frequency. The wide band noise is thus rejected, leading to an increase in signal-to-noise by a factor of 1000.

Some typical results are shown in Figs. 4 and 5. Figure 4 shows the Stark absorption spectrum of formic acid at 200 mTorr pressure. A 170-cm absorption path length was used. The intensity change due to Stark modulation is plotted as a function of the laser frequency. Two lines are present, at about 150 and 1000 G (100 G ~ 140 MHz). By this technique, the linewidths and center frequencies of the resonances could be ascertained to 1:10^7.

Figure 5 compares the Stark modulation spectrum of methyl acetylene with the direct absorption spectrum [3] (no Stark modulation). At about

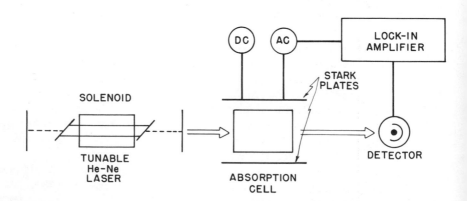

Figure 3. Laser spectrometer, 1964.

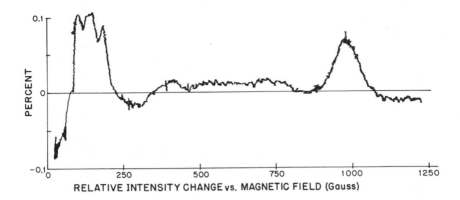

Figure 4. Formic acid (HCOOH), Stark-modulation spectrum near 3.39 μ.
In this trace the dc Stark field is 2.0 esu, the ac Stark field is 4.0 esu,
the sample cell pressure is 0.2 Torr.

900 G one line is seen both via Stark modulation and directly. However, a
second line near 100 G, present in the direct trace (upper trace), is not
observed in the Stark-modulated trace (lower trace). The modulation technique, therefore, actually simplifies the spectrum, emphasizing only those
lines having strong Stark effects.

So here you see the Stark-modulation spectra of infrared molecular
transitions, taken with one of the earliest laser spectrometers. The
Doppler-broadened linewidths of the sample gases, which are clearly
resolved, limit the resolution to a few hundred MHz. Nevertheless, the line
shapes of different gases are quite distinctive and "molecular fingerprints"
such as these might be useful today in analytical applications such as air
pollution studies.

II. LASER-INDUCED LINE NARROWING

Although this was a useful technique, there was the feeling that one could do
better. The reason is because in the optical-infrared region the spectral
response of an ensemble of atoms or molecules moving with a given velocity
is much narrower than the Doppler width. This response is determined by
relaxation processes such as collisional deexcitation and radiative decay.
The resulting linewidth, γ, is referred to as the "homogeneously broadened"
width, since atoms of all velocities relax, on the average, at the same
rate [4]. The net response is built up from the contributions of individual

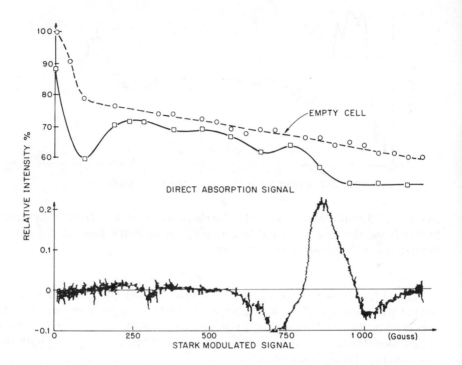

Figure 5.   Methyl acetylene (CH_3C_2H), direct and Stark-modulation spectra near 3.39 μ.   The sample cell pressure is 0.1 Torr.   The upper trace show the direct absorption spectrum (no Stark modulation).   The lower trace show the Stark-modulation spectrum using a dc Stark field of 5.67 esu and an ac Stark field of 11.3 esu.

ensembles over the entire velocity distribution.   Due to the Doppler effect, the response of an ensemble moving with velocity v is offset from the atomi center frequency ω by an amount kv, where k=ω/c is the propagation constant (Fig. 6).   The center frequencies are thus spread over a range ~ ku, where u is the average atomic speed.   At pressures of interest (a few Torr or less) ku ≫ γ.   Accordingly, the contributions from individual velocity ensembles are smeared together, producing a broad Doppler profil in which the information about the spectral response of individual atoms is lost.   Broadening due to the Doppler effect is referred to as "inhomogeneous," since different atomic velocity ensembles resonate at different frequencies.

The lost information can be recovered, however, by means of laser

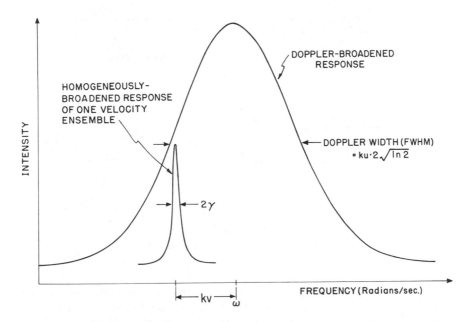

Figure 6.   Doppler-broadened molecular resonance.   For a Maxwellian
velocity distribution of molecules of mass M at temperature T, the full-
width of the intensity profile at half maximum is given by $ku \sqrt{\ln 2}/\pi$ (in
cycles/sec), where $u = (2\kappa T/M)^{1/2}$ and κ is Boltzmann's constant.

saturation techniques.   This is possible because as the laser field traverses
an absorbing or amplifying gaseous medium it produces an atomic beam,
narrow in velocity spread [5].   Consider an intense monochromatic traveling-
wave laser field resonating with a Doppler-broadened transition (Fig. 7).   To
be definite, assume the medium to be in the amplifying phase.   As the laser
beam is amplified, atoms are transferred from the upper energy level of the
transition to the lower level.   However, these changes occur only over a
narrow section of the velocity profile, since only atoms which are Doppler
shifted into resonance with the applied field can couple strongly to it.   Thus,
by virtue of selective saturation, the laser produces a bump on the thermal
velocity distribution of the lower level as measured along the axis of the
laser field, and a notch at the corresponding portion of the thermal velocity
distribution of the upper level (Fig. 7).   These changes are centered at a
velocity $v_0 = (\Omega - \omega)/k$, where Ω is the laser frequency, and extend over a
range $\Delta v \sim \gamma/k$.
     These selective population changes can lead to correspondingly narrow

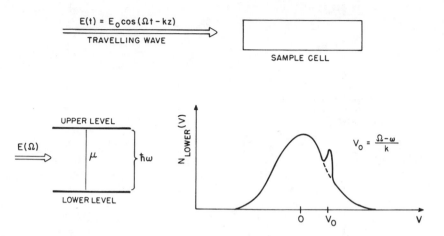

Figure 7.   Laser saturation effect.

spectral resonances.   Consider, for example, the spontaneous emission
arising from a transition formed by the lower level of the saturated transition and a third, lower level [Fig. 8(a)].   Viewed along the axis of the laser
field, the velocity bump will manifest itself as a narrow resonant increase
in emission (change signal) superposed upon the broad Doppler background
signal [Figs. 8(b) and 8(c)].   The spectral width of this change signal will be
of the order of γ and is completely determined by the homogeneous relaxation processes.   The Doppler broadening, whose spectral width is of the
order of ku, is thereby essentially eliminated, making possible a considerable improvement in spectral resolution.

     This narrow resonance is an example of an effect called laser-induced
Doppler line narrowing, which deals with the resonant interaction of intense
laser fields with Doppler-broadened three-level systems [6].   In the above
example the atomic system considered was in the cascade configuration.
Note that similar behavior would occur in a folded configuration, i.e., where
the spontaneous emission arising from the upper level of the saturated transition was studied [Fig. 9(c)].   In the latter case, however, the change signal
would appear as a resonant decrease in the background Doppler profile.   Of
course, if the saturated transition were in the absorbing phase rather than
the amplifying phase, the change signals would be opposite in sign to those
described.

     It should also be noted that observation of the effect is not limited to
spontaneous emission alone: Identical change signals would be observed if
one studied the spectrum of emission (or absorption) at the coupled transition, stimulated by a weak monochromatic probe field collinear with the

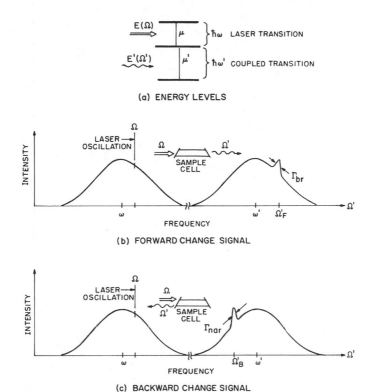

Figure 8.   Laser-induced line narrowing change signals. a) Energy level diagram. b) Forward change signal. c) Backward change signal.

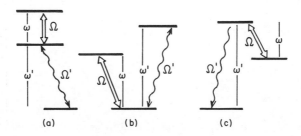

Figure 9.   Energy level configurations.   (a) is a "cascade" configuration, (b) and (c) are "folded" configurations.

377

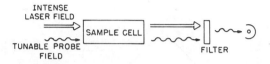

(a) SPONTANEOUS EMISSION VERSION

(b) STIMULATED EMISSION VERSION

Figure 10.   Experimental arrangements for observing laser-induced line-narrowing effects. a) Spontaneous emission version.   b) Stimulated emission version.

laser field (Fig. 10). As will be seen below, the ability to observe the line-narrowing effect both in stimulated emission and spontaneous emission considerably extends the usefulness of the technique.

The laser-induced line-narrowing effect, its principles and applications, are the subject of the present article. Other closely related effects, such as the Lamb dip, enable narrow resonances to be observed in two-level Doppler-broadened systems [7, 8]. These effects, which are based on related manifestations of the saturation effect, will not be discussed here. Further details may be found in other articles in this volume by Letokhov, Hall, Schawlow, and Javan.

The exact location of the change signal follows directly from the Doppler effect. If the emission is viewed in a direction parallel to the propagation direction of the laser field (forward direction) the peak frequency will occur at $\Omega'_F = \omega' + k'v_0$, where ω' is the atomic center frequency of the coupled transition and $k' = \omega'/c$. Thus, in terms of the frequency of the applied laser field,

$$\Omega'_F = \omega' + (\Omega - \omega)\frac{k'}{k} \tag{1}$$

[Fig. 8(b)]. If, on the other hand, the emission is viewed in a direction opposite to the propagation direction of the laser field (backward direction), the peak frequency will occur at $\Omega'_B = \omega' - k'v_0$, so that

$$\Omega'_B = \omega' - (\Omega - \omega)\frac{k'}{k} \tag{2}$$

[Fig. 8(c)].   Therefore, if the laser is tuned by an amount Δ above the atomic center frequency ω, the forward change signal will appear shifted above the corresponding center frequency ω' of the coupled transition by an amount $(k'/k)\Delta$, and the backward change signal will be shifted below ω' by the same amount.   Note the inherent anisotropy of the effect, since forward and backward change signals occur at distinct frequencies (except for $\Delta = 0$).   As will be seen below, this anisotropy can be usefully exploited.

By studying forward and backward change signals together it is possible to determine the atomic center frequency ω' with an accuracy limited only by γ.   In such experiments it is sometimes of interest to utilize a laser field in the form of a standing wave, as could be produced, for example, by reflecting a traveling-wave laser field back upon itself (Fig. 11).   The standing-wave field, at frequency ω, can be decomposed into oppositely directed traveling waves of equal amplitude at the same frequency.   As before, the traveling-wave component propagating in the positive direction will couple to the ensemble of atoms with axial velocity v_0.   Similarly, the oppositely directed traveling-wave component will resonate with the atomic ensemble having axial velocity $-v_0$.   Therefore, in the standing-wave case the forward and backward change signals will appear together at frequencies Ω'_F and Ω'_B, respectively, symmetrically located about ω' (Fig. 11).

The simultaneous appearance of a pair of change signals is not, in itself, surprising.   What is, perhaps, surprising is the fact that these change signals are not identical in shape, the backward signal being taller and narrower than the forward signal [Figs. 8(b), 8(c), and 11].   Similar behavior would occur in a folded system, but the broad and narrow signals would be interchanged.

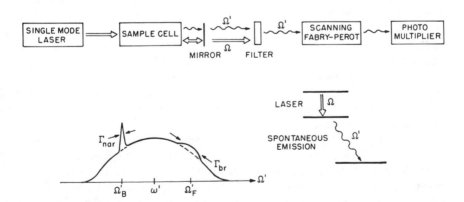

Figure 11.   Standing wave arrangement for observing line-narrowing effects in spontaneous emission.   The laser field is put in the form of a standing wave by means of a partially reflecting mirror.

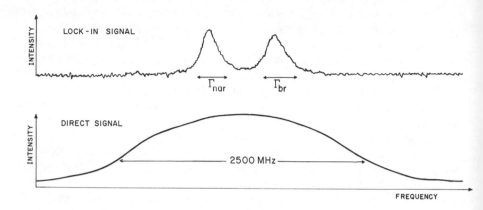

Figure 12.   Experimental traces of standing-wave change signals in spontaneous emission. The experimental arrangement is that of Fig. 11. The
change signals (upper trace) are compared with the direct signal (lower
trace), which consists of the change signals (too small to be seen) superimposed on the broad Doppler profile.

An experimental result using the standing-wave arrangement of Fig. 11
is shown in Fig. 12. In this case the laser transition is the 1.15-μ Ne transition of the He–Ne laser. The frequency spectrum of the spontaneous
emission at 0.61 μ, which is coupled to the lower laser level, is studied by
means of a pressure-scanned Fabry–Perot interferometer. In the lower
trace, which shows the direct spontaneous emission output signal from the
Fabry–Perot, the broad (~ 2500 MHz) Doppler background signal almost
completely obscures the narrow change signals. In the upper trace this
background has been suppressed by means of a lock-in technique. The width
and height differences between forward and backward signals can be clearly
seen. In this example the atomic parameters are such that the asymmetry
is small, but in other cases it can be quite sizeable. The width difference
gives an additional experimental parameter for detailed determination of the
relaxation processes.

This asymmetry in width, initially observed by Holt [9], is the first of
a number of line-shape deformations predicted by the theory. They are due
to the Raman-type processes which occur in a coupled three-level system,
and would not be expected from the preceeding discussion, which was based
on population saturation alone. These processes cause the photons emitted
at the laser transition and the coupled transition to be correlated in frequency. In the cascade system of Fig. 9(a), for example, they lead to the
frequency condition

$$\Omega + \Omega' = \omega + \omega' \tag{3}$$

This equation is the double-resonance condition, requiring that the net energy carried away by a pair of photons simultaneously emitted at frequencies Ω and Ω', respectively, be equal to the separation between upper and lower energy levels. Note that this condition is less restrictive than the corresponding condition for population saturation

$$\begin{aligned}\Omega &= \omega \\ \Omega' &= \omega'\end{aligned} \tag{4}$$

which requires that both of the fields resonate with their respective transitions. Consider an atomic ensemble moving with velocity v. In a coordinate system also moving at v, a laser photon at frequency Ω will appear Doppler shifted to Ω-kv. Therefore, in the moving frame photons at the coupled transition obeying the double-resonance condition (3) will have frequencies close to $\omega+\omega' - \Omega+kv$. As discussed earlier, the frequency of these photons, as observed in the laboratory frame, depends on their relative propagation direction. In the forward direction it is given by

$$\Omega'_{+}(v) = -\Omega + (\omega + \omega') + (k+k')v$$

The corresponding frequency in the backward direction is

$$\Omega'_{-}(v) = -\Omega + (\omega + \omega') + (k-k')v$$

Note that $\Omega'_{+}(v = v_0) = \Omega'_F$ and $\Omega'_{-}(v = v_0) = \Omega'_B$, i.e., for the atomic ensemble with velocity v_0 the frequency condition for Raman-type processes is the same as the population saturation condition. Further note that the velocity dependence of Ω'_{-} is much weaker than that of Ω'_{+}, especially when k and k' are similar in magnitude. In fact, if k' is sufficiently close to k (the condition is that $|k'-k|u \ll \gamma$), Ω'_{-} becomes essentially velocity independent, whereas the velocity dependence of Ω'_{+} remains strong. Therefore, for atoms over the entire velocity distribution Raman-type processes in the backward direction will occur at Ω'_B, spread over a narrow frequency range $\sim |k-k'|u$, whereas in the forward direction they will be spread over a broad range $\sim (k+k')u$, which is about equal to the Doppler width itself. Thus, in the backward direction, where a substantial portion of the atomic velocity distribution contributes near Ω'_B, Raman-type processes will considerably influence the change signal. In the example of Fig. 12 this influence causes a narrowing and an increase in height of the backward change signal. But in the forward direction, where the influence of Raman-type processes is spread over the entire Doppler profile, the change signal is affected to a lesser degree, if at all [10]. As will be seen below, in other cases even more striking distortions of the shape of the backward change signal can occur.

In a folded system such as Fig. 9(b) or 9(c), the frequency condition corresponding to Eq. (3) is the familiar Raman condition,

$$\Omega' - \Omega = \omega' - \omega \tag{5}$$

In this case reasoning similar to that given above shows that the influence of Raman-type processes is most pronounced in the <u>forward</u> direction, so that the forward change signal narrows, whereas the backward signal remains the same [10].

A similar line-shape asymmetry occurs in spontaneous Raman scattering in gases [11]. In these experiments, which study three-level systems formed by two closely spaced levels in the ground state coupled to a common level in an excited state [Fig. 9(c)], an applied pump laser field at frequency Ω can produce Stokes radiation at frequency Ω' by means of double-quantum Raman transitions between the ground-state levels. It is found that the spectral distribution of Stokes radiation scattered into the forward direction is reduced considerably from the Doppler width, while the radiation scattered into the backward direction remains broad in frequency. This asymmetry is also due to the frequency correlation between pump laser and Stokes photons, as in Eq. (5). In this case, however, absorption due to single-quantum transitions is negligible, since the energy separation between the intermediate level and the lower levels is quite large, so that the individual fields are far from resonance with their respective transitions [i.e., Eq. (5) is satisfied but Eq. (4) is not]. In the present case, where the common level is resonant, the occurrence of single-quantum transitions at the coupled transition must be taken into account. Furthermore, the single-quantum transition probability is considerably modified by the presence of an intense field resonating with the laser transition. Interestingly enough, it is found [6] that the expression for the transition rates for both single-quantum and double-quantum processes contain factors which become resonant when the Raman (or double-resonance) condition is fulfilled. The term Raman-type processes, as used above, is intended to describe resonant behavior of this type occurring in both single- and double-quantum processes.

III.   RESULTS OF THE THEORY

The theory of the laser-induced line-narrowing effect has been treated by several authors over the past few years. The stimulated version of the effect, in which both electromagnetic fields are treated classically, has been emphasized in our work at MIT [6, 12–14]. In a complementary approach,

Rautian and his collaborators have treated the spontaneous emission version of the effect, considering the radiation emitted at the coupled transition to be quantized [15]. As mentioned above, the two approaches give equivalent results. A number of other treatments on various aspects of the theory have also appeared [16-22].

The stimulated emission formulation is perhaps the more direct one. In one approach [6, 14, 20] the response of a two-level system to an intense laser field is taken as an unperturbed solution to the problem. The influence of a weak probe field resonating at the coupled transition may then be considered as a small perturbation, which can be conveniently expressed in terms of the unperturbed solution. This approach is advantageous in that the influence of a fully saturating laser field, either in the form of a traveling wave [6, 20] or a standing wave [14] is readily obtainable.

Complete discussions of the theoretical formalism are given elsewhere. The present discussion is limited to a summary of the major results of the theory, with their applications in mind. Portions of the line-shape expressions are quoted below. Complete expressions can be found in Refs. 6, 14, and 23.

A. Traveling-Wave Features

The line-shape effects may be subdivided according to whether the laser field is in the form of a traveling wave or a standing wave. Let us consider the traveling-wave results first. The following discussion is given in terms of a cascade configuration [Fig. 9(a)], where the backward change signal is narrower than the forward one. The same results hold for folded configurations [Fig. 9(b) and 9(c)], except that in this case the narrow change signal occurs in the forward direction and the broad change signal occurs in the backward direction. Note, however, that in both folded and cascade systems the center frequencies of forward and backward change signals occur at Ω_F' and Ω_B', respectively, as given by Eqs. (1) and (2).

In the forward direction, where Raman-type processes tend to be smeared over the entire Doppler profile, the change signal behaves in a manner more or less expected on the basis of population saturation considerations alone [24]. The shape of this change signal is a Lorentzian of full width

$$\Gamma_{br} = \gamma[1 + (2\frac{k'}{k} + 1)Q] \tag{6}$$

(in angular frequency units), where γ is the decay rate of one of the atomic levels (for simplicity, all assumed to be equal here) and Q is the dimensionless saturation parameter

$$Q = \sqrt{1 + (\frac{\mu E}{\hbar \gamma})^2} \qquad (7)$$

where μ is the dipole matrix element of the laser transition and E is the amplitude of the laser field [Figs. 13(a) and 14(a)]. For weakly saturating laser fields ($Q \approx 1$) $\Gamma_{br} \approx 2\gamma(1 + k'/k)$. With increasing laser intensity the change signal increases in size until a limiting magnitude is reached. At the same time the linewidth broadens due to the saturation of the laser transition. The size of the signal is also proportional to the population difference, as measured in the absence of applied fields, between the levels of the laser transition. Accordingly, when these level populations are equal the forward change signal vanishes.

In the backward direction, where Raman-type processes considerably influence the change signal, the situation is more complex, and the line shape depends upon k'/k, the ratio of the frequencies of laser and coupled transitions. For $k' > k$ the line shape is again Lorentzian in shape, of full width

$$\Gamma_{nar} = \gamma[\, 1 + (2\frac{k'}{k} - 1)\, Q] \qquad (8)$$

and again its magnitude is proportional to the population difference between the levels of the laser transition [Fig. 13(b)]. This resonance also broadens with increasing laser saturation. Note, however, that the backward signal is always narrower than the forward signal by $2\gamma Q$. The height of the backward change signal is also somewhat greater than that of the forward signal. An interesting property of these change signals is that their areas (i.e., intensity x frequency) are the same. Note that the width difference between forward and backward change signals is most pronounced when k and k' are comparable in magnitude, as anticipated in the previous discussion of Raman-type processes.

When $k' < k$ the shape of the backward change signal becomes more complex. In this case the change signal is composed of two contributions. One, called the laser-transition change signal, is proportional to the population difference between the levels of the laser transition. The second, called the background change signal, is proportional to the population difference between the levels of the coupled transition. As before, the population differences referred to are measured in the absence of the applied fields.

Consider first the laser-transition change signal. For a weakly saturating laser field the line shape is a Lorentzian of full width equal to 2γ [Fig. 14(b)]. As the laser field intensity increases, however, this shape becomes distorted. For intense laser fields ($S >> \gamma$ the line-shape expression is given approximately by [14, 23]

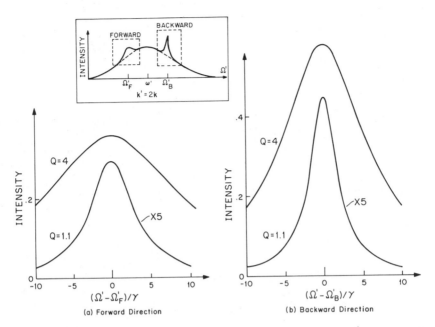

Figure 13. Laser-induced line narrowing change signals, $k' > k$. Both forward (a) and backward (b) signals are shown for small ($Q = 1.1$) and large ($Q = 4$) values of the saturation parameter. The $Q = 1.1$ lineshapes are shown at 5 times normal size. In the example shown $k' = 2k$ and the γ's are all assumed to be equal.

$$\frac{S}{\sqrt{\Delta/\gamma}} \left[\frac{1}{(\Delta - S)^2 + \gamma^2} - \frac{1}{(\Delta + S)^2 + \gamma^2} \right]^{1/2} \tag{9}$$

where

$$\Delta = \Omega' - \Omega'_B$$

and

$$S = \frac{\mu E}{\hbar} \frac{\sqrt{k' (k - k')}}{k} \tag{10}$$

Therefore, the change signal splits symmetrically into two distinctly non-Lorentzian resonances centered at frequencies $\Delta = \pm S$ [Fig 14(b)]. The full width of these resonances is $2/3\gamma$, and their intensity increases as $\sqrt{S/\gamma}$. This line shape is compared to that of the forward change signal in Figs. 14(a) and 14(b). Once again it is found that for a given laser intensity the

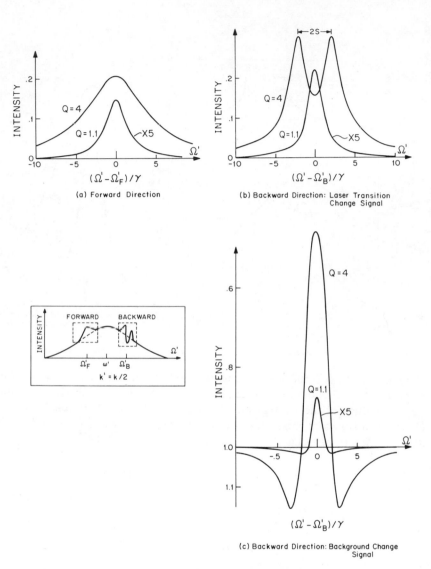

(a) Forward Direction

(b) Backward Direction: Laser Transition
Change Signal

(c) Backward Direction: Background Change
Signal

Figure 14.   Laser-induced line-narrowing change signals, $k' < k$.   Both forward (a) and backward [(b) and (c)] signals are shown for the case $k' = k/2$.   Note in this case that the backward change signal consists of two contributions, (b) the laser-transition change signal and (c) the background change signal.   The splittings in (b) and (c) are manifestations of the high-frequency Stark effect.

386

areas of the two signals are equal.

This splitting is a manifestation of the high-frequency Stark effect [25] and would not be predicted from population saturation considerations. It occurs, essentially, because the intense laser field induces coherent transitions between the levels of the laser transition at a rate $\sim \mu E/\hbar$. Thus, the wave function of the common level and, consequently, the transition probability to the third level, is modulated at this rate. This modulation causes the change signal to split into two. Similar effects are well known in the microwave region, where Doppler broadening is negligible, but the details are different [26].

In the $k' < k$ case there is a second, "background" contribution to the change signal. This line shape is always of one sign at its peak and the opposite sign at the wings, arranged in such a way that the net area is exactly zero [Fig. 14(c)]. As explained above, the magnitude of this change signal component is proportional to the population difference at the coupled transition. Therefore, it will occur even when the level populations of the laser transition are equal. Such an effect is clearly not explainable from population saturation arguments. The fact that the net area under this line shape is zero is a further example of the general rule that the areas of the change signals are always equal to the corresponding areas predicted from population saturation considerations. This rule can be shown to follow from an elementary quantum mechanical analysis of the transition rates of three-level systems interacting with EM fields [27].

In the case of a weakly saturating laser field the background line-shape contribution is given by

$$1 + \frac{1}{2} \frac{S^2 (\Delta^2 - \gamma^2)}{(\Delta^2 + \gamma^2)^2} \tag{11}$$

so that the separation between the two peaks is fixed at $2/3\gamma$. Note that as $S \to 0$ the background change signal vanishes, as it should, leaving only the broad Doppler profile. With increasing saturation the line-shape exression becomes more complicated. The exact expression, valid for all values of S, is given by [14]

$$\text{Im}\left(\frac{\Delta + i\gamma}{\sqrt{S^2 - (\Delta + i\gamma)^2}} \right) \tag{12}$$

where the real part of the square root is chosen to be positive. This resonance is similar in shape to that of the laser-transition contribution, Eq. (9), except that in this case the wings cross the origin and change sign [Fig. 14(c)], as is necessary to produce a net area of zero. As with the laser-

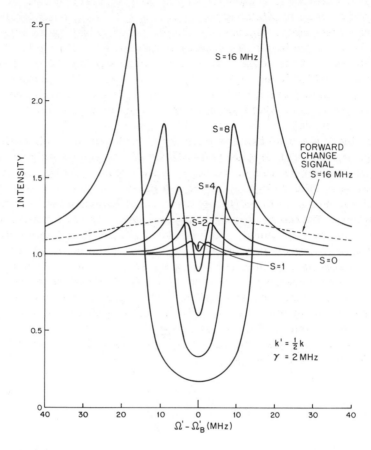

Figure 15.   Line shape of the change signal of a non-degenerate three level system as a function of the parameter S (in MHz) for $Y = 2$ MHz and $k' = k/2$. Notice that in each case the splitting is equal to 2S, and that the linewidth remains constant as S increases.   In the example shown the laser-transition change signal component and the background change signal component are weighted equally.

transition contribution the peaks, which are due to the high-frequency Stark effect, occur at $\Delta = \pm S$.

The net backward line shape is the algebraic sum of the laser-transition and background contributions, weighted according to the relative level population differences.   Since both pairs of peaks occur at the same frequencies, the overall features are quite similar to those of the component line shapes

(Fig. 15): With increasing saturation the line splits symmetrically into two narrow resonances of width $\sim \gamma$, separation 2S, and height increasing as the square root of S. Therefore, with increasing laser intensity the splitting increases linearly with S. Note that the width of the peaks if independent of S and does not broaden with increasing laser field. This shows that for the backward change signal, $k' < k$, the saturation effect manifests itself in the splitting of the line shape and not in the width of its features. This is different from the usual saturation broadening as occurs, e.g., in the forward direction. In Fig. 15 the forward and backward line shapes are compared for the case S = 16.

This unique saturation behavior suggests an interesting spectroscopic identification technique [23] applicable to both cascade and folded configurations (Figs. 16 and 17). Due to M-level degeneracy, an atomic or molecular three-level system actually consists of a number of degenerate three-level subsystems [Fig. 17(a)]. In accordance with angular momentum selection rules, each of these subsystems will have different electric dipole matrix elements. Thus, each subsystem will give rise to a pair of symmetrically located peaks, separated by 2S(M), where S(M), Eq. (10), now depends on the matrix element of the laser transition of the Mth subsystem. Therefore, with increasing saturation the change signal of a molecular system with degeneracy will split into a number of symmetric pairs of peaks. This number is completely determined by the angular momentum quantum numbers of the particular three-level system under study. The relative separation and intensity of the peaks depends on the type of transition involved (P, Q, or R branch), as well as the relative orientation of the laser and coupled E-fields. By analyzing the intensity pattern one can unambiguously determine the angular momentum quantum numbers of the three levels involved.

Such an experiment could be done by subjecting a molecular absorber to an intense laser field and probing a coupled transition with a weak tunable field (Fig. 16). Figure 17 shows the expected line shape calculated for an actual three-level system in NH_3 near 10 μ. In the system chosen the laser transition, which coincides with the P(13) line of the 10-μ N_2O laser at 928 cm^{-1}, is the asQ(8,7) line. The probe transition is the saP(9,7) line at 746 cm^{-1}. Since this system is in the folded configuration, the structure would occur in the forward change signal. Figure 17(a) shows how the NH_3 system can be decomposed into nondegenerate three-level subsytsems. The case shown is for linearly polarized laser and probe fields, E and E', respectively, with E perpendicular to E'. Figure 17(b) shows the contributions from the individual M-components, and Fig. 17(c) gives the resulting line shape. The equal spacing of the peaks is characteristic of a Q-branch laser transition, since in that case the matrix elements, and therefore S(M), are linearly proportional to M. The numbers of peak pairs, eight, immediately sets the J values of the laser transition levels at 8. The concave

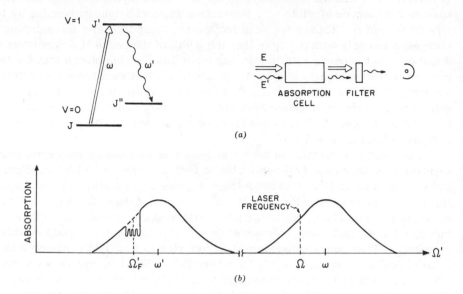

(a)

(b)

Figure 16.   Possible experiment for studying level degeneracy effects.
a) Energy level diagram and schematic of experimental arrangement. b)
Doppler profiles of the coupled transitions. In the case of a folded system,
as in the figure, the change signal will be centered at Ω'_F. The broken line
at Ω indicates the frequency of the pump laser.

intensity pattern is characteristic of a P-branch coupled transition, so that
the J value of the third level must be 9.

Some other possible intensity patterns are shown in Figure 18. The
number of components, their relative spacing and relative intensity, as well
as the different intensity patterns observed when laser and coupled transition
E-fields are perpendicular or parallel, all give information which can lead
to the assignment of the angular momentum quantum numbers involved.

Figure 17.   Intensity pattern for NH_3. In the case shown, laser field and
probe field are polarized at right angles. a) Degenerate sublevels, decomposed into three-level subsystems. b) Contributions from individual M-
components. c) Total line shape for part (b). The inset in (b) shows the
levels involved.

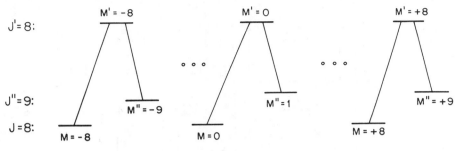

J' = 8:   M' = -8   M' = 0   M' = +8

J" = 9:   M" = -9   M" = 1   M" = +9

J = 8:   M = -8   M = 0   M = +8

(a) Degenerate Sublevels

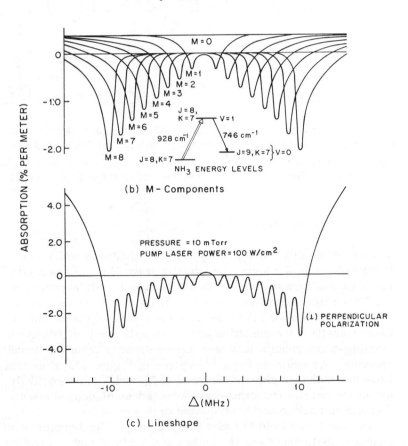

M = 0

0

M = 1
M = 2
M = 3
M = 4
M = 5
M = 6
M = 7
M = 8

-1.0

-2.0

J = 8,
K = 7   V = 1

928 cm⁻¹   746 cm⁻¹

J = 9, K = 7 } V = 0

J = 8, K = 7   V = 0

NH₃ ENERGY LEVELS

(b) M – Components

ABSORPTION (% PER METER)

4.0

2.0

0

-2.0

-4.0

PRESSURE = 10 mTorr
PUMP LASER POWER = 100 W/cm²

(⊥) PERPENDICULAR
POLARIZATION

-10   0   10

△ (MHz)

(c) Lineshape

391

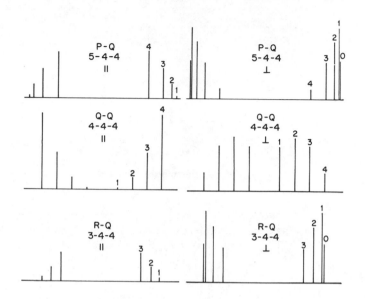

Figure 18. Typical intensity patterns. The figure shows all possibilities when the coupled transition is a Q-branch transition with J = 4. The laser transition can ghen be P, Q, R (rows 1, 2, 3, respectively). The first column shows the case E∥E′. The second column is for the case E⊥E′.

B. Standing-Wave Features

Let us now briefly consider the additional features which occur when the laser field is in the form of a standing wave [14]. As described above, in this case the forward and backward change signals appear together, symmetrically located about the center frequency of the coupled transition. When the laser field is well detuned from the center of the Doppler profile of the laser transition the resulting change signals are well resolved. In this case, standing-wave results may be obtained directly from the traveling-wave analysis. An example for $k=k'$ is given in Figure 19. Note that the two Lorentzian change signals centered at Ω'_F and Ω'_B, respectively, are the broad and narrow traveling-wave resonances discussed above. As expected, there is no background contribution in this case.

As the laser field is tuned to the center of its Doppler profile, the change signals approach each other and begin to merge together. In this case the resulting line shape is not simply the superposition of broad and narrow change signals, except in the limit of a weakly saturating laser field. Additional fine structure develops on the laser-transition change signal. In

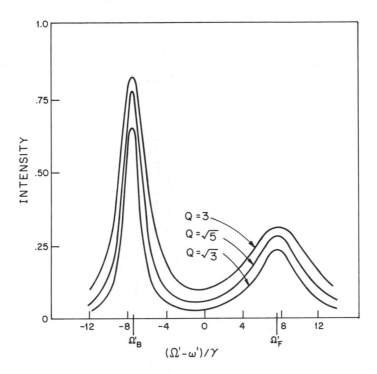

Figure 19. Standing-wave features, laser detuned. The example shown is for a cascade system with the decay rates of both levels of the laser transition taken equal to γ, and the decay rate of the third level taken as $\gamma_1 = \gamma/19$. The laser is detuned by an amount $\Omega-\omega=7.5\gamma$, and $k = k'$. Curves are given for three values of the saturation parameter $Q = \sqrt{3}$, $\sqrt{5}$ and 3. Note that in this case there is no background change signal.

addition, a nonzero background contribution (with zero area) now occurs. An example is shown in Fig. 20 for the same parameters as in Fig. 19. The new line-shape features occur because at resonance the same velocity ensemble of atoms (near $v=0$) can simultaneously interact with both traveling-wave components of the standing-wave field, giving rise to additional resonant behavior. One example of these effects is discussed below. The reader is referred to Ref. 14 for additional discussions.

The condition of tuning the laser field close to the center of its Doppler profile is also the Lamb-dip (central tuning dip) condition. Recent theoretical treatments of the Lamb dip [28, 29] predict that when the central tuning condition is met a fine structure develops at the center of the velocity distri-

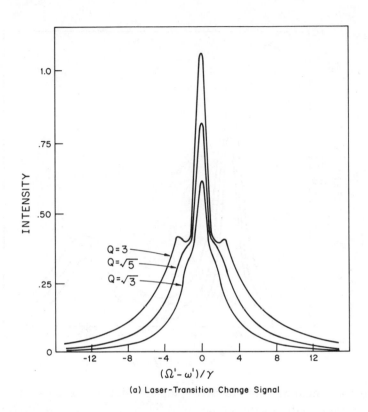

(a) Laser-Transition Change Signal

Figure 20.   Standing-wave features, laser tuned to resonance.   All of the parameters are the same as in Fig. 19, except that now $\Omega = \omega$.   a) Laser-transition change signal.

bution of the level populations of the laser transition [Fig. 21(a)]. This feature is caused by the coherent ringing of slow atoms moving through the spatial nodes of the standing-wave field [29]. The structure is particularly significant when the laser field is intense. However, it does not lead to corresponding fine structure in the Lamb dip. It does, however, manifest itself in the laser-induced line-narrowing change signal. An example is shown in Fig. 21, where the velocity distribution of the common level [Fig. 21(a)] is compared to the line shape at the coupled transition [Fig. 21(b)]. The fine structure is clearly observed on the wings of the central peak. (The central peak itself is due to spatial interference effects [14]). Thus, the experimental observation of curves such as Fig. 21(b) would provide evidence of this population structure.

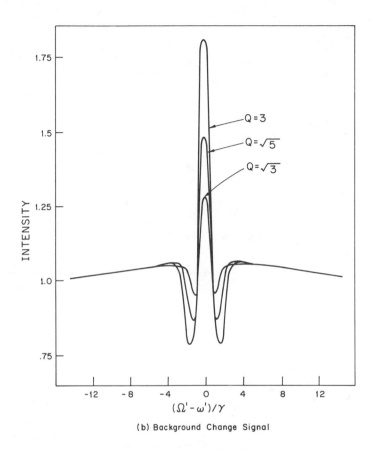

(b) Background Change Signal

Figure 20 b.   Background change signal.   Note that in this case structure
appears in the background change signal also.

C.   Summary

Only a few of the line-shape features described above have been observed as
of yet: The change signals have been observed in both spontaneous
[9,30-35] and stimulated [36-48] emission.   The anisotropic nature [49] of the
line-narrowing effect in the traveling-wave configuration has been established [47,48].   The width asymmetry between broad and narrow change
signals has been confirmed [9,40,43].   The zero-area background signal
produced by a weakly saturating traveling-wave laser field, Eq. (11), has
been observed [42].   The other predictions, including the line-shape distortions due to the high-frequency Stark effect and standing-wave saturation,
await experimental confirmation.

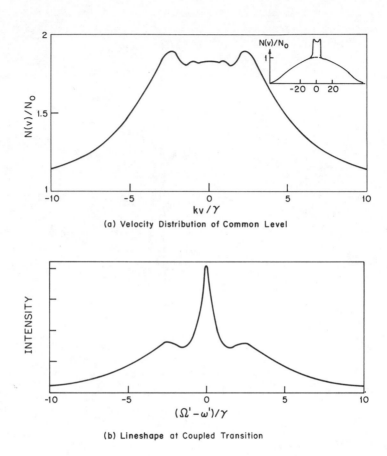

(a) Velocity Distribution of Common Level

(b) Lineshape at Coupled Transition

Figure 21. Standing-wave change signal, fine structure. The parameters are the same as in Fig. 20. a) Plot of spatially averaged population distribution vs. axial velocity of the lower level of the laser transition, subjected to an intense standing-wave field on resonance ($\Omega = \omega$). The inset shows the full axial velocity distribution. b) Laser-induced change signal for the same conditions as in (a). Note that structure similar to the population fine structure occurs at the wings of the narrow central dip of this profile. The central dip itself originates from interference effects.

IV. APPLICATIONS

The narrow resonances described above have been utilized in a number of practical applications [9, 30–48]. The discussions of this section will describe some of the recent activities at MIT. A number of other techniques based on the laser-induced line-narrowing effect have been developed by other research groups. Of these, two deserve special mention. In the studies of a three-level laser by Beterov and Chebotayev [43] an active laser medium, held slightly below threshold, is brought into oscillation by the presence of a laser field resonating at the coupled transition. The resulting laser signals, which occur over narrow spectral regions, have been used to study the width asymmetry and relaxation processes in the helium-neon system. In the "tuned laser differential spectrometry" of Hänsch and Toschek [40-42], the change signals induced in a neon transition in a He-Ne discharge are studied by monitoring the transmission with an external probe laser. One of the most interesting results of this study is the observation of the zero-area background signal of Eq. (11).

Narrow resonances of the type described here are also potentially useful in establishing precise length and frequency standards in the optical-infrared region of the spectrum. A complete discussion of these aspects is given in the paper by Javan [50].

A. Spontaneous Emission Line Narrowing

The first application to be discussed is the spontaneous emission version of the line-narrowing effect, which has been utilized previously to measure linewidths [9, 32, 33] and isotope shifts [30] in neon. The technique has been extended recently to studies of inhomogeneously broadened solids [34, 35]. The present discussion illustrates the technique and presents new measurements of hyperfine structure in Ne^{21}, leading to an accurate value of the neon quadruple moment [31].

The level scheme under study (Fig. 22) consists of the hyperfine sublevels of the $2s_2$, $2p_4$, and $1s_4$ fine-structure levels of Ne^{21}, which form the 1.15-μ ($2s_2$-$2p_4$) and 0.61-μ ($2p_4$-$1s_4$) cascade transitions. Ne^{21}, with nuclear spin $I = 3/2$, has both a magnetic dipole moment, μ_m, and an electric quadrupole moment, Q. Accordingly, hyperfine interactions split each fine-structure energy level (angular momentum J) into a number of hyperfine components, which may be specified by their total angular momentum $F = I + J$. The hyperfine splittings are determined by A and B, the magnetic dipole and electric quadrupole hyperfine constants, respectively (Fig. 22).

Let us first see how the observation of change signals can lead to a knowledge of hyperfine separations. Consider a pair (1, 2) of the closely spaced three-level systems formed by the hyperfine splitting (Fig. 23). Suppose, for simplicity, that the laser field is tuned to the center of the Doppler profile of the laser transition of system No. 1, i.e., $\Omega = \omega(1)$. Then both forward and backward change signals of system No. 1 will appear at $\omega'(1)$, the center frequency of the corresponding coupled transition (Fig. 23).

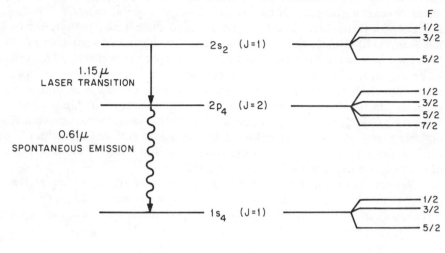

HYPERFINE SPLITTING ENERGIES ARE OF THE FORM:

$$\Delta W_{\text{Magnetic Dipole}} = A \cdot \frac{K}{2}$$

$$\Delta W_{\text{Electric Quadrupole}} = B \cdot \frac{\frac{3}{4} K(K+1) - I(I+1)J(J+1)}{2I(2I-1)J(2J-1)}$$

where $K = F(F+1) - J(J+1) - I(I+1)$

$$W_F = W_J + \Delta W_{\text{Magnetic Dipole}} + \Delta W_{\text{Electric Quadrupole}}$$

Figure 22.   Hyperfine splitting of $1s_4$, $2p_4$ and $2s_2$ levels in Ne[21]. W_F is
the total energy of the hyperfine level, W_J is the energy of the fine-structure
level; A and B are the magnetic dipole and electric quadrupole interaction
constants of level J, respectively; and $K = F(F+1) - I(I+1) - J(J+1)$ where
F, the total angular momentum, can take on the values $F = I+J, I+J-1, \ldots,$
$|I-J|$.

Since the laser field is tuned to $\omega(1)$, it must be somewhat detuned from
$\omega(2)$.   Therefore, the forward change signal of system No. 2 will occur at a
frequency $(k'/k)\Delta$ below $\omega'(2)$, where Δ is the detuning of the laser frequency
below $\omega(2)$.   The backward signal will occur above $\omega'(2)$ by the same amount
(Fig. 23).   Note that the hyperfine splitting at the coupled transition is given
by the separation between the change signals of system No. 1 and the
frequency midway between forward and backward change signals of system

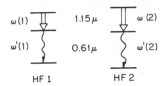

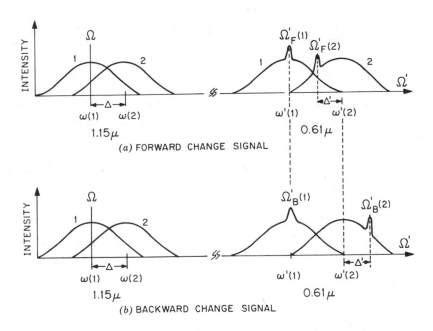

Figure 23. Measurements of closely-spaced structure using the spontaneous emission line-narrowing technique. a) Forward change signal. b) Backward change signal. See text for further details.

No. 2 (Fig. 23). Furthermore, the separation between the forward and backward signals of system No. 2 is a measure of the detuning of the laser field, hence $\omega(1)$, from $\omega(2)$. Therefore, by studying forward and backward change signals together the hyperfine separation at both laser and coupled transitions can be determined.

Actually, it is not necessary for the laser to be tuned to $\omega(1)$, nor is it necessary to study forward and backward change signals at the same time. For a laser field of arbitrary frequency Ω the separation between the change signal pair, obtained from Eqs. (1) and (2), is given by

$$\delta_F = \Omega'_F(1) - \Omega'_F(2) = [\omega'(1) - \omega'(2)] - [\omega(1) - \omega(2)]\frac{k'}{k}$$

$$\delta_B = \Omega'_B(1) - \Omega'_B(2) = [\omega'(1) - \omega'(2)] + [\omega(1) - \omega(2)]\frac{k'}{k}$$

(13)

Therefore, from an experimental determination of δ_F and δ_B, hyperfine splittings at both laser and coupled transitions may be uniquely determined.

The hyperfine interactions in Ne^{21}, which produce 18 such three-level systems, can be analyzed from Eqs. (13). Note that forward and backward patterns contain information on the hfs of both 1.15- and 0.61-μ transitions. Separate observation of forward and backward signals provides two distinct patterns which must be fitted by a single set of parameters. These parameters, the hyperfine A and B constants for each of the three levels, can thus be extracted from the separations between the features of the spectrum. Once the A's and B's are known, μ_m and Q may be obtained from hyperfine-structure theory.

Note the advantages of the traveling-wave approach. If the laser field were in the form of a standing wave, forward and backward change signals would occur together, resulting in a much more complex pattern. Another

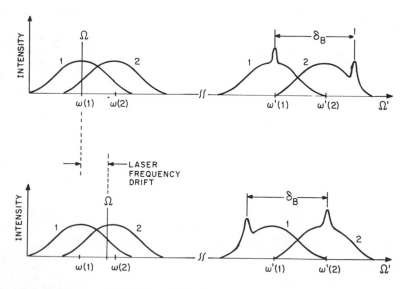

Figure 24. Stability of change-signal intensity pattern. The backward signal of Fig. 23 is shown for two different laser frequencies. Note that although the change-signal pattern shifts in frequency, the separation between the peaks, δ_B, remains fixed.

important feature of the traveling-wave configuration is that the relative positions of the change signals are independent of laser tuning, as can be seen from Eqs. (13), where δ_F and δ_B are not functions of Ω. This is illustrated in Fig. 24, which shows that as the laser frequency and change signal pattern drift the spacings between change signals remain fixed. Therefore, in experiments of this type the laser frequency drift must be small during a measurement, but the absolute frequency of the laser may change between measurements. This is an advantage of the present approach over Lamb-dip techniques, where absolute frequency stability is required.

The experimental setup is shown in Fig. 25. The output beam of an intense stable single-mode 1.15-μ He-Ne laser is focused into an external sample cell containing a Ne^{21} gas discharge at a low pressure ($\sim$ 100 mTorr). The 0.61-μ spontaneous emission from the sample cell, emitted in either the forward or backward direction, is analyzed using a pressure-scanned Fabry-Perot interferometer. A chopper, placed between the laser and the sample cell, enables use of a lock-in amplifier to subtract off the spontaneous emission background and improve the signal-to-noise ratio.

Figure 26 shows typical experimental spectra and the theoretical fit for forward [Fig. 26(a)] and backward [Fig. 26(b)] runs. In each case the upper experimental trace shows the narrow change signal and the lower trace shows the Doppler-broadened background observed without the lock-in amplifier. Note that the hyperfine structure is ordinarily entirely masked by the Doppler widths. The largest change signals in Fig. 26 have intensities of

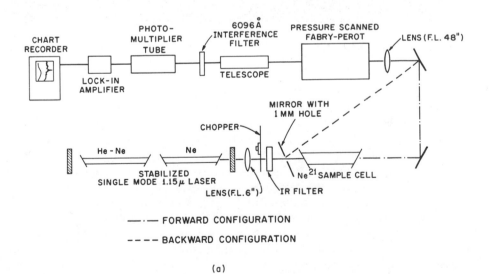

(a)

Figure 25 a. Experimental arrangement for studying Ne^{21} hfs. Note the different configurations for observing forward and backward 0.6 μ radiation.

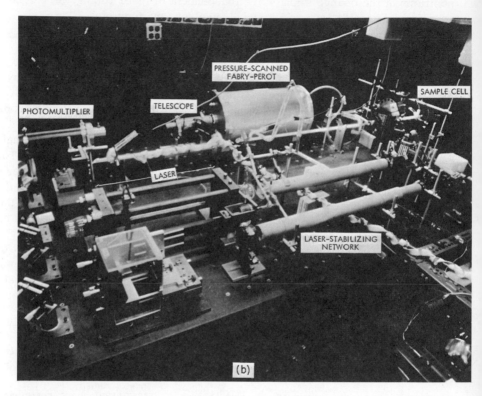

Figure 25 b.   Photograph of experimental arrangement for studying hyperfine structure of Ne[21].

about 1/2% of the intensity of the spontaneous emission background.

It is emphasized that the spectra observed with this technique are richer than ordinary 0.61-μ hyperfine spectra in that our data contain information about both the laser transition <u>and</u> the coupled transition.   The markings on Fig. 26 denote the eight three-level systems producing the observed change signals, the contributions of the remaining ten change signals being negligible.   The two additional features in the forward and backward traces are due to the presence of Ne[20] and Ne[22] in the sample of Ne[21].

The hyperfine B constants were obtained from the best fit of the theory to the data.   (The hyperfine A constants could be calculated from a precise value of μ_m, obtained from NMR measurements [51]).   Then, using hyperfine-structure theory, the quadrupole moment Q was extracted [31].   The

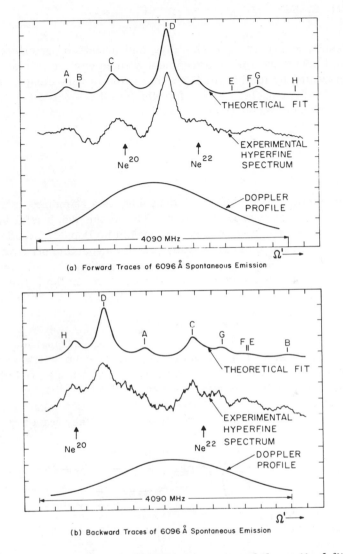

(a) Forward Traces of 6096 Å Spontaneous Emission

(b) Backward Traces of 6096 Å Spontaneous Emission

Figure 26.   Comparison of experimental traces and theoretical fit.   Note that the Doppler profiles for forward and backward traces completely mask the hyperfine structure.   The change signals labeled are the eight most prominent and are associated with three-level cascade systems of the type $F \to F' \to F''$.   Denoting these systems as ordered triplets, they are A = (3/2, 3/2, 1/2);   B = (3/2, 5/2, 5/2);   C = (3/2, 3/2, 3/2);   D = (5/2, 7/2, 5/2); E = (5/2, 5/2, 3/2);   F = (1/2, 1/2, 1/2);   G = (3/2, 5/2, 3/2);   H = (5/2, 5/2, 5/2). For the computer fits to the data we used $\Gamma_{br} = 265$ MHz and $\Gamma_{nar} = 225$ MHz. Of these widths ~ 100 MHz is instrumental;   also, the Doppler width is 2500 MHz.

final result, corrected for shielding and antishielding effects, is
Q = (+ 0. 1029 ± 0. 0075)b, in agreement with an earlier less accurate study
[52]. As noted by Rabi et al. [52], this value is very close to that of the
"almost" mirror nucleus Na^{23}, for which [53] Q = (+ 0. 100 ± 0. 01)b, in
accord with shell model predictions.

B.   Mode Crossing

A second technique, called mode crossing, is a stimulated version of the
laser-induced line narrowing effect. This technique, useful in resolving
closely spaced structure ordinarily masked by Doppler broadening, has been
applied in our laboratories to study fine and hyperfine structure and g-factors
[36, 37, 45] by Brewer to study g-factors in methyl bromide and other gases
[38] and by Johnson and Wolga to study structure in the helium-neon system
[39]. Detailed treatments are given elsewhere [3, 9]. The present discussion summarizes the technique as applied to measuring g-factors in atomic
oxygen.

   In the mode-crossing technique the molecular transitions studied are
folded three-level systems with closely spaced coupled transitions, and the
em fields resonating with laser and coupled transitions are both intense
monochromatic fields, as would be produced by a multimode laser (Fig. 27).
The effect manifests itself as a resonant change in the gain (or attenuation)
of the medium, induced by the applied fields as they traverse the sample

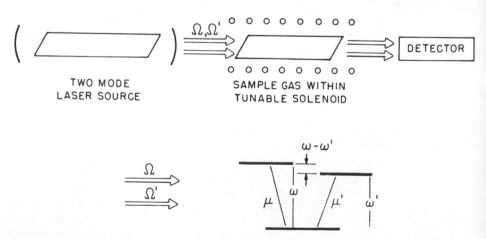

Figure 27. Mode crossing effect, simplified experimental arrangement and
energy level diagram.

cell. The resonance occurs as the separation between the closely spaced levels approaches the frequency separation between the two laser modes. Tuning may be achieved by varying either the laser frequencies (e. g. , by changing the length of the laser cavity), or else the level separation (by Stark or Zeeman effects), but in practice the latter method is usually more convenient.

This resonant behavior is the manifestation of the laser-induced line-narrowing change signal as it appears in transmission, and only occurs when the applied fields can saturate their respective transitions. Since the applied fields are propagating in the same direction and the levels are in the folded configuration, the resonance condition is given by Eq. (1) (forward change signal). Specializing to the case where $k' - k \ll \gamma/u$, this condition becomes

$$\Omega - \Omega' = \omega - \omega' \tag{14}$$

where Ω and Ω' are the frequencies of the applied laser fields. For both fields weakly saturating their respective transitions, the resonant change signal is a Lorentzian of full width $= \gamma + \gamma'$, where γ and γ' are the decay rates of the closely spaced levels [6, 13]. (If $\gamma = \gamma'$, this result agrees with Eq. (8) in the limit $k' \to k$.) The fact that the width of the mode-crossing resonance is independent of the decay rate of the common level is important in systems such as oxygen, where this rate is much larger than those of the "crossing" levels.

An important experimental consequence of the mode-crossing frequency condition, Eq. (14), is that it only depends upon the frequency separation between the laser modes, and not on their absolute frequencies. This makes the effect easily observable with a free-running multimode laser, where the mode spacing remains fixed although the absolute frequencies may wander over a sizeable portion of the broad Doppler profile during the time of observation [Fig. 28 (a)]. This feature is not general and holds only for the forward change signal in the approximation of closely spaced levels. It does not occur for the backward change signal, i. e. , if the two applied fields were incident upon the sample cell in opposite directions. In that case the reso-nant condition would be

$$\Omega + \Omega' = \omega + \omega'$$

which sensitively depends on the absolute frequencies of the applied fields, so that a small instability would cause the modes to drift out of resonance [Fig. 27 (b)]. Consequently, observation of the latter signals would require absolute frequency control as in Lamb-dip experiments. Furthermore, this change signal is broader than the forward signal because in this case the

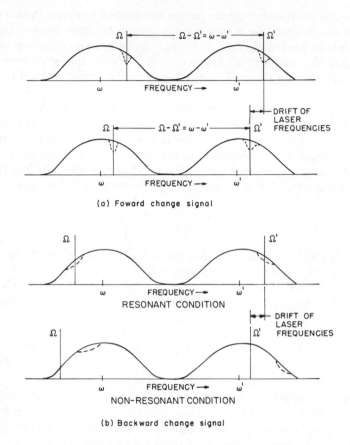

Figure 28. Frequency stability of mode-crossing change signals under laser drift. The dotted lines indicate the regions of gain depletion due to the intense fields. a) Forward change signal. b) Backward change signal. Note that in (a) the resonance condition is maintained under laser drift, whereas in (b) it is not.

decay rate of the common level also contributes to the linewidth. In the limit of weakly-saturating laser fields this change signal is a Lorentzian of full width = $\gamma + \gamma' + 2\gamma''$, where γ'' is the decay rate of the common level. Therefore, when $\gamma'' \gg \gamma, \gamma'$, as is often the case in practice, this resonance will be much broader and less pronounced than the forward change signal.

In the case of a Doppler-broadened laser medium with closely spaced tunable structure, the mode-crossing effect is also observable as a resonant change in the output signal of the laser itself. Note that in this case only the forward change signal is stable against the frequency drift, so that in a free-

TABLE I.   ATOMIC OXYGEN g-FACTORS

Wave length	Transition (Upper Level - Lower Level)	g_{exp}	g_{LS} (Upper Level)
0.84 μ	$3p^3P - 3s^3S$	1.51 ± 2%	3/2
2.89 μ	$4p^3P - 4s^3S$	1.51 ± 2%	3/2
4.56 μ	$4p^3P - 3d^3D$	1.51 ± 2%	3/2

running laser the backward change signal tends to wash out, thus simplifying the observation of the effect.

Mode-crossing resonances were studied in three atomic oxygen laser transitions (Table I). These experiments used a free-running Brewster-angle laser about 3 m long. The laser medium itself was Zeeman tuned by means of a long pair of rectangular coils transverse to the axis of the laser tube. The Brewster-angle end windows were oriented so that the polarization of the laser fields was perpendicular to the magnetic field ($E \perp B$). Fields up to 180 G could be applied over the entire length of the laser tube. The field inhomogeneity was approximately 10%. Adjacent modes of the laser were spaced $c/2L = 49.2$ MHz apart. Mode-crossing signals were observed as resonant changes in the laser output power. To enable a lock-in technique to be used, a small additional modulation field of about 1 G was used. Therefore, the signals observed in the data are actually first derivatives of the change signals.

A typical trace for the 8446 - Å transition ($3p^3P$-$3s^3S$) is shown in Fig. 29. The mode-crossing signals are the resonances observed above about 20 G. (Ignore the large resonance at zero G for the present.) A mode-crossing signal is observed each time the splitting between a pair of upper levels connected to a common lower level approaches the separation between any two axial modes of the multimode laser. For $E \perp B$ the selection rule is $\Delta M = 2$. The g-values can be obtained from the mode-crossing signals using the relationship $2\mu_0 g\Delta B = c/2L$ [obtained from Eq. (14)], in which ΔB is the separation between resonances in units of magnetic field and μ_0 is the Bohr magneton. The measured g-values are listed in Table I. They are in excellent agreement with the values predicted by L-S coupling for the upper levels. [In the three cases studied the upper multiplets are ^3P, for which L-S coupling predicts g-values of 3/2 for both J=2 and J=1 levels. In each case more than one fine-structure level may have contributed to the gain of the laser medium. The present experiments did not resolve possible small variations between g(J = 1) and g(J = 2).] Mode crossings of the lower

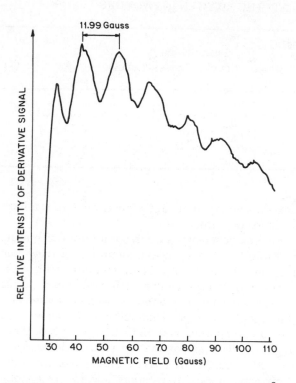

Figure 29. Mode crossing signal in atomic oxygen at 8446 Å.   The output power of a multimode 8446 Å oxygen laser is studied as a function of magnetic field applied to the active medium.   The derivative signal, obtained using a lock-in technique, is shown.

levels were not observed because of instabilities due to laser drift, and also because the linewidths were so broad that the signals overlapped and flattened out.   The broad linewidths are due to the large decay rates of the lower laser levels, as discussed above.

C.   Unidirectional Amplifiers

As a final application of the line-narrowing effect, I would like to describe a device having the property of being able to amplify waves of a given frequency traveling in one direction, but attentuate waves of the same frequency traveling in the opposite direction, i.e., a undirectional laser amplifier [49].   Such a device may be useful in the area of communications and in other applications which require complete isolation of input and output signals. The idea makes use of the inherent anisotropy of the line-narrowing effect in the traveling-wave configuration.

Let us begin by considering the optical pumping of a Doppler-broadening molecular gas by an intense monochromatic traveling-wave field. As an example, consider the molecular system of Fig. 30, which consists of two rotational levels in the ground vibrational state and a rotational level in the first excited vibrational state forming a pair of coupled Doppler-broadened transitions centered at ω and ω', respectively. At thermal equilibrium both transitions will be in the absorbing phase. However, under the appropriate conditions the pump field may bring the coupled transition into the amplifying phase over a narrow range of frequencies.

This approach is similar in some ways to the Raman lasers produced in liquids and solids, where the pump field produces Stokes gain at a coupled transition by means of the stimulated Raman effect [54]. In that case, however, the common level is usually many electron volts above the ground state, so that single-photon transitions are highly nonresonant. In the present case, where the common level is resonant, the gain is considerably modified by direct single-quantum absorption at the coupled transition. Another important distinction is the occurrence of Doppler broadening, which considerably modifies the details of the gain. The unidirectionality is a direct consequence of the Doppler effect.

To see how the line-narrowing effect leads to unidirectional gain, suppose the coupled transition is probed by a weak monochromatic field propagating in a direction parallel to the pump laser field (i. e., in the forward direction). As we know, a resonant decrease in the broad background absorption will occur at Ω'_F, as given by Eq. (1) [Fig. 31(a)]. A weak probe field propagating in the backward direction will also produce a resonant decrease in absorption, but at frequency Ω'_B [Eq. (2)], and if the laser is detuned from the center of its Doppler profile the two resonances will be well

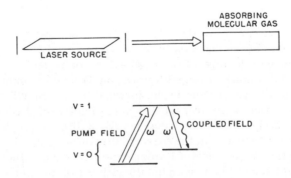

Figure 30. Optical pumping scheme for unidirectional amplifiers.

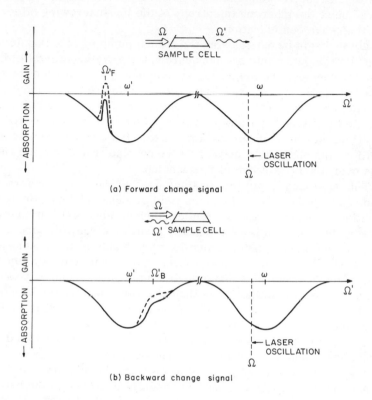

Figure 31.   Absorption profile as influenced by an intense monochromatic field resonating with a coupled transition.   a) Probe field and pump field parallel.   b) Probe field and pump field antiparallel.   The pump field frequency Ω is also indicated.

resolved [Fig. 31(b)].   As the intensity of the saturating field is increased, the magnitudes of the changes in the absorption profile may increase sufficiently to change the sign of the absorption coefficient, resulting in directionally dependent amplification within narrow frequency intervals.   A light wave of frequency Ω'_F propagating in the forward direction will then be amplified, whereas the same wave propagating in the backward direction will be attenuated.   Similarly, a wave at frequency Ω'_B will be amplified in the backward direction but attenuated in the forward direction (Fig. 31).

Due to the occurrence of Raman-type processes, the line shapes of forward and backward change signals will not be the same.   Since we are dealing with a folded level configuration, the backward change signal will be broad and the forward change signal will be narrow. Therefore, the magnitude of the absorption decrease observed in the forward direction will be greater than that observed in the backward direction. In particular, as the pump field intensity is increased, the reversal in the sign of the absorption coefficient first appears for the forward direction at Ω_F', where the magnitude of the effect is largest [Fig. 31(a)]. In this case gain is obtained only for waves propagating in the forward direction, while in the backward direction the coupled transition remains in the absorbing phase within its entire Doppler profile. In fact, as is shown below, it is possible to have a situation where gain can only occur in the forward direction, even in the limit of a fully saturating pump field.

The magnitude of the gain can be quite large for a fully saturating laser field ($Q \gg 1$). For example, in the case $k' \approx k$ [55] it is given by

$$G_F = |\alpha|(p-1) \tag{15}$$

in the forward direction, and by

$$G_B = |\alpha|[(1/3)p-1] \tag{16}$$

in the backward direction. Here, $|\alpha|$ is the magnitude of the linear absorption coefficient of the coupled transition and $p = \delta N/\delta N'$, where δN and $\delta N'$ are the population differences between the levels of the laser and coupled transitions, respectively, in the absence of applied fields. As emphasized above, these expressions include effects due to Raman-type processes, as well as population saturation effects. It is interesting to compare them to the corresponding expression for gain obtained by considering population saturation alone, i.e., by neglecting effects due to Raman-type processes:

$$G_{P.S.} = |\alpha|[(1/2)p-1] \tag{17}$$

In this expression the requirement for net gain is that $p > 2$, which is, of course, identical to the requirement for producing a net population inversion at the coupled transition by fully saturating the laser transition. As can be seen from Eq. (15), the actual requirement for obtaining net gain in the forward direction is that $p > 1$. Therefore, the gain in the forward direction can be considerably larger than that expected from estimates based upon population saturation considerations alone. Also, note that net gain can be achieved even in the absence of a population inversion. This feature is characteristic of Raman-type processes and also occurs in stimulated Raman lasers [54].

As for the backward direction, note that gain can only be obtained for $p > 3$. Accordingly, for $1 < p < 3$, amplification occurs within a narrow band of frequencies centered at Ω_F' for waves propagating in the forward direction, but absorption takes place across the entire Doppler profile of the coupled transition for propagation in the backward direction. It is interesting to note that in the backward direction the influence of Raman-type processes actually serves to reduce the magnitude of the gain compared to the value predicted by population saturation considerations.

Unidirectional amplification may be achieved in several types of systems. For a set of rotational-vibrational transitions in thermal equilibrium gain can only be achieved for $k' > k$. However, in cases where the level populations are not determined by thermal equilibrium, such as in a gas discharge, gain can also be achieved for $k' > k$, provided that $p > 1$. Similar considerations also apply to cascade configurations such as Fig. 9(a), except that in this case maximum gain will be obtained in the backward direction instead of the forward direction.

The unidirectional effects are being studied in hydrogen fluoride gas at room temperature. In these experiments a low pressure ($\sim$ 100 mTorr) HF sample cell is pumped by one of the $v=1 \rightarrow 0$ rotational-vibrational lines of a transverse discharge HF "pin" laser [56]. The output pulses are about 1 μsec long and have peak powers of a few kW. Both the 2.7-μ $P_1(J)$ lines [$(v=1, J-1) \rightarrow (v=0, J)$] and the 2.4-$\mu$ $R_1(J)$ lines [$(v=1, J+1) \rightarrow (v=0, J)$] can be used. An HF energy level diagram is given in Fig. 32. Pumping with the $P_1(J)$ lines [or the $R_1(J-2)$ lines] produces gain at the coupled rotational transitions $(v=1, J-1) \rightarrow (v=1, J-2)$, which fall in the far-infrared (30-250 μ). Pumping with the $R_1(J)$ lines at 2.4 μ should lead to gain at the coupled $P_1(J+2)$ vibrational-rotational transitions at around 2.7 μ. To investigate the unidirectional nature of the gain, the sample cell is placed in a ring cavity (Fig. 33) consisting of two gold-coated mirrors of large radius of curvature and a special flat mirror which is transparent to the pump laser wavelength but reflective at the wavelength of the coupled transition. For far-infrared gain [$P_1(J)$ pumping] a flat made of BaF_2 or a similar salt can be used. For gain at the $P_1(J)$ transitions (near 2.7 μ) a dielectric-coated quartz flat is used. The pump power is coupled in through the flat. A beam splitter in the ring cavity, suitably oriented, couples out a small fraction of the power in either forward or backward direction.

At the time of the Esfahan Symposium (August, 1971) only a small and erratic signal at one of the $P_1(J)$ lines had been observed. Subsequently, while this manuscript was being prepared, strong unidirectional signals have been obtained both in the far [47] and near [48] infrared. In all cases the ratio of forward/backward output intensities is generally in the range 50-100, and for some of the lines near threshold, gain and laser oscillation are observed only in the forward direction. In the case of the far-infrared unidirectional lines, the gains are enormous, typically 2 or 3 per cm, and oscillation is obtained easily without mirrors. In these cases a

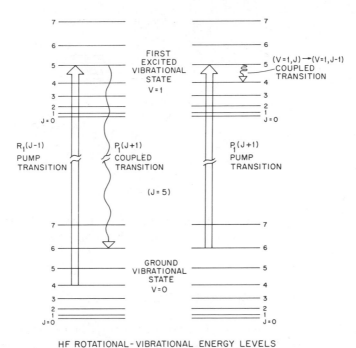

HF ROTATIONAL-VIBRATIONAL ENERGY LEVELS

Figure 32. HF energy level diagram showing pumping schemes for unidirectional amplification at coupled vibrational transitions (left hand side) and coupled rotational transitions (right hand side).

forward/backward intensity ratio of 5–10 is observed.

In the case of the near-infrared lines, the gains are less but still large, typically 10–30% per cm, and a resonant cavity is necessary in order to obtain laser oscillation. Particularly interesting is the case of $P_1(3)$ vibrational transition at 2.61 μ, pumped by the $R_1(1)$ vibrational transition at 2.48 μ. Here, calculations (including level degeneracy) based on population considerations alone predict loss, even for full saturation of the pump transition, whereas a complete analysis including Raman–type processes gain in the forward direction only, in agreement with experiment. This is an example in which amplification is achieved in the absence of a net population inversion, as discussed above.

The unidirectional lines are listed in Tables II and III. Complete details are available elsewhere [47, 48].

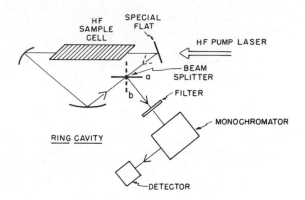

Figure 33. Ring laser cavity configuration. Forward (backward) far-infrared output power is coupled out of the cavity when the beam splitter is in
position a (b).

V. CONCLUSION: PAST, PRESENT, FUTURE

This brings us to the end of our discussions of laser saturation spectroscopy
in coupled Doppler-broadened systems. Although the achievements up to the
present have been impressive, I believe that the most significant developments in this area are yet to come. Reliable tunable monochromatic sources,
which are just now becoming available, will allow the line-narrowing techniques to be routinely applied throughout the infrared, visible and ultraviolet

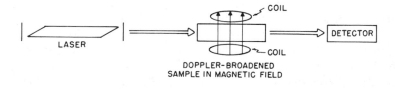

Figure 34. Stimulated Hanle effect, experimental setup and energy level
scheme.

TABLE II. OPTICALLY PUMPED HF VIBRATIONAL TRANSITIONS
[$P_1(J)$ signifies the (v=1, J-1) → (v=0, J) transition;
$R_1(J)$ signifies the (v=1, J+1) → (v=0, J) transition.]

Pump Transition		Laser Transition	
Designation	Wavelength	Designation	Wavelength
$R_1(1)$	2.48 μ	$P_1(3)$	2.61 μ
$R_1(2)$	2.45 μ	$P_1(4)$	2.64 μ
$R_1(3)$	2.43 μ	$P_1(5)$	2.67 μ
$R_1(4)$	2.41 μ	$P_1(6)$	2.71 μ
$R_1(5)$	2.40 μ	$P_1(7)$	2.74 μ

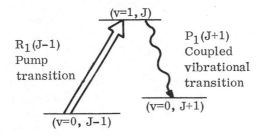

TABLE III. OPTICALLY PUMPED HF ROTATIONAL TRANSITIONS (v=1)
[$P_1(J)$ signifies the (v=1, J-1) → (v=0, J) transition.]

Pump Transition		Laser Transition	
Designation	Wavelength	J_{upper} → J_{lower}	Wavelength
$P_1(2)$	2.58 μ	1 → 0	256.0 μ
$P_1(3)$	2.61 μ	2 → 1	126.5 μ
$P_1(4)$	2.64 μ	3 → 2	84.4 μ
$P_1(5)$	2.67 μ	4 → 3	63.4 μ
$P_1(6)$	2.71 μ	5 → 4	50.8 μ
$P_1(7)$	2.74 μ	6 → 5	42.4 μ
$P_1(8)$	2.78 μ	7 → 6	36.5 μ

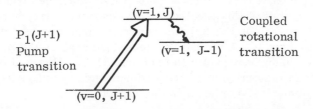

regions, thus opening a new era of atomic and molecular spectroscopy of unprecedented resolution. Resonances as narrow as 1 kHz and even 1 Hz will be attained, enabling ultra-fine level structure to be resolved. Such high resolution will also make possible observation of phenomena such as molecular recoil and transverse Doppler effects. Line narrowing techniques will also be extended to nuclear systems. For example, it should be possible to observe the line-narrowing effects in a coupled three-level system consisting of an optical transition and a nuclear transition sharing a common level. This will make possible high resolution nuclear spectroscopic studies, investigation of isomeric shifts, and extension of frequency measuring techniques into the gamma-ray region of the electromagnetic spectrum.

Finally, to close the circle between past and present, let us consider an improved version of Professor Zeeman's experiment which uses the new laser techniques. The effect is called the stimulated Hanle effect [57]. The experimental requirements and setup are identical to those of the mode-crossing technique, except that only a single laser mode is necessary (Fig. 34). As before, the sample to be studied is a folded Doppler-broadened system with closely spaced tunable structure. The attenuation of the laser field by the sample is studied as a function of the level spacing. As the levels approach one another within their homogeneously broadened widths, a resonant change in transmission is observed. This change occurs because as the closely spaced levels become degenerate the coupling between the atomic system and the applied field increases. The effect can also be viewed as a degenerate mode-crossing resonance, where both applied fields are at the same frequency. As in that case, resonances can be observed only when the laser field saturates its transition.

This effect is closely related to the Hanle effect [58], well known since the 1920's. In the Hanle effect the angular distribution of resonance fluorescence from an atomic system with degenerate levels is found to change as the degeneracy is removed (with respect to the homogeneously broadened width). This change is due to interference between the wave functions of the closely-spaced levels. In the laser experiments the same effect manifests itself, but in stimulated emission (or absorption) rather than fluorescence. In another version of this effect, called level crossing [60], the degenerate condition occurs at a finite value of magnetic field rather than at zero field. A detailed quantum-mechanical analysis of stimulated level crossing and the stimulated Hanle effect will be given elsewhere [57].

An experimental result for the 8446-Å atomic oxygen transition is given in Fig. 35. The tail of this level-crossing resonance already appeared in the mode-crossing trace of Fig. 29, where a very large resonant change signal occurred near zero G. Note that this technique enables the Zeeman sublevels to be resolved using a magnetic field of only 11 G. This is a tremendous improvement in resolution over conventional techniques, as exemplified by Professor Zeeman's experiment in the Introduction.

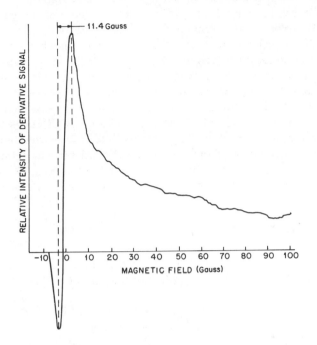

Figure 35.   Stimulated Hanle effect signal observed in atomic oxygen at 8446 Å.   The output power of the 8446-Å oxygen laser is studied as a function of magnetic field applied to the active medium.   The derivative signal, obtained using a lock-in technique, is shown.

So, once again, the homogeneously broadened width has been recovered from a Doppler-broadened system.   This is how the techniques of laser saturation spectroscopy enable us to find a needle in a haystack.

ADDENDUM (OCTOBER, 1972)

The HF optical pumping experiments have recently led to the first observation of Dicke superradiance in an extended, optically thick sample [61].   In these experiments a low pressure (mTorr) cell of HF gas at room temperature is optically pumped from one of the rotational transitions in the ground vibrational state to produce a total inversion at the coupled rotational transition in the first excited vibrational state.   There is no intentional feedback in the apparatus and the effects of unwanted regeneration are minimal.   After a delay of the order of a μsec, an intense burst of radiation is emitted at the frequency of the rotational transition. The envelope of this pulse is highly non-exponential, and ringing often occurs. The emitted radiation is also extremely directional. Furthermore, the effective lifetime of the excited rotational-vibrational state is reduced from its incoherent value by a factor of 10^7. These features are all characteristic of Dicke superradiance, i.e. coherent spontaneous emission. A semiclassical analysis shows that the superradiant lifetime of the state is inversely proportional to the number of excited atoms, and this is observed in the experiments by studying the delay as a function of sample cell pressure. The analysis leads to the conclusion that for a high-gain system the pulse evolution is determined by a single parameter, τ_R, and that inhomogeneous broadening is unimportant. Close agreement between theory and experiment is obtained. A more detailed account will be submitted for publication in Physical Review shortly.

REFERENCES

1. H. P. Zeeman, Phil. Mag. 43, 226 (1897).
2. M. S. Feld, K. Shimoda, M. Kovacs, and C. Shields, Bull. Am. Phys. Soc. 10, 87 (1965).
3. Such direct spectra have also been obtained by H. S. Gerritsen and S. A. Ahmed, Phys. Letters 13, 41 (1964).
4. Strictly speaking, γ is the net decay rate due to homogeneous relaxation mechanisms.
5. An atomic beam produced in this way is different from a conventional atomic beam collimated from an oven. The latter type of beam contains atoms of all axial velocities but very little lateral motion. In contrast, the laser saturation effect selects atoms within a narrow range of axial velocities, but with a wide range of lateral motions.
6. M. S. Feld and A. Javan, Phys. Rev. 177, 540 (1969).
7. W. E. Lamb, Jr., Phys. Rev. 134, A1429 (1964).
8. C. Freed and A. Javan, Appl. Phys. Letters 17, 53 (1970).
9. H. K. Holt, Phys. Rev. Letters 20, 410 (1968).
10. The only deviation in the line shape of this change signal from that predicted from population saturation is a small additional saturation broadening.
11. See, for example, A. Weber, S. P. S. Porto, L. E. Cheesman, and J. J. Barrett, J. Opt. Soc. Am. 57, 19 (1967); and C. J. Schuler, in Progress in Nuclear Energy, ser. 9, Analytical Chemistry, vol. 8, part 2 (Pergamon Press, Oxford, 1968).
12. M. S. Feld, J. H. Parks, H. R. Schlossberg, and A. Javan, Physics of Quantum Electronics, edited by P. L. Kelley, B. Lax, and P. E. Tannenwald. (McGraw-Hill, New York, 1966), p. 567.
13. H. R. Schlossberg and A. Javan, Phys. Rev. 150, 267 (1966).

14. B. J. Feldman and M. S. Feld, Phys. Rev. A 5, 899 (1972).
15. G. E. Notkin, S. G. Rautian, and A. A. Feoktistov, Zh. Eksp. Teor. Fiz. 52, 1673 (1967) [Sov. Phys. JETP 25, 1112 (1967)].
16. H. K. Holt, Phys. Rev. Letters 19, 1275 (1967).
17. J. A. White and R. Bose, Bull. Am. Phys. Soc. 13, 172 (1968).
18. T. Hänsch and P. Toschek, Z. Physik 236, 213 (1970).
19. T. Hänsch and P. Toschek, Z. Physik 236, 373 (1970).
20. T. Ya Popova, A. K. Popov, S. G. Rautian, and R. I. Sokolovskii, Zh. Eksp. Teor. Fiz. 57, 850 (1969); 57, 2254(E) (1970) [Sov. Phys. JETP 30, 466 (1970); 30, 1208 (1970)].
21. A. K. Popov, Zh. Eksp. Teor. Fiz. 58, 1623 (1970) [Sov. Phys. JETP 31, 870 (1970)].
22. R. Bose and J. A. White, Nuovo Cimento 5B, 11 (1971).
23. N. Skribanowitz, M. J. Kelly, and M. S. Feld, Phys. Rev. A 6, 2302 (1972).
24. Population saturation considerations predict a Lorentzian change signal of full width equal to $\Gamma_{p.\,s.} = 2\gamma(1+(k'/k)Q)$. Therefore, Γ_{br} and $\Gamma_{p.\,s.}$ are identical for $Q \approx 1$, but they differ for $Q \gg 1$.
25. S. H. Autler and C. H. Townes, Phys. Rev. 100, 703 (1955).
26. A. Javan, Phys. Rev. 107, 1579 (1957).
27. B. J. Feldman and M. S. Feld (to be published).
28. S. Stenholm and W. E. Lamb, Jr., Phys. Rev. 181, 618 (1969).
29. B. J. Feldman and M. S. Feld, Phys. Rev. A 1, 1375 (1970).
30. R. H. Cordover, P. A. Bonczyk, and A. Javan, Phys. Rev. Letters 18, 730 (1967).
31. T. W. Ducas, M. S. Feld, L. W. Ryan, Jr., N. Skribanowitz, and A. Javan, Phys. Rev. A 5, 1036 (1972).
32. W. G. Schweitzer, Jr., M. M. Birky, and J. A. White, J. Opt. Soc. Am. 57, 1226 (1967).
33. I. M. Beterov, Yu A. Matyugin and V. P. Chebotayev, ZhETF Pis. 10, 296 (1969) [JETP Letters 10, 187 (1969)].
34. A. Szabo, Phys. Rev. Letters 25, 924 (1970); 27, 323 (1971).
35. L. A. Riseberg, Phys. Rev. Letters 28, 786 (1972).
36. H. R. Schlossberg and A. Javan, Phys. Rev. Letters 17, 1242 (1966).
37. G. W. Flynn, M. S. Feld, and B. J. Feldman, Bull. Am. Phys. Soc. 12, 669 (1967).
38. R. G. Brewer, Phys. Rev. Letters 25, 1639 (1970), and in this volume.
39. T. F. Johnson, Jr., and G. J. Wolga, Phys. Letters 27A, 639 (1968).
40. T. Hänsch and P. Toschek, IEEE J. Quantum Electron. QE-4, 467 (1968).
41. T. Hänsch and P. Toschek, IEEE J. Quantum Electron. QE-4, 530 (1968); QE-5, 61 (1969).
42. T. Hänsch, R. Keil, A. Schabert, Ch. Schmelzer, and P. Toschek, Z. Physik 226, 293 (1969).

43. I. M. Beterov and V. P. Chebotayev, ZhETF Pis. Red. 9, 216 (1969) [JETP Letters 9, 127 (1969)].
44. M. S. Feld, B. J. Feldman, and A. Javan, Bull. Am. Phys. Soc. 12, 669 (1967); L. H. Domash, B. J. Feldman, and M. S. Feld, Phys. Rev. A 7, 262 (1973).
45. M. S. Feld, Ph. D. thesis (M. I. T. 1967).
46. J. S. Levine, P. A. Bonczyk, and A. Javan, Phys. Rev. Letters 22, 267 (1967).
47. N. Skribanowitz, I. P. Herman, R. M. Osgood, Jr. , M. S. Feld, and A. Javan, Appl. Phys. Letters 20, 428 (1972).
48. N. Skribanowitz, I. P. Herman, and M. S. Feld, Appl. Phys. Letters 21. 466 (1972).
49. N. Skribanowitz, M. S. Feld, R. E. Francke, M. J. Kelly, and A. Javan, Appl. Phys. Letters 19, 161 (1971).
50. A. Javan, in this volume.
51. J. T. Latourette, W. E. Quinn, and N. F. Ramsey, Phys. Rev. 107, 1202 (1957).
52. G. M. Grosof, P. Buck, W. Lichten, and I. I. Rabi, Phys. Rev. Letters 1, 214 (1958).
53. M. L. Perl, I. I. Rabi, and B. Senitzky, Phys. Rev. 98, 611 (1955).
54. See, for example, B. P. Stoicheff in Quantum Electronics and Coherent Light, Proceedings of the International School of Physics "Enrico Fermi", Course XXXI, edited by C. H. Townes and P. A. Miles (Academic Press, New York, 1964), p. 306.
55. Equation (15) is valid either if $0 < k'-k \ll k$ or if $0 < k-k' \ll k/Q^2$. Equation (16) holds whenever $0 < |k-k'| \ll k$.
56. J. Goldhar, R. M. Osgood, Jr. , and A. Javan, Appl. Phys. Letters 18, 167 (1971); D. R. Wood, E. G. Burkhardt, M. A. Pollack, and T. J. Bridges, Appl. Phys. Letters 18, 112 (1971).
57. A. Sanchez and M. S. Feld, in Proceedings of the Aussois Conference on High Resolution Molecular Spectroscopy, Colloques Internationaux du CNRS, 1973 (to be published).
58. W. Hanle, Z. Physik 30, 93 (1924); Ergeb. Exakt. Naturw. 4, 214 (1925).
59. For a complete discussion of the Hanle Effect see A. C. G. Mitchell and M. W. Zemansky, Resonance Radiation and Excited Atoms (Cambridge University Press, London, 1971).
60. P. A. Franken, Phys. Rev. 121, 508 (1961).
61. N. Skribanowitz, I. P. Herman, J. C. MacGillivray and M. S. Feld, Phys. Rev. Letters 30, 309 (1973).

NONLINEAR INFRARED SPECTROSCOPY

AND COHERENT TRANSIENT EFFECTS

Richard G. Brewer
IBM Research Laboratory
San Jose, California

ABSTRACT

Molecular examples of nonlinear infrared spectra are given where the resolution is of order 10^9. Homogeneous line shapes which are one thousandth of the Doppler width can be sampled using Lamb dip, optical double resonance, and level crossing techniques. The measurements have been extended to two spectral regions, 3.39 μm using a tunable cw He-Ne laser and 9-11 μm with a cw CO_2 laser. By Stark tuning, it has been possible to obtain vibration-rotation line assignments unambiguously, their transition matrix elements, and high-precision dipole-moment values for ground and excited vibrational states. In related transient studies, photon echo and optical nutation have been easily observed in $C^{13}H_3F$ and NH_2D by applying Stark pulses which shift the molecular levels into resonance with cw laser radiation. Theoretical analysis of the nutation effect agrees qualitatively with observation, and echo characteristics closely follow predictions of existing theories. The T_2' pressure dependence, from infrared echo measurements, indicates that $C^{13}H_3F$ relaxes primarily by rotational energy transfer.

I. INTRODUCTION

This occasion marks the tenth anniversary of the gas laser [1], and indeed, this talk will be concerned with spectroscopic applications of gas lasers. My paper will be divided into two parts: (1) Nonlinear infrared spectroscopy and (2) coherent optical transient effects. The first topic will cover the application of Lamb dip, optical double resonance, and level crossing to high-resolution studies of molecular infrared transitions. These are measurements under steady-state conditions. The dynamic or transient experiments form the second subject, and at the time of this meeting, these results are but one month old. I will describe a new technique, which is both simple and versatile, for monitoring photon echoes, optical nutation, and self-induced transparency effects.

Many of the optical examples given here will of course resemble that of NMR and much of the information which can be derived will be similar. It may be useful at the outset to note, however, that these topics are sufficiently broad that the motivation in these studies could follow many different paths. To mention a few examples, molecular structure, relaxation processes, new nonlinear optical effects, coherence phenomena, metrology, geophysical applications, and so on, can all be examined.

II. HISTORICAL NOTE ON NONLINEAR SPECTROSCOPY

Our initial effort in nonlinear spectroscopy commenced at IBM in late 1966, almost four years after Lamb's prediction [2] of a power dip in the tuning characteristic of a gas laser had been proposed and verified experimentally [3,4]. The Lamb-dip experiments of Szöke and Javan [3], and also of McFarlane, Bennett, and Lamb [4] were the first examples of nonlinear spectroscopy using lasers, and related techniques [5,6] such as optical double resonance and level crossing followed. (The literature has grown extensively since and I shall not attempt to give a complete historical account here.) In all of these cases, atomic systems and atomic lasers in particular were examined. The application of these techniques to molecules meanwhile met with greater difficulty even though attempts were made in Professor Javan's laboratory, Professor Shimoda's, ours, and perhaps elsewhere as well. For reasons which are perhaps still not clear, these initial efforts failed, but in 1969, a few papers appeared in which Lamb-dip resonances were observed in I_2 [7], CH_4 [8], NH_2D [9], and SF_6 [10]. The work I will be covering, therefore, spans the last two- to three-year period.

III. COMMENT ON PRECISION AND SENSITIVITY

The nonlinear techniques to be described resemble in many ways some older spectroscopic methods with regard to precision and sensitivity as well as in the richness of detailed information which may be derived. Many similarities can be found with Kastler's optical pumping, with microwave spectroscopy, and perhaps least of all with conventional infrared spectroscopy. Just as in microwave spectroscopy, for example, these nonlinear methods permit high spectral resolution and high detection sensitivity. These properties are directly attributable to the laser source, which resembles the klystron in its high intensity, low noise, and high spectral purity.

The resolution or Q value we are considering here is of order 10^9, i.e., at least 10^3 times greater than what conventional infrared spectroscopy can offer. This value is, of course, the highest resolution to date in the infrared, and it is for this reason that these methods are of interest in stabilizing lasers for use as optical frequency standards. A spectroscopic consequence of this resolution is that it has been possible for the first time to uncover very small splittings in infrared transitions due either to Stark or Zeeman effect or hyperfine structure.

Nonlinear absorption signals as small as 10^{-8} of the incident laser power can now be detected. This sensitivity implies sample gas pressures of a few mTorr in a path length of a few centimeters. It has been possible, therefore, to detect trace isotopic molecular impurities in natural abundance and in individual quantum states, and it should be possible to observe (hot band) transitions between excited vibrational states as well.

IV. LAMB DIP

In molecular gases, the minimum linewidth which may be resolved by conventional infrared spectroscopy is due to the inhomogeneous Doppler broadening. The best resolution which has been obtained in this way is illustrated in the middle spectrum of Fig. 1. These are P(7) multiplet lines of the CH_4 ν_3 band at 3.4 μm, and note that the Doppler width of 260 MHz is almost resolved in each line [11]. For comparison, the entire ν_3 band at a lower resolution of 100 Doppler widths is shown in the top spectrum [12]. However, by means of nonlinear techniques, the homogeneous linewidth which may be one thousandth or less of the Doppler width may be sampled. This is indicated schematically in the lowest spectrum of Fig. 1.

These narrow spectral features cannot be monitored, of course, by

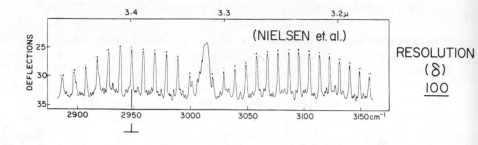

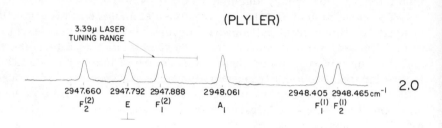

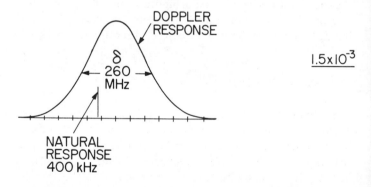

Figure 1. The ν_3 band of CH_4 at 3.3 μm. Top trace: low resolution; the P(7) line group is identified at 2950 cm^{-1} with a tic mark. Middle trace: the same P(7) line group resolved by conventional high-resolution spectroscopy. Bottom trace: the Doppler line shape with a narrow homogeneous line shape superimposed. Resolution is given in each case in terms of the Doppler width $\delta = 260$ MHz.

424

merely saturating a transition. The saturation behavior must exhibit an enhanced response with frequency tuning. Several such methods have been devised, and we shall consider the Lamb dip first. In this case, a standing-wave radiation field from a laser excites a molecular transition. As the laser frequency is tuned through the absorber's Doppler line shape, saturation occurs everywhere but it is enhanced only at the Doppler peak. Those molecules, which have only a transverse velocity to the laser beam, can interact simultaneously with both travelling-wave components. Off the peak, the interaction is with a single travelling wave and is thus reduced.

Our initial observations of molecular Lamb dips were actually made with NH_2D [9], but we shall consider here more recent studies in CH_4 [13]. A Lamb-dip response is shown in the upper trace of Fig. 2 for a ν_3 band transition of CH_4, the E Coriolis component of P(7) (see Fig. 1 also). The spectrum is obtained by tuning a stable single-mode cw 3.39-μm He-Ne laser, with an internal Stark cell, through the Doppler line center. Using an axial magnetic field, the laser frequency may be shifted ~ 3 GHz into near coincidence with the CH_4 transition while a fine frequency sweep is provided by a piezoelectric scan of the cavity length. Derivative line-shape resonances are observed as a change in laser output using weak audio modulation of cavity length and phase-sensitive detection.

The homogeneous linewidth is about 60 kHz HWHM and corresponds to a resolution of $\sim 10^9$. However, with the application of a Stark field, the transition splits into 13 equally spaced and completely resolved components,

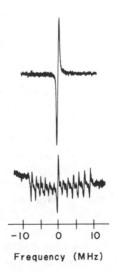

Frequency (MHz)

Figure 2. Lamb-dip spectra of CH_4 in the absence of a Stark field (top) and in the presence of a field of 1660 V/cm (bottom).

all falling within 20 MHz or one-tenth of the Doppler width. The number of lines confirms unambiguously that the upper level is J=6, in agreement with the P(7) assignment. This Stark effect is first order and results from a vibrational anharmonicity which removes the three-fold v_3 vibrational degeneracy--an effect which was predicted by Mizushima et al in 1953 [14, 15].

The magnitude of this induced dipole moment cannot be obtained from the Lamb-dip results as there is no frequency scale. It has been determined [13], however, by an optical double-resonance technique, which we will describe in more detail in the next section. The result is μ = 0.0200 ± 0.0001 D and although it is quite small, it can be determined with considerable precision. Thus, we see that even the observation of a single molecular transition under high resolution can yield a variety of significant results. Such findings will be particularly useful in testing ab initio calculations which are becoming increasingly more accurate and applicable with the aid of large-scale computational methods.

We note that several nonlinear molecular resonances have now been observed in our laboratory within the tuning range of the 3.39-μm laser and that CH_4 is but one example. Such quantities as transition matrix elements [9, 16] and the g value [16] of $^1\Sigma$ molecules have been obtained.

V. OPTICAL DOUBLE RESONANCE AND LEVEL CROSSING

Double-resonance spectroscopy is a familiar subject in the microwave and radio-frequency regions, but it is only recently that the analog has been carried out at optical frequencies. As in all double-resonance effects, two transitions sharing a common level are excited simultaneously by two radiation fields. For the case of two overlapping Doppler-broadened transitions, this interaction will only occur when both fields are Doppler shifted equally, or $\Omega_1 - \omega_1 = \Omega_2 - \omega_2$. Here the optical frequencies are Ω_1 and Ω_2 and the corresponding molecular-level intervals are ω_1 and ω_2. Thus, if the splitting $\omega_1 - \omega_2$ is tuned by the Stark or Zeeman effect, an enhanced nonlinear absorption response will occur for the resonance condition $\Omega_1 - \Omega_2 = \omega_1 - \omega_2$. Note that this is similar to the Lamb dip in that the same velocity group must interact simultaneously with two radiation fields at resonance, but off resonance the nonlinear interaction is with only one field and is therefore reduced. Since only a single narrow velocity group is monitored, its resonance line shape will be homogeneously broadened and again, it may be orders of magnitude narrower than the Doppler width.

Experiments of this type have been carried out on some v_3 band lines of CH_3F in the 10-μm region [17]. This symmetric top has a permanent dipole moment μ and exhibits a first-order Stark effect in an external field E so that the nonlinear resonance condition becomes

$$\Omega_1 - \Omega_2 = 2\mu EK/J(J+1)\hbar \tag{1}$$

Once the rotational assignment (J, K) is made, Eq. (1) permits a dipole-moment determination.

The experimental setup is indicated in Fig. 3. While a multimode cw 3.39-μm He-Ne laser sufficed in the CH_4 studies described earlier, at 10 μm two separate cw CO_2 lasers, each oscillating in single mode, are required. The difference frequency $\Omega_1 - \Omega_2$ is easily determined by monitoring the beat signal with a counter, and the Stark field E can be obtained with a digital voltmeter and a precision measurement of the gap spacing. Many of these and other details can be found in our earlier paper [17].

Recently we have introduced two new features which greatly improve the precision of these double-resonance measurements. First, instead of using two free-running CO_2 lasers, one laser is frequency locked to the other in a servo loop. An error voltage derived from the beat $\Omega_1 - \Omega_2$ and a reference frequency drives the slave laser by piezoelectric cavity tuning. In this way, instabilities and drift in $\Omega_1 - \Omega_2$ no longer limit the precision as this difference frequency can be controlled to 1/300,000. Second, the spectrum is digitized by a computer as it is being taken, and the line position, shape, and width can be derived from a least-squares computer analysis. If necessary,

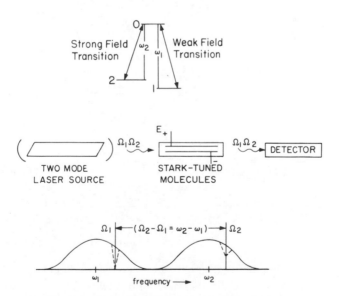

Figure 3. Optical double-resonance technique showing the level structure, experimental arrangement, and the tuning behavior.

the signal-to-noise ratio can be enhanced by multiple sweep time averaging. The present uncertainties are in the determination of line center and in the field measurement.

A CH_3F double-resonance spectrum with these improvements is presented in Fig. 4. The transition assignment, which is determined from the Stark tuning rate, is $(v_3, J, K) = (0, 12, 2) \rightarrow (1, 12, 2)$. The first line to appear is the zero-field level crossing signal which is nothing more than the degenerate case of optical double resonance. The resonance condition is simply $\omega_1 - \omega_2 = 0$ and can be observed with a single frequency source. The next two lines correspond to the resonance condition with finite level splitting. Since the dipole moments of ground and excited states are slightly different, there are two such resonances. It is thus possible to obtain precise dipole-moment values for both ground and excited vibrational states with equal ease. Our current results are summarized in Table I and compare favorably with our limits of uncertainty of $\sim 1/2000$ with earlier microwave [18] and recent molecular beam [19] work.

The laser technique is of course not limited to the ground vibrational or lowest rotational states as is usually the case in microwave studies. In principle, the dipole moments of all six normal modes of CH_3F could be examined and would constitute a severe test of molecular ab initio

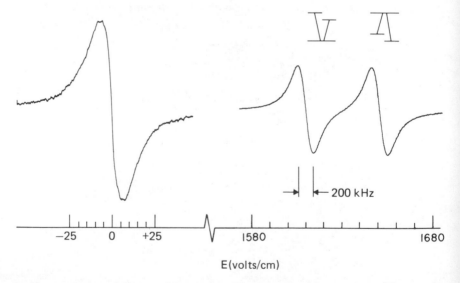

E(volts/cm)

Figure 4. Stark-tuned optical double-resonance spectrum of $C^{12}H_3F$ at 3.1 mTorr pressure. The transition assignment of this v_3 band line is $(J, K = 12, 2 \rightarrow 12, 2)$. Excitation is by two CO_2 lasers oscillating on the P(2 line (9.6-μm band) and with a frequency difference of 39.629 MHz.

TABLE I.   ELECTRIC DIPOLE MOMENT OF $C^{12}H_3F$

V	J	K	Dipole Moment (Debye)	Technique
0	0, 1, 2	0, 1	1.8572 (10)	Microwave spectroscopy [18]
0	1	1	1.85850 (4)	Molecular beam [19]
0	5	4	1.85852 (11)	Molecular beam [19]
0	12	2	1.8596 (10)	Optical double resonance
$v_3=1$	12	2	1.9077 (10)	Optical double resonance

calculations.   Thus, in addition to the v_3 vibrational mode listed in Table I, we are performing similar double-resonance measurements of the v_1 mode using the 3.39-μm He-Ne laser.   Additional studies will be needed to see the small variation of dipole moment with rotational level.

VI.   A MOLECULAR SIDEBAND METHOD

Still another molecular level crossing effect [20] developed by Dr. Luntz of this laboratory may be observed with Stark tuning when a radio-frequency field is applied simultaneously across the Stark plates.   The rf field modulates the molecular wave function, producing molecular sidebands.   With Stark tuning, the sideband and parent levels cross; the resonance condition being the same as in the double-resonance case.   These level crossings have been observed in CH_4; they yield the same dipole moment as obtained with the double-resonance technique, and their intensity is predicted by the Autler-Townes theory [21].   The effect is quite similar to Cohen-Tannoudji's studies of atoms "dressed" by rf magnetic fields [22], and is a useful alternate method for obtaining narrow resonances.

VII.   PHOTON ECHOES AND OPTICAL NUTATION [23]

Let us consider now some of the coherent optical transient effects which we have observed rather simply using a new pulse technique [24].   Previous echo [25-27] and nutation studies [28,29] have required one or more coherent optical pulses, but in the present work the exciting laser radiation is cw and the molecular level splitting is pulsed.   The experimental arrangement is given in Fig. 5 and the principle is illustrated in Fig. 6.   It will be seen that

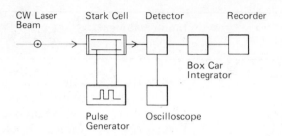

Figure 5. Monitoring technique for observing optical transient signals following one or more Stark pulses.

for molecules such as CH_3F, which exhibit a first-order Stark effect, transient optical signals are readily seen in transmission by electronically gating the optical absorption with a pulsed electric field.

In considering Fig. 6, assume for simplicity a nondegenerate two-level system which exhibits a change in transition frequency of $(\Delta\gamma)E$ when a Stark pulse of amplitude E is applied. Prior to the pulse, molecules of velocity v are continuously excited by radiation of frequency Ω, but when the pulse

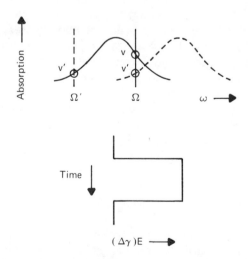

Figure 6. Optical switching behavior of a Doppler-broadened transition to a Stark pulse of amplitude E. The frequency shift is $(\Delta\gamma)E$ where $\Delta\gamma$ is the difference in the gyroelectric ratio of lower and upper levels. The laser frequency is Ω and emission can occur at Ω'.

appears this group is no longer in resonance and it will spontaneously radiate. At the same time, a new velocity group v′ will be switched into resonance. If the pulse period is sufficiently long, this group will exhibit an oscillating macroscopic electric dipole moment and several cycles of damped nutation may be optically monitored. When the pulse terminates, the initial velocity group v will be excited and it too will nutate. Both of these nutation sequences are shown in Fig. 7 for $C^{13}H_3F$. Experimental details are given elsewhere [24]. The transition is $(v_3, J, K) = (0, 4, 3) \rightarrow (1, 5, 3)$ which was assigned recently by the optical double-resonance technique described earlier. In addition, similar optical ringing patterns have been observed [24] for a nondegenerate transition of NH_2D, $(v_2, J, M) = (0, 4, 4) \rightarrow (1, 5, 5)$.

An oversimplified calculation of these nutation signals may be obtained by considering a nondegenerate two-level system (a, b) which is suddenly excited by an optical field $\mathcal{E} = \mathcal{E}_0 \sin \omega t$. If one neglects such quantities as the pulse rise time, propagation effects, the influence of Doppler broadening,

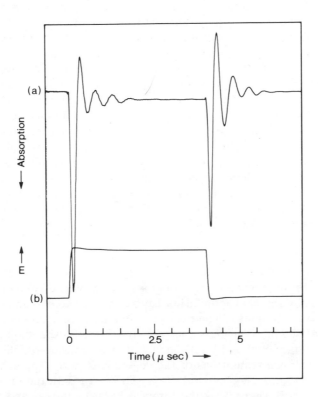

Figure 7. Optical nutation in $C^{13}H_3F$ at 4.8 mTorr pressure. The detector aperture is reduced to a 1-mm diameter. Pulse amplitude: E = 35 V/cm.

and level degeneracy, the problem is then equivalent to transient nutation i:
NMR--a problem which has already been solved [30]. The time-dependent
power absorbed following sudden excitation at time t=0 will be of the form

$$P(t) = (n_a - n_b) \left(y^2 \hbar w_0 / \gamma \right) \sin(2\gamma t) \exp(-t/T_2')$$

where w_0 is the level splitting, $\gamma = \frac{1}{2} [(w - w_0)^2 + 4yy^*]^{1/2}$, and $y = \mu_{ab} \mathcal{E}_0 / 2\hbar$.
The derivation of Eq. (2) follows Ref. 31. The optical ringing pattern of
Eq. (2) will thus resemble Fig. 7, and that of NH_2D, the nutation frequency
being $\Omega = \mu_{ab} \mathcal{E}_0 / \hbar$ at exact resonance.

Consider now the photon-echo experiment. Here, two Stark pulses
produce an optical echo as indicated in the $C^{13}H_3F$ response of Fig. 8. Th
first pulse places the system in a superradiant state which dephases as the
pulse terminates. The dephasing mechanism is Doppler broadening where
the velocity spread is determined by the pulse width τ. At a later time τ_s,
a second pulse of width 2τ reverses the dephasing process and restores the
macroscopic electric dipole moment which emits an echo at time $\sim 2 \tau_s$.

The observed echo differs, however, from the ordinary echo signal
because its frequency is shifted from the laser frequency. This can be see
in Fig. 6. During the pulse, molecules of velocity v' are excited by radiatic
of frequency Ω, but after the pulse they can emit at Ω'. The effect arises
because the identity of the velocity groups excited during the two pulses is
preserved afterward, and the transition frequencies will be different in zer
and nonzero Stark fields. The laser and echo radiation, therefore, produc
a beat signal at the detector because the two beams are collinear and have
the same polarization. Since the laser radiation is stronger than the echo
intensity, a pulsed heterodyne signal appears, and hence the detection sens
tivity is greatly enhanced. This technique, therefore, monitors the echo
amplitude $\mathcal{E}$ due to heterodyne detection, rather than its intensity, and its
decay envelope will be given by

$$\mathcal{E}/\mathcal{E}_0 = \exp(-2\tau_s/T_2')$$

Here, T_2' is the homogeneous part of the transverse relaxation time.

Several echo characteristics have now been tested and compare favoral
with existing theories. The beat phenomenon, for example, was verified by
noting that the beat frequency varies linearly with the Stark pulse amplitude
[compare Figs. 8(a) and 8 (c)], and with a magnitude which corresponds to
the more intense transitions of the degenerate set. The echo delay time is
also in satisfactory agreement, to within $\sim 1\%$, with the predicted value [32
$(2 \tau_s + \tau' - \tau/2)$ where τ_s is the interval between pulses and τ and τ' are the
widths of first and second pulses. The duration of the echo is $\sim \tau$ as
expected [32]. In addition, we have demonstrated that the echo disappears

when only one pulse is present and that its amplitude is a maximum when the first and second pulse areas are $\pi/2$ and π, respectively. Finally, from Eq. (3) the dependence of echo amplitude (extrapolated to $\tau_s = 0$) on molecular density N is verified as being linear, and thus the echo intensity varies as N^2 in accord with the theory of superradiance [33]. It may be of interest that with the available signal-to-noise ratios of 10^3 or larger, 10^{10} molecules/cm^3 or less can yield a detectable echo signal.

The echo measurements for $C^{13}H_3F$ exhibit a T_2' pressure dependence which can be represented by

$$1/\pi T_2' = 250 \text{ kHz} + (31 \text{ kHz/mTorr})p \tag{4}$$

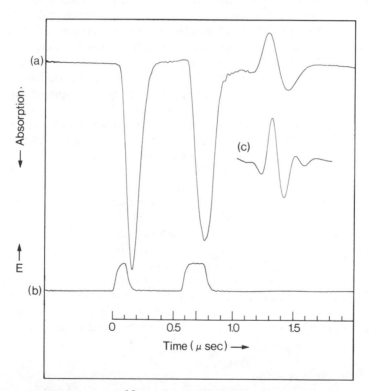

Figure 8.   Photon echo in $C^{13}H_3F$ at 5.2 mTorr pressure.   (a) Optical response to $\pi/2$ and π pulses followed by the echo beat signal.   (b) Corresponding Stark pulses where the interval $\tau_s = 570$ nsec and the pulse widths for τ and τ' are 90 and 180 nsec.   Pulse amplitude: $E = 35$ V/cm.   (c) Echo signal for the case $E = 60$ V/cm.   The other conditions are the same as (a) but note that the beat frequency is $\sim 2\times$ larger.   Detector aperture is not reduced in (a) and (c).

It is valid over at least one decade of pressure (0.5-10 mTorr). Here, the quantity $1/\pi T_2'$ is the equivalent Lorentzian linewidth (full width at half-maximum); the pressure-independent value of 250 kHz reflects the molecular transit time across the laser beam; and the pressure-broadening parameter $2\Delta\nu/p = 31$ kHz/mTorr. Note that echo measurements of T_2' should not be subject to power-broadening effects and that Eq. (4) does not require this correction as in steady-state linewidth determinations [34].

Equation (4) can be compared to the $C^{12}H_3F$ line-broadening parameter obtained from microwave measurements where the principal relaxation mechanism is rotational energy transfer and can be explained in terms of dipole-dipole interactions [35]. The microwave value is $2\Delta\nu/p = 40$ kHz/mTorr for the $J,K = 0,0 \rightarrow 1,0$ transition and is estimated to be 25 kHz/mTorr for the $J,K = 4,3 \rightarrow 5,3$ transition [35]. These number suggest that the rotational relaxation process is dominant for our infrared transition as well and that low-angle elastic scattering (which changes the molecular velocity component along the beam direction) is not. Other mechanisms which contribute to a lesser degree include resonant vibration energy transfer and disruptive binary collisions which bring about significa phase shifts.

It is interesting to note that optical linewidths derived from nonlinear spectroscopic techniques, such as Lamb dip and double resonance, might b highly sensitive to low-angle elastic scattering since only $\sim 1/10^3$ of the Doppler width is sampled. Microwave spectroscopy, on the other hand, monitors the entire Doppler line shape and is insensitive to velocity changi collisions. The echo experiment falls between the two cases since $\sim 1/15$ c the Doppler width is sampled; this may help to explain why elastic scatteri is not of greater importance in our measurements.

The use of a cw laser with pulsed Stark fields is clearly a convenient alternate method for investigating relaxation processes and optical transier phenomena and has the experimental advantage of being simple and versatil Self-induced transparency [36] could be examined as well and by analogy wi NMR, a variety of other experiments such as adiabatic fast passage [37] an multiple-pulse techniques [38] which yield T_1 should be feasible.

REFERENCES

1. A. Javan, W. R. Bennett, Jr., and D. R. Herriott, Phys. Rev. Lette 6, 106 (1961).
2. W. E. Lamb, Jr., Phys. Rev. 134, A1429 (1964).
3. A. Szöke and A. Javan, Phys. Rev. Letters 10, 521 (1963).
4. R. A. McFarlane, W. R. Bennett, and W. E. Lamb, Jr., Appl. Phys Letters 2, 189 (1963).

5. H. R. Schlossberg and A. Javan, Phys. Rev. Letters 17, 1242 (1966); Phys. Rev. 150, 267 (1966).

6. M. S. Feld and A. Javan, Phys. Rev. 177, 540 (1969).

7. G. R. Hanes and C. E. Dahlstrom, Appl. Phys. Letters 11, 362 (1969).

8. R. L. Barger and J. L. Hall, Phys. Rev. Letters 22, 4 (1969).

9. R. G. Brewer, M. J. Kelly, and A. Javan, Phys. Rev. Letters 23, 559 (1969).

10. P. Rabinowitz, R. Keller, and J. T. LaTourrette, Appl. Phys. Letters 14, 376 (1969); F. Shimizu, Appl. Phys. Letters 14, 378 (1969).

11. Kindly provided by E. K. Plyler prior to publication.

12. A. H. Nielsen and H. H. Nielsen, Phys. Rev. 48, 864 (1935).

13. A. C. Luntz and R. G. Brewer, J. Chem. Phys. 54, 3641 (1971), for Lamb dip and double resonance in CH_4. For level crossing in CH_4 see A. C. Luntz, R. G. Brewer, K. L. Foster, and J. D. Swalen, Phys. Rev. Letters 23, 951 (1969).

14. M. Mizushima and P. Venkateswarlu, J. Chem. Phys. 21, 705 (1953).

15. Laser Stark spectroscopy of this CH_4 transition has been examined also by K. Uehera, K. Sakurai, and K. Shimoda, J. Phys. Soc. Japan 26, 1018 (1969); however, the linear absorption technique utilized was limited by Doppler broadening and it was not possible to resolve the Stark pattern or obtain an accurate dipole moment measurement.

16. R. G. Brewer and J. D. Swalen, J. Chem. Phys. 52, 2774 (1970); A. C. Luntz and R. G. Brewer, J. Chem. Phys. 53, 3880 (1970).

17. R. G. Brewer, Phys. Rev. Letters 25, 1639 (1970).

18. P. A. Steiner and W. Gordy, J. Mol. Spectroscopy 21, 291 (1966).

19. S. C. Wofsy, J. S. Muenter, and W. Klemperer, J. Chem. Phys. 55, 2014 (1971).

20. A. C. Luntz, Chem. Phys. Letters (to be published).

21. S. H. Autler and C. H. Townes, Phys. Rev. 100, 703 (1955).

22. C. Cohen-Tannoudji, in this volume; S. Haroche, C. Cohen-Tannoudji, C. Audoin, and J. P. Schermann, Phys. Rev. Letters 24, 861 (1970).

23. The criticism of existing nomenclature which was raised at this meeting by W. E. Lamb is valid in my opinion. The echo phenomenon is not a single-photon event, and optical transitions do not exhibit nutation. In any event, I will not put forward new phrases now when the older ones seem to be so firmly entrenched.

24. R. G. Brewer and R. L. Shoemaker, Phys. Rev. Letters 27, 631 (1971).

25. N. A. Kurnit, I. D. Abella, and S. R. Hartmann, Phys. Rev. Letters 13, 567 (1964); Phys. Rev. 141, 391 (1966).

26. C. K. N. Patel and R. E. Slusher, Phys. Rev. Letters 20, 1087 (1968).

27. B. Bölger and J. C. Diels, Phys. Letters 28A, 401 (1968).

28. G. B. Hocker and C. L. Tang, Phys. Rev. Letters 21, 591 (1968).

29. P. W. Hoff, H. A. Haus, and T. J. Bridges, Phys. Letters 25, 82

(1970).

30. H. C. Torrey, Phys. Rev. 76, 1059 (1949).

31. R. P. Feynman, F. L. Vernon, Jr., and R. W. Hellwarth, J. Appl. Phys. 28, 49 (1957).

32. A. L. Bloom, Phys. Rev. 98, 1105 (1955); S. Fernbach and W. G. Proctor, J. Appl. Phys. 26, 170 (1955).

33. R. H. Dicke, Phys. Rev. 93, 99 (1954).

34. Additional studies will be needed to determine whether or not the pressure-dependent part of Eq. (4) varies with the molecular transit-time effect and hence with beam diameter.

35. J. S. Murphy and J. E. Boggs, J. Chem. Phys. 47, 4152 (1967); our estimate of the correction factor is an interpolation of calculated line-widths versus J, K taken from Fig. 6 of this paper.

36. S. L. McCall and E. L. Hahn, Phys. Rev. Letters 18, 908 (1967); Phys Rev. 183, 457 (1969); Phys. Rev. A2, 861 (1970).

37. N. Bloembergen, E. M. Purcell and R. V. Pound, Phys. Rev. 73, 679 (1948).

38. E. L. Hahn, Phys. Rev. 80, 580 (1950).

HYDROGEN MASER RESEARCH[*]

Norman F. Ramsey
Department of Physics, Harvard University
Cambridge, Massachusetts

I. INTRODUCTION

Atomic hydrogen is the simplest of all atoms to interpret theoretically; as a result, its study has been a fruitful source of new ideas in physics. The most precise studies of atomic hydrogen have been the atomic hydrogen maser experiments on the hyperfine structure of atomic hydrogen in its electronic ground state.

The energy levels of atomic hydrogen in its electronic ground state are shown in Fig. 1. In weak magnetic fields the higher energy $F = 1$ states correspond to the electron and proton spins being parallel while the spins are antiparallel in the $F = 0$ state. With the hydrogen maser the transitions between the various hyperfine energy levels of Fig. 1 have been studied under various conditions.

II. THE HYDROGEN MASER

The atomic hydrogen maser was first developed at Harvard University by Kleppner, Goldenberg, and Ramsey [1-4]. A schematic diagram of the hydrogen maser is shown in Fig. 2. Since ordinary hydrogen that comes in

[*]This work is partially supported by the National Science Foundation and the Office of Naval Research.

437

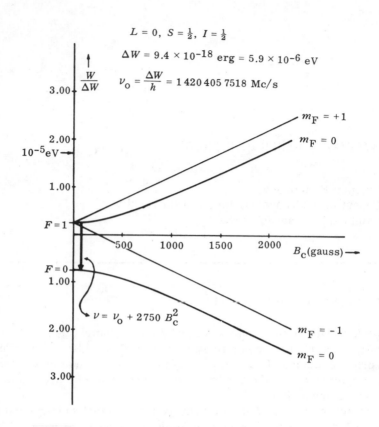

Figure 1.   Atomic hydrogen hyperfine structure showing the dependence of the energy levels on the strength of an external magnetic field.   The heavy arrow indicates the transition used in stable atomic oscillators.

a storage tank is molecular hydrogen (H_2) rather than atomic hydrogen (H), it is first necessary to convert the molecular hydrogen to the atomic form. This is done by admitting the molecular hydrogen into a small quartz or Pyrex bulb which forms the atomic hydrogen source.   This bulb is surrounded by a coil of wire which is excited with a standard radio-frequency oscillator such as is used in a small radio broadcasting station.   The oscillating field from this coil establishes a gas discharge in the bulb similar to the familiar gas discharge in a neon advertising sign.   In this gas discharge the molecules of hydrogen are broken up into atoms, and most of what emerges from a small hole in the source bulb is atomic hydrogen.

The atoms of hydrogen emerge into a vacuum region which is exhausted

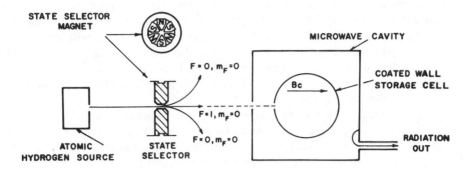

Figure 2. Schematic drawing of atomic hydrogen maser.

to a low pressure (less than 10^{-6} Torr). In such a low-pressure region the atoms will travel in straight lines unless acted upon by force. In particular, some of the atoms will go straight through the state selector magnet, as shown in Fig. 2. The state selector magnet has six poles, half of which are north poles and half south poles, arranged as shown schematically in the cross section in the upper portion of Fig. 2. By symmetry, the magnetic field must be zero on the exact central axis of the magnet. On the other hand, if the atom goes slightly off the central axis, the magnetic field will rapidly increase. Consequently, the energy of the atom will change when it is off axis. If the atom is in the $F = 0$ state, its magnetic energy will decrease as the atom gets farther off the axis as can be seen from Fig. 1. Since atoms prefer to go in the direction where the energy is lower, an atom of hydrogen in the $F = 0$ state which is off the central axis will be subject to a force pulling it still farther off the central axis. Therefore, such atoms will be defocused by the state selector magnet. On the other hand, the energies of the $M_F = + 1$ and 0 states for atoms with $F = 1$ atom increase as the atom gets farther off the axis. Consequently, the force on it pushes it back toward the axis; i.e., it is focused. The dimensions of the magnet are selected in such a fashion that the $F = 1$ atoms are focused on a small hole in a 6-in.-diam Teflon-coated quartz storage bulb.

As a result of the above focusing action on the $F = 1$ state and the defocusing action on the $F = 0$ state, the quartz bulb is dominantly filled with atoms in the higher-energy $F = 1$ state. The bulb is also surrounded by a microwave cavity tuned to the 1420-MHz frequency, characteristic of atomic hydrogen. Since the atoms in the bulb are dominantly in the higher-energy state, this arrangement satisfies all the requirements for maser amplifications. Indeed, such a device can be used as an amplifier. Moreover, if the amplification is sufficient it can also be a self-running oscillator. It is easy to see how oscillation can be established in such a device. If there is a weak noise signal present at the appropriate 1420-MHz frequency, stimulated emission will exceed absorption and the original signal will be amplified to a larger one which will be further amplified. The signal will thus get bigger and bigger up to the point where most of the energy being brought into the bulb by the atoms in the higher-energy state is absorbed. At this condition an equilibrium steady-state oscillation will be established. Although the power in the oscillation is quite weak--about 10^{-12} W--the oscillation is highly stable and is concentrated in an unprecedently narrow frequency band. Consequently the signal can easily be seen despite its low total power. Ordi narily the atoms are stored in the bulb for about one-third of a second, during which time each atom makes over 10, 000 collisions with the wall of the containing vessel.

An electrical coupling loop is inserted into the cavity so that some of the microwaves' oscillatory power can be coupled out for observation. The signal that emerges from such a maser proves to be unprecedentedly stable. This highly stable signal provides the basis for the experiments which will be described in the latter portion of the present report.

A photograph of a hydrogen maser is shown in Fig. 3. The entire device is about 4 ft tall. The vertical cylinders are vacuum pumps, and the large cylinder with a horizontal axis is the tuned microwave cavity. Inside that cylinder is the Teflon-coated quartz bulb containing atomic hydrogen dominantly in the high-energy F = 1 hyperfine state. In normal use the microwave cavity is further surrounded by three successive concentric layers of molypermalloy, which shields the apparatus from the magnetic disturbances in the room.

A photograph of the six-pole focusing magnet used in the hydrogen maser is shown in Fig. 4. The six Alnico magnets are shown in the photograph. The poles of the magnets alternate successively north and south. The atomic beam goes along the axis of the cylinder.

The unprecedentedly high stability of the maser microwave oscillation arises from a combination of four desirable features, all of which contribute to increased stability. These factors include:

(a) The atoms reside in the storage bottle for a much longer period of time than the atoms remain in a normal molecular beam apparatus. Conse- quently the characteristic resonance line is much narrower and the output signal is more stable, since the peak of a narrow line can be much more accurately located than the peak of a broad line. The narrowing of the line is just that to be expected from the Heisenberg Uncertainty Principle. According to this principle, a longer observation time makes possible a narrower resonance and consequently a more stable frequency. The narrow ness of the resonance also diminishes the pulling of the maser frequency by any mistuning of the microwave cavity; in addition the cavity can be accu-

Figure 3.   Photograph of hydrogen maser.

rately tuned by adjusting its tuning to be such that the output frequency is independent of the intensity of the beam of hydrogen atoms [3,5]. It has been shown by Crampton [5] that this method of cavity tuning eliminates the effect of a spin exchange frequency shift except for a small measurable shift of a few parts in 10^{13} due to the change in hyperfine frequency during the time of the collision [5].

(b) The atom is relatively free and unperturbed while radiating, unlike an atom in a resonance experiment using liquids, solids, or gases at relatively high pressure. Consequently, all atoms will have the same characteristic frequency, and the resultant resonance will not be broadened as it would be if it consisted of a superposition of a number of resonance at slightly different frequencies. Unfortunately the atoms of hydrogen are not totally free, since they must collide at intervals with the Teflon-coated wall of the storage bulb. This produces a shift in the maser frequency of about

Figure 4.   Photograph of six pole focusing magnet.

2 parts in 10^{11}. This correction, however, can be experimentally determined by observing the frequency of the output with bulbs of two different
sizes and then by extrapolating to a bulb of infinite diameter as discussed
later in greater detail.

(c) A further advantage of the hydrogen maser is that the first-order
Doppler shift is greatly reduced. With the very-narrow-line characteristic
of the atomic hydrogen maser, the relevant quantity for the Doppler shift is
the ratio of the average velocity of the hydrogen atom to the velocity of light.
Since the hydrogen atoms enter the storage bulb through a small hole, and
then stay in the storage bulb for about 1 sec before finally emerging from
the same hole, the average velocity is zero, or close to zero. Consequently
the first-order Doppler shift is completely negligible. There is a small
second-order Doppler shift due to the relativistic slowing down of any moving
clock or oscillator. Since the second-order Doppler shift depends upon the
velocity squared, it is not averaged to zero while the atom is in the bulb.
On the other hand, the second-order Doppler shift is a correction that can be

exactly calculated if the temperature and hence the mean square velocity of the atom is accurately known.

(d) A final advantage that contributes to the high precision of the atomic hydrogen maser is the low-noise characteristic of maser amplification. Since the amplifying element is an isolated simple atom, there is little opportunity for any extra noise to develop beyond the theoretical minimum noise. As a result, the maser is a very-low-noise amplifier, and the oscillation will be much more stable since the frequency is less likely to drift to a nearby noise peak.

For the above four reasons the hydrogen maser frequency should be very stable. This prediction has been confirmed experimentally as shown in Fig. 5. In the upper portion of that figure, the frequencies of two hydrogen masers are compared approximately once an hour for a period of 24 h. Since one large division on the vertical scale corresponds to a stability of one part in 10^{13} and since the variations are only about one-tenth of a large division, the two hydrogen masers over the period of time observed were stable to one part in 10^{14}. For many experiments this stability is all that is required for the desired measurement. On the other hand, often the observer wishes to know the absolute rate of the oscillation in terms of those of a totally free hydrogen atom. In such cases the correction for the wall shift must be known. Although this can be determined by making measurements with storage bulbs of two different sizes, the necessity for making the correction usually degrades the accuracy and a further degradation of the accuracy of absolute measurements comes from the necessity of measuring the cesium hyperfine structure as the standard of time.

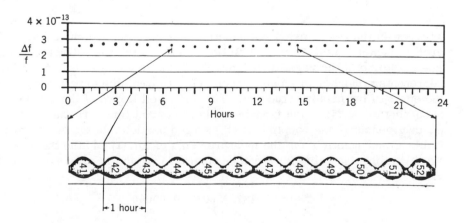

Figure 5. Relative frequency stability of two hydrogen masers.

III.   ATOMIC HYPERFINE SEPARATIONS

One of the first applications of the atomic hydrogen maser was to measure with increased accuracy the separation of the energy levels of atomic hydrogen and its isotopes. Hydrogen has been measured by Crampton, Kleppner, and Ramsey [6], Vessot et al. [7], Peters et al. [7], Kartaschoff [7], and others [7]. All results are consistent with Eq. (7) below. Deuterium and tritium have been measured by Crampton, Kleppner, Robinson, Mathur, and Ramsey [8, 9] with the results in Eqs. (2) and (3).

$$\Delta\nu(H) = 1,420,405,751.768 \pm 0.010 \text{ Hz} \tag{1}$$

$$\Delta\nu(D) = 327,384,352.5222 \pm 0.0017 \text{ Hz} \tag{2}$$

$$\Delta\nu(T) = 1,516,701,470.7919 \pm 0.0071 \text{ Hz} \tag{3}$$

The measurement on atomic hydrogen is based upon the international definition of the second in terms of the hyperfine separation of atomic cesium. All other results in the present report are measured in terms of the hydrogen hyperfine frequency and will be based on the assumption that Eq. (1) is exactly correct.

The theory of the atomic hydrogen hyperfine separation has been extensively considered from the point of view of quantum electrodynamics [10]. The value of $\Delta\nu$ for the hydrogen atom is given theoretically by

$$\Delta\nu = \frac{16}{3}\left(1 + \frac{m}{M}\right)^{-3}\frac{\mu_p}{\mu_o}\,\alpha^2\,cR_\infty\left(1 + \frac{\alpha}{2\pi} - 0.328\frac{\alpha^2}{\pi^2} - (1-\ln 2)\alpha^2 + K\right) \tag{4}$$

where m is the rest mass of the electron and M of the proton, μ_p/μ_o is the proton magnetic moment in Bohr magnetons, α is the fine-structure constant $(\alpha = e^2/\hbar c)$, c is the velocity of light, R_∞ is the Rydberg constant, and K represents a number of small quantum electrodynamic corrections including the effect of the proton structure. The least well known of the constants occurring in Eq. (4) is the fundamentally important fine-structure constant, α, that measures the coupling of all charged particles to the electromagnetic field. Consequently from the measurements of Eq. (1) and the theory of Eq. (4), a precise value for α can be obtained [11] and is consistent with the value

$$\alpha^{-1} = 137.03608 \pm 0.00080 \tag{5}$$

where the listed error is due to uncertainty in the theoretical corrections of Eq. (4), particularly in the correction for the proton structure. As emphasized by Cohen and Du Mond [11] and by others, the value of α given in this experiment differed significantly from that obtained from the early fine-structure measurements [12]. At one time this disagreement was attributed [11, 13] to errors in the corrections to Eq. (4), but further investigations [14-16] have shown that it is difficult to adjust the theory sufficiently to reconcile the two measurements. A completely different measurement [16] based on the Josephson effect in superconductivity has given a value of α in agreement with the result obtained from the maser hyperfine-structure experiments here described. In addition new results on the hydrogen fine structure by Metcalf, Brandenberger, and Baird [17] are compatible with Eq. (5). The situation was clarified by Appelquist and Brodsky [18] who found an error in the sign of some of the higher-order corrections to the theoretical Lamb shift. When this theoretical result is combined with the most recent fine-structure measurements, the value of α inferred from the fine-structure measurements agrees well with the value in Eq. (5) which comes from hyperfine measurements here described and from the Josephson effect. At the present time, in fact, the remarkable situation exists that there is excellent agreement between experiment and all theoretical calculations that depend only on quantum electrodynamics [18, 19]. With the consistency of all these measurements the hyperfine experiments can be reinterpreted as a means to measure $\delta N^{(2)}$, the proton polarizability contribution [16] due to the various excited states and to the internal structure of the proton. From this point of view, the hyperfine measurements on atomic hydrogen show that

$$\delta N^{(2)} = 2.5 \pm 4.0 \text{ ppm} \tag{6}$$

It can be seen from Eqs. (1) and (3) above that the frequency for atomic tritium is approximately the same as for atomic hydrogen. As a result, in this measurement it was easily possible to tune a normal hydrogen maser to operate at the frequency for tritium. Consequently the measurements could be made with high accuracy; in fact, the above quoted number for the hyperfine separation of atomic tritium is probably the most accurate measurement of any quantity of physical significance that has ever been made. The principal problem in the case of the tritium measurement arose from the intense radioactivity of the tritium isotope which required more complicated gas handling procedures.

For the atomic deuterium, however, the frequency differs from that of atomic hydrogen by more than a factor of 4, in which case the wavelengths and the necessary size of the cavity do likewise. Until very recently no one had made a true deuterium maser of this different size. Consequently, a quite different technique had to be used for the first deuterium measurements. If atomic deuterium is admitted into an operating atomic hydrogen maser, the intensity of the signal is reduced by the collisions between the hydrogen and deuterium atoms. The cause of this reduction is the phenomenon of spin exchange, i. e. , when a hydrogen and a deuterium atom are in collision, the electrons of the two atoms are shared and the atoms on emerging may have exchanged electrons with each other and consequently each acquires an electron with a different spin orientation than before the collision. This spin exchange essentially quenches the hydrogen hyperfine radiation, since the spin exchange changes the properties of the atom to such an extent that the original atom is essentially lost for the purposes of the measurement. The effectiveness of the spin exchange quenching, however, depends upon the orientation state of the deuterium atom. Consequently, if an oscillatory field is present at the deuterium hyperfine frequency, the quenching effect of the deuterium is altered and the intensity of the hydrogen maser signal, in turn, is changed. In other words, the strength of the hydrogen maser signal provides a detector for the existence of the atomic deuterium transition, and the atomic deuterium hyperfine frequency can thereby be measured. The first maser measurement on deuterium was made in this way. The accuracy of the measurement was considerably less than that for hydrogen and tritium. However, as discussed in Sec. XI below, a deuterium maser has recently been constructed at Harvard and the value in Eq. (2) is from that deuterium maser which is discussed below.

The experimental values of $\Delta \nu$ for D and T in Eqs. (2) and (3) can also be compared to theory though the theoretical values are less accurate than for H, due to the theoretical dependence on the nuclear structure. The theoretical value [20] for $\Delta \nu$ of D is

$$\Delta \nu_{Dth} = 327,394,000 \pm 33,000 \text{ Hz} \tag{7}$$

in excellent agreement with experiment. In the case of tritium there are the following two different theoretical values of which the first one [21] agrees well with theory while the second one [22] does not

$$\Delta \nu_{Dth} = 1,516,695,000 \pm 8000 \text{ Hz} \quad [21] \tag{8}$$

$$\Delta \nu_{Dth} = 1,516,684,570 \pm 3000 \text{ Hz} \quad [22] \tag{9}$$

Although the H maser is most suitable for measurements on atomic hydrogen, other atoms can also be measured by utilizing the spin-exchange technique described above for the earliest hydrogen maser measurements on atomic deuterium. In this fashion Crampton, Berg, Robinson, and Ramsey [23] showed that the coefficients A of the hyperfine magnetic interaction and B of the electric quadrupole interaction were given by

$$A = 10,450,929.06 \pm 0.19 \text{ Hz} \tag{10}$$

$$B = + 1.32 \pm 0.20 \text{ Hz} \tag{11}$$

The coefficient B is about 10^5 times smaller than the smallest nuclear quadrupole interaction energy ever measured previously. The small size of this interaction energy is primarily due to the almost spherical distribution of the electron charge in the nitrogen atom. The small magnitude of B is consistent with theory but the experimental and theoretical values differ since the latter [24] is

$$B_{th} = 3.4 \pm 1.6 \text{ Hz} \tag{12}$$

IV. HYDROGEN MASER CLOCKS

Because of their high accuracy and stability, the hydrogen masers are useful as time and frequency standards [25]. They have been used, for example, as the time references in many of the long-base-line radio-astronomy interference experiments, especially when the two interfering radio telescopes are unusually far apart. They have also been used in some of the radar-astronomy tests of relativity by the effect of the sun's mass upon radar signals transmitted close to the sun. Kleppner, Vessot, and Ramsey [25] have pointed out that the gravitational red shift of the earth could be measured to an accuracy of about 1 part in 10^5 by comparing ground-based and satellite-borne hydrogen maser clocks.

V. DEPENDENCE OF HYDROGEN HYPERFINE SEPARATION ON ELECTRIC FIELD

Fortson, Kleppner, and Ramsey [26] and Gibbons and Ramsey [27] observed the effect of a strong electrostatic field (Stark shift) on the hydrogen hyperfine separation. The latter authors used a hydrogen maser separated into three regions to only one of which the electric field was applied as shown in Fig. 6. The effect of an electric field had been predicted a number of years ago theoretically, but the theoretical calculation was thought at the time to be irrelevant since the maximum effect expected was a million times less than the accuracy of experimental observations at that time. Hence the calculation was considered to be an interesting mathematical exercise with no significance as to possible experiments. However, with the accuracy of the hydrogen maser, it is possible not only to observe the effect but also to

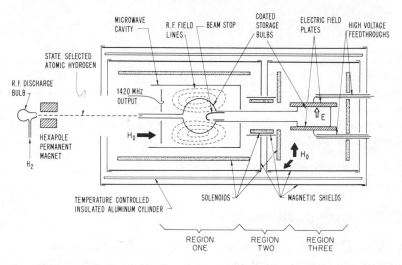

Figure 6.   Three region maser for determining effect of electric field on atomic hydrogen hyperfine separation [26].   The electrostatic field is applied only in region three.

test the theory quantitatively.   The results of the first experiments [26] are shown in Fig. 7.   As can be seen, the shift of frequency was about a part in 10^{11} which was too small to be observable in previous experiments but could be measured accurately with the hydrogen maser.   It is also noteworthy that the shift is proportional to the square of the electrostatic field, as should be the case theoretically with the transition observed.   The dashed line in Fig. 7 corresponds to the theoretical prediction, while the three error bars indicate the experimental observations.   As can be seen, the agreement between

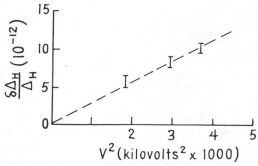

Figure 7.   Dependence of atomic hydrogen hyperfine separation on voltage.

theory and experiment is excellent. In the first experiment, the magnetic and electric fields were parallel to each other. A new experiment was recently complete [27] with observations when the fields were not only parallel to each other but also perpendicular to each other. The results of the two sets of experiments are shown in Table I.

TABLE I.   STARK SHIFT $\delta \Delta \nu \times 10^5 / E^2$ (Hz/esu^2)

Condition	Experiment	Theory	Reference
$\vec{E} \parallel \vec{H}$	-6.7 ± 0.4	-7.625	Fortson et al. [25]
	-7.1 ± 0.7	-7.625	Gibbons et al. [26]
$\vec{E} \perp \vec{H}$	-7.1 ± 0.7	-7.142	Gibbons et al. [26]

VI.   EFFECT OF MAGNETIC FIELD UPON HYPERFINE SEPARATION

In the preceding section we discussed the effect of an electrostatic field in changing the hyperfine frequency. A similar question can be asked about the effect of a strong external magnetic field.

In this case, as in the electrostatic one, a theory had already been developed as to how the energy levels should vary with the strength of an external magnetic field. The Breit-Rabi formula predicts this variation as shown in Fig. 1.

In the presence of a magnetic field a number of different transitions can be measured and compared at different magnetic fields. A study of such transitions has just been completed by Brenner [28] in our laboratory and is currently being prepared for publication. He has been able to test the predictions of the Breit-Rabi formula to an accuracy of 7 parts in 10^9 and finds that the predictions of this theory are in full agreement with experimental measurements.

VII.   PROTON MAGNETIC MOMENT

A particularly significant experiment has been done by Myint, Kleppner, Robinson, and Ramsey [29] and by Winkler, Walther, Myint, and Kleppner [30] by studying the hyperfine structure of hydrogen atoms in a strong magnetic field of 3500 Oe. The transition from a state with m = 0 to a state with m = -1 is primarily dependent upon the magnitude of the magnetic moment of the electron, while the transition from the m = +1 to m = 0 state depends to a considerable degree on the magnetic moment of the proton. Consequently, by measuring the ratio of these two frequencies in the same magnetic field it is possible to obtain an accurate value for the ratio of the magnetic moment of the electron to the magnetic moment of the proton. Although this ratio has been measured in past NMR experiments, all of these experiments have suffered from the necessity of making a shielding correction for the effects of the electrons in the hydrogen molecules. This correction, in turn, was based upon a theory [31] the author developed a number of years ago but which could not previously be checked experimentally. With atomic, rather than molecular, hydrogen, on the other hand, the shielding correction is much simpler and can be easily calculated with the known wave functions for the hydrogen atom. An experiment with atomic hydrogen, therefore, makes possible for the first time a determination of the proton magnetic moment free from any uncertainty due to magnetic shielding.

A schematic diagram of the apparatus used in these experiments is shown in Fig. 8. For experimental convenience in suspending the equipment, the atomic beam in this case runs vertically downward. The atoms of hydro-

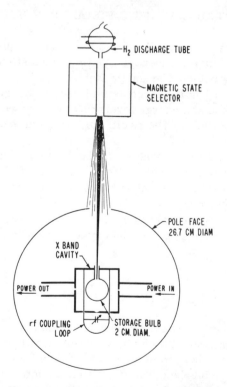

Figure 8. Schematic diagram of high-field hydrogen maser apparatus.

gen are produced in a gas discharge tube shown at the tope of the figure, and
the atoms are focused by a six-pole magnetic state selector into a quartz
storage bottle that is about seven times smaller than the usual 6-in.-diam
hydrogen maser storage bulb. The storage bulb must be much smaller due
to the higher frequency and shorter wavelength of the characteristic radiation. The resonance cavity and the storage bulb are placed between the pole
tips of a magnet at 3500 G. The result of a resonant stimulated emission
associated with the transition indicated by the dashed arrow is shown in Fig.
9. It can be seen from the dashed arrow that the resonance frequency is
approximately proportional to the strength of the external magnet field. It
can also be seen from the recorder trace that the half-width of the resonance
is three parts in 10^8. Consequently, if the magnetic field is unstable by
three parts in 10^8, the resonance will shift by its width and will consequently
tend to be obliterated. As a result, there is a stringent requirement for
unprecedentedly high stability and uniformity of the magnetic fields. Much
of the work in developing this apparatus was devoted to obtaining a suitable
magnetic field. In addition to this high-frequency transition, the transition
between the two energy levels at the top of the diagram in Fig. 10 has also
been observed.

From the ratio of these two frequencies at the same magnetic field, a
value is obtained for the ratio of the magnetic moment $(g_J)_H$, of the hydrogen
atom to the apparent magnetic moment, $(g_p)_H$, of the proton shielded inside
the hydrogen atom. For both of these quantities it is necessary to make a
small known correction to go to the desired ratio μ_e/μ_p of the magnetic
moment of free electron to the magnetic moment of the free proton [31, 32].
By a surprising coincidence the correction factor for both of these quantities
is just the same to the order of α^2 and is equal to $(1 + \frac{1}{3}\alpha^2)$. Consequently

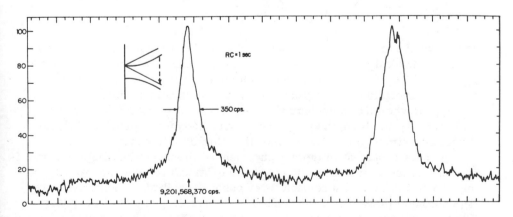

Figure 9. Typical resonance of high-field maser for the transition indicated
by the dashed arrow on figure insert.

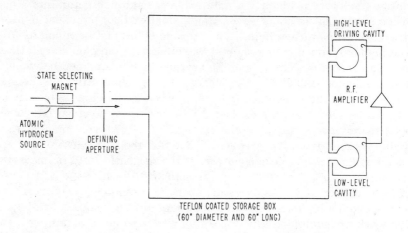

Figure 10.   Schematic diagram of large box hydrogen maser.

to this order the correction factors cancel and the desired magnetic moment ratio is equal to the ratio measured experimentally.   The result for the ratio of the free electron magnetic moment to the free proton magnetic moment is as follows, where K stands for a small higher-order correction:

$$\frac{\mu_e}{\mu_p} = \frac{(1 + \frac{1}{3}\alpha^2)(g_J)_H}{(1 + \frac{1}{3}\alpha^2)(g_p)_H} K = -658.2106827\,(6) \tag{13}$$

From the above ratio and the known value of the magnetic moment of the electron, the value of the magnetic moment of the free proton is found to be μ_p = 0.001 521 032 1 ± 0.000 000 000 9 Bohr magnetons.   This value of the proton magnetic moment agrees fully with the previously measured value from NMR when the molecular shielding constant is obtained from the author's [31] shielding theory.   The result, therefore, provides the first direct experimental confirmation of this shielding theory.

An added interest in the measurement of the proton magnetic moment is its appearance in the derivation of the fine-structure constant from Eq. (4) and the hyperfine measurements previously described.   The derivation depends upon the value of the unshielded magnetic moment of the proton. Since, as discussed earlier, there had been a disagreement between the value of the fine-structure measurements in atomic hydrogen, a possible source of the discrepancy could have been an error in the proton magnetic moment due to an incorrect magnetic shielding constant. The result in Eq. (13) showed that the discrepancy could not be explained in this way. The present agreement between the hyperfine value for α and that obtained at a much later date by the totally different method based on the Josephson effect [16] indicated that no explanation for the hyperfine result was required and that the hyperfine value for α was correct all along.

VIII. ATOMIC MAGNETIC MOMENTS

From the Zeeman transitions ($\Delta F = 0$, $\Delta m \pm 1$) in Fig. 1, the magnetic moment of the hydrogen atom can be measured. The same can be done with deuterium and from the two measurements together the ratio of the hydrogen to deuterium magnetic moment, $g_J(H)/g_J(D)$, can be measured. The first hydrogen maser measurement of this quantity was that of Larson, Valberg, and Ramsey [33] who found

$$g_J(H)/g_J(D) = 1 + (9.4 \pm 1.4) \times 10^{-9} \qquad (14)$$

in approximate but not perfect agreement with the theoretical value

$$g_J(H)/g_J(D)_{th} = 1 + 7.3 \times 10^{-9} \qquad (15)$$

Subsequently Kleppner, Walther, and Phillips [33] found with the high-field maser

$$g_J(H)/g_J(D) = 1 + (7.22 \pm 0.10) \times 10^{-9} \qquad (16)$$

This result is in excellent agreement with theory. Hughes and Robinson [34] with a quite different optical pumping experiment found

$$g_J(H)/g_J(D) = 1 + (7.2 \pm 3.0) \times 10^{-9} \qquad (17)$$

also in excellent agreement with experiment but with a larger error estimate. Larson and Ramsey [36, 37] have also recently found the following preliminary result for tritium:

$$g_J(H)/g_J(T) = 1 + (10.7 \pm 2.0) \times 10^{-9} \qquad (18)$$

The measurements of atomic magnetic moments with the hydrogen maser can be extended to other atoms as well with the spin-exchange technique described in Sec. I. With such a spin-exchange technique, Valberg and

Ramsey [38] have measured for rubidium

$$g_J(Rb^{87})/g_J(H) = 1.0000235855\,(6) \tag{19}$$

in agreement with the even higher precision optical pumping measurements of Hayne, Ensberg, and Robinson [39] and Hughes and Robinson [35].

IX.   LARGE STORAGE BOX HYDROGEN MASER

In several of the previous discussions it has been mentioned that the principal source of uncertainty in the hydrogen maser measurements arises from the necessity of making a wall shift correction for the effect of the wall upon the hydrogen atom when the atom is in the vicinity of the wall. Although this correction can be made by extrapolating results on masers with a different sized storage bulbs, the uncertainty in the determination of the wall correction remains the principal source of uncertainty in many of the maser measurements. Zitzewitz [40,41] and Vessot [42] are undertaking experiments to reduce this uncertainty by finding a wall-coating material that is superior to Teflon. So far, however, Teflon remains the best wall-coating substance.

An alternative means of reducing the effect of the wall shift has been undertaken by Uzgiris in our laboratory. He has constructed a maser with a storage box that is ten times larger in diameter than the normal 6-in.-diam storage bulb. Since the atom will strike the wall only one-tenth as frequently in such a storage box, the wall shift will be ten times less. In addition, a longer total storage time can be arranged, which makes the resonances even sharper.

A schematic diagram of the large storage box hydrogen maser is shown in Fig. 10, and a photograph of the maser inside its triple layer of molypermalloy shielding is shown in Fig. 11. The arrangement of this maser is necessarily different from that of previous hydrogen masers and a number of new principles are involved. In particular the cavity can no longer surround the entire storage box, since the wavelength of the radiation is small compared to the diameter of the storage box. However, an equally narrow resonance is obtained if the cavity in which the maser oscillation occurs surrounds only a portion of the storage box, provided the atoms make a number of transits between the small storage box and the large box before finally exiting through the entrance cavity of the large box. Since the atoms are in the region of the cavity only for a very short period of time, it is necessary to have a higher level of excitation than would occur from simple spontaneous maser oscillation. As a result two cavities are used, each surrounding a small portion of the storage box and about 80 dB of amplifier gain is provided between the two cavities. In this manner the atoms are

Figure 11.   Photograph of large box hydrogen maser showing the storage cylinder, vacuum chamber, and magnetic shielding box.

placed in a superradiant state by the intense oscillations in the high-level driving cavity and are thereby able to produce spontaneous maser oscillation in the low-level cavity, which in turn is amplified to the high-level driving cavity.

Oscillations with this maser were first obtained during the summer of 1967.   From a study of the maser operation during the first few months, it became apparent that two improvements should add greatly to its reliability and its effectiveness.   One was to increase the gain between the two cavities, and the other was to increase the electromagnetic shielding between the two cavities to prevent spontaneous nonmaser oscillation at the higher gain. Both of these changes have now been made, and the expected improvement has occurred.   On the basis of the experience so far, we are optimistic that the large box maser should enable us to improve most of our past measurement accuracies by a factor of 10 and to undertake a number of new measurements that have not been possible previously.

X.   WALL SHIFT MEASUREMENTS

Zitzewitz and Ramsey [40, 41] have studied the Teflon wall shifts at different temperatures with the results shown in Fig. 12.   One interesting result was the observation that at about 80°C the wall shift passes from positive to negative values and consequently vanishes at the crossing temperature.   This feature is of value in maser experiments that are limited by the wall shift. Zitzewitz [40, 41] also found that the abrupt changes with temperature in the slope of the wall shift curve in Fig. 12 are correlated to known phase changes in Teflon.

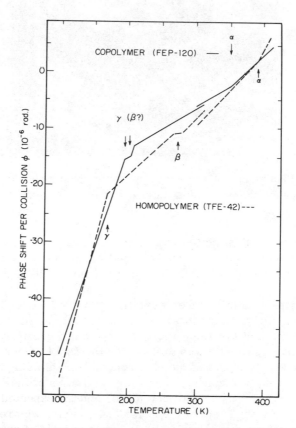

Figure 12.   Experimental phase shift per collision versus temperature.

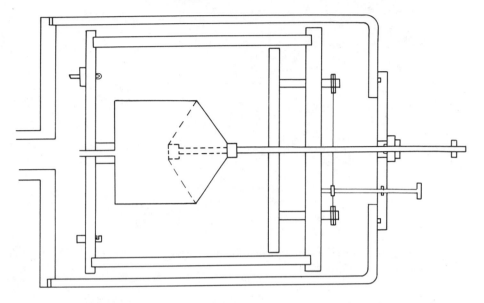

Figure 13.   Deformable bulb H maser [40].   The conical surface can be in the positions indicated by the full or dashed lines.

Until recently all frequency shifts due to wall collisions were measured by extrapolation with the use of Teflon–coated storage bulbs of different diameters.   However, Zitzewitz, Uzgiris, and Ramsey [40] showed that the accuracy of this extrapolation was reduced by the differences in the wall coatings on the different bulbs.   Brenner [43] pointed out that this difficulty could be overcome by the use of a single flexible storage bulb whose volume could be altered while keeping the same surface.   This technique was further developed by Debely [44] who used a storage cylinder, one of whose ends was a thin conical sheet of Teflon which could be in either of the two positions shown in Fig. 13.   With this deformable bulb technique the wall-shift measurements are all made on the same surface.

At present Reinhardt [45] is adapting this defomable bulb technique to the large box hydrogen maser by using the configuration shown in Fig. 14.   An alternative use of the deformable bulb technique has been suggested by Vessot [42] who has proposed operating a hydrogen maser at the temperature where the wall shift vanishes [41] and using the deformable bulb to locate that temperature experimentally, i.e., to operate at the temperature for which the output frequencies are the same in the two different deformable bulb configurations.

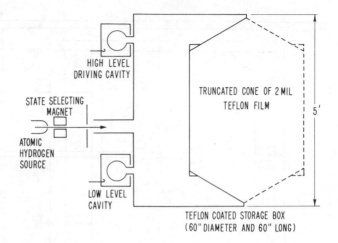

Figure 14. Large storage box H maser with deformable bulb for measurement of wall shift [38].

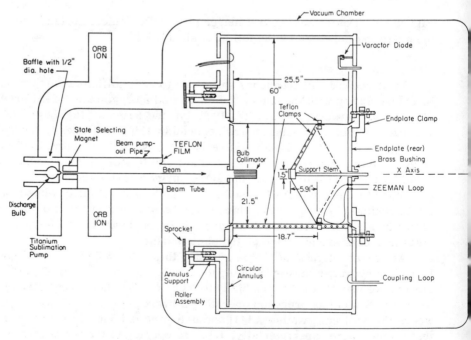

Figure 15. Deuterium maser [40].

XI. DEUTERIUM MASER

Wineland and Ramsey [46] have used the magnetic shield of the large box maser as the shield for a deuterium maser as shown in Fig. 15. A deformable bulb technique was used in this case as well to determine the wall shift. The results of the experiment have already been listed above in Eq. (2).

REFERENCES

1. H. M. Goldenberg, D. Kleppner, and N. F. Ramsey, Phys. Rev. Letters 5, 361 (1960).
2. D. Kleppner, H. M. Goldenberg, and N. F. Ramsey, Phys. Rev. 126, 603 (1962).
3. N. F. Ramsey, Prog. Radio Sci., 7, 111 (1965).
4. D. Kleppner, H. C. Berg, S. B. Crampton, N. F. Ramsey, R. F. C. Vessot, H. E. Peters, and J. Vanier, Phys. Rev. 138, A972 (1965).
5. S. B. Crampton, Phys. Rev. 158, 57 (1967); (private communication, 1971).
6. S. B. Crampton, D. Kleppner, and N. F. Ramsey, Phys. Rev. Letters 11, 338 (1963).
7. R. Beehler, D. Halfoud, R. Harrach, D. Allan, D. Glaze, C. Snider, J. Barnes, R. Vessot, H. Peters, J. Vanier, L. Cutler, and M. Bodily, Proc. IEEE (correspondence) 54, 301 (1966); H. E. Peters, J. Holloway, A. S. Bagley, and L. S. Cutler, Appl. Phys. Letters 6, 34 (1965); R. Vessot, H. Peters, J. Vanier, R. Beehler, D. Halford, R. Harrach, D. Allan, D. Glaze, C. Snider, J. Barnes, L. Cutler, and M. Bodily, IEEE Trans. Instrum. Meas. IM-15, 165 (1966); H. Hellwig, R. F. C. Vessot, M. Levine, P. W. Zitzewitz, H. E. Peters, D. W. Allan, and D. J. Glaze, IEEE Trans. Instrum. Meas. IM-19, 200 (1970).
8. B. S. Mathur, S. B. Crampton, D. Kleppner, and N. F. Ramsey, Phys. Rev. 158, 14 (1967).
9. S. B. Crampton, H. G. Robinson, D. Kleppner, and N. F. Ramsey, Phys. Rev. 141, 141 (1966).
10. S. J. Brodsky and G. W. Erickson, Phys. Rev. 148, 26 (1966); and references listed therein.
11. E. R. Cohen and J. W. M. DuMond, Rev. Mod. Phys. 37, 537 (1965).
12. S. Triebwasser, E. S. Dayhoff, and W. E. Lamb, Jr., Phys. Rev. 89, 98 (1953).
13. S. D. Dress and J. D. Sullivan, Phys. Rev. 154, 1477 (1967).
14. C. Iddings and B. Platzman, Phys. Rev. 115, 919 (1959); C. K. Iddings, Phys. Rev. 138, B446 (1965).

15. H. Grotch and D. R. Yennie, Phys. Rev. (to be published).
16. W. H. Parker, B. N. Taylor, and D. N. Langenberg, Phys. Rev.
    Letters 18, 287 (1967); Rev. Mod. Phys. 41, 375 (1969).
17. H. Metcalf, J. R. Brandenberger, and J. C. Baird, Phys. Rev. Letters
    21, 165 (1968).
18. T. Appelquist and S. J. Brodsky, Phys. Rev. Letters 24, 562 (1970).
19. N. F. Ramsey, Fine and Hyperfine Structure of Atomic Hydrogen, Pro-
    ceedings of International Conference on Precision Measurements, (U. S.
    National Bureau of Standards, U. S. GPO, Washington, D. C., 1971).
20. F. Low and E. Salpeter, Phys. Rev. 83, 478 (1951).
21. R. Avery and R. G. Sachs, Phys. Rev. 74, 433 (1948); 74, 1320 (1948).
22. E. N. Adams, Phys. Rev. 81, 1 (1951).
23. S. B. Crampton, H. C. Berg, H. G. Robinson, and N. F. Ramsey,
    Phys. Rev. Letters 24, 195 (1970).
24. W. W. Holloway, E. Lusiher, and R. Novick, Phys. Rev. 126, 2109
    (1962).
25. D. Kleppner, R. F. C. Vessot, and N. F. Ramsey, Astrophys. Space
    Sci. 6, 13 (1970).
26. E. N. Fortson, D. Kleppner, and N. F. Ramsey, Phys. Rev. Letters
    13, 22 (1964).
27. P. Gibbons and N. F. Ramsey, Phys. Rev. A 5, 73 (1972).
28. D. Brenner, Ph. D. thesis (Harvard University, 1969); Phys. Rev. 185,
    26 (1969).
29. T. Myint, D. Kleppner, N. F. Ramsey, and H. G. Robinson, Phys.
    Rev. Letters 17, 405 (1966).
30. P. F. Winkler, F. G. Walther, M. T. Myint, and D. Kleppner, Bull
    Am. Phys. Soc. 15, 44 (1970); P. F. Winkler, Ph. D. thesis (Harvard
    University, 1971); D. Kleppner (private communication, 1971).
31. N. F. Ramsey, Phys. Rev. 78, 699 (1950).
32. E. H. Lieb, Phil. Mag. 46, 311 (1955).
33. D. J. Larson, P. A. Valberg, and N. F. Ramsey, Phys. Rev. Letters
    23, 1369 (1969).
34. D. Kleppner and F. Walther (private communication, 1971).
35. W. M. Hughes and H. G. Robinson, Phys. Rev. Letters 23, 1209 (1969).
36. D. Larson and N. F. Ramsey (private communication, 1970).
37. D. Larson, Ph. D. thesis (Harvard University, 1970).
38. P. A. Valberg and N. F. Ramsey, Phys. Rev. A 3, 554 (1971).
39. G. S. Hayne, E. S. Ensberg, and H. G. Robinson, Phys. Rev. 174, 23
    (1968); Phys. Rev. Letters, 23, 1209 (1969).
40. P. W. Zitzewitz, E. Uzgiris, and N. F. Ramsey, Rev. Sci, Instrum.
    41, 81 (1970).
41. P. W. Zitzewitz and N. F. Ramsey, Phys. Rev. A 3, 51 (1971).
42. R. F. C. Vessot (private communication,1971)
43. D. Brenner, Bull. Am. Phys. Soc. 14, 443 (1969), J. Appl. Phys. 41,

2942 (1970).

44. P. Debely, Rev. Sci. Instrum. 41, 1290 (1970).

45. V. Reinhardt (private communication, 1971).

46. D. Wineland and N. F. Ramsey, Phys. Rev. A 5. 821 (1972)

SATURATED ABSORPTION LINE SHAPE

J. L. Hall*
Joint Institute for Laboratory Astrophysics
University of Colorado, Boulder, Colorado

There are a number of relevant physical phenomena which will have to be taken into account in a proper theory of the saturated absorption line shape. These include: (1) collisional phase-shift effects [1] and velocity/frequency shifting small-angle collisions [2], (2) power broadening associated with the saturation process [3], (3) z-axis modulation of the saturation parameter [4], (4) the distribution of transverse velocities for a given axial velocity group [4], and, most importantly, (5) coherence-limiting effects due to the finite spatial extent of the light beam [2] and deviations from planar wavefronts [5].

Such a total proper theory does not yet exist to the author's knowledge. Indeed we know of no real saturation theory whatever that is applicable in the low-pressure regime of interest in the standards context (low pressure reduces the pressure-induced frequency offset). Thus we have found it useful to parameterize our observed resonances in terms of a very simple physical "hole-burning" model. It is found that the assumed Lorentzian line shape gives a good representation of the data. Within certain limits, the dependence of the linewidth upon pressure and laser intensity can be well represented by a three-parameter formula of the form suggested by a saturation model [2]. The fact that the three resonance parameters of the model

*Laboratory Astrophysics Division, National Bureau of Standards and Department of Physics and Astrophysics, University of Colorado.

take on systematically different values as one changes the laser spot size gives rise to a very satisfying and essentially quantitative corroboration of our understanding of the relevant physical scaling laws. On the other hand, our values of the phenomenological parameters have no transparent numerical relationship with the phase memory time and transition dipole moment, for example, of a particular molecular velocity group. Our data were obtained to teach us the scaling laws that might lead to more interesting optical-frequency-standard results. The parameterized results are presented here as an aid to communication between experimentalists and theoreticians, and to provide a quasiexperimental testing ground for their calculations.

In the low-pressure (10^{-2} to 10^{-4} Torr) regime of interest in the frequency-standards context, the importance of a realistic treatment of collisional effects is greatly reduced. It is sufficient to use a single pressure-broadening parameter to represent collision-induced transitions out of the resonant velocity interval. Basically the resonance line has become so sharp that there are thousands of resolvable velocity intervals within the Doppler linewidth. Thus a negligible fraction of excited absorbers are returned to the resonant velocity groups by a second collision.

On the other hand, many molecules--perhaps a majority--which originally satisfy the resonance condition in v_z will experience a coherent interaction with the radiation field during their entire transit of the laser beam spot. Thus it is absolutely essential to account for the finite duration of the excitation pulse as the transverse dimension of the spatially bounded laser beam is crossed by the absorbers at their (weighted) thermal velocity.

Since we are interested in the longest phase-coherent interaction time to minimize the resonance linewidth, we use carefully prepared single-spatial-mode excitation beams. The phase variation observed along a straight transverse molecular trajectory will then be sufficiently small if the wavefront radius of curvature is sufficiently great. A suitable phase criterion ($\approx \pm \pi/8$ rad) gives

$$R > b \equiv (2\pi w_o^2/\lambda)^{1/2}$$

where b is the equivalent confocal parameter of the laser beam characterized by Gaussian mode radius w_o. Thus a maximum absorption cell length < 2b is implied for the configuration symmetric around the plane containing the beam waist. One then anticipates a "time-of-flight" linewidth $\Delta v = k(<v>/w_o)$ where $<v>$ is a suitably weighted thermal average velocity and k is a dimensionless constant of order unity.

We may estimate the value of k as follows. For the Gaussian beam profile, about 40% of the energy lies inside an aperture of <u>diameter</u> = w_o. Two degrees of freedom contribute to the transverse velocity, giving

$v_{rms} = (2KT/M)^{1/2} = 5.5 \times 10^4$ cm/sec. The typical "pulse duration" is therefore $\tau \simeq w_o/v_{rms}$ leading to a Lorentz half-width at half maximum, $\Delta\nu_{HWHM} = 1/2\pi\tau$, finally giving k = $1/2\pi$. Thus in this crude model one would expect $w_o\Delta\nu_{HWHM} = v_{rms}/2\pi = 88$ kHz mm.

From the experimental results with single-mode excitation at room temperature (see text below and Table I) we find $w_o\Delta\nu_{HWHM} = (70\pm5)$ kHz mm, corresponding to k = 1/8.

Now the Gaussian time-domain impulse is a self-apodized function, and so the line wings of the frequency-domain response function are free of the oscillations which are usually characteristic of pulse-interrogation resonance problems. Also the nonlinear character of the saturation resonance weights the slow velocity end of the velocity distribution. Thus it is almost surprising that the two values for k are in such close accord.

Naturally if the spatial quality of the beam is poor, a coherent interaction will not be obtained over the entire aperture and k can be more like 1/2 or unity. In fact if there are spatially distinguishable standing-wave fields, one can observe "Ramsey two-cavity fringes". It seems appropriate first to study the pure single-mode case.

The line shapes which we wish carefully to investigate can be as narrow as 23 kHz HWHM, a fractional width of 2.6×10^{-10}. It is obvious that mere mechanical stability will not suffice very long in such a domain! A very powerful kind of spectrometer can be built based on the concept of Frequency-Offset-Locking [6]. We <u>acquire</u> good frequency stability in a laser servo controlled to the apparent top of the saturated absorption peak in CH_4. This excellent stability is then <u>transferred</u> to the powerful laser which illuminates the absorption cell of interest. The reference laser need not be understand-able in the same fundamental terms we hope to achieve in the external absorption case--it need only be (empirically) established to provide a frequency reference of good stability. In Fig. 1 we show the present version of this reference device. The high signal-to-noise ratio, 109 dB for a 1-sec average, should clearly provide some impressive frequency stability if optimally utilized. At present it is found that short-term frequency noise of the basic laser near the modulation frequency is degrading the performance severely [7]. To investigate this point, a nonoptimal device with about the same free-running frequency stability but a much smaller signal-to-noise ratio was heterodyned against one of our "standard CH_4 laser frequency references". [Actually both were beat against an offset laser (No.2) to eliminate the sign ambiguity near zero beat.] In Fig. 2 we show the time-sequence beat frequency data for three averaging times. The 5-MHz basic value has been suppressed. It is clear from the figure that the fractional frequency fluctuation is still decreasing with averaging time near 1 sec. These data and many more have been processed to yield a plot of the Allan Variance as a function of averaging time. See Fig. 3. [The Allan Variance is a fractional measure of the (rms) average frequency difference between

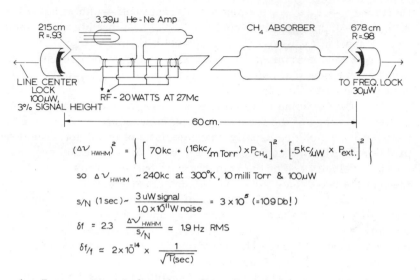

Figure 1.   Laser-saturated methane frequency reference.   This version of the device has a 1-mm mode waist in the absorption cell.   The bore of the gain cell is about 3.2 mm, with $P(Ne^{20}) = 0.3$ Torr, $P(He^3) = 5.0$ Torr.   The connections in the reservoir are for dc-fired Ba getter.

adjacent samples of the input test frequency.   It is widely used in the frequency standards game because it is well behaved in the presence of noise with long time correlations, such as $1/f$ noise.]   In this case the test frequency was the (offset) second beat (audio) between the two (5 MHz) laser beat frequencies.   The frequency noise of the common heterodyne laser No. 2 drops out of this second beat, leaving just the frequency noise of laser No. 1 plus that of laser No. 3.

From Fig. 3 it may be seen that the $\tau^{-1/2}$ averaging law expected for a white-noise process is followed up to averaging times of about 100 sec, where the demonstrated (in-)stability is a matter of 3 or 4 Hz!   Ultimately beyond about 10^4 sec, the stability is degraded by changes in the same systematic errors which limit the present reproducibility to about $1/2 \times 10^{-11}$.   In principle, the short-term frequency stability of the device

Figure 2.   Time-domain frequency stability data.   By taking the audio beat between the beats of the two self-stabilized devices with a common local oscillator, we observe directly the difference of the two stabilized optical frequencies.   The stability is improving with averaging time near 1 sec.   The averaging time in the lower photo was 45 sec.

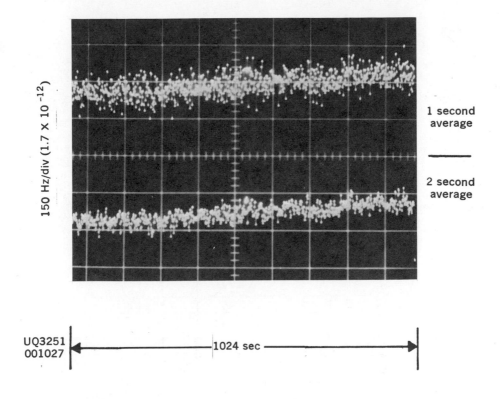

150 Hz/div (1.7 × 10⁻¹²)

1 second
average

2 second
average

UQ3251
001027

|← 1024 sec →|

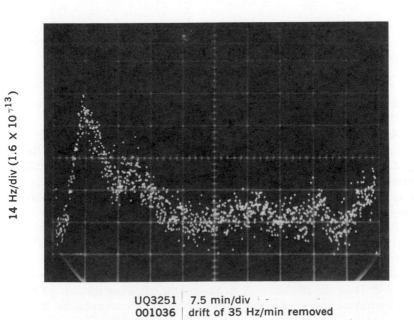

14 Hz/div (1.6 × 10⁻¹³)

UQ3251   7.5 min/div
001036   drift of 35 Hz/min removed

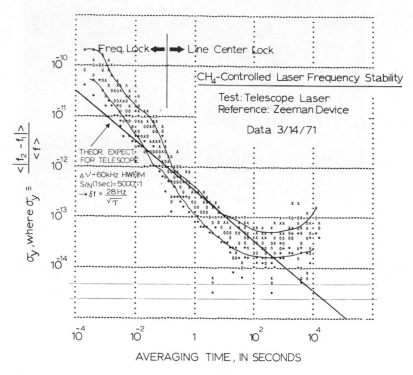

Figure 3.   Allan Variance, calculated from data like that of Fig. 2.   The frequency-lock loop operates for times shorter than 0.1 sec, the line center lock dominates at larger times.

shown in Fig. 1 should be about one decade better than the nonoptimal device tested in these experiments.

As a fundamental standard, the device of Fig. 1 still has important resetability defects due to problems with definition of the base line. However as a frequency reference for laser spectroscopy [8], for geophysical strain seismometry [9], and for precision interferometry [10], the CH_4 frequency reference allows experiments of unprecedented sensitivity, stability, and precision.

For example, in the Frequency-Offset-Locked Laser Spectrometer, illustrated in Fig. 4, we are able to study the saturation resonance itself in delicious detail.   The two frequency-offset-lock loops function very well indeed, transferring essentially the entire available stability of the reference laser (No. 1) to the powerful (7 mW) controlled laser (No. 3).   The use of a local oscillator laser (No. 2) offset 5 MHz red eliminates the troublesome

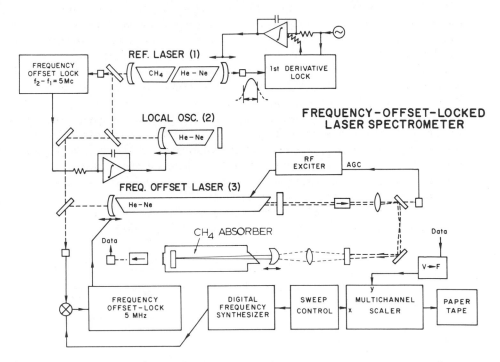

Figure 4.   The Frequency-Offset-Locked Laser Spectrometer.   Laser No. 1 servo stabilized to its own saturation peak provides excellent frequency stability which is transferred--with a 5-MHz red frequency offset--to laser No. 2.   Laser No. 3 is controlled relative to No. 2 by the synthesizer/frequency offset lock loop.   The optical output power of No. 3 is also stabilized.   The cavity containing the CH_4 absorber, servo controlled to track the input wavelength, gives good spatial-mode purity.

region near zero beat.   Using the external cavity saturation signal as a frequency discriminator, we find the variations in the absolute frequency of laser No. 3 to be about 5 kHz peak to peak, with periods typically in the 10- to 100-msec range.   For 1-sec averages, the rms frequency excursion is about 30 Hz.   The fact that the frequency fluctuations can be measured in this way already proves that a more suitable reference device could provide better stability if it is needed.

The laser spectrometer employs an automatic laser intensity control to provide a frequency-independent base line for the absorption signals.   At present we are interested in the effects of mode size and purity, making it

attractive to put the absorbing gas inside an external resonant cavity. The
minor bother of having to servo stabilize the cavity length to track the
applied optical input frequency/wavelength is compensated by the "contrast
amplification" implicit in a properly designed cavity setup. (Basically for a
low-loss cavity, say of finesse ≈ 40 and resonance transmission $\simeq 1/3$, one
can parlay a 1% resonant absorption signal into a 10% fractional change of a
10 times lower basic dc power level. This gambit is obviously helpful in
the frequency reproducibility business.)

In the laser spectrometer of Fig. 4, the digitized resonance line shapes
are obtained directly without frequency modulation and associated derivative
detection. Thus it is appropriate to least-squares fit the data to the Lorentzian function itself. An antisymmetric term proportional to the Lorentz
derivative is useful in lowering the residuals, although its contribution
seldom exceed 1% of the basic signal [11].

A fit typical of all the data to be presented is shown as Fig. 5. The
Lorentz fit becomes even more precise at higher pressure. By now several

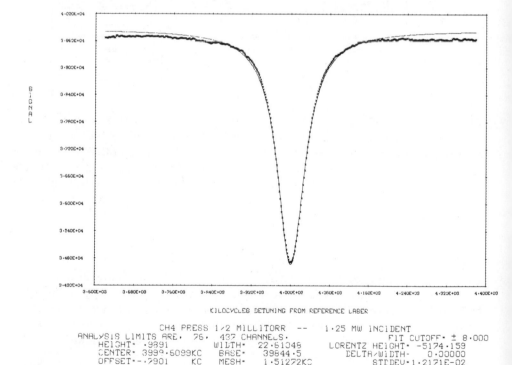

Figure 5. Lorentz fit to typical data from "First FBTK" run in external
120-cm cell. The saturation parameter is 0.38 and the observed linewidth
(HWHM) is 33.6 kHz. The signal as plotted is inverted.

hundred resonance curves have been recorded. Our "eyeball" estimates of
the linewidths and heights have been checked out in a few dozen interesting
cases using the computer least-squares-fit program. As suggested previously, we do not ascribe any fundamental significance to these resonance
parameters; rather they form a convenient media for discussion of the data.

(As an interesting side topic, in Fig. 6 we show a line shape characteristic of low-pressure data: The smooth reference curve is the pure Lorentz
function. Basically, at low pressures, fluctuations in the collision rate from
its average value occasionally allow some slow absorber molecules to interact for quite a long time. At low powers these particles give a dispropor-
tionate and very sharp contribution at line center. It is found that this line
shape can be well reproduced by "peaking" the Lorentzian through subtraction

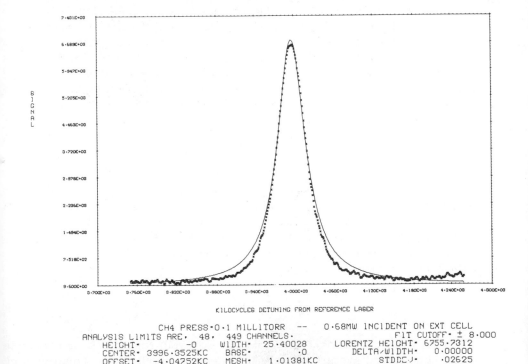

KILOCYCLES DETUNING FROM REFERENCE LASER

CH4 PRESS•0•1 MILLITORR --   0•68MW INCIDENT ON EXT CELL
ANALYSIS LIMITS ARE• 48•   449 CHANNELS•   FIT CUTOFF• ± 8•000
HEIGHT•   -0   WIDTH• 25•40028   LORENTZ HEIGHT• 6755•7312
CENTER• 3996•3525KC   BASE•   •0   DELTA/WIDTH• 0•00000
OFFSET• -4•04752KC   MESH• 1•01381KC   STDDEJ•   •02625

Figure 6. Lorentz fit to 0.1 m Torr data from "First FBTK" run. The
saturation parameter is 0.25. The measured linewidth is 25.6 kHz (HWHM).
Note the characteristic sharpening in the line wings at low power and low
pressure. Although the whole signal represents only about 1.5% of the 0.7%
unsaturated double-pass absorption, the optical isolation is sufficient to
prevent any important curvature of the experimental baseline due to residual
interference effects.

of a second-derivative term, although presumably a Gaussian would be more appropriate at the lowest pressure [12].

This body of data is made vastly more useful by considering the ensemble of resonance widths to depend parametrically upon aperture, laser power, and absorber pressure. As a matter of convenience we will model the resonances by pressure-broadened and power-broadened homogeneous holes "burned" into a Gaussian velocity distribution. Thus we define P_1 as the linewidth <u>intercept</u> at zero pressure and zero power--ultimately to be related by a proper theory to the mode size [13]. In view of the nonlinear nature of the saturation resonance, we will consider only molecules which transit the entire laser spot without important collisions, neglecting those particles of shorter interaction time whose coherence is terminated by collisions. Thus it is reasonable to identify P_1 as the energy "relaxation" rate R, energy being carried out of the laser beam as molecular excited-state occupation probability. The phase interruption rate, γ, is increased beyond R by very-long-range collisions, typically of impact parameter greater than ~ 15 Å. More robust collisions deviate the molecular trajectory, transferring enough z-axis momentum to destroy the saturation resonance condition [2]. These molecules do not contribute very much to the signal due to the intrinsically nonlinear character of the saturated absorption process. We take the phase interruption rate

$$\gamma \text{ (Pressure)} = P_1 + P_2 \times \text{(Pressure)} \tag{1}$$

If μ is the transition dipole moment, the maximum radiative interaction rate may be represented by $\mu E_0/\hbar$, where E_0 is the optical electric field on the laser beam axis due to one running-wave component.

Thus we could define the saturation parameter as $S_0 \equiv (\mu E_0/\hbar)^2/\gamma R$. However, for the present purposes it is more useful to define an <u>effective</u> saturation parameter, S, by the relation

$$S = \left(\frac{\mu}{\hbar}\right)^2 \frac{1}{\gamma R} \ [E^2] \tag{2}$$

where $[E^2]$ is an "effective" electric field (squared) resulting from a suitable three-dimensional average of the laser intensity distribution. For two-level quantum systems interacting with a coherent radiation field, it is known that saturation increases the linewidth by the factor $(1+S)^{1/2}$ [2, 3, 14]. Thus we can model

$$\Gamma \text{ (Pressure, Power)} = \gamma \text{ (Pressure) } (1+S)^{1/2} \tag{3}$$

giving

$$\Gamma^2 \text{(Pressure, Power)} = \gamma^2 \text{(Pressure)} + \gamma \text{ (Pressure)} \left(\frac{\mu}{\hbar}\right)^2 [E^2]/R. \tag{4}$$

The fitting function used was therefore

$$[\Delta\nu_{HWHM}(\text{Pressure, Power})]^2 = [P_1 + P_2 \times (\text{Pressure})]^2$$

$$+ [P_1 + P_2 \times (\text{Pressure})] \times P_3 \times (\text{Power})/P_1 \qquad (5)$$

The great utility of this gambit may be surmised from Figs. 7 and 8 which are essentially two perpendicular slices through the three-dimensional surface of linewidth versus pressure and power. In both figures the "+" character indicates the experimental linewidth, the "0" is the data corrected for the finite value of the orthogonal abscissa. Thus in Fig. 7 we have a set of power-broadening experiments at five values of the pressure. Note all results at fixed power are projected onto essentially the same point by the

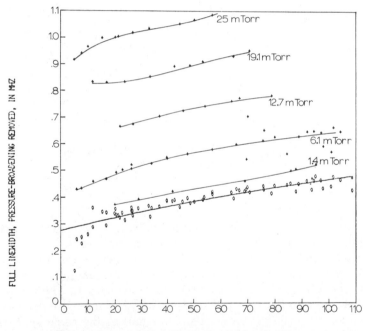

TRANSM POWER IN UW (MULT X4402 CM-2 FOR CENTRAL INTENSITY)
FOLDED CAVITY (WO=.490 MM) FORM 4,PG 69(3) 8/14/71

Figure 7. Projection onto pressure = 0 plane of experimental linewidth surface. The measured points are plotted "+". For the corrected data, plotted as "0", Eq. (5) has been used to remove the effect of nonzero pressure in the experiment. Equation (5) is plotted as the smooth line through the "0" 's. See text regarding points at low powers.

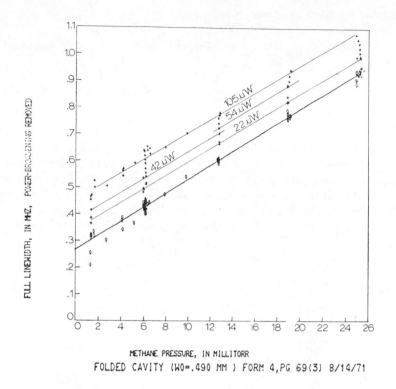

FOLDED CAVITY (WO=.490 MM) FORM 4,PG 69(3) 8/14/71

Figure 8.   Experimental line width surface projected onto power = 0 plane. Equation (5) was used to correct for finite measurement intensity. See caption of Fig. 7.

assumed pressure correction. Similarly in Fig. 8 we have experimental linewidth at several power levels versus methane pressure. The saturation correction is sufficient to essentially eliminate the power broadening [12].

A pressure-dependent T_1 model gave, within the statistical limits, the same "best-value" parameters as our assumed model, with a fairly negligible improvement in the residuals. Several other intensity dependences were investigated, but none gave residuals as small as the assumed form.

The output parameters for data taken at five apertures are presented in Table I. The half-linewidth, P_1, varies from 850 to 23 kHz for a larger-than-corresponding change in the laser spot size. The data are presented in historical order to allow the reader to assess for himself the significance of spatial mode purity. For example the first two experiments, EUZ and First FBTK, differ only subtly in the focusing and recollimation adjustments. The EUZ data are presented only to illustrate the significance of the spatial

TABLE I. (All entries are ordinary frequency, not radian frequency.)

Run ID	Beam[a] Radius=w	P_1 (kHz)	P_2 (kHz/mTorr)	P_3/P_1 (kHz/mW)	P_1^2/P_3 (mW)	$w P_1$ (kHz mm)
EUZ[b]	5 mm	29.3 ± 2.3	12.66 ±0.17	8.9 ± 2.0	3.3 ±1.0	145
First FBTK[c]	5 mm	23.28 ± 0.48	9.92 ±0.21	9.10 ± 0.85	2.56 ±0.29	116
Small spot[d]	85 μ	846 ±45	10.27 ±0.91	635 ± 162	1.33 ±0.41	72
First cavity[e] total data	0.706 mm	104.0 ± 5.6	15.7 ±0.4	59.4 ± 6.0	1.75 ±0.27	73
Second cavity[e]	0.490 mm	141.3 ± 2.3	13.26 ±0.09	148.0 ± 5.5	0.95 ±0.05	70

[a]For $1/e^2$ intensity.
[b]Passive beam expansion.
[c]Same as footnote a but better alignment.
[d]Focused spot.
[e]Length-stabilized resonant cavity, $F \sim 40$.

aspects of these experiments. Because of this sensitivity it is difficult to assess the systematic error effects in the entries in Table I: The errors listed are standard-deviation random errors only. Even so, we think that there is a meaningful peak in the pressure-broadening coefficient, P_2, near apertures of 1/2 mm. A detailed interpretation of this effect in terms of a collision pseudopotential would be facilitated if we could obtain reliable pressure-shift data as well, but the data at the larger apertures are just too uncertain. (The shift is only $\sim$ 75 Hz/mTorr.)

In Table I we report P_3/P_1 (rather than P_3) as it is probably a more useful number. The modest cross correlation implicit in the least-squares procedure affects the error estimate.

In the last column of Table I, the deviation upward from $w_0 P_1$=70 kHz mm is perfectly correlated with other kinds of knowledge regarding the spatial quality of the laser beam (tolerance of the auto collimation condition, size of the interference fringes with reduced optical decoupling, etc).

In the next-to-last column of Table I, one may see that, as expected, the concept of a "saturation intensity" in lifetime-limited (atomic) cases maps into an analogous "saturation power" for long-lived absorbing particles whose radiative interaction time is dominated by the transit-time effect.

Corresponding data on the signal size, rather than width, are available from the experiments described. However a number of interesting new factors are operative in this case, and we will defer discussion of the mode-size scaling results to a later publication.

In this paper we show that large amount of saturation data can be represented by a simple model. We hope that the availability of these results and demonstration of the time-of-flight scaling effect will encourage someone to develop a real theory applicable in this interesting limit.

ACKNOWLEDGMENTS

In the course of this work a number of colleagues have made important contributions. Throughout the last several years the productive and pleasant collaboration with R. L. Barger has been appreciated and enjoyed. Some of the early data were taken in collaboration with E. E. Uzgiris. As usual, P. L. Bender has made a number of useful suggestions and instructive criticisms. We thank J. Levine for applying some of his computer expertise to our Lorentz curve–fitting problem. The cheerful and competent computing help of Mrs. P. Kunasz is appreciated.

REFERENCES

1. B. L. Gyorffy, M. Borenstein, and W. E. Lamb, Jr., Phys. Rev. 169 340 (1968).
2. J. L. Hall, in Lectures in Theoretical Physics, edited by K. T. Mahanthappa and W. E. Britten (Gordon and Breach, New York, 1971).
3. T. Hänsch and P. Toschek, IEEE J. Quantum Electron. QE-4, 530 (1968); See also Ref. 14.
4. S. Stenholm and W. E. Lamb, Jr., Phys. Rev. 181, 618 (1969); W. K. Holt, Phys. Rev. A 2, 233 (1970); B. J. Feldman and M. S. Feld, Phys. Rev. A 1, 1375 (1970); J. Shirley (private communication).
5. V. S. Letokhov, Zh. Eksp. Teor. Fiz. 56, 1748 (1969) Sov. Phys. JETP 29, 937 (1969); S. N. Bagaev, L. S. Vasilenko and V. P. Chebotaev, "Shape of Lamb Dip in Molecular Gas at Low Pressure," Preprint #15, USSR Academy of Science Siberian Division, Institute of Semiconductor Physics [translated at JILA by E. Weppner].
6. R. L. Barger and J. L. Hall, Phys. Rev. Letters 22, 4 (1969).

7. The author is indebted to G. Kramer of PTB for discussion of this idea.

8. For example, see E. E. Uzgiris, J. L. Hall, and R. L. Barger, Phys. Rev. Letters 26, 289 (1971).

9. H. S. Boyne, J. L. Hall, R. L. Barger, P. L. Bender, J. Ward, J. Levine, and J. Faller, in Laser Applications in the Geosciences, edited by J. Gauger and F. F. Hall (Western Periodicals, North Hollywood, Calif., 1970), p. 215. Also see, J. Levine and J. L. Hall, J. Geophys. Res. 77, 2595 (1972).

10. R. L. Barger and J. L. Hall, in Proceedings of the International Conference on Precision Measurements and Atomic Constants, Teddington, Sept. 1971 (to be published). See also R. L. Barger and J. L. Hall, "Precision Wavelength Measurement of the Methane 3.39 μm Saturated Absorption Line by Laser-Controlled Interferometry"(to be published).

11. This small asymmetry is found to be strongly saturation-dependent and is thought to be a manifestation of the "recoil momentum" effect. See A. P. Kol'chenko, S. G. Rautian, and R. I. Sokolovskii, Zh. Eksp. Teor. Fiz. 55, 1864 (1968) [Sov. Phys. JETP 28, 986 (1968)].

12. In all of the data presented here we have substantially eliminated the regime of anomalously narrow lines near zero power and zero pressure: different physical processes are operative there.

13. As already noted, at very low powers and pressures the line is formed mostly by very slow particles. However P_1 here is to be associated with the same average transverse velocity that is important for the line formation in the regime of moderate pressure and power broadening (say > 20%). Thus we give an intercept definition for P_1 and for the present modeling purpose do not take the fit very seriously just near the origin.

14. R. Karplus and J. Schwinger, Phys. Rev. 73, 1020 (1948).

PRECISION INTERFEROMETRIC WAVELENGTH COMPARISON

N. A. Kurnit
Department of Physics, Massachusetts Institute of Technology
Cambridge, Massachusetts

ABSTRACT

Developments in the technology for measuring absolute wavelengths in the infrared using a scanning Michelson interferometer are described. The ratio of the infrared wavelength to the wavelength of a He-Ne secondary standard can be obtained to an accuracy approaching 1 part in 10^8 by using a computer to extract the relative phases of the infrared and red fringes, thereby giving fractional fringe accuracy.

The development of stable lasers had made possible, particularly in combination with nonlinear resonant saturation techniques, a wide range of high-precision and high-resolution experimental advances, many of which have been described at this conference. This is by way of a progress report on some work on precision wavelength measurements in the infrared which is part of a long-range program in our laboratory aimed at high-precision spectroscopy, the exploration of means of achieving frequency and wavelength standards in the infrared, and the establishment of a connection between the time and length standards. The latter goal is being pursued through the measurement of the frequency and wavelength of the same laser transition to yield a precise determination of the speed of light.

These wavelength measurements utilize an evacuated scanning Michelson interferometer first used by Daneu, Hocker, Javan, Rao, and Szőke of MIT and Zernike of the Perkin-Elmer Corporation for measurement of H_2O and D_2O far-infrared wavelengths [1]. By counting the number of fringes of a 633-nm He-Ne laser corresponding to an integral number of infrared fringes, the ratio of the infrared wavelength to the He-Ne wavelength could be determined to an accuracy of one part in 10^6 in a scan lasting only a few minutes. Further improvements in accuracy required the measurement, as discussed below, of fractional fringes.

At the time, a preliminary determination of the speed of light was made [1, 2], and an accurate determination by measurement of the frequency and wavelength of the 84-μ D_2O transition appeared feasible. Subsequent extension of laser harmonic frequency mixing techniques to the 10-μ region and even shorter wavelengths [3, 4], coupled with the higher power, better detectors, and smaller diffraction effects associated with these shorter wavelengths, made CO_2 a more attractive candidate for this measurement. This paper will be limited to a discussion of the progress in CO_2 wavelength measurements, performed in collaboration with Monchalin, Kelly, Thomas, Javan, Szőke and Zernike [5].

The experimental arrangement is shown in Fig. 1. The two lasers whose wavelengths are to be compared are each isolated from the interferometer by a polarizer and quarter-wave plate, sent through beam expanders, and are then combined by means of a beam splitter at the entrance to the interferometer. This instrument is constructed with a fixed flat mirror in the short arm and a front surface mirror corner-cube in the long arm. The corner-cube is suspended from two guides and can be pulled in either direction at constant velocity through a distance of 0.7 m by a motor driven taut string.

The use of a corner-cube has two major advantages, both of which result from the fact that the beam always returns parallel to itself: (a) the phase and visibility of the fringes are not affected by small angular deflections of the corner-cube which may occur during translation, and (b) flat fringes are obtained only if the light is directed normal to the flat mirror, thus ensuring that the two beams whose wavelengths are to be compared traverse the interferometer along identical paths. Since a misalignment of one of the lasers by an angle δ results in a path-length difference proportional to $1-\cos\delta$, the relative error this introduces into the measurements is $\delta^2/2$. The infrared laser can be aligned perpendicular to the flat mirror to within 10^{-4} rad by spreading the interference pattern to one fringe over a diameter of 50 mm; the red laser can be aligned to within one-quarter of this tolerance. The error due to laser misalignment can thus be held to 5 parts in 10^9.

The paths followed by the two beams through the interferometer are shown in more detail in Fig. 2. The NaCl beam splitter has a wedge angle $\alpha = 9$ mrad in order to provide angular discrimination against unwanted

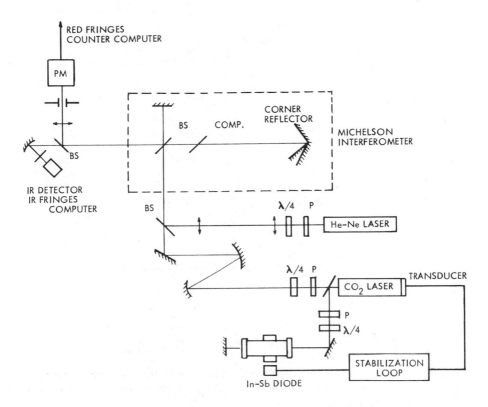

Figure 1. Scanning Michelson interferometer and associated optics for comparison of CO_2 and He-Ne wavelengths.

reflections. A similar NaCl wedge is oriented as shown, with surface B parallel to surface A, and serves to compensate for the angular deflection as well as temperature variations in path length introduced by the beam splitter. The wedge angles of the beam splitter and compensator are equal within 0.5 mrad, which is sufficient to make the two beams parallel to within 2.5×10^{-5} rad. Complete thermal compensation is not possible because the beams reflected by the flat mirror and the corner-cube pass through slightly different thicknesses of NaCl since the corner-cube inverts the returned beam. Moreover, approximately ten times more CO_2 power passes through the beam splitter than through the compensator. Measurement of temperature changes after the interferometer has been allowed to come to thermal equilibrium indicate that thermal drift during the time taken for a complete scan should not alter the difference in optical path length

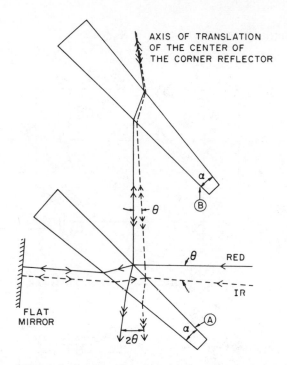

Figure 2.   Paths of the infrared and red beams in the interferometer. Parallelism of the two beams is achieved by adjusting the input direction of each to obtain flat fringes.

traversed by the red and infrared beams by more than 1 part in 10^9 of the total path.

The He-Ne laser is a Perkin-Elmer model 5800 with a thermally stabilized Invar cavity. It can be locked to 1 part in 10^8 at the center of the Lamb dip, and, after sufficient warmup, will maintain this accuracy during the course of a run with the feedback loop open. The latter mode of operation is preferred because the frequency modulation of the laser required for closed-loop operation results in a modulation of the interferometer fringes. The wavelength of this laser has been measured against the ^{86}Kr standard by means of a Fabry-Perot interferometer, yielding a vacuum wavelength of 0.6329915 μm [6]. This wavelength will have to be reverified because of the possible shift of the Lamb dip with tube age.

The CO_2 laser is one constructed in our laboratory utilizing a stable Invar four-rod cavity design. The laser is locked to the center of the

Doppler profile of a CO_2 absorber in a low-pressure CO_2 absorption cell (Fig. 1) by monitoring the 4.3-μ $00^{\circ}1 - 00^{\circ}0$ fluorescence as a standing-wave field is tuned through resonance [7]. When the laser is tuned to the center of the Doppler-broadened transition, both traveling-wave components interact with the same set of atoms and the nonlinear saturation results in a dip in the fluorescence over a width determined by natural, collision, power, and transit-time broadening. The output of the InSb detector which monitors the fluorescence is fed to a lock-in amplifier with a 1-sec time constant. The derivative signal obtained by modulating the laser by ±300 kHz at a 100-Hz rate as the laser is tuned across the resonance is shown in Fig. 3. The width of the dip is 1.5 MHz and the laser has been locked to the center with a stability of 1 part in 10^8 of the laser frequency. This technique is particularly useful for establishing frequency and wavelength standards in the 10-μ region since it is applicable to any CO_2 laser line. Since a low-pressure (< 100 mTorr) absorber is used, pressure shifts are small. It is expected that considerable improvement in frequency stability can yet be realized by means of this technique.

In order to obtain the maximum possible accuracy from the interferometric measurements, it is necessary, as already mentioned, to measure fractional fringes. Since 2.1×10^6 red fringes are counted in one scan, an accuracy of 1 part in 10^8 requires that we be able to split a red fringe to 1 part in 50 and an infrared fringe to nearly 1 part in 1000. This task is not as formidable as it sounds since we can utilize as many fringes as necessary to obtain the required accuracy. A better way of stating what must be accomplished is that the phase of the infrared fringes must be determined relative to the first and last red fringe zero crossings to an accuracy of slightly better than 10^{-2} rad.

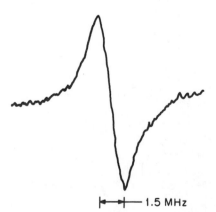

|←—•—|— 1.5 MHz

Figure 3.   Derivative of the 4.3-μ fluorescence signal used to stabilize the CO_2 laser.

For this purpose, the interferometer is tied to a PDP-12 computer which is used to sample the infrared fringes and store them on magnetic tape. The computer is triggered on negative-going red fringe zero crossings and samples the infrared fringe amplitude for 1000 red fringe crossings at the beginning of a scan (after the corner-cube has been brought up to constant velocity) and six more times at intervals of 300,000 red fringes.

A fraction of one sample of data is shown in Fig. 4(a). The use of the red fringes to trigger the computer ensures that fluctuations in the mirror velocity do not result in distortion of the recorded fringes. As a function of red fringe number, the infrared fringes appear as a sine wave whose frequency is given by the ratio of red to infrared wavelengths. Since this number is known to a high degree of accuracy (better than 1 part in 10^6) from earlier measurements, it is a simple process to fit the data to a sine wave and extract the phase of the fringes. (Linear and quadratic variation of sine-wave amplitude and centerline are also allowed for in this fitting process, but these are found to be negligible.)

The deviation between the data and the fitted sine wave is shown, greatly expanded, in Fig. 4(b). The rms deviation is 0.6% in this example and is indicated on the figure. A close inspection of the noise shows components at harmonics of the fringe frequency, and these can be clearly exhibited by Fourier analysis. These harmonic components arise from detector and amplifier nonlinearities as well as from the presence of a small amount of feedback into the laser. The latter can result in a pulling of the laser frequency which gives rise to a possible systematic error. The present reproducibility of our results to ±2 parts in 10^7 is believed to be limited largely by this effect. Sufficient decoupling of the lasers from the interfer-

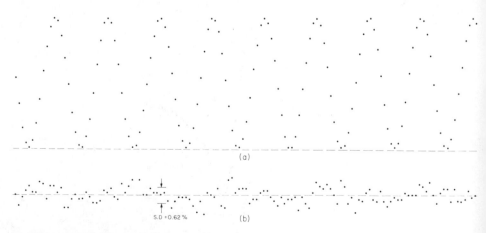

Figure 4.   (a) Infrared fringes from interferometer. (b) Expanded difference between fringes and fitted sine wave (0.6% rms deviation).

ometer should permit this source of error to be reduced below 1 part in 10^8. Within a given run, internal consistency between different samples has shown this method to be capable of extracting the wavelength ratio to better than 1 part in 10^8.

For a beam of finite cross section A, the measurement of absolute wavelengths to a fractional precision approaching λ^2/A requires an accurate knowledge of the effect of diffraction on the measured propagation constant. Due to the long wavelength used (4 mm), the diffraction correction in the presently accepted speed of light determination [8] had to be known to within a few percent in order to obtain the stated accuracy of 3 parts in 10^7. One of the major advantages of measuring wavelengths in the 10-μ region is that the diffraction correction can be reduced to ~ 1 part in 10^8, which is of the order of the ultimate accuracy imposed by the reproducibility of the present wavelength standard.

The spatial dependence of the field of a TEM_{00} Gaussian beam with a 1/e radius w_0 at the beam waist (z=0) can be approximated by [9, 10]

$$E_{00}(r, z) = \exp(-r^2/w_0^2) \exp[-i\left(k - \frac{\lambda}{\pi w_0^2} + \frac{r^2}{w_0^2}\frac{\lambda}{\pi w_0^2}\right) z] \qquad (1)$$

provided that z is in the near field, i.e., $z \ll \pi w_0^2/\lambda$. The latter condition has been used to replace arctan $(\lambda z/\pi w_0^2)$ by $\lambda z/\pi w_0^2$ and $w^2(z)$ by w_0^2. Under the experimental conditions $2w_0$ = 25 mm at 0.633 and either 10.5 mm or 25 mm at 9.3 μm, $\pi w_0^2/\lambda$ is, respectively, 800 m, 9.3 m, or 55 m, which all fulfill the requirement that the output of the interferometer remains in the near field. The fringes formed by the interference of the waves coming from the flat mirror and the corner-cube will vary as $\cos[2k'(r)\Delta z]$ where $k'(r)$ is related to the propagation constant $k=2\pi/\lambda$ for a plane wave in free space by

$$k'(r) = k - \frac{\lambda}{\pi w_0^2} + \frac{r^2}{w_0^2}\frac{\lambda}{\pi w_0^2} \qquad (2)$$

A detector placed at r=0 measures a k which is too small by $\lambda/\pi w_0^2$. If the detector sees the whole beam, the effect of the last term in Eq. (2), when averaged over the intensity distribution, is to decrease this difference by a factor of 2, to $\lambda/2\pi w_0^2$. The fractional difference $(k-k')/k=\lambda^2/4\pi^2 w_0^2$ is, for the example given above, 6×10^{-11} at 0.633 μm, 8×10^{-8} at 9.3 μm with a 10.5-mm beam, and 1.35×10^{-8} at 9.3 μm with a 25-mm beam. Corrections to these numbers due to a possible small admixture of other modes and the presence of narrow cracks between the corner-cube mirrors have been analyzed and found to be small [10].

The recent work in perfecting these wavelength-measuring techniques has concentrated on the measurement of the CO_2 R(12) $00^{\circ}1 - 02^{\circ}0$ transition at 9.3 μm since the frequency of this transition can be measured by mixing with the third harmonic of the 28-μm water vapor laser plus a microwave difference frequency [3]. Similar frequency multiplying steps connect the H_2O laser transition with the Cs frequency standard, thus providing an absolute frequency measurement of the R(12) CO_2 transition [11]. We obtain as a preliminary value of the wavelength

$$\lambda_{R(12)} = 9.317\ 247\,(2)\ \mu m \quad (\pm 2 \text{ in } 10^7)$$

Together with the measured frequency [3]

$$\nu_{R(12)} = 32.176\ 085\,(18)\ THz \quad (\pm 6 \text{ in } 10^7)$$

this yields a value for the speed of light,

$$c = \nu\lambda = 299\ 792\ 550\,(240)\ m/sec \quad (\pm 8 \text{ in } 10^7)$$

Several other groups are presently close to obtaining a value for the speed of light to an accuracy of ~ 1 part in 10^8, the limit imposed by the reproducibility of the ^{86}Kr wavelength standard. Once a value to this accuracy is agreed upon, it will be possible to dispense with the wavelength standard and reference all length measurements to the frequency standard. Frequency measurements are inherently more accurate because of the reproducibility of the frequency standard and because they do not depend upon the quality of optical components.

Absolute wavelength measurements in the infrared nevertheless will remain a useful spectroscopic tool. Measurements to an accuracy of 1 part in 10^6 or 10^7 can be done fairly readily, and an accuracy of 1 part in 10^8 appears to be achievable. Frequency measurements, while basically more accurate, are often more time consuming and require different lasers and/or klystrons for different wavelength regions. The use of a scanning interferometer such as described here permits rapid laser wavelength measurements to an accuracy not previously obtainable in the infrared. A grating spectrometer with a resolution of 1 part in 10^6 at $\lambda = 10$ μm requires a 10-m grating; a Fabry-Perot interferometer of the same length can give perhaps a hundredfold improvement in resolution, but requires careful analysis of the effects of diffraction and mirror phase shift. Rapid developments in the area of tunable lasers promise to make precision spectroscopy possible throughout the infrared region. A scanning Michelson interferometer such as described here can help to provide a set of accurately measured lines to serve as bench marks for absolute wavelength measurements.

ADDENDUM (OCTOBER, 1972)

Precision measurements of the frequency [12] and wavelength [13] of the 3.39 μm He–Ne laser have recently provided a value for the speed of light [14] to an accuracy limited by the Kr wavelength standard: c = 299 792 456.2(1.1) m/sec. In the course of these measurements, Evenson et al. [12, 14] measured the frequency of the R(10) $00^\circ 1 - 02^\circ 0$ CO_2 transition to be $\nu_{R(10)}$ = 32.134 266 891 (24) THz. As in our experiment the CO_2 laser was stabilized at the center frequency of the nonlinear saturation resonance of the type initially observed by Freed and Javan [7]. This reduces the value of $\nu_{R(12)}$ quoted above to 32.176 079 5 THz, yielding a value of c = 299 792 490 (60) m/sec from our wavelength measurements. Baird, Riccius, and Siemsen [15] have performed CO_2 wavelength measurements by measuring the wavelength of sidebands placed on a 633-μm He–Ne laser by nonlinear mixing with a CO_2 laser in a proustite crystal. Our value of $\lambda_R(12)$ is in agreement within experimental error with their tabulated value calculated from measurements on several lines together with the known [16] rotational constants. In this experiment also, the laser is stabilized at the peak of the nonlinear saturation resonance. Together with Evenson's frequency measurements, they obtain a value of c = 299 792 460 (6) m/sec. This accuracy of ±2 in 10^8 is achieved by averaging over a number of wavelength determinations with a scatter of ~±1 in 10^7. An independent measurement of the speed of light by Bay, Luther, and White [17] by an interferometric determination of the ratio of an optical frequency to the frequency of a microwave signal used to generate optical sidebands yields c = 299 792 462 (18) m/sec.

REFERENCES

1. V. Daneu, L. O. Hocker, A. Javan, D. Ramachandra Rao, A. Szöke and F. Zernike, Phys. Letters 29A, 319 (1969).
2. L. O. Hocker, J. G. Small and A. Javan, Phys. Letters 29A, 321 (1969).
3. V. Daneu, D. Sokoloff, A. Sanchez, and A. Javan, Appl. Phys. Letters 15, 398 (1969); K. M. Evenson, J. S. Wells, and L. M. Matarrese, Appl. Phys. Letters, 16, 251 (1970).
4. D. R. Sokoloff, A. Sanchez, R. M. Osgood, Jr., and A. Javan, Appl. Phys. Letters 17, 257 (1970).

5. J.-P. Monchalin, A. Javan, N. A. Kurnit, A. Szőke, and F. Zernike, Bull. Am. Phys. Soc. 16, 1402 (1971). These experiments are continuing with the added collaboration of M. J. Kelly and J. E. Thomas.
6. G. Koppelman, MIT Laser Group Internal Report.
7. C. Freed and A. Javan, Appl. Phys. Letters 17, 53 (1970).
8. K. D. Froome, Proc. Roy. Soc. (London) A247, 109 (1958). This and other measurements are discussed in: B. N. Taylor, W. H. Parker, and D. N. Langenberg, Rev. Mod. Phys. 41, 375 (1969).
9. H. Kogelnik and T. Li, Appl. Optics 5, 1550 (1966).
10. J.-P. Monchalin, M. S. thesis (MIT, 1971).
11. Such a frequency multiplying chain is presently being constructed in our laboratory. See the paper by A. Javan in this volume.
12. K. M. Evenson, G. W. Day, J. S. Wells, and L. O. Mullen, Appl. Phys. Letters, 20, 133 (1972); J. D. Cupp, B. L. Danielson, G. W. Day, L. B. Elwell, K. M. Evenson, D. G. McDonald, L. O. Mullen, F. R. Petersen, A. S. Risley, and J. S. Wells, Conference on Precision Electromagnetic Measurements, Boulder, Colorado, June, 1972 (IEEE, New York, 1972), p. 79; K. M. Evenson, J. S. Wells, F. R. Petersen, B. L. Danielson, and G. W. Day, Appl. Phys. Letters 22, 192 (1973).
13. R. L. Barger and J. L. Hall, Conference on Precision Electromagnetic Measurements, Boulder, Colorado, June, 1972 (IEEE, New York, 1972) p. 76; R. L. Barger and J. L. Hall, Appl. Phys. Letters 22, 196 (1973).
14. K. M. Evenson, J. S. Wells, F. R. Petersen, B. L. Danielson, G. W. Day, R. L. Barger, and J. L. Hall, Phys. Rev. Letters 29, 1346 (1971).
15. K. M. Baird, H. D. Riccius, and K. J. Siemsen, Opt. Commun. 6, 91 (1972).
16. T. J. Bridges and T. Y. Chang, Phys. Rev. Letters 22, 811, (1969).
17. Z. Bay, G. G. Luther, and J. A. White, Phys. Rev. Letters 29, 189 (1972).

B. Spectroscopy of Gases

EXPERIMENTAL TESTS OF THE QUANTUM THEORY OF MOLECULAR HYDROGEN

G. Herzberg
Division of Physics, National Research Council of Canada
Ottawa, Canada

I. INTRODUCTION

One of the early and most significant successes of wave mechanics was the demonstration (by Hylleraas, Heitler, London, and others) about 45 years ago that, unlike the Bohr theory, it can account for the ground and excited states of two-electron systems (He and H_2). To be sure, it was quite a long time before a really precise quantitative test was accomplished. It was only 13 years ago that a fairly definitive study [1] of the ground state of He and several excited states of Li^+ was completed [2]. The ground state of the latter still remains to be done.

The study of He and Li^+ was made more interesting, after the earlier semiquantitative comparisons, by the advent of quantum electrodynamics and the possibility of establishing for the first time a Lamb shift in a two-electron system. The study of H_2 and its isotopes is still in progress [3-5] and will form the subject of this lecture since it has at least a slight connection to the main theme of this meeting: H_2 has been found to provide the shortest wave-length laser yet developed.

One of the original aims of our work on H_2, HD, and D_2 was the determination of the Lamb shifts in these two-electron systems. We did not realize at the time that these shifts would be as small as later theoretical work has indicated. The Lamb shift of the H_2 molecule differs only slightly from that of two H atoms and is far from the much larger shift of the united atom, that is, He.

II.   EARLY WORK ON H_2 AND H_2^+

A.   Theory

The earliest calculations of the dissociation energy of the H_2^+ ion were carried out by Burrau [6] in 1927.   They were followed by the work of Guillemin and Zener [7], Teller [8], Hylleraas [9], Jaffe [10], Sandeman [11] and Johnson [12].   These studies led to the value

$$D_0^0 (H_2^+) = 21345 \pm 20 \text{ cm}^{-1}$$

which, until a few years ago, was the accepted theoretical value.

The ground state of the H_2 molecule and its dissociation energy was first calculated by Heitler and London [13] in 1927.   Their work was refined by Wang [14], Rosen [15], Weinbaum [16], and finally, by the monumental work of James and Coolidge [17] in 1933, carried out before the development of modern computers.   It resulted in the value

$$D_0^0 (H_2) = 36104 \pm 100 \text{ cm}^{-1}$$

which, until 1960, was considered to be the best theoretical value.

The ionization potential of the H_2 molecule follows from $D(H_2)$ and $D(H_2^+)$ from the simple equation

$$\text{I. P. } (H_2) = \text{I. P. } (H) + D_0^0 (H_2) - D_0^0 (H_2^+) \tag{1}$$

and, with the values for $D_0^0(H_2)$ and $D_0^0(H_2^+)$ given and
I. P. (H) = 109678.8 cm^{-1}, becomes

$$\text{I. P. } (H_2) = 124438 \pm 100 \text{ cm}^{-1}$$

Another important early theoretical result was the calculation of the repulsive state of H_2 resulting from two H atoms when their spins are parallel, by Heitler and London [13] and James, Coolidge, and Present [18]. As was first suggested by Winans and Stueckelberg [19] this state represents the lower state of the continuous spectrum of hydrogen in the visible and near-ultraviolet regions, a suggestion that has been confirmed in detail by the calculations of Coolidge, James, and Present [20, 21].

B.   Experimental

Witmer [22] in 1926 was the first to give a fairly good value for the dissociation energy of the neutral H_2 molecule from the emission spectrum of the H_2 molecule in the vacuum ultraviolet, simplified by the addition of argon. Dieke and Hopfield [23] used the edge of a continuous absorption in the vacuu ultraviolet to determine the dissociation energy, a method that was greatly refined and made more precise by Beutler [24]. From the potential diagram, Fig. 1, it is clear that the dissociation energy is given by

$$D_0^0(H_2) = \nu_{lim} - E(H, n=2) \tag{2}$$

where ν_{lim} is the observed limit of continuous absorption for zero rotation (J=0) and $E(H, n=2)$ is the excitation energy of the H atom in the n=2 state. The value that Beutler obtained was

$$D_0^0(H_2) = 36116 \pm 6 \ cm^{-1}$$

The ionization potential of the hydrogen molecule was first obtained by an extrapolation of a very short Rydberg series by Richardson [25] and later, more precisely, from preionization and predissociation data by Beutler and Jünger [26]. They obtained

$$I.P. (H_2) = 124429 \pm 13 \ cm^{-1}$$

If this value is combined with the experimental value for $D_0^0(H_2)$ according to Eq. (1) one obtains for the H_2^+ ion

$$D_0^0(H_2^+) = 21366 \pm 15 \ cm^{-1}$$

This was the experimental situation until 1960 when Herzberg and Monfils [27] repeated Beutler's work for the three isotopes H_2, HD, and D_2 with much higher resolution and obtained the values given in the second column of Table I. By adding the zero-point energies the values for D_e

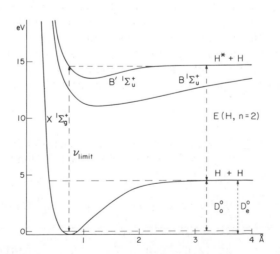

Figure 1.   Potential energy diagram of H_2 explaining the relation of the absorption limit to the dissociation energy of the ground state. In addition to the ground state $^1\Sigma_g^+$ the two lowest $^1\Sigma_u^+$ states are shown. The observed absorption continuum corresponds mainly to a transition to the B' $^1\Sigma_u^+$ state.

TABLE I.   DISSOCIATION ENERGIES (1960)

	$D_0^0 (cm^{-1})$	$D_e^0 (cm^{-1})$
H_2	36113.6 ± 0.3	38292.3 ± 0.5
HD	$36399._9 \pm 1._0$	$38290._3 \pm 1._5$
D_2	36743.6 ± 0.5	38290.8 ± 0.7
H_2^∞	...	38287.0 ± 0.8
calc.	(Kolos and Roothaan)	38286.9

(dissociation energies referred to the equilibrium position) in the third column of Table I were obtained which, when extrapolated to infinite mass, yield the value in the fourth row of Table I.   At the same time Kolos and Roothaan [28] had carried out new calculations of the potential function of H_2 assuming fixed nuclei and had obtained the value given in the last row of Table I, which agrees in a very satisfactory way with the experimental value

III.   RECENT WORK ON THE DISSOCIATION ENERGIES OF H_2, HD, AND D_2

Unfortunately, as the theoretical calculations were improved (Kolos and Wolniewicz [29]) a systematic discrepancy arose since, by the variation principle, the theoretical energies upon improvement of the approximation go down, that is, the dissociation energies increase. The new theoretical work produced, moreover, directly values for D_0^0 rather than D_e.   The discrepancies between theory and experiment that arose in this way are shown in Table II.

TABLE II.   OBSERVED AND CALCULATED
DISSOCIATION ENERGIES (1968)

	D_0^0 Observed [27]	D_0^0 Calculated [29]
H_2	36113.6 ± 0.3	36117.4
HD	$36399._9 \pm 1.0$	36405.2
D_2	36743.6 ± 0.5	36748.0
	cm^{-1}	cm^{-1}

In order to ascertain the cause of these discrepancies a new study of the far-ultraviolet spectrum of H_2, HD, and D_2 was undertaken (Herzberg [4]) with two important improvements: (1) The spectra were taken with our 10-m vacuum spectrograph instead of the 3-m instrument used for the 1960 measurements, leading to an improvement in resolving power by a factor of 3, and (2) the spectra were taken at liquid-nitrogen temperature in order to remove many disturbing and overlapping lines arising from rotational levels other than J=0.

Figures 2, 3, and 4 show sections of the new spectra for H_2, D_2, and HD. In H_2 there are two J=0 lines which overlap the absorption limit. These had been unresolved in the previous work which therefore had given too small a value for the limit. In D_2 in the previous work the J=0 limit had not been observed and it had been necessary to extrapolate from the J=1, J=2 limits. In the new work, at liquid-nitrogen temperature the J=0 limit of D_2 is clear

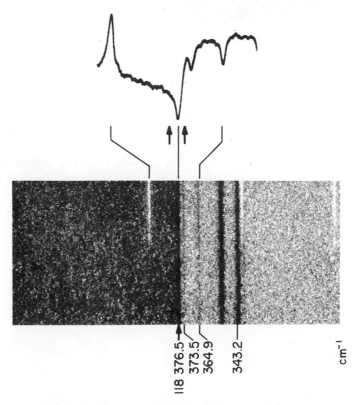

Figure 2. The $J''=0$ absorption limit of H_2 near 844.8 Å. The limit is marked by an arrow in the spectrogram. In the enlarged photometer curve at the top it must lie between the two vertical arrows.

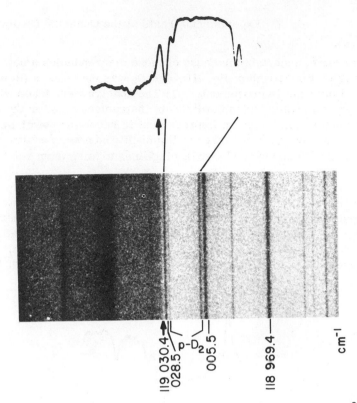

Figure 3. The $J'' = 0$ absorption limit of D_2 near 840.1 Å.

TABLE III. CALCULATED AND OBSERVED DISSOCIATION
ENERGIES OF H_2, HD, AND D_2 (1970)

	Theor.[a]	Obs.[4]
$D_0^0(H_2)$	36117.9	<36118.3 >36116.3
$D_0^0(HD)$	36405.5	36406.6 36405.8
$D_0^0(D_2)$	36748.2 cm^{-1}	36748.9 ± 0.4 cm^{-1}

[a]Including nonadiabatic corrections according to Bunker [49].

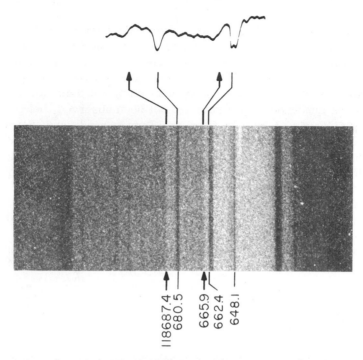

Figure 4.   The $J''=0$ absorption limits of HD near 842.6 Å. The two limits are marked by arrows.

and not overlapped by any line and can be accurately measured.   In HD two limits are observed which correspond to dissociation at the limit into either H + D* or D + H*.

In Table III the new dissociation energies obtained from the observed limits after a very small correction for the rotational barrier (of the order of 0.2 cm^{-1}) are compared with the latest theoretical values.   There is still a very small difference between theory and experiment but it is not much larger than the experimental error (± 0.4 cm^{-1}) and an improvement of the theoretical calculations would now make a change in the right direction.   It may be mentioned that the predicted Lamb shift is only -0.2 cm^{-1} which is well within the estimated error limits of the observations and therefore cannot be determined.

IV.   VIBRATIONAL AND ROTATIONAL LEVELS
OF THE GROUND STATE

The Raman data of Stoicheff [30] still give the best values for the first two rotational constants B_0 and B_1, as well as for the first vibrational quantum $\Delta G_{1/2}$ of H_2, HD, and D_2. The quadrupole spectrum of H_2, first investigat by Herzberg [31] and later studied under higher resolution by Rank and his collaborators [32, 33], supplies corresponding constants for v=2 and v=3. For D_2 the quadrupole spectrum has not been observed, but recently Bredo and I [34] have determined from the Lyman bands and Werner bands in the ultraviolet the rotational constants and the vibrational intervals for a the vibrational levels of the ground state with an accuracy comparable to th of Stoicheff. This study was similar to an earlier study by Herzberg and Howe [35] of the higher vibrational levels of H_2 not covered by the study of the Raman and quadrupole spectra.

For HD, apart from the Raman spectrum, the dipole infrared spectrum has been studied by Durie and Herzberg [36], giving rotational and vibration al constants up to v=4. Figure 5 gives as an example the 4-0 band of the

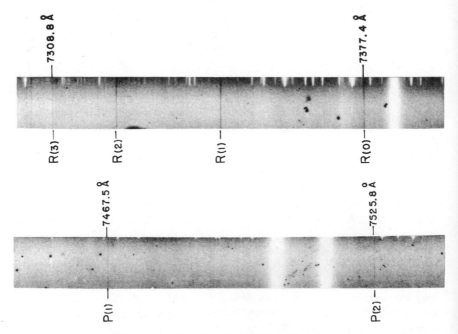

Figure 5.   4-0 rotation-vibration band of HD.   The absorbing path was 1000 m.   The spectrum was taken in the first order of a 6.5 m grating.

infrared spectrum of HD. The higher vibrational intervals have been observed with less accuracy from resonance series observed in mixtures of argon and HD by Takezawa, Innes, and Tanaka [37].

Kolos and Wolniewicz [29], in addition to the dissociation energies, have also derived from the theoretical potential function, values for the vibrational intervals $\Delta G_{v+1/2}$ of H_2, HD, and D_2. In addition, they have derived expectation values of r^{-2} (in atomic units) which are readily converted to predicted values of the rotational constants

$$B_v = \frac{h}{8\pi^2 c\mu} \left[\frac{1}{r^2} \right]_{average}$$

In Tables IV and V the observed ΔG values are compared with the theoretical ones for H_2 and D_2, respectively. In all cases there is a systematic discrepancy which for low v for H_2 is of the order of 0.8 cm^{-1}, for D_2 of the order of 0.4 cm^{-1}. In Figure 6 the deviations are represented graphically, including the data for HD.

Another way of presenting this discrepancy is to plot the differences

TABLE IV.   VIBRATIONAL QUANTA IN THE
GROUND STATE OF H_2 [35]

v	$\Delta G_{v+1/2}$	
	Obs.	Theor.
0	4161.14	4162.06
1	3925.98	3926.64
2	3695.24	3696.14
3	3468.01	3468.68
4	3241.56	3242.24
5	3013.73	3014.49
6	2782.18	2782.82
7	2543.14	2543.89
8	2292.96	2293.65
9	2026.26	2026.81
10	1736.66	1737.13
11	1414.98	1415.54
12	1049.18	1048.98
13	621.96	620.16

TABLE V.   VIBRATIONAL QUANTA IN THE
GROUND STATE OF D_2 [34]

v	$\Delta G_{v+1/2}$ Obs.	Theor.
0	2993.56[a]	2993.96
1	2873.6[b]	2874.82
2	2757.18	2757.79
3	2642.20	2642.40
4	2528.08	2528.21
5	2414.52	2414.75
6	2301.20	2301.45
7	2187.51	2187.72
8	2072.61	2072.98
9	1956.14	1956.46
10	1836.87	1837.24
11	1714.00	1714.48
12	1586.39	1586.83
13	1452.52	1452.84
14	1310.84	1311.23
15	1159.27	1159.59
16	995.14	995.67
17	815.43	815.66
18	615.72	615.60
19	391.15	390.59
20	141.69	

[a]From Stoicheff [30].
[b]From Takezawa, Innes, and Tanaka [37].

between the observed and calculated energies of the levels (rather than their intervals). This is done in Figure 7. As the curves show, the magnitude of the discrepancy first increases with v but then fairly suddenly turns and decreases. According to Poll and Karl [38] the discrepancy between observed and calculated vibrational levels is of the right order of magnitude to be accounted for by the effect of nonadiabatic corrections, but a quantitative calculation of these effects has not yet been made.

In Figure 8 the differences between the observed and calculated rotational constants are plotted. Again, for H_2 and D_2 the observed values are systematically lower than the calculated values. The effect is small and just outside the limit of accuracy of the observations. For HD the scatter of

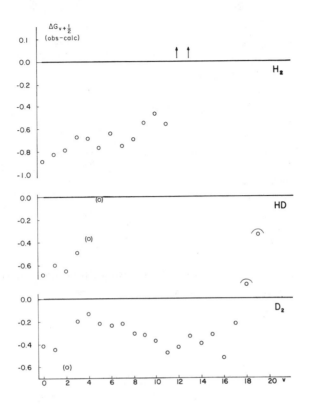

Figure 6.   Deviations of observed vibrational quanta, $\Delta G_{v+1/2}$, of H_2, HD, and D_2 from those obtained from theory.

the observations is greater than the expected effect would be.   It must be assumed that the slight systematic differences for H_2 and D_2 (of the order of 0.005 cm^{-1}) are again due to the effect of nonadiabatic corrections.

In the study of the Lyman and Werner bands in ordinary discharges rotational levels are observed up to about J=10 and often only to lower J values. However, in a study of absorption spectra of flash discharges in H_2 we have been able to observe [39] very high rotational levels for a number of vibrational levels of the H_2 molecule in its ground state.   In Table VI, as an example, for $v'' = 5$ and $v'' = 6$ the observed rotational levels are compared with those calculated by Le Roy [40] on the basis of the Kolos–Wolniewicz potential.   The differences between observation and theory are small and change slightly with increasing rotational quantum number.   The major part of the differences corresponds to the shifts of the $v'' = 5$ and $v'' = 6$ levels, as given by Fig. 7.   The slight variation in a given vibrational level is the same

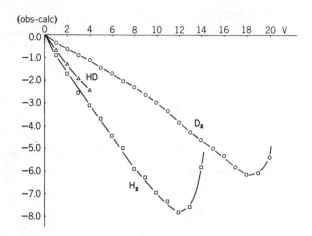

Figure 7. Deviations of observed vibrational levels G_v of H_2, HD and D_2 from those obtained from theory.

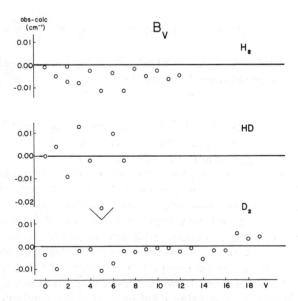

Figure 8. Deviations of observed rotational constants B_v of H_2, HD and D_2 from those obtained from theory.

TABLE VI.   OBSERVED [39] AND CALCULATED [40] ROTATIONAL
LEVELS IN THE v = 5 AND 6 VIBRATIONAL LEVELS OF THE
GROUND STATE OF H_2

J	v = 5			v = 6		
	Obs.	Calc.	Δ	Obs.	Calc.	Δ
0	18492.1	18495.9	−3.8	21505.8	21510.4	−4.6
1	18581.8	18585.7	−3.9	21589.8	21594.4	−4.6
2	18760.4	18764.3	−3.9	21757.0	21761.7	−4.7
3	19026.1	19030.0	−3.9	22005.5	22010.2	−4.7
4	19376.6	19380.1	−3.5	22333.2	22337.6	−4.4
5	19807.1	19811.1	−4.0	22735.7	22740.3	−4.6
6	20315.1	20318.9	−3.8	23210.4	23214.5	−4.1
7	20895.0	20899.1	−4.1	23751.2	23755.7	−4.5
8	21542.7	21546.4	−3.7	24354.6	24358.8	−4.2
9	22251.3	22255.4	−4.1	25013.8	25018.6	−4.8
10	23017.5	23020.6	−3.1	25725.1	25729.4	−4.3
11	23832.0	23836.1	−4.1	26480.9	26485.7	−4.8
12	24692.3	24696.2	−3.9	27277.0	27281.6	−4.6
13	25590.4	25594.8	−4.4	28105.8	28111.1	−5.3
14	26522.1	26526.2	−4.1	28963.5	28968.4	−4.9
15	27479.6	27484.5	−4.9	29841.7	29847.5	−5.8
16	28459.9	28463.9	−4.0	30737.5	30742.4	−4.9
17	29453.7	29458.8	−5.1	31641.1	31647.0	−5.9
18	30459.0	30463.5	−4.5	32549.8	32555.1	−5.3
19	31466.6	31472.1	−5.5	33454.2	33460.2	−6.0
20	32474.3	32478.8	−4.5	34349.7	34355.2	−5.5
21	33471.7	33477.4	−5.7	35226.2	35232.4	−6.2
22	34456.0	34461.4	−5.4	36076.8	36082.3	−5.5
23	35417.3	35423.2	−5.9	26886.7	36892.7	−6.0

effect that causes the slight differences between observed and calculated B
values mentioned earlier.   The last observed rotational levels are close to
the last stable rotational level of each set.   The one for $v'' = 6$ lies above the
dissociation limit (but is below the centrifugal barrier).

Another way of comparing theory and experiment is by way of the equilibrium constants r_e, B_e, and ω_e.   There are, however, two difficulties that
limit the significance of such a comparison:

(1) H_2, HD, and D_2 are peculiar in that there is both in the B_v and the

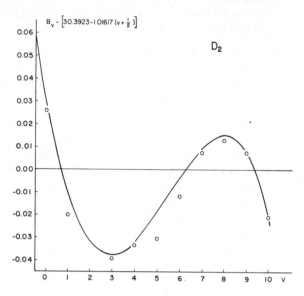

Figure 9.   Deviations of observed B_v values of D_2 from a linear B_v curve. The curve represents the theoretical values of Kolos and Wolniewicz.

ΔG curve a strong positive curvature for low v values: This is illustrated for B_v in Fig. 9 in which the deviations of the observed B_v values of D_2 from a linear relation are shown.   It is clear that an extrapolation of the observed B_v values to v = -1/2, that is, to the effective B_e value, is beset with considerable uncertainty, and the same situation arises for ω_e obtained by extrapolating ΔG to v = -1/2.

(2)   There are a number of corrections that must be applied to the apparent B_e and ω_e values (following Dunham [41], now usually called Y_{01} and Y_{10}, respectively) before they can be compared with the theoretical values, namely, (a) the Dunham corrections which are a result of the finer interaction of rotation and vibration, (b) the adiabatic, and (c) the nonadiabatic corrections due to the interaction of rotation and electronic motion.

Following Bunker [42] we give in Tables VII and VIII the theoretical values of r_e, B_e, and ω_e, as well as the theoretical values from which the Dunham corrections, the adiabatic and finally the nonadiabatic corrections have been subtracted.   These last values should be equivalent to the observed effective values given in the last row of Tables VII and VIII.   Here it must be noted that the calculations are somewhat uncertain since they depend on a development of the Kolos–Wolniewicz potential into a power series, a

TABLE VII. CALCULATED AND OBSERVED ROTATIONAL CONSTANTS

	H_2	HD	D_2
Theory			
r_e		— 0.74140 Å —	
B_e	60.893	45.677	30.462
Y_{01}^{BO}	60.881	45.670	30.459
Y_{01}^{a}	60.847	45.651	30.451
Y_{01}^{na}	60.842	45.646	30.449
Exper.			
$Y_{01} \sim B_e$ 60.862		45.660	30.444

TABLE VIII. CALCULATED AND OBSERVED VIBRATIONAL CONSTANTS

	H_2	HD	D_2
Theory			
ω_e	4405.4	3815.5	3115.9
Y_{10}^{BO}	4404.4	3814.6	3115.4
Y_{10}^{a}	4402.3	3813.5	3114.8
Y_{10}^{na}	4401.3	3812.8	3114.4
Exper.			
$Y_{10} \sim \omega_e$ 4403.4		3814.0	3115.5

development that introduces an appreciable uncertainty in the corrections, and also it must be realized that the experimental values for B_e and ω_e are much less accurate than the B_v and ΔG values. If that is kept in mind, the comparison between theory and experiment in Tables VII and VIII must be considered as quite satisfactory.

If one disregards all corrections and uses the apparent B_e values at their face values one obtains "observed" r_e values of 0.74139 Å and 0.74156 Å for H_2 and D_2, respectively, which should be compared with the theoretical value of the minimum of the Kolos-Wolniewicz potential of 0.74140 Å.

V.   IONIZATION POTENTIALS AND DERIVED QUANTITIES

A.   Rydberg Series

In the spectra of atoms with a $s^2 \, {}^1S$ (or $s \, {}^2S$) ground state there is only one Rydberg series, namely, $n^1P - {}^1S$ (or $n^2P - {}^2S$). In a molecule with a $\sigma^2 \, {}^1\Sigma$ ground state we expect correspondingly two Rydberg series, namely, $n^1\Pi - {}^1\Sigma^+$ and $n^1\Sigma^+ - {}^1\Sigma^+$. Two such series have indeed been observed for the hydrogen molecule, but until recently only the lowest members had been found.

Near the Rydberg limit the spectrum of hydrogen is extremely complicated because, for each value of the principal quantum number n, in each of the Rydberg series we have a whole band system with its widely spaced vibrational and rotational structure. A very considerable simplification of the spectrum can be obtained if it is observed at liquid-nitrogen temperature particularly if para-hydrogen is used, since then only the transitions with $J'' = 0$ occur. Under these conditions we did observe two series of lines for each value of the vibrational quantum number v' of the upper state [3, 5]. For $v' = 1$ this is shown in Fig. 10. It turns out, however, that the two series of a given v' have slightly different limits because one of them $(n^1\Sigma)$ can only correlate with the $N = 0$ level of the ion, while the other $(n^1\Pi)$ can only correlate with $N = 2$ of the ion, and these two rotational levels of the ion are approximately 180 cm^{-1} apart. (N is the quantum number of the total angular momentum apart from spin).

In Fig. 11 the two series of levels are shown schematically, together with the continuous energy region that extends beyond each of them. It is immediately clear from Fig. 11 that the levels of the Π series which lie above the limit of the Σ series are subject to preionization, and indeed Fig. 10 shows that the corresponding lines are broad and show apparent emission in the way that was first found by Beutler [43] for xenon and explained by Fano [44].

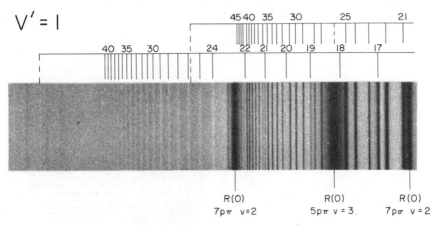

Figure 10. Rydberg series of para-hydrogen for $v' = 1$. The two series corresponding to $N = 0$ and $N = 2$ are marked. The latter appears, above the limit of the former, as an apparent emission series. The three strong lines marked at the bottom belong to different v' Rydberg series.

In addition to the preionization above the first ionization limit there are perturbations between accidentally coinciding levels of the two series below the limit. For example, the n=20 level of the second series (Fig. 11) has about the same energy as the n=34 level of the first series. These perturbations lead to irregularities in the series, which are clearly shown in Fig. 10. They must be fully understood before a reliable extrapolation of the series limit is possible. Fano [45] has developed an elegant theory for these perturbations. Figure 12(a) shows a comparison of the observed series plotted as a difference against an ordinary Balmer series (R/n^2). The full lines are calculated from the Fano formulas. One sees clearly the perturbed places in the $^1\Sigma$ series.

We have also observed the spectrum of ortho-hydrogen, that is, lines with $J'' = 1$. The upper levels of the Q(1) lines form only a single series and are not perturbed. This is seen in Fig. 12(b) which shows a plot for the Rydberg Q(1) series. To be sure, there is one perturbation in this series: it is due to interaction with a Q(1) level of a different vibrational series. Such perturbations with $\Delta v \neq 0$ occur also in the para-Rydberg series and complicate matters still further. They have been fully analyzed in a paper soon to be published [5].

The absorption spectra at low temperature had to be taken at a fairly high pressure of helium (40 mm). This pressure was necessary because the source of the continuous background required it and because the hydrogen (of

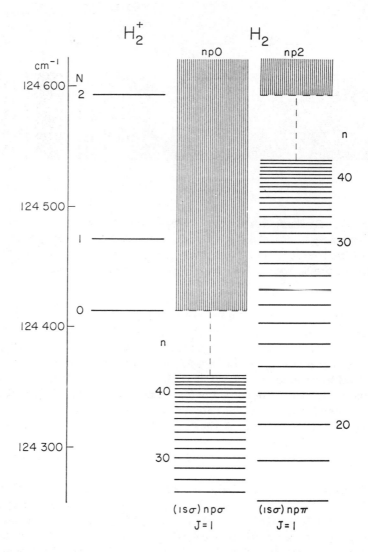

Figure 11.   Energy levels of the Rydberg electron near the ionization limit ($v' = 0$) and correlation with the rotational levels of H_2^+.   The vertical hatching indicates a continuous energy range.

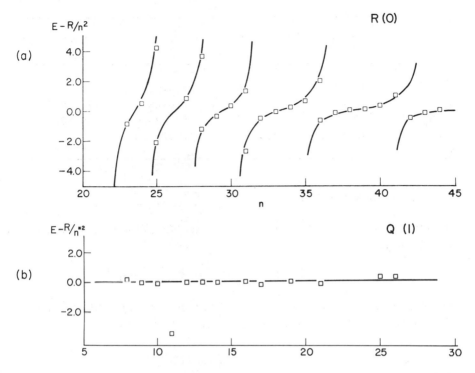

Figure 12.   Deviation of R(0) and Q(1) Rydberg series of H_2 from R/n^2 and $R/n*^2$ respectively for $v'= 1$.

low pressure) had to be retained in the low-temperature absorption tube between light source and slit without solid windows. On account of the fairly high pressure of He there is a shift of the lines of the Rydberg series. The magnitude of this shift was established by comparison with pictures of para-hydrogen at room temperature when the hydrogen was at extremely low pressure in the body of the spectrograph. In this way it was found that under the conditions of the low-temperature experiments the shift is about 1 cm^{-1} independent of n for n>10. The magnitude of the shift is very closely the same as that observed in atomic Rydberg series at the same density. Already in 1934 Fermi [46] had given a very satisfactory theoretical account of this shift.

The various ionization limits obtained for v=0, 1, 2, and 3 and corrected to zero pressure are summarized in Table IX. The N = 1 limits are referred to the $J''= 0$ level, as are the others, even though the observed N = 1 Rydberg limits are at longer wavelengths because they originate from the $J''= 1$ level.

TABLE IX. LIMITS OF RYDBERG SERIES ABOVE $v''=0$, $J''=0$

v	N = 0	N = 1	N = 2	N = 4
0	124417.2	124476.0	124591.5	
1	126608.4	126664.2	126773.6	127152.2
2	128672.6	128724.8	128828.0	129185.7
3	130613.4	130662.3	130760.9	

B. Ionization Potential

The Rydberg limit for N=0 and v=0 corresponds to the ionization potential of the H_2 molecule, that is,

$$\text{I. P.}_{\text{obs.}} (H_2) = 124417.2 \pm 0.4 \text{ cm}^{-1}$$

which may be compared with the latest theoretical value (Hunter and Pritchard [47], Jeziorski and Kolos [48], Bunker [49])

$$\text{I. P.}_{\text{theor.}} (H_2) = 124417.3 \text{ cm}^{-1}$$

It is seen that the agreement is most satisfactory. The predicted Lamb shift (which is included in the theoretical value given) is 0.3_3 cm^{-1}. In view of the estimated error of the observed value the presence of a Lamb shift cannot be experimentally established from the present observations.

It is interesting to note that the ionization potential of D_2 recently determined by Takezawa [50] gives equally good agreement only after a pressure shift correction is made, similar to the one for H_2. One then finds

$$\text{I. P.}_{\text{obs.}} (D_2) = 124745.6 \pm 0.6 \text{ cm}^{-1}$$

while the theoretical value is

$$\text{I. P.}_{\text{theor.}} (D_2) = 124745.2 \text{ cm}^{-1}$$

C. The Dissociation Energy of the Molecular Ion

From the ionization potentials and the dissociation energies given earlier we obtain, according to Eq. (1), the following values for the dissociation energies of the ions:

$$D_{0\ \text{obs.}}^{0}\ (H_2^+) \leq 21379.9 \pm 0.4\ \text{cm}^{-1}$$

$$D_{0\ \text{obs.}}^{0}\ (D_2^+) = 21711.9 \pm 0.6\ \text{cm}^{-1}$$

These values may be compared with the theoretical values

$$D_{0\ \text{theor.}}^{0}\ (H_2^+) = 21379.3\ \text{cm}^{-1}$$

$$D_{0\ \text{theor.}}^{0}\ (D_2^+) = 21711.6\ \text{cm}^{-1}$$

which are considered to be accurate to ± 0.2 cm^{-1}. That the agreement is again very satisfactory is not surprising since $D_0^0(H_2^+)$ is determined by $D_0^0(H_2)$ and I. P. (H_2) and for both of the latter very good agreement between theory and experiment was found. The theoretical Lamb shift for $D_0^0\ (H_2^+)$ is only 0.2 cm^{-1}.

D. Rotational Levels of H_2^+

The series limits given in Table IX supply several rotational and vibrational levels of H_2^+. These are compared with theoretical values of Hunter and Pritchard [47] in Table X. It is seen that again the agreement between theory and experiment is very satisfactory. The observed rotational and vibrational levels are not sufficient to obtain reliable values for the equilibrium constants of H_2^+ because higher-power terms in the energy formulas, which cannot be determined from the few available levels, are quite significant in such a light molecule. However, using the standard formulas one obtains from the observed energy levels an equilibrium internuclear distance of 1.0525 Å, while the theoretical value is 1.05687 Å.

VI. EXCITED ELECTRONIC STATES

From the analysis of the Lyman and Werner bands one obtains, of course, in addition to the energy levels of the ground state those of the excited states. The comparison with the theoretical values of Kolos and Wolniewicz [51] is less satisfactory than for the ground state because the excited states are subject to much greater nonadiabatic corrections which have not yet been calculated, and also the adiabatic corrections are large and have been

TABLE X.   OBSERVED AND CALCULATED
VIBRATIONAL AND ROTATIONAL
LEVELS IN THE GROUND STATE
OF H_2^+

	Obs.	Calc.
$\Delta G_{1/2}$	2191.2	2191.2
$\Delta G_{3/2}$	2064.2	2063.9
$\Delta G_{5/2}$	1940.8	1941.0
F(1)-F(0)		
$v = 0$	58.8	58.2
$v = 1$	55.8	55.2
$v = 2$	52.2	52.2
$v = 3$	48.9	49.3
F(2)-F(0)		
$v = 0$	174.3	174.2
$v = 1$	165.2	165.1
$v = 2$	155.4	156.2
$v = 3$	147.5	147.6

calculated only for the upper state of the Lyman bands. We shall not pursue this matter further in this review.

It is, however, of interest, in connection with the subject of this Symposium, to report briefly on some absorption experiments in flash discharges carried out a few years ago [52]. In the bottom strips of Fig. 13 are shown two sections of absorption spectra in the visible region of flash discharges through hydrogen; they are compared with emission spectra of the same flash discharge (top strips) and emission spectra of an ordinary ac hydrogen discharge (middle strips). It is seen that the absorption spectrum consists of a large number of absorption lines which are readily identified as the following well-known transitions in the visible region: j $^3\Delta_g \leftarrow$ c $^3\Pi_u$, i $^3\Pi_g \leftarrow$ c $^3\Pi_u$, g $^3\Sigma_g^+ \leftarrow$ c $^3\Pi_u$, that is, the c $^3\Pi_u$ excited state which lies about 12 eV above the ground state is the lower state of these absorption lines. This state is a metastable state and this fact

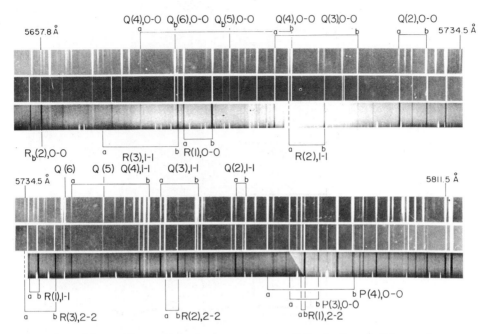

Figure 13.   Absorption and emission spectra of H_2 in the visible region. Only two small sections of the spectrum are shown.   The bottom strips are the absorption spectrum of the flash discharge, the top strips the emission spectrum of the same discharge.   The center strips show for comparison the spectrum of an ordinary ac discharge.

may account for the result that in spite of its high energy a sufficient number of molecules are formed in this state during a flash to make the observation of absorption possible.

A characteristic feature of the absorption spectrum is that the relative intensities in the Λ doublets of the first two systems are quite anomalous. Under normal conditions such as in an ordinary hydrogen discharge (middle strip in each section of Fig. 13) there is an alternation such that alternately the shortward and longward component of the Λ doublet is the stronger one. But in absorption in the R and P branches for both even and odd N the longward components (b) are by far the strongest, while for the Q branches the shortward components (a) are by far the strongest; this intensity relation is reversed in flash emission (upper strips in each section of Fig. 13).

The explanation of this anomalous behavior is readily found in a predissociation of the lower c $^3\Pi_u$ state into the b $^3\Sigma_u^+$ state corresponding to

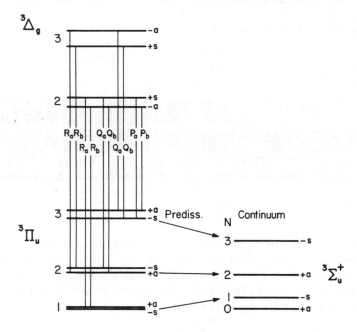

Figure 14.   Energy level diagram for the first lines in the j $^3\Delta_g$ - c $^3\Pi_u$ bands of H_2 showing allowed predissociation to the continuum of the b $^3\Sigma_u^+$ state. In the continuum at the right there is no energy difference for different N values.

normal atoms. As shown by Fig. 14, according to the selection rules $\Delta J = 0$, $+ \not\leftrightarrow -$, this predissociation can only affect one Λ component of the $^3\Pi_u$ state. The lines corresponding to this component are missing or very weak in absorption. In emission the intensity should not be affected by the predissociation and this is found to be the case in a normal hydrogen discharge, but in a flash discharge when there is a strong excitation of the upper levels an inversion of population arises for those transitions that end on the predissociated levels, since these predissociated levels are very rapidly depopulated

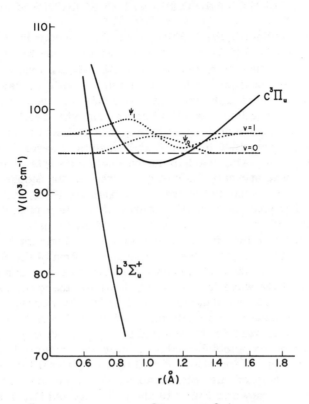

Figure 15.   Potential curves of the c $^3\Pi_u$ and b $^3\Sigma_u^+$ states of H_2. The vibrational wave functions of the lowest two vibrational levels of the c $^3\Pi_u$ state are indicated by dotted lines.

by predissociation.   Thus the high intensity in emission of the lines that are weak in absorption must be ascribed to an incipient laser action for these transitions.   Indeed, further in the infrared we have seen one line that completely burnt out the photographic plate, a line that was independently observed by Bockastén, Lundholm, and Andrade [53] and others.

     That a predissociation between the c $^3\Pi_u$ and b $^3\Sigma_u^+$ state is possible is shown by the potential diagram of Fig. 15 which shows that the two potential functions in the critical region come fairly close together, so that the Franck-Condon factor is appreciable.   It should be mentioned that this predissociation has also been observed by Lichten [54] in a study of beams of excited c$^3\Pi_u$ hydrogen molecules.

VII.   PREDICTED AND OBSERVED CONTINUOUS SPECTRA

As already mentioned, the ordinary continuous spectrum of hydrogen extending from 5000 to 1700 Å is well known to be due to the transition a $^3\Sigma_g^+$ - b $^3\Sigma_u^+$. The lower state is the Heitler-London repulsive state arising from two hydrogen atoms with parallel spin. Coolidge, James, and Present [20, 21] have calculated the intensity distribution in the continuum from first principles and have found excellent agreement with the observed distribution.

Recently a new continuum of hydrogen was observed in our laboratory which extends from 1600 to 1400 Å and consists of several broad maxima, as shown in Fig. 16. After various unsuccessful attempts to account for this continuum it was shown by Dalgarno, Herzberg, and Stephens [55] that this continuum corresponds to the transition from the B $^1\Sigma_u^+$ state to the continuum of the X $^1\Sigma_g^+$ ground state; in other words, that it is part of the Lyman system of hydrogen. If one were to calculate the intensity distribution of the continuous wave function by δ functions one would not obtain such a fluctuating continuum. It is only when the maxima and minima of the wave functions in the continuum are taken into account that this spectrum can be accounted for.

In Fig. 17 the wave functions are shown for the upper state v'= 9 and for a lower state in the continuum. It is clear that by varying the energy in the continuum a fluctuating intensity distribution is obtained, as shown in Fig. 18. Such fluctuations were first predicted by Condon [56] in his second paper on the Franck-Condon principle and have often been referred to as Condon diffraction bands. If the effect of different vibrational levels in the upper state is superimposed one obtains surprising agreement between theory and observation, as shown in Figs. 19 and 20 for H_2 and D_2, respectively.

The calculations have been carried out only for the lowest rotational

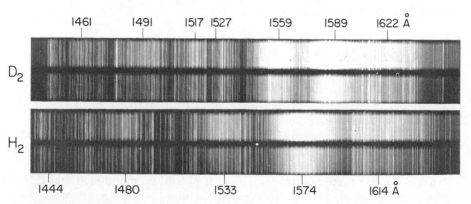

Figure 16.   Emission spectra of H_2 and D_2 in the region 1650–1440 Å. Two spectra with different exposure times are shown for both H_2 and D_2. The positions of the intensity maxima of the continuum are indicated.

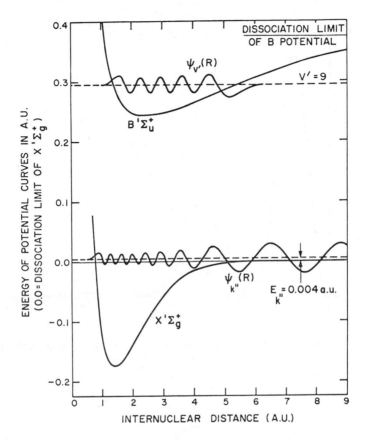

Figure 17. Wave functions of the $v'= 9$ level in the B $^1\Sigma_u^+$ state and of the continuum in the X $^1\Sigma_g^+$ ground state of H_2 for E = 0.004 a.u.   The potential curves in the two states are included.

level $J'=0$. In this case each vibrational level gives rise to a distribution as given in Fig. 18 with a sharp, short wavelength limit. The observed limits are much less sharp, presumably because several rotational levels are overlapped. If this spectrum could be observed under conditions when only one rotational level is effective, very sharp limits should be found which could serve as an independent check of the value of the dissociation energy.

Since for the new continuum there is a population inversion, the lower state being continuously depopulated by dissociation, one may ask whether this transition lends itself to the construction of a tunable laser in the vacuum ultraviolet, an extension of the discrete laser of Hodgson [57] based on the

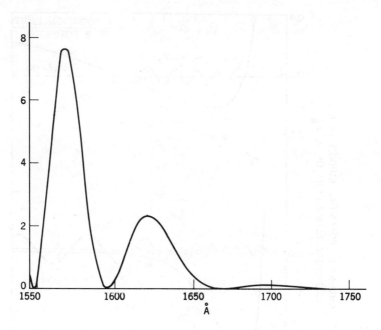

Figure 18.   Theoretical intensity distribution in the partial emission continuum arising from the $v' = 9$ level of the B $^1\Sigma_u^+$ state.

TABLE XI.   SUMMARY OF AGREEMENT BETWEEN
THEORY AND EXPERIMENT

I. P.	within	2 ppm
$D_0^0(H_2, HD, D_2)$	within	10 ppm
$D_0^0(H_2^+, D_2^+)$	within	10 ppm
B_v	within	120 ppm
B_e	within	250 ppm
r_e	within	150 ppm
ΔG	within	200 ppm
ω_e	within	500 ppm

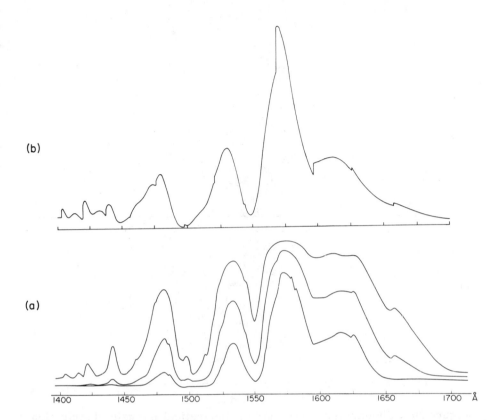

Figure 19.   Photometer curves (a) and theoretical intensity distribution (b) in the H_2 continuum between 1700 and 1400 Å. In (a) three curves corresponding to different exposure times are shown in order to bring out both the weak and the strong features.   The theoretical curve in (b) refers to zero rotation.

same band system in the discrete region.

VIII.   CONCLUSION

I believe that the preceding discussion shows that on the whole there is excellent agreement of the observed data on H_2, HD, and D_2 with theoretical data based on ab initio calculations.   In Table XI the quality of the agreement for various constants is summarized.   The remaining very small discrep-

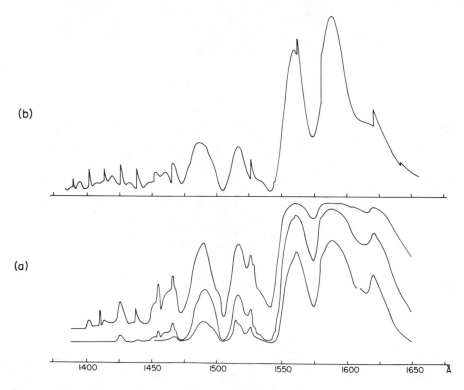

Figure 20.   Photometer curves (a) and theoretical intensity distribution (b) in the D_2 continuum between 1700 and 1400 Å. See caption of Fig. 19.

ancies must be ascribed partly to nonadiabatic corrections, as mentioned earlier, and partly to error of measurement.

REFERENCES

1. G. Herzberg, Proc. Roy. Soc. 248A, 309 (1958).
2. G. Herzberg and H. R. Moore, Can. J. Phys. 37, 1293 (1959).
3. G. Herzberg, Phys. Rev. Letters 23, 1081 (1969).
4. G. Herzberg, J. Mol. Spectry. 33, 147 (1970).
5. G. Herzberg, and Ch. Jungen, J. Mol. Spectry. 41, 425 (1972).
6. Ø. Burrau, K. Danske Vid. Selskab. 7 (No. 14), (1927).
7. V. Guillemin and C. Zener, Proc. Nat. Acad. 15, 314 (1928).
8. E. Teller, Z. Phys. 61, 458 (1930).

9. E. A. Hylleraas, Z. Phys. 71, 739 (1931).
10. G. Jaffé, Z. Phys. 87, 535 (1934).
11. I. Sandeman, Proc. Roy. Soc. Edinb. 55, 72 (1935).
12. V. A. Johnson, Phys. Rev. 60, 373 (1941).
13. W. Heitler and F. London, Z. Phys. 44, 455 (1927).
14. S. Wang, Phys. Rev. 31, 579 (1928).
15. N. Rosen, Phys. Rev. 38, 2099 (1931).
16. S. Weinbaum, J. Chem. Phys. 1, 593 (1933).
17. H. M. James and A. S. Coolidge, J. Chem. Phys. 1, 825 (1933); 3, 129 (1935).
18. H. M. James, A. S. Coolidge and R. D. Present, J. Chem. Phys. 4, 187 (1936); 6, 730 (1938).
19. J. G. Winans and E. C. G. Stueckelberg, Proc. Nat. Acad. 14, 867 (1928).
20. A. S. Coolidge, H. M. James, and R. D. Present, J. Chem. Phys. 4, 193 (1936).
21. H. M. James and A. S. Coolidge, Phys. Rev. 55, 184 (1939).
22. E. E. Witmer, Phys. Rev. 28, 1223 (1926).
23. G. H. Dieke and J. J. Hopfield, Z. Phys. 40, 299 (1926); Phys. Rev. 30, 400 (1927).
24. H. Beutler, Z. phys. Chem. B 29, 315 (1935).
25. O. W. Richardson, Molecular Hydrogen and its Spectrum (Yale Univ. Press, New Haven, 1934).
26. H. Beutler and H. O. Jünger, Z. Phys. 101, 304 (1936).
27. G. Herzberg and A. Monfils, J. Mol. Spectry. 5, 482 (1960).
28. W. Kolos and C. C. J. Roothaan, Rev. Mod. Phys. 32, 219 (1960).
29. W. Kolos and L. Wolniewicz, J. Chem. Phys. 49, 404 (1968).
30. B. P. Stoicheff, Can. J. Phys. 35, 730 (1957).
31. G. Herzberg, Can. J. Res. 28A, 144 (1950).
32. O. Fink, T. A. Wiggins and D. H. Rank, J. Mol. Spectry. 18, 384 (1965).
33. J. V. Foltz, D. H. Rank, and T. A. Wiggins, J. Mol. Spectry. 21, 203 (1966).
34. H. Bredohl and G. Herzberg, Can. J. Phys. (to be published)
35. G. Herzberg and L. L. Howe, Can. J. Phys. 37, 636 (1959).
36. R. A. Durie and G. Herzberg, Can. J. Phys. 38, 806 (1960).
37. S. Takezawa, F. R. Innes, and Y. Tanaka, J. Chem. Phys. 46, 4555 (1967).
38. J. D. Poll and G. Karl, Can. J. Phys. 44, 1467 (1966).
39. G. Herzberg and B. J. McKenzie (to be published).
40. R. J. Le Roy, J. Chem. Phys. 54, 5433 (1971).
41. J. L. Dunham, Phys. Rev. 41, 721 (1932).
42. P. R. Bunker, J. Mol. Spectry. 35, 306 (1970).

43. H. Beutler, Z. Phys. 93, 177 (1935).
44. U. Fano, Nuovo Cimento 12, 156 (1935); Phys. Rev. 124, 1866 (1961).
45. U. Fano, Phys. Rev. A 2, 353 (1970).
46. E. Fermi, Nuovo Cimento 11, 157 (1934).
47. G. Hunter and H. O. Pritchard, J. Chem. Phys. 46, 253 (1967).
48. B. Jeziorski and W. Kolos, Chem. Phys. Letters 3, 677 (1969).
49. P. R. Bunker, J. Mol. Spectry. 42, 478 (1972).
50. S. Takezawa, Abstr. Columbus Meeting on Molecular Spectroscopy, June 1971.
51. W. Kolos and L. Wolniewicz, J. Chem. Phys. 48, 3672 (1968).
52. G. Herzberg, Sci. of Light 16, 14 (1967).
53. K. Bockastén, T. Lundholm, and O. Andrade, J. Opt. Soc. Am. 56, 1260 (1966).
54. W. Lichten, Phys. Rev. 120, 848 (1960); 126, 1020 (1962); Bull. Am Phys. Soc. 7, 43 (1962).
55. A. Dalgarno, G. Herzberg, and T. L. Stephens, Astrophys. J. 162, L49 (1970).
56. E. U. Condon, Phys. Rev. 32, 858 (1928).
57. R. T. Hodgson, Phys. Rev. Letters 25, 494 (1970).

HIGHLY EXCITED STATES OF THE HELIUM ATOM*

W. E. Lamb, Jr. , D. L. Mader,† and W. H. Wing
Department of Physics, Yale University
New Haven, Connecticut

We may begin by asking the question "What role has the helium atom played
in the development of the laser?" Among the possible answers are that
Maiman [1] was introduced to microwave-optical spectroscopy of the helium
triplets in his Ph.D. research at Stanford University in 1952-1956, and in a
few years invented the ruby laser. Also, Javan [2] used collisional transfer
of energy from helium metastables to neon atoms in the first gas laser. We
have heard [3] in this conference that laser operation at 96 μ occurs between
the 3^1P_1 and 3^1D_2 states of helium. Now, however, the situation is about
to reverse. As should be clear by the end of the lecture, in the future
lasers will play an important part in high-precision spectroscopy of the
helium atom.

We may distinguish two kinds of atomic fine structure. Relativistic
fine structure involves spin effects (and quantum-electrodynamic level
shifts). Examples occur with H, He, ..., Na, etc. The doublet structure
of the sodium D lines falls in this category. This kind of fine structure has

---

*Research sponsored by the Air Force Office of Scientific Research under
Grant No. F44620-71-C-0042.
†Present Address: Columbia Astrophysics Laboratory, Columbia University,
New York, New York 10027.

been studied extensively in recent years using atomic beams, bottles, optical-pumping cells, etc. The second kind of fine structure is more nearly a pure electrostatic effect. It causes, for example, the separation of 3P and 3S in sodium and the yellow color of the D lines. The high Rydberg series of any atom (molecule) gives information on screening of the nuclear charge(s) by inner core electrons, as well as on exchange and core-polarization effects. We will be concerned today with the electrostatic fine structure of the helium atom.

Figure 1 shows the energies of some states of the helium system. The ground state has two 1s electrons. The first excited states are about 20 eV higher. One of these, $1s2p^1P_1$ (or 2^1P_1), emits the strongest helium resonance line, of wavelength 584 Å. The ionization limit lies at about 25 eV. In today's lecture we will be discussing microwave transitions between high Rydberg states of the helium atom which lie between 20 and 25 eV above the ground state. We got into this interesting field by accident, and some of this story will be told more or less as it happened to us.

Therefore, let us first go higher in the energy diagram, Fig. 1. Beginning at 65 eV we reach excited states of singly-ionized helium, which is a hydrogen-like system. Leventhal and some of us at Yale [4] were studying the fine structure of the n = 3 state, which is caused by relativistic and quantum-electrodynamic effects. Figure 2 shows this fine structure and the notation used to denote the states and their magnetic sublevels. The energies being sought were the level shift S and the doublet separation ΔE. The transition frequencies used ranged up to about 40 GHz.

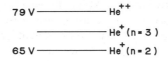

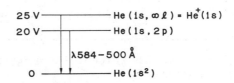

Figure 1. Energy levels of the helium atom and ion. States with two excited bound electrons are not shown.

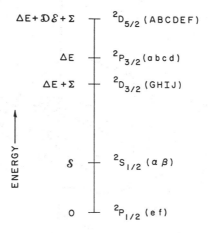

Figure 2.   Fine structure of n = 3, He$^+$. The Zeeman sublevels belonging
to each fine-structure level are shown in order of decreasing magnetic
quantum number. The energy levels are not drawn to scale. The P-state
doublet separation ΔE is about 52 GHz, the S-state level shift $\mathcal{S}$ is 4.2 GHz
and the P-D level shift Σ is 85 MHz.

The apparatus used in this work is shown in Fig. 3. Helium gas at a
pressure of a few millionths of an atmosphere is bombarded by an electron
beam of about 1 mA current and a square-centimeter area. The visible and
ultraviolet (uv) light emitted by the excited helium ions is brought by a
hollow ellipsoidal light pipe to a photomultiplier detector. Microwaves are
applied through a waveguide to the excited ions. A uniform magnetic field
is also applied. Whenever Zeeman tuning brings the frequency of an
allowed ionic transition close to the microwave frequency, transitions are
induced between the corresponding fine-structure levels. This causes a
change in the amounts of visible and uv light emitted in decay to lower states,
and hence changes the photodetector current, provided the detector responds
selectively to one of the decay lines and the initial excitation produces a
population difference. This type of apparatus is known colloquially as a
"bottle machine". Modulation of the microwave (rf) power at an audio
frequency and lock-in detection of the photocurrent greatly improves the
signal-to-noise ratio. The detector shown here is sensitive to a fairly
narrow band of radiation around the wavelength 1640 Å emitted in the transitions from n = 3 to n = 2 of He$^+$. There are no atomic-helium lines within
the passband of the combined filter and photomultiplier tube. The detector
can be changed so as to be sensitive to other wavelengths.

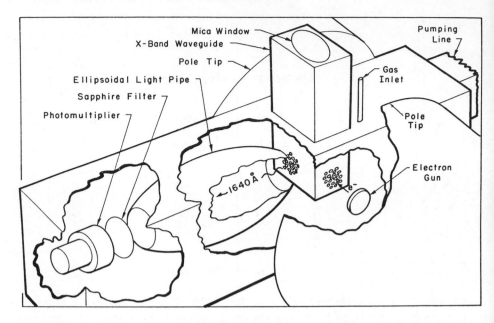

Figure 3. Apparatus used for fine-structure measurements. The photo-multiplier and filter shown respond selectively to the He$^+$ Hα transition (n = 3 → n = 2) at 1640 Å.

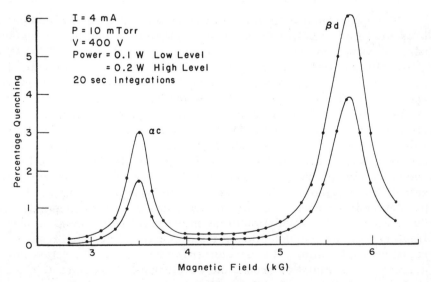

Figure 4. Resonance curves for transitions αc and βd in n = 3 of He$^+$, $3^2S_{1/2} - 3^2P_{3/2}$, at two microwave power levels. The frequency is 40 GHz.

526

Figure 4 shows an example of resonance curves obtained with this apparatus for two of the high-frequency transitions $3^2S_{1/2} - 3^2P_{3/2}$ of He^+. The analysis of such data went well, and there were no unpleasant surprises in the determination of a good value for the zero-field energy splitting $\Delta E - S$.

It was much harder to work with the lower-frequency resonances $3^2S_{1/2} - 3^2P_{1/2}$. Figure 5 shows transition αe of n = 3 with additional structure arising from αf of n = 3 and the electron-cyclotron resonance on the low-magnetic field side of the main peak, as well as transition αe of n = 4 on the high-field side. Transitions with n > 3 can be seen because such states are also excited by the electron bombardment, and produce 1640-Å light through cascade processes to n = 3. Since the fine-structure frequencies decrease with increasing n, these higher-n resonances overlap more closely the low-frequency n = 3 resonances than they do the high-frequency ones shown earlier. Nevertheless, it turned out that the main $3\alpha e$ resonance could be analyzed very well without taking into account the satellite structure shown in the figure. This led to what should have been a highly precise value of the electrodynamic level shift $S = \nu(3^2S_{1/2} - 3^2P_{1/2})$.

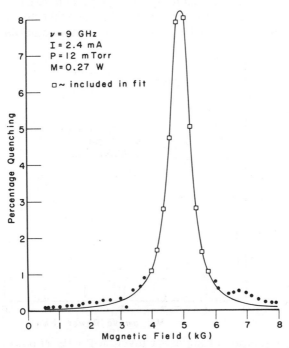

Figure 5. Resonance curve for transition αe in n=3 of He^+, $3^2S_{1/2} - 3^2P_{1/2}$, taken under typical operating conditions for precision data runs. The low-field satellites are attributed to $3\alpha f$ and the electron cyclotron resonance, while that on the high-field side is $4\alpha e$.

It was found that the result did not depend much on pressure or bombarding current. However, an unexpected dependence on radio-frequency power was found, as shown in Fig. 6. At very low power, the apparent level shift began to fall rapidly away from the anticipated value. This effect was far larger than could be explained by rf Stark shifts, which, in any case, should be linear in rf power.

In an attempt to understand this phenomenon, the resonances were studied at very low rf power. Each point in Fig. 7 required an hour of running time. Contributions from n = 5, 6, and 7 states of He$^+$ are seen from an analysis of the data. The high-n states have long lifetimes and large electric dipole matrix elements, so their transitions are saturated by what is a low rf power level for 3αe. The low-power drop-off of Δ in Fig. 6 can be fully understood by taking into account the presence of the 7βc resonance. No trace of 7βc could be seen in the earlier figure, for which the rf power was many times higher, because there 7βc was so power broadened that it added only a substantially constant baseline to 3αe.

Figure 8 shows how the amount of total detected light and radio-frequency signal depended on electron-bombardment energy. The latter showed a very clear threshold at the expected energy of about 73 eV. However, a small amount of light persisted down to 20-30 eV. One plausible explanation of this is that uv radiation from atomic helium transitions was

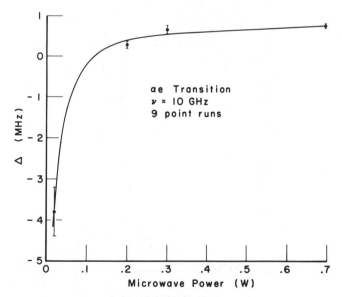

Figure 6.   Variation of apparent level shift with rf power level. Δ is the difference between the observed level shift and a predicted value, and is obtained from computer fits of a Lorentzian function to data like that shown in Fig. 5.

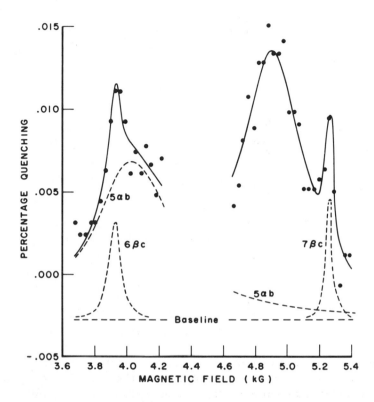

Figure 7.   Very-low-power panoramic taken at 9 GHz with 0.0005 W incident power.   Each point represents 1 h of signal averaging.   The solid
curve is a theoretical fit.   The main resonance 3αe, centered at 4.9 kG, has
a height about 0.1% of its saturated height.   Other resonances are shown as
dashed lines with their bases on a fitted baseline of -0.0035% quenching.

being converted by fluorescence in the sapphire filter, or elsewhere, to
radiation near 1640 Å to which the photodetector could respond.

    At this point we shifted to a study of the n = 2 state of He$^+$.   To avoid
the high-n difficulties we worked at a bombardment energy not too far above
the 65-eV n = 2 threshold so that the higher states of He$^+$ would not be
excited.   For this study we had to change to a broadband ultraviolet photodetector sensitive to light of wavelength $\lesssim$ 1200 Å so that the Lyman-α
radiation 2P - 1S at 303 Å could be detected.   To our surprise, we still
found all sorts of unexpected resonance structure at very low rf power.
Furthermore, this structure persisted even when the bombardment energy

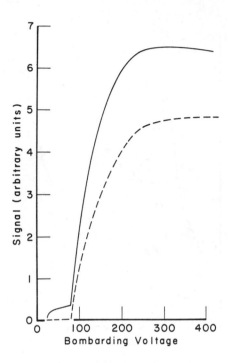

Figure 8.   Excitation curves normalized to unit bombarding current.   The lock-in signal (dashed line) vanishes below the threshold of about 73 eV, but the photomultiplier current has a small plateau down to 20 eV.

was lowered below the 65-eV threshold energy for exciting He^+.   In fact, its threshold was about the same 20-30 eV as that of the total-light "foot" seen by the earlier 1640-Å detector.   Hence, these strange resonances could no longer be attributed to He^+, but rather to transitions in the two-electron helium atom [5].

Figure 9 shows one of many recorder traces of these resonances.   The form depended drastically on the microwave frequency, and also on pressure, bombarding current, and rf power level.   Clearly the complexity of the pattern makes any analysis very difficult.

Figure 10 shows the energy levels $1sn\ell$ of He for $n \geq 2$.   The 1^1S_0 ground state is way below the bottom of the figure.   The energy levels fall into two groups: singlet (or para) and triplet (or ortho).

The energy depends most strongly on the principal quantum number n. The singlet-triplet difference comes from electrostatic repulsion of the two electrons via Heisenberg's exchange interaction.   The dependence on the

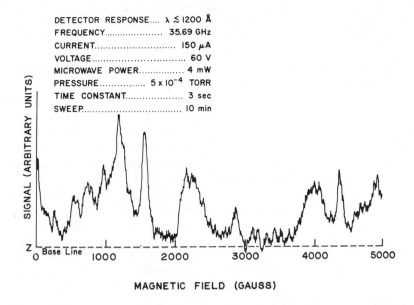

DETECTOR RESPONSE.... λ ≲ 1200 Å
FREQUENCY.................... 35.69 GHz
CURRENT........................... 150 μA
VOLTAGE................................ 60 V
MICROWAVE POWER............... 4 mW
PRESSURE................ 5 x 10⁻⁴ TORR
TIME CONSTANT.................... 3 sec
SWEEP.................................... 10 min

SIGNAL (ARBITRARY UNITS)

Z₀ Base Line   1000   2000   3000   4000   5000

MAGNETIC FIELD (GAUSS)

Figure 9.   Example of resonances attributed to high-n states of He I.

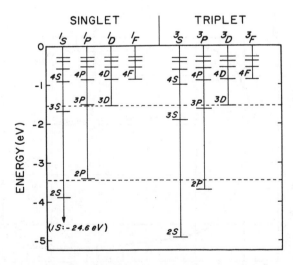

SINGLET   TRIPLET

^1S ^1P ^1D ^1F ^3S ^3P ^3D ^3F

ENERGY (eV)

$(1S: -24.6\,eV)$

Figure 10.   Excited states of the helium atom.   The dotted lines indicate two of the hydrogenic levels for Z = 1.   Energies in eV are measured from the helium ionization limit (=1S of He⁺).

531

orbital angular-momentum quantum number L also arises from electrostatic forces between the two electrons, and is described by the use of an effective quantum number $n^* = n + \delta$ instead of the integer n, where the (usually negative) quantity δ is called a "quantum defect."

On a much smaller scale, the triplet states have a fine structure caused by spin-spin and spin-orbit magnetic interactions.

We shall be especially interested in the case of large n. Our group certainly does not have a monopoly on the study of high-n states. This is illustrated by mention of some work done by others which deals with high-n states in various atoms. Thus, by level-crossing studies, Descoubes [6] got results up to n = 9 for the triplet fine-structure splittings of helium.

Other workers have dealt with much higher n values. Herzberg [7] has told us of the Rydberg series for molecular hydrogen. The principal series of sodium has been seen in absorption spectroscopy [8] to n = 79. Even when a perturbing gas at a pressure of one atmosphere is present, the lines can still be seen clearly resolved beyond n = 30. Under these conditions there are about 10,000 perturbing atoms inside the alkali electron's charge cloud. Figure 11 shows [9] how the resulting pressure shifts at one atmosphere depend on n for various perturbing gases. Fermi [10] has given a theory of these shifts in which the itinerant electron moves in a medium with an effective index of refraction for matter waves. For such large orbits it clearly does not matter much whether the parent atom is an alkali or helium.

Figure 12 shows [9] the corresponding linewidths, also for a pressure

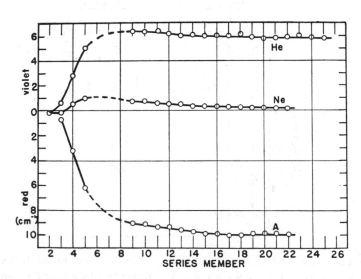

Figure 11. Shifts of the Cs principal series lines perturbed by 1-atm pressures of He, Ne, and Ar (Ref. 9).

of one atmosphere. Since one wave number corresponds to a frequency of 30,000 MHz, these shifts and widths would scale down to a fraction of a megahertz at the pressures used in our work on the high helium levels, to be described later.

Jenkins and Segrè [11] used a Berkeley cyclotron magnet to study high-series alkali lines (n ≤ 36) in the presence of a very strong magnetic field. They were able to detect the quadratic Zeeman effect, which is unimportant for low n values. They had a helium buffer-gas pressure of 1% of an atmosphere. Similar more recent work with barium by Garton and Tomkins [12] has reached n ~ 75.

In the last few years, the radio astronomers [13] have become the unchallenged high-n champions. Transitions from n to n - 1, n - 2, ..., etc. for very high n values of a hydrogenlike system lie in the microwave frequency range. In the radio recombination spectrum obtained from some nebulae, lines with values of n as high as 225 have been detected. An example of such work by Churchill and Mezger [14] is shown in Fig. 13. Because of reduced mass effects, transitions in hydrogen, helium, and carbon are clearly resolved.

We now turn to the real subject of our lecture: the high Rydberg states of atomic helium. The lowest few of these states were shown in Fig. 10.

Figure 14 gives the definition of "quantum defect" more precisely. The penetration of the outer $(n\ell)$ electron into the 1s core changes the integral quantum number n to an effective one: $n^* = n + \delta$, where the quantum defect $|\delta|$ depends strongly on the ℓ value. For $n \gg 1$, $|\delta|$ is found to be nearly independent of n. (For sodium, $|\delta_P| \sim 0.85$ for large n.)

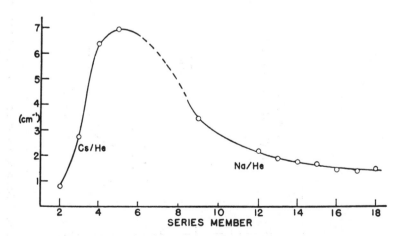

Figure 12.   Half-widths of members of the principal series of Cs and Na perturbed by helium for a pressure of 1 atm (Ref. 9).

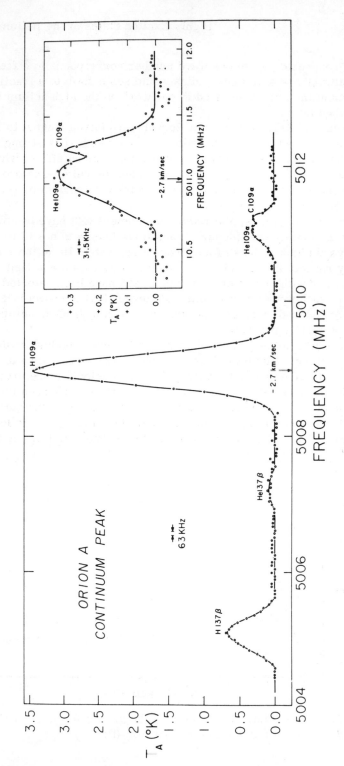

Figure 13. Broadband spectrogram of the 109α region of the spectrum. The frequency resolution is 63.0 kHz for the broadband spectrogram and 31.5 kHz for the narrow-band spectrogram centered on the He 109α line and inserted in the upper right-hand corner of the figure. (Churchill and Mezger, Ref. 14. 109α and 137β denote the transitions n = 110 → 109 and n = 139 → 137, respectively.)

If n >> 1

Inner ~ hydrogen with z = 2
Outer ~ hydrogen with z = 1

Binding energy of outer electron

$T(1s, n\ell) = -\dfrac{R}{(n+\delta)^2}$ "Rydberg Series"

δ = "quantum defect"
~ independent of n $\left\{ \delta = \delta_\infty + \dfrac{c_1}{n^2} + \cdots \right.$

$|\delta|$ = strongly decreasing func. of ℓ

Figure 14.   Schematic representation of high-n states of the singly excited helium atom, with indication of the quantum defect.

Figure 15 will remind older physicists how core penetration was discussed [15] in the days of Bohr orbits and the old quantum theory. Nowadays, quantum mechanics is used to describe how the departure from a pure monopole Coulomb field removes the "accidental" hydrogenic ℓ degeneracy.

Very extensive tables (Moore [16] and Martin [17]) of the U.S. National Bureau of Standards allow the quantum defects $|\delta|$ for helium to be calculated as a function of n and ℓ. The results are shown in Fig. 16. The fall-off for very large n probably comes from the use of a slightly incorrect ionization potential for the helium atom.

Table I gives smoothed $|\delta|$ values, and a listing of the various theoretical contributions to $|\delta|$ for D states. Similar tables could be given for S, P, F, ... states.

We now explain how resonances like $15^1D - 15^1P$ can be detected by our apparatus. Figure 17 shows a possible mechanism. Both the 15P and 15D states are excited by electron bombardment, with 15P the more heavily populated. The atom then decays with emission of radiation. Most of the P states go directly to the ground state 1^1S_0 producing the $15^1P_1 - 1^1S_0$ resonance line at 506 Å. The D states have to cascade down, mostly passing through 2^1P_1 and then emitting the principal resonance line $2^1P_1 - 1^1S_0$ at 584 Å. Our detector probably responds about equally well to 584-Å radiation and to 506 Å. However, the helium gas in the light pipe scatters into its

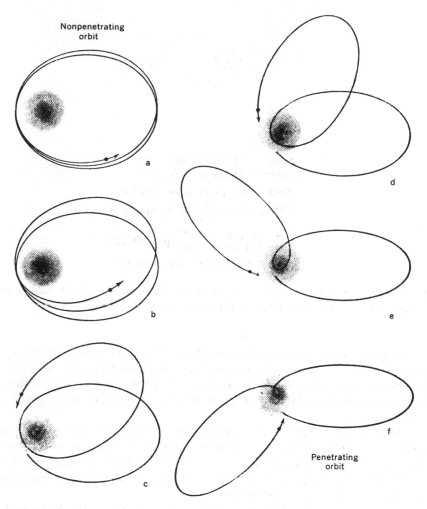

Figure 15.   Effect of core penetration on an orbital electron according to the old quantum theory. The rate of precession of the elliptical orbit increases with increasing core penetration (White, Ref. 15).

walls the 584-Å $n = 2$ resonance radiation produced by 15D decay much more strongly than it does the 506-Å $n = 15$ radiation from 15P decay. Hence the detector will sense a microwave-induced transfer of atoms from the 15P state to the 15D state as a decrease in light intensity.

Table II shows miscellaneous relations which are of importance in working out the properties of the high-n states. For these states the usual

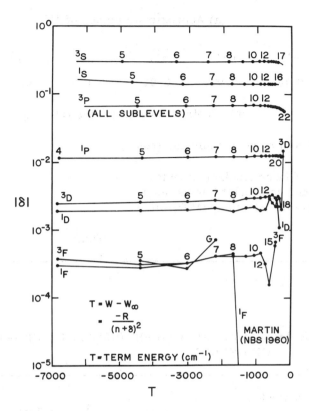

Figure 16.   Summary of quantum defect values for helium, obtained from NBS atomic energy tables (Martin, Ref. 17).   The data for G states were taken under conditions of strong Stark mixing, and have no fundamental significance.

selection rules for allowed transitions are easily broken down.   Even for the unperturbed atom, spin-orbit interaction weakens the singlet-triplet classification.   For F states and states of higher L value, there is little meaning to the resultant spin quantum number [18].   In addition, states of high n are very sensitive to Stark shifts and mixing in electric fields.   An appreciable electric field is induced by the motion of the atom at right angles to a static magnetic field.   The presence of a magnetic field has been almost essential in our work to control the electron bombardment, and even more so to allow us to sweep through broad resonances by Zeeman tuning (which is technically easier than changing the microwave frequency).

TABLE I.

(a) Approximate Values of $|\delta|$

n^3S	n^1S	n^3P	n^1P	nD	nF	nG
0.30	0.14	0.07	0.01	0.003	0.0003	0.00005

(b) Contributions to $|\delta|$ for D States

Polarization of core by outer electron	0.0024
Exchange	0.0007
Screening of inner electron by outer	0.0002
Spin-spin and spin-orbit interactions	0.0001
Other (mass polarization, QED, etc.)	0.00002
Total	~0.003

TABLE II. MISCELLANEOUS PROPERTIES OF HIGH-n STATES
OF HELIUM

Excitation rates	
Decay rates $\Big\} \propto 1/n^3$	
Level spacings	
Number of sublevels	$\propto n^2$
Number of matrix elements	$\propto n^4$
Mean atomic radii	
Electric dipole matrix elements ($\Delta n = 0$) $\Big\} \propto n^2$	
Stark shifts	$\propto \mathcal{E}^2 n^4$
Motional field broadening	$\propto [(\bar{v}/c)\mathcal{H}]^2 n^4$
Zeeman effect:	$H_L = \mu_0 \mathcal{H} \cdot (g_L \vec{L} + g_S \vec{S})$

"Strong-field" electric-dipole rf transitions: Slopes of 1.4 MHz/G

Quadratic Zeeman effect: $\quad H_Q = \dfrac{e^2 \mathcal{H}^2}{8mc^2} r^2 \sin^2\theta$

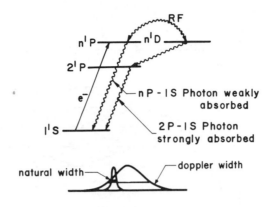

For a stationary atom: $\sigma \sim \dfrac{\lambda^2}{2\pi}$ within natural width

For a collection of N moving atoms:

$$\sigma_{eff} \sim N\sigma \left(\frac{\Delta\nu_{natural}}{\Delta\nu_{doppler}} \right) \sim \frac{\Delta\nu_{2P}/n^3}{\left(\frac{\overline{v}}{c}\right)\left(\frac{c}{\lambda}\right)}$$

At 10^{-3} Torr : $\ell_{2P} \approx 1mm$

$\ell_{20P} \approx 1 meter$

Figure 17.   Possible mechanism for the observation of transitions $n^1P - n^1D$ in highly excited states of helium.   This depends on strong self-reversal of the low members of the principal series by the helium gas in the light pipe shown in Fig. 3.

A third breakdown of usual selection rules comes from multiple-quantum transitions which occur with quite weak microwave fields.   Figure 18 shows some of the possibilities for transitions with L and m_J changing by more than ± 1.

Despite the great complexity of the resonance pattern shown earlier (Fig. 9), it is possible to make some analysis as follows: A sequence of such curves is run at closely spaced frequencies. Figure 19 shows an example of this procedure.   There is a veritable forest of resonances. Nevertheless, with sufficient imagination, it is possible to find paths through the jungle, some of them quite straight.   The various peaks are connected with straight lines which seem to extrapolate back to zero magnetic field to one of a number of discrete frequencies.

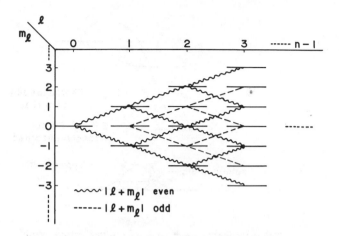

Figure 18.   Breakdown of $\Delta L = \pm 1$, $\Delta M = \pm 1, 0$ selection rules for a singlet state in the presence of a magnetic and transverse electric field. Weak-field transitions occur between adjacent coupled vertices. In strong enough transverse fields, transitions will occur between every pair of states for which there is a coupling path; the states fall into two isolated groups according to the parity of $|\ell + m_\ell|$. If strong electric-field components parallel to $\vec{B}$ are also present, every transition can occur.

Figure 20 shows a summary of this kind of analysis. Each of the horizontal shaded regions was explored and any straight lines found were extrapolated as above. This led to fairly plausible zero-field frequencies of 81.8, 70.2, 60.7, 57.0, 47.6, 46.0, 37.6, 29.4, 24.0, 20.0, 16.0, and 0.0 GHz.

Figure 21 summarizes the existing spectroscopic data on singlet states. A more complicated pattern applies to the triplets.

Table III lists, for several of the transition frequencies just mentioned, all possible identifications that can be made using the energy levels given in the NBS spectroscopic tables [17]. No attention is paid to selection rules, and only in some of the cases does one find a very plausible assignment. It should be noted, however, that if an unequivocal assignment can be made, one will have determined an atomic splitting with an order of magnitude more precision than conventional spectroscopy can give.

In the last few months, two substantial improvements in the apparatus have been introduced. A visible response photomultiplier and a low f-number monochromator are used to monitor light of a particular helium

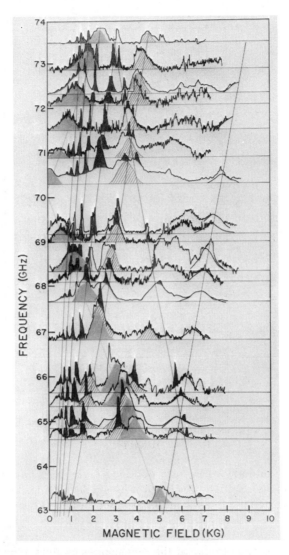

Figure 19. A family of resonance curves of the sort shown in Fig. 9 for a number of microwave frequencies in the range 63–74 GHz. Some shaded peaks are connected by straight lines which are extrapolated back to zero magnetic field.

541

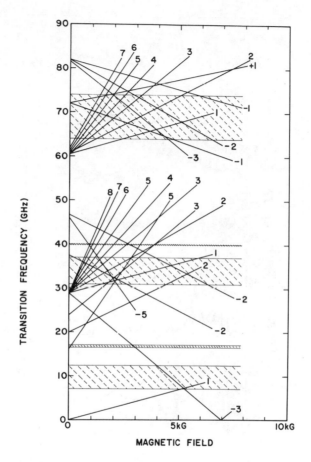

Figure 20.   Summary of results obtained from diagrams like Fig. 19.   The frequency ranges explored are shown by shaded bands.   The transition lines are numbered by the nearest integral slope in units of μ_0/h.

transition.   The complexity of the resonances is thus greatly reduced. Furthermore, since the individual resonances are quite narrow, it is now possible to work in zero magnetic field with varying rf frequency.   This has made it possible to avoid the complexities of the Zeeman splitting and the motional Stark mixing.   The problem now is to find a tunable microwave source with the right frequency, but fortunately very little power is required. Harmonic generation sources have so far been adequate.

TABLE III.   POSSIBLE IDENTIFICATIONS OF SOME
OF THE MORE PROMINENT EXTRAPOLATED ZERO-
FIELD TRANSITION FREQUENCIES OF FIG. 20.   NO
NOTICE OF USUAL SELECTION RULES HAS BEEN
TAKEN.

Observed Resonance (GHz)	Possible Identifications (from 1960 NBS tables[a])	
81.8	80.9	$17^3D - 17^3P$
	82.1	$17^3D - 17^1P$
	83.9	$10^3F - 10^1P$
70.2	70.5	$11^1D - 11^1P$
	70.8	$19^1P - 19^3P$
	72.7	$6^1F - 6^3D$
	72.7	$6^3F - 6^3D$
	70.1	$6^{1,3}G - 6^1D$
	68.1	$5^{1,3}H - 5^3D$
	69.6	$18^1D - 18^3P$
60.7	59.7	$20^3P - 20^1P$
	58.5	$12^3D - 12^1P$
	63.0	$11^3F - 11^1P$
	59.1	$7^3G - 7^3D$
	59.1	$4^1D - 4^3D$
47.6	48.0	$20^3D - 20^3P$
	46.2	$13^3D - 13^1P$
	48.0	$12^3F - 12^1P$
	46.2	$7^{1,3}G - 7^1D$
46.0	45.9	$7^{1,3}F - 7^3D$
	46.2	$7^{1,3}G - 7^1D$
	46.2	$13^3D - 13^1P$

[a]Reference 17.

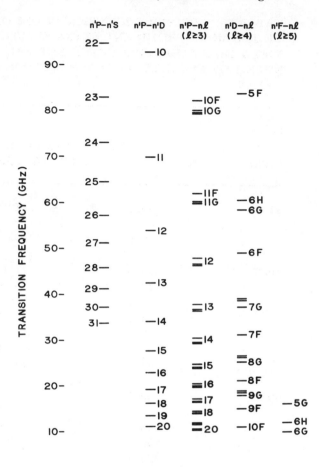

Figure 21.   Expected frequencies for various helium singlet transitions. For S, P, and D states, term energies were reconstructed from the spectroscopic data via Seaton's quantum defect fit (Ref. 20). For F and states of higher L, Edlén's polarization model was used (Ref. 21). The great complexity of the resonance curves of Fig. 19 comes from the Zeeman splitting of these energy levels and the breakdown of selection rules.

Consider the work with the $7^1D_2 - 2^1P_1$ line at 4009 Å. The structure of the n = 7 state as given by the spectroscopic data is shown in Fig. 22. With monochromatic detection, in zero magnetic field, we have obtained the resonances shown in Fig. 23. In weak magnetic field, due to Zeeman mixing of J states the other two 7^3F states (having J = 2 and 4) also appear. Unsaturated linewidths are about 2 MHz. These are somewhat greater than

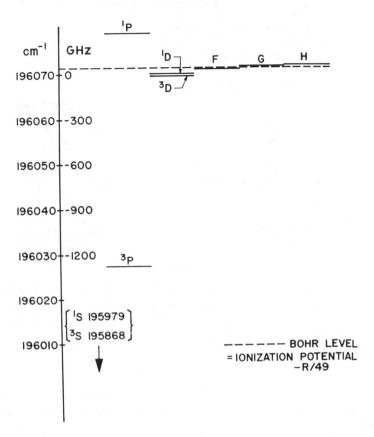

Figure 22.   Fine structure of the n = 7 states of He I according to the compilation of Ref. 17.   The G and H states should lie below the hydrogenic (Bohr) level.   Their altered position here is probably due to the combination of a slightly incorrect ionization poetential and the strong Stark mixing conditions for these data.

the natural width, and presumably still have Stark and plasma contributions.

Table IV gives a comparison of the results from Fig. 23 with values obtained from conventional spectroscopy.

The calculation of high-precision helium energy levels (Hylleraas, Pekeris, Schwartz, ...) has concentrated on states of low quantum number (although Accad, Pekeris, and Schiff [19] have given results for 15^1S and 17^3S, and for P states to n = 5).   The more approximate calculations [20] of quantum defects have an accuracy comparable only to that of conventional

7^{1}D-7F RESONANCES SEEN IN 7^{1}D-2^{1}P LIGHT
(4009 Å) OF HELIUM

CURRENT................. 200 μA
VOLTAGE..................... 60 V
PRESSURE.. 2.5 x 10^{-3} TORR
MAGNETIC FIELD..... ≲ 0.1 G
MICROWAVE POWER.... 1 μW

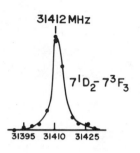

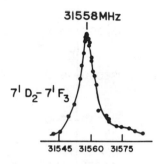

FREQUENCY   (MHz)

Figure 23.   Zero-field resonances in 7^{1}D - 2^{1}P light obtained by tuning the microwave oscillator. The right-hand peak corresponds to the microwave transition 7^{1}D$_2$ - 7^{1}F$_3$. The other peak involves the intercombination transition 7^{1}D$_2$ - 7^{3}F$_3$ which can occur because spin-orbit coupling mixes singlet and triplet states of the same J.

spectroscopic experimental methods. To exploit fully our potential accuracy, it will be necessary to make much more accurate theoretical calculations for the high helium states.

Using monochromatic detection, we have already seen F - D transitions in n = 6, 7, 8, 10, and 11, and 20D - 20P transitions. It will be interesting to see how far up in n we can go. Certainly not as far as the radio astronomers, but perhaps far enough to see the n → n ± 1 transitions as high as n = 40.

The development of tunable lasers will make it possible to work with lower series members for which microwave techniques are difficult due to the high frequency required. Two-step excitation from the ground state to a metastable state by electron bombardment, followed by laser excitation of a particular high nP state, should make it possible to obtain much simpler spectra and sharper lines for higher n values.

It is apparent that we are on the edge of a whole new field of spectroscopy which might be called "microwave quantum-defect spectroscopy" or "microwave Rydberg spectroscopy." We are by no means limited to helium atoms, since all electronic systems, atomic and molecular, have highly excited Rydberg states.

TABLE IV.   SUMMARY OF EXPERIMENTAL WORK
ON THE 7^1D - 7F INTERVALS OF He^4 I.

Source	Interval (GHz)	
Optical data (NBS 1960-Martin[a])	33	
D state smoothed by quantum defect fit (Seaton[b]) + F state polarization theory (Edlén[c])	30.8	
Present work (preliminary)	$7^1D_2 - 7^1F_3$ 31.558	$7^1D_2 - 7^3F_3$ 31.412

[a]Reference 17.
[b]Reference 21.
[c]Reference 22.

Note added in proof: More recent work is listed in Refs. 23 and 24.

REFERENCES

1.   W. E. Lamb, Jr. and T. H. Maiman, Phys. Rev. 105, 573 (1957).
2.   A. Javan, Phys. Rev. Letters 3, 87 (1959).
3.   A. Javan, in this volume.
4.   D. L. Mader, M. Leventhal, and W. E. Lamb, Jr., Phys. Rev. A 3, 1832 (1971). Similar work on n = 4 of He$^+$ has been reported by R. R. Jacobs, K. R. Lea, and W. E. Lamb, Jr., Phys. Rev. A 3, 884 (1971).
5.   A preliminary report of these observations has been made by W. H. Wing, D. L. Mader, and W. E. Lamb, Jr., Bull. Am. Phys. Soc. 16, 531 (1971).
6.   J. P. Descoubes, Physics of the One- and Two-Electron Atoms, edited by F. Bopp and H. Kleinpoppen (North-Holland Publ. Co., Amsterdam, 1969), p. 341.
7.   G. Herzberg, in this volume.
8.   E. R. Thackeray, Phys. Rev. 75, 1840 (1949).

9.   This figure is taken from a review article by H. Margenau and W. W. Watson, Rev. Mod. Phys. 8, 22 (1936) and is based on work reported by C. Füchtbauer in Physik. Z. 35, 975 (1934).

10.   E. Fermi, Nuovo Cimento 11, 157 (1934).

11.   F. A. Jenkins and E. Segrè, Phys. Rev. 55, 52 (1939).

12.   W. R. S. Garton and F. S. Tomkins, Astrophys. J. 158, 839 (1969).

13.   See the review article by P. G. Metzger, Ref. 6, p. 801.

14.   E. Churchill and P. G. Mezger, Astrophys. Letters 5, 189 (1970).

15.   H. E. White, Introduction to Atomic Spectra (McGraw-Hill Book Co., New York, 1934).

16.   C. E. Moore, Atomic Energy Levels, National Bureau of Standards Circular No. 467 (U. S. GPO, Washington, D. C., 1949), Vol. I.

17.   W. C. Martin, J. Res. Natl. Bur. Std. (U.S.) A 64, 79 (1960).

18.   G. Araki, Proc. Phys. Math. Soc. Japan 19, 128 (1937); R. W. Mires and R. M. Parish, Bull. Am. Phys. Soc. 15, 1303 (1970).

19.   Y. Accad, C. L. Pekeris, and B. Schiff, Phys. Rev. A 4, 516 (1971).

20.   H. A. Bethe and E. E. Salpeter, Quantum Mechanics of One- and Two-Electron Atoms (Springer-Verlag, Berlin, 1957); H. A. Bethe, Handbuch der Physik, edited by H. Geiger and K. Scheel (Springer, Berlin, 1933), Vol. 24/1.

21.   M. J. Seaton, Proc. Phys. Soc. 88, 815 (1966).

22.   B. Edlén, Encyclopedia of Physics, edited by S. Flügge (Springer-Verlag, Berlin, 1964), Vol. 27, Secs. 20 and 33.

23.   W. H. Wing and W. E. Lamb, Jr., Phys. Rev. Letters 28, 265 (1972).

24.   W. H. Wing, K. R. Lea and W. E. Lamb, Jr., Atomic Physics, Vol. 3, edited by S. J. Smith and G. K. Walters (Plenum Press, N.Y., 1973), p. 119.

HIGH–RESOLUTION NONLINEAR SPECTROSCOPY OF MOLECULAR

VIBRATIONAL RESONANCES IN GASES*

F. De Martini
Istituto di Fisica "G. Marconi"
Universita di Roma, Italy

ABSTRACT

An account of the theory of the line shape of optically active resonances in gases is given including the effects of Doppler broadening, Dicke narrowing, and pressure broadening. Furthermore two different methods of high–resolution nonlinear spectroscopy applied to the study of the $Q_{01}(1)$ vibrational-rotational resonance in hydrogen are described. The results of these experiments are compared with the corresponding data obtained by spontaneous Raman spectroscopy and are critically discussed on the basis of the theory of the line shape previously outlined.

I. INTRODUCTION

In the present work, we consider the application of some nonlinear optical effects to the spectroscopy of Raman–active vibrational resonances in gases. In particular we shall deal with the study of the line profiles of these

---

*Work supported by C. N. R.

549

resonances in a thermodynamical condition of the gas in which the collision-narrowing (or Dicke-narrowing) process [1] and the dephasing collision-broadening effect determine competitively the shape of the resonance lines.

We shall present first a brief account of the theory of the line shapes in gases on the basis of some recent works on the subject [1-3] and then we shall proceed by describing two experimental methods of high-resolution nonlinear spectroscopy based on the stimulated Raman scattering (SRS) process [4]. These methods have been applied so far to the study of the profile of the $Q_{01}(1)$ vibrational resonance in hydrogen gas. As it will be clarified later in the paper, these methods can be of general interest and of far reaching application in molecular and solid-state spectroscopy. In particular the use of the currently available tunable dye lasers makes possible the application of these methods to the study of all Raman-active resonances in material media.

II.   COLLISION NARROWING OF RAMAN-ACTIVE VIBRATIONAL LINES

The process of motional narrowing is well known in nuclear magnetic resonance [5]. It can be simply explained by considering the effect of the phase fluctuations of the precessing spins due to the fluctuations of the local magnetic field that each molecule of the gas experiences in its motion. These phase fluctuations, that are responsible for the inhomogeneous broadening of the resonance line, may be thought to arise from "soft" adiabatic collisions with the spins surrounding the molecule [6,7].

A similar situation is present in a gas of molecules when optical levels are involved. In this case the mechanism that is responsible for the inhomogeneous broadening is generally the Doppler effect for which the observer detects, owing to the molecular motion, a change $\delta\omega = (\bar{v}/c)\omega_0$ of the frequency ω_0 emitted by each molecule in its own frame. $\bar{v}$ is the component of the average velocity of the particle in the direction of the observer. In the time interval τ_D between two collisions the observer experiences a phase jump of the radiation coming by a single molecule equal to $\delta\varphi \simeq \delta\omega\tau_D$. After n such intervals the mean square dephasing $\overline{\Delta\bar{\varphi}^2}$ will be

$$\overline{\Delta\bar{\varphi}^2} = n\delta\varphi^2 = n\left(\frac{\bar{v}\,\omega_0}{c}\right)^2 \tau_D^2 \tag{1}$$

The number of intervals n in a time t is simply $n = t/\tau_D$. If we take T_2 as the time for the radiation emitted from a group of molecules to get about one radian out of step we find, after (1)

$$1 = \frac{T_2}{\tau_D} \left(\frac{\bar{v}}{c}\right)^2 w_0^2 \tau_D^2 \tag{2}$$

or, for the ensemble of molecules,

$$\frac{1}{T_2} = 4\pi^2 \frac{\bar{v}\ell}{\lambda_0^2} \sim 4\pi^2 \frac{D}{\lambda_0^2} \tag{3}$$

In (3) we have written the gas diffusion coefficient $D = kT/m\beta \sim \bar{v}\ell$ in terms of the average molecular speed $\bar{v}$, of the mean free path 1, and of the coefficient of dynamical friction of the gas β [8]. To a first approximation we can set β^{-1} equal to the mean free time τ_D between deflecting collisions [9]. We note that the shorter τ_D (that is, the more rapid is the motion) the narrower is the resonance.

The expression of the linewidth $1/T_2$ reported in Eq. (3) is equal to the one calculated by Dicke [1] by a more rigorous statistical calculation involving the use of the Einstein diffusion probability function [8]. The results of the Dicke's analysis have some interesting features:

1. The linewidth (3) is generally much narrower, in normal thermodynamical conditions of the gas, than the linewidth $\Delta w_D = 2 w_0 (2kT \ln 2/mc^2)^{1/2}$ of the Gaussian line that accounts for the usual Doppler broadening.

2. The lineshape of the resonance line of the emitted radiation is Lorentzian despite the evident inhomogeneous character of the broadening.

We should consider at this point that the theory given by Dicke in Ref. 1 does not consider in a satisfactory way the other processes that contribute, in actual cases, to the shape of the resonance line, namely, the effect of Doppler (Gaussian) broadening in the condition of high gas rarefaction and the dephasing effect due to "hard" collisions that becomes overwhelming at high density. In order to take into account the overall process in a consistent way, we need to reformulate the statistical approach to the problem. We shall do this here on the basis of the recent theory of Galatry [2] that may be well adapted to the problem of the Raman scattering.

We consider the intensity spectrum $I(w)$ of the radiation emitted by a gas of free electric dipoles. By considering, in a first instance, only the Doppler effect, the autocorrelation function of the field may be written in the following form:

$$F(\tau) \equiv \frac{1}{2\pi} \int_{-\infty}^{+\infty} I(w) \exp(-iw\tau) \, dw = \lim_{T \to \infty} \frac{1}{T} \int_{-T}^{+T} \exp\left(-\frac{2\pi i}{\lambda_0}[x(t_0+\tau)-x(t_0)]\right) dt_0 \tag{4}$$

The usual way of evaluating the last integral of (4) in the theory of pressure broadening is to apply the ergodic hypothesis, i.e., to replace the average over time by an average over all possible paths x in the time τ_D [10]. We can use the impact approximation (collisions of negligible duration) with collisions distributing the free paths in accordance with a Poisson law with mean time τ_D. The autocorrelation function may be written in that case as follows:

$$F(\tau) = \langle \sum_{m=0}^{\infty} \frac{1}{m!} \left(\frac{\tau}{\tau_D}\right)^m \exp(-\tau/\tau_D) \; P_m(v_o, \Delta v_1, \Delta v_2, \cdots \tau_1, \tau_2, \cdots /\tau)$$

$$\times \exp\left\{-\frac{2\pi i}{\lambda} [\Delta x(v_o, \Delta v_1, \Delta v_2, \cdots \tau_1, \tau_2)]_m\right\} dv_o \cdots d\tau_i \rangle \tag{5}$$

where $[\;]_m$ denotes $\Delta x = x(t_o + \tau) - x(t_o)$ for a path containing m collisions as function of velocity v_o at time t_o, the m velocity changes Δv_i and the m durations τ_i ($\Sigma_i \tau_i = \tau$). P_m is the conditional probability density of the various parameters in a path containing m collisions. Breene [11] and Galatry [2] have evaluated the expression of $F(\tau)$ by making use of the Maxwellian distribution for the set of velocities $\{v_o\}$ and by introducing for the set of statistical spatial variables $\{\Delta x\}$ and the "diffusion" probability function due to Chandrasekhar [9]:

$$W(\Delta x/v_o, \tau) = (A/\tau)^{1/2} \exp\{-A[\Delta x - v_o [1 - \exp(-\beta\tau)]/\beta]^2\} \tag{6}$$

with

$$A = \left(\frac{m\beta^2}{2KT}\right) [2\beta\tau - 3 + 4 \exp(-\beta\tau) - \exp(-2\beta\tau)]$$

The autocorrelation function for the field may be evaluated now in closed form and for a gas of equal molecules of mass m it is

$$F(\tau) = \exp\left\{-\frac{4\pi^2 kT}{m\beta^2\lambda_o^2} [\beta\tau - 1 + \exp(-\beta\tau)]\right\} \tag{7}$$

We can now take into account the dephasing effects due to the "hard" collisions. By making the hypothesis of no statistical correlation between the ensemble $\{\Delta x\}$ and the one of the phase jumps due to collisions $\{\Delta\varphi\}$, the phase autocorrelation function is found to be equal to [10, 11]

$$F'(\tau) = \exp\left\{-\frac{\tau}{\tau_L} [(1-A) + iB]\right\} \tag{8}$$

where τ_L is the mean free time between the Poisson distributed collisions that disturb the phase and

$$A + iB = \int_0^{2\pi} \exp(i\Delta\varphi) p(\Delta\varphi) d(\Delta\varphi) \tag{9}$$

The factor $(1-A)/\tau_L$ is responsible for the pressure broadening (collision limited lifetime) while B/τ_L represents the shift of the center line due to pressure. $p(\Delta\varphi)$ is the probability distribution of the phase jumps due to collisions.

The final overall correlation function is the product of expressions (7) and (8) and its Fourier transform gives the profile of the resonance line according to (4).

By inspection of (7) we note that the quantity $\beta\tau \sim \tau/\tau_D$ strongly affects the shape of the line. For a highly rarefied gas (mean free path larger than the wavelength of the emitted radiation) expression (7) may be well approximated by $F(\tau) \simeq \exp[-(2\pi^2 kT/m\lambda_0^2)\tau^2]$ leading to the well-known Doppler Gaussian lineshape. For larger gas densities both Doppler effect and pressure broadening are present and the correlation function becomes

$$F(\tau) \simeq \exp\left[-\left(\frac{(1-A)}{\tau_L} + \frac{4\pi^2 D}{\lambda_0^2}\right)\tau - i\frac{B}{\tau_D}\tau\right] \tag{10}$$

The corresponding line has a Lorentzian profile:

$$I(\omega) = \frac{(1-A)/\tau_L + 4\pi^2 D/\lambda_0^2}{(\omega-\omega_0-B/\tau_L)^2 + [(1-A)/\tau_L + 4\pi^2 D/\lambda_0^2]^2} \tag{11}$$

We verify that the width of the line is determined competitively by the homogeneous broadening effect of the inelastic collisions (pressure broadening) and by the inhomogeneous Dicke-narrowing effect that is proportional to $\tau_D = Dm/kT$. By calling $\alpha \equiv (\tau_L/\tau_D) \geq 1$ the inverse of the fraction of collisions that are active in dephasing the field, we can verify that the minimum linewidth is reached at a gas pressure corresponding to the following value of the coefficient of dynamical friction: $\beta = 4\pi^2\alpha kT/(1-A)m\lambda_0^2$. The minimum value of the linewidth $\Delta\omega$ may be shown to be proportional to the width of the Gaussian Doppler linewidth at low density $(\Delta\omega)_D$ and inversely proportional to the square root of α:

$$\Delta\omega = \left(\frac{2(1-A)}{\ln 2}\right)^{1/2} \frac{\Delta\omega_D}{\sqrt{\alpha}} \tag{12}$$

When each collision is a "hard" collision that effectively limits the lifetime of the optical level, $\alpha = 1$ and the narrowing is not present, being always $A \ll 1$. In general this is the case of the optical transitions that possess a dipole moment. In that case the larger probability of the dephasing process is generally due to the effect of the long-range field interactions between the particles. On the other hand if the optical transition is somewhat "screened" against the perturbations due to collisions, as for some infrared inactive molecular vibrational-rotational states, the effect may be observable and even quite large. The first evidence for the effect of Dicke narrowing has been reported by Rank and Wiggins [12] for the S(1) line of the 1-0 quadrupole band of hydrogen. The same effect has been studied for the $Q_{01}(1)$ rotational line of H_2 by Lallemand et al. [13] and by Murray and Javan [14] by a Raman scattering technique. The work [13] that reports the application of a high-resolution nonlinear spectroscopy will be considered in some detail in Sec. III. The work [14] deals with a spontaneous Raman scattering experiment. A similar work has been reported by Cooper et al. [15] for the $S_0(0)$ and $S_0(1)$ lines of H_2 and by May et al. [16] for the vibrational Q lines in N_2 and CO.

The application of the spontaneous Raman spectroscopy to the study of molecular spectra in gases at low pressure involve in general severe problems of detection requiring the use of photon counting techniques. It seems therefore justified to devote Sec. III to the description of two techniques of nonlinear SRS spectroscopy that have been applied to the study of the $Q_{01}(1)$ resonance in H_2. As we shall see, the obvious advantage which is a consequence of the high intensity of the scattered beams is the possibility of obtaining information on molecular parameters that are inaccessible to the spontaneous spectroscopy. This is due to the possibility offered by the high intensity of the laser fields of creating in the medium high-order nonlinear effects corresponding to various dynamical schemes of interactions.

III.   METHODS OF HIGH-RESOLUTION NONLINEAR
SPECTROSCOPY IN HYDROGEN

The first experiment we want to present is due to Lallemand et al. [13] (Fig. 1). A ruby laser pulse (~ 30 nsec duration), after passing through a test cell (1) creates a strong SRS in the oscillator (2). The coherent Stokes beam emitted in backward direction is made to interact in (1) with the primary laser beam inducing in that cell a coherent molecular excitation at the difference frequency $\omega_L - \omega_S$. The Stokes frequency ω_S may be slightly varied by changing the pressure of the gas in the oscillator (2). This effect of frequency tuning is linear with pressure for gas densities less than ~ 80 Amagats. It is due to the shift of the vibrational Q resonance of the H_2 molecule due to the perturbation of the intermolecular potential induced by collisions.   This effect, first studied by May et al. [17] is present at various degrees for all vibrational-rotational resonances of molecules in a gas.   As far as the spectroscopical application of this effect is concerned we should consider that the stimulated emission occurs at the center of the spontaneous Raman line [4] and that the stimulated Stokes line is in general extremely narrow (of the order of the laser linewidth).   In reference [13] the Stokes gain of the hydrogen amplifier (1) has been measured as a function of $\Delta\omega = \omega_L - \omega_S$ for various pressures of that cell.   This has been done by measuring the increase of the Stokes intensity I in the amplifier by means of two photomultipliers.   The ratio of the Stokes intensities at the ends of the cell of length L is given by

$$I_{out}/I_{in} = \exp\left(\frac{4\pi\omega_S}{cn_S}\, \chi''(\omega)\, |E_L|^2 L\right) \tag{13}$$

where the laser flux density is given by $(c/2\pi n_S)\, |E_L|^2$.

We see that, by varying $\Delta\omega$ we can sweep over the resonant imaginary part of the Raman susceptibility $\chi(\omega) = \chi'(\omega) + i\chi''(\omega)$ and determine its linewidth.   In Fig. 1 the plot of the linewidth of $\chi''(\omega)$ as function of the gas pressure of the oscillator is reported.   It shows the marked effect of collision narrowing relative to the dynamics of the backward-wave interaction

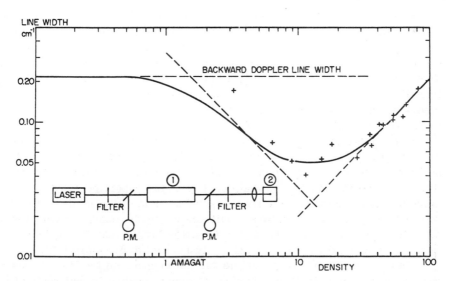

Figure 1.   Determination of the width of the $Q_{01}(1)$ vibrational line of H_2 as a function of pressure by means of a Raman amplifier cell for <u>backward</u> Stokes light.   (Lallemand et al. [13]).

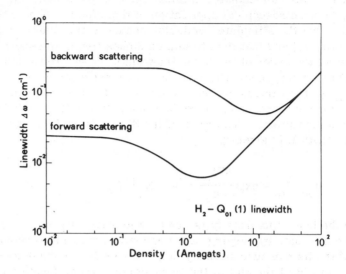

Figure 2.   Plots of the width of the $Q_{01}(1)$ vibrational line of H_2 as a function of gas density for forward and backward Raman scattering.

taking place in the Raman amplifier.   Owing to simple kinematical considerations on the Doppler effect we can verify that the linewidths corresponding to the Dicke narrowing region for the backward-wave Raman interaction are larger by a factor $\sim (\omega_L + \omega_S / (\omega_L - \omega_S)$ than the ones corresponding to the forward scattering.   This is shown in the plots of Fig. 2 valid for the $Q_{01}(1)$ transition of H_2 at $300^\circ K$.   These plots have been drawn on the basis of the theory we have outlined in Sec. II by assuming, to a first approximation, Lorentzian line shapes in the narrowing region.

Quite unfortunately in Ref. (13) the profiles of the Raman lines in the narrowing region are not reported.   This is due, very likely, to the lack of a sufficiently precise experimental information on $\chi''(\omega)$ owing to the exponential functional dependence on that quantity of the experimental data.   Furthermore we should consider the inaccuracy intrinsic of the method of comparing two large signals differing by a small amount at the ends of a very low gain amplifier, and the fact that, at gas densities larger than 70-80 Amagats, the induced Raman shift is no more linearly related to pressure [17].

On the other hand the experimental information on the line shapes in the narrowing region is of large physical interest because in that region the effect of inhomogeneous Doppler broadening, leading to a Gaussian line at very low pressures, becomes comparable with the effect of homogeneous pressure broadening corresponding to a Lorentzian line.   Furthermore it should be very interesting to check the validity of the theory reported in Sec. II which anticipates that the effect of Dicke narrowing still contributes to a Lorentzian profile.   In this respect also the recent experimental work of Buijs and Gush [18] on the electric field induced absorption in H_2 does not lead to results sufficiently precise in spite of the highly sophisticated spec-trometer they use.

A very sensitive and somewhat indirect method for the determination of the line shapes has been recently applied to the study of the $Q_{01}(1)$ resonance of the H_2 molecule by the author and co-workers [19].   In that experiment the collision-narrowing region is explored in the <u>forward</u> direction.   This leads one to consider the narrow widths of the lines $\chi''(\omega)$ corresponding to the lower plot of Fig. 2.   Furthermore in that experiment we do not reach direct information on $\chi''(\omega)$ but on the modulus of the Raman susceptibility $(|\chi'(\omega)|^2 + |\chi''(\omega)|^2)^{1/2} = |\chi(\omega)|$.   The results require in that case some mathematical processing in order to reach information on $\chi''(\omega)$ and $\chi'(\omega)$.   We point out that this last quantity has been so far inaccessible to the methods of "spontaneous" molecular spectroscopy.   The determination of the curve $|\chi(\omega)|$ instead of $\chi''(\omega)$ presents some substantial advantages in view of the exact knowledge of the line profile and, consequently, of the physical processes affecting the resonance.   The "resonant" curve $|\chi(\omega)|$ extends over a much larger frequency domain than $\chi''(\omega)$ owing to the dispersive part of the susceptibility $\chi'(\omega)$.   Furthermore, the Kramers-Krönig expressions that relate $\chi'(\omega)$ to $\chi''(\omega)$ critically determine the shape of $|\chi(\omega)|$ on the basis of the exact analytical frequency dependence of $\chi'(\omega)$ and $\chi''(\omega)$ [4].   For instance a Gaussian $\chi''(\omega)$ line shape leads to an expression of $|\chi(\omega)|$ that is very different from the one corresponding to a Lorentzian profile.

The experimental setup is shown in Fig. 3.   The ruby laser worked on a single axial mode with a linewidth $\simeq 0.008$ cm^{-1} and it induced a SRS process in the oscillator A.   In A the pressure of a gaseous mixture of H_2+He

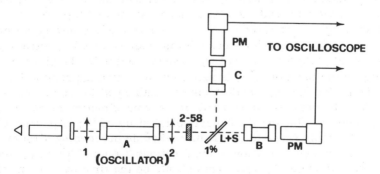

Figure 3.   Schematic diagram of the experimental apparatus for the study of the $Q_{01}(1)$ vibrational line profile by coherent <u>forward</u> scattering.

could be varied over about an order of magnitude to provide the Stokes
frequency tuning in analogy with Ref. 13. The laser + Stokes beams emerging
from the oscillator were superimposed in the test H_2 gas cell B and in the
reference gas cell C. That last cell was filled with nitrogen at atmospheric
pressure. Two photomultipliers detected the light signals at frequency
$\omega_{AS} = (2\omega_L - \omega_S)$ generated in the cells owing to the four-photon nonlinear
interactions [20, 21]. The nonlinear source polarization at frequency
$(2\omega_L - \omega_S)$ created in the two cells may be written in the form [4]:

$$P(\omega_{AS}) = \chi^{(3)}(\omega_{AS}, \omega_L, \omega_L, -\omega_S)\, E(\omega_L)E(\omega_L)E^*(\omega_S) \qquad (14)$$

where $\chi^{(3)}$ is the third-order nonlinear susceptibility of the two gases at
frequency ω_{AS}. In cell B, $\chi^{(3)}$ coincides with the (resonant) Raman susceptibility of H_2, $\chi_k(\omega) = \chi'(\omega) + i\chi''(\omega)$, owing to the value of ω_S emitted by the
oscillator A. In cell C $\chi^{(3)}$ is the nonresonant third-order susceptibility
that may be taken as a frequency-independent real quantity. If the Raman
gain in B is much smaller than the inverse of the interaction length of the
process, the anti-Stokes intensity detected at the end of that cell is given, in
terms of the laser and Stokes intensities I_L and I_S, by the expression:
$I(\omega) = |\chi_k(\omega)|^2 I_L^2 I_S$. The same expression is valid for the process taking p
place in C but with the constant $\chi^{(3)}$ of N_2 in place of $\chi_k(\omega)$. In Fig. 4 the
ratio $I_{H_2}(\omega)/I_{N_2}(\omega)$ is plotted vs frequency for (Research grade) H_2 at 20 atm
and 300°K.

We can verify in Fig. 2 that the thermodynamical conditions of the gas
corresponding to Fig. 4 for forward scattering correspond to a line shape
that is competitively determined by the effect of motional narrowing and by
the collision broadening process. Furthermore, owing to the value of β, we
should expect a Lorentzian frequency dependence of $\chi''(\omega)$ according to Eq.
(11). However in that case, owing to Kramers-Krönig relations, the function
$|\chi_k(\omega)|^2$ should be Lorentzian, in contrast with the shape reported in Fig. 4.
In addition, we note that the experimental shape of $|\chi_k(\omega)|^2$ is not symmet-
rical showing a shift of the center of mass of the line toward lower frequencies. This effect, which agrees with the experimental spectra $\chi''(\omega)$ reported
by May et al. [16] for N_2 and CO, is shown better by the plots of Fig. 5. In
that figure the function $\chi''(\omega)$ is assumed, to a first approximation, to be a
Lorentzian line having a half-intensity width equal to the one reported in
Ref. 18. The corresponding first-order approximation function $\chi'(\omega)$ is
easily determined using the experimental shape of $|\chi_k(\omega)|^2$ reported in
Fig. 4. We could proceed by Kramers-Krönig processing $\chi'(\omega)$ to obtain the
second-order approximation of the functions $\chi''(\omega)$ and $\chi'(\omega)$ by a further use
of the experimental $|\chi_k(\omega)|$. This should be the first step of an interative
process leading, by successive approximations, to the real shapes of $\chi'(\omega)$
and $\chi''(\omega)$.

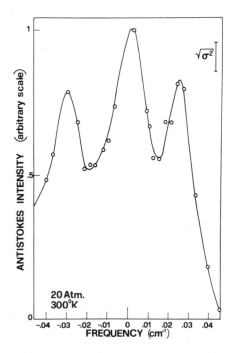

Figure 4.   Plot of the anti-Stokes intensity versus the frequency shift (Δw–w_o).   The conditions of the H_2 gas are p=20 atm, T=300°K.

In order to extract direct spectroscopic information from the curve $|X_k(w)|^2$ one should determine experimentally the phase of the anti-Stokes field $\varphi(w) = \arctan X'(w)/X''(w)$ by simply detecting on the cathode of a photomultiplier the interference of the anti-Stokes fields emitted by the cells B and C.   That additional phase information, combined with the corresponding one for the modulus of $X_k(w)$, leads immediately to $X'(w)$ and $X''(w)$.

We have described in that section two nonlinear techniques in which the method of coherent molecular excitation [22] is applied to a work of high-resolution molecular spectroscopy.   Obviously, in the present case the use of a pressure-tuned SRS hydrogen oscillator has been of great help in the investigation of the $Q_{01}(1)$ resonance of the H_2 molecule.   However we should consider that the use of the now available tunable dye lasers makes possible the general application of the same (or similar) nonlinear methods in molecular and in solid-state spectroscopy without restriction on the system under study and on the resonance excited.

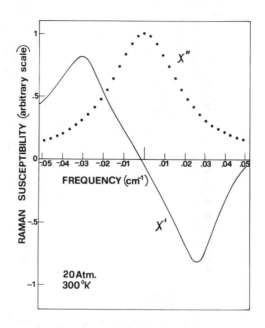

Figure 5.   Plot of the real and imaginary parts of the Raman susceptibility versus ($\Delta\omega - \omega_0$) and corresponding to the conditions of Fig. 4. (See text).

REFERENCES

1. R. H. Dicke, Phys. Rev. 89, 472 (1953); J. P. Wittke and R. H. Dicke Phys. Rev. 103, 620 (1956).
2. L. Galatry, Phys. Rev. 122, 1218 (1961).
3. H. M. Foley, Phys. Rev. 69, 616 (1946); J. I. Gersten and H. M. Foley, J. Opt. Soc. Am. 58, 933 (1968).
4. Bloembergen, Nonlinear Optics (W. A. Benjamin, New York, 1965).
5. E. M. Purcell, R. V. Pound, and N. Bloembergen, Phys. Rev. 70, 986 (1946).
6. C. P. Slichter, Principles of Magnetic Resonance (Harper and Row, New York, 1963).
7. A. Abragam, The Principles of Nuclear Magnetism (Oxford at the Clarendon Press, Oxford, England, 1961).
8. J. Jeans, An Introduction to the Kinetic Theory of Gases (Cambridge at the University Press, Cambridge, 1962).
9. S. Chandrasekhar, Rev. Mod. Phys. 15, 1 (1943).
10. M. Margenau and M. B. Lewis, Rev. Mod. Phys. 31, 569 (1959).

11. R. G. Breene, Rev. Mod. Phys. 29, 94 (1957).
12. D. H. Rank and T. A. Wiggins, J. Chem. Phys. 39, 1348 (1963).
13. P. Lallemand, P. Simova, and G. Bret, Phys. Rev. Letters 17, 1239 (1966); P. Lallemand and P. Simova, J. Mol. Spectroscopy 26, 262 (1968).
14. J. Murray and A. Javan, J. Mol. Spectroscopy 29, 502 (1969).
15. V. G. Cooper, A. D. May, E. H. Hara, and H. F. Knapp, Can. J. Phys. 46, 2019 (1968).
16. A. D. May, J. C. Stryland, and G. Varghese, Can. J. Phys. 48, 2331 (1970).
17. A. D. May, V. Degen, J. C. Stryland, and H. L. Welsh, Can. J. Phys. 39, 1769 (1961); A. D. May, G. Varghese, J. C. Stryland, and H. L. Welsh, Can. J. Phys. 42, 1058 (1964).
18. H. L. Buijs and H. P. Gush, Can. J. Phys. 49, 2366 (1971).
19. F. De Martini, G. P. Giuliani, and E. Santamato, Opt. Commun. 5, 126 (1972).
20. R. W. Minck, R. W. Terhune, and W. G. Rado, Appl. Phys. Letters 3, 181 (1963).
21. W. G. Rado, Appl. Phys. Letters 11, 123 (1967).
22. F. De Martini, Nuovo Cimento 51B, 16 (1967).

NONLINEAR SPECTROSCOPY OF MOLECULES

Tadao Shimizu
Department of Physics, University of Tokyo
Tokyo, Japan

ABSTRACT

Several experiments on the nonlinear effects of molecules recently performed in our laboratories are introduced, after a brief historical review of the nonlinear spectroscopy of molecules is presented.

I. INTRODUCTION

In the conventional spectroscopy of atoms and molecules, the frequency of the spectral line is determined by the Bohr condition and the absorption signal intensity is proportional to the intensity of the incident radiation. However, a variety of nonlinear effects, such as a saturation in absorption, a shift of the resonance frequency, a broadening and splitting of the spectral line, a two-photon absorption, a double-resonance effect and so on, become appreciable as the intensity of incident radiation increases. These nonlinear behaviors of molecules had been observed only in the radio and microwave frequency regions, before the laser light sources became available.

Radio and microwave spectroscopy have been in a favorable situation for observing nonlinear effects because strong and coherent radiation sources are available in these frequency regions. The "strength" of the coherent oscillator may be expressed in terms of flux density of the photons in a proper propagation mode. For example, we can easily obtain an output power of 100 mW/cm^2 at 1 cm wavelength from a typical klystron oscillator. The photon flux density is

$$F = P/h\nu = 5 \times 10^{21} \text{ photons/sec cm}^2$$

which is large enough to cause a saturation effect in absorbing molecules at moderate pressure. The saturation becomes evident when the condition

$$(\bar{\mu}E/\hbar)\,\tau \gtrsim 1$$

is satisfied, where $\bar{\mu}$ is the dipole matrix element between the two energy levels of the molecule, τ is the relaxation time, and E is the field strength. This condition can be rewritten by the equivalent expression [1]

$$\alpha F \gtrsim (N_2-N_1)/\tau$$

where N_2-N_1 is the difference in populations on the two levels 2 and 1, and α is the absorption coefficient. When the pressure of molecules is 1 mTorr, N_2-N_1 is of the order 10^{10}/cm^3 and τ is 10^{-5} sec. Then we obtain

$$\alpha F \gtrsim 10^{15}/\text{scc cm}^3$$

The output power of the klystron can saturate the molecular transition with an absorption coefficient as small as 2×10^{-7} cm^{-1}. The power required to obtain the same photon flux increases, being proportional to the frequency of radiation. In the case of infrared radiation, a power of several tens of Watts per cm^2 in a single mode is necessary, which may be available only when lasers are used.

Broadening of a molecular spectral line due to the saturation effect was first observed in ammonia by Townes [2] in 1946. The first observation of a two-photon transition in a molecule was the electric resonance experiment on a RbF molecular beam at a frequency of several megahertz performed by Hughes and Grabner [3] in 1950. The radio frequency-optical double-resonance experiment in atoms was initiated by Bitter [4] and Brossel and Kastler [5] in 1949, while the double-resonance effect in a molecule was first observed in the ammonia beam maser at microwave frequency by Shimoda and Wang [6] in 1955. Since the saturation effect in three-level systems of molecules was theoretically discussed by Javan [7] in 1957, a variety of experiments have been carried out. In an earlier stage, for example, investigations on OCS spectral lines by Battaglia, Gozzini, and Polacco [8], and Shimoda [9] and on HCOOH spectral lines by Yajima [10] were reported.

Since the advent of lasers, spectroscopic studies of nonlinear properties of molecules have become possible also in the infrared and visible wavelength regions. Nonlinear spectroscopic methods such as the Lam dip, mode crossing, and transient saturation have been applied to molecular spectroscopy.

Several spectroscopic studies on the nonlinear effects in molecules recently performed in our laboratories will be discussed.

II. MICROWAVE TWO-PHOTON ABSORPTION

The simultaneous absorption of two microwave photons of identical energy by a molecule was observed by Oka and Shimizu [11]. A highly sensitive double-resonance technique was successfully employed to detect the weak two-photon transition. The experiment was performed in several three-level systems of CD_3CN and PF_3 molecules. The largest signal was observed in the three-level system consisting of the J = 2, 1, and 0 levels of CD_3CN. Absorption due to the transition J = 1 ← 0 at 15.716 GHz was turned into emission in the presence of pumping radiation with exactly half the frequency of the J = 2 ← ← 0 transition. Since the dipole moment of CD_3CN is large (3.92 D) and the J = 1 level lies close to the intermediate state of the two-photon transition (7.85 GHz in the frequency difference), the microwave power of 20 W is large enough to produce "population inversion" between the J = 1 and J = 0 levels. The saturation behavior of the two-photon transition was well explained by a modified Karplus-Schwinger formula [11].

III. INFRARED-MICROWAVE DOUBLE-PHOTON ABSORPTION

Although laser pumping is a powerful method in molecular spectroscopy, it is usually difficult to find a good coincidence between the frequencies of laser and absorption lines. The frequency of a tunable microwave photon can be added to that of the laser photon by utilizing the nonlinear property of molecular absorption. Laser spectroscopy with high resolution and high accuracy thus becomes applicable in a wider frequency range for a variety of molecular transitions.

The infrared-microwave double-photon absorption was observed first in the $v_2 = 1$, J = 4, K = 4(s) ← $v_2 = 0$, J = 4, K = 4(s) dipole-forbidden transition of $^{15}NH_3$ with the P(15) N_2O laser radiation [12] and later in several tens of other infrared transitions of ammonia. As shown in Fig. 1(b) the frequency of the infrared allowed transition is about 0.3 GHz higher than the laser frequency. Strong absorption of laser radiation was observed in the presence of microwave power of 20 W at the sum frequency of 23.05 GHz (J = 4, K = 4 inversion frequency) and 0.3 GHz. A saturation dip method would make the double-photon absorption signal free from Doppler width limitations at the infrared frequency. The absolute frequencies of infrared

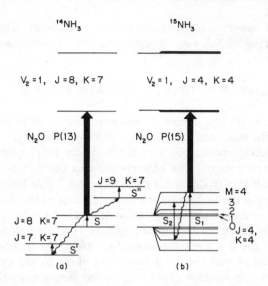

Figure 1. Schematic energy levels of ammonia molecules involved in the infrared-microwave double resonance with N_2O laser.

laser lines are accurately measured.   The frequency of the infrared absorption line can be accurately determined by measuring the microwave frequency in the double-photon transition.

IV.   INFRARED-MICROWAVE DOUBLE RESONANCE

The absorption intensity of the ground state $J = 8$, $K = 7$ inversion line of NH_3 was changed up to 800% in the presence of the P(13) N_2O laser radiation [13], the frequency of which was in resonance with that of the $v_2 = 1$, $J = 8$, $K = 7(s) \leftarrow v_2 = 0$, $J = 8$, $K = 7(a)$ infrared transition within 10 MHz [Fig. 1(a)].   The changes in the absorption signals were also observed in the $J = 9$, $K = 7$ and the $J = 7$, $K = 7$ inversion transitions.   These are due to collisional transfer of the nonequilibrium population of the pumped level $J = 8$, $K = 7(a)$ to the $J = 9$, $K = 7(s)$ and the $J = 7$, $K = 7(s)$ levels.   The probability and the selection rules of the collision-induced transitions have been extensively studied by Oka by the method of microwave-microwave double resonance [14].   The infrared pumping is hundreds of times more efficient than the microwave pumping, because in the former case the population of the upper level of the pumped transition is negligibly small.   A ten times larger change in the signal due to the collision-induced transition was observed by the infrared laser pumping than that by the microwave pumping.

A large change in the absorption intensity of the $J = 4$, $K = 4$ microwave transition was also observed in the presence of the P(15) N_2O laser radiation when the frequency of infrared transition $v_2 = 1$, $J = 4$, $K = 4(s)$ ← $v_2 = 0$, $J = 4$, $K = 4(a)$ was tuned into the laser frequency by applying a Stark field of 4.565 kV/cm [Fig. 1(b)].

A double-resonance signal was observed in the three- and four-level systems of H_2CO and HDCO by Takami and Shimoda [15]. The 5_1, 4_1, and 3_1 transitions were pumped and saturated by the microwave power and the $v_5 = 1$, $6_{0,6}$ ← $v_5 = 0$, $5_{1,5}$ infrared transition at 2850.633 cm^{-1} was monitored by the Zeeman-tuned 3.5-μm He-Xe laser (Fig. 2). The absorption cell was placed inside the laser cavity. When the infrared radiation was strong and the saturation at the infrared transition occurred, the double-resonance signal was observed to reverse its sign at low pressure. The effect was well explained by a theory of double resonance in the case when both radiations were strong [1].

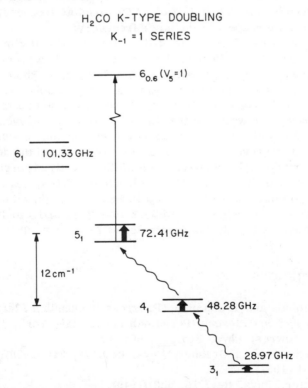

Figure 2. Schematic energy levels of formaldehyde molecule involved in the infrared-microwave double resonance with He-Xe laser.

ADDENDUM (SEPTEMBER, 1972)

Infrared-microwave two-photon spectroscopy has been extensively studied. Several tens of infrared-microwave two-photon absorption lines of $^{14}NH_3$ and $^{15}NH_3$ molecules were observed by using infrared lasers in the 10-μm wavelength region [16]. A signal of the inverse Lamb dip in the two-photon process, which was proposed in this text, was also recently observed in the forbidden transitions of the $^{15}NH_3$: $\nu_2\,[^qQ_-(4,4)]$ and the $^{14}NH_3$: $\nu_2[^qQ_+(5,4)]$ by using the P(15) N_2O laser line and the R(6) CO_2 laser line, respectively [16]. The frequency-tunable microwave radiation was "added" to the laser radiation, so that the sum or the difference of the laser frequency and the microwave frequency came into resonance with the absorption line frequency. The infrared radiation was in a standing-wave mode, but the microwave radiation was not necessarily a standing wave.

Although the observed dip width was still much larger than the collisional width at a gas pressure of several mTorr, it was as narrow as 0.8 MHz. This technique will become very useful in high-resolution infrared spectroscopy and in determination of the absolute frequency of infrared transitions with respect to the laser frequency.

Since the first observation of the saturation dip at a microwave frequency frequency [17], the method has been developed into a precise spectroscopic study of the microwave transitions. The Doppler width of a microwave line is very small and of the order of several tens of kilohertz. The diameter and length of the absorption cell should be large, so that the radiation is a well-defined plane wave and the time of interaction between the radiation and the molecule, which traverses the microwave beam, is sufficiently long.

Several improvements in the absorption tank and the detecting system have been made. The narrowest dip width recently observed is 0.78 kHz in the $J = 3 \leftarrow 2$ transition of the OCS molecule at 36,488,812.8 kHz [18]. The pressure and microwave power dependences of the dip width were also investigated in the ranges of 1 to 30 x 10^{-5} Torr and 5 to 40 nW/cm^2, respectively. A broadening parameter of 70 Hz/10^{-5} Torr was obtained.

REFERENCES

1. K. Shimoda and T. Shimizu, Progress in Quantum Electronics, edited by Sanders and Stevens (Pergamon Press, New York, 1972).
2. C. H. Townes, Phys. Rev. 70, 665 (1946).
3. V. Hughes and L. Grabner, Phys. Rev. 79, 314 (1950); 79, 829 (1950); 82, 561 (1951).
4. F. Bitter, Phys. Rev. 76, 833 (1949).
5. J. Brossel and A. Kastler, Compt. Rend. 229, 1213 (1949).
6. K. Shimoda and T. C. Wang, Rev. Sci. Instr. 26, 1148 (1955).

7. A. Javan, Phys. Rev. 107, 1579 (1957).
8. A. Battaglia, A. Gozzini, and E. Polacco, Nuovo Cimento 14, 1076 (1959).
9. K. Shimoda, J. Phys. Soc. Japan 14, 954 (1959).
10. T. Yajima, J. Phys. Soc. Japan, 16, 1594 (1961); 16, 1709 (1961).
11. T. Oka and T. Shimizu, Phys. Rev. A 2, 587 (1970).
12. T. Oka and T. Shimizu, Appl. Phys. Letters 19, 88 (1971).
13. T. Shimizu and T. Oka, Phys. Rev. A 2, 1177 (1970).
14. T. Oka, J. Chem. Phys. 47, 13 (1967); 47, 4852 (1967); 47, 5410 (1967); 48, 4919 (1968); 49, 3135 (1968); 49, 4234 (1968).
15. M. Takami and K. Shimoda, Japan. J. Appl. Phys. 10, 658 (1971).
16. S. M. Freund and T. Oka, VII international Quantum Electronics Conference, Montreal, 1972.
17. C. C. Costain, Can. J. Phys. 47, 2431 (1969).
18. C. C. Costain and T. Shimizu (to be published).

C. Spectroscopy of Condensed Matter

BRILLOUIN AND RAMAN SPECTROSCOPY WITH LASERS

B. P. Stoicheff
Department of Physics, University of Toronto
Toronto, Canada

ABSTRACT

With the advent of the laser and with recent improvements in spectroscopic
and detection techniques, light scattering has once again become an important and flourishing field of research.   These new developments in experimental techniques have produced increased sensitivity and resolution and
have made experimental work easier and more efficient.   They have helped
to bring to fruition many significant experiments which earlier were extremely difficult or impossible to do.   In Brillouin scattering it is now possible to
precisely measure frequency shifts and linewidths as a function of scattering
angle and to investigate their dependence on temperature and pressure and,
in particular, their behavior in the neighborhood of critical points.   Laser
Raman scattering has been especially valuable in the study of solids and a
variety of new and important Raman-type spectra have been observed.
Spectra have been excited with radiation of wavelength up to 10.6 μ;   resonance Raman spectra with $\Delta n = 9$ have been reported;   and high resolution with
widths of a few MHz have been shown to be possible.   These and other results
of recent investigations in Raman and Brillouin scattering in gases, liquids,
solids, and plasmas are briefly reviewed.

INTRODUCTION

The laser, with its remarkable properties of high intensity, monochromaticity, and directionality, has stimulated a surge of activity in Raman and Brillouin spectroscopy. Just ten years ago the situation was very different; only a handful of spectroscopists knew of Brillouin scattering, and research in Raman spectroscopy was considered by many to be passé. In the years prior to the laser, research in light scattering was carried out at only a few laboratories, and notably by Fabelinskii in the U.S.S.R., Krishnan in India, Mathieu in France, Rank in the U.S.A., and Welsh in Canada. A review of the accomplishments in Brillouin scattering in this pre-laser period has been given by Fabelinskii [1], and in Raman spectroscopy by the author [2].

Since the first reports of laser-excited Raman spectra [3,4] in 1961 and Brillouin spectra [5-7] in 1964, there has grown a voluminous literature on the subject of light scattering. It will be the purpose of this paper to summarize the highlights of these recent investigations in Brillouin and Raman spectroscopy and to briefly outline the advances in experimental techniques.

Another flourishing branch of light scattering is Rayleigh scattering. Especially important are measurements of linewidths in the vicinity of critical points and of critical mixtures of liquids using the exquisite resolution provided by light-beating spectroscopy. This interesting subject will not be pursued here and the reader is referred to several published reviews [8-11].

I. BRILLOUIN SPECTROSCOPY

A. Brief Résumé

When a beam of monochromatic light passes through a transparent medium, a small part of the light is scattered in all directions, some at the incident frequency and some with changed frequency. It is customary to distinguish the scattered light of changed frequency as of two types. The scattered light with relatively large frequency shifts (independent of scattering angle) caused by rotational and vibrational transitions of individual molecules in fluids, and by optical phonons and other excitations in solids, is the well-known Raman radiation. The scattered light with small frequency shift (varying with scattering angle) caused by thermal fluctuations in the medium is the Brillouin radiation [12].

The Brillouin spectrum of a fluid is represented in Fig. 1: it consists of Stokes and anti-Stokes components, shifted from the Rayleigh line at ν_0, downward and upward, respectively, by the frequency of thermal sound waves or phonons, ν_s. The frequency shift is given by the Brillouin equation

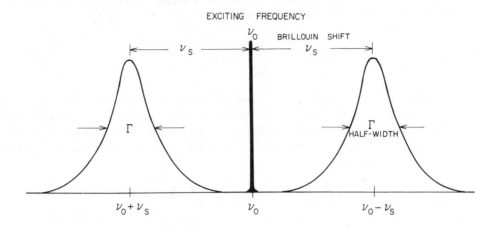

Figure 1.   Diagram of the predicted Brillouin spectrum for a fluid.

$$\Delta\nu = \nu_s = \pm 2\nu_0 \frac{V}{c}\, n\, \sin\frac{\theta}{2} \tag{1}$$

where n is the refractive index of the medium, and θ the scattering angle. If the scattering angle θ is well defined, $\Delta\nu$ can be used to determine the sound wave velocity V. Also, the widths Γ of the Brillouin components are a measure of the damping or attenuation of the waves. Finally, by varying the angle θ from the forward to the backward direction, the velocity and attenuation of acoustic waves in a range of frequencies from approximately 10^7 to 10^9 Hz may be investigated. Thus, in effect, we can study "acoustical spectroscopy" by light scattering. Not only can we investigate possible dispersion relations by this means, but also obtain information on molecular motions and of physical properties of media which affect the velocity or attenuation of these thermal or acoustic waves.

The intensity of scattered light is derived from a consideration of the fluctuation of the dielectric constant $\delta\epsilon\,(\bar{r}, t)$ which is usually written

$$\delta\epsilon\,(\bar{r}, t) = \left(\frac{\partial\epsilon}{\partial\rho}\right)_T \delta\rho\,(\bar{r}, t) + \left(\frac{\partial\epsilon}{\partial T}\right)_\rho \delta T(\bar{r}, t) \tag{2}$$

The first term describes the variation in dielectric constant with density (ρ) at constant temperature; it arises from propagating density fluctuations and results in the Brillouin components. The second term describes the variation in dielectric constant with temperature at constant density; it arises from the decay of temperature concentrations or entropy and results in the Rayleigh component. Finally, the intensity of the Brillouin components is given (apart from a numerical constant) by

$$I_s = \frac{I_0 \nu_0^4}{r^2} V_0 \rho^2 \left(\frac{\partial \epsilon}{\partial \rho}\right)^2 kTK_s \qquad (3)$$

Here I_0 is the incident intensity, V_0 the volume of the scattering medium at a distance r from the observer, T the absolute temperature, and K_s the bulk compressibility. From thermodynamic arguments, Landau and Placzek showed that the ratio of intensity of the Rayleigh to the Brillouin components in fluids is given by

$$\frac{I_c}{2I_B} = \frac{C_p}{C_v} - 1 = \gamma - 1 \qquad (4)$$

where C_p and C_v are the specific heats at constant pressure and volume, respectively.

Detailed derivations of the Brillouin equation and scattering intensity for fluids have been given by Rytov [13], Mountain [14], Van Kampen [15], Cummins [11], Benedek and Greytak [16] amongst others, and for crystals by Benedek and Fritsch [17]. Also, reviews on Brillouin scattering have been published recently by Chu [10], Cummins [11], Fleury [18], Krishnan [19], McIntyre and Sengers [20] and by the author [21].

B.   Experimental Techniques

A typical apparatus for Brillouin scattering is shown in Fig. 2. The narrow light beam from a laser is incident on the scattering medium. The scattered light, at angle θ, is collected through a small aperture and analyzed by a pressure-scanned Fabry-Perot interferometer. The resulting interference ring system is projected on a screen. A pinhole at the center of the ring system transmits light to a photomultiplier; the signal is amplified and recorded.

At the present time, the He-Ne laser radiation at 6328 Å is most widely used for excitation of Brillouin spectra. Frequency-stabilized He-Ne lasers are commercially available which emit up to 0.5 mW in a line of width $\ll 1$ MHz and are stable to 1 MHz for a day. More powerful He-Ne lasers are also standard items with emission in a line of width up to 1 GHz or 0.03 cm^{-1}. More recently Ar-ion lasers have become available with powers of ~ 1 W at 4880 or 5145 Å; although this power is emitted in a single longitudinal mode, stability still remains a problem for the highest-resolution

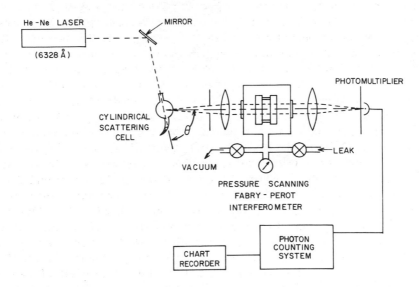

Figure 2. A typical apparatus for Brillouin spectroscopy of liquids.   (After Stegeman et al. [22]).

work.

Detailed accounts of apparatus and technique have been published by Benedek and Greytak [16], and Stegeman et al. [22]. The use of spherical plate interferometers [23] for higher resolution and stability has been described, as well as of conical lenses [16] and of cylindrical reflectors [24] for efficient collection of scattered light at specific angles θ. Piezoelectric scanning of the interferometer has also been used in conjunction with various signal-averaging techniques [25].

In place of the interferometer or other optical instruments, Lastovka and Benedek [26] used the heterodyne technique of light-beating spectroscopy to detect thermal Brillouin scattering at 30 MHz for small-angle scattering. A completely different concept, although also suggested by Brillouin [12] in 1922, making use of injected sound in the medium and the examination of scattered laser light by superheterodyne techniques, has been described by Gordon and Cohen [27].

C. Spectra of Gases

From the brief review of the origin of Brillouin scattering given in Sec. I A, it is obvious that a system of free particles will not give rise to a Brillouin spectrum: rather, it produces a Doppler broadened Rayleigh line. In other words, if the mean free path of the particles is longer than the phonon wavelength thermal or acoustic waves cannot propagate in the medium. However, as the density of the medium is increased, the mean free path becomes shorter and shorter, thermal waves begin to propagate and the broad Rayleigh line is resolved into the Rayleigh-Brillouin triplet. This evolution of Brillouin scattering is shown in Fig. 3 from a study of gaseous H_2, HD, and D_2 at densities ranging from 0.5 to 104 amagat, by May and co-workers [28]. They noted good agreement of the observed spectra with computed spectra at high densities where the hydrodynamic theory of Mountain [14] is applicable, and at low densities (where the ratio of molecular collision frequency to the frequency of sound ~ 1) where the kinetic theory of Sugawara and Yip [29] is applicable. However, neither theory is in agreement with experiment at intermediate densities, and the authors propose that this is evidence of a coupling between relaxation of the rotational states and thermal conduction.

The validity of the hydrodynamic and kinetic theories was also checked by Greytak and Benedek [23], who carried out experiments on Ar, Xe, N_2, CO_2, and CH_4 at 1 atm pressure, and varied the wavelength of the thermal waves being analyzed by changing the scattering angle from the forward to the backward direction. Some of their results are shown in Fig. 4. Scattering in the forward direction probes waves of small k vector (or long wavelength), and at a pressure of 1 atm CO_2 exhibits the Brillouin components: conversely, scattering in the backward direction probes waves of short wavelength, and in CO_2 gives rise to the broad spectrum shown in the lower part of Fig. 4. Such experiments form important tests of present kinetic theories

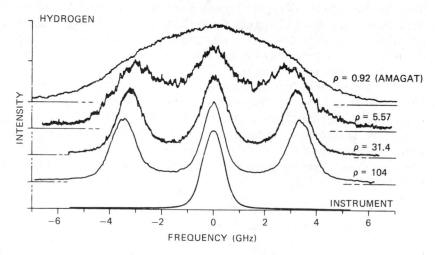

Figure 3.   Rayleigh and Brillouin spectra of gaseous H_2 at various densities. (After Hara, May, and Knaap [28].)

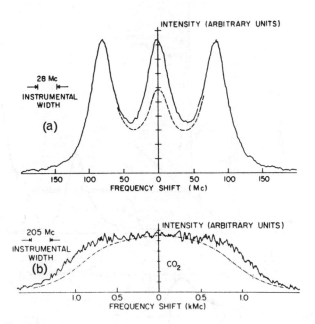

Figure 4.   Brillouin spectra of CO_2 gas at 1 atm: (a) at a scattering angle of 19.6°;   (b) at angle of 169.4°.   (After Greytak and Benedek [23].)

and of the Boltzmann equation.

Perhaps the most interesting results have been obtained from studies of Xe and CO_2 in the neighborhood of their critical points.   Figure 5 represents the Brillouin spectrum of Xe at four temperatures along the critical isochore investigated by Cannell and Benedek [30].   Here it is readily seen that there is a marked velocity dispersion and increase in linewidth or attenuation close to the triple point.   In addition, there is a growth of intensity between $\omega = 0$ and the Brillouin component attributed to the growth of a diffusive mode as $T-T_c$ goes to zero.   An analysis based on Mountain's theory of a fluid whose bulk viscosity relaxes with a single relaxation time has yielded the observed temperature dependence of the relaxation time, the bulk viscosity, the specific heat ratio C_p/C_v, the correlation range, and the compressibility and C_p-C_v at $k = 0$, $\omega = 0$.   Cummins and Swinney [31] have measured the sound velocity in xenon on the critical isochore and found that the sound dispersion agrees with the Kawasaki theory.   Earlier measurements on CO_2 near the critical point also showed considerable sound dispersion [32] and increase in attenuation [33] as T_c is approached.   However, comparison with theory is more difficult because of the internal degrees of freedom of the CO_2 molecule.

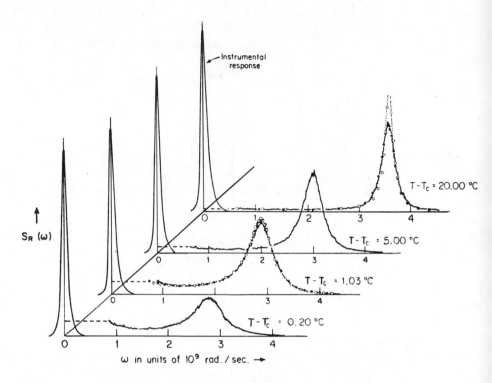

$S_R(\omega)$

ω in units of 10^9 rad./sec. $\rightarrow$

Figure 5.   Brillouin spectra of Xe at various temperatures along the critical isochore.   (After Cannell and Benedek [30].)

Further investigations in the critical region, including measurements of Rayleigh and Brillouin scattering, will be valuable for testing recent theories of the critical state.

    Another interesting problem is the behavior of the spectra of gaseous mixtures at various concentrations.   Such an experiment has been reported by Bloembergen [34] and colleagues for SF_6 and He.   With increasing concentration of He, the Rayleigh–Brillouin triplet of pure SF_6 broadens rapidly, the Brillouin shift increases, and the intensity of a broad component (centered at $\omega = 0$) associated with concentration fluctuations rises markedly, to such an extent that it dominates the spectrum at 70% He concentration.   This behavior of the mixture is explained in terms of strong coupling between the diffusive concentration mode and the sound wave, and by taking into account the internal degrees of freedom of SF_6.   Good agreement was obtained between observations and calculations by assuming that the mixture is in the hydrodynamic regime.

D. Spectra of Liquids

The spectra of liquids have been investigated more than of gases and solids, mainly because of the higher intensity of scattering of liquids. The first experiments with laser excitation were concerned with observing the spectra and measuring the shifts and linewidths, usually at 90° scattering angle. This was soon extended to 90° and 180° by Chiao and Fleury [35] in order to observe possible relaxation effects, and to estimate relaxation frequencies for a variety of liquids. More recently, several detailed studies have been carried out on the variation of frequency shifts and linewidths with scattering angle from the forward to the backward direction.

A series of studies was undertaken by the author and his colleagues [22, 36] to determine the usefulness of Brillouin spectroscopy in evaluating acoustic properties of liquids, and to assess the accuracy achievable with this method. Three different liquids were chosen for these studies: CCl_4 which has a relaxation of the vibrational specific heat in the frequency region available to Brillouin spectroscopy, CS_2 whose relaxation frequency is known to be below 100 MHz from acoustic measurements, and ethyl ether whose relaxation frequency is estimated to be an order of magnitude above that of CCl_4.

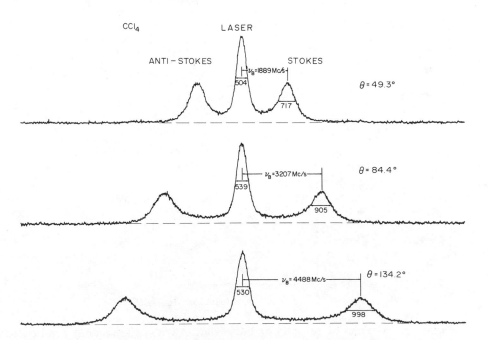

Figure 6.   Brillouin spectra of liquid CCl_4 at $20^\circ C$, taken at various scattering angles. (After Stegeman, et al. [22]).

Brillouin spectra of CCl_4 (at 20°C) recorded at scattering angles of 49°, 84°, and 134° are shown in Fig. 6. The changes of frequency shift and of linewidth with scattering angle are readily apparent. Their measurement led to the determination of velocities and attenuations of thermal waves in the frequency range of 1.6–4.7 GHz. A large dispersion (>10%) and corresponding change in attenuation was found (Fig. 7), characteristic of a thermal relaxation of the total internal specific heat. A broad new component [37] between the Rayleigh and Brillouin lines was observed for the first time (Fig. 6), which is also characteristic of a relaxation process. This "relaxation" line is centered at $\omega = 0$, it has a breadth of the order of the relaxation frequency,

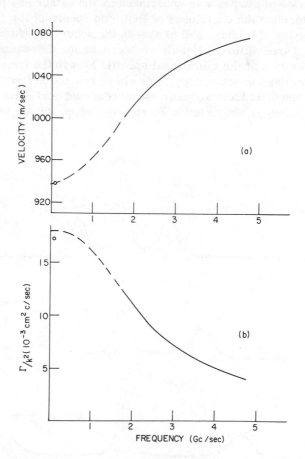

Figure 7.   Graphs of velocity and attenuation versus frequency for liquid CCl_4 at 20°C.   (After Stegeman et al. [22].)

and is due to isotropic scattering as are the Rayleigh and Brillouin components. Analyses of these data were carried out in two ways: one based on a simple extension of acoustic equations to light scattering, and the other, on Mountain's theory [14] of light scattering in the region of a thermal relaxation with a single relaxation time. It was found that the observed spectra and derived data agree with Mountain's theory in all respects, including the modification of the Landau-Placzek ratio [Eq. (4)] to $I_c/(2I_B+I_M) = \gamma - 1$ (where I_M is the intensity of the "relaxation" line). Similar results were found by Carome et al. [38].

Improved techniques and a single-mode laser were used for the investigations [36] with CS_2 and ethyl ether. In ethyl ether the phonon velocity and attenuation were constant in the frequency range studies (0.8-4.1 GHz), and agreed with acoustic measurements at 100 MHz. Also, the observed intensity ratio $I_c/(2I_B) = 0.38\pm0.02$ agreed with the simple ratio of Eq. (4), namely, $\gamma-1 = 0.37$. These results are in agreement with a relaxation frequency >10 GHz. In CS_2, a considerable decrease in attenuation was observed in the range 1.7-6.1 GHz. The data were analyzed according to Mountain's single relaxation theory and yielded values of the vibrational specific heat $C_I = 4.00$ cal/mole$^\circ$C, and relaxation frequency $1/2\pi\tau = 85$ MHz, which are in good agreement with acoustic measurements. In this example, the expected "relaxation" line is superimposed on the Rayleigh component. Therefore instead of $I_c/(2I_B) = \gamma-1 = 0.55$, Mountain's theory gives $(I_c + I_M)/(2I_B) = [(C_p-C_I)/(C_v-C_I)]-1$, the calculated value of 0.83 being in good agreement with the measured value 0.81 ± 0.05.

An extremely difficult experiment has been carried out by St. Peters, Greytak, and Benedek [39] who measured the velocity and attenuation of thermal waves in superfluid He. Their results are shown in Fig. 8; there is dispersion in the sound velocity and a dramatic increase in attenuation near T_λ, along with a subsidiary peak in the attenuation near 1.5°K, but without corresponding dispersion. These observations appear to be in qualitative agreement with recent theories of weakly interacting phonons and rotons. Woolf, Platzman, and Cohen [40] used light scattering from transducer-generated sound waves to measure the velocity and attenuation in liquid He and obtained similar results.

Brillouin scattering from the monatomic liquids argon and neon has been studied by Fleury and Boon [41]. They found that the hypersonic ($\sim$ 3 GHz) velocities in argon decrease linearly from 850 m/sec at 85°K to 742 m/sec at 100°K, but depart slightly from low-frequency (1 MHz) data obtained under the same thermodynamic conditions. This small negative dispersion is in qualitative agreement with theoretical calculations which take into account the frequency dependence of the transport coefficients. In neon the hypersonic velocity decreases (not quite linearly) from 620 m/sec at 24.9°K to 508 m/sec at 32°K. Comparison with ultrasonic data (at 30 and 1.3 MHz)

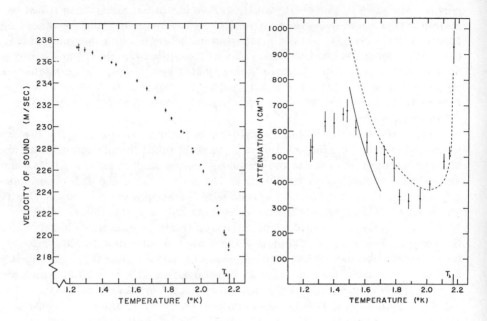

Figure 8.   Graphs of velocity and attenuation versus temperature for liquid He.   (After St. Peters, Greytak, and Benedek [39].)

indicates that the hypersonic velocity is always lower by 0.5 to 1.5%, a discrepancy which is larger than the experimental errors.   The frequency dependence of the transport properties does not account for this difference and Fleury and Boon suggest that quantum effects may be responsible.

All of the Brillouin spectra discussed above are associated with longitudinal waves and consist of isotropic scattering.   It has long been known that for liquids consisting of molecules with nonspherical polarizability, these spectra are superimposed on a depolarized component known as the Rayleigh wing: it is centered at zero frequency, is very broad, sometimes extending as far as 10 or even 100 cm^{-1}, and is believed to originate from molecular rotations in liquids.   Recently, Starunov, Tiganov, and Fabelinskii [42], and Stegeman and Stoicheff [43] have observed a sharp depolarized component of only a few GHz breadth, centered at the exciting frequency, in the spectra of liquids composed of anisotropic molecules.   Typical spectra are shown in Fig. 9 for various states of polarization of the incident and scattered light. The I_V^V spectrum arising from isotropic scattering shows the expected sharp Brillouin components due to longitudinal waves, the sharp Rayleigh line, and

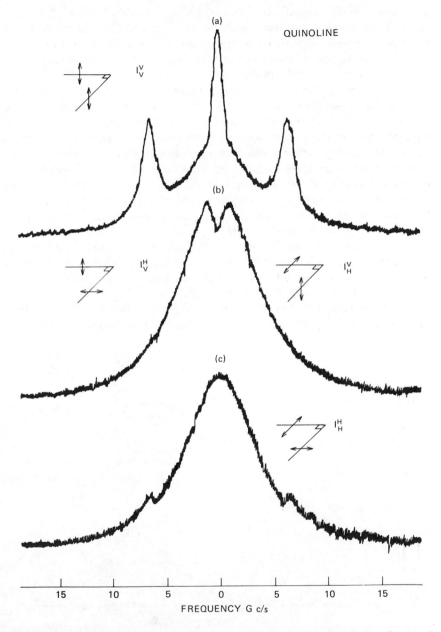

Figure 9. Spectra of liquid quinoline showing polarized and depolarized components. The intensity of (a) is attenuated by a factor of 2. (After Stegeman and Stoicheff [43].)

an intense broad component. The I_V^H and I_H^V spectra consist of a pronounced doublet, symmetrically displaced about the exciting frequency, and having a peak separation of the order of the breadth of the doublet. Finally, the I_H^H spectrum consists of a single broad component of Lorentzian shape with a weak Brillouin doublet in the wings.

The observed line shapes and intensities of the I_H^V and I_H^H spectra, as well as the dependence of the doublet splitting on $\sin(\theta/2)$ are in quantitative agreement with theories of Leontovich [44], Rytov [13], and Volterra [45] on light scattering from orientational fluctuations of anisotropic molecules caused by heavily damped shear waves. Distinct doublet spectra have been observed for more than 20 different liquids composed of anisotropic molecules, but having widely varying viscosities, dipole moments, and molecular structures. Analyses of the spectra have given values of shear wave frequencies, relaxation times ($\sim 10^{-11}$ sec), and of shear moduli ($\sim 10^8$ dyn/cm^2). A brief review of experimental data and comparison with recent theories is given in Ref. 46. An especially interesting investigation on the temperature dependence of the doublet was carried out by Fabelinskii, Sabirov, and Starunov [47] on salol whose viscosity ranges from 10^{-3} to 10^9 P over the temperature interval 130 to -70°C.

Surface waves or "ripplons" on liquid surfaces have recently been studied by light scattering. In this process the component of momentum

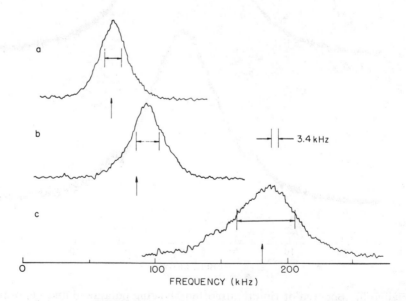

Figure 10. Ripplon spectra of different frequencies from liquid methanol. (After Katyl and Ingard [48].)

along the surface as well as the energy are conserved leading to a Brillouin doublet of frequency $w_0 \pm w_s$, where w_s is the "ripplon" frequency. Surface waves are rapidly damped by shear viscosity and it is found that for wavelengths shorter than a critical value $\Lambda_c = 1.16\ \pi\eta^2/\rho\sigma$, the surface vibrations are overdamped leading to nonoscillatory decay and to Rayleigh line broadening. (Here η is the shear viscosity, ρ the liquid density, and σ the surface tension.) Only for wavelengths longer than Λ_c do oscillatory vibrations or "ripplons" occur; for most common liquids "ripplon" frequencies ($w_s/2\pi$) are less than 100 MHz. Scattering from underdamped and overdamped surface waves in several liquids has been investigated by Katyl and Ingard [48], and in the neighborhood of the liquid-vapor phase transition in CO_2 by Bouchiat and Meunier [49]. Figure 10 shows experimental "ripplon" spectra of different frequencies (and scattering angles) from liquid methanol obtained with a 5-mW argon-ion laser and an optical heterodyne spectrometer. The line shapes are Lorentzian with half-intensity widths $\Delta w = 2\tau$ dependent on frequency, with the "ripplon" lifetime given by $\tau = \rho/2\eta q^2$.

E.   Spectra of Solids

Considerable work on Brillouin spectra of solids has recently been reported. While the main emphasis has been on the determination of elastic constants and damping of phonons, there have been several interesting studies of ferroelectric phase transitions and of the phonon bottleneck using Brillouin scattering.

The spectra of isotropic solids such as glass and fused quartz contain two pairs of Brillouin components, one pair associated with longitudinal waves and the other with transverse waves. In general, the spectra of simple crystals exhibit three pairs of components, while birefringent crystals exhibit more complex spectra [50].

The theory of Brillouin scattering from cubic crystals has been analyzed by Benedek and Fritsch [17], and tested by them with the spectra of KCl, KI, and RbCl. They measured the frequency shifts (10 to 15 GHz) for longitudinal and "mixed" longitudinal-transverse acoustic phonon branches as a function of the direction of propagation. Thus they deduced the velocities of the phonons, and were able to determine the elastic constants C_{11}, C_{12}, and C_{44} for each crystal. Their values were in good agreement with ultrasonic measurements at 10 MHz, indicating an absence of dispersion over the region 10 MHz to $\sim$ 10 GHz.

More recently, Brillouin spectroscopy has been used by the author and his co-workers to determine values of the elastic constants of rare-gas single crystals. Only a few values are presently available, based on ultrasonic and neutron-scattering experiments. The advantage of the light-scat-

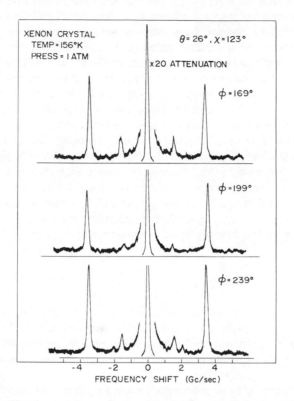

Figure 11. Brillouin spectra of a xenon single crystal at various orientations φ about a vertical axis. (After Gornall and Stoicheff [51]).

tering method is that the size of the crystal need only be ~ 1 mm³, and single crystals of such small dimensions are relatively easily grown. Spectra of xenon single crystals [51] have been investigated (Fig. 11) leading to the first determination of the elastic constants of xenon. Preliminary results for neon single crystals have just been reported [52], and experiments with argon and krypton are in progress.

Attenuation of longitudinal phonons in the solids α-quartz, fused silica, and calcium fluoride has been measured by Durand and Pine [25] using a single-mode laser, a multiscanned confocal Fabry-Perot interferometer, and a multichannel analyzer. By observing the backward scattered spectra, they measured widths of ~30 MHz for phonon frequencies of ~30 GHz. The corresponding attenuation length is ~ 60 μ which is not measurable by conventional ultrasonic techniques.

In a study of the ferroelectric crystal triglycine sulfate (TGS) Gammon

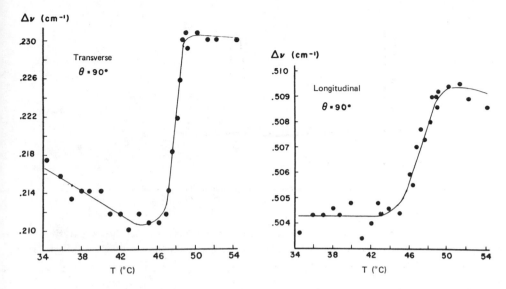

Figure 12.   Brillouin shifts of longitudinal and transverse components near
the phase transition in triglycine sulfate. (After Gammon and Cummins [53].)

and Cummins [53] found a marked dispersion with temperature of the hypersonic velocity in the neighborhood of the phase transition (49° C). The
Brillouin shifts for transverse waves increased by $\sim 10\%$ and those for longitudinal waves by $\sim 1\%$ as shown in Fig. 12.   The dispersion follows the usual
form for single relaxation processes which in this crystal is interpreted as
a coupling of acoustic phonons to polarization fluctuations.   From the dispersion temperatures for phonons of different frequencies the relaxation time
τ was found to depend on temperature as $\tau = A(T-T_c)^{-1}$, in accordance with
theory (although the constant A is smaller by an order of magnitude than the
ultrasonic value).   An even more dramatic discontinuity with temperature
was observed by Brody and Gammon [54] in the ferroelectric crystal potassium dihydrogen phosphate (KDP).   They measured a frequency change in the
transverse Brillouin components from ~ 5.3 GHz at 140°K to ~ 0.6 GHz at
the phase transition temperature $T_c = 122^\circ$K.   The frequency rapidly
increased to ~ 5.5 GHz at 120°K in the ferroelectric phase.   The results for
$T > T_c$ agree with the model that there is piezoelectric coupling of the acoustic shear mode to the ferroelectric soft mode observed by Kaminow and
Damen [55] in the Raman spectrum.   The order-disorder transition in
ammonium chloride (NH_4Cl) has also been investigated by Brillouin scattering, by Lazay et al. [56].

Brya, Geschwind, and Devlin [57] used Brillouin scattering to observe a

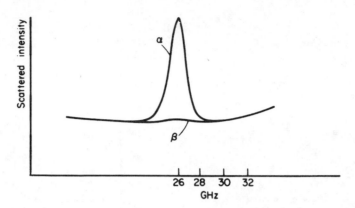

Figure 13.   Brillouin scattering in MgO with Ni^{2+} impurity, (α) with and (β) without resonant microwave radiation incident on the crystal.   (After Brya, Geschwind, and Devlin [57].)

large nonequilibrium distribution of acoustic phonons caused by a "phonon bottleneck" in MgO when the spin system of the Ni^{2+} paramagnetic impurtiy is excited with resonant microwave radiation. This result is shown in Fig. 13 Without microwave radiation and with the single crystal of MgO immersed in liquid He at $2^{\circ}K$, the Brillouin spectrum in the frequency range of ~ 25 to 34 GHz is given by the essentially flat curve labelled β. Included in this frequency range are the frequencies 25.6 and 31.3 GHz which correspond to transverse phonons of the same acoustic branch travelling in crystallographically inequivalent directions for the particular geometry of the experiment.   When the crystal is irradiated with cw microwave radiation at 25.6 GHz and the Ni^{2+} spin system is saturated, the excited spins relax to the lattice phonons and excite acoustic phonons within the frequency bandwidth of the microwave resonance.   However, because these phonons cannot transport their energy rapidly enough to the bath, the phonons remain excited above their equilibrium value, and form the "phonon bottleneck." The resulting Brillouin spectrum is given by curve α: it shows a dramatic increase in intensity in the region of 25.6 GHz over the thermal scattering at $2^{\circ}K$ (curve β), and corresponds to a temperature of $\sim 60^{\circ}K$.   The fact that there is no observed intensity change at 31.3 GHz indicates the selectivity of the bottleneck process.

   Before ending this section, mention should be made of stimulated Brillouin scattering [57, 1] and especially of its use to measure phonon lifetimes in the range 10^{-7} to 10^{-9} sec.   Experiments have been carried out on CS_2 [58], crystal quartz [59], and on liquid helium [60-61] in the vicinity of the λ point.

II. RAMAN SPECTROSCOPY

A. Introduction

Several books on Raman spectroscopy have been published recently including "Laser Raman Spectroscopy" by Hendra and Gilson [62], and the multi-authored compilations edited by Szymanski [63], and by Anderson [64]. For reviews of recent Raman studies with laser excitation, the reader is referred to the chapters in Ref. 63 by Schrotter and by Koningstein, and to Ref. 64 for a chapter by Hathaway on instrumentation and techniques. Much of the credit for the rapid advance of laser Raman spectroscopy and for developments in experimental technique belongs to Porto, and many of his contributions will be described in the sections below dealing with spectra of gases, liquids, and solids.

The classic review of the theory of Raman scattering is the paper by Placzek [65] written in 1934. He treated the rigorous theory, the polarizability theory as well as selection rules for rotational and vibrational spectra, all illustrated by many examples of Raman spectra of gases and liquids. Only in 1964 did a comparable treatment of the Raman spectra of solids appear, by Loudon [66]. A more recent and comprehensive review of the theory of Raman scattering in crystals has been given by Cowley in Ref. 64.

B. Spectra of Gases

The use of laser excitation has paved the way for great improvements in high-resolution Raman spectroscopy--as discussed by Weber et al. [67]. Not only is this due to the monochromaticity of laser radiation but also because scattering in the forward direction will effectively minimize the Doppler breadths associated with light scattering. An example [68] of the improvement possible is indicated in Fig. 14. Here are shown profiles of a rotational line of H_2 gas for both forward and 90° scattering. At 90° scattering, the full width at half maximum intensity is 0.15 cm^{-1}, corresponding to the Doppler breadth at 20°C. In the forward direction, the observed linewidth is reduced to 0.04 cm^{-1}, which is the total instrumental width (including the laser and Fabry-Perot linewidths) of the experiment. While this represents an improvement of almost a factor of 10 over the effective resolution achieved in prelaser work (~ 0.3 cm^{-1}), appreciable further improvement should be possible with the most recently available single-frequency lasers.

A general improvement in scattering intensity has been achieved by a careful consideration of the illumination of a gas sample in the focal region of a laser beam, and of the light-gathering optics used to collect the scattered light. Thus Barret and Adams [69] have been able to record photoelectrically

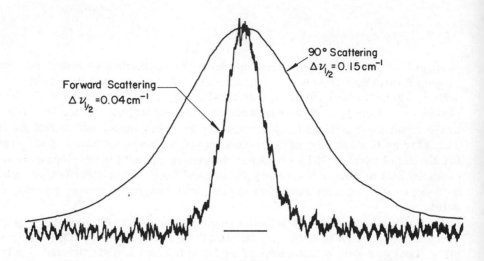

Figure 14.   Profiles of the S(1) rotational line of H_2 gas at 2 atm pressure in forward and $90°$ scattering.   (After Clements and Stoicheff [68].)

the rotation-vibration spectra of gaseous O_2, N_2, and CO_2 from only 10^{11} molecules in volumes of $\sim 10^{-7}$ cm^3.

   In 1953, Dicke [70] discussed the narrowing of Doppler-broadened lines when the density of a gas is increased so that the mean free path of the atom or molecules is much less than the wavelength of the emitted radiation.   He showed that under these conditions the line would have a Lorentzian profile with full width at half-intensity maximum given by $\Delta \nu = k^2 D_0 / \rho \pi$, where k is the wave vector of the emitted light, D_0 is the coefficient of self-diffusion, and ρ is the density.   The narrowing has been observed in the microwave and infrared regions and its effect on the gain in stimulated Raman emission has been measured by Lallemand et al. [71].   Recently, detailed studies of the dependence of linewidths of rotational and rotation-vibrational Raman lines in H_2 on gas density have been reported by Cooper et al. [72] and Murray and Javan [73].   Figure 15 shows graphs of this dependence for the Q(1) Raman line of H_2 for forward and backward scattering.   For both directions of scattering, the Doppler breadth rapidly decreases and reaches a minimum as the density increases;   the breadth then increases linearly with density as collision broadening becomes effective at the higher density.   The data at high densities thus give the pressure broadening coefficient, while the data at low densities give the diffusion coefficient.   The results obtained to date are in agreement with the theory of Dicke and with calculated values of pressure broadening coefficients;   however, the diffusion coefficient determined

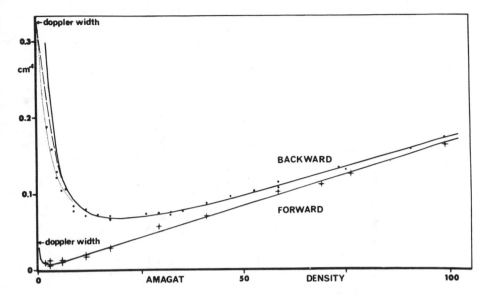

Figure 15. Measured linewidths of Q(1) rotation–vibration line of H_2 gas at various densities, for forward and backward scattering. (After Murray and Javan [73].)

from the rotational linewidths is larger than the value obtained from gas kinetic measurements, and is possibly an indication that collisions which modify the translational motion and collisions which lead to line broadening may be partially correlated.

It has long been known that Rayleigh scattering and Raman scattering (for totally symmetric vibrations) of isotropic molecules such as CCl_4 in the liquid and gaseous states is slightly depolarized. Recently, there have been several theoretical and experimental contributions to this field. Levine and Birnbaum [74] have discussed a binary collision model and have applied it to the noble gases. During a collision, there is a distortion of the electronic structure of the colliding atoms resulting in an anisotropic polarizability for the pair. This additional polarizability scatters light into a broad frequency band whose width is determined by the duration of the collision and whose intensity varies as the square of the density, and decreases exponentially with frequency. McTague and Birnbaum [75] have observed such collision-induced scattering in gaseous Ar and Kr (Fig. 16) and have confirmed the above predictions. Thibeau, Oksengorn, and Vodar [76] have measured the pressure-induced depolarization scattered by compressed noble gases, and have explained their results in terms of a cage model and the statistical

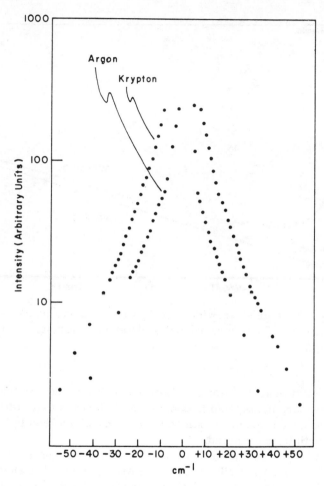

Figure 16.   Scattered light intensity versus wave number for argon and krypton.   (After McTague and Birnbaum [75].)

fluctuations of the internal electric field at each atom due to its irregular environment.   Further experiments and refinements in the theories have bee published by Lallemande [77] and by Gersten [78].   Observations of depola- rized spectra of liquid argon [79] and xenon [80] have also been reported.   It should be noted that collision-induced light scattering is the analogue of far- infrared pressure-induced absorption first observed by Kiss and Welsh [81] in 1959, and such studies will complement each other in the search for suit- able intermolecular potentials.

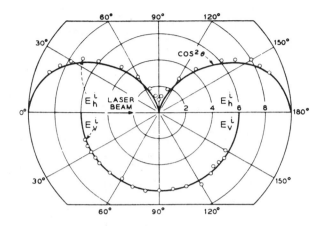

Figure 17.   Angular dependence of the Raman scattering for the 992–cm^{-1} vibrational band of C_6H_6.   (After Damen, Leite, and Porto [82].)

C.   Spectra of Liquids

The angular dependence of Raman scattering has been accurately measured by Damen, Leite, and Porto [82].   Their results for the totally symmetric stretching vibration (A_{1g}) of C_6H_6 at 992 cm^{-1} are shown for the two polarizations of the incident beam in Fig. 17.   For the incident electric field vector perpendicular to the plane of observation (E_v^i) the intensity distribution is constant with scattering angle θ.   For the electric field vector parallel to the observation plane (E_h^i) the intensity distribution follows a cos^{2}θ law.   Both results (and others for bands of e_{2g} symmetry) are in agreement with Placzek's [65] predictions.   A method of absolute measurements of Raman scattering cross sections for liquids has been developed and used by Kato and Takuma [83].   They make a direct comparison of the intensity of scattered light with that of a blackbody, by transmitting both through the same optical path (including a double monochromator) alternatively by means of a rotating sector disk.   Measurements were made on CCl_4, CS_2, benzene, toluene, and nitrobenzene with standard deviations of ∼ 3%.   Linewidth measurements using a Fabry–Perot interferometer have been reported [84] for the liquids O_2, N_2, CS_2, benzene, and toluene.   For N_2 and O_2 the full widths at half–intensity are surprisingly small, being 0.067 and 0.117 cm^{-1}, respectively.

One of the most exciting observations is the Raman spectrum of rotons in superfluid He^4 recently reported by Greytak and Yan [85].   Figure 18 shows experimental traces of the spectrum at 1.16 and 2.14°K, just below

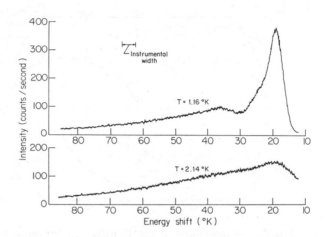

Figure 18.   Raman spectra of liquid He^4 at a scattering angle of $90°$, at two temperatures below the λ point.   (After Greytak and Yan [85].)

T_{λ}.   The main feature of the spectrum at $1.16°$K is the sharp asymmetric line with a maximum corresponding to an energy shift of $18.5 \pm 0.5°$K (or 13 cm^{-1}).   At higher temperatures the line broadens rapidly and at $2.14°$K is barely distinguishable from the broad background present at all temperatures.   The spectrum is highly depolarized and its total intensity is 1.6×10^{-3} of the Brillouin spectrum.   The line occurs at twice the energy required for excitation of a single roton, and is interpreted as arising from the excitation of two rotons having momenta $\vec{k}$ and $\vec{k}'$ of almost equal magnitude, but of opposite directions.   In this way, momentum is conserved since $\vec{k}+\vec{k}'$ is equal to the very small momentum of the optical photon, and energy is conserved, since the scattered photon suffers a loss corresponding to twice the roton energy.   The occurrence of this spectrum in light scattering was first proposed by Halley [86].   Accurate measurements [85] at $1.2°$K have shown that the two-roton line occurs at an energy shift which is less than twice the energy of a single roton at that temperature.   This result is explained by the existence of a two-roton bound state having a binding energy of $\sim 0.4°$K.   Raman scattering has thus been shown to be a valuable probe for the study of elementary excitations in superfluid helium, and has provided information on roton energies and lifetimes which complements that obtained by inelastic neutron scattering.

D. Spectra of Solids

The use of laser excitation has had a profound effect on investigations of Raman scattering in solids. This is due to the high intensity of laser radiation, to its directionality, and to the ease with which the radiation may be linearly polarized (with a rejection ratio of $\sim 10^{-5}$ for radiation which is orthogonally polarized). Polarization measurements give information on the components of the change of polarizability tensor which are nonzero for a specific mode of vibration, and thus lead to an unambiguous designation of the symmetry species of vibrational modes. And measurements of angular dependence of frequency and intensity give information of the effect of momentum conservation on Raman frequencies and selection rules.

An example of the effect of the state of polarization of the incident and scattered light on Raman scattering, taken from the work of Porto and his colleagues [87], is shown in Fig. 19 for MnF_2. This crystal belongs to the point group D_{4h}: it contains two molecules per unit cell and therefore has 15 vibrational modes. Four of these modes are Raman active having symmetries A_{1g} (with α_{XX}, α_{YY}, $\alpha_{ZZ} \neq 0$), E_g (with α_{YZ}, α_{XZ}, α_{ZY}, $\alpha_{ZX} \neq 0$), B_{1g} (α_{XX}, $\alpha_{YY} \neq 0$), and B_{2g} (with α_{XY}, α_{YX}, $\neq 0$). The experiment was carried out on a carefully oriented single crystal of MnF_2, and in order to examine a specific component α_{ij} of the scattering tensor, the incident light was polarized in the i direction and scattered light polarized in the j direction was observed. Thus the symmetries of the observed Raman bands in Fig. 19 were readily determined. In (a) is shown the A_{1g} band at 341 cm^{-1}; in (b) the E_g band at 247 cm^{-1}; in (c) again the A_{1g} band at 341 cm^{-1}, and the B_{1g} band at 61 cm^{-1}; and in (d) the B_{2g} band at 476 cm^{-1}. Similar investigations have been carried out by Porto et al. [87] on the crystals TiO_2, MgF_2, FeF_2, ZnO, calcite, and others.

Crystals which have inversion symmetry do not exhibit first-order Raman spectra, but may have intense second-order or two-phonon Raman spectra. Fleury and Worlock [88] have induced first-order Raman scattering in the cubic crystals $KTaO_3$ and $SrTiO_3$ by an applied electric field which removes the inversion symmetry of the crystals. The relatively intense second-order Raman spectrum of $KTaO_3$ (without electric field) is reproduced in Fig. 20(a). The much weaker first-order scattering was detected by applying an alternating electric field at 210 Hz and observing the scattered light modulated synchronously at twice the applied field frequency. The observed spectrum is shown in Fig. 20(b). It consists of a single band whose intensity increased as the square of the applied field, and whose frequency decreased with decreasing frequency. It was identified as the "soft" or "ferroelectric" mode. A similar Raman band observed in $SrTiO_3$ showed striking electric field shift (from 11 to 43 cm^{-1} for an increase in field from 400 to 12,000 V/cm, with the crystal at 8°K) and a splitting with increased electric field.

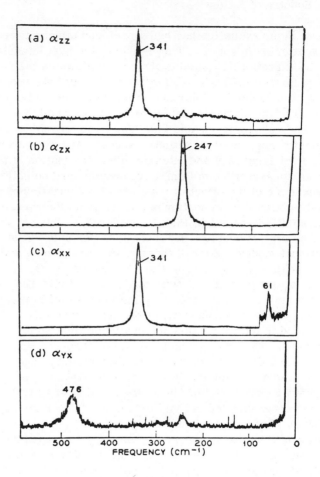

Figure 19.   Raman spectrum of MnF$_2$ obtained for various states of polarization of the incident and scattered light. (After Porto, Fleury, and Damen [87].)

Raman scattering has been used to investigate a variety of phase transitions.   The temperature dependence of the "soft" mode frequency has helped to clarify the interaction of optical and acoustic modes in the ferroelectric transitions of KDP [55] and SrTiO$_2$ [88].   The order-disorder transition (at T$_\lambda \sim 235^\circ$K) in NH$_4$Br has been shown by Wang [89] to have a marked effect on the intensity of a polarized Raman band at 56 cm^{-1}:   the integrated intensity is a maximum at T$_\lambda$, decreases rapidly with decreasing temperature, and is essentially zero at 150°K.   This behavior is in quantitative

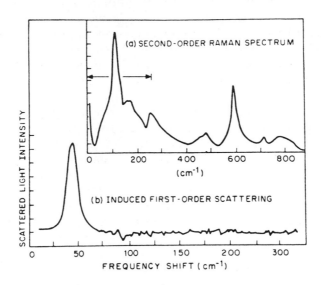

Figure 20.   Spectrum of light scattered from KTaO$_3$ at 77 K.   (a) Intrinsic second-order spectrum without an external electric field. (b) Electric-field-induced spectrum.   (After Fleury and Worlock [88].)

agreement with the observed specific heat anomaly, and suggests a decay of short-range ordering of NH$_4^+$ ions below T$_\lambda$.   Crystalline quartz is known to undergo a transition at 573°C from a low-temperature α phase with symmetry D$_3$ to a high-temperature β phase with symmetry D$_6$.   One of the Raman-active modes of A$_1$ species in the α phase (at 147 cm^{-1} at room temperature) is thought to play a fundamental role in this phase transition.   The observed Raman band [90] is found to grow in intensity, and to shift to lower frequencies according to the relation $\nu^2 \sim |T-T_c|^{0.45}$, as the α-β transition is approached, in general agreement with the thermodynamic theory of second-order phase transitions.

Parker, Feldman, and Ashkin [91] have developed a technique for light scattering from opaque surfaces and have observed Raman bands of Si, Ge, of the monatomic metals Be, Zn, Mg, Bi, and of the alloy AuAl$_2$.   Their method made use of an argon-ion laser with the sample placed in the laser cavity and a polished surface of the sample acting as a third cavity mirror. The frequencies of the observed bands are in good agreement with neutron-scattering data, but are more precise, and the Raman linewidths have led to the first determination of the lifetimes of optical phonons in these materials.

Following the first observations of electronic Raman spectra in crystals by Hougen and Singh [92] using mercury arc excitation, there is now considerable activity in this branch of light scattering with lasers. Spectra of rare-earth ions arising from transitions between Stark levels (split by the crystal field) have been observed by Koningstein and his colleagues [93]. The dependence of frequencies on applied magnetic field and of linewidths on temperature have been used to distinguish the electronic Raman bands from those due to lattice vibrations of the host crystal.

A rich variety of new Raman spectra has been observed recently. They include scattering from F centers in NaCl and KCl [94]; from spin waves or one-magnon modes in FeF_2 [95] and CoF_2 [96]; from two-magnon modes in MnF_2 [97], and coupled two-magnon excitations in K_2NiF_4 [98]; from polaritons in GaP [99], ZnO [100], quartz [101], and GaAs [102]; from the mixing of one- and two-phonon processes in $BaTiO_3$ [103]; from plasmons in GaAs [104]; and from single particle excitations of electrons in InSb in the presence of a magnetic field [105]. Discussions of many of these new Raman spectra are included in a review of Cowley in Ref. 64, and in reviews by Patel [106] and Mooradian [107]. The reader is also referred to the published proceedings of two recent international conferences on light scattering in solids, edited by Wright [108] and Balkanski [109].

Here, mention will only be made of a few highlights of recent experiments with semiconductors. Leite and Porto [110] have observed resonant enhancement of Raman scattering for one-, two-, three-, and four-phonon processes in CdS with excitation by 4880 Å which lies near the absorption edge of CdS at $77°K$. An extension of these experiments has led to the observation of up to the ninth-phonon process (Fig. 21) in CdS with excitation at 4579 Å and with resonance at the scattered photon energy. More recently, a magnetic field has been shown to enhance the scattering from single-phonon processes in CdS. The results of Damen and Shah [111] are presented in Fig. 22 where a factor of ~ 30 increase in Raman intensity is observed to occur at ~ 90 kG. This increase is shown to arise by tuning the band-gap

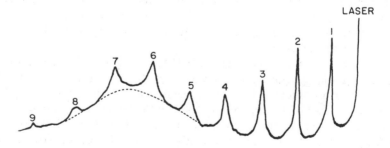

Figure 21.   Resonant Raman spectrum of CdS at ~ 300 K with excitation at 4579 Å. (After Leite, Scott, and Damen [110].)

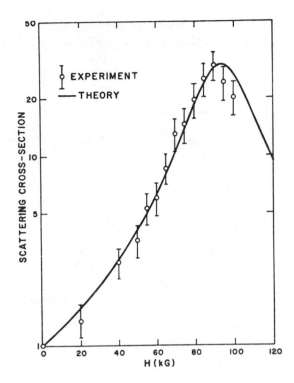

Figure 22.   Raman scattering intensity in CdS and its variation with magnet-ic field--a demonstration of magnetic-field-induced resonance.   (After Damen and Shah [111].)

energy into resonance with the incident photon by application of the magnetic field.

The longest wavelength radiation used for Raman scattering to date is 10.6 μ from the CO_2 laser.   With this radiation, Slusher, Patel, and Fleury [112] studied the Raman scattering from electrons in InSb in the presence of a magnetic field.   They observed transitions between Landau levels ($\Delta \ell = 1$, $\Delta \ell = 2$) and between spin states ($\Delta S = 1$, $\Delta \ell = 0$) of the Landau levels, and noted the effect of the magnetic field.   Spectra at two different magnetic fields are shown in Fig. 23.   It is seen that the Raman shifts of all three bands increase with increasing field, and that the spin-flip ($\Delta S = 1$, $\Delta \ell = 0$) band is more intense and narrower than the $\Delta \ell = 1$, $\Delta \ell = 2$ bands.   Further investigation of this new Raman spectrum led to the development of the tuna-ble spin-flip Raman laser by Patel and Shaw [113, 106].

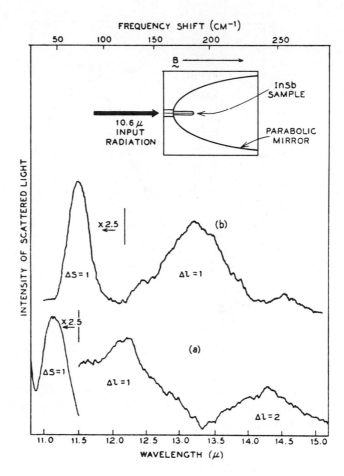

Figure 23.   Raman spectrum from electrons in InSb in the presence of a magnetic field (a) of ~ 26 kG, and (b) of ~ 37 kG.   The inset is a schematic view of the experiment.   (After Slusher, Patel, and Fleury [112].)

E.   Nonlinear and Stimulated Raman Scattering

With the availability of extremely intense radiation from pulsed ruby lasers, many exciting experiments in nonlinear optics have been performed, including some in light scattering.   These have been described in reviews by Fabelinskii [1] and Bloembergen [114], among others.   Three experiments which show promise as useful techniques for the study of matter will be discussed here.

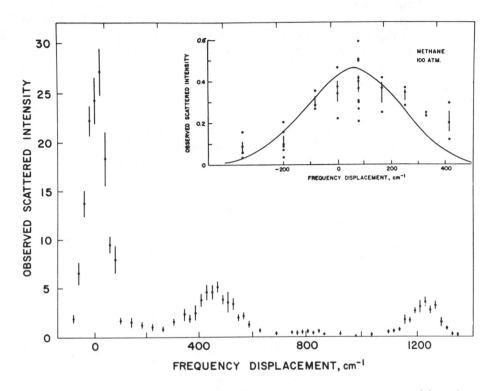

Figure 24. Nonlinear elastic and inelastic scattering spectrum of fused quartz excited with ruby laser radiation at 6940 Å. The inset shows the observed and calculated nonlinear rotational spectrum of CH_4 gas. (After Terhune, Maker, and Savage [115], and Maker [116].)

Terhune, Maker, and Savage [115] found that when a beam of light from a giant-pulse ruby laser is focused in liquids and solids, scattered radiation at nearly twice the laser frequency is obtained. Only about 10^{-13} of the incident radiation is scattered and its intensity varies as the square of the incident intensity. Both elastic and inelastic scattering with frequency displacements characteristic of the medium is produced. An example of nonlinear scattering in fused quartz is shown in Fig. 24. Maker [116] has developed the technique further and has succeeded in obtaining spectra of gases, which are 100-fold less intense.

The dipole moment of a molecule in an electric field is given by $P = \mu + \alpha E + \beta E^2/2 + \gamma E^3/6 + \ldots$, where μ is the permanent dipole moment. When the electric field is that of a light wave, $E = E_0 \sin \omega t$, then P will have components oscillating at ω, 2ω, and 3ω as follows: $P(\omega) = \alpha E + \gamma E^3/8$; $P(2\omega) = \beta E^2/4$; $P(3\omega) = \gamma E^3/24$. Since the intensity of scattering depends on P^2, a measurement of the intensity at 2ω and 3ω gives directly values of β^2 and γ^2, respectively. Similarly, for inelastic (or hyper-Raman) scattering, values of the derivatives of the second- and third-order polarizability with respect to the normal coordinate of the vibration may be determined. Thus, one of the important applications of nonlinear light scattering is in the evaluation of higher-order polarizabilities and their derivatives. Also, selection rules for hyper-Raman spectra will generally differ from those for normal Raman scattering, and will lead to new spectra and new information on molecular structure. For example, the nonlinear spectrum of fused quart (Fig. 24) differs significantly from the known Raman spectrum, and is similar to the infrared spectrum in the 400 and 800 cm^{-1} regions. Also, since CH_4 has spherical polarizability, it does not exhibit a pure rotational Raman spectrum, the selection rule being $\Delta J=0$; however, in the hyper-Raman effect the corresponding selection rule is $\Delta J=0$, ± 1, ± 2, ± 3, and a pure rotational spectrum is observed (Fig. 24). Finally, measurements of the spectral widths of elastic second-harmonic spectra of numerous liquids have led to the determination of orientational relaxation times [116], and these values compare well with those obtained by nuclear magnetic resonance and dielectric relaxation.

Jones and the author [117] have shown that when a molecular medium is irradiated simultaneously by intense monochromatic light of frequency ν_0 and by a continuum of higher frequencies, absorption occurs from the continuum at frequencies $\nu_0 + \nu_M$, where ν_M is a Raman-active frequency of the medium. The absorption lines at $\nu_0 + \nu_M$ arise from the modulation of the electric-field-induced dipole moment by the molecular frequencies ν_M, and

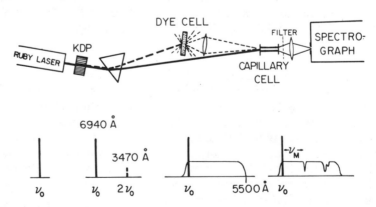

Figure 25. Schematic diagram of experimental arrangement for inverse Raman spectroscopy and corresponding spectra. (After McLaren and Stoicheff [118].)

are the analog of stimulated Raman emission at $\nu_0 - \nu_M$. This "inverse Raman effect" has been observed in liquids and solids. A method [118] for photographing the absorption spectra is shown in Fig. 25.

Radiation at ν_0 is produced by a Q-switched ruby laser, and the required continuum is obtained from the normal fluorescence of a liquid dye excited with second-harmonic radiation (at $2\nu_0$) generated by a KDP crystal. The laser beam and the continuum radiation are then trapped by total internal reflection in a glass capillary containing the scattering liquid. In this way, temporal and spatial coincidence of both light beams is ensured. Spectra have been photographed with single pulses having duration times of ~ 40 nsec. More recently, Alfano and Shapiro [119] have used a different method of generating the continuum, and have pushed the time duration to a few pico-seconds. Inverse Raman spectra may be useful in the study of short-lived species and of transient phenomena including Stark and Zeeman effects with pulsed electric and magnetic fields.

During the process of stimulated Raman emission a substantial popula-tion of molecules in the upper vibrational level may be produced. Ducuing and DeMartini [120] have probed this coherently excited population by observing the time variation of the spontaneous anti-Stokes emission, excited by a second light pulse which is delayed with respect to the initial pulse. In this way, they have measured the vibrational lifetime of gaseous H_2 in the pressure range 20 to 60 kg/cm^2. More recently, with the development of

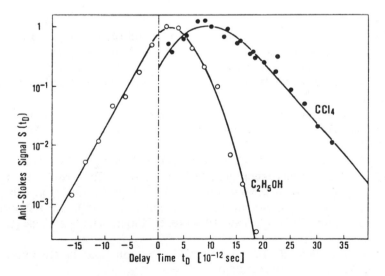

Figure 26.   Measured anti-Stokes signal versus delay time for ethyl alcohol and carbon tetrachloride, used for the direct determination of molecular dephasing times. (After von der Linde, Laubereau, and Kaiser [122].)

mode-locked lasers emitting picosecond pulses, this technique has been used to determine directly the very short lifetimes and dephasing times of optical phonons in condensed matter. Alfano and Shapiro [121] have measured the lifetime of the $1086-cm^{-1}$ optical phonon in calcite at various temperatures. They obtained values of 8.5 psec at $297°K$ and 19.1 psec at $100°K$. Kaiser and co-workers [122] have measured the dephasing time of the coherently excited vibration at $459 \ cm^{-1}$ in carbon tetrachloride. Their measurements are shown in Fig. 26, from which a lifetime of 4.1 psec was deduced. The above exploratory experiments have shown the importance of this new technique for the direct determination of lifetimes and dephasing times of excited states.

CONCLUSION

The many examples cited here of the application of laser sources to light scattering have clearly demonstrated the importance of these remarkable light sources to this field of research, and in addition, the value of Brillouin and Raman spectroscopy in furthering our knowledge of the structure and properties of matter.

REFERENCES

1. I. L. Fabelinskii, Molecular Scattering of Light, (Nauka Press, Moscow, 1965) [English transl. by Plenum Press, New York, 1968.]
2. B. P. Stoicheff, in Methods of Experimental Physics, Vol. 3 Molecular Physics, edited by D. Williams (Academic Press, New York, 1962), p. 111.
3. S. P. S. Porto and D. L. Wood, in Symposium on Molecular Structure and Spectroscopy, Columbus, Ohio, 1961; J. Opt. Soc. Amer. 52, 251 (1962).
4. B. P. Stoicheff, Bull. Can. Assoc. Physicists 17, 11 (1961); in Proc. Xth Colloq. Spectroscopicum Internationale, edited by E. R. Lippincott and M. Margoshes (Spartan Books, Washington, 1963), p. 399.
5. D. I. Mash, V. S. Starunov, E. V. Tiganov, and I. L. Fabelinskii, Zh. Eksp. Teor. Fiz. 47, 783 (1964) [Sov. Phys. JETP 20, 523 (1964)]. (1964)].
6. G. B. Benedek, J. B. Lastovka, K. Fritsch, and T. Greytak, J. Opt. Soc. Amer. 54, 1284 (1964).
7. R. Y. Chiao and B. P. Stoicheff, J. Opt. Soc. Amer. 54, 1286 (1964).
8. G. B. Benedek, in Polarization Matiere et Rayonnement, Livre de Jubile en l'honneur du Professeur A. Kastler, (Presses Universitaires de

France, Paris, 1969), pp. 49–84.

9. H. Z. Cummins and H. L. Swinney, in Progress in Optics VIII, edited by E. Wolf (North-Holland, Amsterdam, 1971), pp. 135–197.

10. B. Chu, Ann. Rev. Phys. Chem. 21, 145 (1970).

11. H. Z. Cummins, Proceedings of the International School of Physics, Enrico Fermi, Course 42, edited by R. J. Glauber (Academic Press, New York, 1969), p. 247.

12. L. Brillouin, Ann. Phys. 17, 88 (1922).

13. S. M. Rytov, Zh. Eksp. Teor. Fiz. 33, 166 (1957); 33, 514 (1957); 33, 669(1957)[Sov. Phys. JETP 6, 130(1958); 6, 401 (1958); 6, 513 (1958)].

14. R. D. Mountain, Rev. Mod. Phys. 38, 205 (1966); J. Res. Nat. Bur. Stand. 70A, 207 (1966).

15. N. G. Van Kampen, Proceedings of the International School of Physics, Enrico Fermi, Course 42, edited by R. J. Glauber (Academic Press, New York, 1969), p. 235.

16. G. B. Benedek and T. Greytak, Proc. IEEE 53, 1623 (1965).

17. G. B. Benedek and K. Fritsch, Phys. Rev. 149, 647 (1966).

18. P. A. Fleury in Physical Acoustics, edited by W. P. Mason (Academic Press, New York, 1970).

19. R. S. Krishnan, in Raman Effect, edited by A. Anderson (Marcel Dekker Inc. New York, 1971), Vol. I, p. 343.

20. D. McIntyre and J. V. Sengers, in Physics of Simple Liquids, edited by H. N. V. Temperley, J. S. Rowlinson, and G. S. Rushbrooke, (North-Holland, Amsterdam, 1968), p. 449.

21. B. P. Stoicheff, in Molecular Spectroscopy, edited by P. Hepple (Elsevier, Amsterdam, 1968), p. 261.

22. G. I. A. Stegeman, W. S. Gornall, V. Volterra, and B. P. Stoicheff, J. Acoust. Soc. Amer. 49, 979 (1971).

23. T. Greytak and G. B. Benedek, Phys. Rev. Letters 17, 179 (1966).

24. G. M. Searby and G. W. Series, Opt. Commun. 1, 191 (1969).

25. G. E. Durand and A. S. Pine, IEEE J. Quantum Electron QE-4, 523 (1968); D. Jackson and E. R. Pike, J. Phys. E. 1, 394 (1968).

26. J. B. Lastovka and G. B. Benedek, in Physics of Quantum Electronics, edited by P. L. Kelley, B. Lax, and P. E. Tannenwald, (McGraw-Hill, New York, 1966), p. 231.

27. E. I. Gordon and M. G. Cohen, Phys. Rev. 153, 201 (1967).

28. E. H. Hara, A. D. May, and H. F. P. Knaap, Can. J. Phys. 49, 420 (1971).

29. A. S. Sugawara and S. Yip, Phys. Fluids 10, 1911 (1967).

30. D. S. Cannell and G. B. Benedek, Phys. Rev. Letters 25, 1157 (1970).

31. H. Z. Cummins and H. L. Swinney, Phys. Rev. Letters 25, 1165 (1970).

32. R. W. Cannell and G. B. Benedek, Phys. Rev. Letters 19, 1467 (1967).

33. N. C. Ford, Jr., K. H. Langley, and V. G. Puglielli, Phys. Rev. Letters 21, 9 (1968).

34. W. S. Gornall, C. S. Wang, C. C. Yang, and N. Bloembergen, Phys. Rev. Letters 26, 1094 (1971).
35. P. A. Fleury and R. Y. Chiao, J. Acoust. Soc. Amer. 38, 1057 (1965).
36. S. Gewurtz, W. S. Gornall, and B. P. Stoicheff, J. Acoust. Soc. Amer. 49, 994 (1971).
37. W. S. Gornall, G. I. A. Stegeman, B. P. Stoicheff, R. H. Stolen, and V. Volterra, Phys. Rev. Letters 17, 297 (1966).
38. E. F. Carome, W. H. Nichols, C. R. Kunsitio-Swyt, and S. P. Singal, J. Chem. Phys. 49, 1013 (1968).
39. R. L. St. Peters, T. J. Greytak, and G. B. Benedek, Opt. Commun. 1, 412 (1970).
40. M. A. Woolf, P. M. Platzman, and M. G. Cohen, Phys. Rev. Letters 17, 294 (1966).
41. P. A. Fleury and J. P. Boon, Phys. Rev. 186, 244 (1969); J. Phys. (Paris) 33, C1-19 (1972).
42. V. S. Starunov, E. V. Tiganov, and I. L. Fabelinskii, ZhETF Pis. Red. 5, 317 (1967) [JETP Letters 5, 260 (1967)]; I. L. Fabelinskii and V. S. Starunov, Appl. Opt. 6, 1793 (1967).
43. G. I. A. Stegeman and B. P. Stoicheff, Phys. Rev. Letters 21, 202 (1968); Phys. Rev. A7, 1160 (1973).
44. M. A. Leontovich, Izv. Akad. Nauk. SSSR Ser. Fiz. 5, 148 (1941) [J. Phys. (USSR) 4, 499 (1941)].
45. V. Volterra, Phys. Rev. 180, 156 (1969).
46. G. D. Enright, G. I. A. Stegeman, and B. P. Stoicheff, J. Phys. (Paris) 33, C1-207 (1972).
47. I. L. Fabelinskii, L. M. Sabirov, and V. S. Starunov, Phys. Letters 29A, 414 (1969).
48. R. H. Katyl and U. Ingard, Phys. Rev. Letters 19, 64 (1967); 20, 248 (1968).
49. M. A. Bouchiat and J. Meunier, Phys. Rev. Letters 23, 752 (1969).
50. S. M. Shapiro, R. W. Gammon, and H. Z. Cummins, Appl. Phys. Letters 9, 157 (1966); V. Chandrasekharan, Proc. Indian Acad. Sci. A33, 183 (1951); J. Phys. 26, 655 (1965).
51. W. S. Gornall and B. P. Stoicheff, Solid State Commun. 8, 1529 (1970); Phys. Rev. B 4, 4518 (1971).
52. B. P. Stoicheff, W. S. Gornall, H. Kiefte, D. Landheer, and R. A. Mc-Laren, Light Scattering in Solids II, edited by M. Balkanski (Flammarion Sciences, Paris, 1971), p. 450.
53. R. W. Gammon and H. Z. Cummins, Phys. Rev. Letters 17, 193 (1966).
54. E. M. Brody and H. Z. Cummins, Phys. Rev. Letters 21, 1263 (1968).
55. I. P. Kaminow and T. C. Damen, Phys. Rev. Letters 20, 1105 (1968).
56. P. D. Lazay, J. H. Lunacek, N. A. Clark, and G. B. Benedek, Light Scattering in Solids I, edited by G. B. Wright (Springer-Verlag, New York, 1969).

57. W. J. Brya, S. Geschwind, and G. E. Devlin, Phys. Rev. Letters 21, 1800 (1968).

58. R. Y. Chiao, C. H. Townes, and B. P. Stoicheff, Phys. Rev. Letters 12, 592 (1964).

59. D. Pohl, M. Maier, and W. Kaiser, Phys. Rev. Letters 20, 366 (1968).

60. G. Winterling and W. Heinicke, Phys. Letters 27A, 329 (1968).

61. W. Heinicke, G. Winterling, and K. Dransfeld, Phys. Rev. Letters 22, 170 (1969).

62. P. J. Hendra and T. R. Gilson, Laser Raman Spectroscopy (Wiley-Interscience, London, 1970).

63. H. A. Szymanski, Raman Spectroscopy (Plenum Press, New York, 1967 and 1970), Vols. 1 and 2.

64. A. Anderson, The Raman Effect (M. Dekker Inc., New York, 1971).

65. G. Placzek, Handbuch der Radiologie, edited by E. Marx, (Akademische Verlagsgesellschaft, Leipzig, 1934), Vol VI, Part 2, p. 205: See also G. W. Chantry, in Ref. 64, p. 49.

66. R. Loudon, Adv. Phys. 13, 423 (1964); 14, 621 (1965).

67. A. Weber, S. P. S. Porto, L. E. Cheeseman, and J. J. Barrett, J. Opt. Soc. Amer. 57, 19 (1967).

68. W. R. L. Clements and B. P. Stoicheff, J. Mol. Spectr. 33, 183 (1970).

69. J. J. Barrett and N. I. Adams, III, J. Opt. Soc. Amer. 58, 311 (1968).

70. R. H. Dicke, Phys. Rev. 89, 472 (1953).

71. P. Lallemand, P. Simova, and G. Bret, Phys. Rev. Letters 17, 1239 (1966).

72. V. G. Cooper, A. D. May, E. H. Hara, and H. F. P. Knaap, Can. J. Phys. 46, 2019 (1968).

73. J. R. Murray and A. Javan, J. Mol. Spectr. 29, 502 (1969).

74. H. B. Levine and G. Birnbaum, Phys. Rev. Letters 20, 439 (1968).

75. J. P. McTague and G. Birnbaum, Phys. Rev. Letters 21, 661 (1968).

76. M. Thibeau, B. Oksengorn, and B. Vodar, Compt. Rend. 265B, 722 (1967); J. Phys. (Paris) 29, 287 (1968).

77. P. M. Lallemand, Phys. Rev. Letters 25, 1079 (1970).

78. J. I. Gersten, Phys. Rev. A 4, 98 (1971).

79. J. P. McTague, P. A. Fleury, and D. B. Du Pre, Phys. Rev. 188, 303 (1969).

80. W. S. Gornall, H. E. Howard-Lock, and B. P. Stoicheff, Phys. Rev. A 1, 1288 (1970).

81. Z. J. Kiss and H. L. Welsh, Phys. Rev. Letters 2, 166 (1959); D. R. Bosomworth and H. P. Gush, Can. J. Phys. 43, 729 (1965).

82. T. C. Damen, R. C. C. Leite, and S. P. S. Porto, Phys. Rev. Letters 14, 9 (1965).

83. Y. Kato and H. Takuma, J. Opt. Soc. Amer. 61, 347 (1971).

84. W. R. L. Clements and B. P. Stoicheff, Appl. Phys. Letters 12, 246 (1968).

85. T. J. Greytak and J. Yan, Phys. Rev. Letters 22, 987 (1969); T. J. Greytak, R. Woerner, J. Yan, and R. Benjamin, Phys. Rev. Letters 25, 1547 (1970).

86. J. W. Halley, Phys. Rev. 181, 338 (1969); M. Stephen, Phys. Rev. 187, 279 (1969).

87. S. P. S. Porto, P. A. Fleury, and T. C. Damen, Phys. Rev. 154, 522 (1967); T. C. Damen, S. P. S. Porto, and B. Tell, Phys. Rev. 142, 570 (1966); S. P. S. Porto, J. A. Giordmaine, and T. C. Damen, Phys. Rev. 147, 608 (1966).

88. P. A. Fleury and J. M. Worlock, Phys. Rev. Letters 18, 665 (1967); 19, 1176 (1967).

89. C. H. Wang, Phys. Rev. Letters 26, 1226 (1971).

90. S. M. Shapiro, D. C. O'Shea, and H. Z. Cummins, Phys. Rev. Letters 19, 361 (1967).

91. J. H. Parker, Jr., D. W. Feldman, and M. Ashkin, Phys. Rev. 155, 712 (1967); Phys. Rev. Letters 21, 607 (1968).

92. J. Hougen and S. Singh, Phys. Rev. Letters 10, 406 (1963); Proc. Roy. Soc. (London) A277, 193 (1964).

93. J. A. Koningstein and G. Schaack, J. Opt. Soc. Amer. 60, 755 (1970); 60, 1110 (1970).

94. J. M. Worlock and S. P. S. Porto, Phys. Rev. Letters 15, 697 (1965).

95. P. A. Fleury, S. P. S. Porto, L. E. Cheesman, and H. J. Guggenheim, Phys. Rev. Letters 17, 84 (1966).

96. R. M. Macfarlane, Phys. Rev. Letters 25, 1454 (1970).

97. P. A. Fleury, S. P. S. Porto, and R. Loudon, Phys. Rev. Letters 18, 658 (1967).

98. P. A. Fleury and H. J. Guggenheim, Phys. Rev. Letters 24, 1346 (1970).

99. C. H. Henry and J. J. Hopfield, Phys. Rev. Letters 15, 964 (1965).

100. S. P. S. Porto, B. Tell, and T. C. Damen, Phys. Rev. Letters 16, 450 (1966).

101. J. P. Scott, L. E. Cheesman, and S. P. S. Porto, Phys. Rev. 162, 834 (1967).

102. C. K. N. Patel and R. E. Slusher, Phys. Rev. Letters 22, 282 (1969).

103. D. L. Rousseau and S. P. S. Porto, Phys. Rev. Letters 20, 1354 (1968).

104. A. Mooradian and A. L. McWhorter, Phys. Rev. Letters 19, 849 (1967); A. Mooradian and G. B. Wright, Phys. Rev. Letters 16, 999 (1966).

105. R. E. Slusher, C. K. N. Patel, and P. A. Fleury, Phys. Rev. Letters 18, 77 (1967).

106. C. K. N. Patel, in this volume.

107. A. Mooradian, in this volume.

108. G. B. Wright, Light Scattering Spectra of Solids (Springer-Verlag, New

York, 1969).

109. M. Balkanski, Light Scattering in Solids (Flammarion Sciences, Paris, 1971).

110. R. C. C. Leite and S. P. S. Porto, Phys. Rev. Letters 17, 10 (1966): R. C. C. Leite, J. F. Scott, and T. C. Damen, Phys. Rev. Letters 22, 780 (1969).

111. T. C. Damen and J. Shah, Phys. Rev. Letters 26, 249 (1971).

112. R. E. Slusher, C. K. N. Patel, and P. A. Fleury, Phys. Rev. Letters 18, 77 (1967).

113. C. K. N. Patel and E. D. Shaw, Phys. Rev. Letters 24, 51 (1970).

114. N. Bloembergen, Am. J. Phys. 35, 989 (1967); see also P. Lallemand, Ref. 64, Chap. 5.

115. R. W. Terhune, P. D. Maker, and C. M. Savage, Phys. Rev. Letters 14, 681 (1965).

116. P. D. Maker, in Physics of Quantum Electronics, edited by P. L. Kelley, B. Lax, and P. E. Tannenwald (McGraw-Hill, New York, 1966), p. 60; Phys. Rev. A 1, 923 (1970).

117. W. J. Jones and B. P. Stoicheff, Phys. Rev. Letters 13, 657 (1964).

118. R. A. McLaren and B. P. Stoicheff, Appl. Phys. Letters 16, 140 (1970).

119. R. R. Alfano and S. L. Shapiro, Chem. Phys. Letters 8, 631 (1971).

120. F. DeMartini and J. Ducuing, Phys. Rev. Letters 17, 117 (1966).

121. R. R. Alfano and S. L. Shapiro, Phys. Rev. Letters 26, 1247 (1971).

122. D. von der Linde, A. Laubereau, and W. Kaiser, Phys. Rev. Letters 26, 954 (1971).

NEW DEVELOPMENTS IN THE RAMAN SPECTROSCOPY OF SOLIDS*

Aram Mooradian
Lincoln Laboratory, Massachusetts Institute of Technology
Lexington, Massachusetts

ABSTRACT

A discussion is given of some recent developments in the experimental techniques and applications of Raman spectroscopy. Some of the new techniques discussed include optical filters, detectors, image tubes, and laser sources. Applications include electronic materials analysis and continuously operating spin-flip lasers.

I. INTRODUCTION

Raman spectroscopy has developed very rapidly in the past few years with the wide spread availability of low-cost reliable lasers (especially argon-ion lasers). Scattering from a large number of new excitations has been observed in a variety of different materials. A number of scattering phenomena have been well understood and are now being used in materials characterization and device development. The present discussion will deal with recent developments in the techniques of laser Raman spectroscopy of solids and their present and future applications to materials characterization and

---

*This work was sponsored by the Department of the Air Force.

device application.   Emphasis here [1] will be primarily on the Raman spectroscopy of semiconductors and its applications.

II. SOME NEW EXPERIMENTAL TECHNIQUES

A. Optical Filters

One of the most severe problems in Raman spectroscopy has been the rejection of stray laser light in the spectrometer which tended to obscure weak Raman lines.   There have been a number of attempts [1] to overcome this problem using optical and electronic techniques.   One of the most successful methods yet devised uses an absorptive gas filter of iodine [2] to selectively reject the single-frequency 5145 Å light from an argon-ion laser used as the exciting source.   The absorption spectrum of iodine vapor at about $50^{\circ}C$ shown in Fig. 1 consists of a large number of electronic absorption lines with a Doppler limited width of about 500 MHz and a separation of about 1 cm^{-1}. When this filter is placed between the sample and the spectrometer, only the laser light is strongly attenuated while the relatively broad Raman lines have a negligible attenuation.   This technique was first used in Brillouin spectroscopy and requires that the argon-ion laser operate in a particular longitudi-

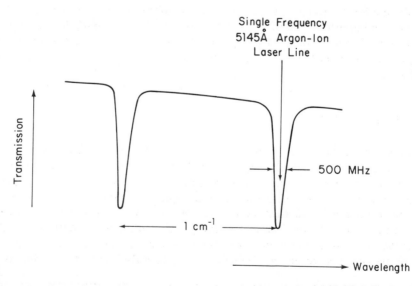

Figure 1.   Electronic absorption of iodine vapor near the 5145-Å argon-ion laser line (not to scale).

nal mode that is coincident with an electronic absorption line.

With the ever increasing development and use of tunable lasers (see below), it should be possible to lock the frequency of a tunable laser to any of an enormous number of sharp absorption lines in gases. The same gas could then be used as a wide field of view narrow–band absorptive rejection filter for the Raman collection optics. For even higher laser frequency stability, it would be possible to lock the laser to the Lamb dip of the gas. Figure 2 shows a possible experimental arrangement for such a system. Some of the uses of this scheme could be in resonance scattering studies and for use with large-f-number compact spectrometer systems for field use. Another application might include use in remote Raman radar systems where the laser would be locked to the absorption line of a gaseous pollutant and the resonance Raman or resonance fluorescence could be detected. Time gating, of course, would be used to reject the laser in such a system.

B. Detectors

The low efficiency of many Raman processes has always posed the problem of developing sensitive detectors over a broad band, especially in the near infrared. Highly sensitive photomultiplier tubes in the visible and near infrared have existed for some time. However, for detecting near–infrared Raman light which occurs when a visible laser scatters from a substance with a large Raman frequency or when a near–infrared laser (i.e., Nd:YAG) is used as the exciting source, a number of new cessiated alloy semiconductor materials (such as $Ga_xIn_{1-x}As$) have been successfully used to improve

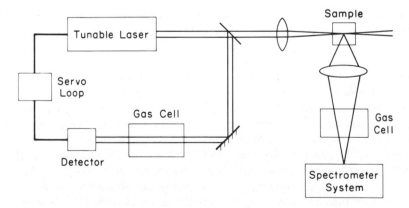

Figure 2. Schematic setup used to lock the wavelength of a tunable laser to the narrow absorption line of a gas. The same gas absorption line is used in the wide field of view filter to reject the laser line.

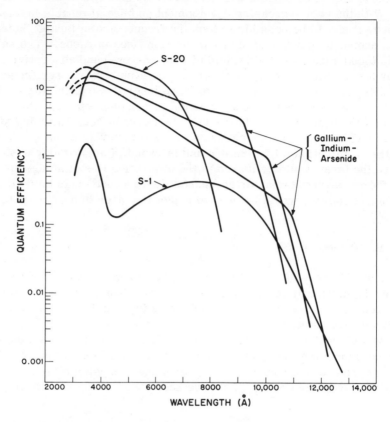

Figure 3.   Quantum efficiency of some commercially available photomultiplier tubes as a function of wavelength.

detector performance in the near infrared.   Figure 3 shows the response
characteristics of a few such commercially available devices.   Because the
semiconductor elements are opaque in their sensitive region, the emitted
photoelectrons are collected from the front surface.   While cooling such
tubes does reduce the dark count to levels desirable for Raman spectroscopy
there does occur a significant shift of the response fall-off to higher energy
as the band gap increases with lower temperature in $Ga_xIn_{1-x}As$.

Another recent development in Raman spectroscopy is the use of image
intensifier tubes to increase the detection sensitivity [3].   The image-tube
system replaces the photographic plate and can record the entire Raman
spectrum continuously.   In contrast to the old techniques of photographic film
the advantages of the image-tube technique is the linearity of response, the

real-time readout, and a greater range of response than photographic film. In contrast to using a simple photomultiplier tube and scanning the spectrometer wavelength, the image-tube system can achieve a sensitivity increase of two orders of magnitude by recording the same wavelength range all at once. There is an additional advantage of electronic control of wavelength coverage and resolution. Figure 4 shows a schematic arrangement of such a system. The light from the dispersed spectrum on the photoemitter of the image intensifier causes electrons to be emitted. The emitted electrons are accelerated and then strike a phosphor which emits many more photons than were originally incident. The image dissector tube then scans and stores the image on the phosphor. Photon-counting techniques can be employed in the image tube for maximum sensitivity. In such a system, the signal will grow out of the noise since the noise increases as the square root of the integration time while the signal grows linearly. Note that the spectral response of the image intensifier tube can be anywhere including the near infrared but that the phosphor and the image dissector tube response is only in the most sensitive part of the visible. The image-tube system must be used with a spectrometer that can display the entire dispersed spectrum, thus excluding the use of a usual double spectrometer.

Avalanche photodiodes may some day replace the photomultiplier tubes as sensitive low-frequency detectors if the multiplication factors can be increased enough to overcome amplifier noise. Their major advantages include high quantum efficiency (about 50%), infrared response, and compact size.

C. Laser Sources

A large number of fixed-frequency lasers have successfully been used as a

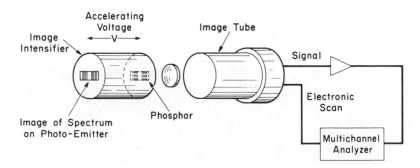

Figure 4. Arrangement of image intensifier system for recording of Raman spectra.

Raman excitation source. While the most popular fixed-frequency laser has been the argon-ion laser, the Nd:YAG laser has been particularly useful for scattering from semiconductors which are transparent in the near infrared.

One of the more exciting recent developments in laser spectroscopy has been tunable lasers. Tunable lasers are particularly important for the study of resonant Raman effects and high-resolution spectroscopy. Dye lasers and parametric oscillators can span the range from the near ultraviolet to the near infrared and are commercially available. The use of these devices in the near future will be made to provide careful studies of resonant effects in various materials.

Semiconductor injection lasers have been made to operate from about 6000 Å to over 35 μ in the infrared by a suitable choice of laser material. Injection lasers made from GaAs, $Ga_{1-x}As_xP$, and $Ga_{1-x}Al_xAs$ are commercially available at relatively low cost and can operate with enough (several milliwatts) power in a single axial mode to provide a temperature-tunable source for Raman spectroscopy [1]. External frequency control optics on injection lasers might provide even more single-frequency power which would be tunable over a few hundred Å for each alloy composition at a fixed temperature [4].

A recent advance [5] in the technology of semiconductor laser devices has been achieved by using large arrays of commercially available GaAs diode lasers to optically pump a rod of InSb. The InSb rod (0.2 × 0.2 × 10 mm) was placed on a heat sink between two linear arrays of GaAs diode lasers (25 diodes per array) capable of producing nearly 10 W of average power at ~ 8500 Å (see Fig. 5). The short wavelength pump light is absorbed on opposing surfaces of the crystal and the excited carriers then diffuse into the bulk of the crystal providing nearly uniform excitation within the sample volume. Laser action occurred in the InSb in a large number of axial modes near 5.2 μ with the entire package at liquid helium temperature. The InSb laser threshold occurred almost immediately after the onset of laser action

Figure 5. Photograph of GaAs diode laser array used to pump a rod of InSb. The dimensions of the rod corresponds to a unity Fresnel number.

in the GaAs arrays (less than 1 A for the series-wired arrays near liquid helium temperature). Present indications are that such a device is capable of power conversion efficiencies at low temperature of a few percent. Some of the advantages of such a device are simplicity of fabrication, high average power compared to presently available diode lasers of the same material (a few hundred milliwatts from this InSb device presently seems possible), and the simplicity of operation characteristic of diode lasers. The wavelength range of such a device can in principle be extended by using a number of other semiconductor crystals in place of the InSb, such as $Hg_xCd_{1-x}Te$, $Pb_xSn_{1-x}Te$, $Pb_xS_{1-x}Se$, etc. Single-frequency operation has yet to be achieved to provide a usable source for spectroscopy studies.

III. APPLICATIONS OF RAMAN SPECTROSCOPY

A. Materials Analysis

Raman spectroscopy has always been used most extensively as an analytic tool. With the use of lasers as a Raman excitation source, light scattering from a large class of new materials and new excitations has been observed and understood. Many of the new class of phenomena can readily be used in the analysis of materials. A particularly interesting example of such a system is the electronic properties of semiconductor crystals.

Raman scattering from electronic excitations in semiconductors has been studied in great detail [1]. From these results it is now possible to use the scattering properties of an electron gas in, say, GaAs to make fairly accurate determinations of the electron density as a function of the position within the crystal. Figure 6 shows a series of Raman spectra of GaAs with various doping levels taken at room temperature using a Nd:YAG laser operating at 1.06 μ. The scattering from both the single particle and collective mode excitations has been accurately calibrated relative to the intensity of the TO phonon mode which is not affected by the presence of electrons in the crystal. A careful comparison of the variation of electron scattering efficiency has been made with Hall data. From these results and the experimental accuracy of the Raman data, it is possible to determine the electron concentration in GaAs with an absolute accuracy of between two and three significant figures. This determination can be made at any point within an ingot of GaAs without the need of any external electrical leads or any destruction of the sample. Reasonably good optical surfaces are necessary to maintain the image quality within the sample. It is possible with the state-of-the-art instrumentation to detect electron concentrations in GaAs to less than $10^{12}cm^{-3}$ in a probe volume 50 μ in diameter by 1 mm long using a 5-W CW

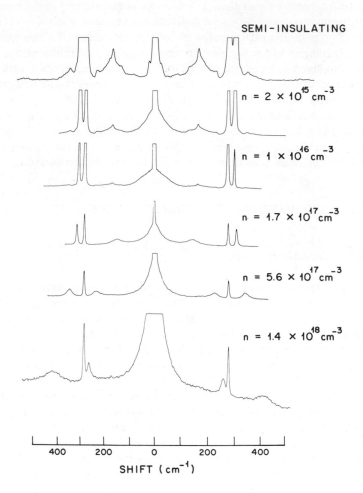

Figure 6.   Raman spectra of n–type GaAs samples at room temperature taken using a 1.064-μ Nd:YAG laser. Stokes components are to the right. There is an arbitrary gain for each of the traces.

Nd:YAG laser at 1.06 μ. In the concentration range $< 10^{12}$cm^{-3} to 10^{16}cm^{-3} the single particle electron scattering would serve as the monitor and for concentrations over 10^{16}cm^{-3} scattering from the collective (plasmon-phonon) modes would be the best indicator. Monitoring of electron concentrations well above room temperature would also be possible. A similar type of probing of CdTe, InP, AlSb, ZnTe, ZnSe, CdS, CdSe, etc., and alloys of these materials is also possible. However, GaAs is the only material whose

electron scattering has been carefully measured and compared with theory
[1].   Because of the ever increasing use and production of GaAs and related
electronic semiconductor crystals for optoelectronic devices, such nonde-
structive testing techniques may become quite useful in materials evaluation.

Electronic Raman scattering from impurity states in semiconductor
crystals has also been observed [6,7] in which the Raman frequency shift
corresponds to the energy difference between the ground state and one or
more excited states.   Figure 7 shows a Raman spectrum of cadmium-doped
$(p \sim 1 \times 10^{16}\,cm^{-3})$ GaAs taken at 1.6°K using a Nd:YAG laser.   The sharp
peak at $\sim 200\ cm^{-1}$ corresponds to a transition from the ground state to an
excited state which is about half the ionization energy for this acceptor (as
determined from Hall data).   The continuum near the laser line (as well as
continua near the main excited state lines) has been interpreted by Henry,
Hopfield, and Luther [6] in terms of an acoustic phonon sideband associated
with an electronic transition.   This continuum scattering near the laser line
persists and even increases somewhat in intensity as the temperature
increases up to room temperature.   At 300°K, this scattering has been interpreted as intravalence-band single-particle hole scattering [1].   Recently,
Auyang, Wolff, and Lax [8] have proposed a scattering process in p-type III-
V semiconductors in which a hole makes a transition between the degenerate
(at k = 0) light and heavy hole bands.   It is not clear why the single-particle
hole scattering should persist down to 1.6°K if indeed all of the free holes
are frozen out onto isolated acceptor states.   Since the linewidth of single-
particle electron (hole) scattering is proportional to the energy of the pump

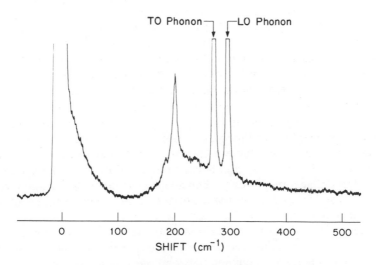

Figure 7.   Electronic Raman scattering from cadmium impurities in GaAs at
1.6°K.   Hole concentration at room temperature is $1 \times 10^{16} cm^{-3}$.

photon it might be possible to distinguish this type of mechanism by using a number of different pump frequencies from a Nd:YAG laser or even a tunable laser.

It is interesting to note that there is an appreciable broadening of the electronic Raman lines as the concentration (and hence overlap of the impurity states) increases. Despite this, it is possible to characterize the net acceptor concentration, $N_A - N_D$, up to 10^{19} cm^{-3} in GaAs. The lower limit of detection is $< 10^{13}$ cm^{-3}. Again the concentration is determined in the bulk by comparing the relative intensity to the optical phonon mode intensity. Scattering from Cd, Zn, Mg, and Mn acceptors has been observed in GaAs, and a number of different donors and acceptors in silicon. These measurements usually require low temperature while the hole concentration measurements by Raman scattering can be made at room temperature or higher.

B.   Device Applications

A number of interesting devices using stimulated Raman scattering have recently been demonstrated. Some of these Raman lasers can be continuously tuned (such as the spin-flip Raman laser and the polariton Raman laser) over an appreciable range in the infrared region. In the past, most Raman lasers have required high peak pump power (on the order of several kilowatts for liquids such as nitrobenzene and CS_2) to achieve threshold. The first continuously operating Raman laser of any kind has recently been reported [9] using a cw carbon monoxide laser to pump spin-flip excitations in InSb. Thresholds [10] below 50 mW of pump power within the uncoated crystals were measured. In contrast, the first reported observation of stimulated spin-flip scattering [11] used a Q-switched CO_2 laser to get thresholds of a few hundred watts within the crystal. The main reason for such a low Raman threshold is the use of excitation frequencies near the energy band gap of InSb where there occurs an enormous resonance enhancement to the scattering efficiency. The experimentally measured resonance enhancement is shown compared [12] to theory in Fig. 8. Because of the increased absorption near the band gap, the net Raman gain did not vary more than a factor of about 5 for pump wavelengths between 5.1 and 6.5 µ. The phenomenon of scattering enhancement near a band-gap resonance occurs generally in most solids and more cw Raman oscillators can be expected in the future by near resonant pumping.

Figure 9 shows a recorder spectrum taken of a cw spin-flip Raman laser both below (lower trace) and above (upper trace) threshold. Note that the upper trace has no gain change between the laser and the first Stokes peak. For this case of collinear stimulated scattering, pump depletion is clearly evident. Conversion efficiencies in excess of 75% with cw output power in

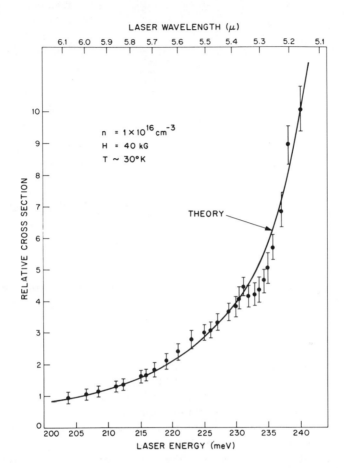

Figure 8. Resonant enhancement of $\Delta\ell = 0$ spontaneous spin–flip scattering in InSb.

excess of 1 W have been reported [10]. Because of such low threshold power densities near resonance, saturation occurs from pump depletion rather than spin saturation, thus allowing the high conversion efficiencies. A CO TEA laser with a few hundred watts in a single line has also been used [13] to excite as many as three stimulated Stokes components in InSb with comparable efficiency in the first Stokes component.

Figure 10 shows the fine–tuning characteristics of a cw spin–flip laser operating in a collinear geometry. Both mode hopping and mode pulling are observed and have been accounted for in terms of index and gain peak varia-

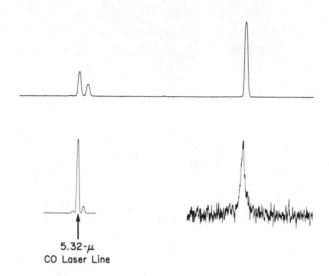

5.32-μ
CO Laser Line

Figure 9.   Recorder traces of spontaneous (lower trace) and cw–stimulated (upper trace) collinear spin–flip scattering in InSb for H = 35 kG.   Note there is no scale change between the laser and the stimulated line.

tion with magnetic field.   Oscillation occurred only in a single axial mode since the spontaneous linewidth [14] for collinear scattering ($\lesssim$3–5 GHz) is less than the mode spacing in this case ($\sim$ 6. 2 GHz).   The fine frequency tuning characteristics appear to make the spin–flip Raman laser a promising though relatively complicated device for high resolution and ultrahigh resolution (Lamb dip) spectroscopy in the future.

Because of the relatively low thresholds for the cw spin–flip scattering in InSb, it might be possible to pump such a device with a semiconductor laser.   The GaAs diode laser array pumped InSb rod described in Sec. II C is a possible candidate in which to achieve a spin–flip Raman laser pumped collinearly by the band–gap laser radiation.   In such a case the pump frequency would always remain as close as possible to resonance.   For the case of InSb, the band–gap radiation and the spin–flip output would tune away from the zero magnetic field laser energy at about equal rates in opposite directions.

ADDENDUM (SEPTEMBER, 1972)

An InSb spin–flip Raman laser pumped continuously by a carbon monoxide

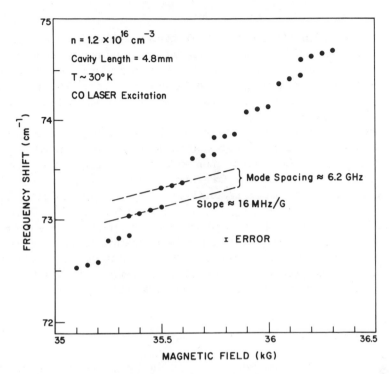

Figure 10.   Frequency dependence of single axial mode output of cw spin–flip Raman laser as a function of magnetic field.

laser near 5.2 μ has produced Stokes pulses of $< 10^{-9}$ sec in duration when an electric field of a few V/cm was applied or removed [15] from the sample. This effect is interpreted in terms of a transient Raman gain which occurs when electrons are heated or allowed to cool in the lower spin state of the conduction band.   Because of the nonparabolicity of the InSb conduction band, the spin–flip frequency of an electron is dependent upon its kinetic energy, ϵ, along the magnetic field.

$$\omega_s = \omega_{so}(1 - 2\epsilon/\epsilon_g)$$ (1)

where ω_{so} is the spin frequency at the bottom of the band and ϵ_g is the band gap energy.   The fractional change of the spontaneous Raman frequency for a Maxwellian distribution having an electron temperature, T_e, and neglecting drift effects is

$$\langle (\omega_{so} - \omega_s)/\omega_{so} \rangle = kT_e/\epsilon_g \qquad (2)$$

For electric fields of a few V/cm, frequency shifts of several GHz can occur which are comparable or larger than the unloaded cavity mode frequencies of the spin-flip laser. A transient pulse occurred whenever the spontaneous frequency was swept through one of the cavity modes. This effect can be used to provide tunable, fast ir pulses or an FM laser with only a few tens of millivolts of drive.

The spontaneous linewidths of spin-flip scattering have been measured [16] in InSb with extremely high resolution using small-signal gain in a Raman amplifier. One carbon monoxide laser at frequency ω_1 was used as a pump and a second laser with frequency ω_2 was used as a probe. The line-shape was measured with a resolution of a few tens of kilohertz (field stability) as the spin frequency, ω_s, was tuned through $\omega_1 - \omega_2$. At $T = 1.5^\circ K$, linewidths as small as 100 MHz were observed in samples with electron concentrations in the range $10^{14} - 10^{15}$ cm^{-3} in magnetic fields as low as several hundred gauss. Detailed line-shape studies are important in understanding the spin-flip line broadening mechanisms and characterizing the operation of the InSb spin-flip Raman laser such as tuning and threshold.

REFERENCES

1. A more detailed account of some of the techniques and physical interpretation of the Raman spectroscopy of semiconductors appears in the article by A. Mooradian, Laser Handbook, edited by F. T. Arrechi and E. O. Schultz-DuBois (North-Holland Publishing Company, Amsterdam, 1972).
2. G. E. Devlin, J. L. Davis, L. Chase, and S. Geschwind, Appl. Phys. Letters 19, 138 (1971).
3. M. Bridoux, Revue d'Optiques 46, 389, (1967); C. M. Savage and P. D. Maker, Appl. Opt. 10, 965 (1971); S. Mende, Appl. Opt. 10, 829 (1971); C. Hilsum, S. Fray, and C. Smith, Solid State Commun. 7, 1057 (1969).
4. H. D. Edmonds and A. W. Smith, IEEE J. Quantum Electron. 6, 356 (1970).
5. A. Mooradian and J. Rossi, Lincoln Lab. Solid State Research Report 2, 16 (1972).
6. C. H. Henry, J. J. Hopfield, and L. C. Luther, Phys. Rev. Letters 17, 1178 (1966).
7. G. B. Wright and A. Mooradian, in Proceedings of Ninth International Conference on Physics of Semiconductors (Publishing House, Nauka, Moscow, 1968) Vol. II; Phys. Rev. Letters 18, 608 (1967).

8. Y. C. Auyang, P. A. Wolff, and B. Lax, Bull. Am. Phys. Soc. 16, 442 (1971).
9. A. Mooradian, S. R. J. Brueck, and F. A. Blum, Appl. Phys. Letters 17, 481 (1970).
10. S. R. J. Brueck and A. Mooradian, Appl. Phys. Letters 18, 229 (1971).
11. C. K. N. Patel and E. D. Shaw, Phys. Rev. Letters 24, 451 (1970).
12. S. R. J. Brueck and A. Mooradian, Phys. Rev. Letters 28, 161 (1972).
13. H. Fetterman, W. Barch, and A. Mooradian, M. I. T. Lincoln Laboratory Solid State Research Report (1971:3), p. 22.
14. S. R. J. Brueck, Ph. D. thesis (MIT, 1971).
15. A. Mooradian, S. R. J. Brueck, E. J. Johnson, and J. A. Rossi, Appl. Phys. Letters 21, 482 (1972).
16. S. R. J. Brueck and A. Mooradian (to be published).

RAMAN SPECTRA OF SOLIDS INVOLVED IN

ORDER-DISORDER PHASE TRANSITIONS

W. G. Harter and S. P. S. Porto
University of Southern California
Los Angeles, California

Theoretical discussions [1, 2] of second-order phase transitions in solids give the impression that a crystal lattice attains a higher or lower symmetry when it is passing through the transition. Usually there is a low-symmetry phase which is called the <u>ordered phase</u> because there is some crookedness or ordering in the lattice at a given direction which is often associated with some sort of dipole moment along that direction. In the other phase which is called the <u>disordered phase</u> it is claimed that the lattice loses the asymmetry or ordering it had before and thereby attains a higher space-group symmetry. It is claimed that the symmetry group H of the ordered phase is a subgroup of the symmetry group G of the disordered phase [2]. Landau [3] has derived criteria which tell whether certain G(H) group-subgroup pairs could be involved in a second-order phase transition. For example, any pair in which G has three times the number of elements of H is forbidden, while a pair in which G has twice the size of H is allowable.

It is fairly certain that an example of a second-order phase transition is found in $NaNO_2$ [4]. In the high-symmetry phase $NaNO_2$ is thought to have a crystalline form that is schematized [5] in Fig. 1(a) with symmorphic symmetry $G = D_{2h}^{25}$, while in the low-temperature phase, Fig. 1(b), the symmetry is $H = C_{2v}^{20}$ which is half as big as $G = D_{2h}^{25}$. These ideas about the form of $NaNO_2$ come from x-ray crystallography studies [5].

It has been assumed that once we are given information such as the above along with appropriate lattice constants one could perform, using one or more

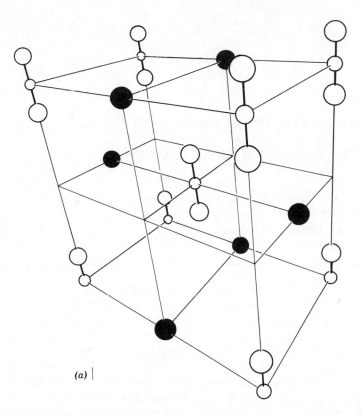

(a)

Figure 1(a). Hypothetical crystal structures of NaNO$_2$ deduced by x-ray diffraction. (a) Disordered phase has D$_{2h}^{25}$ symmetry. Sodium atoms (black objects) are centered between nitrite sites.

of existing models, a lattice dynamics calculation for either phase and reproduce the phonon spectrum with reasonable success. Short of this one could use the irreducible representations of the group G or H depending upon which phase was studied, to label the modes of small oscillation, and thereafter derive selection rules for infrared and Raman light scattering. Indeed the rules would necessarily be more selective in the high–temperature phase since we are assuming it has the higher symmetry G. Those ideas could be thoroughly tested in the Raman and infrared experiments of Hartwig et al. [6] in which the spectra of NaNO$_2$ were obtained at many different temperatures at either side of the λ point (164°C). At temperatures below T$_\lambda$ the k = 0 phonon spectra matched well with the expected modes which we have

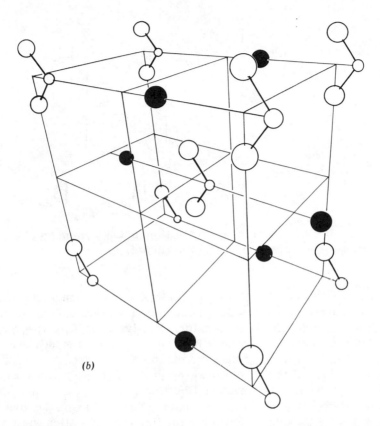

(b)

Figure 1(b). Ordered phase has C_{2v}^{20} symmetry. Nitrite configurations are isoceles triangles, and sodium is moved off center.

classified by irreducible representations of little group C_{2v}. Eight of the twelve modes gave visible Raman lines.

Above T_λ the symmetry is supposedly D_{2h}^{25} and among the $k = 0$ modes possible in a crystal of the form shown in Fig. 1(a), only those labeled A_{1g}, B_{1g}, and B_{2g} by the little group D_{2h} (Fig. 2) should be Raman active. However our spectra continued to show at least six Raman lines clearly, even as the temperature was raised 100 degrees above T_λ. Just below T_λ two close lying lines became degenerate and were not resolved thereafter as the temperature was raised.

We conclude, in agreement with Hartwig et al. , that $NaNO_2$ cannot have the form pictured in Fig. 1(a) at any temperature except in an average sense. The x-ray picture is an average structural picture, and hence misleading.

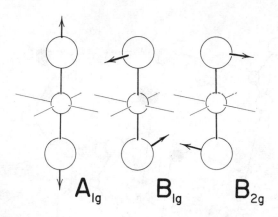

Figure 2.   If $NaNO_2$ had the configuration of D_{2h}^{25} symmetry, only the three k = 0 modes shown here would be Raman active.

Any theoretical model that starts with this picture cannot be expected to give correct analysis of the lattice dynamics or the electronic structure.

The NO_2 triangle does not deform easily; in fact the frequencies for NO_2 internal modes are greater by a factor of 5 than modes that mostly involve relative motion of the NO_2 and the Na. Therefore it appears that the correct picture of the high-symmetry phase of $NaNO_2$, at least for optical-frequency experiments, involves libration of the NO_2 triangles, each one capable of changing direction depending more or less on the direction of its neighbors. If the dispersion curve for this type of motion were fairly flat, Fourier combinations of various "libration waves" of different wavelength could be made that consisted of fairly large microdomains of low group velocity within which phase and amplitude of libration were fairly uniform. If also the average period of libration was much lower than phonon lifetimes, a Raman process could occur in an environment that was <u>not much different than that of the low-symmetry phase.</u>

These conditions may occur in $NaNO_2$ and in room-temperature $CaCO_3$ and $NaNO_3$; they surely occur for the disordered phases of the amonium halides, but there are many solids for which the situation and the resulting Raman spectrum will be very different. We believe that order-disorder occur with similar character in $SrTiO_3$, $BaTiO_3$, TiO_2, VO_2, and $KTaO_3$. In those crystals the microdomain structure may disappear almost completely and the spectral lines will show a large broadening because conservation of k is not necessary. In other solids, such as KDP, there is evidence that the tunneling frequencies are comparable to the optical phonon frequencies and a vibronic interaction takes place. In this case we expect to find extreme broadening at low frequencies and in a way the Raman spectrum of the disordered phase of KDP should resemble the low–frequency Raman spectrum of glasses.

In short, in any of these solids which undergo phase transitions, x–ray structural data or any structural information gained by averaging over times greater than 10^{-9} sec will undoubtedly be of no analytical value. Instead the structure of these solids must be deduced from Raman scattering, infrared absorption, or other means in which $\tau < 10^{-9}$ sec.

It is perhaps not surprising to find that "average pictures" are often meaningless for dynamic analysis of a solid in the disordered phase. For example, in the alloy of Cu and Zn, the high–symmetry phase is supposed to consist of a perfect body–centered–cubic lattice, in which the probability of finding a Cu or a Zn atom is equal at all lattice points. Clearly then one cannot then say that CuZn has O_h^9 symmetry in the ordinary sense.

REFERENCES

1. L. D. Landau and E. M. Lifshitz, Statistical Physics (Addison–Wesley, Reading, Mass., 1958).
2. J. L. Birman, Ferroelectricity: General Motors Symposium, edited by E. F. Weller (Elsevier, New York, 1967).
3. Ref. 1, p. 437.
4. S. Sawada, S. Nomura, S. Fujii, and I. Yoshida, Phys. Rev. Letters 1, 320 (1958).
5. I. Shibuya, J. Phys. Soc. Japan. 16, 490 (1961).
6. W. Hartwig, E. Wiener–Avnear, and S. P. S. Porto, Phys. Rev. 5, 79. 1972.

LASER MAGNETOREFLECTION SPECTROSCOPY[*]

M. S. Dresselhaus
Department of Electrical Engineering, and Center for Materials
Science and Engineering, Massachusetts Institute of Technology
Cambridge, Massachusetts

ABSTRACT

The basic principles involved in magnetoreflection spectroscopy are reviewed and the utility of laser sources for these investigations is discussed. Specific applications are made to the study of (1) interband Landau level transitions for the investigation of electronic-energy-band structure, (2) the optical Shubnikov-de Haas effect to establish its characteristic features and physical mechanisms, and (3) far-infrared transitions involved in cyclotron resonance, electron spin resonance, polaron effects in cyclotron resonance, shallow impurity states in semiconductors, and electron paramagnetic resonance.

LASER MAGNETOREFLECTION SPECTROSCOPY

There are several basic reasons why magneto-optical studies of solids are significant to the field of solid-state physics on one hand, and to the field of

---

[*]Work supported in part by the Advanced Research Projects Agency.

laser physics on the other. Because of the large Landau level separations and large spin splittings which occur in solids at modest magnetic fields, the magnetic-energy-level structure provides a basis for the fabrication of lasers, tunable over a wide frequency range. Furthermore, the resonant property of the density of states in a magnetic field permits optical transitions to be studied under resonant conditions. Thereby an accurate determination can be made of the electronic-energy-band parameters which characterize the energy-band structure.

We review in Sec. I the fundamental concepts which dominate magnetooptical studies of solids. Application of magnetooptic techniques and lasers to the study of interband Landau level transitions and energy-band structure of solids is made in some detail in Sec. II by considering the problem of the four-coupled π bands in graphite. The application to the optical Shubnikov-de Haas effect is considered in Sec. III. Finally, in Sec. IV we briefly review a few of the applications that have been made of far-infrared lasers to studies of magnetic energy levels in solids, using techniques such as cyclotron resonance, electric dipole excited electron spin resonance, polaron effects in the cyclotron resonance, shallow impurity levels in semiconductors, and electron paramagnetic resonance.

I. INTRODUCTION AND GENERAL THEORY

The eigenstates of a free electron in a magnetic field are plane-wave states for motion along the magnetic field and harmonic oscillator states for motion in the plane perpendicular to the magnetic field. The corresponding electronic energy eigenvalues for the magnetic field directed along the z direction,

$$E_{n,s}(k_z) = \frac{\hbar^2 k_z^2}{2m} + \hbar \omega_c (n + \frac{1}{2} + s) \tag{1}$$

are expressed in terms of the three quantum numbers n, s, and k_z, which are, respectively, the harmonic oscillator index, the electron spin $= \pm 1/2$, and the component of the wave vector along the magnetic field. These energy levels are completely degenerate in the quantum number k_y, denoting the component of the wave vector in the plane perpendicular to H_z. For each value of n, the magnetic energy levels have a simple parabolic k_z dependence. At a fixed k_z value, the magnetic energy level spacing is $\hbar \omega_c$, where ω_c is the cyclotron frequency $\omega_c = eH/mc$. With an electron g factor of 2, this magnetic-energy-level spacing of $\hbar \omega_c$ is maintained, whether or not the electron spin is considered explicitly, as is shown in Fig. 1. The equality of the spin splitting $g\mu_B H$ (where μ_B is the Bohr magneton) and the orbital

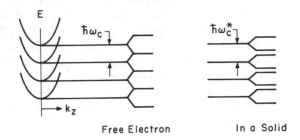

Figure 1. Electronic energy levels in a magnetic field for free electrons and for electrons in a solid.

level separations $\hbar w_c$ ensures that the spin-up magnetic energy level with index n is exactly degenerate with the spin-down level with index (n+1).

In a solid, the electronic energy levels are no longer the simple free-electron states given by Eq. (1); however, in the effective mass approximation, these eigenvalues can be represented in an approximate way by rewriting Eq. (1) in terms of two effective mass quantities and an effective g factor

$$\mathcal{E}_{n,s}(k_z) = \frac{\hbar^2 k_z^2}{2m_z^*} + \hbar w_c^* (n + \frac{1}{2} + \frac{1}{2} g_{eff} s) \qquad (2)$$

in which m_z^* is the effective mass component along the magnetic field and is associated with plane-wave motion in this direction, and the effective cyclotron frequency is $w_c^* = eH/m_c^* c$. Since the cyclotron effective mass, which dominates the electronic motion in the plane perpendicular to the magnetic field, is small for many semiconductors (e. g., $m_c^* = 0.014m$ for the conduction electrons in InSb), sizeable Landau level separations, $[\mathcal{E}_{n+1,s}(k_z) - \mathcal{E}_{n,s}(k_z)] = \hbar w_c^*$, can be achieved with modest magnetic fields. It is this property of the magnetic energy levels in a solid that is exploited in producing magnetically tuned lasers.

Introduction of the effect of electron spin in a solid is a more complicated matter than for the free electron and the Landau level spacing $\hbar w_c^*$ will in general be different from the spin splitting $g_{eff} \mu_B H$ as is seen in Fig. 1. If the spin-orbit interaction in the solid is small, then the effective g factor assumes approximately the free-electron value of 2. Thus, for a material like graphite, which for $\vec{H} \parallel c$ axis has a Landau level separation about 20 times larger than the free-electron separation of $\hbar w_c$, the spin splitting

$g_{eff}\mu_B H$ is nevertheless very close to the free-electron value of $2\mu_B H$. On the other hand, for the situation of two tightly coupled bands and large spin-orbit interaction (the energy gap is small compared with the spin-orbit interaction energy), the effective g factor assumes the value of $g_{eff} = 2m/m_c^*$. In this case, the Landau level spacings and spin splittings are equal as they are in the free-electron situation, except that the tightly coupled bands tend to have small effective masses and large effective g factors. In general, the Landau level spacing and spin splitting for a solid will be unequal, as illustrated in Fig. 1, with the Landau level separation governed by the effective mass tensor for an electron in band j

$$\left(\frac{1}{m_j^*}\right)_{\alpha\beta} = \frac{\delta_{\alpha\beta}}{m} + \sum_\mu \frac{v_{j\mu}^\alpha v_{\mu j}^\beta + v_{j\mu}^\beta v_{\mu j}^\alpha}{\mathcal{E}_j(0) - \mathcal{E}_\mu(0)} \tag{3}$$

while the spin splitting is governed by the effective g factor

$$\left(g_j^{eff}\right)_{\alpha\beta} = 2\delta_{\alpha\beta} + \frac{m}{i}\sum_\mu \frac{v_{j\mu}^\alpha v_{\mu j}^\beta - v_{j\mu}^\beta v_{\mu j}^\alpha}{\mathcal{E}_j(0) - \mathcal{E}_\mu(0)} \tag{4}$$

where $\mathcal{E}_j(0)$ and $\mathcal{E}_\mu(0)$ are the band energies at the extrema in bands j and μ. Each of these tensors differs from the free-electron values $\delta_{\alpha\beta}/m$ and $2\delta_{\alpha\beta}$ because of the coupling of band j to other electronic energy bands μ through the velocity matrix element $\vec{v}_{j\mu}$. It is the different dependence of these two tensors on the velocity matrix elements that is responsible for the different values of the Landau level separations and spin splittings in solids. The interband coupling which is responsible for the correction terms to the free-electron g factor produces a mixing of the spin-up and spin-down states away from the band extrema, thereby making possible an electric dipole transition between a spin-split doublet, a situation which is forbidden in the free-electron case. Not only are the effective mass tensors and effective g factors anisotropic, but the effective g factor for electrons can be of either sign, as opposed to the free-electron case where it is energetically favorable for the electron spin to align antiparallel to the applied magnetic field.

The occurrence of small effective masses and of correspondingly large Landau level separations provides one reason for the interest in the magnetic-energy-level structure of solids for laser applications. A second attractive property of the magnetic-energy-level structure is the resonant behavior of optical transitions in a magnetic field. The probability for making optical transitions can be understood in a general way from Fermi's Golden Rule

$$W_{ij} = \frac{2\pi}{\hbar} \, \rho\,(E) \, |\mathcal{H}'_{ij}|^2 \tag{5}$$

where the perturbation Hamiltonian for the interaction between the electron and the electromagnetic field is $\mathcal{H}' = -\,(e/mc)(\vec{p} \cdot \vec{A})$. Although this perturbation couples different electronic states, the matrix element $\mathcal{H}'_{ij}$ does not generally exhibit the kind of resonant behavior that can occur in the density of states. In the absence of a magnetic field, the density of states merely exhibits a threshold at the band edge $\mathcal{E}_c$

$$\rho\,(\mathcal{E}, 0) \sim (\mathcal{E} - \mathcal{E}_c)^{1/2}; \quad \mathcal{E} > \mathcal{E}_c$$
$$= 0 \quad\quad\quad ; \quad \mathcal{E} < \mathcal{E}_c \tag{6}$$

On the other hand, the density of states in a magnetic field

$$\rho\,(\mathcal{E}, H) = \frac{1}{\pi^2} \, \frac{eH}{\hbar c} \sum_{n, s} \left(\frac{\partial \mathcal{E}_{n, s}\,(k_z)}{\partial k_z} \right)^{-1} \tag{7}$$

contains a singularity at the extremum of each magnetic subband, labelled by the Landau level index n. A graphical display of the threshold behavior of the density of states in zero magnetic field $\rho\,(\mathcal{E}, 0)$ is given by the continuous curve in Fig. 2, which is to be contrasted with the resonant behavior of the

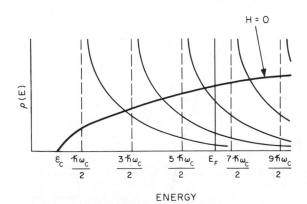

Figure 2. Electron density of states versus energy for a simple parabolic energy band. The dark solid curve for the density of states in zero magnetic field exhibits a threshold at the band extremum. The contribution to the density of states in a magnetic field for each energy subband exhibits a singularity at the extremum of the subband.

density of states in a magnetic field $\rho(\mathcal{E}, H)$, also shown in this figure.   The singularities in the magnetic density of states give rise to resonant behavior in a number of observables, thereby allowing accurate measurements to be made of resonant fields.   Resonant transitions between Landau levels in the same energy band (intraband transitions) and in different energy bands (interband transitions) are widely exploited in studies of the energy-band structure of solids.

To present the principles underlying these resonant intraband and inter-band transitions in solids, the simplified model of Eq. (2) is assumed for the magnetic energy levels for two noninteracting and nondegenerate energy bands with energy extrema at $\vec{k}=0$. Neglecting the electron spin, these energy bands in zero magnetic field are represented in Fig. 3 by the parabolic curves for both the occupied valence band $\mathcal{E}_-$ and the partially occupied conduction band $\mathcal{E}_+$

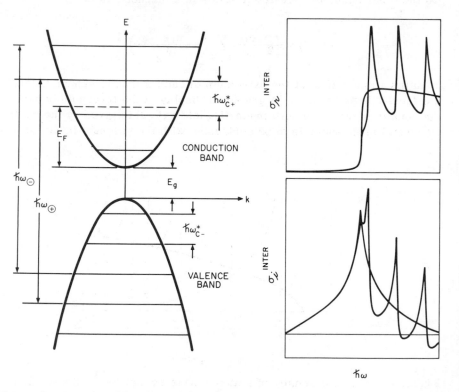

Figure 3.   Interband Landau level transitions in a solid following the selection rule $\Delta n = \pm 1$. Resonances in the real and imaginary parts of the inter-band conductivity are shown.

$$\mathcal{E}_+ = \frac{E_g}{2} + \frac{\hbar^2 k^2}{2m_+^*}$$

and   (8)

$$\mathcal{E}_- = -\frac{E_g}{2} - \frac{\hbar^2 k^2}{2m_-^*}$$

Since the magnetic density of states exhibits singularities at the extrema of the magnetic energy subbands, the magnetic energy levels at $k_z=0$ are of major interest and the energy of Landau levels at $k_z=0$ are indicated by horizontal lines on Fig. 3. De Haas-van Alphen type resonances occur in any observable depending on the density of states whenever a Landau level extremum passes through the Fermi level E_F. Intraband Landau level transitions are excited at constant k_z by photons of energy $\hbar\omega = \hbar\omega_{c+}^* = \hbar e H/m_+^* c$. Because of the exclusion principle, this cyclotron resonance excitation can be made for all k_z values for which there is an occupied electron state in one Landau level and an unoccupied electron state in the adjacent Landau level. The intensity for these cyclotron resonance transitions is large on the basis of this simple model because transitions can occur for a large range of k_z values. That is, the Landau level separation is $\hbar\omega_{c+}$ for all k_z values around the extremum. According to Fig. 3, contributions to the cyclotron resonance signal are made from $n=2 \rightarrow n=3$ for k_z values near $k_z=0$, while contributions from lower levels such as $n=1 \rightarrow n=2$ are made for larger k_z values.

In addition to the intraband cyclotron resonance transitions, interband Landau level transitions also occur, but these transitions occur at much higher photon energies. In this case, an electron in an occupied Landau level in the valence band absorbs a photon $\hbar\omega$ to make a transition to some unoccupied state in the conduction band, subject to suitable selection rules for this process. For a strictly noninteracting two-band model, the selection rule $\Delta n=0$ would apply, whereby states n_v in the valence band would only couple to states $n_c=n_v$ in the conduction band. Of particular interest to us here is the case of interband Landau level transitions in graphite, where the selection rules are $\Delta n = \pm 1$, and these selection rules are indicated in Fig. 3. For this to be a resonant process, interband transitions must occur at singularities in the joint density of states which are found at $k_z=0$. The insets at the right-hand side of Fig. 3 show that at zero magnetic field, there is a threshold for the interband conduction process in σ_r^{inter} consistent with the exclusion principle discussed above. Corresponding to this threshold in the real part of the interband conductivity σ_r^{inter}, structure is also found in the imaginary part of the interband conductivity σ_i^{inter}, and this is a consequence of the Kramers-Kronig relation. In the presence of a magnetic field, resonances appear in both the real and imaginary parts of the conductivity whenever the photon energy equals the energy difference at $k_z=0$ between some occupied valence level and an unoccupied conduction level which are coupled by the electromagnetic interaction. As a consequence of these resonances in the conductivity, resonances also appear in observables such as the reflectivity.

The analysis of these interband Landau level transitions can provide a determination of the effective mass (and also effective g factors when the electron spin is included) not only for the electron and hole carriers, but also for bands that are not occupied. Furthermore, such studies provide the most accurate determination of the energy band gaps in solids. Let us, for example, consider the simple parabolic band model of Eq. (8) and the magnetic-energy-level structure which this model implies. If no selection rules were operative and transitions could occur from any valence state to any conduction state, then there would be a resonant absorption of electromagnetic radiation whenever the photon energy is equal to

$$\hbar\omega = \mathcal{E}_+(n_+) - \mathcal{E}_-(n_-) = E_g + \hbar\omega_{c+}^*(n_++1/2) + \hbar\omega_{c-}^*(n_-+1/2) \tag{9}$$

If the selection rule of $n_+ - n_- = \pm 1$ is operative, as it is in graphite, the sense of circular polarization which couples n_- to $[n_+ = (n_-+1)]$ gives a series of resonant absorptions which can be labelled by n_- = integer, where the resonant photon energy is given by

$$\hbar\omega_\ominus \equiv \mathcal{E}_-^{res}(n_-) \equiv \mathcal{E}_+(n_-+1) - \mathcal{E}_-(n_-) = E_g + \hbar\omega_{c+}^* + (n_-+1/2)(\hbar\omega_{c+}^* + \hbar\omega_{c-}^*) \tag{10}$$

Thus, at a constant magnetic field, the separation in energy of neighboring resonances is just the sum of the Landau level separations in the valence and conduction bands which is sometimes written in terms of the reduced cyclotron frequency ω_{cr}^*

$$\mathcal{E}_-^{res}(n_-+1) - \mathcal{E}_-^{res}(n_-) = \hbar\omega_{c+}^* + \hbar\omega_{c-}^* \equiv \hbar\omega_{cr}^* \tag{11}$$

and this separation is illustrated in Fig. 4. In this figure and in Eqs. (9)-(10), it is seen that all the resonant photon energies approach the energy gap E_g as $H \rightarrow 0$. By using such a fan chart, the analysis of the interband Landau level transitions for simple parabolic bands yields the sum of the Landau level separations of the two bands as well as the energy gap between them.

To find the Landau level separation in each of the bands individually, we must compare the resonant photon energies for the two senses of circular polarization. If we denote by $\mathcal{E}_\mp^{res}(n_+)$ the resonant photon energy for the

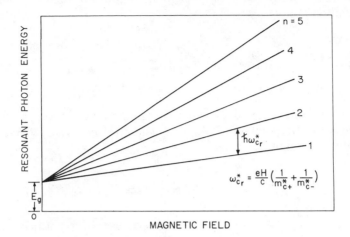

Figure 4. Fan chart of resonant magnetic fields as a function of photon energy for simple parabolic bands for one sense of circular polarization. Such diagrams are analyzed to yield the band gap and band curvatures of the valence and conduction bands.

other sense of circular polarization which couples a state n_- in the valence band to a state $n_+ = (n_- - 1)$ in the conduction band, then the resonant photon energy for the nth transition in each series differs by

$$\mathcal{E}_+^{res}(n) - \mathcal{E}_-^{res}(n) = \hbar\omega_{c-}^* - \hbar\omega_{c+}^* \tag{12}$$

so that from Eqs. (11) and (12) we can obtain $\hbar\omega_{c-}^*$ and $\hbar\omega_{c+}^*$ separately.

The energy-band structure in actual solids is more complex than the simple parabolic band model given by Eq. (8). The importance of the magneto-optic techniques is that they provide a powerful tool for the investigation of complicated energy-band structures as will be demonstrated in our discussion of the graphite problem. Energy bands do not follow the parabolic band model because of the coupling that exists between bands. The simplest coupled band model that can be devised consists of two bands that are strongly coupled to each other, but are not coupled at all to any other bands. This model is called the coupled two-band model and is applicable to some narrow-gap semiconductors and semimetals with small effective masses. Because of their small effective masses, these materials have energy level separations that can be tuned over a wide frequency range with a magnetic field.

For the case of the two-band model, we can write the energy $\mathcal{E}_i(\vec{k})$ at a wave vector $\vec{k}$ away from the energy extremum in terms of Brillouin–Wigner perturbation theory as

$$\mathcal{E}_i(\vec{k}) = \mathcal{E}_i^o + \sum_j' \frac{|\mathcal{H}'_{ij}|^2}{\mathcal{E}_i(\vec{k}) - \mathcal{E}_j^o} \tag{13}$$

where $\mathcal{E}_i^o$ is the energy at the band extremum and $\mathcal{H}' = \vec{k} \cdot \vec{p}/m$ is effectively a perturbation which relates $\mathcal{E}_i(\vec{k})$ to $\mathcal{E}_i^o$ and couples band i to band j through the matrix element

$$\mathcal{H}'_{ij} = \frac{\vec{k}}{m} \cdot \langle i|\vec{p}|j \rangle \tag{14}$$

If there are only two bands that are coupled, then Eq. (13) results in two coupled equations that can be solved exactly to yield

$$\mathcal{E}_i(\vec{k}) = \frac{1}{2}(\mathcal{E}_i^o + \mathcal{E}_j^o) \pm \frac{1}{2}\left[(\mathcal{E}_i^o - \mathcal{E}_j^o)^2 + |\mathcal{H}'_{ij}|^2\right]^{1/2} \tag{15}$$

This equation is more conveniently expressed by writing the energy separation at the band extremum as the energy gap E_g and taking the zero of energy halfway between the two bands at the band extremum. This gives

$$\mathcal{E}_{\pm}(\vec{k}) = \pm \frac{1}{2}\left[E_g^2 + 2E_g \hbar^2 (\vec{k} \cdot \overleftrightarrow{\alpha} \cdot \vec{k})/m\right]^{1/2} \tag{16}$$

where $\overleftrightarrow{\alpha}$ is the reciprocal effective mass tensor. The magnetic energy levels for the two band model at $k_z = 0$ are given by

$$\mathcal{E}_{\pm}(n, s) = \pm \frac{1}{2}\left[E_g^2 + 4E_g(n + \frac{1}{2} - s)\,\hbar\omega_c^*\right]^{1/2} \tag{17}$$

where n is the Landau level index and $s = \pm 1/2$ is the spin quantum number. Inspection of Eq. (16) shows that the energy levels of the coupled two–band model have a nonparabolic $\vec{k}$ dependence, which becomes increasingly nonparabolic as we move away from the band extremum. The magnetic-energy-level structure for the two–band model is also interesting. Provided that the spin–orbit interaction is much larger than the energy gap, we find an equality between the Landau level spacing $\hbar\omega_c^*$ and the spin splitting $g_{eff}\mu_B H$ which results in a degeneracy

$$\mathcal{E}_{\pm}(n, s = -1/2) = \mathcal{E}_{\pm}(n+1, \ s = 1/2) \tag{18}$$

very similar to what occurs for the free electron. When the coupling to other bands is considered, g_{eff} is no longer equal to $2m/m_c^*$ and this degeneracy is lifted. By studying the dependence of the interband Landau level transitions on photon energy and magnetic field, accurate values can be deduced for the effective mass tensor and the energy gap separating the two tightly coupled bands.

II. APPLICATION TO STUDIES OF ENERGY-BAND STRUCTURE OF SOLIDS--GRAPHITE

The utilization of laser sources in studies of the magneto-optical spectrum of solids has significantly advanced our understanding of the energy-band structure of these materials. To illustrate the techniques, we consider here the case of graphite where the laser sources have provided the resolution necessary for quantitative study of the complicated magnetoreflection spectrum which characterizes the resonant Landau level transitions in the low quantum limit. To show the power of the technique, it is appropriate to consider a material like graphite which has a fairly complicated energy-band structure arising from the strong coupling between four energy bands.

The earliest magnetoreflection studies in graphite did not use laser sources. Even though these experiments were carried out under low resolution conditions, they did serve to map out many aspects of the problem [1, 2]. Magnetoreflection resonances occur between magnetic energy levels associated with the four π bands of graphite which determine the Fermi surface and infrared properties of this material. The energy-band structure is described by an effective Hamiltonian $\mathcal{K}$ for four coupled bands, consistent with the crystal symmetry and valid for regions of k space around the edges HKH and H′K′H′ of the Brillouin zone where the Fermi surface is located [3, 4],

$$\mathcal{K} = \begin{bmatrix} E_1 & 0 & H_{13} & H_{13}^* \\ 0 & E_2 & H_{23} & -H_{23}^* \\ H_{13}^* & H_{23}^* & E_3 & H_{33} \\ H_{13} & -H_{23} & H_{33}^* & E_3 \end{bmatrix} \tag{19}$$

in which the various terms are given by

$$E_1 = \Delta + \gamma_1 \Gamma + \frac{1}{2} \gamma_5 \Gamma^2$$

$$E_2 = \Delta - \gamma_1 \Gamma + \frac{1}{2} \gamma_5 \Gamma^2$$

$$E_3 = \frac{1}{2} \gamma_2 \Gamma^2$$

$$H_{13} = \frac{1}{\sqrt{2}} (-\gamma_0 + \gamma_4 \Gamma) \sigma \exp(i\alpha) \qquad (20)$$

$$H_{23} = \frac{1}{\sqrt{2}} (\gamma_0 + \gamma_4 \Gamma) \sigma \exp(i\alpha)$$

$$H_{33} = \gamma_3 \Gamma \sigma \exp(i\alpha)$$

and

$$\Gamma = 2 \cos\pi\xi$$

In writing these equations, the wave vector is measured from the center of the edge of the Brillouin zone and is expressed in dimensionless polar coordinates in terms of the lattice constants a and c

$$\xi = c\kappa_z / 2\pi$$

$$\sigma = \frac{1}{2} \sqrt{3} \, a \, |\kappa|,$$

and

$$\exp(i\alpha) = (\kappa_y - i\kappa_x) / |\kappa| \qquad (21)$$

where $|\kappa| = (\kappa_x^2 + \kappa_y^2)^{1/2}$. Once the seven parameters γ_0, γ_1, γ_2, γ_3, γ_4, γ_5, and Δ are specified, the energy levels are determined in much the same way that the specification of the effective mass tensor and the energy gap gives the energy levels of the coupled two-band model (see Sec. I). Furthermore, once the energy bands are known, a direct determination can be made of the Fermi surface, which dominates the transport properties of the solid. With the low-resolution magnetoreflection experiment, it was possible to determine values for the parameters γ_0 and Δ and the ratio γ_0^2/γ_1.

The initial motivation for the laser magnetoreflection studies in graphite was to provide a more complete and accurate determination of the band parameters of the energy-band model. Inspection of the graphite magnetic-energy-level structure for $\vec{H} \parallel \vec{c}$ (Fig. 5) reveals why a high-resolution magnetoreflection study is expected to be fruitful. First of all, it is seen that the Landau level spacing at high quantum number about the energy extrema at the K point is different in the valence and conduction bands; this effect is governed by the parameter γ_4. Second, in the low quantum limit, the Landau level spacing becomes quite irregular; this irregularity is

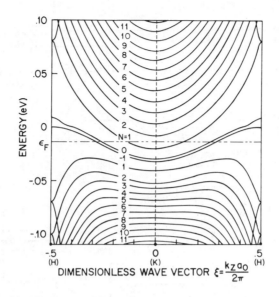

Figure 5. Landau level contours of graphite at a field of 50 kG. The Fermi level is indicated by the horizontal line.

largely governed by the parameter γ_3. As it turned out, these high-resolution measurements yielded other exceedingly useful information about the energy-band structure of graphite.

The high-resolution magnetoreflection experiments in graphite were carried out using a cw gas laser system which provided laser lines from 3.39 to 21.752 μm using mixtures of various gases. Since the magnitude of the magnetoreflection resonances is generally only about 1% of the reflectivity, stringent requirements are imposed on the stability of the laser output power. The schematic diagram of the experimental setup in Fig. 6 indicates the feedback network used for laser output stabilization as well as the detection system which proved satisfactory. Circularly polarized radiation was provided by a Fresnel rhomb. To satisfy the resonant condition $\omega_c \tau \gg 1$, it was necessary to make the measurements at high magnetic fields and low temperatures.

Traces were taken of the dependence of the reflectivity on magnetic field at a number of photon energies. The magnetoreflection spectrum taken at a photon energy of $\hbar\omega = 0.0693$ eV is shown in Fig. 7 for the two senses of circular polarization. Analysis of such traces shows that the interband Landau level transition excited with ⊖ polarization from a valence state n to a conduction state (n+1) occurs at a lower magnetic field than the

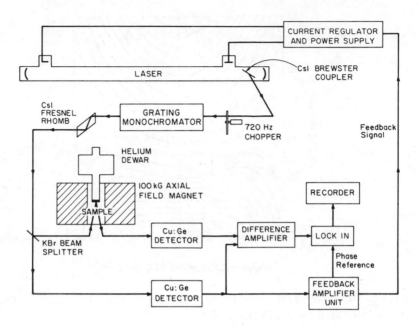

Figure 6. Block diagram of laser magnetoreflection experiment.

transition from a valence state (n+1) to a conduction state n excited by the
other ⊕ sense of circular polarization. From this result, the signs of
several of the Slonczewski-Weiss band parameters are determined; namely,
that $Y_2<0$, $\Delta>0$, $Y_4>0$, and $E_F<0$ [5]. The fact that Y_2 is negative implies
that the carriers at point K in the Brillouin zone must be <u>electrons.</u>
Previous to this high-resolution magnetoreflection study, the carrier assignment of locating the holes at K was widely accepted.

It would appear that the study of interband Landau level transitions
would be a rather indirect method for the identification of the sign of
carriers. A more direct measurement would appear to be the cyclotron
resonance measurement using circularly polarized radiation. In the cyclotron resonance experiment, we would expect the ⊖ polarization to excite
electrons and the ⊕ polarization to excite holes. Cyclotron resonance
measurements using circularly polarized radiation had, in fact, been
carried out for graphite [6], but the results for a given polarization did not
yield a single resonance line that could be identified with either electrons or
holes. Instead, a rather complicated spectrum was observed as is seen in
Fig. 8; this spectrum was not satisfactorily interpreted until the high-resolu-

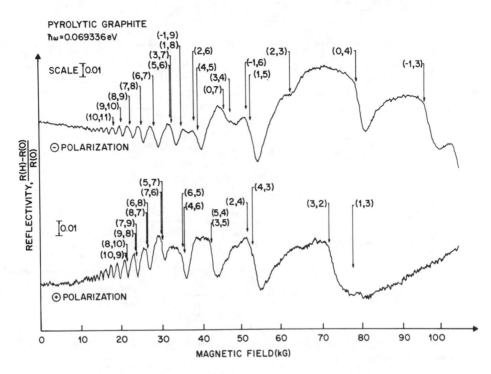

Figure 7. Experimental recorder traces obtained in the laser magnetoreflection experiment in graphite. Resonant magnetic field values predicted from the Slonczewski-Weiss band model are indicated by arrows. For each transition the Landau level indices for the valence and conduction levels involved are indicated by the first and second numbers in parenthesis, respectively. The field direction and geometry are such that electron cyclotron resonance would occur for the traces labelled "⊖ Polarization" and hole cyclotron resonance for those labelled "⊕ Polarization" (Ref. 14).

tion magnetoreflection study was made and the correct carrier assignment was established [5]. This carrier assignment for the electron at K is also preferred in considering (1) the filling of electron states on the nearly free-electron energy-band picture and (2) the observation that the minority carrier in graphite behaves like a hole upon boronation [7]. Subsequently, additional evidence has been presented [8] in support of this carrier assignment: (1) studies of Shubnikov-de Haas oscillations in the Hall effect and thermopower [9], (2) studies of the effect of neutron damage on the Shubnikov-de Haas periods in graphite [10], (3) studies of galvanomagnetic properties [11, 12].

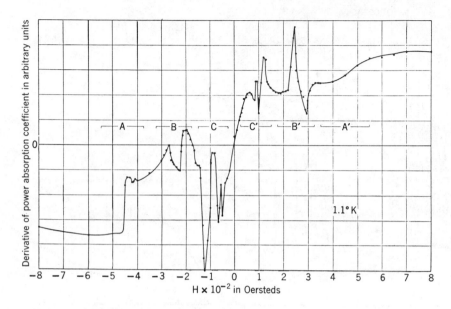

Figure 8.   Cyclotron resonance spectrum for graphite using circularly polarized radiation (Ref. 6).

Although the initial motivation for the study of the difference in Landau level spacing in the conduction and valence bands was to determine an accurate value for the Slonczewski-Weiss band parameter γ_4, the actual study of this effect led to the location of electrons and holes in the Brillouin zone and the turning of the bands upside-down with respect to the Fermi level.   Likewise, the initial motivation for the detailed identification of the spectrum in the low quantum limit was to obtain a quantitiative determination of the Slonczewski-Weiss band parameter γ_3.   Here again, the actual study provided other significant information with far-reaching consequences.   To illustrate this point, let us examine the spectrum of Fig. 7 in the low quantum limit.   Consider, for example, the spectrum for $\ominus$ polarization.   Whereas the transitions at low magnetic fields appear to be regularly spaced and to adhere to the selection rule for "allowed" transitions for the $\ominus$ polarization [level n in the valence band to level (n+1) in the conduction band], the structure at high magnetic fields is irregular and a breakdown in the selection rule occurs.   In fact, the intensity for the "allowed" transition from a valence state n=2 to a conduction state n=3 is considerably weaker than the "forbidden" transition from the conduction state n=0 to another

conduction state n=4. The strength of the "forbidden" transitions are inti-
mately connected with the magnitude of Y_3. If Y_3 were identically zero, the
"forbidden" transitions would have zero intensity. Therefore, by comparing
the intensity for the "forbidden" and "allowed" transitions, an accurate value
for Y_3 can be deduced. Such an analysis was, in fact, carried out [13, 14] and
it was found that the value of Y_3 was too large ($Y_3 = 0.29$ eV) to permit the
use of any form of perturbation theory in the calculation of the magnetic-
energy-level structure in graphite. Consequently, a large matrix technique
was devised to handle the large number of interacting magnetic energy states
for the coupled π bands. This new approach to the magnetic-energy-level
structure of graphite has important consequences in the interpretation of all
experimental data relevant to the electronic properties of graphite in a
magnetic field. Confirmation of the large magnitude of Y_3 and the inapplica-
bility of perturbation theory for the calculation of the magnetic-energy-level
structure in graphite has been provided by recent work on cyclotron resonance [15] and the Shubnikov-de Haas effect in the extreme quantum limit [16].

   With regard to the original aim of the high-resolution magnetoreflection
experiment, a detailed determination of the Slonczewski-Weiss band parameters was also carried out [13, 14]. Once the energy-band structure is
specified [Eqs. (19) and (20)], the Fermi surface can be calculated (see Fig.
9). From the energy bands, we can predict the cyclotron effective masses
for the carriers as well as the extremal Fermi surface cross-sectional
areas that are measured in a de Haas-van Alphen type experiment. With a
quantitative value now available for Y_3, the warping of the electron orbits

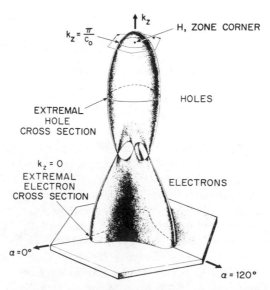

Figure 9. Graphite Fermi surface model.

about the K point can be found for the first time.

Of the various experimental techniques that are presently used to provide information about the energy-band structure of solids, the study of interband Landau level transitions is one of the most powerful techniques. The application of this technique to the study of the energy-band structure of graphite represents one of the most fruitful applications of this technique.

III. APPLICATION TO STUDIES OF THE OPTICAL SHUBNIKOV-DE HAAS EFFECT--ANTIMONY

Whereas the resonances associated with intraband and interband Landau level transitions involve the actual excitation of an electron from one state to another through the absorption of electromagnetic radiation, there are resonant processes in solids which are due to kinematic effects. An example of a resonant kinematic process is the Shubnikov-de Haas effect, whereby in an increasing magnetic field, magnetic-energy-level extrema pass through the Fermi level.

The observable in this case is the electrical conductivity which is proportional to the density of states at the Fermi level. Since the density of states in a magnetic field exhibits singularities at the extrema of each magnetic subband (see Fig. 2), resonances in the electrical conductivity are found whenever the extremum of a magnetic subband passes through the Fermi level. For parabolic noninteracting energy bands, the resonance condition can be simply written as

$$E_F = \hbar w_c^* (n + \frac{1}{2}).$$

(22)

For this simple model, these resonances are periodic in $1/H$ with a period

$$H_n^{-1} - H_{n-1}^{-1} = e\hbar / E_F m_c^* c$$

(23)

in which H_n is the resonant field at which the nth Landau level passes through the Fermi level. Since the resonant magnetic fields in the Shubnikov-de Haas effect do not depend on the photon energy, this effect can be readily distinguished from the optically excited Landau level transitions for which the resonant magnetic fields are strongly frequency dependent.

The importance of the Shubnikov-de Haas and de Haas-van Alphen-type phenomena to the study of the energy-band structure of solids arises from the fact that for a general Fermi surface, these de Haas-van Alphen periods are inversely proportional to the extremal Fermi surface cross-sectional areas [17] and the temperature dependence of the de Haas-van Alphen amplitudes can be analyzed to yield values for the cyclotron effective mass [18]. Such measurements have, in fact, proven to be invaluable for the mapping out of the Fermi surfaces in solids. Although the observation of the Shubnikov-de Haas effect at optical frequencies presents a more difficult experimental problem than at dc or audio frequencies, nevertheless, this approach is attractive insofar as the electrical conductivity σ as expressed by the Drude theory,

$$\sigma = ne^2\tau/m^*(1-i\omega\tau) \qquad (24)$$

becomes insensitive to scattering and relaxation processes in the limit $\omega\tau \gg 1$. The hope was that by studying the Shubnikov-de Haas effect at both optical and low frequencies, it would be possible to distinguish between resonant contributions from the carrier density n and from the relaxation time τ.

The first observation of the optical Shubnikov-de Haas effect was made in antimony using conventional optical sources [19]. In Fig. 10, we see small oscillations superimposed on a large magnetic-field-dependent reflectivity. These small oscillations were identified with the optical Shubnikov-de Haas effect because (1) the magnetic fields at which these oscillations occurred were found to be independent of photon energy, (2) the oscillations were periodic in $1/H$, and (3) the values of the observed periods were in good agreement with de Haas-van Alphen periods measured in antimony by other techniques for the binary, bisectrix, and trigonal orientations of the magnetic field.

Having identified this effect, a line-shape calculation was carried out to gain insight into the dominant physical mechanism. In this calculation, it was assumed that the Shubnikov-de Haas resonances originated in the free-carrier contribution to the conductivity through resonances in the carrier density and in the density of states in a magnetic field. The magnitude of the resonances in the optical reflectivity was predicted to be small, except near the plasma frequency where the real part of the dielectric constant is very small. Near the plasma frequency, a small change in the dielectric constant due to the optical Shubnikov-de Haas mechanism could produce a measurable change in the optical reflectivity. Examples of magnetoreflection line shapes predicted from this model are shown in Fig. 11. In this figure, it is seen that for frequencies below the plasma frequency ω_p, the resonant magnetic fields occur at a reflectivity minimum; on the other hand, for $\omega > \omega_p$, the resonant magnetic fields are identified with reflectivity maxima. The resonant frequencies are independent of photon energy. Also of interest is the predicted frequency dependence of the magnitude of the magnetoreflection resonances, shown in Fig. 12; in this figure, the magnitude of the resonances is plotted as a positive quantity for the reflectivity minima, but as a negative quantity for the reflectivity maxima.

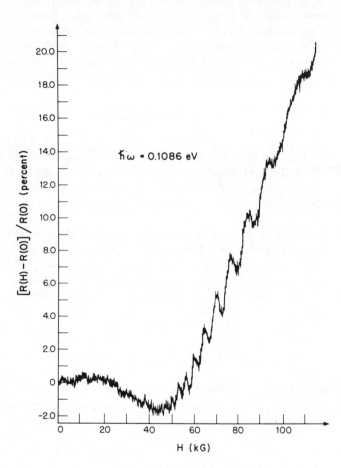

Figure 10.   Recorder trace showing the fractional change in the reflectivity of antimony plotted as a function of magnetic field for $\vec{H} \parallel$ binary axis.   The photon energy is 0.1086 eV and the sample temperature is $\sim 25^\circ$K (Ref. 19). The small oscillations are identified with the optical Shubnikov-de Haas effect.

In order to study the physical mechanisms behind the optical Shubnikov-de Haas effect, a much more detailed study of the magnetoreflection line shape was undertaken utilizing the high resolution possible with laser sources [20].   As in the case of studies of the interband Landau level transitions, the introduction of a laser source into the study of the optical Shubnikov-de Haas effect resulted in the observation of a host of unexpected

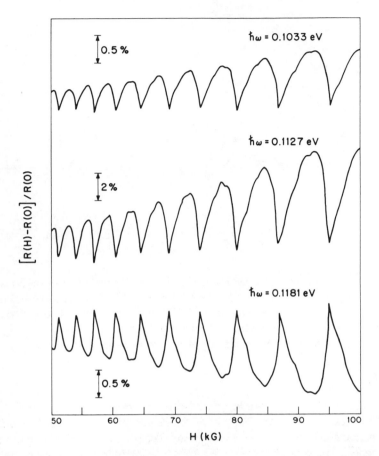

Figure 11. Calculated reflectivity spectrum for antimony at the photon energies 0.1033 eV, 0.1127 eV, and 0.1181 eV assuming T=0° K for $\vec{H}$ || trigonal axis. For the top two traces, the passage of a Landau level through the Fermi surface is associated with the reflectivity minima, while for the bottom trace, the reflectivity maxima correspond to the passage of a Landau level through the Fermi surface (Ref. 20).

phenomena. Shown in Fig. 13 is a magnetoreflection trace taken on an anti-mony crystal at T = 1.89° K for the magnetic field directed along a binary direction using a neon laser line at 10.978 μ. This trace differs very significantly in several respects from the trace in Fig. 10 taken at about the same photon energy and with the same orientation of the magnetic field:
(1) The amplitudes of the Shubnikov-de Haas resonances are larger by more

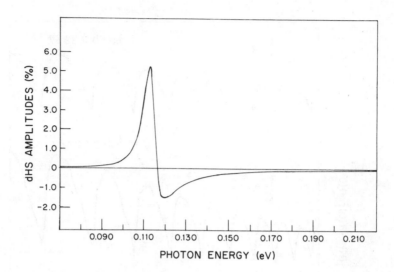

Figure 12.   Calculated fractional change in the reflectivity associated with the passage of the n=11 Landau level through the Fermi surface as a function of photon energy, for antimony with $\vec{H}\,\|$ trigonal axis.   The positive and negative values indicate that the passage of a Landau level through the Fermi surface is associated with a reflectivity minimum and maximum, respectively (Ref. 20).

than one order of magnitude and (2) the nonresonant background magnetoreflectivity is absent.   Perhaps even more significant is the fact that these resonances are not confined to a small frequency range about the plasma frequency, but are observed with large amplitude over a wide frequency range.   This is evident in the results of Fig. 14, taken on another antimony crystal, but with the same magnetic field orientation, $\vec{H}$ along a binary direction.   The upper trace, taken with a xenon line at 18. 500 μ, and the lower trace taken with a xenon line at 5. 5739 μ, both yield resonances with large amplitudes at frequencies where the analysis of Fig. 12 would indicate the resonances to have negligible amplitudes.   The presence of large amplitude resonances in the reflectivity at photon energies well above and below $\hbar\omega_p$ is a feature which is common to all high-quality samples.   It is also found that for such samples the passage of a magnetic-energy-level extremum through the Fermi level is associated with either reflectivity minima or maxima, depending on the sample, and that no line-shape reversal occurs at the plasma energy $\hbar\omega_p$.   Although none of these phenomena can be understood on the basis of the simple theory of the optical Shubnikov-de Haas effect based on resonances in the carrier density (Figs. 11 and 12), the

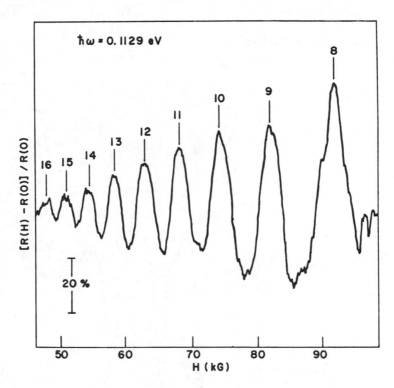

Figure 13. Experimental magnetoreflection trace obtained for antimony at $\hbar\omega = 0.1129$ eV (neon laser line) and T = 1.89°K with $\vec{H}\parallel$ binary axis and polarization $\vec{E}\parallel$ bisectrix axis. The resonances are identified with an optical Shubnikov-de Haas mechanism (Ref. 20).

observed phenomena must be identified with a Shubnikov-de Haas type mechanism, since the observed resonances are independent of photon energy, as shown in Fig. 15, and the observed periods are in good agreement with the best values available for the electron periods in antimony [21]. The fact that the magnitude of the optical Shubnikov-de Haas effect does not decrease above the plasma frequency indicates that the physical mechanism must involve interband processes, since the intraband contribution to the dielectric constant ϵ_1 falls off as $1/\omega^2$ (the Drude tail). On the other hand, measurements made at high temperatures (T $\gtrsim 20$°K) or on samples of inferior quality can be described quite well in terms of the simple calculation based on resonances in the carrier density.

At low temperatures and on high-quality samples, the optical Shubnikov-

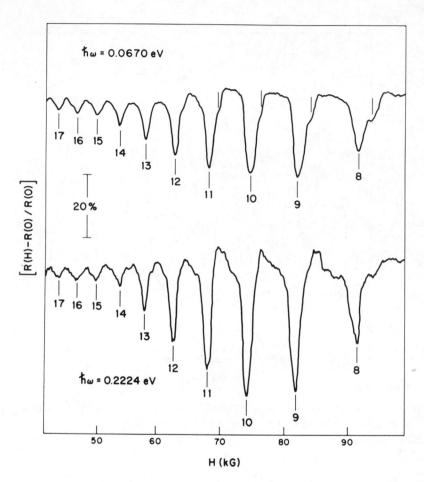

Figure 14.   Experimental magnetoreflection traces obtained for antimony at $\hbar\omega = 0.0670$ eV and 0.2224 eV (Xenon laser lines) for $T = 1.96°$K, $\vec{H} \parallel$ binary axis and polarization $\vec{E} \parallel$ bisectrix axis (Ref. 20).

de Haas effect exhibits a number of other interesting features.   For example, when the experiment is carried out with the magnetic field along a bisectrix direction, resonances due to both electrons and holes are observed.   In Fig. 16, taken with $\vec{H} \parallel$ bisectrix axis, the reflectivity maxima are identified with electron resonances while the reflectivity minima are identified with holes. It is of special interest to observe that the resonant amplitudes associated with the holes increase much more rapidly with increasing magnetic field than do the resonances associated with the electrons.   If the Shubnikov-de

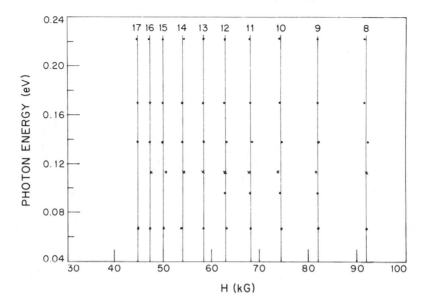

Figure 15.   Photon energy of Shubnikov-de Haas resonances versus magnetic field for antimony with $\vec{H} \parallel$ binary axis.   The frequency independence of these resonances is demonstrated (Ref. 20).

Haas mechanism should involve the interband conductivity through a relaxation process to the Fermi level, then this optical Shubnikov-de Haas mechanism could provide a method for studying the relaxation to electron and hole states independently.

The question of line-shape reversals is a complicated one.   Some samples do not exhibit any line-shape reversals, either as a function of photon energy or magnetic field (see Figs. 13 and 14).   Other samples show Shubnikov-de Haas resonances for a particular period at both reflectivity minima and reflectivity maxima, with line-shape reversals occurring as a function of photon energy.   On some samples oriented with the magnetic field along the trigonal direction, several line-shape reversals have been observed as a function of magnetic field, but at a single photon energy.

Although the laser was initially introduced to facilitate the detailed study of the magnetoreflection line shape, the actual use of the laser source resulted in the discovery of a new Shubnikov-de Haas mechanism operative at low temperature and on high-quality samples.

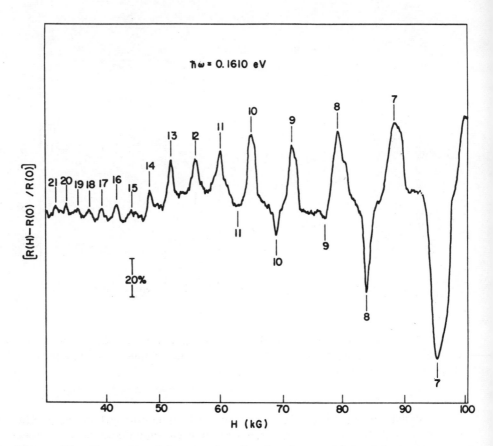

Figure 16.   Experimental magnetoreflection trace obtained for antimony at $\hbar\omega = 0.1610$ eV (neon laser line), $T = 1.95°K$, $\vec{H} \parallel$ bisectrix axis and the polarization $\vec{E} \parallel$ binary axis.   The reflectivity maxima are associated with electrons and the minima with holes.

IV.   APPLICATION OF FAR-INFRARED LASERS

Lasers have proven to be especially useful in the study of a variety of phenomena in solids occurring in the far-infrared region of the spectrum. For one thing, lasers have provided sources of reasonable intensity in an intensity-starved region of the spectrum.   It has been recognized for some time, that the cyclotron resonance technique, which had yielded such valuable information on the effective masses in semiconductors, could be applied to a greater variety of materials, if the measurements could be made at frequencies higher than the standard microwave frequencies. In order for cyclotron resonance to be observable, it is necessary for an electron to complete a cyclotron orbit before scattering; this can be expressed by the condition $\omega_c^* \tau > 1$. If the relaxation time for a given material is limited by the greatest purity and crystal perfection which can be realized with the present technology, then it is only by going to higher frequencies that the resonance condition $\omega_c^* \tau > 1$ can be satisfied. By going to higher frequencies, the resonances are also pushed to higher magnetic fields, where an increased signal is expected because the density of states in a magnetic field is proportional to H. Higher frequencies are furthermore attractive for studying the irregular spacing of the magnetic energy levels in the low quantum limit (called quantum effects in semiconductors). To study such quantum effects in semiconductors, it is necessary to confine the thermal population to only a very small number of magnetic energy levels; this condition can be written as $\hbar \omega_c^* / kT \gg 1$. By going to higher frequencies, quantum effects can be studied over a wide range of temperatures.

Far-infrared lasers have, in fact, been applied to the study of cyclotron resonance in a large number of semiconductors: Ge, Si, InSb, GaSb, Te, $Hg_x Cd_{1-x} Te$, CdS, CdSe, CdTe and ZnO [22]. For a number of these materials, the relaxation time is sufficiently short so that the higher–resonance frequency is essential for satisfying the resonance condition $\omega_c^* \tau > 1$. The higher resolution of the laser sources has also been exploited in studying quantum effects in semiconductors like Ge and InSb.

Of greater interest to our present discussion, however, are the new types of phenomena which can now be studied with the availability of far-infrared laser sources. One such application is to the study of electric dipole transitions in the electron spin resonance (ESR) of semiconductors. For a free-electron system, an electric dipole transition is strictly forbidden between the spin-split doublet of a magnetic energy level. In a solid, however, such transitions can occur with measurably large intensity. Basically, these transitions are possible in a solid because we can always expand the energy bands at a point $(\vec{k}_0 + \vec{k})$ in terms of the energy bands at a nearby point $\vec{k}_0$, using the method of $\vec{k} \cdot \vec{p}$ perturbation theory. In so doing, each state at $\vec{k}_0 + \vec{k}$ is expressed in terms of the complete set of states at $\vec{k}_0$, which for most practical applications is taken at an energy extremum. Even if the transition between two states is strictly forbidden at $\vec{k}_0$, the admixture at a point $(\vec{k}_0 + \vec{k})$ of other $\vec{k}_0$ states causes a breakdown in this selection rule. Thus, in solids, electric dipole transitions can be made between a spin-up and a spin-down state. Electric-dipole excited electron spin resonance has been observed in InSb using a far-infrared $H_2O-D_2O-SO_2$ pulsed laser source [23]. To interpret the magnetic-energy-level transitions which occur away from band extrema, line-shape studies are necessary. Because of the relative sharpness of these resonances (a linewidth of 80 G or 0.16 cm^{-1} for a 118.6-μ H_2O laser excitation energy), it is attractive to use a laser source for this application. Thus, the far-infrared lasers are now having a large impact on the determination of both the cyclotron effective mass and effective g factors in semiconductors.

The study of cyclotron resonance phenomena in the far infrared has an additional attraction insofar as matching the excitation frequency with the natural longitudinal optical phonon frequencies in solids. In polar semiconductors, the electronic motion is perturbed by the lattice vibrations which give rise to polarization effects. This electron-phonon coupling is enhanced when the Landau level separation is close in energy to the optical phonon frequency. Such interactions between the electronic motion and the lattice modes have been observed in the polar semiconductor CdTe using HCN, DCN, H_2O, and D_2O lasers, operating at 337, 195, 172, and 119 μ, respectively [24]. On the basis of this work it has been possible to verify the general Fröhlich theory describing this electron-phonon interaction, and to obtain a numerical value for the coupling constant governing this interaction.

Far-infrared lasers have also had an impact on the study of the impurity problem in semiconductors. From an elementary point of view, it is possible to describe the shallow donor impurity levels in a semiconductor in terms of hydrogenic levels through the effective mass approximation, replacing the free-electron mass by the effective mass, and the electron charge e by an effective charge $e/\sqrt{\epsilon}$, screened by the dielectric constant ϵ. Although this theory was not strictly applicable to the ground state donor levels of the semiconductors silicon and germanium for which the theory was originally developed [25], it did appear to work very well for the donor levels in GaAs. However, by examining these donor levels under higher-resolution conditions, a multiplicity of ground-state levels have been studied with considerable success using high magnetic fields and far-infrared laser sources [26]. Such ground-state splittings arise from the small differences in the Coulomb potential in the vicinity of the various kinds of donor impurity ions. These differences in the Coulomb potential are important only when the donor electron is close to the impurity ion site itself; the effect is thus called the central cell correction. Magnetic fields tend to localize the donor electron on the donor impurity site, thereby increasing the central cell correction; here laser sources are useful for resolving the small splittings.

Although electron paramagnetic resonance (EPR) studies have in the past been carried out at microwave frequencies, there are distinct advantages in performing such experiments with far-infrared laser sources. If the ground-state multiplet spans a range of say 10-50 cm^{-1}, microwave frequencies are only of limited value in studying the Zeeman pattern for these states. It is only by going to higher frequencies that a sufficient number of transitions can be excited between these magnetic energy levels to determine the parameters of the spin Hamiltonian describing the field dependence of this multiplet structure [27]. It is clear that far-infrared lasers will provide a valuable tool for the investigation of electron paramagnetic resonance spectra and a host of other phenomena in solids.

REFERENCES

1. M. S. Dresselhaus and J. G. Mavroides, IBM J. Res. Develop. 8, 262 (1964).
2. M. S. Dresselhaus and J. G. Mavroides, Carbon 4, 465 (1966).
3. J. C. Slonczewski and P. R. Weiss, Phys. Rev. 109, 272 (1958).
4. J. W. McClure, Phys. Rev. 108, 612 (1957).
5. P. R. Schroeder, M. S. Dresselhaus, and A. Javan, Phys. Rev. Letters 20, 1292 (1968).
6. J. K. Galt, W. A. Yager, and H. W. Dail, Phys. Rev. 103, 1586 (1956).
7. D. E. Soule, IBM J. Res. Develop. 8, 268 (1964).
8. J. W. McClure, Proceedings of the Conference on the Physics of Semimetals and Narrow-Gap Semiconductors, Dallas, 1970, edited by D. L. Carter and R. T. Bate (Pergamon, New York, 1971) p. 127.
9. J. A. Woollam, Phys. Letters 32A, 115 (1970).
10. J. D. Cooper, J. Woore, and D. A. Young, Nature 225, 721 (1970).
11. V. V. Kechin, Sov. Phys. Solid State 11, 1448 (1970).
12. I. L. Spain, Proceedings on the Conference on the Density of States, Gaitersburg, National Bureau of Standards Special Publication 323 (1969), p. 717.
13. P. R. Schroeder, Ph. D. thesis (Massachusetts Institute of Technology, 1969).
14. P. R. Schroeder, M. S. Dresselhaus, and A. Javan, Proceedings of the Conference on the Physics of Semimetals and Narrow-Gap Semiconductors, Dallas, 1970, edited by D. L. Carter and R. T. Bate (Pergamon, New York, 1971) p. 139.
15. H. Suematsu and S. Tanuma, J. Phys. Soc. Japan 33, 1619 (1972); H. Ushio, T. Uda, Y. Uemura, J. Phys. Soc. Japan 33, 1551 (1972).
16. J. W. McClure and J. A. Woollam (private communication), and J. Woollam, Phys. Rev. Letters 25, 810 (1970).
17. L. Onsager, Phil. Mag. 43, 1006 (1952).
18. A. H. Kahn and H. P. R. Frederidse, Solid State Physics, Vol. 9, edited by F. Seitz and D. Turnbull (Academic, New York, 1959), p. 257.
19. M. S. Dresselhaus and J. G. Mavroides, Solid State Commun. 2, 297 (1964).
20. F. P. Missell, Ph. D. thesis (Massachusetts Institute of Technology, 1971).
21. L. R. Windmiller, Phys. Rev. 149, 472 (1966).
22. K. J. Button, B. Lax, and C. C. Bradley, Phys. Rev. Letters 21, 350 (1968); K. J. Button, G. Landwehr, C. C. Bradley, P. Grosse, and B. Lax, Phys. Rev. Letters 23, 14 (1970); K. J. Button, B. Lax, and

D. R. Cohn, Phys. Rev. Letters 24, 375 (1970).

23. B. D. McCombe and R. J. Wagner, Phys. Rev. B 4, 1285 (1971).

24. J. Waldman, D. M. Larsen, P. E. Tannenwald, C. C. Bradley,
D. R. Cohn, and B. Lax, Phys. Rev. Letters 23, 1033 (1969).

25. W. Kohn, Solid State Physics, Vol. 5, edited by F. Seitz and
D. Turnbull (Academic, New York, 1957), p. 257.

26. H. R. Fetterman, D. M. Larsen, G. E. Stillman, P. E. Tannenwald,
and J. Waldman, Phys. Rev. Letters 26, 975 (1971).

27. R. S. Rubins and H. R. Fetterman, Bull. Am. Phys. Soc. 16, 522
(1971).

D. Tunable Sources and Their Applications

SPECTROSCOPY WITH TUNABLE LASERS IN THE VISIBLE REGION

A. L. Schawlow
Stanford University
Stanford, California

ABSTRACT

The several kinds of tunable lasers and some methods for making them really monochromatic are reviewed. Gas lasers, and particularly ion lasers tunable over several thousand MHz, have been used by Hänsch, Levenson, and Sorem for saturation spectroscopy on a number of lines of molecular iodine. The absorption cell is external to the laser. It is found that the quadrupole coupling parameter is nearly independent of the molecular rotation and vibration state, but the magnetic hyperfine structure constant is large for states near the dissociation energy. For broader tunability dye lasers have been investigated. Holzrichter and Macfarlane used a tuned flashlamp-pumped dye laser to generate excitons and magnons in antiferromagnetic MnF_2. The growth and relaxation of these excitations is observed through the magnetization produced in the crystal. Several experiments have been done with dye lasers pumped by a pulsed nitrogen laser. These include amplification of an entire image with good resolution, preserving wavelength, obtained by Hänsch and Varsanyi. Laser action has been obtained for several dyes in gelatin. Ways have been found by Hänsch to make a pulsed tunable dye laser extremely monochromatic, and coherent beams no wider than 7 MHz have been generated. Hänsch and Shahin have applied this laser, with the external cell method of saturation spectroscopy, to resolve hyperfine structures in the atomic sodium D lines. By delaying the probe pulse after saturation, the restoration of equilibrium has been studied both with and without a buffer gas. In the latter case, a narrow transmission band remains for at least several hundred nanoseconds after the saturating pulse.

Lasers are still very primitive devices. The only thing that saves us most of the time is that, being physicists, we are not assigned a task to do with the lasers but are rather able to do whatever can be done with the available lasers. We have been quite resourceful in adapting to the limitations of the laser and doing what is possible with the existing ones. But we are however now entering a time, I think, when some of these limitations are being overcome, and in particular, the limitation on tunability. (I should restrict my discussion to the visible and near-visible region. Others at this symposium have discussed the problems in the infrared.) As an example of some of the difficulties, a few years ago we wanted to do some work on photochemistry and the only way we could do it was to find a molecule which had lines coinciding with an available laser. At that time there was the ruby laser, which we could tune over about 9 cm^{-1} by changing the temperature. To match that ruby laser we had to use bromine which turned out, after some years, to be a most unfortunate choice from the point of view of photochemistry but that was all we could do with the available lasers [1,2]. Similar restrictions are encountered with other lasers which have a rather narrow tuning range, and the only way that we could use them for spectroscopy was to find something which happened to coincide with their particular laser lines.

Let us first discuss an experiment that we have done recently in that way. We used available argon and krypton ion lasers and found a molecule which happened to coincide with their output wavelengths, in particular, the iodine molecule. The method used in these experiments, which were done by Hänsch, Levenson, and Sorem [3,4] is a method of saturation spectroscopy. The technique used was developed out of earlier work by Hänsch and Toschek [5] and was recently introduced in different context by Bordé and by Smith and Hänsch [6]. A scheme of the experiment is shown in Fig. 1. The advantage of an argon or krypton ion laser is twofold. It has quite a lot of power so that we can get saturation outside the laser. Also the spontaneous emission lines are rather broad, so that we can tune over perhaps 5000 MHz. The light from the laser is split into two parts at the beam splitter. One part, which we may call the saturating beam, is chopped. It goes through the iodine cell and bleaches its way through. The other part, which we call the probe beam, goes through in almost exactly the opposite direction. In

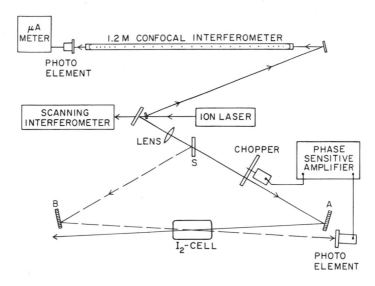

Figure 1. Saturation spectrometer for iodine hyperfine structures.

Fig. 1, the angle between the beams is exaggerated, as it is of the order of 1 mrad. If these beams affect the same molecules, then the saturating beam will bleach a path for the probe beam and the signal at the photodetector will increase. Thus, as the saturating beam is chopped, the light reaching the photodetector is modulated. Now this happens, of course, only if the two beams are seeing the same molecules. That happens most simply if the laser is tuned exactly to the center of the resonance line so that its output is absorbed by molecules which are going neither to the left nor to the right, but only transversely to the beam. For those molecules, neither beam is Doppler shifted and the two beams are in resonance with the same molecules. So, in that way, as is familiar in saturation spectroscopy, we get rid of the Doppler broadening. One of the advantages of this method is that it is convenient to work outside the laser, and particularly so if you want to work in a magnetic field. Although it will not be discussed here, we have recently begun work on the Zeeman effect of these saturation resonances in iodine.

The laser is made to oscillate in a single axial mode by a quartz etalon, thermostatically stabilized. The frequency is scanned by a piezoelectric drive on one end mirror. As the scan rate may not be constant, it is calibrated by passing some light from the laser through a 1.2-m confocal interferometer.

In this experiment, we are looking at single rotational lines in the vibrational bands of the transition from the ground $^1\Sigma_{0g}$ to an electronically excited state $^3\Pi_{0u}$. There are an enormous number of lines in these bands, but we can study only those few absorption lines which happen to coincide with one of the wavelengths produced by one of our lasers. Fig. 2 indicates the kind of hyperfine structure we may expect to find. The hyperfine splittings are primarily due to the quadrupole moment of the $I_2{}^{127}$ nuclei and secondarily to a magnetic hyperfine interaction. There are two nuclei which have spin 5/2 for the ordinary isotopes of iodine. The hyperfine interaction Hamiltonian is

$$\mathcal{H}_{hfs} = \sum_{i=1}^{2} eQq \frac{3(\vec{I}_i \cdot \vec{J})^2 + \frac{3}{2}\vec{I}_i \cdot \vec{J} - |\vec{I}|^2 |\vec{J}|^2}{2J(2J-1)\,I_i(2I_i-1)} + C\,\vec{I}_i \cdot \vec{J}$$

Each nucleus couples independently to the field gradient and the energy levels of the two nuclei simply add because there is very little coupling between the nuclei directly. For one nucleus you will have states with $m_I = \pm 1/2, \pm 3/2,$ and $\pm 5/2$. In an axially symmetric field gradient the level spacings are in the ratio 1 to 2 so that the energies are 0, 1, and 3. If you have two such nuclei, you just simply add up the energies in all possible combinations and the third column of Fig. 2 shows what you get. There are five equally spaced levels and then a double spacing. This is what you would get if there were no magnetic couplings, no magnetic $\vec{I} \cdot \vec{J}$

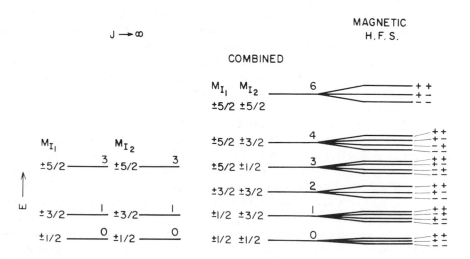

Figure 2.   Expected hyperfine level pattern from quadrupole and magnetic interactions (for states with large J).

term, and also if the rotational angular momentum J were infinite.   Actually, in some of these lines observed it is pretty close to it.   We have observed lines with J values as high as 117.   That is very large.   If the $\vec{I} \cdot \vec{J}$ term is small but appreciable, these levels split further, as shown in the last column, producing triplets and quartets.   The selection rules are extremely simple in that the nuclear orientations do not change in the transition, so that the pattern of the lines is just the same as the pattern of the levels.   In other words, we have an upper state with the sort of pattern shown in Fig. 2 and a lower state with the same kind of pattern.   Each level in the upper state connects with the corresponding level of the lower, so that the line pattern is just the same as the level pattern.   No crossover transitions are observed.   Thus only the difference of the quadrupole couplings for the two electronic states can be observed in most lines.   Hyperfine structures of iodine lines coincident with the 6328-Å helium-neon laser line have been observed and analyzed by Hanes and Dahlstrom [7, 8].

Fig. 3 shows the observed hyperfine structures of two molecular iodine

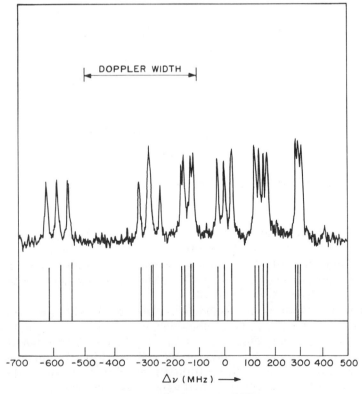

Figure 3 (a).   Hyperfine structure of I_2^{127} P(117) 21-1 line.

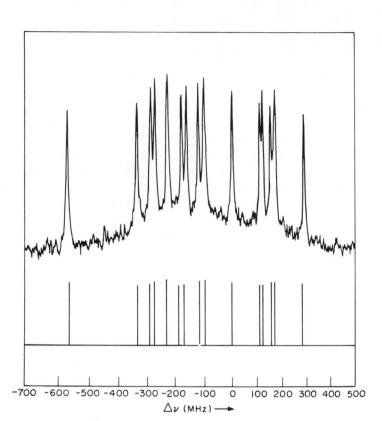

Figure 3 (b).   Hyperfine structure of $I_2{}^{127}$ R(78) 40–0 line.

lines, where the J values of the lower state are respectively odd [Fig. 3(a)] and even [Fig. 3(b)]. In the latter case, the Pauli exclusion principle prevents states where both nuclei have the same quantum numbers. Then all triplets become singlets and so the number of components in the hyperfine pattern is reduced from 21 to 15.

It is worth noting that some individual components are strong, well-resolved, and have a width of about one part in 10^8. It should be fairly easy to lock the laser onto one of the components with a precision of about 1% of the linewidth, so that we would have a convenient laser standard in the green or in the yellow reproducible within a part in 10^{10}. This modulation method is quite sensitive and gives a good signal-to-noise ratio.

By using the argon and krypton lasers and all the available lines, and also using the radioactive iodine as well as the normal one, we can get

measurements of the quadrupole coupling for a number of lines. It is found
that the difference in eQq varies by only about 100 out of 2400 MHz over the
range of lines observed. The $\vec{I} \cdot \vec{J}$ coupling, though, does vary quite considerably. In Fig. 4, the magnetic hyperfine interaction constant is plotted
against the energy of the upper state above the bottom of the $^3\Pi_0$ band. It is
expressed as $G = CI/\mu$, in order to correlate data obtained with two different

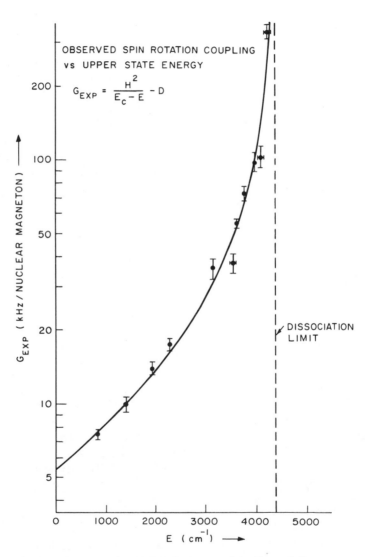

Figure 4.   Magnetic hyperfine interaction as a function of the energy of the
upper state above the bottom of the $^3\Pi_0$ band.

isotopes which have different spin and nuclear moments. Near the dissocia-tion limit, the $\vec{I} \cdot \vec{J}$ coupling becomes very large, most likely because it comes about largely from mixing in another state which is close to that limit.

At low J values you can do something about separating the quadrupole couplings in the upper and lower state. The experiments show that the quadrupole coupling does not change much with either rotational or vibra-tional states. Thus, it is reasonable to assume that q_M, the field gradient along the molecular axis, is constant within a given electronic state. The measured quadrupole coupling is $eq_J Q$, where q_J is the corresponding quantity averaged over the rotation and so oriented along J. They are related by $q_J = q_M / [2 + (3/J)]$. It follows that by measuring the difference in quadrupole coupling for two known lines, we can solve for q_M in the upper and lower electronic states. Although the errors are large, we find $eQq = - 2700 \pm 400$ MHz for the ground electronic state of I_2^{127} and $eQq = - 800 \pm 400$ MHz for the $^3\Pi_0$ electronic state.

It is, however, somewhat frustrating that, in these experiments, we are able to study only those iodine lines that happen to lie near an argon or krypton laser line, and we would like, in particular, to look at some very low J values. The lowest J value that we have found so far is 10. We looked therefore at more broadly tunable lasers. So far, none of them is quite monochromatic enough to do justice to the fine structures in the iodine spec-trum. However, parametric oscillators and dye lasers both provide tunable sources of coherent visible light. For both of them, progress is rapid, but at any frequency, extreme monochromaticity is not easily achieved when the device can oscillate over a wide range. Most of the visible can now be covered with tunable parametric oscillators and certainly a large part of the infrared [9]. One such device, based on the designs of Harris and available commercially, starts with a YAG:Nd laser which can oscillate at several wavelengths near 1 μ, followed by a lithium iodate doubler, and a lithium niobate parametric converter. These lasers give pulses usually about 1 μsec in length with repetition rates up to a thousand per second. That is a fairly low duty cycle but for many experiments the high repetition rate makes it seem essentially continuous. The tuning range is a few hundred angstroms for each laser line. In all there is continuous tuning range of several thousand angstroms in the long-wavelength part of the visible region. With other versions, you can go further into the infrared.

We have, however, in our own work used instead what are to us rather simpler devices, the various kinds of dye lasers. The methods used for refining the monochromaticity and stability are largely applicable to other sources of coherent light.

A dye laser that was constructed by Holzrichter [10, 11] in our labora-tory consisted of a cylindrical cell containing an organic liquid dye, with external mirrors. It was pumped by straight flashlamps which had enlarged sections at the ends in order to provide a reservoir for pressure relief because these are flashed with a short intense current pulse. The circuitry was all coaxial to give a fast rise time. The laser produced pulses about 1 μsec long and the peak power was a few kilowatts with sodium fluorescein dye. It was tuned by a diffraction grating which replaces one of the end mirrors. For finer tuning, the output mirror is replaced by a quartz etalon, and sometimes two etalons of thickness 0.5 mm and 2 mm. The laser output was thus reduced from a band of several hundred cm^{-1} to about 0.1 cm^{-1} with little loss of power.

This laser was used for an experiment in solid-state physics and in particular for studies of ions in crystals, for which the oscillator strengths are typically low. Thus a fairly large amount of pumping energy is needed to saturate any sort of an absorption. That is one reason for using a flash-lamp-pumped laser. This particular one gave only a few millijoules but it is worth noting that Bradley has been able to get energy outputs of as much as 2 j in 1 μsec out of flashlamp-pumped dye lasers, which is a power of several megawatts [12].

In the experiment of Holzrichter and Macfarlane, the laser is tuned to 5420 Å and used to excite excitons and magnons in manganese fluoride. Manganese fluoride is a crystal which is well known to be antiferromagnetic at low temperatures. That is, there are two sublattices with their spins arranged antiparallel. With the laser light you can raise some of the manganese ions, through a rather sharp absorption line at 5420 Å, to excited states and thereby change the spin. However, since this is a concentrated crystal the transitions are not strictly attributable to single ion transitions, but rather produce fairly narrow Frenkel excitons. The excitation is observed in this experiment through its magnetization detected by a pickup coil placed around the crystal. This is quite a difficult experiment and required a semiconductor preamplifier in the liquid helium to get a fast rise time, followed by additional amplifiers. Fig. 5 shows the apparatus.

By stressing this crystal, which has been extensively studied, the two exciton lines can be split so that the laser can be tuned to excite spins on either the one sublattice or the other sublattice. Moreover, the sharp exciton line is accompanied by spin-wave sidebands and if the laser is tuned a few angstroms, it can pump the spin-wave sideband. In that case, an exciton is produced on one sublattice and a magnon on the other sublattice. It was thus possible to measure relaxation times for both excitons and magnons.

This flashlamp-pumped laser can be refined and I believe that both the parametric oscillator and the flashlamp-pumped laser can be made as mono-chromatic as you need it to be, but you have to work for it. One problem is that the liquid in the dye cell becomes optically inhomogeneous under the intense heating of the flashlamp.

676

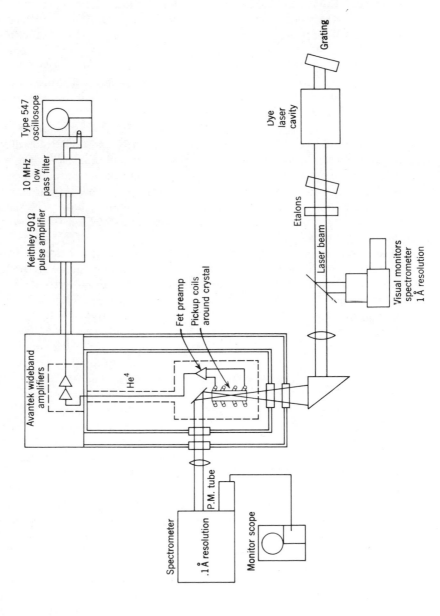

Figure 5. Apparatus for laser photomagnetism studies.

In many respects it is a lot easier to work, if you can, with a nitrogen laser. Nitrogen lasers are very effective for pumping dye lasers. They emit as a pump source in the ultraviolet at 3371 Å. The power output of one typical nitrogen laser is about 100 kW in a pulse which lasts about 10 nsec, or about as long as the excited states of typical dye molecules. It can be focused with a cylindrical lens to a line on the surface of the cell containing a dye, and many materials will lase under this kind of excitation [13]. I have been telling people for years that anything will lase if you hit it hard enough. If it has not lased yet, you have not hit it hard enough. This dye laser gives us a chance to prove our point. For example, following a suggestion of Mrs. F. Jurian, Tincture 99, Merthiolate (trade mark Eli Lilly and Company) was found to lase when pumped by the nitrogen laser.

To tune the laser you can put a diffraction grating at one end of the dye cell and a partially reflecting mirror at the output. As an alternative, an acousto-optic filter can be used for tuning the laser [14]. A wide variety of dyes have been discovered in the last few years and more are coming out all the time, so that laser action can be obtained at nearly any wavelength in the visible and near-visible region of the spectrum. There are some dyes which have tremendous tuning ranges. Methyl-umbelliferone, which forms an exciplex, fluoresces very strongly and will lase [15] from about 4000 Å in the violet, almost ultraviolet, right up into the green, beyond 5500 Å. So it is really a spectacular thing to turn this grating and see the color change so dramatically. Other dyes are used for other wavelength regions.

You can also take this dye laser output and pump other substances by end pumping. For example, when a blue beam from a nitrogen-pumped dye laser is directed into a single drop of fluorescein, a highly directional green beam emerges. This occurs since an extremely large gain is produced along the path traversed by the exciting beam, because of the large oscillator strength of the dye molecule transition.

With nitrogen lasers and with nitrogen-pumped liquid dye lasers as pumps, we have been able to obtain laser action from dyes in gelatin [16]. This provides further confirmation of our suspicion that nearly any substance will provide laser action if excited violently enough. It also provides the first edible laser material, if nontoxic dyes like fluorescein are used. Gelatin has long been used as an optical medium, particularly for color filters, and two commercial filters, Wratten 22 and 29, have been made to lase when pumped by a nitrogen laser. Moreover, gelatin is the base for most photographic films, so that complex film structures can be formed photographically. Thus Kogelnik and Shank [17] have shown that structures with periodic variation of refractive index, produced photographically in the gelatin medium, can provide selective reflection to tune laser action in the dyed gelatin.

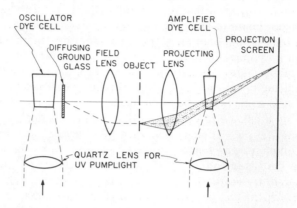

Figure 6.   Arrangement for image amplification.

Hänsch and Varsanyi have shown that nitrogen-pumped dye lasers can provide high-gain high-resolution amplification of an entire image [18].   The device is not merely an image intensifier but a true amplifier by stimulated emission.   Figures 6 and 7 show the arrangement.   The source of light to be amplified is an oscillator cell which contains dye pumped by the nitrogen laser.   It has to be rather bright, but not actually a laser.   The light from the small spot source is collimated and passed through an object, which is a transparency, and then is focused by a projection lens onto the image screen.   The amplifying cell is inserted between the projection lens and the screen, but at a place where the light has a small cross-section diameter. The dye is concentrated enough so that the pump light only gets in about a tenth of a millimeter so that the amplifying is done by a small volume of liquid on the edge of the cell, about a tenth of a millimeter in cross section. Even though the length of the dye cell is only 1.3 mm, the image is amplified by a factor of 100 to 1000 and the resolution is nearly diffraction limited for the 1/10-mm aperture.   This high resolution is quite remarkable in view of the very high pumping intensity from one side of the cell.   But refractive index changes from thermal expansion cannot occur before the short laser pulse ends, because the rate of expansion is limited by the velocity of sound.   This kind of an amplifier is not a low light level device because any maser amplifier produces a lot of spontaneous emission, as has been discussed by many people.   However, it could perhaps be made useful for large-screen displays.   It has the remarkable property, in contrast to image intensifiers, that it does preserve wavelengths.   The bandwidth of the rhodamine 6G dye that we used here is about 200 Å, but other dyes could

Figure 7.   Photograph of image amplifier.   The source is outside the picture to the left.   The amplified image can be seen on the right.

give greater bandwidths with reduced gain.

The absence of thermal blurring in the image amplifier suggested that
a well-collimated beam could be obtained from a dye laser cell.   Such a
beam would give sharply tuned reflection from a diffraction grating, whereas a range of wavelengths would be reflected for a beam broadened by
refractive inhomogeneities.   Hänsch [19] has shown that the effects of divergence from diffraction in the dye cell can be reduced and better collimation
at the grating can be attained, by placing a telescope between the dye cell
and the grating, as shown in Fig. 8.   Moreover, the telescope illuminates a
larger area on the grating, which also enhances the resolution.   With this
arrangement a linewidth of less than 0.1 Å is obtained.   A tilted quartz
Fabry-Perot etalon, 6 mm thick with broadband reflecting coatings on each

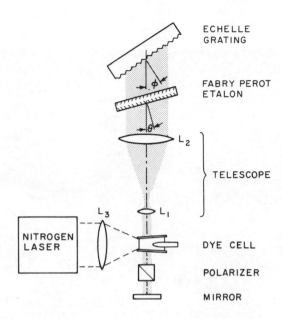

ECHELLE
GRATING

FABRY PEROT
ETALON

L_2

TELESCOPE

L_3 L_1

NITROGEN
LASER

DYE CELL

POLARIZER

MIRROR

Figure 8.   Monochromatic tunable dye laser.

side, placed between the telescope and gratings, further narrows the output to about 300 MHz, or 0.01 cm^{-1}. This is less than the Doppler width of most spectral lines in gases. Fine tuning is obtained by tilting the etalon.

One of the things you have to remember in trying to make a pulsed dye laser really monochromatic is that the pulse is rather short. It is only about 10 nsec from the nitrogen laser and the dye really only gives good gain for about 5 nsec or so. Therefore, light only travels about 5 ft in that time and you simply cannot use a high-Q resonator structure that requires light to travel back and forth over a long distance because there just is not time enough for it during the pulse. And so rather than try and select modes here very well, Hänsch has used a structure which provides good wavelength selection essentially by putting in a highly filtering spectrograph into the laser. With rhodamine 6G, the gain is so high that the output mirror can have very low reflectivity, and the shortness of the pulse makes it nearly useless to increase the resonator Q for regenerative line narrowing. This dye laser is really a very simple device, but it will produce an output stable within 1/100 Å with pulses up to 100 times a second, and do this for hours on end.

However this is not yet narrow enough for high-resolution saturation spectroscopy. Therefore, to do saturation spectroscopy, Hänsch has added to this a small passive confocal resonator with piezoelectric scanning. It filters to select a portion of the output. When you add the confocal resonator you do throw away some of the light, but the peak output power is about 1 W. To scan this device over a thousand megacycles or so, the grating is first set to the desired line, then the etalon is slowly tilted for the next finer tuning. The final fine tuning of the piezoelectric drive on the confocal resonator is synchronized by deriving its driving voltage from a potentiometer geared to the motor which turns the micrometer screw to tilt the etalon. This is a very simple method, but extremely effective, giving a linewidth of 7 MHz tunable over any desired range of about 2000 MHz. Incidentally, the confocal resonator lengthens the pulse from 5 nsec to 30 nsec by storing light in itself. With this device it has been possible, for the first time, to do high-resolution saturation spectroscopy experiments in an atomic vapor [20].

Figure 9 shows the apparatus used by Hänsch and Shahin for saturation spectroscopy of the D lines of sodium. As in the experiments of Hänsch, Levenson, and Sorem on iodine, a probe beam goes through the absorption cell one way and the saturating beam goes in the opposite direction. However, because the laser output fluctuates from pulse to pulse, an additional probe beam is used, which does not pass through the saturating region, and a differential amplifier corrects for fluctuations in the intensity of the laser beam. The sodium vapor used in this experiment is at a rather low pressure, about 10^{-7} Torr, but since sodium is a very strong absorber and has high oscillator strength, a power of a fraction of 1 W is sufficient.

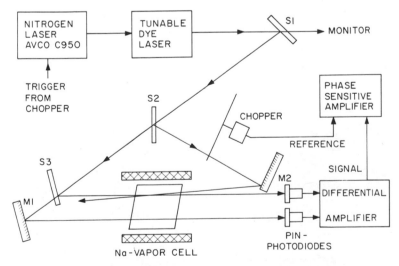

Figure 9.   Saturation spectrometer for sodium.

Hyperfine structures in sodium have been studied extensively, and the ground-state $^2S_{1/2}$ splitting is 1772 MHz. The $^2P_{1/2}$ has a spacing of 192 MHz and that of the $^2P_{3/2}$ is enough smaller so that it is not resolved even in these experiments.

Fig. 10 shows the hyperfine structure of the D_2 line, $^2P_{3/2} \leftarrow ^2S_{1/2}$, observed by this method. On this scale, the other D line would be about 90 m away, and so it is evident that the dispersion and resolution are very high. It will be noted that there is an inverted bump, corresponding to enhanced rather than decreased absorption halfway between the two peaks. With this particular apparatus we are able to put to good use the fact that the laser is pulsed to investigate this phenomenon. The probe beam can be delayed by an optical delay line until after the saturating beam has gone. Figure 10(b) shows the saturated absorption when the probe pulse is delayed 700 nsec. The fluorescent lifetime is 20 nsec so that excited states have decayed long before the probe pulse arrives, and yet the bleaching remains, about half as much as it was originally. The reason for this is that after the

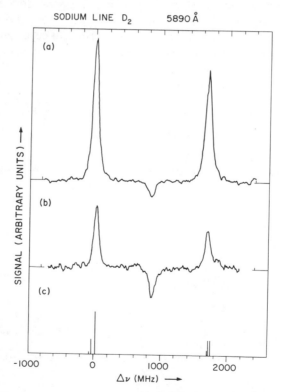

Figure 10.   (a) Saturation spectrum of the sodium D_2 line without delay; (b) Like (a), but with a probe delay of 700 nsec; (c) D_2 hyperfine transitions.

saturating beam excites atoms from one hyperfine level of the $^2S_{1/2}$ state, the excited atoms fluoresce and make transitions partly to their original level but also in large part to the other hyperfine level of the ground state. Thus, the population in the first hyperfine level, for atoms with the particular velocity, remains reduced and consequently the absorption is reduced, until those atoms move out of the beam and are replaced by others.

This remanent spectral bleaching provides a very narrow filter which could have a fractional band width of perhaps 10^{-8}, or with better substances maybe even 10^{-10}. Moreover a saturation filter could be made to transmit an entire image. So it is possible to imagine this device, or something like it, as a kind of a velocity-gated viewer through which you could view a scene, illuminated by a laser searchlight which also bleaches the filter. Then you would be able to see light which comes back only from stationary objects. If the objects are moving you would not see them. Of course, you could offset the filter and see moving objects.

It can be seen in Fig. 10 that the enhanced absorption dip between the two peaks is stronger when the probe pulse is delayed. This dip is observed

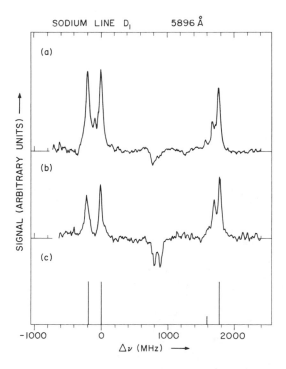

Figure 11.   (a) Saturation spectrum of the sodium D_1 line without delay; (b) Like (a), but with a probe delay of 56 nsec; (c) D_2 hyperfine transitions.

when the laser, being tuned halfway between the transitions from the two ground-state levels, cannot interact with atoms having velocities perpendicular to the laser beams. However, there are some atoms which are moving with a component of velocity along the laser beam such that, to them, the saturating beam frequency appears to coincide with one of the transitions and the probe appears to coincide with the other. Then the probe beam will sense the enhanced absorption resulting from the optical pumping process just described.

Fig. 11 shows the line, $^2P_{1/2} \leftarrow {}^2S_{1/2}$, i.e., the sodium D_1 line. The splitting, 192 MHz, of the state is clearly resolved. Between each pair of upper-state components, there is a crossover transition, again caused by molecules which have the proper component of velocity along the beams. The relative intensities are as they should be from the selection rules. The saturation method rather exaggerates the differences, as the signal is proportional to the square of the oscillator strength.

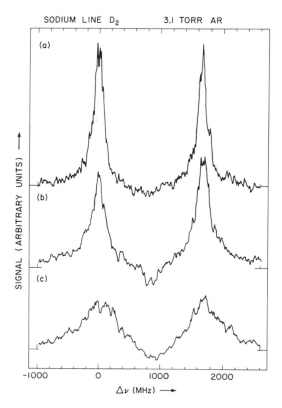

Figure 12.   Pressure-broadened line shapes with 3.1 Torr of argon. (a) Without delay, (b) 38-nsec delay, (c) 100-nsec delay.

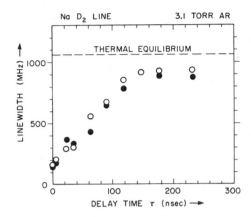

Figure 13.   Linewidth of sodium hyperfine components as a function of probe delay time, with 3.1 Torr of argon.

With this apparatus it is now possible to study how the atoms restore their velocity distribution. The saturating pulse bleaches a few atoms which have nearly zero component of velocity in the direction of the beam. Thus these atoms are labeled so that their recovery can be studied. The kind of collisions needed to restore the absorption are those which redistribute the velocity distribution. Harder collisions, or more of them, are needed than, say, for microwave relaxation of polar molecules because the collisions have to restore the actual velocity distribution as the spin states are rather insensitive to collisions. Weak, rather distant, collisions are the commonest ones. It takes quite a lot of them to restore the equilibrium in velocity. One can tell something about the hardness of them from what is known about interatomic forces. But to investigate them experimentally, a cell was made with a buffer gas (3.1 Torr of argon) and, as expected, the line broadens as the delay increases. Fig. 12 shows the line shapes with various probe delays. Fig. 13 shows the variation of linewidth with delay time. It can be fitted from a diffusion theory which assumes that no one collision makes a large change in the velocity direction of an atom. The comparison is complicated by other sources of line broadening. It does look, however, that this method can be used to study the collision processes in the atoms.

Continuous-wave dye lasers, also exist [21]. When pumped by an argon ion laser, various dyes have given continuous laser operation from approximately 5200 Å to 7000 Å. Single-mode operation has been achieved [22]. Ultimately, it should be possible to make continuous dye lasers with extremely high stability. However, the amplification available from the dye is less than in pulsed operation, and tuning elements must be carefully designed to have low losses. Meanwhile, pulsed tunable lasers now exist that can be made monochromatic enough to do saturation spectroscopy on any line from any gas, atomic or molecular, anywhere near the visible region. Because the pulses are so short, the states observed do not have to be long-lived states. The states may last a few nanoseconds. In all, tunable lasers have become very powerful tools with many interesting scientific applications.

REFERENCES

1. W. B. Tiffany, H. W. Moos, and A. L. Schawlow, Science 157, 40 (1967).
2. W. B. Tiffany, J. Chem. Phys. 48, 3019 (1968).
3. T. W. Hänsch, M. D. Levenson, and A. L. Schawlow, Phys. Rev. Letters 26, 16 (1971).
4. M. S. Sorem, M. D. Levenson, and A. L. Schawlow, Phys. Letters 37A, 33 (1971).
5. T. W. Hänsch and P. Toschek, IEEE J. Quantum Electron. QE-4, 467 (1968).
6. P. W. Smith and T. W. Hänsch, Phys. Rev. Letters 26, 29 (1971); C. Bordé in Proceedings of the Sixth International Quantum Electronics Conference, Kyoto, Japan, 1970 (to be published).
7. G. R. Hanes and C. E. Dahlstrom, Appl. Phys. Letters 14, 362 (1969).
8. M. Kroll, Phys. Rev. Letters 23, 631 (1969).
9. S. E. Harris, Proc. IEEE 57, 2096 (1969).
10. J. F. Holzrichter and A. L. Schawlow, Ann. N. Y. Acad. Sci. 168, 703 (1970).
11. J. F. Holzrichter, R. M. Macfarlane, and A. L. Schawlow, Phys. Rev. Letters 26, 946 (1971).
12. D. J. Bradley, Phys. Bull. 21, 116 (1970).
13. J. A. Myer, C. L. Johnson, E. Kierstead, R. D. Sharma, and I. Itzkan, Appl. Phys. Letters 16, 3 (1970).
14. D. J. Taylor, S. E. Harris, S. T. K. Nieh, and T. W. Hänsch, Appl. Phys. Letters 19, 269 (1971).
15. C. V. Shank, A. Dienes, A. M. Trozzolo, and J. A. Myer, Appl. Phys. Letters 16, 405 (1970).
16. T. W. Hänsch, M. Pernier, and A. L. Schawlow, IEEE J. Quantum Electron. QE-7, 45 (1971).
17. H. Kogelnik and C. V. Shank, Appl. Phys. Letters 18, 152 (1971).
18. T. W. Hänsch, F. Varsanyi, and A. L. Schawlow, Appl. Phys. Letters 18, 108 (1971).

19.   T. W. Hänsch, Appl. Opt. <u>11</u>, 895 (1972).
20.   T. W. Hänsch, I. Shahin and A. L. Schawlow, Phys. Rev. Letters <u>27</u>, 707 (1971).
21.   O. G. Peterson, S. A. Tuccio, and B. B. Snavely, Appl. Phys. Letters <u>17</u>, 245 (1970).
22.   M. Hercher, Opt. Comm. <u>3</u>, 346 (1971).

SPIN-FLIP RAMAN LASERS

C. K. N. Patel
Bell Telephone Laboratories
Holmdel, New Jersey

ABSTRACT

This paper reviews some of the recent work in the field of Raman lasers. In particular, we will emphasize the very recent developments in the field which has led to the operation of the spin-flip Raman laser at magnetic fields as low as 400 G. This development allows one to obtain a tunable spin-flip Raman laser radiation without having to resort to a superconducting magnet for the magnetic field. We will also briefly summarize some of the applications of the spin-flip Raman laser in areas such as high-resolution spectroscopy, transient infrared spectroscopy, pollution detection using the spin-flip Raman laser, and further nonlinear optics using the spin-flip Raman laser.

I. INTRODUCTION

There are a number of different possible techniques of producing tunable laser radiation in the infrared portion of the spectrum. These include using parametric oscillators, tunable diode lasers, and the spin-flip Raman lasers. In the present discussion we will concentrate only on the spin-flip Raman laser without any attempt being made to compare the three different techniques. The motivation for obtaining tunable spin-flip Raman laser leads us back to the time about 1964-1965 when the CO_2 lasers and the CO

689

lasers were undergoing very vigorous investigations. In 1965 cw powers in excess of 100 W were reported from a CO_2 laser and Q-switched powers in excess of 10 kW were obtained [1]. The CO laser was also shown to be capable [2] of producing a large amount of cw as well as Q-switched power output. Starting from this point the question was raised, how can we convert the fixed frequency output from these molecular lasers into tunable laser radiation which would increase the usefulness by a significant amount? Of the various possibilities we chose to explore the idea of Raman scattering and it has led to the development of the spin-flip Raman laser [3] as we know today.

In particular, Raman scattering from electrons in the presence of a magnetic field is the situation that will be considered from a number of different scattering processes. The spin-flip Raman scattering which results from a process involving the inelastic scattering of an input photon at frequency ω_0 to a frequency ω_s given by

$$\omega_s = \omega_0 - g\mu_B B \tag{1}$$

led to the development of the spin-flip Raman laser. Here ω_0 is incident photon frequency, g is the g value of electrons, μ_B is the Bohr magneton, and B is the magnetic field. Of course, Eq. (1) describes only the first-Stokes Raman scattering. We will also describe the anti-Stokes, second-Stokes, and third-Stokes spin-flip Raman lasers.

Section II describes the early results on spontaneous spin-flip Raman scattering [4] along with some of the theoretical background [5, 6] leading to the first experiment of the tunable spontaneous raman scattering. No attempt is made to give a complete review of the spontaneous work but only a very brief introductory description is given.

In Sec. III we briefly review the development and operation of the spin-flip Raman laser and describe some of the operating characteristics of the spin-flip Raman lasers. Here again details will not be given because recently a rather complete description of the operation of the spin-flip Raman laser has appeared [7]. We will touch only upon those aspects which either were not dealt with in the complete paper or those aspects which need to be reemphasized. In this section we will also discuss the linewidth considerations for the output for the spin-flip Raman laser and we will show that we have extended our measurement techniques to a point where we believe that the linewidth of the spin-flip Raman laser is considerably narrower than the Doppler width of an absorption line used to monitor the spin-flip Raman laser radiation.

In Sec. IV we will describe some of the very recent experiments on low-field operation of the spin-flip Raman laser where we have extended the tuning range of the spin-flip Raman laser to magnetic fields as low as 400 G.

This operation is obtained using a very low carrier concentration InSb sample and it leads us to the conclusion that for some applications a super-conducting magnet may not be necessary to obtain tunable Raman laser radiation in the infrared.

In Sec. V we will discuss some of the applications of the spin-flip Raman laser and show that the spin-flip Raman laser has a considerable number of applications at the present time as well as a very large future. In the last section we will conclude by pointing to some of the future developments which will be useful and which appear to be possible. These will extend the tuning range as well as the usefulness of the spin-flip Raman lasers.

II. SPONTANEOUS RAMAN SCATTERING FROM ELECTRONS IN SEMICONDUCTORS

When looking for processes which give rise to tunable Raman scattering, i.e., a Raman-scattering process in which the Raman-scattered frequency is changeable can be best visualized by looking at Fig. 1 which shows a simple picture of what happens during a process of inelastic scattering of a photon in a medium. A photon at frequency ω_0 having a polarization vector ϵ_0 and a propagation vector $\vec{k}_0$ is incident on a material shown in the circle. We look for scattered light at some angle θ and at frequency ω_S having a polarization vector ϵ_S and a propagation vector $\vec{k}_S$. In a process which involves Raman scattering the output frequency ω_S is different from the incident frequency ω_0. The energy conservation requirement tells us that in the material in which the inelastic Raman scattering is obtained, an energy equal to $\hbar\omega$ given by

$$\hbar\omega = \hbar\omega_0 - \hbar\omega_S \tag{2}$$

$\omega = \omega_0 - \omega_S$ ENERGY TRANSFER

$\underset{\sim}{q} = \underset{\sim}{k}_0 - \underset{\sim}{k}_S$ MOMENTUM TRANSFER

Figure 1. Schematic of Raman scattering.

should be absorbed. This implies that the material in which Raman scattering is obtained must have an elementary excitation having an energy equal to $\hbar\omega$ as dictated by the energy conservation requirement. There is also the momentum requirement which dictates that a momentum k given by

$$\vec{k} = \vec{k}_0 - \vec{k}_s \tag{3}$$

should be taken up by the elementary excitation responsible for the inelastic light scattering. It is easy to see that in order to obtain Raman-scattered frequency, ω_s, which is tunable we need a process or a medium which has elementary excitations whose frequency ω can be easily changed. Electron gas in a semiconductor is one such convenient medium. Electron gas in a semiconductor possesses two distinct types of elementary excitations. The first arises from the collective oscillation of the electrons, i.e., plasma oscillations, or the plasmons. The plasma frequency, ω_p, is given by

$$\omega_p = (4\pi ne^2/m^*\epsilon_0)^{1/2} \tag{4}$$

where n is the electron or free-carrier concentration, e is the electron charge, m^* is the effective mass of the carriers in the semiconductor, and ϵ_0 is the dielectric constant. The other set of variable frequency elementary excitations of the electron gas arise from the single-particle excitations, with or without magnetic field. The first process, namely, the collective oscillations, easily leads to a Raman scattering yielding tunable Raman frequency, however, this process requires that in order to change the Raman frequency the carrier concentration be changed which is not very convenient. Thus for the purposes of the present discussion we will restrict ourselves only to the single-particle excitations of the electron gas.

In 1966, Wolff suggested [5] that single-particle excitations of electrons in a magnetic field should be capable of giving rise to sizeable inelastic light scattering in a semiconductor. He suggested that in the presence of a magnetic field the electron in a semiconductor such as InSb should give rise to a Raman-scattering process in which the scattered frequency ω_s is given by

$$\omega_s \approx \omega_0 - 2\omega_c \tag{5}$$

ω_c is the cyclotron frequency of the electrons given by $\omega_c = eB/m^*c$, and c is the velocity of light. To see this a little more clearly we show Fig. 2 which illustrates the energy-level diagram of electrons in InSb in presence of the magnetic field (at the bottom of the electron band, i.e., at $k_z = 0$ where k_z is the z-directed momentum of the electrons in InSb with magnetic field being along the z direction). The electrons are quantized in Landau levels where the Landau level quantum number, ℓ, equals 0, 1, 2, 3, etc. Each one of the Landau levels is split into two spin substates because of the removal of the spin degeneracy. Wolff suggested that process marked (a) should have a large cross section and this process is tunable because the cyclotron frequency is tunable by changing the magnetic field. Subsequently, Yafet [6] proposed that the process shown by (b) should also have a large cross section. This process is the one in which the Landau level quantum number of the electron is not changed but only the spin state is changed. Raman scattering is described by Eq. (1). Experimentally a third process shown as (c) corresponding to

$$\omega_s = \omega_0 - \omega_c \tag{6}$$

was observed in addition to the processes (a) and (b) when Slusher, Patel, and Fleury reported [4] the first results of spontaneous Raman scattering from electrons in InSb in the presence of a magnetic field.

Figure 3 shows one of these results of spontaneous Raman scattering with the incident photons at a wavelength of $10.6 \ \mu$ (using the CO_2 laser). Here we see the output spectrum of spontaneous light scattering at two

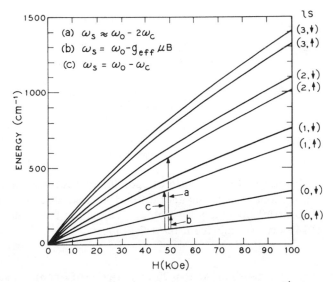

Figure 2.   Energy-level diagram of electrons in InSb at $\vec{k}_z = 0$ where the magnetic field is directed along the z direction. The energy levels are plotted from the bottom of the conduction band in zero magnetic field as a function of the magnetic field.

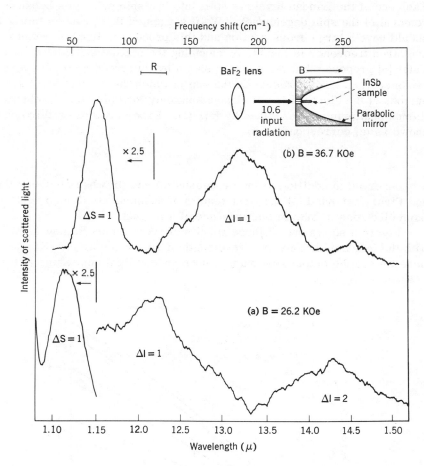

Figure 3.   Spectrometer analysis of spontaneous Raman scattering from electrons in presence of the magnetic field in InSb as a function of wavelength for two different magnetic fields. The pump frequency corresponds to the 10.5915-μ transition from the CO_2 laser.

different magnetic fields. At B $\approx$ 26 kG, three different peaks are observed. These are identified as the spin-flip Raman scattering, $\Delta\ell = 1$ and $\Delta\ell = 2$ Raman scattering. As the magnetic field is increased to 37 kG, all three peaks are seen to move to longer wavelengths indicating the tunable nature of the present Raman scattering. Three points should be mentioned.

1. $\Delta l = 2$ process has the largest tunability and the spin-flip process has the smallest tunability.

2.   The Raman-scattering cross section for the spin-flip Raman process appears to be considerably larger than that for the $\Delta l = 1$ and $\Delta l = 2$ process.

3.   The linewidth of the spin-flip Raman process appears to be considerably narrower than that for the $\Delta l = 1$ and $\Delta l = 2$ processes.

Figure 4 shows the tuning characteristics of the spontaneous Raman scattering for electrons in InSb. We see that using modest magnetic fields of the order of 50-100 kG large tuning range is obtained for all the three processes. The cross sections for scattering for the three processes are approximately 10^{-24} cm^2 sr^{-1} for the $\Delta l = 1$ and the $\Delta l = 2$ processes, and

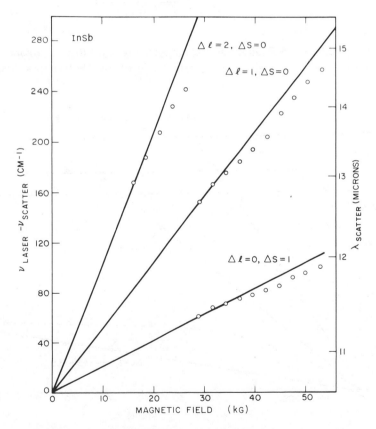

Figure 4.   Tuning characteristics of spontaneous Raman scattering from electrons in InSb as a function of magnetic field.

about 10^{-23} cm^2 sr^{-1} for the spin–flip Raman process. These cross sections are per electron and are in agreement with the calculated values [5, 6].

Subsequently, Raman scattering was also investigated in InAs [7] and PbTe [8]. Spin–flip Raman scattering as well as the $\Delta l = 1$ and $\Delta l = 2$ processes were observed. Tuning characteristics for spontaneous Raman scattering from InAs and PbTe are shown in Figs. 5 and 6, respectively.

However, the cross section for Raman scattering in InSb appears to be about the largest we have seen to date. In order to go from the spontaneous Raman scattering to stimulated Raman scattering one has to calculate the amount of Raman gain one would observe in a particular Raman process. Equation (7) shows the Raman gain in terms of various parameters:

$$g_S = \frac{16\pi^2 c^2 (S/l\,d\Omega)}{\hbar\omega_s^3 n_p^2 (\bar{n}+1)\,\Gamma}\,I \qquad (7)$$

where $S/l\,d\Omega$ is the Raman-scattering efficiency which is proportional to the number of carriers and the single-particle cross section (as discussed in

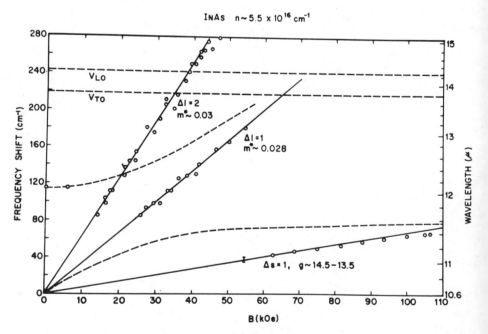

Figure 5.   Tuning characteristics of spontaneous Raman scattering from electrons in InAs as a function of magnetic field.

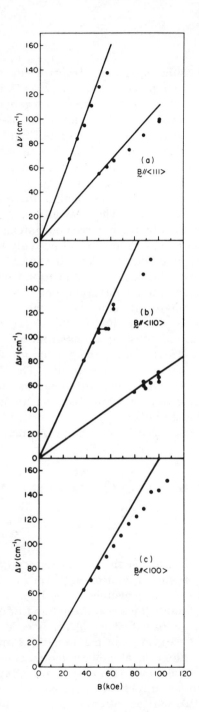

Figure 6.   Tuning curves for spontaneous spin-flip Raman scattering from electrons in PbTe as a function of magnetic field for three different orientations of the magnetic field with respect to the PbTe crystalline axes.

Ref. 9), Γ is the spontaneous SFR linewidth, n_p is the refractive index at pump frequency, I is the intensity of the pump radiation, and $(\bar{n}+1)$ is the Boltzmann factor (see Ref. 9). It can be seen that to obtain a large gain one needs a large scattering cross section and a small linewidth. The measured spontaneous Raman cross sections are the largest for the spin-flip process, and the measured linewidth for the spin-flip process, Γ_{SFR}, appears to be less than 1 cm^{-1}. The $\Delta \ell = 1$ and the $\Delta \ell = 2$ processes yield a spontaneous Raman linewidth of 50-100 cm^{-1}. The reason for the difference between the two linewidths arises from the particular relaxation mechanism for the different levels involved in the Raman scattering. The linewidths for $\Delta \ell = 1$ and $\Delta \ell = 2$ are essentially determined by the electron collisional relaxation time in InSb since the Landau levels relax very effectively by electron collisions. On the other hand, electron collisions are notably poor in reversing the spin of the electrons, and thus the spin relaxation time is considerably longer than the electron momentum relaxation time. It is this fact that gives us the linewidth for the spin-flip Raman process which is considerably narrower than that for the $\Delta \ell = 1$ and the $\Delta \ell = 2$ processes. When one looks at details, one finds that the spin-flip Raman linewidth is essentially given by the nonparabolicity effects, i.e., the g value for the electrons which varies as a function of energy of the electrons. Thus in spite of the smallest tuning range, we see that the spin-flip process appears to be more suited for obtaining stimulated Raman scattering or a Raman laser because of its large gain. One should refer to Ref. 9 for details of the discussion of Raman gain and the losses that have to be overcome.

III.   CHARACTERISTICS OF THE SPIN-FLIP RAMAN LASER

A.   Stokes Spin-Flip Raman Laser

Introduction of appropriate numbers [3] in Eq. (7) shows that Raman gain $\approx 10^{-5}$I cm^{-1}, where I is the intensity of pump radiation in W cm^{-2}. Thus using 10.6μ power ~ 1 kW as obtained from a Q-switched CO_2 laser even a mild focusing gives rise to very large Raman gains and stimulated Raman scattering should be possible. In 1969, Patel and Shaw [3] reported the first spin-flip Raman laser. Since the first observation of the spin-flip Raman laser, progress has been rapid and the latest developments include cw-operating spin-flip Raman lasers, high peak power spin-flip Raman lasers, and spin-flip Raman lasers operating on the anti-Stokes processes as well as on higher-order Stokes processes. We will not attempt to review all the spin-flip Raman laser details that have been published to date but touch upon only those which appear to be common to all the different spin-flip Raman

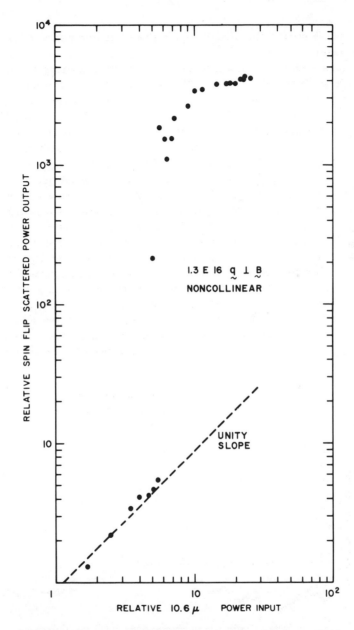

Figure 7. Spin-flip Raman-scattered output at $\lambda = 11.7$ μ as a function of pumped power at 10.6 μ (in arbitrary units) for n–InSb with $n_e = 1.3 \times 10^{16}$ cm^{-3}, $T = 18°K$, and B = 40 kG. The maximum pump power is 1.0 kW inside of sample and the maximum SFR laser output is 100 W peak.

lasers and describe the central operating characteristics of the spin–flip Raman lasers.

Figure 7 shows the input versus output characteristics for the spin–flip Raman laser pumped at 10.6μ [3]. The Raman–scattered power is measured at a wavelength of 11.7μ from an InSb sample having a carrier concentration of 1.3×10^{16} cm^{-3} in a magnetic field of ~ 45 kG. At low pump power intensities we see a linear region of output versus input indicating essentially a spontaneous region where the linewidth of the output scattered light is still ≈ 2 cm^{-1} (see Fig. 8) and the relative intensity of the output scattered light is about 10^{-6} of the input radiation. As we increase the pump intensity we see a very sharp increase in the output power for a very small increase in the input power. This point is identified as a point at which the Raman gain becomes greater than the losses in the InSb sample. We should point out that in these experiments the Raman laser consisted of an InSb sample in a magnetic field with no high reflectivity coatings on the sample. This means that the reflectivity was approximately 36% for the Raman cavity. Even with such a poor cavity the Raman gain exceeds the Raman losses and stimulated Raman scattering is obtained. The maximum Q–switched power output from the CO_2 laser used in this experiment was about 1 kW and spin–flip Raman laser power output was approximately 100 W. By using higher pump powers, outputs in excess of 1 kW have now been

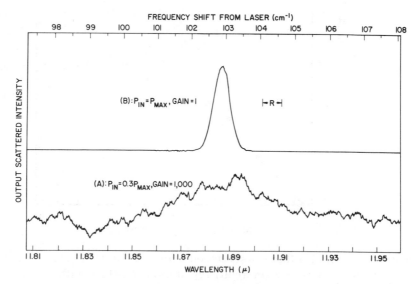

Figure 8.   Spectral analysis of SFR–scattered power output below and above stimulated emission threshold (the gain in the figure refers to the amplifier gain) for n–InSb, R is the spectrometer resolution.

obtained [10].   In addition to the observation of increased Raman-scattering efficiency above the threshold for obtaining a spin–flip Raman laser, we should also look at the linewidth above and below threshold.   Figure 8 shows the spectral analysis of the Raman-scattered power output below and above threshold.   We see that in going from below SFR laser threshold to above, the linewidth of the output radiation is narrowed considerably.   In Fig. 8 the stimulated emission linewidth is limited by the spectrometer resolution. We have attempted to measure the linewidth of the spin–flip Raman laser using a spectrometer but that gives only a upper limit of ~ 0.03 cm^{-1} limited by the spectrometer resolution.   In Sec. V we will show that we have been able to resolve the linewidth of the spin–flip Raman laser to ~ 0.006 cm^{-1} using molecular absorption in gases.   The linewidth of the output radiation is extremely narrow and calculations [9] yield a linewidth which is of the order of a few Hz.

Figure 9 shows the tuning characteristics of the spin–flip Raman laser pumped at 10.6 μ where we see a tuning curve extending from about 10.9 μ to around 13 μ using a magnetic field ~ 100 kG.   This is the overall tuning characteristics and it says nothing about the fine–tuning characteristics of the spin–flip Raman laser.   The laser action due to the spin–flip Raman process occurs in the InSb sample with specular reflection from the surfaces which form the laser cavity.   Figure 10 shows the Raman laser cavity modes that are separated by $c/2n\ell$, where n is the refractive index of InSb.   For a cavity length, $\ell \approx 2$ mm the separation of the cavity modes is ~ 0.6 cm^{-1}. The spontaneous SFR line having a width of ~ 1 cm^{-1} shown underneath

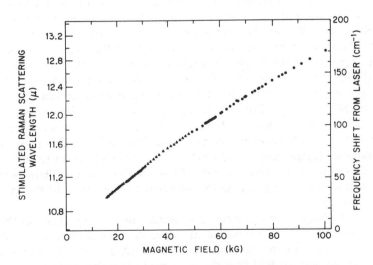

Figure 9.   Overall tuning curve for the spin–flip Raman laser pumped with the CO$_2$ laser at 10.591 μ.

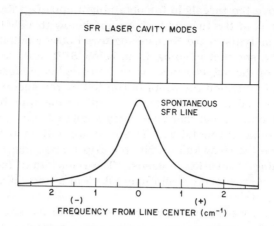

SFR LASER CAVITY MODES

SPONTANEOUS
SFR LINE

2   1   0   1   2
(−)   (+)

FREQUENCY FROM LINE CENTER (cm⁻¹)

Figure 10.   Cavity modes and spontaneous SFR emission line for assumed cavity length of 2 mm and InSb SFR spontaneous linewidth of 1 cm^{-1}.

indicates that as we sweep the spontaneous line by changing the magnetic field, the laser output frequency should jump from one cavity mode to the next cavity mode. Experimentally we attempted to see how the output frequency tunes. Two distinct observations should be reported.

    1. At very high pumped intensities, e.g., pump intensities about a factor of 20 above the threshold pump power, essentially no mode jumping was observed. Figure 11 shows the measurements of the output frequency using a spectrometer. We have subtracted out the linear portion of the tuning curve from the output frequency and have plotted the error as a function of SFR laser frequency. We see that the maximum tuning error from linearity is ≤ 0.03 cm^{-1}. What is more interesting is that the tuning error does not show a periodicity of the cavity mode separation shown by two vertical bars in the figure. If there were any mode jumping effects the error should be periodic in the cavity mode spacing. The error period is considerably larger and is ~ 2 cm^{-1}. This periodicity is traced to the periodic error in the drive screw of the spectrometer. Thus even the 0.03 cm^{-1} tuning error may not be the inherent tuning error of the spin–flip Raman laser.

    2. The tuning characteristic at pump power very close to the threshold for spin–flip Raman laser action are considerably different. Here the output of the spin–flip Raman laser shows considerable variation as the magnetic field is tuned and we find that there is mode jumping and discrete frequency tuning. But this happens only very close to the spin–flip Raman laser threshold.

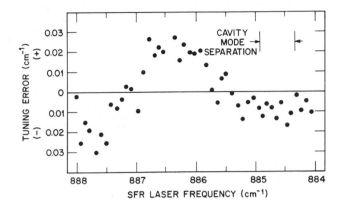

Figure 11.   Tuning error of the output frequency of the SFR laser as a function of expected SFR laser frequency.   The tuning error is obtained by subtracting out from the actual SFR laser frequency the expected tuning characteristics known from the previously measured g values.

It is not unreasonable to expect that the tuning behavior will be different at threshold and considerably above threshold.   We should point out that this behavior is not inconsistent with the fact that we have a very low Q cavity because of the reflectivity of only about 36%, and a spontaneous spin-flip Raman linewidth which is ≤ 1 cm^{-1}.   Some frequency pulling should occur even at very high pump powers, but at the present time measurements have not been possible.   The argument against superradiant spin-flip Raman scattering instead of the SFR laser are given in Ref. 9.

Another operating characteristic of the spin-flip Raman laser is a minimum magnetic field below which the SFR laser does not oscillate. This magnetic field is governed by the quantum limit consideration discussed in Ref. 9 and reproduced in Fig. 12 where we show the minimum magnetic field versus the carrier concentration.   The solid line is the theory and the dots are the experimental points.   This requirement arises from the fact that we must have the upper spin level out of the Fermi sea in order to make a spin-flip Raman laser oscillate.   This argument which has been experimentally verified shows that to operate a spin-flip Raman laser at lower magnetic fields one should use lower carrier concentration.   There is, however, a problem with decreasing the carrier concentration, viz., the Raman gain is proportional to the carrier concentration and thus the threshold for spin-flip Raman laser oscillation will go up in trying to obtain a low field tuning. We will discuss this in somewhat greater detail in Sec. IV.

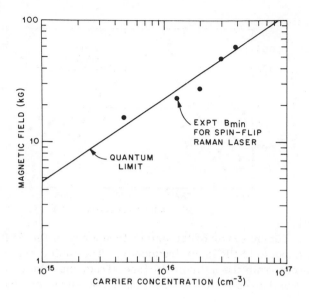

Figure 12.   The lowest magnetic field to obtain stimulated SFR scattering as a function of carrier concentration in InSb (T ≈ 20°K, $\vec{q}_e \perp \vec{B}$). The solid line shows the theoretically calculated magnetic field necessary to reach the quantum limit.

B.   Anti-Stokes Spin-Flip Raman Laser

Corresponding to every Stokes process there is an anti-Stokes Raman-scattering process.   Under suitable and proper experimental conditions the anti-Stokes Raman-scattering process should also be seen in stimulated emission.   (See Ref. 11 for details.)   Experimentally the anti-Stokes SFR laser in InSb is depicted in Fig. 13 which shows the spectrum of the Stokes and anti-Stokes spin-flip Raman laser pumped at 10.6 μ.   The SFR laser action now occurs at two frequencies one below and one above the pump frequency.   The output frequency of the anti-Stokes spin-flip Raman laser is given by

$$\omega_{as} = \omega_o + g\mu_B B \tag{8}$$

The anti-Stokes spin-flip Raman laser tuning is given in Fig. 14.   The anti-Stokes spin-flip Raman laser is tunable from 9.2 μ to 10 μ.   In some of the recent experiments [10], Aggarwal et al. have reported an anti-Stokes spin-flip Raman laser tunable to somewhat shorter wavelengths.

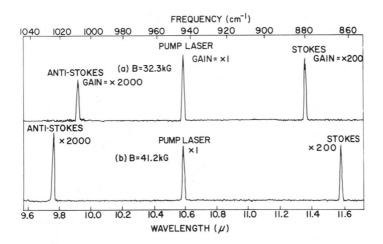

Figure 13.   Spectrometer analysis of the output from InSb sample at B = 33.32 kG and B = 41.2 kG showing the stimulated anti-Stokes SFR scattering (left) stimulated Stokes SFR scattering (right) and the pump laser (center) wavelengths.   The widths of the spectral lines are spectrometer resolution limited.   The Stokes and the anti-Stokes line positions appear asymmetrical with respect to the laser because the spectrometer sweep is linear in wavelength.

C.   cw Spin-Flip Raman Laser

The threshold for the 10.6 μ pumped CO_2 SFR laser is [9] approximately 100 W.   In order to decrease this threshold to a point where cw spin-flip Raman laser action is possible, one must either increase the gain or reduce the losses.   A technique for increasing the Raman gain is the resonant enhancement of the scattering cross section, given in Eq. (9)

$$\text{Enhancement} \approx \left(\frac{E_g^{'2}}{E_g^{'2} - (\hbar \omega_o)^2} \right)^2 \tag{9}$$

where E_g' is the effective band gap and $\hbar \omega_o$ is the pump photon energy.   The use of a pump frequency very close to the band gap of InSb (e.g., using a CO laser [2] as the pump to obtain radiation at 5.3 μ which is very close to the band gap of InSb) should give rise to a considerably larger cross section for

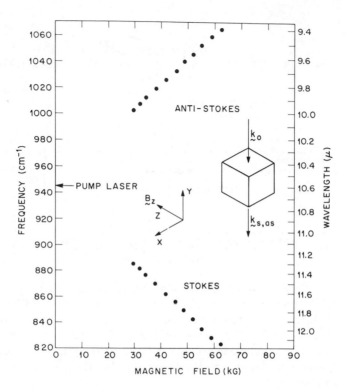

Figure 14.   Tuning of the anti-Stokes and the Stokes spin-flip Raman lasers in InSb as a function of the magnetic field when pumped at 10.5915 μ.   (The inset shows the experimental geometry.)

the spin-flip Raman scattering.   Without going through the details it can be seen that this scattering enhancement can be quite large.   Mooradian et al. [12] showed that this enhancement is more than enough to produce a cw spin-flip Raman laser.   Subsequently, it was also shown that the enhancement of the cross section is sufficient to give rise to the second as well as the third stimulated Stokes scattering [13].   Figure 14 shows the operation of the cw spin-flip Raman laser at 5.3 μ using the CO laser as a pump [13], where we see the first- as well as second-Stokes spin-flip Raman lasers.   Figure 15 shows the tuning characteristics for the first- and second-Stokes cw spin-flip Raman laser. Under ideal conditions the third-Stokes SFR laser has been observed under cw operation.   The maximum cw spin-flip Raman laser output on the first Stokes has been ~1 W corresponding to ~70% conversion from pump to the first-Stokes radiation.   Figure 16 shows the tunability of a first- and second-Stokes SFR laser using the cw CO laser as a pump.

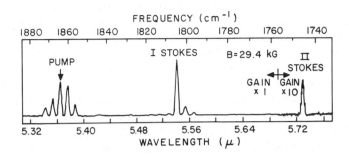

Figure 15.   Spectrum of the output from an SFR laser in n-InSb pumped with a cw carbon monoxide laser showing the pump wavelength, first- and second-Stokes SFR laser wavelength.

D.   Miscellaneous

In all of the above experiments, the optimum electron concentration for maximum spin-flip Raman laser output was seen to be around 1.3×10^{16} cm^{-3}.   Reference 9 gives arguments about why an optimum carrier concentration should exist.   In order to go to lower magnetic fields, one should use lower carrier concentration samples which will be dealt with in the next section.   It should also be pointed out that second-Stokes spin-flip Raman laser action has also been recently [10] obtained at 10.6 μ.

IV.   LOW-FIELD SPIN-FLIP RAMAN LASER [14]

It can be seen [15] that the CO_2 laser as well as the CO laser output can be made to occur on a number of frequencies separated by ~ 2 cm^{-1} for the CO_2 laser and ~ 4 cm^{-1} for the CO laser.   This implies that in order to obtain a continuously tunable laser source the maximum tunability that is required from the spin-flip Raman laser is only ~ 4 cm^{-1}.   This corresponds to a magnetic field of only ~ 2 kG.   Thus if it is possible to extend the minimum magnetic field range from around 14 kG described [9] in Sec. III to lower magnetic fields, a permanent magnet or an electromagnet should yield nearly the same tuning range as that can be obtained using superconducting magnets. What is done here is that one tunes the SFR laser from one laser line to the next laser line at which point one switches the pump frequency and starts the tuning curve again.   Thus it is important to investigate the low-field operating characteristics.   Recently we have obtained the low-field spin-flip Raman laser at magnetic fields as low as 400 G [14].   This was obtained using an InSb sample having an electron concentration of around 1×10^{15} cm^{-3}.

This electron concentration is more than a factor of 10 lower than the optimum required for producing maximum spin–flip Raman laser output as described in Sec. III.   Most of the experiments were carried out at 5.3 μ using the CO laser as the pump in the low–field spin–flip Raman laser experiments given below.   The enhancement of cross section which facilitates the cw spin–flip Raman laser is given by Eq. (9).   The maximum enhancement obtainable is determined by the phenomenlogical relaxation time, τ, although it is not introduced formally in the Eq. (9).   To the best of the author's knowledge nobody has yet put the relaxation time formally but it must appear in the enhancement and the maximum enhancement would be determined by τ.   By lowering the electron concentration the electron collision time in InSb can be considerably increased.   In particular, the relaxation time for electrons in InSb with an electron concentration of about 1×10^{15} cm^{-3} is approximately three times longer than that in the InSb sample having a carrier concentration of 1×10^{16} cm^{-3}.   Thus at least phenomenologically the enhancement would be approximately a factor of 3^4 larger in the 1×10^{16} cm^{-3} sample.   Thus the reduction in carrier

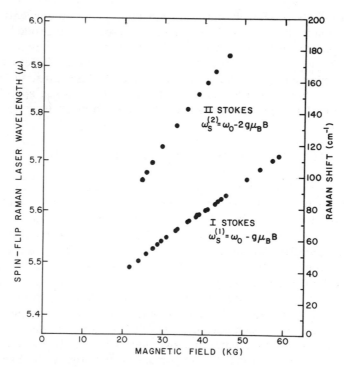

Figure 16.   Tunability of the cw first- and second–Stokes SFR laser in InSb obtained with a pump at 5.3648 μ from a cw CO laser.

concentration will be more than offset by the increase in the maximum enhancement. In addition to the improvement in the maximum enhancement of cross section, the increased electron collision time also affects the spin relaxation time, and there will be a decrease in the spontaneous spin–flip Raman scattering linewidth [14]. A lower spontaneous SFR linewidth can result from the lower carrier concentration material since the Fermi energy is considerably smaller and thus the contribution of the nonparabolicity to the linewidth will also be smaller. It is estimated that the increase in the Raman–scattering efficiency coming from increased enhancement and the decrease in the spontaneous Raman scattering linewidth will more than make up for the lower electron concentration. Experimentally this is observed. Using an electron concentration of 1×10^{15} cm^{-3}, we observed SFR laser at magnetic fields as low as 400 G with only about a factor–of–2 increase in the threshold for stimulated spin–flip Raman scattering when compared with the

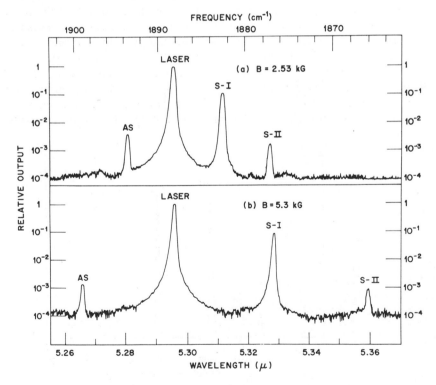

Figure 17.   Spectrum of a low magnetic field spin-flip Raman laser at magnetic fields of 2.6 and 5.2 kG. Note that we are observing the cw anti-Stokes SFR laser at 5.3 μ for the first time.

threshold for the 1×10^{16} cm^{-3} sample. The threshold for stimulated SFR scattering using the 5.3-μ pump radiation and electron concentration of 1×10^{16} cm^{-3} is reported [16] to be 50 mW.

Figure 17 shows the SFR laser operation at two different magnetic fields, 2.5 kG and 5.3 kG. We see the first-Stokes, second-Stokes, and now for the first time cw anti-Stokes spin-flip Raman laser oscillation. It should be pointed out that we are operating under cw conditions. Subsequent to the work which is shown in Fig. 17, we have improved the experimental conditions and used a small 4 in. Varian electromagnet to obtain the low-field operation. We observed that extremely high conversion efficiencies can also be obtained even at low magnetic fields. Figure 18 shows the operation of the spin-flip Raman laser at the magnetic field of 1141 G. This field is obtained by using a permanent magnet. There are no filters or no polarizers which discriminate against any of the radiation and we see that the power put on the first-Stokes spin-flip Raman laser is $\approx$ 1 W. There is $\approx$ 1 W of unused pumped radiation coming through and there is approximately 300 mW of anti-Stokes SFR laser. There is about 10 mW of the second-Stokes spin-flip laser radiation out. This experiment was carried out under true cw-operating conditions.because now the sample is immersed in the helium so a considerably better heat transfer can be obtained.

Figure 19 shows the low-field tuning characteristic of the spin-flip Raman laser. (For details of this work see Ref. 14.) It can be seen that now all that is necessary is a small electromagnet or a permanent magnet capable of magnetic fields of the order of 2-3 kG, to obtain a tunable spin-

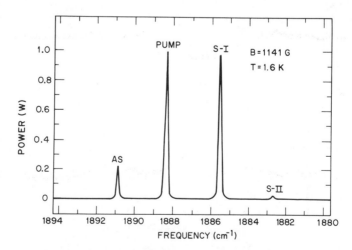

Figure 18.   Low magnetic field SFR laser at 1141 G showing very high conversion efficiency from the pumped radiation at 5.3 μ to the SFR laser. The magnetic field of 1141 G is obtained using a permanent magnet.

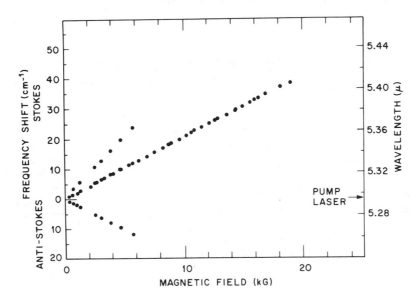

Figure 19.   Low magnetic field SFR laser tuning curves using the $P_{9-8}(12)$ transition of the cw CO laser as the pump.

flip Raman laser over a very wide frequency range by switching the pump frequency every time one runs out of a magnetic field. This will allow us to design experiments where the size of the magnets is important. In Fig. 19 we see a temperature of $1.6°K$. However, we have also observed the low-field spin-flip Raman laser action at temperatures as high as $20-25°K$ indicating that liquid helium is not a necessary requirement, and a closed-cycle cooler which takes us down to $20°K$ should be more than adequate.

The low-field spin-flip Raman laser has also been operated at $10.6\ \mu$ using the Q-switched CO_2 laser as the pump [17]. Here, however, no advantage is obtained from the increased enhancement yet the advantage from the narrower linewidth still exists. This has allowed us to obtain the spin-flip Raman laser at low fields using only about 5 times higher pump power than that required for optimum carrier concentration in InSb samples. The details of the low-field $10.6-\mu$ pumped spin-flip Raman laser will be published elsewhere [17].

V.   APPLICATIONS OF THE SPIN-FLIP RAMAN LASER

The extremely narrow linewidth and the wide tuning range (see the summary

712   C. K. N. Patel

TABLE I. SPIN–FLIP RAMAN LASER.

Pump ⟹ ⊠ InSb ⟹ Tunable SFR laser

Pump	Threshold power	Power output		Tunability (100 kG)	Remarks
		Pulsed	Cw		
CO_2 laser ~10.6 μ	~100 W	$> 10^3$ W		$9.0\,\mu - 14.6\,\mu$	I, II Stokes Anti-Stokes
CO laser ~5.3 μ	~0.2 W	~20 W	~0.5 W	$5.2\,\mu - 6.2\,\mu$	I, II, III Stokes Anti-Stokes

Measured linewidth $\lesssim 0.03$ cm^{-1}
Measured linearity of tuning 1:3 × 10^4

as in Table I) gives us a very versatile tool of coherent radiation which has a very large number of applications. Some of these applications are listed below:

1. ir Spectroscopy
    (a) Conventional spectroscopy
    (b) Time-resolved spectroscopy
2. Pollution detection
3. Local oscillator in communications and radar
4. Further nonlinear optics
    (a) SHG and other
    (b) Difference-frequency mixing to generate tunable far ir in the 100-μ range.

We have investigated a few of these applications and we briefly describe these below.

A. Spectroscopy

Using the 10.6-μ pumped SFR laser, we investigated the absorption spectrum of NH_3 in the region around 12 μ. The experimental setup is shown in Fig. 20. The magnetic field reading from the magnetic field sensor is converted into frequency by tuning curve given in Fig. 9.

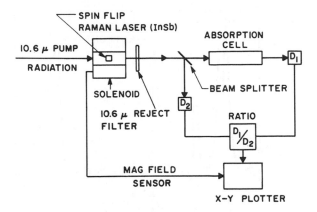

Figure 20.   Experimental setup for using the tunable SFR laser as an infrared spectrometer.

Figure 21 shows the absorption spectrum of ammonia obtained using the SFR laser spectrometer shown in Fig. 20, as well as the best spectrum of ammonia that is available to date.   Comparison between the two indicates the superiority of the SFR laser as an infrared spectrometer.   It should be pointed out that the spectrum that is given in Fig. 21 using the spin–flip Raman laser as a source is not the best obtained to date.   The problem that one has to watch out for when trying to obtain extremely high resolution results is that the tuning of the SFR laser is very rapid.   The tuning rate is approximately 70 MHz/G.   The linewidths of absorption lines in the infrared are typically in the range 50-100 MHz which means that a very precise control of the magnetic field is necessary to obtain highest resolution possible from the SFR laser spectrometer.   Figure 22 shows the absorption spectrum of water vapor at a frequency of 1885.24 cm^{-1} obtained using the low–field cw SFR laser [18] at two different air pressures.   We see the complete difference between the two spectra.   The observed broadening of the water vapor line at 760 Torr is in very good agreement with that calculated [19].   At low pressure we measure a linewidth of ~ 195 MHz.   It should be pointed out that all of the 195 MHz comes from the Doppler width of ~ 100 MHz, and a pressure broadening of ~ 130 MHz.   So the linewidth of the SFR laser is indeed considerably narrower than 195 MHz, and there is no reason to doubt that it is any wider than what is given by the Schawlow–Townes limit.   We are at the present time in the process of measuring the true linewidth of the spin–flip Raman laser using heterodyne techniques [18].

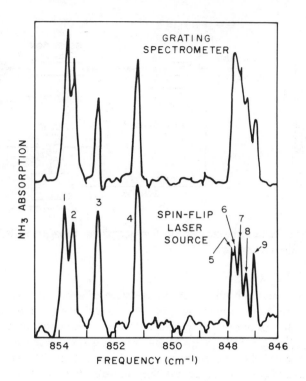

Figure 21.   Comparison of the absorption spectrum of ammonia taken with (a) SFR laser, (b) conventional grating spectrometer (see Ref. 19 for details as well as for the identification of the transition).

B.   Transient Infrared Spectroscopy

To date it has not been possible to carry out time dependent spectroscopy, i.e., study time-dependent populations of molecular species in the infrared because of the low intensity of the sources available until the present time. The Q-switched CO_2 laser pumped spin-flip Raman laser has completely changed all of this and it should allow us to measure the transient populations by monitoring the infrared absorption and thereby allow us to trace out chemical reactions of transient species of molecules. In particular molecular ions and molecular free radicals should be investigated. The experimental setup to carry out this experiment is shown in Fig. 23 which is very similar with that used for doing conventional spectroscopy shown in Fig. 20, with one difference. Now the absorption cell has a xenon flashlamp

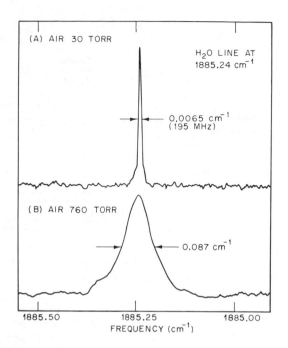

Figure 22.   Absorption spectrum of H_2O line in air at 1885.24 cm^{-1} at (a) p_{air} = 30 Torr and (b) p_{air} = 760 Torr.

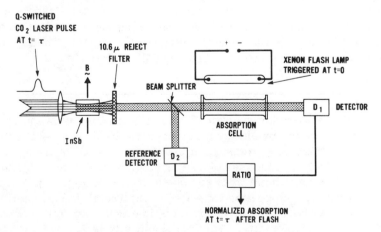

Figure 23.   Experimental setup for using the tunable SFR laser in transient infrared spectroscopy.

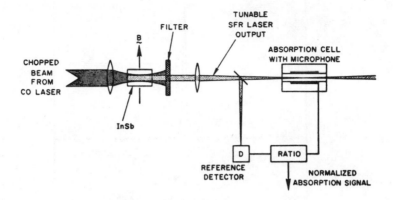

Figure 24. Experimental setup for air pollution detection by opto-acoustic absorption using a spin-flip Raman laser.

around it which is triggered at time $t - 0$ while the Q-switched pulse from the spin-flip Raman laser appears at $t = \tau$ after the flash. Thereby we monitor the absorption spectrum of the gas at $t = \tau$. We can monitor a given species by sitting at one wavelength (i.e., a given magnetic field for the SFR laser) or by measuring the entire spectrum at a given $t = \tau$. The time delay τ can be changed at will. Experiments show that this is possible and at the present time we have very preliminary results on the dissociation of H_2CO by electronic flash. We have been monitoring the excited-state absorption of HCO which happens to be in the range of 852 cm^{-1}. These experiments and details are under investigation and will be reported at a later date [20].

Figure 25. Absorption spectra of various gas samples in the range 1815 to 1825 cm^{-1} obtained by changing the magnetic field from 30 to 40 kG. All spectra are taken at a pressure of 300 Torr. (a) Calibration spectrum of 20 ppm NO in N_2. Lines numbered 1, 5, 6, 8, and 11 are due to NO (see Ref. 21). The remaining lines are due to H_2O (see Ref. 19). Vertical sensitivity 1.0. Integration time 1 sec. (b) The absorption cell noise. Vertical sensitivity is 100, integration time is 4 sec. (c) Room air at $21°C$ and 30% relative humidity. Vertical sensitivity is 1. Integration time is 1 sec. (d) Air sample from Route 22 near Plainfield, N.J. Vertical sensitivity is 1. Integration time is 1 sec. (e) Automobile exhaust. Vertical sensitivity is 0.4, Integration time is 1 sec.

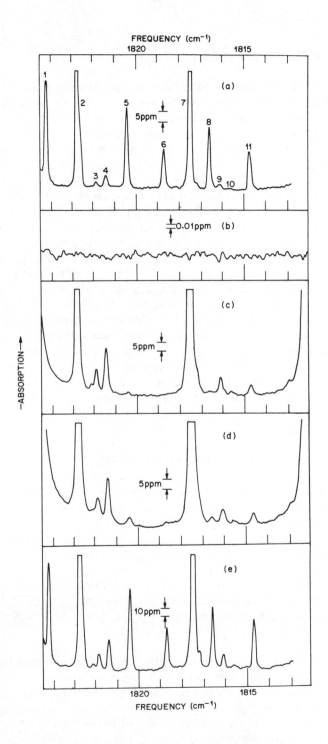

717

C.   Pollution Detection

Recently we have used the SFR laser for measuring the nitric oxide concentrations in air [21].   It is seen that using a setup shown in Fig. 24 we can measure very low concentrations of a number of different pollutants in air. The setup consists of a spin-flip Raman laser as a tunable laser source and an acousto-optic absorption cell [22] which allows us to measure very low absorption in the gas.   The details of this setup are given in Ref. 21, and will not be repeated here.   However, in Fig. 24 we show the results of this experiment where we have been able to measure nitric oxide concentrations as low as 0.01 parts per million with an integration time of the order of 1 sec.   This experiment was carried out using a spin-flip Raman laser power output of about 10 mW.   Since the acousto-optic detection technique is essentially a calorimetric technique, an increase in the pump power would directly results in improvement in the ultimate detection sensitivity.   With the presently available cw spin-flip Raman laser output of the order of 1 W, we hope to lower the minimum detectable limits for NO from $1:10^8$ to $\sim 1:10^{10}$.   These experiments will be reported in the near future.   We see from Fig. 25 that the pollution setup using the spin-flip Raman laser and the acousto-optic detection allows us not only to measure the very low concentrations but also the very high concentrations [21].   Since this technique is essentially an infrared spectroscopy technique, it is applicable to just about any pollutant that has its absorption spectrum in the region from 5 to 6.5 μ and from 9 to 14.6 μ.

VI.   CONCLUSIONS AND THE FUTURE

The future of the spin-flip Raman laser appears to be bright both for technological developments and for new applications of the spin-flip Raman laser.   In both of these we have barely scratched the surface.   The future possibilities in the SFR lasers in extending the tuning range and improving the operating characteristics are listed below.

1.   InSb pumped with 14-μ HF laser ~ 14-20 μ tuning;
2. $Hg_{1-x}Cd_xTe$ pumped with CO_2 laser at 10.6 μ, HF laser at 14 μ ~ 9-20 μ (cw possible because of gap variation with X; large g value).

In addition the range of tunability can be considerably increased by using nonlinear optics and some of these experiments are listed in Table II. These indicate that we should be able to extend the tuning range from around 3 μ to in excess of 1000 μ by suitable combinations of pump frequency and the tunable Raman laser frequency.   Some of these experiments are being pursued at the present time.

TABLE II.   FUTURE (Together with Nonlinear Optics)

Primary Sources		Resultant Tunability		
Tunable Source	Fixed-Frequency Source			
1. SFR CO_2	H_2O			
9.0-14.6 μ	27.97 μ	DIFF:	13.3 - 30.5	μ
		SUM:	6.76 - 9.6	μ
2. SFR CO_2	SHG			
9.0-14.6 μ			4.5 - 7.3	μ
3. SFR CO_2	CO			
9.0-14.6 μ	5.0- 6.7 μ	SUM:	3.2 - 4.6	μ
4. SFR CO	SHG			
5.3- 6.5 μ			2.65 - 3.25	μ
5. SFR CO	CO_2			
5.3- 6.5 μ	9.2-10.8 μ	DIFF:	3.56 - 22.8	μ
6. SFR CO_2	CO_2			
9.0-14.6 μ	9.2-10.8 μ	DIFF:	30.0 - 1000.0	μ

ADDENDUM (SEPTEMBER, 1972)

The low-field operation of the SFR lasers described in the text has now been extended to lower fields. Lowest field at which SFR laser has been operated is ~ 200 G using a $n_e \approx 1 \times 10^{15}$ cm^{-3} InSb sample. Problems arising from carrier freeze-out in the low carrier concentration samples at fields higher than 40 kG have been observed. However, these problems do not interfer with the low-field operation which allows operation of the SFR laser with small electromagnets rather than a superconducting solenoid. The case of control of magnetic field and the resettability of the electromagnet has allowed us to carry out a number of new experiments. These include very high resolution spectroscopy, Lamb-dip spectroscopy, linewidth measurements of the SFR laser by heterodyning, lowering of the minimum detectability in the pollution detection, etc. These will be outlined briefly:

1. Linewidth measurements [18]: Output from an SFR laser has been heterodyned with the power output from a cw gas laser to obtain an rf beat whose observed width was ~1 kHz. The rf spectrum analyzer resolution was 1 kHz indicating that the linewidth of the SFR laser is ≪1 kHz in agreement with a calculated width of ~1 kHz.

2. High-resolution spectroscopy [23]: Both absorption spectroscopy as well as modulation spectroscopy has shown that the linewidth of ≪1 kHz

allows ultrahigh resolution spectroscopy. Modulation of the SFR laser magnetic field allows us to measure small abosrption using modulation spectroscopy techniques. (i) Pressure broadening of H_2O lines with various buffer gases has been studied for the first time. (ii) Λ doubling of NO has been observed in modulation spectroscopy near 5.3 μ. (iii) Saturation spectroscopy has allowed us to measure linewidths smaller than the Doppler-broadened linewidths of gaseous absorbers. Lamb-dip width of ~50-100 kHz has been measured.

   3. Pollution detection: Previous studies had shown that acousto-optic detection technique in conjunction with a SFR laser source allows detection of NO concentration as low as 10 ppb in air using a 4-sec time constant. With the low-field operation of the SFR laser, we have improved [24] the detection capability to $\lesssim$1 ppb with a 1-second integration time.

   4. Far-ir generation: The electron spins in InSb have been used to generate far ir in the 100-μ range by difference-frequency mixing [25]. It is expected that the SFR laser itself will be generating the far ir at a frequency $\nu_{far\ ir} = g\mu_B B$. Thus the SFR laser is expected to give tunable near-ir power output as well as tunable far-ir power output, considerably increasing the versatility and usefulness of the SFR laser.

REFERENCES

1.   C. K. N. Patel, P. K. Tien, and J. H. McFee, Appl. Phys. Letters 7, 290 (1965).
2.   C. K. N. Patel, Appl. Phys. Letters 7, 246 (1965).
3.   C. K. N. Patel and E. D. Shaw, Phys. Rev. Letters 24, 451 (1970).
4.   R. E. Slusher, C. K. N. Patel, and P. A. Fleury, Phys. Rev. Letters 18, 77 (1967).
5.   P. A. Wolff, Phys. Rev. Letters 16, 225 (1966).
6.   Y. Yafet, Phys. Ref. 152, 858 (1966).
7.   C. K. N. Patel and R. E. Slusher, Phys. Rev. 167, 413 (1968).
8.   C. K. N. Patel and R. E. Slusher, Phys. Rev. 177, 1200 (1969).
9.   C. K. N. Patel and E. D. Shaw, Phys. Rev. B3, 1279 (1971).
10.   R. B. Aggarwal, B. Lax, C. E. Chase, C. R. Pidgeon, D. Limpert, and F. Brown, Appl. Phys. Letters 18, 383 (1971).
11.   E. D. Shaw and C. K. N. Patel, Appl. Phys. Letters 18, 215 (1971).
12.   A. Mooradian, S. R. J. Brueck, and F. A. Blum, Appl. Phys. Letters 17, 481 (1970).
13.   C. K. N. Patel, Appl. Phys. Letters 18, 274 (1971).
14.   C. K. N. Patel, Appl. Phys. Letters 19, 400 (1971).
15.   C. K. N. Patel in Lasers, edited by A. K. Levine (M. Decker, New York, 1968), Vol. 2, p. 1.

16. S. R. J. Brueck and A. Mooradian, Appl. Phys. Letters 18, 229 (1971).
17. C. K. N. Patel (to be published).
18. C. K. N. Patel, Phys. Rev. Letters 28, 649 (1972).
19. W. S. Benedict and R. F. Calfee, Line Parameters for 1.9 and 6.3 μ Water Vapor Bands, ESSA Professional Paper 2 (U.S. Department of Commerce, Washington, D. C., 1967).
20. C. K. N. Patel (to be published).
21. L. B. Kreuzer and C. K. N. Patel, Science 173, 45 (1971).
22. L. B. Kreuzer, J. Appl. Phys. 42, 2934 (1971).
23. C. K. N. Patel, Proceedings of the III Rochester Conference on Coherence (to be published).
24. C. K. N. Patel (to be published).
25. V. T. Nguyen and T. J. Bridges, Phys. Rev. Letters 29, 359 (1972); T. L. Brown and P. A. Wolff, Phys. Rev. Letters 29, 362 (1972).

TUNABLE SEMICONDUCTOR LASERS AND THEIR

SPECTROSCOPIC USES[*]

P. L. Kelley and E. D. Hinkley
Lincoln Laboratory, Massachusetts Institute of Technology
Lexington, Massachusetts

ABSTRACT

A discussion is given of the properties of semiconductor lasers together
with results of some spectroscopic studies made with these lasers.

I. INTRODUCTION

Tunable lasers are useful devices in basic research as well as in other
fields where spectroscopic methods can be applied. There are presently
four types of coherent sources which show appreciable tuning: these are
(1) optical parametric oscillators [1], (2) dye lasers [2], (3) spin-flip Raman
lasers [3], and (4) tunable semiconductor lasers. We will confine ourselves
in this report to tunable semiconductor lasers; for a discussion of the other
three types the reader is referred to the pertinent references as well as to
other articles in this volume.

---

[*]This work was sponsored by the Department of the Air Force and the
Environmental Protection Agency.

II.   CHARACTERSTICS OF SEMICONDUCTOR LASERS

Recombination-radiation semiconductor lasers operate by stimulating emission across a band gap.   Population inversion is achieved by electron injection across the band gap either with current (diode), by optical pumping, or by electron-beam excitation.   Infrared semiconductor laser materials in the 1-30-μm region include binary compounds such as InAs, InSb, GaSb, PbS, PbSe, PbTe, and pseudobinary alloys such as $Pb_{1-x}Sn_xTe$, $Pb_{1-x}Sn_xSe$, $PbS_{1-x}Se_x$, $Pb_{1-x}Cd_xS$, $InAs_{1-x}P_x$, $In_{1-x}Ga_xAs$, and $GaAs_xSb_{1-x}$.   In the visible and near-infrared region materials such as CdS_xSe_{1-x}, $GaAs_{1-x}P_x$, and $Al_xGa_{1-x}As$ can be used.   In what follows we discuss lead-salt diode lasers, as most of the high-resolution tunable laser experiments have been performed with these devices.

The lead-salt diode lasers are made from vapor-grown single crystals of semiconducting material [4] by cleaving the crystals into rectangular parallelepipeds (each containing a p-n junction) of approximate overall dimensions 0.12 by 0.05 by 0.03 cm.   Low-resistance Ohmic contacts are formed, and the laser is then mounted onto the cold finger of a cryogenic Dewar.   Liquid-helium temperatures are presently required for continuous laser operation, with pulsed operation possible at liquid-nitrogen temperature (77°K).   The infrared emission frequency of the laser is, within the spontaneous emission bandwidth, determined by the refractive index of the semiconductor and the physical length of the cavity.

Devices made in this manner have produced up to 1 mW of continuous (cw) tunable narrow-line radiation at liquid-helium temperature, and 10 W pulsed at 77°K.   The nominal infrared frequency of these lasers is determined by the energy gap, which is set by the chemical composition [5].   (Some compositional tuning ranges are shown in Fig. 1.) Once a diode laser has been fabricated, further control of its energy gap is possible by changing pressure [6], temperature [7], or magnetic field [8].   (For example, PbSe diodes have been pressure tuned from 8 to 22 μm.)

Fine tuning within a single cavity mode can also be accomplished by changes in pressure, temperature, or magnetic field--through the refractive-index variation with these parameters, which determines the cavity mode frequency.   In the spectroscopic experiments described below the fine tuning was achieved by changing the diode current [9], which changes the heating rate and hence the temperature.   A single laser can be tuned in this manner over 40 cm^{-1} in continuous bands up to 2 cm^{-1} wide; but because of a change in laser operation from one cavity mode to another, only about one-half of this tuning range is covered.

The measured [10] linewidths of these lasers have been found to be in essential agreement with predictions [11] of the fundamental (quantum-phase-noise-limited) linewidth.   This was observed by heterodyning emission from a stable CO_2 laser with that from a single-mode diode laser operating at

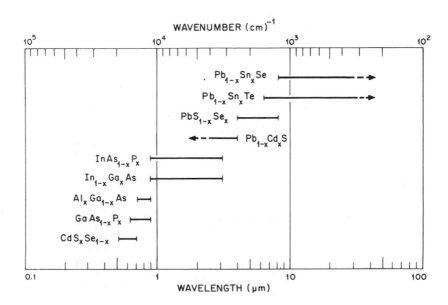

Figure 1. Compositional tuning ranges of some semiconductor alloys.

10. 6 μm, and observing a Lorentzian line shape representing the diode laser power spectrum (see Fig. 2). The width of this Lorentzian line varies inversely with laser output power. The diode shown in Fig. 2 was operating at 240 μW. There are two reasons why this fundamental linewidth can be readily observed, compared with the difficulty of detecting it in gas lasers: (1) the Q of the diode cavity and the power output are such as to give a fundamental width which is wide compared to that which one would normally expect from a gas laser; (2) the small physical size of the laser cavity makes it very stable against acoustic and temperature fluctuations.

The diode laser linewidth is obviously more than adequate for Doppler-limited absorption spectroscopy in this region of the infrared where absorption lines are of the order of 50 MHz (~ 0.002 cm^{-1}) wide. Conventional infrared absorption spectroscopy with blackbody sources and spectrometers typically gives considerably less resolution (approximately 900 MHz at best) and sensitivity. The laser's substantially greater power per unit spectral interval (compared to a blackbody) simplifies detection at high resolution. Figure 3 compares diode laser and grating spectrometer scans of the sP(1, 0) line of the ν_2 band of NH$_3$. Note that with the diode laser one is observing the Doppler-limited lineshape--by comparison, a very large

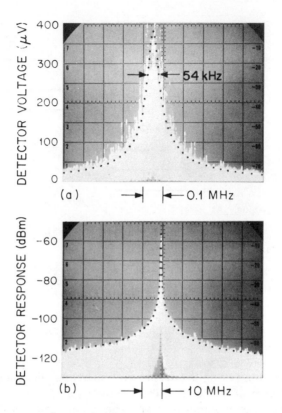

Figure 2. Spectrum-analyzer display of beat note between a 240-μW single-mode PbSnTe diode laser and the P(14) line of a stable CO_2 gas laser. The i.f. bandwidth is 10 kHz.

grating would be required to fully resolve this line by conventional techniques. It should be pointed out that greater resolution (~ 0.03 cm^{-1} in this wavelength range) has been obtained in the laboratory with grating instruments, but at the expense of considerable difficulty; and that Fourier transform methods are also being developed to alleviate this problem. With the high power levels and narrow linewidths available from diode lasers it may also prove feasible to use nonlinear saturation resonance techniques to observe homogeneous widths within the inhomogeneous Doppler profiles of molecular gases.

By going to external cavities with grating control, it may be possible to obtain continuous tuning over wider ranges (i.e., mode hopping may be avoided) and greater frequency stabilization in the pulsed mode. (The diodes

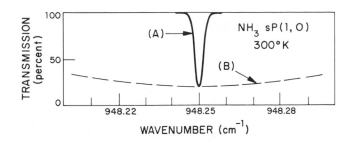

Curve	Instrument	Resolution	$p \times L\,(T\text{-}cm)$
(A)	Diode Laser	Linewidth	$0.05\,T \times 10\,cm$
(B)	Grating Spectrometer	$0.2\ cm^{-1}$	$5\,T \times 200\,cm$

Figure 3.   Experimental comparison between a tunable $Pb_{0.88}Sn_{0.12}Te$ diode laser scan and a diffraction grating scan of the $sP(1,0)$ line of the ν_2 band of NH_3 at room temperature.

mode chirp under pulsed operation at about +20 MHz/nsec, which makes them at present undesirable for some spectroscopic applications.) Wider tuning ranges are to be expected because external control can utilize most of the broad luminescence bandwidth of about a hundred wave numbers. Furthermore, pressure tuning over wide wavelength ranges becomes more practical if stable frequency operation is available in the high-temperature pulsed mode. It has also been suggested that frequency tuning and control could be achieved by piezoelectrically adjusting the laser cavity length. Improvements in diode technology and the use of optical pumping are expected to yield greater output powers.

III. SPECTROSCOPIC APPLICATIONS

We now discuss some of the infrared molecular spectroscopic studies which have been carried out with cw tunable lead-salt lasers which were compositionally tuned to the desired wavelength regions.

A. Sulfur Hexafluoride (SF_6)

Portions of the ν_3 band of SF_6 have been studied with diode lasers [12], and a representative scan in the vicinity of the P(16) CO_2 laser line is shown in Fig. 4. While the complex spectra are not understood in detail, they have

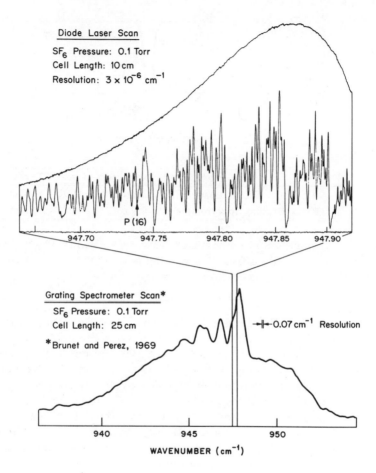

Figure 4.   Infrared spectrum of ν_3 band of SF_6. (Bottom) Grating spectrometer absorption curve of the ν_3 band of sulfur hexafluoride at room temperature (adapted from the curve of Brunet and Perez [14]); SF_6 pressure,
0.1 Torr; cell length, 25 cm. (Top) Diode laser transmission scan of a
segment of this band near the P(16) CO_2 laser line; SF_6 pressure, 0.1 Torr;
cell length, 10 cm.

been correlated with band contour calculations and estimates of the effect of
second-order Coriolis splitting [13].   The results are illustrative of several
points concerning tunable laser spectroscopy.   For example, the laser
spectra can be compared with those obtained with a grating spectrometer, as
is shown in the figure.   The indicated position of the P(16) CO_2 line was

obtained by observing the beat signal between CO_2 laser emission and that from a diode laser that was sent through an absorbing cell and tuned around the CO_2 laser line. The spectrum analyzer displayed a portion of the SF_6 absorption spectrum with the zero beat giving the position of the P(16) laser line. While this method accurately calibrates a dozen or so of the lines in the SF_6 spectrum which occur within 1 or 2 GHz of the known laser line, the wavelengths for SF_6 lines beyond this region cannot be as accurately established because of possible nonlinearities in the tuning rate of these lasers. Tuning linearity can, of course, be checked with a Fabry-Perot. Calibrated diode laser scans of SF_6 have also been obtained in the vicinity of the P(14), P(18), P(20), and P(22) CO_2 laser transitions. The results show overlapping Doppler lines near the laser lines which makes difficult the interpretation of observations of self-induced transparency [15] and photon echoes [16] in SF_6 using the P(20) transition of CO_2.

B. Sulfur Dioxide (SO_2)

The ν_1 band of SO_2 in the 8.6 µm region has also been studied with PbSnTe diodes [17]. These results illustrate how tunable lasers can be used to determine band centers when reference spectra are available. A theoretical SO_2 spectrum has been calculated in Ref. 17, using a previously described method [18], based on accurate microwave values [19] of the rotational constants of both the ground and first excited state of the ν_1 mode. The theoretical spectrum was then compared with the experimental spectra (see Fig. 5, for example) and the transitions identified. The SO_2 transitions are unlabeled in the figure. However, as an example, we point out that the large peak just to the left of the aR(11, 11) NH_3 line is the transition $18_{1, 17} \leftarrow 18_{2, 16}$ where the notation J_{K_A, K_C} is used for the rotational quantum numbers. The frequencies of these SO_2 transitions were determined relative to absorption lines of accurately known NH_3 transitions [20]. A value of the frequency of the ν_1 band center was found to be 1151.71 ± 0.01 cm^{-1} which can be compared with the experimental value of 1151.38 cm^{-1} obtained by Shelton et al. [21]. The uncertainty of ± 0.01 cm^{-1} is due in part to the uncertainty in the ammonia calibration spectra and in part due to the intercomparison of the ammonia and SO_2 spectra. A more careful statistical analysis should yield a more accurate value.

C. Nitric Oxide (NO)

Because of its unpaired electron nitric oxide has an interesting infrared spectrum. Using a PbSSe diode laser observations [22] have been made of Λ doubling as well as the strong Zeeman effect of this molecule; recently,

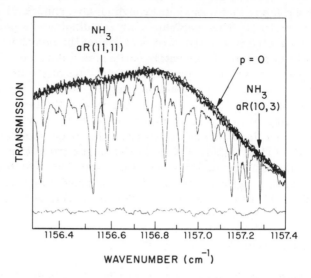

Figure 5.   Diode laser scans of SO_2 and NH_3 absorption lines at room temperature. SO_2 scans were taken at 0.1, 1, and 10 Torr, NH_3 at 1 Torr. The
absorption cell length was 30 cm. (From Ref. 17).

other experimental studies [23] have been made of nuclear hyperfine structure in the Q branch and magnetic rotation. Figure 6 shows laser scans of
both the $R(15/2)_{1/2}$ and the $R(15/2)_{3/2}$ transitions. Note that the Λ doubling
is distinctly observable in the transition involving $^2\Pi_{1/2}$ states. The Doppler
width is 128 MHz, in agreement with theory. The predicted Λ-doublet
separation, using microwave data [24] is 324 MHz while the infrared observations give 318 MHz (the difference is within experimental error). The
results for the Zeeman splitting of the $R(15/2)_{3/2}$ level is shown in Fig. 7.
The circles represent experimental data while the solid lines represent
theory [25]. The derivative spectrum in Fig. 7(b) was taken by modulating
the diode current (frequency-modulating the laser).

D. Monodeuterated Ammonia (NH_2D)

There is a large linear Stark effect in the ground vibrational state levels
$a4_{0,4}$ and $s4_{1,4}$, due to the strong mixing of these nearby levels. Here, s(a)
refers to the symmetric (antisymmetric) inversion state. Stark tuning
through CO_2 laser lines of several ν_2 R-branch transitions involving these
ground-state levels has been observed [26-28]. Tunable laser spectroscopy

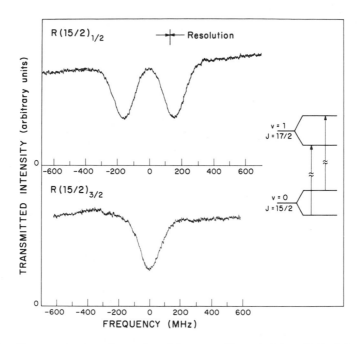

Figure 6.   Transmission through a 10-cm cell containing NO at 0.2 Torr pressure versus frequency (a) for the $R(15/2)_{1/2}$ line and (b) for the $R(15/2)_{3/2}$ line.   The energy level diagram at the right shows schematically the Λ type doubling and the transitions involved.   (From Ref. 22).

was carried out in the vicinity of 949.4 cm^{-1} and the transitions in the Stark multiplet $s5_{1,4} \leftarrow a4_{0,4} + s4_{1,4}$ were seen [29] in modulated (see Fig. 8) and unmodulated electric fields.   Those transitions whose frequency increases with increasing electric field had been previously observed with the P(14) line of the CO_2 laser.   The transitions whose frequency decreases with increasing electric field become relatively weaker as the electric field decreases since the admixture of $a4_{0,4}$ is decreasing.   In zero field the transition $s5_{1,4} \leftarrow s4_{1,4}$ is forbidden while the transition $s5_{1,4} \leftarrow a4_{0,4}$ is allowed.

IV.   CONCLUSION

There are a large number of fundamental spectroscopic applications of tunable lasers as a general class, and of tunable semiconductor lasers in

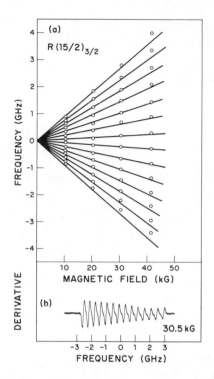

Figure 7.   Zeeman spectra of the R(15/2)$_{3/2}$ line of NO;   (a) transition
frequencies versus magnetic field, experimental results appear as circles,
theory as solid lines; (b) recorder trace of a Zeeman derivative spectrum
at 30.5 kG.   Data taken with 1.5 Torr NO in a 5-cm cell.   (From Ref. 22).

particular.   It is too early to predict which tunable coherent sources will
turn out to be the most useful for a given spectroscopic purpose in a particular wavelength range.   Much more confidence can be expressed in the view
that tunable lasers will have impact not only on a wide variety of areas in
fundamental research but also on more routine problems in chemical,
biological, and environmental analyses.

ADDENDUM (SEPTEMBER, 1972)

A number of results pertaining to the spectroscopic uses of tunable semiconductor lasers have recently been obtained.
    With optical pumping lasing bandwidths of up to 6% of the center

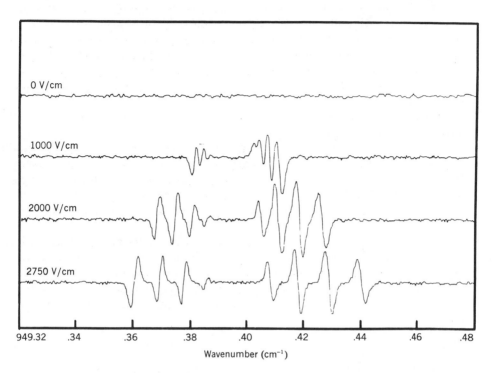

Figure 8. Diode laser scans of Stark-modulated spectra of NH_2D at various dc Stark fields. Stark modulation field was 2.5 V/cm, and the modulation frequency was 20 kHz. The 30-cm cell contained 0.2 Torr NH_2D. (From Ref. 29.)

frequency have been observed in PbSSe [30]. With an external cavity for frequency control it is hoped that a reasonably stable optically pumped semiconductor laser can be constructed.

The tuning characteristics of PbSSe diode lasers in a magnetic field have been studied [31]. The fine tuning of a laser mode with field was found to be highly nonlinear. Gross tuning with magnetic field and fine tuning with current has proven useful for spectroscopy.

High-resolution (Doppler-limited) time-resolved spectroscopy of cesium [32] and excited xenon [33] have been carried out with a pulsed 77°K GaAs laser. In the case of xenon the laser was coarse tuned by pressure. The spectroscopy was carried out with the frequency swept output occurring during a laser pulse. Optical pumping of cesium has also been achieved using a GaAs laser [34].

With a cw PbSSe diode laser operating in the vicinity of 5.3 μm peak gain (loss) coefficients and line shapes have been measured for several transitions in a CO laser amplifier [35]. The gain (loss) coefficients have been studied as a function of discharge current and partial pressure of the constituent gases in the discharge tube mixture. Improved values of the relative line positions have been obtained for some of the observed transitions.

Transitions in the 6.3-μm ν_2 band of water vapor have also been studied with PbSSe diode lasers operating near 5.3 μm [36]. Several transitions involving states of high rotational energy have been observed to have widths 2 to 4 times narrower than previously predicted for sea level air. Errors in relative position up to several linewidths have also been found. These discrepancies have important implications for the infrared properties of the atmosphere, in particular the transmission characteristics for monochromatic sources.

Collisional ("Dicke") narrowing of two of these water vapor transitions has also been observed. Xenon, argon, and nitrogen were used as buffer gases [37]. As pressure was increased lines narrowed (and absorption increased) from the low-pressure Doppler values, had a minimum in the 50–200 Torr range, and were dominated by pressure broadening above a few hundred Torr.

REFERENCES

1. S. E. Harris, Proc. IEEE 57, 2096 (1969); R. G. Smith in Laser Handbook, edited by F. T. Arrechi and E. O. Schultz-DuBois (North-Holland, Amsterdam, 1972).
2. B. B. Snavely, Proc. IEEE 57, 1374 (1969); M. Bass, T. F. Deutch, and M. F. Weber, in Lasers, edited by A. K. Levine and A. J. deMaria (Dekker, N. Y. 1971), Vol. 3, p. 269; D. J. Bradley, Proc. Electrooptical Systems Conference, Brighton, England (to be published).
3. C. K. N. Patel and E. D. Shaw, Phys. Rev. Letters 24, 451 (1970); A. Mooradian, S. R. J. Brueck, F. A. Blum, Appl. Phys. Letters 17, 481 (1970); C. K. N. Patel and E. D. Shaw, Phys. Rev. B 3, 1279 (1971); S. R. J. Brueck and A. Mooradian, Appl. Phys. Letters 18, 229 (1971).
4. A. R. Calawa, T. C. Harman, M. Finn, and P. Youtz, Trans. Met. Soc. Amer. Inst. Mining Eng. 242, 374 (1968).
5. J. O. Dimmock, I. Melngailis, and A. J. Strauss, Phys. Rev. Letters 16, 1193 (1966); J. F. Butler, A. R. Calawa, and T. C. Harman, Appl. Phys. Letters 9, 427 (1966); J. F. Butler and T. C. Harman, Appl. Phys. Letters 12, 347 (1968); for a review, see I. Melngailis, J. Phys. Colloq. C-4, Suppl. No. 11-12, C4-84 (1968); T. C. Harman, J. Phys.

Chem. Solids 32, Suppl. 1, 363 (1971).

6. J. M. Besson, J. F. Butler, A. R. Calawa, W. Paul, and R. H. Rediker, Appl. Phys. Letters 7, 206 (1965); J. M. Besson, W. Paul, and A. R. Calawa, Phys. Rev. 173, 699 (1968).

7. E. D. Hinkley, C. Freed, and T. C. Harman, Solid State Research Report, MIT Lincoln Laboratory (1968:3), p. 19; T. C. Harman, A. R. Calawa, I. Melngailis, and J. O. Dimmock, Appl. Phys. Letters 14, 333 (1969).

8. J. F. Butler and A. R. Calawa, in Physics of Quantum Electronics, edited by P. L. Kelley, B. Lax, and P. E. Tannenwald (McGraw-Hill, New York, 1966), p. 458; A. R. Calawa, J. O. Dimmock, T. C. Harman, and I. Melngailis, Phys. Rev. Letters 23, 7 (1969).

9. E. D. Hinkley, T. C. Harman, and C. Freed, Appl. Phys. Letters 13, 49 (1968).

10. E. D. Hinkley and C. Freed, Phys. Rev. Letters 23, 277 (1969).

11. A. L. Schawlow and C. H. Townes, Phys. Rev. 112, 1940 (1958); for modifications of the Schawlow-Townes expression see, R. D. Hempstead and M. Lax, Phys. Rev. 161, 350 (1967) and references cited therein.

12. E. D. Hinkley, Appl. Phys. Letters 16, 351 (1970).

13. E. D. Hinkley, D G. Sutton, and J. I. Steinfeld, paper presented at the 25th Symposium on Molecular Structure and Spectroscopy, Ohio State University, 1970 (unpublished).

14. H. Brunet and M. Perez, J. Mol. Spectroscopy 29, 472 (1969).

15. C. K. N. Patel and R. E. Slusher, Phys. Rev. Letters 19, 1019 (1967).

16. C. K. N. Patel and R. E. Slusher, Phys. Rev. Letters 20, 1087 (1968).

17. E. D. Hinkley, A. R. Calawa, P. L. Kelley, and S. A. Clough, J. Appl. Phys. 43, 3222 (1972).

18. W. S. Benedict, S. A. Clough, L. Frenkel, and T. E. Sullivan, J. Chem. Phys. 53, 2565 (1970).

19. G. Steenbeckeliers, Ann. Soc. Sci. Bruxelles 82, 331 (1968).

20. K. N. Rao, J. B. Curtin, and P. Yin (to be published); the ammonia transitions are identified from J. S. Garing, H. H. Nielsen, and K. N. Rao, J. Mol. Spectroscopy 3, 496 (1959).

21. R. D. Shelton, A. H. Nielsen, and W. H. Fletcher, J. Chem. Phys. 21, 2178 (1953).

22. K. W. Nill, F. A. Blum, A. R. Calawa, and T. C. Harman, Chem. Phys. Letters 14, 234 (1972).

23. F. A. Blum, K. W. Nill, A. R. Calawa, and T. C. Harman, Chem. Phys. Letters 15, 144 (1972).

24. J. J. Gallagher and C. M. Johnson, Phys. Rev. 103, 1727 (1956); J. J. Gallagher, F. D. Bedard, and C. M. Johnson, Phys. Rev. 93, 729 (1954).

25. G. C. Dousmanis, T. M. Sanders, and C. H. Townes, Phys. Rev. 100, 1735 (1955).

26. R. G. Brewer, M. J. Kelly, and A. Javan, Phys. Rev. Letters 23, 559 (1969).
27. R. G. Brewer and J. D. Swalen, J. Chem. Phys. 52, 2774 (1970).
28. M. J. Kelly, R. E. Francke, and M. S. Feld, J. Chem. Phys. 53, 2979 (1970).
29. R. E. Francke, M. S. Feld, and E. D. Hinkley (unpublished).
30. A. Mooradian, A. J. Strauss, and J. A. Rossi, IEEE J. Quantum Electron. 9, 347 (1973).
31. K. W. Nill, F. A. Blum, A. R. Calawa, and T. C. Harman, Appl. Phys. Letters 21, 132 (1972).
32. S. Siahatgar and U. E. Hochuli, IEEE J. Quantum Electron. 5, 295 (1969).
33. A. S. Pine, C. J. Glassbrenner, and J. A. Kafalas, IEEE J. Quantum Electron. 9, No. 8 (1973).
34. G. Singh, P. DiLavore, and C. O. Alley, IEEE J. Quantum Electron. 7, 196 (1971).
35. F. A. Blum, K. W. Nill, A. R. Calawa, and T. C. Harman, Appl. Phys. Letters 20, 377 (1972).
36. F. A. Blum, K. W. Nill, P. L. Kelley, A. R. Calawa, and T. C. Harman, Science 177, 694 (1972).
37. R. S. Eng, A. R. Calawa, T. C. Harman, P. L. Kelley, and A. Javan, Appl. Phys. Letters 21, 303 (1972).

EMISSION PROCESSES AND OPTICAL PUMPING

NONEQUILIBRIUM DISTRIBUTIONS OF MOLECULAR STATES IN INTERSTELLAR SPACE[*]

C. H. Townes
Department of Physics, University of California
Berkeley, California

I. INTRODUCTION

Interstellar space contains a number of different components, each with its own characteristics and its own degree of excitation, often far from thermodynamic equilibrium. These components include:

1. Stellar radiation, with a spectral distribution corresponding to a temperature of $10^4 \, ^\circ K$, but diluted by a factor of about 10^{14} on the average. Locally, its intensity may be much greater, especially in regions near new stars, where the intensity may be exceedingly high.
2. The isotropic microwave radiation, or "relict" radiation. This has approximately the spectral distribution and energy density of blackbody radiation at $2.7^\circ K$. It is believed to be the remains of radiation associated with the initial explosion of the universe, going back specifically to the first few million years of the universe's lifetime. The total energy density of this radiation is approximately the same as that of stellar radiation.
3. Cosmic rays, including gamma rays. These have various but

[*]Work partially supported by the National Aeronautics and Space Administration Grant NGL 05-003-272.

uncertain origins, possibly coming from supernova explosions, from pulsars, or from white dwarfs. Their total energy density is comparable with that of components (1) and (2).

4. X rays. X rays are emitted locally by a variety of objects. They may also originate from inverse Compton scattering of the radiation field by the high-energy particles in space.

5. Gases, including plasmas. The most abundant atomic species is hydrogen, with average density about $10^{-1}/cm^3$ within the galaxy and less than $10^{-6}/cm^3$ in intergalactic space. The abundance of helium is about 15% that of hydrogen. Carbon, oxygen, and nitrogen have abundances in the range 10^{-3} to 10^{-4} that of hydrogen; other elements are still less abundant.

6. Dust particles, with typical size about 0.1 μ. The composition of these particles is not clearly known. However, some evidence indicates that they are largely composed of graphite and silicates, with other elements and molecules either admixed or frozen on the surface. The average mass density of dust is approximately 10^{-2} that of hydrogen, which means that the average number of dust particles per unit volume is about 10^{-13} that of hydrogen atoms.

II. INTERSTELLAR CLOUDS

Interstellar gases have frequently been considered to be largely composed of atoms or ions, and to have a density of about one atom per cm^3. The ultraviolet radiation from stars can ionize such gases, producing electrons of temperatures which are characteristic of stellar temperatures, that is, about 10^4 °K. However, there are also substantial amounts of neutral atomic hydrogen and it is now clear that there are large quantities of molecular hydrogen. In fact, the gas can be roughly described as being in one of two more-or-less stable conditions. One is a very dilute gas of density of about 1/10 atom per cubic centimeter, which tends to be at a rather high temperature and frequently ionized. The other stable state is represented by a cloud of much denser gas. Once the gas density increases substantially, collisions allow it to cool more effectively. This overcomes the heating by ultraviolet and other ionizing radiation, so that the gas cools and contracts further. As the cloud contracts, it intercepts less of the energetic radiation and contracts further, possibly into clouds of density as high as 10^4 to 10^6 atoms/cm^3. Under such conditions, the dust is usually dense enough to make what is generally recognizable as a dark cloud, or dust cloud, obscuring stellar light beyond it.

Dust clouds have not until rather recently received much attention, partly because their darkness does not allow many of the observations with visible light that are normal to astronomy. However, microwave and infra-red radiation emanate from such clouds, and by comparison with optical

Figure 1.   The "Horsehead Nebula", a dark cloud in Orion with illumination from an ionized region behind it.

waves such radiation can penetrate them easily, because of the small size ($\sim 0.1\ \mu$) of the dust particles.   Figure 1 is a picture of a dust cloud called the Horsehead Nebula, in Orion.   It is a dark dust cloud illuminated by a bright nebula associated with a highly ionized region.   Such dust clouds are quite common, and sometimes very extensive.   Figure 2 shows dark clouds in an exploding galaxy.   Dust clouds are apparently typical not only of cool quiet regions but also as condensates following violent explosions.   Figure 3 shows the Trifid Nebula within our own galaxy, split into three parts by dark dust lanes.   The Milky Way, the disk of our own galaxy, is full of dust clouds, as may be seen from Fig. 4.   This figure represents a montage of photographs of the Southern Milky Way.   An almost continuous band of dark clouds in various irregular shapes is seen throughout the Milky Way.   Such clouds are not only common; they contain a substantial fraction of the entire galactic material.

Figure 2.   An exploding galaxy, NGC 3034, showing dark dust clouds.
(200-in. telescope, Mt. Palomar.)

Thus, dust clouds are important in themselves as a common phenomenon
of our universe and a substantial part of its mass. Furthermore, they
represent a stage in the dynamic processes which shape other galactic
phenomena. Some of the dust may have originated in stars or explostions.
Perhaps all of it did. In turn, condensations of gas and dust are the origin
of a number of stars which may heat up and disperse the clouds that gave
them birth. Furthermore, it is in the dust clouds that one finds most
molecules which exist in interstellar space.

IV.   MOLECULES IN INTERSTELLAR SPACE

Molecules have been known in interstellar space since about 1940, when they
were identified from optical spectra taken by already well-established

Figure 3.   The Trifid Nebula in our own galaxy, split by three intersection dust "lanes".   (200-in. telescope, Mt. Palomar.)

techniques.   Three molecular free radicals, each of which are rare on earth, were discovered at about that time--CH, CH$^+$, and CN.   Much later, in 1963, the radical OH was found by its microwave spectrum [1].   These discoveries, and the supposition that gases in interstellar space are very rarefied and subject to ionizing radiation, led to the general expectation that

Figure 4.   The Southern Milky Way, shown by a montage of separate photographs.   Many dark clouds are seen throughout the Milky Way.   (Courtesy Bart Bok.)

molecules in interstellar space are both rare and simple. However, it is clear that dust clouds should provide the most hospitable location for molecules, since gases in these clouds tend to be denser than in other interstellar regions, and furthermore the dust provides shielding from ionizing and dissociating ultraviolet radiation. In 1968 a search of such clouds revealed the first two polyatomic molecules, NH_3 [2] and H_2O [3]. Both have convenient spectra in the microwave region; the wavelengths are, futhermore, long enough to penetrate such clouds easily. Shortly afterwards, the organic molecule H_2CO was similarly discovered [4]. Since then, the number and complexity of molecules has continued to increase more and more rapidly, as indicated in Fig. 5, which shows the accelerating buildup in numbers of species over the last several years. The total number of molecular species which have been identified in interstellar space is approximately 20, and it is quite clear that many others will be found as further searches are conducted and increasing sensitivity is achieved in the microwave and infrared regions. Many of the molecular lines are of intensity

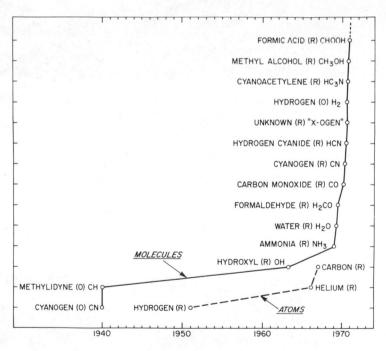

Figure 5. Discovery of interstellar molecules as a function of time, showing the very rapid increase which has occurred within the last several years. Atoms detected by radio astronomy are also indicated. (Prepared November 1, 1970, courtesy E. Lilley.)

sufficiently high to be rather easily detected.   Their rapid detection during
these last several years is perhaps primarily because their existence and
importance was realized, but it has also been associated with an increasing
availability of millimeter-wave techniques and antennas. The present list of
molecules, their isotopic species, and their frequencies are given in Table I.
     Energy levels of the lower rotational states of formaldehyde, H_2CO, are
shown in Fig. 6.   All the transitions indicated by an arrow have been
observed, although the one most commonly seen is the 1_{01}-1_{11} transition.
The formaldehyde levels are separated into two groups, depending on the
symmetry around the molecular axis.   Rotation of the molecule around its
axis by 180° simply interchanges its two hydrogen atoms.   Thus, the
symmetry with respect to hydrogen atoms is analogous to that of the hydrogen molecule, and formaldehyde exists in both ortho and para forms, as
does H_2.   On the left-hand side of Fig. 6 one sees the energy levels of paraformaldehyde, and on the right-hand side those of orthoformaldehyde.   These
two sets of levels represent two more or less completely separable gases.
In interstellar space, transitions between them rarely occur, possibly taking
place only when the molecules are absorbed on dust particles.

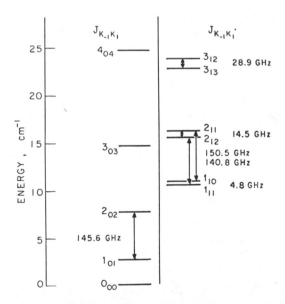

Figure 6.   Energy levels of the lower rotational levels of H_2CO, showing
two different series associated with hydrogen nuclear spins parallel
(orthoformaldehyde) and antiparallel (paraformaldehyde).   The right-hand
series of levels are those of orthoformaldehyde, and left-hand series are
for paraformaldehyde.   Transitions so far observed from interstellar H_2CO
are indicated by arrows.

TABLE I. OBSERVED MICROWAVE RESONANCES OF INTERSTELLAR MOLECULES. A "yes" indicates that the hyperfine structure has been detected (after Rank et al. [5]).

Date of Discovery	Molecule	Rotational Quantum Numbers	Transition Type	ν (GHz)	Hyperfine Structure	Emission or Absorption
1963	^{16}OH $^2\Pi_{3/2}$	$J=3/2$	Λ doublet	1.665	Yes	E, A
1969		$=5/2$	Λ doublet	6.030	Yes	E
1970		$=7/2$	Λ doublet	13.448		E
1968	^{16}OH $^2\Pi_{1/2}$	$J=1/2$	Λ doublet	4.765	Yes	E
1969		$=5/2$	Λ doublet	8.135		E
1966	^{14}NH$_3$ para	$J=3/2$	Λ doublet	1.637	Yes	A
1968		$(JK)=1,1$	inversion doublet	23.694		E
	para	$=2,2$	inv. doublet	23.722		E
	ortho	$=3,3$	inv. doublet	23.870		E
	para	$=4,4$	inv. doublet	24.139		E
	ortho	$=6,6$	inv. doublet	25.056		E
1969	H$_2$^{16}O ortho	$J_{K_{-1}K_1}=5_{23}-6_{16}$	rotational	22.235		E
1969	H$_2^{12}$C^{16}O ortho	$J_{K_{-1}K_1}=1_{11}-1_{10}$	rotational	4.83	Yes	A
		$=2_{12}-2_{11}$	rotational	14.488		A
		$=3_{13}-3_{12}$	rotational	28.974		A
1971	H$_2^{12}$C^{16}O ortho	$=1_{11}-2_{12}$	rotational	140.839		E
	H$_2^{12}$C^{16}O para	$=1_{01}-2_{02}$	rotational	145.603		E
	H$_2^{13}$C^{16}O ortho	$=1_{10}-2_{11}$	rotational	150.598		E
1969	ortho	$=1_{11}-1_{10}$	rotational	4.593		A
1970	^{12}C^{16}O	$J=0-1$	rotational	115.271	None	E
1971	^{13}C^{16}O	$J=0-1$	rotational	110.201		E

Year	Molecule	Transition	Type	Frequency	Detected	E
1971	$^{12}C^{18}O$	J=0–1	rotational	109.782		E
1970	$^{12}C^{14}N$	J=0–1	rotational	113.49	Yes	E
1970	$H^{12}C^{14}N$	J=0–1	rotational	88.631		E
1970	$H^{13}C^{14}N$	J=0–1	rotational	86.339		E
1970	X ogen (unknown – possibly HCO$^+$)	J=0–1	rotational	89.1090		E
1970	$H^{12}C_3^{14}N$	J=0–1	rotational	9.100	Yes	E
1970	$^{12}CH_3^{16}OH$	$J_{K_{-1}K_1}=1_{11}-1_{10}$	rotational	0.843		E
1971	$^{12}CH_3^{16}OH$	(JK)=4,1–4,2	rotational	24.933		E
		=5,1–5,2	rotational	24.959		E
		=6,1–6,2	rotational	25.018		E
		=7,1–7,2	rotational	25.124		E
		=8,1–8,2	rotational	25.294		E
1970	$H^{12}C^{16}O^{16}OH$	$J_{K_{-1}K_1}=1_{11}-1_{10}$	rotational	1.6388		E
1971	$^{12}C^{32}S$	J=2–3	rotational	146.969	None	E
1971	$^{28}Si^{16}O$	J=2–3	rotational	130.246	None	E
1971	$^{12}CH_3^{12}C_2H$ ortho	(JK)=4,0–5,0	rotational	85.457		E
1971	$H^{14}N^{12}C^{16}O$	$J_{K_{-1}K_1}=3_{03}-4_{04}$	rotational	87.425		E
		$=0_{00}-1_{01}$	rotational	21.981		E
1971	$^{16}O^{12}C^{32}S$	J=8,9	rotational	109.463		E
1971	$^{12}CH_3^{12}C^{14}N$ ortho	(JK)=5,0–6,0	rotational	110.463		E
	para	=5,1–6,1	rotational	110.381		E
	para	=5,2–6,2	rotational	110.375		E
	ortho	=5,3–6,3	rotational	110.364		E
	para	=5,4–6,4	rotational	110.349		E
	para	=5,5–6,5	rotational	110.330		E
1971	X$_2$ (unknown – possibly HCN)		rotational	90.665		E
1971	$^{14}NH_2H^{12}C^{16}O$	$J_{K_{-1}K_1}=2_{11}-2_{12}$	rotational	4.619	Yes	E
1971	CH_3HCO	$J_{K_{-1}K_1}=1_{10}-1_{11}$	rotational	1.065		E

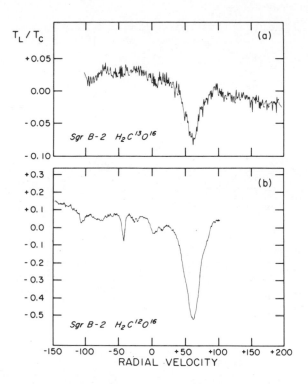

Figure 7. Absorption due to the $1_{01}01_{11}$ formaldehyde transition in Sagittarius B2. Resonances due to both $H_2{}^{12}CO$ and $H_2{}^{13}CO$ are shown. The frequency scale is converted to radial velocity in accordance with the Doppler formula. [Courtesy B. Zuckerman et al., Astrophys. J. <u>157</u>, L167 (1969).]

    Figure 7 shows an observed formaldehyde resonance. This particular resonance, the $1_{01}-1_{11}$ transition, is typically seen as an absorption line in the continuum radiation which orignates from some ionized region on the other side of a gas cloud. The resonance is plotted with intensity along the ordinate and radial velocity along the abscissa. Actually, what is measured is intensity and frequency. Since the molecular frequency is known from laboratory work, any deviation from this laboratory frequency is interpreted as a Doppler shift, representing a radial velocity with respect to the earth along the line of sight. Figure 7 also shows an absorption spectrum due to the isotopic species of formaldehyde which contains carbon 13 rather than carbon 12. Its Doppler velocity distribution naturally agrees with that of the more abundant isotopic species, and its intensity is weaker because of the lesser abundance of carbon 13.

Molecules are thus found throughout our galaxy, some detected by the absorption of a microwave continuum, and some by emission from their own excited states. By far the largest number of molecules are concentrated in the plane of our galaxy, as indicated in Fig. 8. This figure shows the distribution of molecular observations in the celestial sphere. In this figure, each dot represents a location where molecules have been detected. In most cases, the points far from the direction of the galactic plane represent clouds which are rather close to us. Thus they may not actually be very far from the galactic plane, while still being at a rather large angle with respect to this plane. The occurrence of molecules is seen to be widespread, and rather similar to the distribution of the remainder of galactic matter.

The estimated total column density of various molecules in the direction of the Orion Nebula and of the galactic center are listed in Table II. Most of these molecules are, in a sense, quite rare. Their abundance is typically 10^{-6}–10^{-8} that of hydrogen. Furthermore, the total column density corresponds to a thickness of about one hundredth of a millimeter of gas at normal

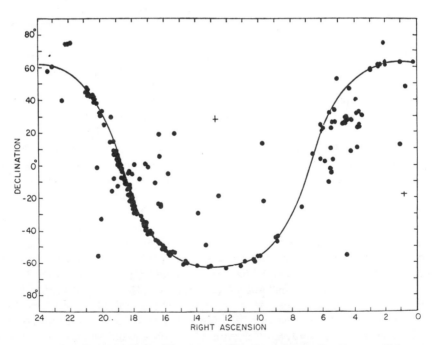

Figure 8. The distribution of known molecular concentrations within our galaxy. Each point indicates the direction for observed microwave molecular absorption or emission lines. The solid curve represents the galactic plane, near which most molecules are concentrated (Rank et al. [5]).

temperature and pressure. On the other hand, the enormous area of each cloud means that even with this small column density, the total mass of an individual molecular species present in a cloud may be as large as that of the sun.

IV. THE USE OF OBSERVED SPECTRA OF INTERSTELLAR MOLECULES

The presence and characteristics of molecular spectra in interstellar space tell us, of course, something about the composition of our galaxy. The presence of molecules in interstellar space is also pertinent to biology and the origins of life, because it is just such molecules as are observed that have generally been thought to be the raw material for the creation of life, and to be produced in primitive planetary atmospheres by natural processes, such as ultraviolet excitation or lightning strokes. While the same molecular combination of atoms which occur in an interstellar cloud is unlikely to persist during the formation of a star and its planets, the presence of these molecular species in interstellar space indicates that they are readily formed by natural processes. Hence, the materials generally assumed to be necessary for the initiation of life can be expected to be commonplace in primitive planetary atmospheres. Such molecules are also important in affecting the characteristics of the clouds in which they exist, even though they are few in number. By comparison with atomic species or hydrogen molecules, they are quite effective in radiative cooling, and probably play a substantial role in determining the temperatures of some clouds.

The most important aspect of the discovery of a rich variety of molecular spectra is that they now provide diagnostic probes deep into the interior of clouds with which to examine conditions and processes within them, much as nuclear magnetic and quadrupole resonances have provided the physicist with probes with which to diagnose the structure and behavior of matter in the laboratory. Observations of the various states of excitation of molecules and their relative abundances tell us much about the cloud densities, temperatures, and constituents. The spectral shapes of molecular resonances yield information on the cloud velocities, including both turbulence and systematic velocities which may correspond, for example, to rotation of these dark nebulae. The isotropic relict radiation directly affects molecular excitation in many cases; thus relative populations of molecular states give us insight into the intensity of the isotropic radiation, particularly in the millimeter and far-infrared region where direct measurements are very difficult. Finally, relative abundances of the isotopic resonances of molecules give indications of the origin of materials from which the molecules are formed, or of the clouds in which they are found.

The famous carbon-oxygen-nitrogen cycle for production of nuclear energy would, if it were dominant, produce a ratio of about 3 to 1 for the abundance of carbon 12 to that of carbon 13. On earth, that ratio is about 90. There is no known reason why the terrestrial ratio should be widespread, since it presumably represents some random mixture of material produced by the carbon-oxygen-nitrogen burning cycle in stars and that from other sources. However, present indications are that this terrestrial ratio is remarkably widespread in interstellar material. In some areas of space, this ratio may be somewhat less than 90, but in no interstellar region can we be sure that it approaches the 3 to 1 ratio produced by the carbon-oxygen-nitrogen cycle and, in almost all cases where the ratio can be surely established, it is remarkably similar to that on earth.

V. SOME INTERPRETATION OF MOLECULAR SPECTRA

Many of the interesting phenomena connected with molecular spectra in interstellar space are associated with nonequilibrium conditions. Some of the more spectacular results of these nonequilibrium conditions are naturally occurring interstellar masers. However, a wide variety of other interesting situations occur as the result of competing mechanisms of excitation and deexcitation.

Spontaneous emission rates for microwave molecular transitions range from about 10^{-3} to 10^{-10} sec. Stimulated emission due to the isotropic relict radiation is of the same order, or one or two powers of 10 faster. The rate of molecular collision for typical clouds of interstellar space is in the same general range. Thus, the molecular states are usually warmed by intermolecular collisions occurring at kinetic temperatures of a few tens to a few hundreds of degrees, and cooled by radiation into a radiation field of only a few degrees. Since rates for the heating and cooling are frequently comparable, but of various relative magnitudes and often involving more than one step, a wide variety of excitation conditions exist, including population inversion and refrigeration. In some cases, there may also be infrared radiation from stellar or nebulae sources which does the exciting, with collisions and radiation at other frequencies providing the deexcitation.

In some clouds, the rate of collision may be much slower than the rate of transitions induced by the relict radiation, and thus molecules come to equilibrium with this radiation. In such cases, molecular line intensities cannot differ noticeably from the intensity of the isotropic 2.7°K radiation. Thus, even a large abundance of molecules may give no detectable emission radiation. They may, however, be detected by absorption if an intense emission source is beyond the molecules.

The fact that molecular emission lines are observed at all sets a lower limit on the rate of collision, and hence on the cloud density. The density of particles in a cloud which would give collision rates just equal to the radiative transition rates is [5]

$$n = \frac{64\pi^4 |\mu|^2}{3h\lambda^3 \sigma v [1-\exp(-h\nu/kT)]} \tag{1}$$

While the rate of collision need not necessarily be just equal to the radiative rate in order for a line to be observed, it cannot be very much less than this, so that expression (1) gives a rough lower limit to the molecular density. For typical electric dipole matrix elements, this density is approximately $n = 3 \times 10^3/\lambda^3$, where λ is the wavelength in centimeters. Thus, observation of centimeter-wavelength radiation indicates particle densities greater than $3 \times 10^3/cm^3$ and observation of millimeter line densities as large as perhaps $10^6/cm^3$. Since hydrogen atoms observed in the dark cloud regions have much smaller densities than these, and yet hydrogen must be the most common element present, observation of molecular emission appears to give rather direct evidence for the presence of molecular hydrogen of densities $10^3-10^6/cm^3$. A density of $10^6/cm^3$ may represent approximately an upper limit to the cloud density because of the possibility of gravitational collapse. For a cloud of 10^6 hydrogen molecules/cm^3, gravitational collapse can occur in about 30,000 years. Thus such a cloud's lifetime would be short unless there were a rapid rotation or other mass motion to prevent such a collapse. Very likely, many of the observed clouds are collapsing locally and thus forming new stars. It is also likely that the clouds are quite nonuniform and turbulent as a result of self-gravitation; this turbulence probably produces much of the rather broad linewidth that is found in some of the more massive clouds.

The relative abundance of various types of molecules is indicated in Table II. CO is typically by far the most abundant species other than H_2. The relative abundance of other molecules tends to decrease as their stability against dissociation decreases and their complexity increases. Figure 9 shows the relative stability of various molecular bonds. Clearly, CO is exceptionally stable, and simple molecules displaying most of the more stable bonds shown in Fig. 9 have been found. Such molecules as N_2 and O_2 have not yet been detected, almost surely only because they do not have intense microwave or infrared transitions.

The velocities of various molecules detected in one given direction are displayed in Fig. 10. It is evident that most of the molecules have very similar velocity distributions. Since they coincide both in velocity and in direction, it is presumed that they coexist in the same clouds. By contrast, the velocity of hydrogen atoms can be seen to be quite different from those of molecules in the same direction, indicating that the majority of hydrogen atoms do not coexist with the molecules; this is consistent with the conclusion that the clouds containing heavier molecules are quite dense and most of the hydrogen in them is in molecular form.

The case of ammonia, whose energy levels are shown in Fig. 11,

TABLE II. APPROXIMATE COLUMN DENSITIES OF MOLECULES FOUND IN THE DIRECTIONS OF THE ORION NEBULA AND OF SAGITTARIUS (SOURCE SAGITTARIUS B2) (after Rank et al.[5]).

Molecule	Column density in Sag. B2 (molecules/cm^2)	Column density in Orion (molecules/cm^2)	Comment
H_2	$\geq 10^{22}$	$\sim 2 \times 10^{23}$	indirect determination
OH	$>5 \times 10^{16}$	?	T assumed $29°K$ for Sag. B2 primarily maser radiation from Orion
CO	$\sim 10^{19}$	$\sim 10^{18}$	optically dense clouds, hence column density quite uncertain
CN	not detected	$\sim 10^{15}$	T assumed $50°K$
CS	$\sim 10^{14}$	$2 \times 10^{13} - 5 \times 10^{14}$	
SiO	$\sim 4 \times 10^{13}$	not detected	T assumed $30°K$
H_2O	?	?	maser radiation
HCN	not determined	$\sim 10^{15}$	T assumed $20°K$ for Orion
OCS	$\geq 3 \times 10^{15}$	not detected	
NH_3	$\geq 10^{17}$	not detected	T assumed $35°$
H_2CO	$\sim 2 \times 10^{15}$	$\sim 3 \times 10^{14}$	T assumed $3°K$
NHCO	not determined	not detected	
HC_3N	$\sim 2 \times 10^{16}$	not detected	T assumed $50°$
HCOOH	$10^{13} - 3 \times 10^{15}$	not detected	line interfered with by $O^{18}H$ transitions
CH_3OH	?	$\sim 5 \times 10^{16}$	size of clouds unknown. possibly maser radiation from Sag. B2, very small cloud in Orion
CH_3CH	$\sim 2 \times 10^{14}$	not detected	
CH_3C_2H	not determined	not detected	
$X_1{}^a$	not detected	$\sim 10^{15}$	frequency 89,190 MHz, rough estimate only of abundance
$X_2{}^b$	not detected	not detected	frequency 90,665 MHz found in sources W 51 and DR 21
NH_2HCO	not determined	not detected	

[a]unknown called x–ogen
[b]unknown

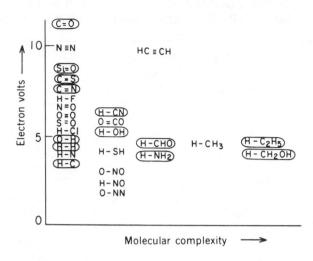

Figure 9.   Dissociation energies of various molecular bonds.   Those cases enclosed in capsules have already been found in interstellar space (Rank et al. [5]).

provides a number of interesting examples of important excitation and relax-ation processes.   Microwave emission corresponding to transitions between the inversion-doubled levels of ammonia has so far been found for the rotational states designated by 1,1; 2,2; 3,3; 4,4; 6,6; and 2,1.   The two quantum numbers in such designations correspond to J and K respectively, the total rotational angular momentum and its component along the symmetry axis.   For most of these pairs of levels, the radiative relaxation time is of such value that the molecules which provide collisional excitation must have a density of $10^3/cm^3$ or greater.   All of the first five states are metastable, in that radiative transitions can change the value of the quantum number K only very slowly or not at all and each of these represents the lowest J value for a given value of K.   However, the occurrence of the 2,1 inversion transition is surprising in that these levels are not metastable, and decay in the rather short time of 10^3 sec to the 1,1 level.   The excitation of these levels thus implies molecular densities as high as about $10^7/cm^3$.

Figure 10.   Microwave spectra of molecules and of hydrogen atoms from the direction of the center of our galaxy.   The frequency scale has been converted into a radial velocity scale by the Doppler formula.   The velocities of most molecular resonances are in general agreement, that of hydrogen atoms (H I) is different (Rank et al. [5]).

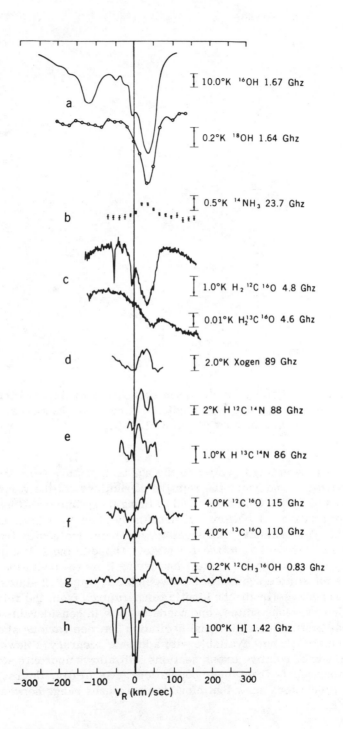

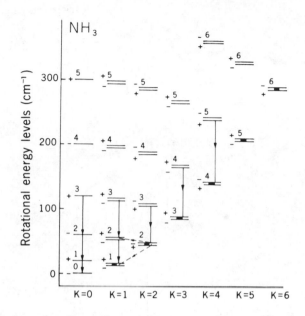

Figure 11. Energy levels of the rotation-inversion spectrum of ammonia. Positive and negative signs indicate parities of the states, and arrows some of the dipole transitions (Rank et al. [5]).

It is tempting to estimate the gas temperature from the relative intensities, and hence the relative populations, of the various metastable states of NH_3. Collisions tend to bring these into equilibrium. However, their relative populations are not simply given by a Boltzmann factor for each. For example, collisions may transfer molecules from the 2, 2 to the 2, 1 state as well as to the 1, 1 state. Those in the 2, 1 state almost immediately decay to the 1, 1 state, thus giving it more population than would be the case for simple equilibrium between the 1, 1 and 2, 2 states. Thus, in order to properly estimate the kinetic temperatures from the relative populations in these various states, one needs to know in considerable detail the relative cross section for collisional excitation between various states--numbers which are not now available with adequate accuracy. However, a rough estimate of relative cross sections still allows approximate temperature measurements from these populations. Generally, the ratio of the 1, 1 and 2, 2 populations show that cloud temperatures range between about $20°$ K and $100°$ K.

VI. VARIOUS TYPES OF NONEQUILIBRIUMS

If one compares the population of the 3, 3 states with that of the 1, 1 states, again a temperature may be obtained, but one which is not consistent with that obtained from the 1, 1 and 2, 2 levels. This is to be expected since levels with K a multiple of 3 are not readily transformed by collisions into those with K not a multiple of 3. This is because the spins of the three hydrogens in ammonia are aligned parallel, giving a total spin of three-halves when K is a multiple of 3, and aligned in such a way that the total spin is 1/2 when K is not a multiple of 3. Thus, these two separate classes of K values correspond to orthoammonia and paraammonia and are not readily interconverted by collision. Probably, the equilibrium between these two types of states is established either by absorption of the gas on the dust grain or by exchange collisions with atomic hydrogen, each of which can be estimated to occur in a time of about 10^6 years. Hence, examination of the relative populations in the 1, 1 and 2, 2 states should reflect the present kinetic temperature of the gas in the cloud while the ratios of populations in the 1, 1 and 3, 3 states should reflect a temperature at a time about 10^6 years ago. It is quite typical that measurement of every new molecular line gives new information on the temperature, excitation, and relaxation processes, because none represents thermal equilibrium. Measurement of the relative populations of a number of paraammonia states gives, for example, information on the average temperature and various higher moments of the temperature when the latter is not uniform.

A striking example of nonequilibrium is given by the spectrum of methyl cyanide shown in Fig. 12. There, the various possible K values appropriate to the transition $J = 5 \leftrightarrow 6$ are all present with the exception of $K = 2$. For some unknown reason, this particular state appears to be very much depopulated. A good explanation for this phenomenon has not yet been found.

Another striking case of nonequilibrium is the absorption due to formaldehyde in dark clouds. Ordinarily, formaldehyde lines are observed in the absorption due to clouds of gas in front of a source of continuum microwaves. However, even when such a source of microwave continuum is not present, an antenna pointed at a dark cloud will frequently detect an absorption line at the formaldehyde frequency. The formaldehyde must in these cases be absorbing some of the isotropic relict radiation, which has a temperature of 2.7°K, and hence have an excitation temperature lower than this. In some cases, this excitation temperature is as low as 0.8°K, considerably colder than either the kinetic temperature or the radiation temperature. A pump providing refrigeration of the H_2CO levels is clearly at work. There are two primary suggestions as to what this pump may be. Thaddeus and Solomon[7] have proposed a radiation mechanism based on a small deviation of the isotropic radiation from that of a blackbody. Such a deviation and a

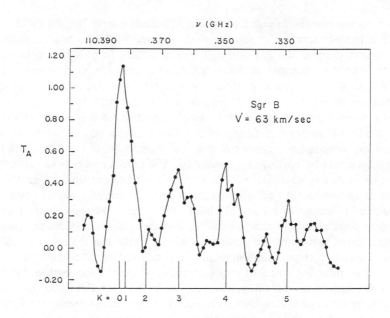

Figure 12.   The spectrum of CH_3CN observed from interstellar molecules in Sagittarius B2.   This represents the $J = 5 \longleftrightarrow 6$ transition, and various values of K (Solomon et al. [6]).

cyclic series of transitions among the four lowest levels of orthoformaldehyde shown on the right-hand side of Fig. 6, can in fact increase the population in the lowest orthoformaldehyde state and thus "refrigerate" the transitions between the lowest pair of levels.   Another mechanism, proposed by Townes and Cheung [8], involves collisions.   This mechanism will be explained in a little more detail, since it has some generality and may also apply to the OH maser.

Figure 13 is a diagram of the formaldehyde molecule.   The two hydrogen atoms are quite light by comparison with carbon and oxygen atoms along the molecular axis.   Most of the collisions to which the molecules are subjected are those due to hydrogen molecules, the dominant species in the interstellar gas.   If these hydrogen molecules are moving in the plane of the formaldehyde molecule and strike either one of its hydrogen atoms, the result is an end-over-end rotation of the molecule about an axis perpendicular to its plane.   On the other hand, if the hydrogen molecule approaches the formaldehyde in a direction perpendicular to the molecular plane and collides with one of the hydrogen atoms, the molecule is not rotated end over

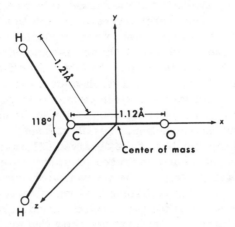

Figure 13. A diagram of the geometry of the H_2CO molecule.

end, but instead is rotated primarily about its C-O axis, because of the very
light weight of hydrogen atoms compared to the masses of carbon and oxygen.
Thus, when hydrogen molecules collide with formaldehyde from random
directions, the lower rotational levels, corresponding to rotations about an
axis perpendicular to the molecular plane, are preferentially excited. A
classical calculation shows that such considerations give an average angular
momentum about an axis perpendicular to the plane greater than that about
an axis in the plane by a factor of about 2. Such excitation corresponds to
putting the molecules predominantely in the lower of the two K-type doublets
of excited rotational states, shown in Fig. 6. The lower of the K-type
doublets for high J values decays by radiation to the lower of the J=1 levels,
thus "refrigerating" this doublet pair. A semiclassical calculation gives
approximately the observed temperature for this transition. However, more
rigorous calculations using quantum mechanics and strong coupling theory
for the collisions are needed, and are underway.

VII. INTERSTELLAR MASERS

As indicated above, perhaps the most striking examples of nonequilibrium
are the interstellar masers produced by water and by the free radical OH.
These amplify and radiate intensely. The mechanisms of excitation and the
nature of these sources has been much discussed, although they still remain
rather uncertain. It seems very likely that a number of the OH masers are
excited by infrared radiation. Other OH masers show characteristics and

are in sources which make such infrared pumping seem less likely. They may be pumped by a collisional mechanism of a type similar to that proposed for formaldehyde. The energy levels of OH are closely analogous to those of formaldehyde, its Λ-type doublets corresponding to the K-type doublets of formaldehyde. The Λ doublets of OH are associated with its unpaired electron, which is off-axis, rather than the light off-axis hydrogen atoms occurring in formaldehyde. By a mechanism similar to that discussed above for formaldehyde, collisions between atomic hydrogen and OH produce a pumping which provides an excess population in the upper of the two Λ-type doublets, and thus maser action. Possibly, such a mechanism may be important in a large number of the observed OH masers. By contrast with atomic hydrogen, molecular hydrogen interacts more strongly with the two paired electrons in π orbits in the OH radical than with the unpaired electrons. The result is that collisions between molecular hydrogen and OH overpopulate the lower of the two doublets, giving a refrigerating action just as in formaldehyde. Observations show that the Λ doublet levels of OH in large dark clouds, where molecular hydrogen should predominate over atomic hydrogen, are at very low effective temperatures. These low temperatures may possibly correspond to refrigeration by such a collision mechanism, although many astronomers have accepted the view that the very low OH temperatures in some dark clouds, of the order of $3°K$, are in fact the real kinetic temperatures of the clouds.

While the collisional mechanism discussed above has not yet been rigorously calculated for very low J values, certain collisional mechanisms can clearly be shown to produce maser action. Figure 14 shows the excitation temperatures of various levels of a diatomic molecule which result from the combination of collisions and reradiation due to both spontaneous and stimulated processes. At very low gas density, radiative processes dominate and the molecular excitation simply corresponds to the temperature of the isotropic radiation. At very high densities, collisions dominate and excitation temperatures correspond to the kinetic temperature of the gas. For intermediate cases, temperatures differ substantially from either one, and inversion is not improbable. In this case, however, the amplification by maser action cannot produce temperatures much higher than that of the relict radiation, because of saturation effects. Figure 15 shows calculated excitations for additional types of collisions.

Although the OH and H_2O masers show the most spectacular characteristics, there may be other types of masers in space. Various other molecular lines do have somewhat odd intensities and may be associated with population inversion. One such case which suggests maser action is the radiation of methyl alcohol. Barrett et al. [10] have detected a number of methyl alcohol lines in the 1-cm region from a very localized point in Orion. While their relative intensities correspond approximately to temperature distribution, there appear to be some deviations, and since the radiations

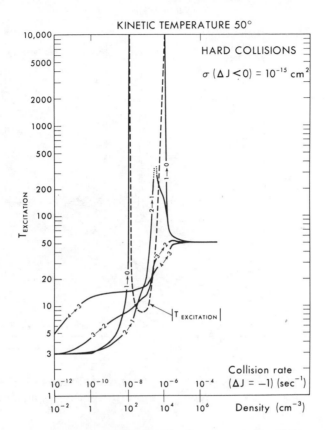

Figure 14.   Excitation temperatures of various rotational transitions of CO
in the presence of hard (short-range) collisions and radiative transitions,
including those induced by the isotropic "relict" radiation (Goldsmith [9]).

appear to come with some intensity from a rather localized point, they may
represent maser action.

The OH and H_2O masers can be positively identified as such from a
variety of striking characteristics.   First, there is the intensity, or more
specifically the radiation temperature.   These masers produce antenna
temperatures in an antenna of about 100-ft diameter which range up to
$5000°K$.   Interferometry between two antennas separated by intercontinental
distances show that their size is, however, enormously smaller than the
antenna beam widths.   Some of these masers are still unresolved by long
baseline interferometers with antennae as far separated as the territories of

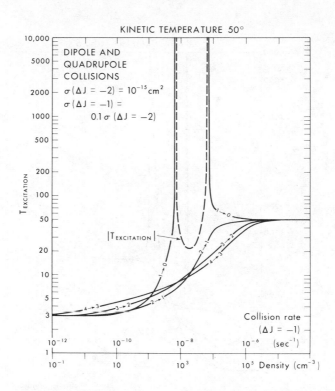

Figure 15. Excitation temperatures of rotational transitions of CO in the presence of long-range collisions of two different types and radiation transitions (Goldsmith [9]).

the Soviet Union and of the United States. Such small size and large intensity indicate effective radiation temperatures at the surface of these objects of 10^{13} °K or more. This radiation is, of course, confined to the very narrow bandwidth of the water line. Such high temperatures cannot occur in other degrees of freedom of the radiating system without completely dissociating the water. Hence, maser amplfication must be involved. There are other telltale signs of maser action. One is polarization. OH masers are frequently almost completely circularly polarized. Water masers are not generally so highly polarized, but some have been found to be about 20% linearly polarized. Furthermore, intensities of these systems vary rather rapidly, sometimes by one order of magnitude within a time span of one week. This, as well as long baseline interferometry, indicates that the source of the radiation is quite small, and hence the radiative temperatures very high. It is particularly easy to understand why fluctuations in intensity are prominent if maser amplification in the sources is large. Since the radiation is amplified exponentially within a cloud, a relative small fractional change in the cloud's characteristics can result in a large fractional change in the final amplified power.

An additional property of masers which has recently been examined with some care is the statistics of the radiation. It is common for maser and laser systems to pulse; in fact it is frequently difficult to prevent pulsing. So far, no characteristic pulses have been detected. The radiation from a saturated maser system differs in another way from normal thermal emission. Its variation in intensity is substantially less than that of blackbody radiation; in the ideal case of a heavily saturated oscillator there is almost no variation in intensity. Recently, Evans et al. [11] have examined the statistical distribution of intensities from such systems and found interstellar masers do in fact have a Gaussian distribution of intensity, just as does random thermal radiation.

General characteristics of some of the more completely measured OH and H_2O maser systems are shown in Table III. The specific numbers are not precise for any one actual frequency of the maser radiation from W49 or Orion, but rather are typical values for some of the more intense frequencies. The total energy emitted at the molecular resonance by one of these masers is frequently comparable to the total radiative energy of the sun, 10^{33} erg/sec. This assumes that emission is isotropic, and hence of the same intensity in all directions as is detected at the earth. Their size can be judged by the angular size measured in a long baseline microwave interferometer, and an estimated distance to the object obtained from a variety of astronomical considerations. This gives linear sizes for the water masers of the order of the size of the earth's orbit, about 10^{13} cm, which is comparable with the size of a large star. OH masers characteristically appear somewhat larger. As indicated above, such a small size and such a large intensity implies exceedingly high effective temperatures. It also implies rather high microwave field strengths at the surface of the object, which induce rather rapid transitions and hence require fast pumping rates. This pumping may be due to radiation, which in itself must then be quite intense, or it may be due to collisions. Collisional pumping requires densities in the range 10^{10}–10^{16} molecules/cm^3 for adequate pumping speeds. Such densities imply total masses in the clouds which are not much less than the solar mass.

Without specific knowledge of the pumping mechanisms, a good deal can still be concluded about the nature of such masers. Some preliminary results of a study underway by Evans et al. [12] will be discussed below.

A question frequently raised is whether the apparent size is in fact the actual size of the amplifying medium, since with coherent amplification it is possible that radiation begins at very localized regions, and then is amplified by a very extensive cloud without change in the shape of the wavefront.

TABLE III. TYPICAL CHARACTERISTICS OF INTENSE MASER–AMPLIFIED OH AND H$_2$O LINES FROM THREE ASTRONOMICAL SOURCES.[a]

	OH			H$_2$O		
	W49	W3	Orion	W49	W3	Orion
Intensity	10^{30} erg/sec	3×10^{28}	2×10^{26}	10^{32}	10^{30}	2×10^{28}
Linewidth	5 kHz	2	10	40	80	50
Size						
Angular	$\sim \dfrac{1}{5 \times 10^6}$ rad	$\dfrac{1}{5 \times 10^7}$	$\lesssim \dfrac{1}{5 \times 10^7}$	$? \dfrac{1}{6 \times 10^9}$	$\lesssim \dfrac{1}{3 \times 10^8}$	$? \dfrac{1}{3 \times 10^8}$
Linear	$\sim 10^{16}$ cm	$\sim 2 \times 10^{14}$	$\lesssim 3 \times 10^{13}$	$\sim 3 \times 10^{13}$	$\lesssim 3 \times 10^{13}$	$\sim 7 \times 10^{12}$
Apparent temperature	$\sim 10^{11}$	$\sim 10^{13}$	$\geq 10^{12}$	$\sim 10^{15}$	2×10^{13}	$\sim 10^{13}$
Nominal gain (T/10^3)	$\sim e^{18}$	$\sim e^{23}$	e^{21}	$\sim e^{28}$	e^{24}	$\sim e^{23}$
Field strength	$\sim 10^{-4}$ esu	$\sim 10^{-5}$	4×10^{-6}	$\sim 4 \times 10^{-3}$	3×10^{-4}	$\sim 3 \times 10^{-4}$
Transition rate	~ 10/sec	$\sim 10^3$	10^2	$\sim 6 \times 10^6$	10^5	$\sim 3 \times 10^4$
Nominal mol. density	10^{11}/cm^3	10^{13}	10^{12}	6×10^{16}	10^{15}	3×10^{14}
Nominal mass	$1/10\ M_\odot$	10^{-1}	3×10^{-5}	60	3×10^{-1}	10^{-4}

[a]The intensity given assumes the source is isotropic. The linewidth is the half-width at half-maximum intensity. Angular sizes are measured by long baseline interferometry and linear sizes deduced from this and distances estimated from other astronomical information. Apparent temperatures are determined from the intensity and size. The nominal gain assumes radiation initially of 10^3 °K is amplified. The field strength is that at the surface of the amplifying cloud, assuming it has the measured size. Transition rates are those due to this field strength. Nominal molecular densities are those which would give the nominal masses.

Another interesting question is whether these maser systems are saturated. It seems possible to conclude that their apparent size is probably not very much smaller than the actual size of the amplifying medium. Considerations which lead to these conclusions, and a general model of the masers, follow.

Assume that there is a cloud which amplifies microwaves initiated by some thermal source. This source would hardly be much hotter than $1000°K$. Hence, since at least for some of the most intense water masers, the radiation emerges from the amplifying medium with an effective temperature of about $10^{14}\,°K$, it must be amplified by a factor of about 10^{11} if the system represents simple traveling–wave amplification. This immediately raises the question whether or not there may be feedback of some type, or waves traveling through the medium in other directions. There does, in fact, appear to be a type of feedback connected with molecular scattering. Unless there are large chunks of material with relatively sharp discontinuities present within the amplifying cloud, there seem to be no mechanisms which are likely to provide feedback except resonant scattering of the molecules themselves. This type of scattering must occur if there are molecules present which are responsible for amplification. The cross section for resonant scattering of radiation has an upper limit of $\lambda^2/2\pi$, when the radiation frequency is exactly at the molecular resonance and the molecule is undisturbed. Assuming that the molecule is disturbed at a rate $1/\tau$ and that the Doppler width of a molecule is $\Delta\nu$, the actual cross section for resonant scattering can be shown to be

$$\sigma = \frac{16\pi\mu^4\tau}{9\hbar^2\lambda^4\Delta\omega} \tag{2}$$

As the wave travels through the medium, it scatters backward a certain amount of radiation which is then amplified as the latter travels backwards through the cloud. The ratio of total intensity of radiation thus emerging within a solid angle $\Delta\Omega$ to the intensity of radiation emerging after amplification of the radiation from the initial source can be shown to be

$$r = \frac{n\sigma e^{\gamma L}\Delta\Omega}{8\pi\gamma} \tag{3}$$

where n is the density of molecules in the two states involved in the transition, γ is the gain per unit length, and L is the thickness of the amplifying cloud. The total gain of the cloud due to the molecules is $g = e^{\gamma L}$ where

$$\gamma = \frac{4\pi\mu^2\Delta n}{3\hbar\lambda\Delta\omega} \tag{4}$$

Δn is the difference in the number of molecules per unit volume in the upper and lower states. From these two relations one can conclude that the total intensity of radiation scattered backwards and amplified is greater than the radiation assumed to be traveling forward unless there is an upper limit on the gain through the amplifying medium given by

$$g \leq \frac{3}{4} \frac{\hbar \lambda^3}{\mu^2 \tau} \frac{\Delta n}{n} \frac{4\pi}{\Delta\Omega} \tag{5}$$

Saturation does not change this equation, since it decreases Δn and decreases the time τ between interruptions proportionately. If the medium is assumed simply to amplify the radiation from some small source, the total known gain and expression (5) set a lower limit on the interruption rate $1/\tau$. This is a lower limit on the molecular density in the cloud, assuming collisions are important in either the excitation or relaxation of the molecular levels. The resulting density is so high that the total mass would be prohibitive unless the cloud is rather small, namely, comparable with the actual linear sizes deduced from the measurement of angle and known distance. Such a conclusion does not represent a precise relationship, but rather a qualitative one which can be expected to be usually correct within one or two orders of magnitude.

Unless the amplifying cloud is very small and very dense, Eq. (5) shows that there is, in fact, so much feedback that the amplifying molecules must saturate to prevent the gain from becoming infinite. Thus the amplifying cloud is very likely a cloud which is not very much larger than the measured size, scattering waves in all directions through the cloud, each of which is amplified. A laser with this type of incoherent feedback has been studied both theoretically and experimentally by Ambartsumian, Basov, Kriukov, and Letokhov [13]. It is known to give Gaussian–type radiation with the same statistics as thermal noise, as has been found for these interstellar masers. It also emits more or less isotropically, and with the same bandwidth produced by a traveling-wave system, even though feedback is present. The bandwidth in this case is given by

$$\Delta\nu = \frac{\Delta\omega}{2\pi} / [\ln(g/\sqrt{2})]^{1/2} \tag{6}$$

This gives a bandwidth reduction of a factor of about 5, which is consistent with the observed frequency width of laser emission and reasonable velocities of individual molecules. Saturation and important amounts of incoherent feedback do not necessarily occur in all interstellar masers, some of which are much less intense than those indicated in Table III, and many of which are not yet measured in size. Furthermore, there are a number of possible variations of this model, even though it seems very likely to be correct for the more intense masers.

Interstellar masers may, of course, be discussed as heat engines. Since they continuously produce energy, there must be both a heat source and a heat sink, of different temperatures. The heat source may be either the kinetic energy of molecules, fed in through collisions, or it may be the radiation field. The heat sink could likewise be either one of these two. While for most of the masers observed, the molecular kinetic energy appears to be the most likely heat source and radiation the cooling mechanism, it is not necessary to make a precise specification of the source and sink for present discussion. In the case of both OH and H_2O, no pumping mechanism has so far been imagined which does not involve the release of one far-infra-red quantum for each microwave quantum. For collisional pumping, colli-sions excite OH to an upper rotational state, which then decays to the ground rotational state, where maser action occurs. Infrared pumping like-wise requires the release of one quantum of rotational energy, of wavelength about 100 μ. Consider now how such radiation can be emitted from the maser as a whole. If it is emitted from the surface of the cloud, the ratio of the effective infrared temperature T_{ir} at the surface is given, in the Rayleigh-Jeans limit, by

$$T_{ir} = (\frac{\lambda_{ir}}{\lambda_M})^2 T_M \tag{7}$$

where T_M is the effective microwave temperature and λ_{ir} and λ_M are the infrared and microwave wavelengths, respectively. Thus, if 100-μ radia-tion from the cloud cools the system, its effective temperature at the surface of the maser would be of the order of $10^9 \, °K$. Since nothing else within the system is so hot, such temperatures cannot, of course, really correspond to the heat sink, unless one wishes to assume maser action at the 100-μ resonance, as well as at microwave frequencies. The latter assumption causes still other complications. Thus, it appears impossible for 100-μ resonant radiation to escape. It must be absorbed internally. Dust particles within the cloud appear to be the only likely possibility for an appropriate heat sink, and in fact can serve this function well. The far-infrared quanta can be absorbed ny these particles, then reradiated in a broad band of frequencies, which would require only a modest surface temp-erature, or temperatures for the dust particles of a few hundred degrees. This type of consideration can be used correspondingly for the heat source, if the molecular excitation which produces pumping is due to infrared radia-tion. From Eq. (7) such infrared radiation must be generated internally within the amplifying medium unless its wavelength is quite short. Other-wise, exorbitantly high temperatures would again be required to produce it. Thus, one can conclude that both the heat source and the heat sink are probably internal to the system, unless there is some special geometry which gives a very large surface to volume ratio for the cloud.   This does not exclude an infrared star surrounded by a shell of gas which is amplifying; the star might be considered internal in such a case, or the surface to volume ratio of the shell very high.   Another reasonable model for such maser systems may be simply clouds of gas, possibly contracting into a protostar, of which some part has dust grains and resulting nonequilibrium conditions which allow maser action.   When these critical conditions are not fulfilled, the system can no longer amplify as a maser, and radiation of the amplified frequencies is hence subject to wide fluctuations in intensity.

There have been many additional considerations about the mechanisms and characteristics of interstellar masers, and a number of suggestions of other masers which may possibly occur.   Space does not permit inclusion of many of these interesting discussions and proposals.   However, most of the details of such systems are yet unsettled and still remain for the astronomer, the spectroscopist, the molecular physicist, and those in quantum electronics to investigate and clarify.

REFERENCES

1. S. Weinreb, A. H. Barrett, M. L. Meeks, and J. C. Henry, Nature 200, 829 (1963).
2. A. C. Cheung, D. M. Rank, C. H. Townes, D. C. Thornton, and W. J. Welch, Phys. Rev. Letters 21, 1701 (1968).
3. A. C. Cheung, D. M. Rank, C. H. Townes, D. C. Thornton, and W. J. Welch, Nature 221, 626 (1969).
4. L. E. Snyder, D. Buhl. B. Zuckerman, and P. Palmer, Phys. Rev. Letters 22, 679 (1969).
5. D. M. Rank, C. H. Townes, and W. J. Welch, Science 174, 1083 (1971).
6. P. M. Solomon, K. B. Jefferts, A. A. Penzias, and R. W. Wilson, Astrophys. J. 168, L107 (1971).
7. P. Thaddeus and P. M. Solomon, Bull. Am. Astron. Soc. 2, 218 (1970).
8. C. H. Townes and A. C. Cheung, Ap. J. 157, L103 (1969).
9. P. Goldsmith, Astrophys. J. 176, 597 (1972).
10. A. H. Barrett, P. R. Schwartz, and J. W. Waters, Astrophys. J. 168, L101 (1971).
11. N. J. Evans II, R. E. Hills, O. E. H. Rydbeck, and E. Kollberg (to be published).
12. N. J. Evans II, A. C. Cheung, M. W. Werner, and C. H. Townes (to be published).
13. R. V. Ambartsumian, N. G. Basov, P. G. Kriukov, and V. S. Letokhov, Progress in Quantum Electronics (Pergamon Press, New York, 1970), Vol. 1, pp. 105-197.

OPTICAL PUMPING IN WEAK DISCHARGES

J. Brossel
Laboratoire de Spectroscopie Hertzienne, Ecole Normale Supérieure
Paris, France

I.   INTRODUCTION

Many problems in laser physics are closely related to those which have been studied extensively over the years in optical pumping.   My purpose is not to describe them all; nor do I intend to make a detailed review of the field, to be fair in a way, to give credit to everyone, or to be complete in any manner. I will limit myself to the description of a few experiments actually being carried out at the Ecole Normale Supérieure, on <u>conventional</u> optical pumping in weak discharges in rare-gas mixtures.   To be more specific, we are interested among other things in the properties of the 2^3S_1 metastable state of helium and in its behavior in different kinds of collisions against other atomic species.   This kind of problem, of course, attracted a lot of attention when people first started work on gas lasers in order to obtain oscillation.   But before I start on this, I would like to mention briefly that other problems bearing close relations to the laser field are actually explored in our laboratory.

Some of them will be described in a communication by Cohen-Tannoudji. They deal with transverse optical pumping of atoms "dressed" with non-resonant radio-frequency photons.   It is shown how atomic levels are perturbed by the coupling with an intense coherent radio-frequency field (what we call the "dressing" of the atom).   It seems to us that quite a few of the results which have been obtained in this way might be valuable in the physics of lasers.

In a very different direction we are interested also in optical pumping
with coherent light (that is to say with a laser source): The gas mixture
under study is excited in a discharge tube which is placed inside, or outside,
the laser cavity (of course, one is limited to atoms for which laser oscillation is possible). One looks at the fluorescence of this tube in the presence
and in the absence of laser excitation. One operates in a static longitudinal
or transverse magnetic field which does not interfere with the working of the
laser itself. This kind of pumping--in the case of the linear response, at
low enough laser intensities--allows one to do experiments that are very
similar to those one does in ordinary optical pumping, i.e., with a conventional light source--Hanle effect, Landé g factors, lifetime and collision
cross-section measurements, etc. This is particularly true for lasers
having a great number of modes inside the Doppler width.

But the specific characters of the laser source--monochromaticity of
the modes--which can be incoherent or locked, great intensity, etc., lead to
effects which have no equivalent in conventional optical pumping, for
instance, the existence of modulations or saturations which have a resonating
behavior. Those effects have been described by many people. They have
been the subject of detailed and very complete theories and lead to beautiful
applications. Many of these are described in papers in this volume.
Accordingly, I am limiting myself to the particular question I mentioned at
the beginning and which I analyze now.

II.   OPTICAL PUMPING IN HELIUM

This paper will deal with pure ^{3}He and ^{4}He and ^{3}He-^{4}He mixtures (Fig. 1).
I will only briefly mention at the end the case of the other rare gases.

Optical pumping in this case operates in the following way [1,2]. A
weak discharge is maintained in helium at a pressure of about 1 Torr. This
produces in a steady-state manner a low population of 2^3S_1 helium metastables. These can be oriented by optically pumping (in a static field $\vec{B}_0$)
this level with the 10,830-Å ($2^3P \rightarrow 2^3S_1$) circularly polarized line. As we
shall see there are many ways to monitor this orientation optically. One of
them is optical absorption on the pumping beam at 10,830 Å. The magnetic
resonance between the Zeeman sublevels of 2^3S_1 (or of its hfs sublevels F
and F' in the case of ^{3}He) can easily be observed. One can study in this way
the longitudinal and transverse relaxation produced by collisions (and determine their angular dependence). One can also measure the structure of this
state, for instance its Landé g factor and the way it is being altered by
different perturbations.

This analysis of the data leads to results which have been described by
people who discovered the field [1,2] and which appear puzzling at first
sight. I will mention only those which are of interest here.

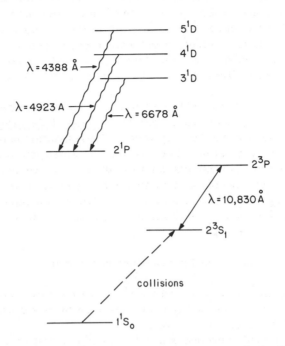

Figure 1.   Energy-level scheme of the helium atom. Optical pumping of the 2^3S_1 metastable level, populated by a weak rf discharge, is produced by the $\lambda = 10,830$ A line.

In pure ^{4}He, the magnetic resonance between the Zeeman sublevels of 2^3S_1 (g = 2) can be very narrow indeed. In the absence of any trace of ^{3}He it is essentially determined by the 2^3S_1 lifetime, i.e., the diffusion time to the walls, when the pumping beam intensity and the intensity of the discharge are sufficiently low (most authors think that metastable atoms are destroyed in a collision against the wall). As a consequence, the Landé g factor of $(^4$He $- 2^3S_1)$ can be obtained with great precision. We describe, below, a measurement of this kind; its interest comes from the fact that we are dealing in this case with a simple atomic system which can be calculated from first principles with an excellent approximation.

In pure ^{3}He, on the other hand, the resonances pertaining to the F = 3/2 and F' = 1/2 hfs sublevels of 2^3S_1 are much broader than the one of ^{4}He in identical conditions. Their width has now a linear dependence on the ^{3}He pressure. It is clearly determined by collisions against ^{3}He atoms in the 1^1S_0 ground state, i.e., by the so-called <u>metastability exchange</u> collisions.

Very clearly, metastability exchange exists in collisions between two

^{4}He atoms, but they do not appear in any way. One has to admit, then, that in those collisions all the information of electronic character which is carried by the incoming (^{4}He - 2^3S_1) metastable atom is transferred to the outgoing one. In fact, electron clouds are exchanged during the process.

We will describe, later on, the properties of metastability exchange collisions between ^{3}He and ^{4}He. I will just mention here that an isotope relaxes the other.

Another very remarkable fact is observed in ^{3}He when the 2^3S_1 metastable state is oriented: There exists a <u>nuclear orientation in the 1^1S_0 ground state</u>, and it can be very large. Its relaxation depends on many factors and can be very slow. Because the concentration of metastable 2^3S_1 atoms is very low in weak discharges, the time which is necessary to obtain the full ground-state nuclear orientation can be very long. Nonetheless, the 2^3S_1 and 1^1S_0 orientations are strongly coupled, and the 1^1S_0 NMR is observed on optical transitions in which the 2^3S_1 level is involved.

A.   Extension of the Franck-Condon Principle

We look now into the reasons why a nuclear orientation appears in (^{3}He - 1^1S_0) because of metastability exchange when there exists an orientation in the (^{3}He - 2^3S_1) metastable level.

Optical pumping produces an electronic orientation of J in 2^3S_1 together with a nuclear orientation of I in this state (essentially via the $a\vec{I}\cdot\vec{J}$ hfs interactions in 2^3P and 2^3S_1). As we shall see the values of metastability exchange cross sections ($\sim 10^{-16}$ cm^2) indicate that the process is very short and lasts for a time θ of about 10^{-13} sec (the time necessary for an He atom, at thermal velocities, to travel by about 1 Å). The interaction which is responsible for metastability exchange is an electrostatic one. <u>It does not act directly on the nuclear spin</u> (nor does it on the electronic spin). If $\mathcal{K}$ is the corresponding Hamiltonian the evolution of the atomic system during the collision is described for $t < \theta$ by $\exp(-i\mathcal{K}t)$.

The evolution due to the static Hamiltonian H acting on $\vec{I}$ directly (Zeeman and hfs interactions) is completely negligible during the collision if H$\theta \ll 1$; this is <u>always</u> the case, in the above circumstances, for times of the order of 10^{-13} sec: during the metastability exchange collision itself the nuclear spin remains unaffected [3, 4]; the nuclear orientation which was present in 2^3S_1 goes fully into the 1^1S_0 state in which the atom is brought. On the other hand, between successive collisions the evolution of $\vec{I}$ is governed by the static Hamiltonian H.

This is a very general situation. In fact, it is the rule in very short collisions (10^{-13} sec) taking place in the gas phase, when the dominant interaction is the electrostatic one.

One understands, in this way, that the orientations of 1^1S_0 and 2^3S_1 are not the only ones to exist and to be coupled in an isotropic discharge in optically pumped ^{3}He. The discharge takes place in a gas where a nuclear orientation $\langle \vec{I} \rangle$ exists in the ground state. Those atoms are raised to different excited states by collisions with electrons and ions which are present in the discharge (note that, in the pressure range used, the directions of their velocities are isotropic in space). Quite clearly this electrostatic excitation process belongs to the class which was described above; during the excitation process itself $\langle \vec{I} \rangle$ is fully transferred to all excited states by an isotropic discharge. During the lifetime τ of the excited state, the dominant Hamiltonian acting on $\vec{I}$ is, as a rule, the $a\vec{I}\cdot\vec{J}$ interaction. The corresponding evolution will be appreciable if $a\tau \gg 1$. The nuclear orientation $\langle \vec{I} \rangle$ will then be transformed in part into electronic orientation $\langle \vec{J} \rangle$ in this state. As is well known, the existence of such an electronic orientation produces a circular polarization of the lines emitted from this state. This allows optical detection of the different orientations which are present in all states where $a\tau > 1$. Even more interesting is the fact that one can detect on the same lines the orientation which is present on other levels, and which happen to be strongly coupled to the orientations of the level from which the transition under study starts. This is particularly true for the ground- and metastable-state orientations. For instance, the corresponding resonances appear on the 6678-Å and 5876-Å lines (and others!). This method for monitoring the 1^1S_0 and 2^3S_1 resonances [5] is very useful and often gives a signal-to-noise ratio which is far better than the one obtained in absorption of lines starting from the ground or metastable states themselves.

Another example for which whatever we have said applies is "charge exchange" between a neutral atom A and an ionized one B^+

$$A + B^+ \rightarrow A^+ + B$$

This reaction allows one to obtain an oriented A^+ ion if one starts with a neutral atom A on which a nuclear orientation exists. It is in this way, we believe, that one obtains the oriented $^3\mathrm{He}^+$ ion in weak discharges of optically oriented ^{3}He [6,7].

As a result of all the facts we have described it appears that in such a discharge, all excited states of the neutral atom (for which $a\tau > 1$), the ground state of the $^3\mathrm{He}^+$ ion, and presumably its excited states as well are oriented and all these orientations are strongly coupled to the metastable- and ground-state orientations. This multiplies the possibilities of the detection.

The above properties have been discovered in the course of the study of many relaxation processes in the gas phase. Bender [8] was the first to suggest that the nuclear spin might be unaffected in collisions of very short duration, and this simplifies the problem considerably. The theory can be worked out in the case of the even isotope ($I = 0$). For $I \neq 0$, the results so derived can be used directly. One has to recouple $\vec{I}$ and $\vec{J}$ during the time which elapses during two collisions. The relaxation times in this $I \neq 0$ case (population, orientation, alignment of each hfs level, transfer rates between hfs levels, etc.) can all be expressed in terms of those found for the $I = 0$ case. As an example, one can mention in this respect the many experiments performed on the relaxation of the 6^3P_1 level of Hg in collisions with rare gases [9] or in Holtsmark collisions [10]. In the same class of problems falls the question of the influence of the nuclear spin on (electronic) spin exchange collisions [11].

The above remarks can be extended without major difficulties to the electron spin $\vec{S}$ but, in this case, the condition $H'\theta < 1$ is much more difficult to meet (H' is the <u>static</u> Hamiltonian acting on $\vec{S}$), in particular because of the magnitude of the Bohr magneton and of the spin orbit interaction. It is very likely that the same comments apply to Penning collisions [12].

It seems legitimate then to talk about an extension of the Franck-Condon principle [13] derived first for molecules: during a sudden perturbation in which electrons only are directly concerned (absorption or emission of electromagnetic radiation, collisions, etc.), the coordinates of the nuclei are unchanged because of the great mass difference of the electron and of an atomic nucleus.

Indeed all we have said does apply to the process of absorption of the light coming from a conventional light source, and to the process of spontaneous emission, when the transition is of the electric-dipole type (it does not affect $\vec{S}$ or $\vec{I}$ directly). For spontaneous emission the duration of the process is extremely small (of the order of optical frequencies, 10^{-15} sec). Nuclear orientation is totally transferred to the lower state by spontaneous emission. On the other hand, the absorption process involves times of the order of $1/\Delta$, Δ being the width of the wave train exciting the atom. Clearly laser excitation does not obey the above criteria.

Another remark should be made. Everything so far has been described on ^{3}He in terms of atomic orientations. It does seem that situations might arise where similar situations would exist for alignments which can be produced by pumping with linearly polarized light.

Finally, it might be worthwhile noting that in a static field $\vec{B}_0$ large enough to decouple partly $\vec{I}$ and $\vec{J}$, a nuclear orientation $\langle \vec{I} \rangle$ may very well induce an electronic <u>alignment</u> of $\vec{J}$ besides an electronic orientation. This appears as a <u>linear polarization</u> of the lines emitted by the isotropic discharge. This fact has been used to measure the hyperfine structures of a few excited states of ^{3}He [15].

III. A STUDY OF METASTABILITY EXCHANGE RELAXATION

In Sec. IV, we will describe the measurement of the Landé g factor of
(^{4}He - 2^3S_1). In those experiments we determine the value of the static field
$\vec{B}_0$ by measuring the NMR frequency of ^{3}He in its ground state 1^1S_0. We
make use, then of rf discharges in ^{3}He - ^{4}He mixtures. In this situation the
(^{4}He - 2^3S_1), (^{3}He - 2^3S_1), and (^{3}He - 1^1S_0) orientations are strongly coupled by metastability exchange. This coupling determines the <u>linewidth</u>, i.e.
the precision of the measurement. It can affect also, as we shall see, the
<u>position</u> of the resonances, in 2^3S_1 and 1^1S_0. Because of this, we studied in
some detail metastability exchange relaxation in pure ^{4}He and ^{3}He and in
^{4}He-^{3}He mixtures. Before we present our results, we will describe the
experimental setup and the observed resonances.

A. Experimental Setup

The experimental setup is shown in Fig. 2. The helium lamp L is followed
by a 1.1-μ filter and by a circular polarizer P_1. The cell containing ^{3}He
(or ^{4}He,..., etc.) is excited by a weak rf discharge and placed in a longitudinal magnetic field $\vec{B}_0$. It is irradiated by a radio-frequency field $\vec{B}_1$
cosωt, linearly polarized at right angles to $\vec{B}_0$. In line with this, we find a
rotating quarter-wave plate and a linear polarizer P_2. This is just a σ^+ or
σ^- circular analyzer depending on the position of the slow axis of the $\lambda/4$
plate. One obtains in this way a modulated signal at twice the frequency of
the rotation of the $\lambda/4$ plate, the amplitude of which is proportional to the
difference of the σ^+ and σ^- intensities emitted by the cell, i.e., to the
amount of orientation present in the atomic state under study. A filter F_2

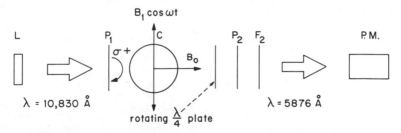

Figure 2. The experimental setup. L: helium lamp--P_1 circular
polarizer; C: discharge cell containing helium; P_2: linear polarizer;
F_2: optical filter passing 5876 Å; PM: phototube.

isolates the line λ on which the detection is made, and is followed by the phototube PM. Most of our measurements were made at $\lambda = 5876$ Å ($3^2D - 2^2P$) and 6678 Å ($3^1D - 2^1P$). Fig. 3 shows a spectrum, for g values between 4 and 3/4, obtained with ^{3}He containing ^{4}He as a slight impurity [7]. One can see two resonances corresponding to the hfs levels $F = 3/2$ and $F' = 1/2$ of (^{3}He - 2^3S_1).

At $g = 2$, there is a very weak resonance which is much more intense in pure ^{4}He, where its width is much reduced and determined, as we said, by the diffusion time to the walls. It corresponds to the 2^3S_1 metastable state of ^{4}He, and it is the line on which we have studied the behavior of (^{4}He - 2^3S_1) under metastability exchange collisions with ^{3}He. Because the corresponding Landé g factor is 2, one might wonder whether this resonance would not be produced by polarized free electrons eventually present in the discharge (in fact, electrons are very hot in the discharge and would have a much broader resonance; it has been observed, though, but in post discharges [14,15]). Besides, the precision measurements that we report in Sec. IV and in which we determine the position of this resonance give an unambiguous way to identify it and exclude the possibility that it might be due to free electrons.

Finally one sees in Fig. 3, at $g = 1$, a resonance due to the ground

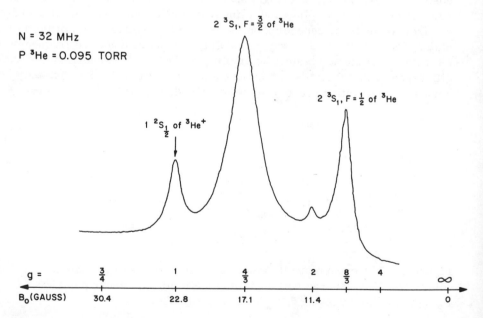

Figure 3. Resonances observed at 32 MHz, in ^{3}He containing a small impurity of ^{4}He. The resonance at $g = 2$ is the (^{4}He - 2^3S_1) resonance.

state $1^2S_{1/2}$ of the $^3He^+$ ion. It has been identified [7] by looking at its behavior at higher field values when a slight decoupling of $\vec{I}$ and $\vec{J}$ occurs. Under those circumstances the Zeeman transitions $(1 \to 0)$ and $(0 \to -1)$ (within the F = 1 hfs level of $1^2S_{1/2}$) have different frequencies. Indeed, as seen in Fig. 4 this line splits in two components; from the observed position of these components one can deduce the associated hfs constant. It coincides to better than 10^{-4} with the well-known [16] hfs constant of $(^3He^+, 1^2S_{1/2})$. The identification is then unambiguous.

One also sees, in Fig. 4, the very narrow and intense lines corresponding to the nuclear magnetic resonance of ^3He in its 1^1S_0 ground state with which we calibrate the field $\vec{B}_0$. These are produced by a different field than the one inducing the paramagnetic resonances $(^3He^+ - 1^2S_{1/2})$, $(^3He - 2^3S_1)$, $(^4He - 2^3S_1)$ that we have just described. It is produced by separate coils wound around the cell and excited by a different generator. The NMR transitions of $(^3He - 1^1S_0)$ are observed at several frequencies obtained by multiplying the basic frequency (stable to 10^{-9}) of a continuously running quartz controlled oscillator.

The width of the $^3He^+$ ion resonance increases linearly with the ^3He pressure: The lifetime of the ground state $(1^2S_{1/2})$ of $^3He^+$ is limited by charge exchange collisions with $(^3He - 1^1S_0)$. The corresponding cross-section has been measured; the associated relaxation is under study.

In the experiments described in Sec. IV, we operate on $^4He - ^3He$ mixtures where the ^3He pressure is quite low (0. 01 Torr). The ^4He pressure on the other hand is in the range 0. 5-1 Torr. This is in order to get a good signal to noise ratio. ^3He is there just to calibrate the field. One finds (we

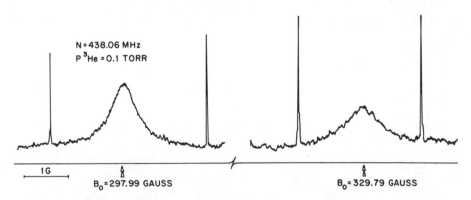

Figure 4.   The splitting of the $(^3He - 1^2S_1)$ resonance in high fields. Because of the $\vec{I} \cdot \vec{J}$ decoupling the transitions $(m_F \to m'_F)$, $(1 \to 0)$ and $(0 \to 1)$ have different frequencies. The sharp and intense resonances correspond to the nmr of ^3He in its ground state. They are used to calibrate the field and are induced by a separate rf circuit.

will come back to this later on) that the width of (^{4}He - 2^3S$_1$) has a linear dependence on ^{3}He pressure. We have to accept a compromise: We cannot take the ^{3}He pressure too low if we want to observe the (^{3}He - 1^1S$_0$) resonance in good conditions. Fig. 5 shows a typical (^{4}He - 2^3S$_1$) resonance (notice that the NMR (^{3}He - 1^1S$_0$) peaks used to calibrate the field are quite narrow under the conditions of the experiment).

B.   Metastability Exchange Relaxation

In the first part, we will deal with the problem in pure ^{3}He. We call $< \vec{F} >$ (and $< \vec{F'} >$) the orientation present in the hfs level $F = 3/2$ (and $F' = 1/2$) of (^{3}He - 2^3S$_1$). Similarly $< \vec{I} >$ is the nuclear orientation in the ground state (1^1S$_0$). The theoretical analysis we have made is based on a model which has been used already by previous workers in the field [3, 4] and which has been justified from first principles [17]. We suppose the following:

1.   At the end of the very short collision process, the outgoing metastable atom has the same electronic density matrix as the incoming one before the collision. There is a transfer of electronic clouds from one nucleus to the other. This is due to spin conservation during a very short electrostatic interaction which does not operate on $\vec{S}$ directly.

2.   Similarly, during the collision, the nuclear spins of the two nuclei remain unaffected.

Table I indicates the evolution of the density matrices in the ground state 1^1S$_0$ and in the metastable (2^3S$_1$) state, due to metastability exchange alone. Immediately after the collision there is no correlation between the

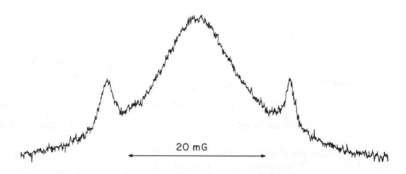

Figure 5.   The (^{4}He - 2^3S$_1$) resonance with the (^{3}He - 1^1S$_0$) markers.

TABLE I. [a]

	Ground state atom	Metastable atom
Before collision	ρ_g	ρ_m
After collision	$\rho_g' = \mathrm{Tr}_e\, \rho_m$	$\rho_m' = \rho_g \otimes \mathrm{Tr}_n\, \rho_m$
After hyperfine coupling	$\rho_g'' = \rho_g'$	$\rho_m'' = \sum_F P_F\, \rho_m'\, P_F$

[a] This table summarizes the density matrix evolution of two ^{3}He atoms undergoing a metastability exchange collision. Tr_e and Tr_n refer respectively to partial tracing over the electronic and nuclear variables; P_F is the projection operator onto the F hyperfine level.

nuclear and spin variables. Finally, in the time interval between collisions, the $a\vec{I}\cdot\vec{J}$ hfs interaction operates in 2^3S_1 and the hyperfine coherences are destroyed (a is large in comparison to the width of 2^3S_1). From this the following equations are derived [18]:

$$\frac{d}{dt}\langle\,\vec{F}\,\rangle = -\frac{1}{\tau}\langle\,\vec{F}\,\rangle + \frac{1}{\tau}\left(\frac{5}{9}\langle\,\vec{F}\,\rangle + \frac{10}{9}\langle\,\vec{F}'\,\rangle + \frac{10}{9}\langle\,\vec{I}\,\rangle\right)$$

$$\frac{d}{dt}\langle\,\vec{F}'\,\rangle = -\frac{1}{\tau}\langle\,\vec{F}'\,\rangle + \frac{1}{\tau}\left(\frac{2}{9}\langle\,\vec{F}'\,\rangle + \frac{1}{9}\langle\,\vec{F}\,\rangle - \frac{1}{9}\langle\,\vec{I}\,\rangle\right) \qquad (1)$$

$$\frac{d}{dt}\langle\,\vec{I}\,\rangle = -\frac{1}{T}\langle\,\vec{I}\,\rangle + \frac{1}{T}\left(\frac{1}{3}\langle\,\vec{F}\,\rangle - \frac{1}{3}\langle\,\vec{F}'\,\rangle\right)$$

τ and T are the average times between two metastability exchange collisions in the metastable 2^3S_1 state and in the ground state, respectively; n and N being the corresponding populations, one has $n/\tau = N/T$. Eqs. (1) show that orientations $\langle\,\vec{F}\,\rangle$ and $\langle\,\vec{F}'\,\rangle$ are directly coupled, besides being coupled to the ground-state orientation $\langle\,\vec{I}\,\rangle$. This fact is a result of electronic spin conservation in the collision. Part of the electronic spin orientation remains on F, another part being transferred from F'. As a result, F and F' Zeeman coherences have a life time which is different, longer than if electron orientation transfer did not exist and longer than τ. There is a strong and direct coupling between $\langle \vec{F} \rangle$ and $\langle \vec{F'} \rangle$ with an associated time constant of the order of τ.

Eqs. (1) allow a complete analysis of metastability exchange relaxation in F and F' levels in pure ^{3}He as well as for the NMR in 1^1S_0, in particular as far as linewidths and resonance frequencies are concerned.

The theoretical ratio of the widths of the F and F' resonances (in cps) deduced from (1) is $\Delta\nu_F/\Delta\nu_{F'} = 4/7 = 0.572$. The ratio of the measured widths [19] in gauss is equal to 1.16 ± 0.02. Taking into account the Landé g's of F and F', its value, when converting into cycles, is 0.579 ± 0.012. The agreement is excellent.

Fig. 6 shows that the widths of the F and F' resonances are indeed proportional to ^{3}He pressure. From the slopes of the lines, one can deduce the same value of τ for the F and F' levels, i.e., the same metastability exchange cross section, $\sigma = (7.6 \pm 0.4) \times 10^{-16}$ cm^2. This value is in very good agreement with the determination of Colegrove et al. [20] deduced from the broadening of the F = 3/2 level, if it is corrected by a factor of 9/4 present in Eqs. (1) and by another (1.3) allowing for the same definition of

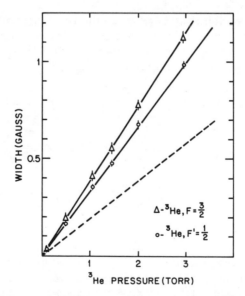

Figure 6.   The linear dependence of the F = 3/2 and F' = 1/2 Zeeman resonances of (^{3}He – 2^{3}S$_1$) on ^{3}He pressure in pure ^{3}He (metastability exchange collisions). Taking the experimental slope of the F = 3/2 curve one obtains from Eqs. (1) in the text the straight line corresponding to the F' = 1/2 level. Experimental points are plotted on the diagram. The dotted line is obtained for the F' = 1/2 level in an incorrect theory where one forgets the transfer of electronic coherence in metastability exchange collisions.

the cross section. Our result agrees also with the one, $(7.9 \pm 2) \times 10^{-16}$. Greenhow [21] deduced from the broadening of the NMR line of the ground state (the lifetime of $\langle \vec{I} \rangle$ in 1^1S_0 is determined by metastability exchange collisions, because transverse orientations rotate at very different frequencies ω_f and ω_m in 1^1S_0 and 2^3S_1).

The same kind of analysis can be applied to ^{3}He - ^{4}He collisions and leads to the interpretation of metastability exchange broadening of 2^3S_1 in the ^{3}He - ^{4}He mixture [22] [Eqs. (1) do not apply then]. In this kind of measurement it is important to choose the proper pressures to use when determining the ratio of the magnetic moment of (^{4}He - 2^3S_1) to the nuclear moment of ^{3}He (Sec. IV). The model we have adopted for metastability exchange has the following consequence: (^{3}He - ^{3}He) and (^{3}He - ^{4}He) cross-sections must be the same. Neglecting the small correction (5%) due to the different thermal velocities of ^{3}He and ^{4}He, this implies that τ should be the same for ^{3}He - ^{3}He and ^{3}He - ^{4}He relaxations.

Fig. 7 (corresponding to a total pressure of 0.1 Torr) gives (full lines) the theoretical result in ^{3}He - ^{4}He mixtures [18,19]; it shows how the widths of the different resonances in 2^3S_1 vary with ^{3}He pressure. In Fig. 7(a) the widths are expressed in cps. In Fig. 7(b), they are converted into gauss. In Fig. 7(b), the experimental points have been plotted (the velocity

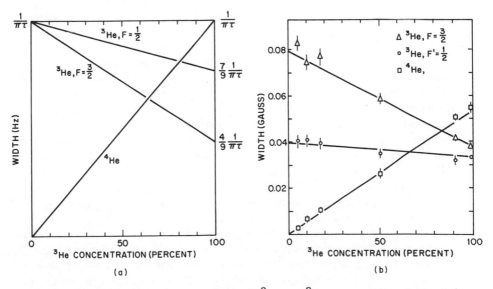

Figure 7. Dependence of the widths of (^{3}He - 2^3S_1), F = 1/2 and F = 3/2 resonances, and of the (^{4}He - 2^3S_1) in ^{3}He - ^{4}He mixtures. In (a), the widths are in cycles/sec. In (b), the same as (a), but widths in gauss. Experimental points are plotted in (b).

correction due to the mass difference between ^{3}He and ^{4}He has been taken into account). Here again the agreement is quite good.

One sees that <u>in pure ^{4}He</u>, $\Delta\nu = 0$: Metastability exchange is not a broadening factor. As soon as one adds ^{3}He a broadening proportional to ^{3}He pressure appears: In a ^{3}He - ^{4}He collision, the <u>electronic</u> orientation goes from (^{4}He - 2^3S_1) to (^{3}He - 2^3S_1). The transverse transferred orientation then precesses at the Larmor frequency which pertains to (^{3}He - 2^3S_1) and which is very different of the one belonging to (^{4}He - 2^3S_1). When a new ^{3}He - ^{4}He metastability exchange collision brings it back on ^{4}He, it has lost, on the average, all phase relation with the orientation which was kept on ^{4}He, as soon as $[\omega_m(^3He) - \omega_m(^4He)]\tau \gg 2\pi$. When this condition is met (and it is, in our experiments), the first ^{4}He - ^{3}He collision led to a complete loss of transverse ^{4}He orientation. This sets the value of its lifetime.

On the other hand, for ^{4}He as in impurity in ^{3}He, every ^{4}He - ^{3}He collision destroys the transverse ^{4}He orientation; the width is then $1/\tau$ [Fig. 7(a)].

The same kind of reasoning can be <u>applied to ^{3}He</u>. In pure ^{3}He, $\langle \vec{F} \rangle$ and $\langle \vec{F}' \rangle$ relax at the rates resulting from Eqs. (1). A very small amount of ^{4}He increases those rates for the same reason we have just given. On the other side, if one takes ^{3}He at a very low concentration in ^{4}He, every collision is against an ^{4}He atom and results in a loss of coherence. $\langle \vec{F} \rangle$ and $\langle \vec{F}' \rangle$ have the same relaxtion time τ. Fig. 7(b) expresses the same facts on the widths of the lines when measured in gauss: In all cases an isotopic species relaxes the other.

The results we have just described on the relaxation of the 2^3S_1 level by metastability exchange are not the only ones one can expect. Circulation of Zeeman coherences between F and F' on one side and between those levels and the ground state on the other will also <u>produce a shift</u> of the 2^3S_1 resonance. In fact, at the $\vec{B}_0$ field values at which one operates this coherence circulation (it is field dependent) is very small and the width of the 2^3S_1 line is too great to detect the corresponding shift. In the experiments described in Sec. IV on the measurement of the Landé g factor of (^{4}He - 2^3S_1), the effect is negligible.

We briefly report now a result pertaining to this kind of effect: The shift of the NMR of ^{3}He in its ground state produced by the coupling of 1^1S_0 and 2^3S_1 due to metastability exchange [23,24].

Of course, in pure ^{3}He, the existence of Pauli's exclusion principle should induce one to take great precautions in the presentation of the results. In spite of those difficulties, we give in a few words what we believe is the essentially correct interpretation of the NMR shift. The transverse ground-state (nuclear) magnetization rotates at ω_f; when a collision raises the atom to 2^3S_1, $\vec{I}$ and $\vec{J}$ get coupled via the $a\vec{I}\cdot\vec{J}$ interaction, so that the nuclear spin precesses at ω_m around the static field. One has $\omega_m \gg \omega_f$. At the next metastability exchange collision bringing the atom back to the ground state the nuclear orientation falls back at an angle $(\omega_m - \omega_f)\tau$ to the nuclear orientation which has remained in the ground state. As a whole the resulting ground-state orientation rotates at a frequency which is slightly different from ω_f. This effect is completely analogous in nature to the type of light shift which is due to real transitions [25]; it does not exist in zero field $(\omega_m - \omega_f = 0;$ the dephasing is zero), nor does it in high fields, when $(\omega_m - \omega_f)\tau \gg 2\pi$, so that, on the average, the circulation of coherence is zero. Figure 8 shows the effect [26]: the position of the NMR resonance of $(^3He - 1^1S_0)$ is shifted more and more when the intensity of the discharge, i.e., the 2^3S_1 concentration, is increased. The true position of the resonance is at the abscissa 10. One can see on this figure the broadening of the NMR at increasing metastable concentrations. We have already explained the origin of this effect.

Figure 9 shows the variations of the shift and of the width of the ground state NMR, due to metastability exchange, at different field values. One should be conscious of the fact that in these experiments these effects have this magnitude because the intensity of the discharge was rather large, and mainly because the static field was quite low $(B_0 \simeq 10^{-3}$ G). It appears that the shift behaves as expected. In Fig. 9, the full lines are theoretical and deduced from Eqs. (1). The experimental points fit very well those curves.

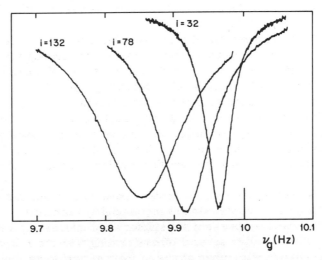

Figure 8.   Dependence of the $(^3He - 1^1S_0)$ nuclear magnetic resonance on metastability exchange collisions. Each curve is taken at a given level i of the discharge, i.e., at a given 2^3S_1 metastable atom concentration. Shift and width of the line increases at increasing number of metastability exchange collisions. The undisplaced position of the ^3He NMR is marked at abscissa 10.

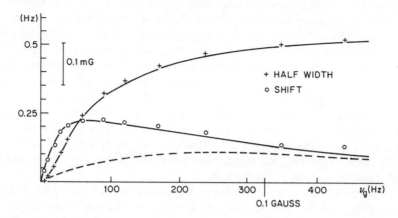

Figure 9.   Dependence of the shift and of the width of the NMR line on the static field $\vec{B}_0$ amplitude.

We will make the following remarks:

1.   As mentioned earlier, in experiments described in Sec. IV, the NMR shift is negligible.

2.   The theory presented here takes into account coherence circulation between F and F' and between both levels and the ground state.   As a consequence the curve giving the shift versus B_0 is not a dispersion curve but the sum of two dispersion curves [27].   Similarly the curve giving the width is not a Lorentzian but the sum of two Lorentzians.

3.   When one does not take into account the transfer of electronic orientation in metastability exchange, one is lead to a considerable error in evaluating coherence circulation.   The real shift, as computed here, is, at low field values, 9 times greater than what one (wrongly) finds in this way.

All the results we have presented above do indicate that we have an essentially correct description of metastability exchange.

These experiments seem satisfactory and interesting for several reasons.   Coming after several others dealing with the relaxation of excited states in collision with other atoms in their ground state (the disorientation is produced by the electrostatic interaction again), they are another example of a successful relaxation study in a weak discharge, i.e., in conditions which are not particularly simple because so many parameters are involved. Moreover, the kind of processes we have analyzed in ^{3}He - ^{4}He mixtures have every chance to exist in other rare gases or rare-gas mixtures.   The nuclear orientation of ^{21}Ne has already been observed [28], and also the one of metastable species in Ne, Ar, Xe [29] and recently krypton. It seems obvious that many problems can be studied in those mixtures which are the basic element of many gas lasers.

IV.   DETERMINATION OF THE LANDÉ g FACTOR IN (^{4}He - 2^3S$_1$)

Using the methods already described we have measured the ratio α of the resonating frequency of (^{4}He - 2^3S$_1$) to the NMR frequency (^{3}He - 1^1S$_0$). We find the following value [30]:

$$\alpha = \frac{\nu_J}{\nu_I} = \frac{1}{2}\frac{\mu_J \ (^4\text{He} - 2^3\text{S}_1)}{\mu_I \ (^3\text{He} - 1^1\text{S}_0)} = 864.02392 \pm 0.00006$$

Fig. 10 gives a histogram of about 100 independent measurements. The rms deviation Δ of this distribution corresponds to about 1% of the width of the line. The quoted error bar is twice the standard deviation Δ', $\Delta' = \Delta/\sqrt{n}$, n being the number of measurements. At this stage, the essential question concerns the existence of possible systematic errors. Many checks have been made in order to detect them. We will just mention those we thought

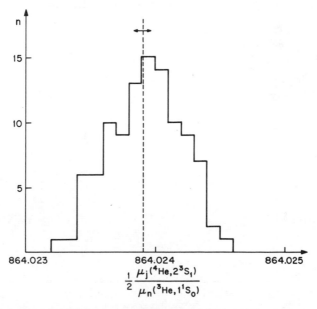

Figure 10.   The histogram of the measurements of ratio α.

of, but we will not discuss them in any detail.

1. <u>Light shifts, due to real and virtual transitions</u> [25]. They might displace the 2^3S_1 resonance. The first shift is field dependent and can be safely calculated; it is quite negligible at 150 G where the circulation of coherence is completely quenched. The second changes sign when one goes from σ^+ to σ^- pumping or when one reverses the field. Its value was about 9 cycles at a resonance frequency of the order of 450 Mc/sec. The ways to eliminate those effects are well known and have been used to get the above value of α.

2. <u>Influence of the discharge intensity</u>. It determines the number of metastable and excited atoms. Because of the coupling between the different orientations existing in a great number of levels, coherence circulations might exist which would shift the $[^4\text{He} - 2^3S_1)$ resonance and the ^3He NMR as well. These effects are expected to be completely negligible in the present case, and indeed no effect of the discharge level has been observed.

3. <u>Magnitude of the static field</u>. We also found that the magnitude of the static field $\vec{B}_0$ had no influence on the value of α. One might think that this could change the coupling between angular momenta and alter the effects due to coherence circulation. On the other hand, the independence of α on B_0 also indicates that the Bloch–Siegert shift (theoretical value $\sim 10^{-9}$) is quite negligible.

4. <u>Influence of the ^3He - ^4He pressure.</u> The values of α are indepen- of ^3He - ^4He pressures (see Fig. 11). The $(^4\text{He} - 2^3S_1)$ resonance might be

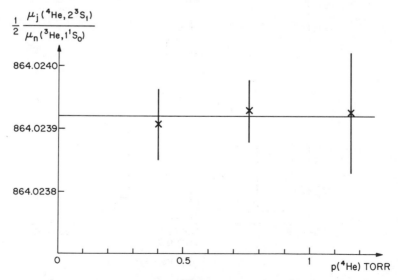

Figure 11. The value of α is independent of ^4He pressure.

displaced by metastability exchange collisions against (^{3}He – 1^1S_0). The very low ^{3}He concentration and the field values used in this experiment lead to undetectable values. As we have seen, a similar situation holds for the NMR of ^{3}He.

5. Static field inhomogeneities. The orientations of (^{4}He – 2^3S_1) and of (^{3}He – 1^1S_0) have a very different repartition within the volume of the discharge cell.

The relaxation time of the NMR is very long and it takes many collisions against the walls to destroy the ^{3}He nuclear orientation which is then uniform in the cell. On the other hand, most authors admit that metastable 2^3S_1 atoms are destroyed in collisions with the walls. Moreover metastable atoms are produced by the discharge. When it is not uniform within the volume of the cell (if one produces it with localized electrodes for instance), the density of metastable atoms greatly varies from one point to another and the diffusion modes which appear may vary considerably depending on conditions. Accordingly, ^{3}He nuclei and 2^3S_1 atoms see a very different "average" field when $\vec{B}_0$ is not uniform. The value of α one gets will depend on many factors: For instance on the position of the electrodes producing the discharge, on the position of the cell in the static field, on the dimensions of the cell, etc. These effects can be very large and can be observed easily. Great precautions were taken to make sure that they were not present and that they do not affect the quoted uncertainty. The measurements were made in Helmholtz coils 60 cm in diameter. The field was explored with a proton resonance. The values found for α are the same in cells 15 and 30 mm in diameter, whatever the position of the exciting electrodes. The same value of α is again obtained if the cell is displaced by 1 cm in three directions in the immediate neighborhood of the geometrical center of the setup.

As a result, it appears that when one takes proper precautions, precision measurements can be made in weak discharges. Their interest in the present case comes from the fact that they deal with a relatively simple atomic system which is calculable with a high degree of precision. This seems to make the experiment worthwhile.

In order to justify this we will make the following remarks. Drake, et al. [31] have measured (by the atomic beam resonance method) the ratio of the Landé g factor of (^{4}He – 2^3S_1) to the Landé g factor of atomic hydrogen in the ground state

$$x_{exp} = \frac{g_J(^4He - 2^3S_1)}{g_J(^1H, 1^2S_{1/2})} = \frac{1}{2}\frac{\mu_J(^4He - 2^3S_1)}{\mu_J(^1H, 1^2S_{1/2})} = 1 - (23.3 \pm 0.8) \times 10^{-5}$$

They also made a theoretical estimate [32] of the same quantity

$$x_{the} = 1 - (23.3 \pm 1) \times 10^{-6}$$

assuming, among other things, that the anomalous moments of the two electrons in He are additive.

Williams and Hughes [33] have measured y, the ratio of the nuclear moment of (^{3}He - 1^1S_0) to the proton moment in molecular hydrogen

$$y = \frac{\mu_J \, (^3He - 1^1S_0)}{\mu_P \, (H_2)} = 0.76178685 \pm 8 \times 10^{-8}$$

when one knows z

$$z = \frac{\mu_J \, (^1H - 1^2S_{1/2})}{\mu_P \, (H_2)}$$

the ratio α we have measured can be written as

$$\alpha = \frac{1}{2} \frac{\mu_J \, (^4He - 2^3S_1)}{\mu_J \, (^3He - 1^1S_0)} = \frac{xz}{y}$$

There exist two independent and precise measurements [34, 35] of z which are in excellent agreement. We can then compare our measurements with previous results. I will not report here the full discussion, but the most logical way to use our value of α seems to combine it with existing values of y and z to obtain a new experimental value of x which is more precise than the one of Drake and Hughes. We obtain in this way

$$x = 1 - (21.6 \pm 0.5) \times 10^{-6}$$

the quoted uncertainty being essentially due to the uncertainty on z. The difference between this value and x_{exp} obtained by Drake and Hughes is 7 times the standard deviation of their measurement (i.e., twice their error bar). Our value is not in agreement with the theoretical estimate of x_{the}. It might be interesting to understand why.

The precision of the measurement of α can be somewhat improved by reducing the temperature at which the discharge is produced (by immersion in liquid air, or liquid helium). But to improve the precision on x one will have to get better measurements of z.

Another way might prove more interesting.   It would be to observe simultaneously in the discharge the (^{4}He - 2^3S_1) resonance and other resonances of simple and calculable systems such as (^{4}He$^+$ - $1^2S_{1/2}$) (we have already observed the ^{3}He$^+$ ion [7]) or free polarized electrons (already seen in post discharges [14,15]), or in helium–hydrogen mixtures, the ground-state resonance of atomic hydrogen (^{1}H - $1^2S_{1/2}$), or the proton resonance in H_2, or even maybe the free proton resonance.   One may hope that different orientations might appear in those species which would be coupled to those we have observed in helium.   One should remember in this respect the beautiful experiments on Penning collisions [12].

It seems to us that it is fair to say that conventional optical pumping in weak discharges does lead to some measurements on simple atomic systems for which the precision can be rather high.   The comparison with highly developed theories of those systems will, no doubt, be useful and interesting.

REFERENCES

1.   F. D. Colegrove and P. A. Franken, Phys. Rev. 119, 680 (1960).
2.   F. D. Colegrove, L. D. Schearer, and G K. Walters, Phys. Rev. 132, 2561 (1963)
3.   L. D. Schearer, thesis (Rice University, 1965).
4.   R. B. Partridge and G. W. Series, Proc. Phys. Soc. 88, 983 (1966).
5.   M. Pavlovic and F. Laloe, J. Phys. 31, 173 (1970). F. Laloë, Thesis (Paris, 1969).
6.   S. D. Baker, E. B. Carter, D. O. Findley, L. L. Hatfield, G. C. Phillips, N. D. Stockwell, and G. K. Walters, Phys. Rev. Letters 20, 738 (1968).
7.   M. Leduc and F. Laloë, Opt. Comm. 3, 56 (1971).
8.   P. L. Bender, thesis (Princeton University, 1956).
9.   J. P. Faroux and J. Brossel, C. R. Acad. Sci. (Fr.) 261, 3092 (1965); 262B, 41 (1966); 262B, 1385 (1966). J. P. Faroux, thesis (Paris, 1969).
10.   A. Omont, J. Phys. (Paris) 26, 26 (1965); thesis (Paris, 1967).
11.   F. Grossetête, J. Phys. (Paris) 25, 383 (1965); thesis (Paris, 1967).
12.   L. D. Schearer, Phys. Rev. Letters 22, 629 (1969). L. D. Schearer and W. C. Holton, Phys. Rev. Letters 24, 1214 (1970).
13.   C. Cohen-Tannoudji, Comments on Atomic and Molecular Physics 2, 24 (1970); A. Kastler, Lectures given at the Summer School of Physics, Izmir, Turkey (1969).
14.   H. G. Dehmelt, Phys. Rev. 109, 381 (1958); J. Phys. Radium 19, 866 (1958).

15.   L. D. Schearer, Phys. Rev. 171, 81 (1968).
16.   H. A. Schuessler, E. N. Fortson, and H. G. Dehmelt, Phys. Rev. 187, 5 (1969).
17.   R. A. Buckingham and A. Dalgarno, Proc. Roy. Soc. A 213, 506 (1952).
18.   J. Dupont-Roc, F. Laloë, and M. Leduc, J. Phys. (Paris) 32, 135 (1971).
19.   J. Dupont-Roc, M Leduc, and F. Laloë, Phys. Rev. Letters 27, 467 (1971).
20.   F. D. Colegrove, L. D. Schearer, and G. K. Walters, Phys. Rev. 135, A353 (1964).
21.   R. C. Greenhow, Phys. Rev. 136, 660 (1964).
22.   R. Byerly, thesis (Rice University, 1967).
23.   L. D. Schearer, F. D. Colegrove, and G. K. Walters, Rev. Sci. Instr. 35, 767 (1964); H. G. Dehmelt, Rev. Sci. Instr. 35, 768 (1964).
24.   A. Dönszelmann, thesis (Amsterdam, 1970); Physica 56, 138 (1971).
25.   J. P. Barrat and C. Cohen-Tannoudji, J. Phys. Radium 22, 443 (1961); C. Cohen-Tannoudji, Ann. Phys. 7, 423 (1962); 7, 469 (1962).
26.   J. Dupont-Roc, C. R. Acad. Sci. (Fr.) 273B, 45 (1971).
27.   J. Dupont-Roc, C. R. Acad. Sci. (Fr.) 273B, 282 (1971).
28.   M. Leduc, F Laloë, and J. Brossel, C. R. Acad. Sci. (Fr.) 271B, 342 (1970).
29.   L. D. Schearer, Phys. Letters 28A, 660 (1969); Phys. Rev. 180, 83 (1969).
30.   M. Leduc, F. Laloë, and J. Brossel, J. Phys. (Paris) 33, 49 (1972).
31.   C. W. Drake, V. W. Hughes, A. Lurio, and J. A. White, Phys. Rev. 112, 1627 (1958).
32.   W. Perl and V. W. Hughes, Phys. Rev. 91, 842 (1953).
33.   W. L. Williams and V. W. Hughes, Phys. Rev. 185, 1251 (1969).
34.   M. Than Myint, D. Kleppner, N. F. Ramsey, and H. G. Robinson, Phys. Rev. Letters 17, 405 (1966).
35.   E. B. D. Lambe, thesis (Princeton University, 1959).

TRANSVERSE OPTICAL PUMPING AND LEVEL CROSSINGS IN FREE AND "DRESSED" ATOMS

C. Cohen-Tannoudji
Laboratoire de Physique de l'Ecole Normale Supérieure,
Université de Paris, Paris, France

I. INTRODUCTION

One of the important characteristics of optical pumping is to provide the possibility of preparing an atomic system in a coherent superposition of Zeeman sublevels [1]. For a $J = 1/2$ angular momentum state, it is equivalent to say that the magnetization $\vec{M}_0$ introduced by the pumping light is not necessarily parallel to the static magnetic field $\vec{B}_0$ (as is usually the case when $\vec{M}_0$ is determined only by the Boltzmann factor in $\vec{B}_0$).

Using such a "transverse" pumping, one can observe level crossing signals in atomic ground states. The width of the observed level crossing resonances may be extremely small as I will show in the first part of this paper. Some possible applications to the detection of very weak magnetic fields will be described.

I will then study the modifications which appear on optical pumping signals when the atoms are no longer free but interacting with nonresonant radio-frequency (rf) photons. These interactions may be visualized in terms of virtual absorptions and reemissions of rf quanta, leading to some sort of "dressing" of the atom by the surrounding quanta. The Zeeman diagram of the "dressed atom" is more complex than the one of the corresponding free atom. The level crossings which were present on the free atom are considerably modified. New level crossings appear. All these effects can be

791

studied by optical pumping techniques as I will show in the second part of this paper.

It may appear surprising to quantize a rf field which is essentially classical and, effectively, all the effects I will describe could be understood in a classical way. I think however that the quantization of the rf field introduces a great simplification in the theory as it leads to a time-independent Hamiltonian for the whole isolated system atom + rf field, much easier to deal with than the time-dependent Hamiltonian of the classical theory. In particular, all the higher-order effects such as multiple quanta transitions, Bloch-Siegert type shifts, Autler-Townes splitting, etc., appear clearly on the Zeeman diagram of the dressed atom. Consequently, this approach could perhaps be generalized to the study of some of the nonlinear phenomena observable with intense laser light.

II. LEVEL CROSSING RESONANCES IN ATOMIC GROUND STATES AND DETECTION OF VERY WEAK MAGNETIC FIELDS

It is well known that the resonance radiation scattered by an atomic vapor exhibits resonant variations when the static field is scanned around values corresponding to a crossing between two Zeeman sublevels of the excited state (Hanle effect – Franken effect) [2]. The width of these resonances is the natural width of the excited state, not the Doppler width. They give useful informations about this state, such as lifetimes, g factors, hyperfine structure [3], etc.

Similar resonances, with a considerably smaller width, can be observed in atomic ground states [4]. To simplify, we will consider a $J = 1/2$ angular momentum state (the calculations could be easily generalized to higher J's). The pumping beam is circularly polarized and propagates along the Ox direction, perpendicularly to the static field $\vec{B}_0$, which is parallel to Oz (Fig. 1).

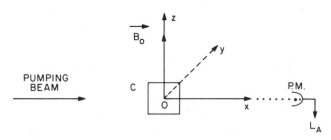

Figure 1. Schematic diagram of the experimental arrangement for the observation of level crossing resonances in atomic ground states. $\vec{B}_0$: static field; P. M.: photomultiplier measuring the absorbed light L_A; C: resonance cell.

As a result of the optical pumping cycle, angular momentum is transferred from the incident quanta to the atoms contained in the resonance cell C.   Let $\vec{M}$ be the total magnetization of the vapor.   It is easy to derive the following equation of evolution for $\vec{M}$:

$$\frac{d}{dt}\vec{M} = \frac{\vec{M}_o - \vec{M}}{T_P} - \frac{\vec{M}}{T_R} + \gamma\vec{M} \times \vec{B}_o \tag{1}$$

The first term represents the effect of optical pumping: if this process was the only one, after a certain amount of time T_P (pumping time), all the spins would be pointing in the Ox direction, producing a saturation magnetization $\vec{M}_o$ parallel to Ox; the second term describes the thermal relaxation process (T_R: relaxation time) due to the collisions against the walls of the cell; the third term, the Larmor precession around $\vec{B}_o$ (γ is the gyromagnetic ratio of the ground state).

Equation (1) looks like the well-known Bloch's equation.   But here, $\vec{M}_o$ is not along $\vec{B}_o$ and its direction is imposed by the characteristics of the pumping beam.

Defining

$$M_{\pm} = M_x \pm iM_y \tag{2}$$

$$\frac{1}{\tau} = \frac{1}{T_P} + \frac{1}{T_R} \tag{3}$$

$$M'_o = M_o \frac{\tau}{T_P}$$

one gets from (1)

$$\frac{d}{dt}M_z = 0 \tag{4a}$$

$$\frac{d}{dt}M_{\pm} = \frac{M'_o}{\tau} - \frac{M_{\pm}}{\tau} \mp i\gamma B_o M_{\pm} \tag{4b}$$

Note that the source term, M'_o/τ, proportional to M_o, appears only in Eq. (4b) relative to the transverse components of the magnetization (transverse pumping).

The steady-state solution of (4) is readily obtained and can be written as

$$M_z = 0 \tag{5a}$$

$$M_{\pm} = \frac{M'_o}{1 \pm i\gamma\,\tau\,B_o} \tag{5b}$$

which gives

$$M_Z = 0 \tag{6a}$$

$$\frac{M_x}{M_o'} = \frac{1}{1 + (\gamma \tau B_o)^2} \tag{6b}$$

$$\frac{M_y}{M_o'} = \frac{-\gamma \tau B_o}{1 + (\gamma \tau B_o)^2} \tag{6c}$$

It follows that M_x and M_y undergo resonant variations when B_o is scanned around 0 (Fig. 2).

These variations result from the competition between optical pumping which tends to orient the spins along the Ox axis and the Larmor precession around $\vec{B}_o$. The critical value of the field, ΔB_o, for which the two processes have the same importance is given by

$$\Delta B_o = 1/\gamma \, \tau \tag{7}$$

ΔB_o is the half-width of the resonances of Fig. 2, which can be detected by monitoring the absorbed or reemitted light, the characteristics of which (intensity, degree of polarization) depend on M_x, M_y, M_z. For example, the photomultiplier P.M. of Fig. 1 detects the absorbed light L_A which is proportional to M_x.

It is possible to get modulated signals by adding a high–frequency rf field $\vec{B}_1 \cos \omega t$ parallel to $\vec{B}_o$ (high frequency means nonadiabatic modulation: $\omega \gg 1/\tau$). The rate equations in the presence of $\vec{B}_1 \cos \omega t$ can be exactly solved [5]. One finds that the zero–field level crossing resonances appear

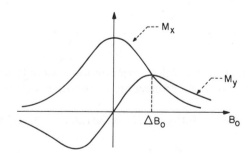

Figure 2.   Variations of the steady–state values of M_x and M_y with B_o. ΔB_o: half-width of the resonances.

also on modulations at the various harmonics $p\omega$ of ω ($p = 1, 2, 3 \ldots$). For example, the ω component of M_x is given around $B_0 = 0$ by

$$M_x(\omega) = M'_0 \, J_0\!\left(\frac{\gamma B_1}{\omega}\right) J_1\!\left(\frac{\gamma B_1}{\omega}\right) \frac{\gamma \tau B_0}{1+(\gamma \tau B_0)^2} \; \sin \omega t \qquad (8)$$

where J_0 and J_1 are the Bessel functions of order 0 and 1. It varies with B_0 as a dispersion curve. The possibility of using selective amplification and lock-in detection techniques with such a signal, increases considerably the signal-to-noise ratio.

Let us calculate the order of magnitude of ΔB_0 for ^{87}Rb which has been experimentally studied [5]. In paraffin-coated cells [6] without buffer gas, T_R (and consequently τ) is of the order of 1 sec; γ is equal to 4.4×10^6 rad sec^{-1} G^{-1}, so that $\Delta B_0 \simeq 10^{-6}$G. Clearly, with such a small width, one has to operate inside a magnetic shield in order to eliminate the erratic fields present in the laboratory (of the order of 10^{-3}G). Five concentric layers of mu-metal (1 m long, 50 cm in diameter, 2 mm thick) have been used for that purpose, providing sufficient protection.

Figure 3 shows an example of the level crossing resonance observed on the modulation at ω of the absorbed light, i.e., on the signal corresponding to theoretical expression (8) ($\omega/2\pi = 400$ Hz). The time constant of the detection is 3 sec. We get a 2-μG width and a signal-to-noise ratio of the order of 3000. It is therefore possible to detect very weak magnetic fields (less than 10^{-9}G) as it appears in Fig. 4 which shows the response of the signal to square pulses of 2×10^{-9} G amplitude around $B_0 = 0$.

Such a high sensitivity is sufficient to measure the static magnetization of very dilute substances. Suppose one places near the ^{87}Rb cell another cell (6 cm in diameter) containing ^{3}He gas at a pressure of 3 Torr (Fig. 5).

The ^{3}He nuclei are optically pumped [7] by a ^{3}He beam B_2, to a 5% polarization. One calculates easily that the oriented ^{3}He nuclei produce at the center of the ^{87}Rb cell (6 cm away) a macroscopic field of the order of 6×10^{-8} G. This field is sufficiently large to be detected on the ^{87}Rb level crossing signal obtained on the B_1 beam.

The experiment has been done [8] and Fig. 6 shows the modulation of the ^{87}Rb signal due to the free precession of the ^{3}He nuclear spins around a small magnetic field applied only on the ^{3}He cell, perpendicular to the directions of B_1 and B_2 (the Larmor period is of the order of 2 min). One sees that one can follow the free decay of the ^{3}He magnetization during hours and hours until it corresponds to only 5×10^{13} oriented nuclei per cm^3. This magnetostatic detection presents many advantages compared to the other optical or radioelectric methods. It could be generalized to other cases (for example to optically pumped centers in solids or to weakly magnetized geological samples). Other spatial or biological applications

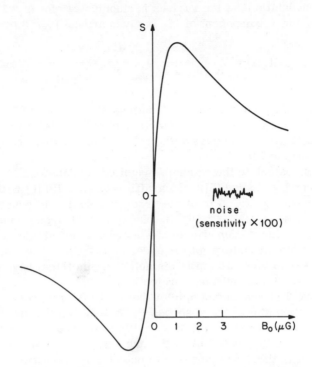

Figure 3.   Zero-field level crossing resonance in the ground state of ^{87}Rb observed on the modulation at $\omega/2\pi$ of the absorbed light L_A ($\omega/2\pi = 400$ Hz). The time constant of the detection is 3 sec. For measuring the noise, the sensitivity is multiplied by a factor 100.

could be considered as it is now possible, by recent improvements [9], to record simultaneously the three components of the small magnetic field to be measured.

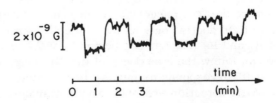

Figure 4.   Test of the sensitivity of the magnetometer: variations of the signal when square pulses of 2 × 10^{-9} G are applied to the resonance cell.

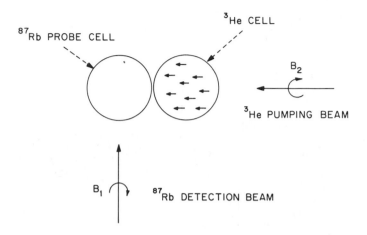

Figure 5.   Detection by the ^{87}Rb level crossing resonance of the magnetic field produced at a macroscopic distance (6 cm) by optically pumped ^{3}He nuclei.   Schematic diagram of the experimental arrangement.

III.   OPTICAL PUMPING OF "DRESSED" ATOMS

To interpret the various resonances which appear in optical pumping experiments performed on atoms interacting with strong resonant or nonresonant rf fields, we will try to develop the following general idea [10, 11]: the light of the pumping beam is scattered, not by the free atom, but by the whole system--atom + rf field in interaction--which we will call the atom "dressed" by rf quanta. Plotted as a function of the static field B_0, the Zeeman diagram of this dressed atom exhibits a lot of crossing and anticrossing points; as for a free atom, the light scattered by such a system undergoes resonant variations when B_0 is scanned around these points. It is therefore possible to understand the various resonances appearing in optical pumping experiments in a very synthetic way. Furthermore, the higher-order effects of the coupling between the atom and the rf field may be handled in a simple way, by time-independent perturbation theory. In some cases (as for example for the modification of the g factor of the dressed atom), the effect of the coupling may be calculated to all orders.

   In the absence of coupling, the energy levels of the whole system are labelled by two quantum numbers, one for the atom (we will take a two-level system, $J = 1/2$), and the other for the field. We will call $|\pm\rangle$ the Zeeman sublevels of the $J = 1/2$ atomic state; their energy in the presence of a

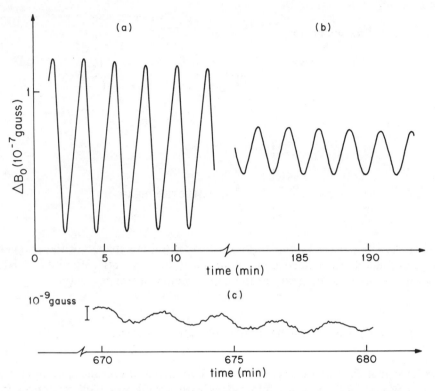

Figure 6. Magnetostatic detection of the Larmor precession of ^{3}He nuclei: (a) just after optical pumping has been stopped; (b) 3 h later; (c) 11 h later (the polarization is now $P \simeq 5 \times 10^{-4}$ and corresponds to 5.3×10^{13} oriented nuclei per cm^3).

static field B_0 parallel to Oz is $\pm \omega_0/2$ ($\omega_0 = -\gamma B_0$; we take $\hbar = 1$). Let $|n\rangle$ be the states of the rf field corresponding to the presence of n quanta and consequently to an energy $n\omega$ (ω is the pulsation of the rf field). The states of the combined system atom + field (without coupling) are the $|\pm, n\rangle$ states with an energy $\pm \omega_0/2 + n\omega$. They are plotted on Fig. 7 versus ω_0. One sees that a lot of crossing points appear for $\omega_0 = 0$, ω, 2ω, 3ω, ... The effect of the coupling V between the atom and the rf field is important at these points. We will first study this coupling in a simple and exactly soluble case, the one of a rotating rf field, perpendicular to $\vec{B}_0$ (this situation leads also to exact solutions in the classical theory).

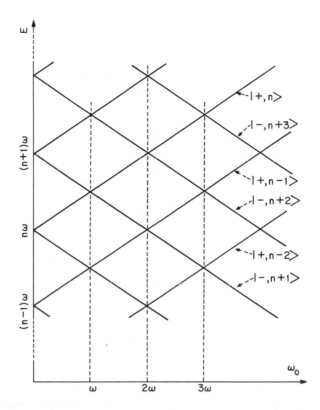

Figure 7. Energy levels of the combined system "atom + rf field" in the absence of coupling.

A. Rotating rf Field Perpendicular to $\vec{B}_0$

The unperturbed states are coupled two by two by V. For example, the $|-,n+1\rangle$ state is coupled only to $|+, n\rangle$ and the other way

$$|-,n+1\rangle \longleftrightarrow |+,n\rangle$$

The physical meaning of such a selection rule is very clear. Each circularly polarized rf quantum carries an angular momentum +1 with respect to Oz (σ^+ photons; we suppose a right circular polarization) and the two states coupled by V must have the same total angular momentum: $-1/2 + (n+1) = + 1/2 + n.$

800   C. Cohen-Tannoudji

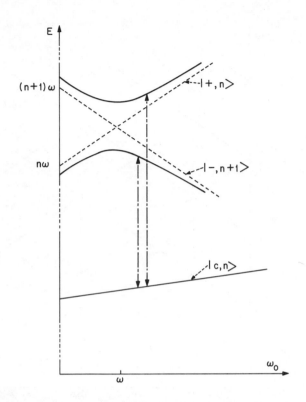

Figure 8. "Anticrossing" resulting from the coupling between the two states $|+, n\rangle$ and $|-, n+1\rangle$. The dotted arrows represent transitions between the two anticrossing levels and a third level (Autler-Townes splitting).

Because of this coupling, the two unperturbed states which cross for $\omega_0 = \omega$ (dotted lines of Fig. 8) repel each other and form what is called an "anticrossing" (full lines of Fig. 8). The minimum distance between the two branches of the hyperbola is obtained for $\omega_0 = \omega$ and is proportional to the matrix element of V between the two unperturbed states. It is possible to show that this matrix element v is proportional to $\sqrt{n+1}$ and may be related to the amplitude B_1 of the classical rf field (more precisely, v is proportional to $\omega_1 = -\gamma B_1$). As n is very large, this matrix element does not change appreciably when n is varied inside the width Δn of the distribution $p(n)$ corresponding to the rf field (for a coherent state [12], $\Delta n \ll n$). Therefore, the anticrossings corresponding to the couples of unperturbed states, $(|-, n+2\rangle, |+, n+1\rangle), (|-, n\rangle, |+, n-1\rangle), \ldots$, have the same characteristics

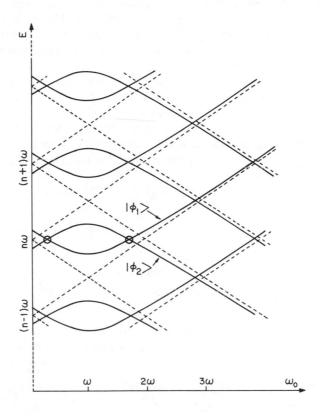

Figure 9. Energy levels of the combined system "atom + rf field" in the presence of coupling. The rf field is circularly polarized and perpendicular to the static field $\vec{B}_0$.

as the one of Fig. 8 (they are deduced from it by a simple vertical translation) and we obtain finally the Zeeman diagram of Fig. 9.

What kind of information can be extracted from this diagram? First, the anticrossings of Fig. 9 reveal the existence of the magnetic resonance occuring for $w_0 = w$. If one starts from the state $|-, n+1 \rangle$, the system is transferred by the coupling V to the other state $|+, n \rangle$ (transition $|- \rangle \rightarrow |+ \rangle$ by absorption of one rf quantum). More precisely, the system oscillates between these two states with an efficiency maximum at the center of the anticrossing (where the mixing between the two unperturbed states is maximum) and at a frequency corresponding to the distance between the two branches of the hyperbola (this is nothing but the well-known Rabi precession). If one looks at the frequencies of the transitions joining the two anticrossing levels of Fig. 8 to a different atomic level $|c,n\rangle$ (not resonantly coupled to the rf field), one finds a doublet (Autler-Townes effect) [13]; the distance between the two components of the doublet and their relative intensities are very simply related to the energies and wave functions of the two anticrossing levels.

A lot of crossing points appear also on the energy diagram of Fig. 9. Let us focus on the two crossings indicated by circles on this figure. The zero-field level crossing of the free atom is shifted by the coupling V; a new level crossing appears near $\omega_0 = 2\omega$ and can be optically detected in transverse optical pumping experiments. The argument is the following: let $|\varphi_1\rangle$ and $|\varphi_2\rangle$ be the two perturbed crossing levels; we have

$$|\varphi_1\rangle = -\sin(\theta/2)|+,n-1\rangle + \cos(\theta/2)|-,n\rangle$$

$$|\varphi_2\rangle = \sin(\theta/2)|-,n+1\rangle + \cos(\theta/2)|+,n\rangle \tag{10}$$

where

$$\tan\theta = \frac{-\gamma B_1}{\omega_0 - \omega} \tag{11}$$

$|\varphi_1\rangle$ and $|\varphi_2\rangle$ contain admixtures of the $|+,n\rangle$ and $|-,n\rangle$ states which correspond to the same value of n so that they can be connected by J_x

$$\langle\varphi_2|J_x|\varphi_1\rangle = \cos^2(\theta/2)\langle n|n\rangle \langle +|J_x|-\rangle$$

$$= (1/2)\cos^2(\theta/2) \neq 0 \tag{12}$$

It is therefore possible to introduce by optical pumping a transverse static magnetization at this crossing point and to get a level crossing signal of the same type as the one described in the first part of this paper (for the other crossings of Fig. 9: $\omega_0 = 3\omega, 4\omega, 5\omega, \ldots$, J_x has no matrix elements between the two crossing perturbed levels, and the level crossings are not detectable).

When the intensity of the rf field (i.e., n) is increased, the distance between the two branches of the hyperbola of Fig. 8 increases and the two crossings of Fig. 9 (indicated by circles) shift towards $\omega_0 = \omega$. These effects appear clearly on Fig. 10 which represents the two corresponding level crossing resonances observed on ^{199}Hg (J = 1/2) [14]. Each curve of figure 10 corresponds to a different value of the amplitude of the rf field (measured by the dimensionless parameter $\omega_1/\omega = -\gamma B_1/\omega$).

Finally, it can be seen in Fig. 9 that, for $\omega_0 = 0$, the Zeeman degener-

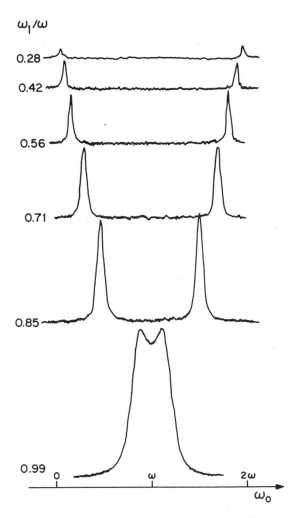

Figure 10. Level crossing resonances observed on ^{199}Hg and corresponding to the two-level crossings indicated by circles on Fig. 9. Each curve corresponds to a different value of the dimensionless parameter ω_1/ω.

acy of the free atom is removed by the coupling with the nonresonant circularly polarized rf field. One can show [15] that, for $\omega_1/\omega \ll 1$, the effect of this coupling is equivalent to that of a fictitious static field $\vec{B}_f$ perpendicular to the plane of the rf field and proportional to $\omega_1^2/\gamma\omega$

$$B_f = \omega_1{}^2/\gamma\omega \qquad (13)$$

In the case of an alkali atom such as ^{87}Rb which has two hyperfine levels in the ground state, $F = 2$ and $F = 1$, with two opposite g factors ($\gamma_2 = -\gamma_1$), it follows from (13) that the two fictitious fields B_{f_2} and B_{f_1} corresponding to $F = 2$ and $F = 1$ are opposite. Therefore, the position of the Zeeman sublevels in the presence of rf irradiation (and in zero static field) is the one shown on Fig. 11(b); it may be compared to the position of the Zeeman sublevels in a true static field producing the same Zeeman separation [Fig. 11(a)]. It follows immediately that the hyperfine spectrum is completely different in a true static field and in the fictitious fields B_{f_1} and B_{f_2} associated with the rf field: we obtain experimentally [15] three $\Delta m_F = 0$ and four $\Delta m_F = \pm 1$ different lines in the first case [Figs. 12(a) and 13(a)]; one $\Delta m_F = 0$ and two $\Delta m_F = \pm 1$ different lines in the second case, as for hydrogen [Figs. 12(b) and 13(b)].

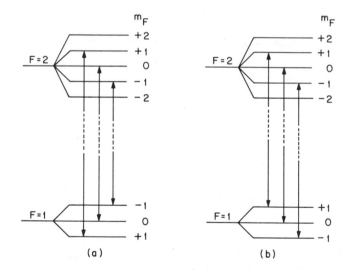

(a)   (b)

Figure 11. Zeeman sublevels in the ground state of ^{87}Rb atoms, (a) in the presence of a true static field $\vec{B}_0$, (b) in the presence of a circularly polarized rf field with $\vec{B}_0 = 0$; the two fictitious static fields $\vec{B}_{f_1}$ and $\vec{B}_{f_2}$ describing the effect of this rf field inside the $F = 1$ and $F = 2$ hyperfine levels are opposite. The arrows represent the three $\Delta m_F = 0$ hyperfine lines which have different frequencies in case (a) and which coincide in case (b).

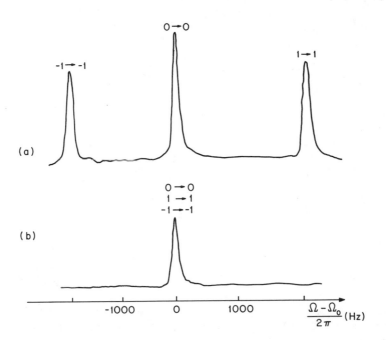

Figure 12. $\Delta m_F = 0$ hyperfine transitions observed on ^{87}Rb atoms, (a) in a true static field $\vec{B}_0$ [see Fig. 11(a)], (b) in the presence of a circularly polarized rf field with $\vec{B}_0 = 0$ [see Fig. 11(b)].

B. Linear rf Field Perpendicular to $\vec{B}_0$

We suppose now that the rf field has a linear polarization, perpendicular to $\vec{B}_0$ (Fig. 14). Such a linear field can be decomposed into two σ^+ and σ^- rotating components. It is equivalent to say that each of the rf quanta has no definite angular momentum with respect to Oz : this angular momentum may be either +1 (σ^+ component) or -1 (σ^- component).

Consequently, the unperturbed $|-,n+1\rangle$ state is now coupled, not only to $|+,n\rangle$ (absorption of a σ^+ photon), but also to $|-,n+2\rangle$ (stimulated emission of a σ^- photon); similarly, the $|+,n\rangle$ state is coupled not only to $|-,n+1\rangle$, but also to $|-,n-1\rangle$

$$
\begin{array}{ccc}
|+,n\rangle & \xleftrightarrow{\;\;\sigma^+\;\;} & |-,n+1\rangle \\[4pt]
\sigma^- \updownarrow & & \updownarrow \sigma^- \\[4pt]
|-,n-1\rangle & & |+,n+2\rangle
\end{array}
\tag{14}
$$

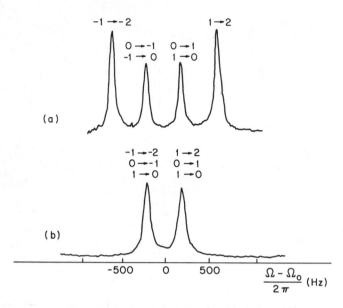

(a)

(b)

Figure 13. $\Delta m_F - \pm 1$ hyperfine transitions observed on ^{87}Rb atoms, (a) in a true static field $\vec{B}_0$, (b) in the presence of a circularly polarized rf field with $\vec{B}_0 = 0$.

These additional couplings which were not present in the previous case (pure σ^+ rf field) are nonresonant for $\omega_0 = \omega$ (the two $|-, n-1\rangle$ and $|+, n+2\rangle$ states do not have the same energy as the two crossing unperturbed levels). They displace however these two crossing levels (from the position indicated by dotted lines in Fig. 15 to the one indicated by interrupted lines) so that the center of the anticrossing $\omega_0 = \omega$ (full lines of Fig. 15) is now shifted by a quantity δ towards $\omega_0 = 0$.   This shift δ is nothing but the well-known Bloch-Siegert shift which is immediately evaluated in this formalism by elementary second-order perturbation theory (the derivation of this shift is much more

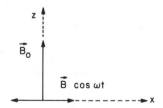

Figure 14. Orientation of the static and rf fields.

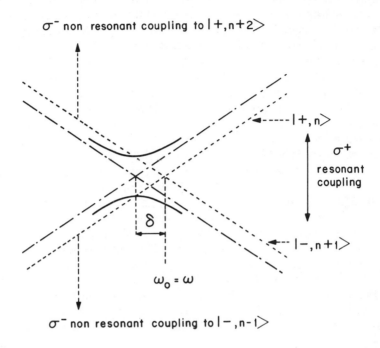

σ^- non resonant coupling to $|+,n+2\rangle$

σ^+
resonant
coupling

$|+,n\rangle$

$|-,n+1\rangle$

δ

$\omega_o = \omega$

σ^- non resonant coupling to $|-,n-1\rangle$

Figure 15.   Origin of the Bloch–Siegert shift observed when the rf field has a linear polarization : nonresonant couplings (dotted arrows) are induced by the σ^- component of the rf field and displace the two anticrossing levels.

elaborate in classical theory) [16].   We must also notice that the two unperturbed levels which cross for $\omega_o = -\omega$ (for example $|-,n+1\rangle$ and $|+,n+2\rangle$) are now coupled by the σ^- component of the rf field.   This gives rise to a new anticrossing near $\omega_o = -\omega$ (symmetrical to the first one with respect to $\omega_o = 0$).

Moreover, one can easily show that each "odd" crossing $\omega_o = (2p+1)\omega$ (with $p = +1, +2, \ldots$) appearing in Fig. 7 becomes now anticrossing.   Let us consider for example the case of the two levels $|-,n+2\rangle$ and $|+,n-1\rangle$ which cross for $\omega_o = 3\omega$

$$|+,n-1\rangle \xleftrightarrow{\ \sigma^+\ } |-,n\rangle \xleftarrow{\ \sigma^-\ } |+,n+1\rangle \xleftrightarrow{\ \sigma^+\ } |-,n+2\rangle$$

$$\sigma^- \updownarrow \qquad\qquad\qquad\qquad\qquad\qquad\qquad \sigma^- \updownarrow \qquad (15)$$

$$|-,n-2\rangle \qquad\qquad\qquad\qquad\qquad\qquad |+,n+3\rangle$$

As shown in (15), they are coupled by V, not directly, but through two intermediate states. It follows that the crossing $w_0 = 3w$ becomes a "third-order anticrossing" (which is also shifted towards $w_0 = 0$, as the $w_0 = w$ anticrossing, as a consequence of nonresonant couplings).

The even crossings $w_0 = 2pw$ (p = 0, 1, 2, ...) of Fig. 7 remain however true crossings (which are also shifted for the same reason as before towards $w_0 = 0$). The argument is the following: for $w_0 = 2pw$, the two crossing levels (for example, $|+,n\rangle$ and $|-,n+2p\rangle$) differ by 2p quanta. The absorption of an even number of quanta (σ^+ or σ^-) cannot provide the angular moment +1 necessary for the atomic transition $|-\rangle \rightarrow |+\rangle$. This excludes any direct or indirect coupling between the two crossing levels.

Finally, through these simple arguments, we get the shape of the Zeeman diagram represented in full lines in Fig. 16 and which is symmetrical with respect to $w_0 = 0$ (the crossing $w_0 = 0$ is not shifted as in the previous case). All the various resonances observable in optical pumping experiments appear in a synthetic way in this diagram.

To the various anticrossings of Fig. 16 are associated the magnetic resonances involving one or several rf quanta [17]. For example, near $w_0 = 3w$, we have a resonant oscillation of the system between the two states $|-,n+2\rangle$ and $|+,n-1\rangle$ which correspond to resonant transitions between $|-\rangle$ and $|+\rangle$ with absorption of three rf quanta. The shift and the rf broadening of the resonances are simply related to the position of the center

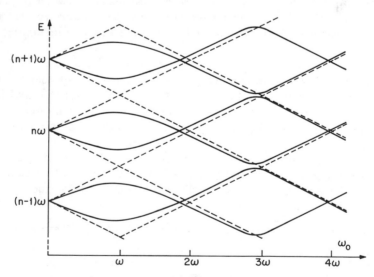

Figure 16.   Energy levels of the combined system "atom + rf field" in the presence of coupling. The rf field is linearly polarized and perpendicular to the static field $\vec{B}_0$.

of the anticrossing and to the minimum distance between the two branches of
the hyperbola. It is also clear that an Autler–Townes splitting must appear
near these higher-order anticrossings.

All the even crossings of Fig. 16 can be detected in transverse optical
pumping experiments : in the perturbation expansion of the two crossing
perturbed levels, one can find unperturbed states with the same value of n
so that J_x can connect the two crossing levels. Figure 17 shows for example the $\omega_0 = 4\omega$ level crossing resonance observed on ^{199}Hg atoms [18].
(This resonance does not appear with a pure σ^+ rf field). Each curve corresponds to an increasing value of the rf amplitude. The Bloch–Siegert type
shift appears very clearly. Such resonances are sometimes called "parametric" resonances or "coherence" resonances as they do not correspond
to <u>real</u> absorptions of one or several rf quanta by the atomic system.

So far, we have explicitly treated the effect of the coupling V as a
perturbation. It is possible to follow qualitatively what happens in the
neighborhood of $\omega_0 = 0$ when the amplitude of the rf field is increased. The
first crossings $\omega_0 = +2\omega$ and $\omega_0 = -2\omega$ shift more and more towards $\omega_0 = 0$.
The separation between the two branches of the anticrossing $\omega_0 = \omega$ increases
more and more. It follows that the slope of the two levels which cross for
$\omega_0 = 0$, i.e., the g factor of the dressed atom, gets smaller and smaller.

More precisely, it is possible to find exactly the eigenstates of the
Hamiltonian $\mathcal{H}_{rf} + V$ which represents the energy of the system in zero
static field [19] ($\mathcal{H}_{rf}$ is the energy of the free rf field) and to treat the
Zeeman term $\mathcal{H}_{at} = \omega_0 J_z$ as a perturbation. This treatment, which takes
into account the effect of the coupling to <u>all orders</u>, gives the slope of the
levels as a function of the dimensionless parameter ω_1/ω. One finds [20]

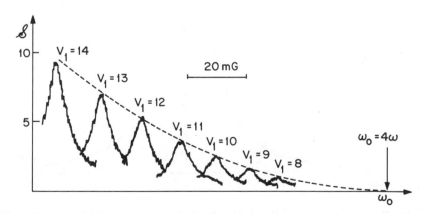

Figure 17. Level crossing resonances observed on ^{199}Hg atoms and
corresponding to the level crossing occuring near $\omega_0 = 4\omega$ in Fig. 16 (V_1 is
the rf voltage, proportional to ω_1).

that the g factor of the dressed atom, g_d, is related to the g factor of the
free atom, g, by the expression

$$g_d = gJ_0(\omega_1/\omega) \tag{16}$$

where J_0 is the zeroth-order Bessel function. This effect can be important.
For example, for all the values of ω_1/ω corresponding to the zero's of J_0,
the dressed atom has <u>no</u> magnetic moment. This modification of g due to
the nonresonant coupling with a filled mode of the electromagnetic field may
be compared to the well-known g-2 effect (anomalous spin moment of the
electron) due to the coupling with the vacuum electromagnetic fluctuations.

If the slope of the levels near $\omega_0 = 0$ is reduced, the width of the zero-
field level crossing resonance discussed in the first part of this paper must
increase. We have observed such a broadening on ^{199}Hg atoms (Fig. 18).

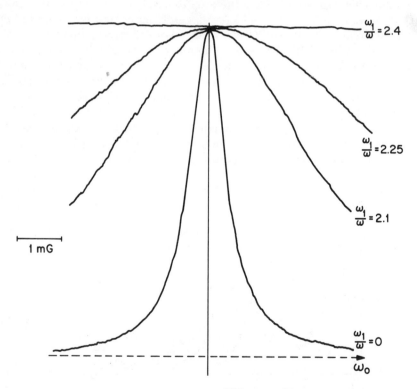

Figure 18.   Zero-field level crossing of ^{199}Hg "dressed" atoms. The rf
field is linearly polarized. The width of the curves is inversely proportional to g_d; it becomes infinite for the value of ω_1/ω corresponding to the
first zero of J_0 [see Eq. (16)].

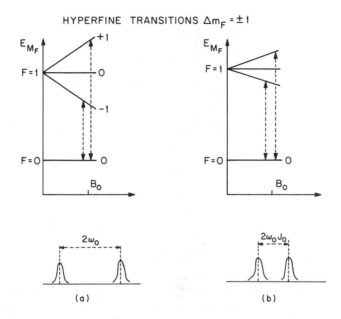

Figure 19. $\Delta m_F = \pm 1$ hyperfine transitions of hydrogen atoms. (a) Free H atoms in a static field B_0; (b) "dressed" H atoms in the same static field B_0. The splitting between the 2 lines is reduced by a factor J_0.

Each curve of Fig. 18 corresponds to a given value of w_1/w. One sees that for $w_1/w = 2.4$ (first zero of J_0), the width of the level crossing resonance becomes infinite. This modification of the g factor has also important consequences on the hyperfine spectrum of hydrogen and alkali atoms. Figure 19 shows the splitting S_0 between the two hyperfine $\Delta m_F = \pm 1$ transitions of H; S_0 is proportional to w_0 [in weak magnetic fields; See Fig. 19(a)]. If we "dress" the atom by a nonresonant linear rf field, the slope of the $F = 1$ sublevels decreases and the splitting S between the two $\Delta m_F = \pm 1$ transitions is reduced by a factor $J_0(w_1/w)$ [see Fig. 19(b)]. The same effect exists also for alkali atoms. We have already mentioned that the two g factors of the $F = 2$ and $F = 1$ hyperfine levels are opposite. As J_0 is an even function, the reduction of the slope of the sublevels is the same in both hyperfine levels. It follows that the splitting between the four $\Delta m_F = \pm 1$ hyperfine transitions is reduced as in the hydrogen case. The four lines coalesce for all the zero's of J_0. This appears clearly in Fig. 20 which represents the evolution of the observed hyperfine spectrum of ^{87}Rb atoms interacting with a nonresonant linear rf field of increasing amplitude [21].

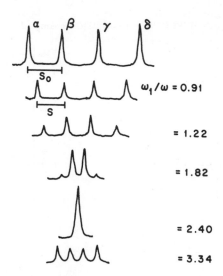

Figure 20.   Evolution of the four $\Delta m_F = \pm 1$ hyperfine transitions of ^{87}Rb atoms "dressed" by a nonresonant linear rf field of increasing amplitude. Experimental results.   For $\omega_1/\omega = 2.4$ (first zero of J_0), the four lines coalesce.

Figure 21 shows the comparison between the experimentally determined ratios S/S_0 measured on H and ^{87}Rb atoms, and the theoretical variations of the Bessel function J_0.

This possibility of changing continuously the g factor of an atom may provide interesting applications.   It has been used, for example, to reduce the effect of static field inhomogeneities on the width of the hyperfine lines. Figure 22(a) shows the hyperfine line $F = 2$, $m_F = 0 \longleftrightarrow F = 1$, $m_F = 0$ of ^{87}Rb broadened by an applied static field gradient.   One observes [10] a narrowing of the line [Fig. 22(b)] when the magnetic moment of the ^{87}Rb atoms is reduced when interacting with a nonresonant linear rf field.

Another application [22] is to allow a coherence transfer between two atomic levels with different g factors: by changing the two g factors through the coupling with a nonresonant linear rf field, one can match the Larmor frequencies in the two atoms and make the coherence transfer possible in nonzero magnetic fields.

I hope that these few example will have proven the versatility of optical pumping techniques and the usefulness of concepts such as the one of dressed atoms.   It would be interesting to see if they could be generalized to other fields of research.

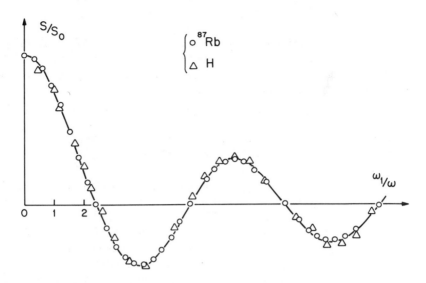

Figure 21.   Plot of the ratio S/S_0 versus ω_1/ω.   The experimental points for
^{87}Rb and H fit into the same theoretical curve.

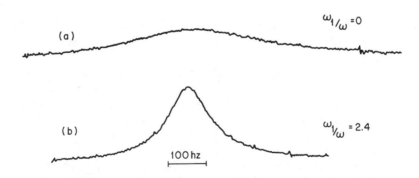

Figure 22.   (a) Recording of the F = 2, $m_F = 0 \longleftrightarrow$ F = 1, $m_F = 0$ hyperfine
line of ^{87}Rb atoms broadened by an applied static field gradient.   (b) Narrowing of the line when the atoms are "dressed" by a nonresonant linear rf
field which cancels the g factor in both hyperfine levels.

REFERENCES

1. For a review article on Optical Pumping, see J. Brossel, in Quantum Optics and Electronics, Les Houches, 1964, (Gordon and Breach, New York, 1965). A. Kastler and C. Cohen-Tannoudji, Progress in Optics, edited by E. Wolf (North-Holland, Amsterdam, 1966), Vol. 5, p. 1.
2. W. Hanle, Z. Phys. $\underline{30}$, 93 (1924); F. D. Colegrove, P. A. Franken, R. R. Lewis, and R. H. Sands, Phys. Rev. Letters $\underline{3}$, 420 (1959).
3. See the review article by Zu Putlitz in Atomic Physics, Proceedings of the First International Conference on Atomic Physics (Plenum Press, New York, 1969), p. 227.
4. Level crossing resonances in atomic ground states have first been observed on ^{111}Cd and ^{113}Cd. J. C. Lehmann and C. Cohen-Tannoudji, Compt. Rend. $\underline{258}$, 4463 (1964).
5. J. Dupont-Roc, S. Haroche, and C. Cohen-Tannoudji, Phys. Letters $\underline{28A}$, 638 (1969); C. Cohen-Tannoudji, J. Dupont-Roc, S. Haroche, and F. Laloe, Revue Phys. Appliquee $\underline{5}$, 95 (1970); $\underline{5}$, 102 (1970).
6. M. A. Bouchiat and J. Brossel, Phys. Rev. $\underline{147}$, 41 (1966).
7. For optical pumping of ^{3}He see F. D. Colegrove, L. D. Schearer, and G. K. Walters, Phys. Rev. $\underline{132}$, 2567 (1963).
8. C. Cohen-Tannoudji, J. Dupont-Roc, S. Haroche, and F. Laloe, Phys. Rev. Letters $\underline{22}$, 758 (1969).
9. J. Dupont-Roc, Revue Phys. Appliquée 5, 853 (1970); J. Phys. (Paris) $\underline{32}$, 135 (1971).
10. For a general and detailed review on the properties of "dressed" atoms, see S. Haroche, thesis (Paris, 1971) [Ann. Phys. (Paris) $\underline{6}$, 189 (1971); $\underline{6}$, 327 (1971).
11. C. Cohen-Tannoudji and S. Haroche, Compt. Rend. $\underline{262}$, 37 (1966); J. Phys. (Paris) $\underline{30}$, 125 (1969); $\underline{30}$, 153 (1969). See also the article in Polarisation, Matiere et Rayonnement, edited by the French Physical Society (Presses Universitaires de France, Paris, 1969). C. Cohen-Tannoudji, Cargese Lectures in Physics, Vol. 2, edited by M. Lévy (Gordon and Breach, New York, 1967).
12. R. J. Glauber, Phys. Rev. $\underline{131}$, 2766 (1963); $\underline{131}$, 2788 (1963).
13. S. H. Autler and C. H. Townes, Phys. Rev. $\underline{100}$, 703 (1955); C. H. Townes and A. L. Schawlow, Microwave Spectroscopy (McGraw-Hill, New York, 1955), p. 279.
14. M. Ledourneuf, These de troisieme cycle (Universite de Paris, 1971).
15. M. Ledourneuf, C. Cohen-Tannoudji, J. Dupont-Roc and S. Haroche, Compt. Rend. $\underline{272}$, 1048 (1971); $\underline{272}$, 1131 (1971). Let us mention that the effect of a nonresonant circularly polarized optical irradiation can also be described in terms of fictitious static fields. See C. Cohen-Tannoudji and J. Dupont-Roc, Phys. Rev. $\underline{5}$, 968 (1972).

16. F. Bloch and A. Siegert, Phys. Rev. $\underline{57}$, 522 (1940).
17. J. M. Winter, thesis (Paris, 1958); Ann. Phys. (Paris), $\underline{4}$, 745 (1959).
18. C. Cohen-Tannoudji and S. Haroche, Compt. Rend. $\underline{261}$, 5400 (1965).
19. N. Polonsky and C. Cohen-Tannoudji, J. Phys. (Paris) $\underline{26}$, 409 (1965).
20. C. Cohen-Tannoudji and S. Haroche, Compt. Rend. $\underline{262}$, 268 (1966). See also Ref. 10
21. S. Haroche, C. Cohen-Tannoudji, C. Audoin, and J. P. Schermann, Phys. Rev. Letters $\underline{24}$, 861 (1970).
22. S. Haroche and C. Cohen-Tannoudji, Phys. Rev. Letters $\underline{24}$, 974 (1970).

MACROSCOPIC APPROACH TO EFFECTS OF RADIATIVE

INTERACTION OF ATOMS AND MOLECULES*

V. A. Alekseev, A. V. Vinogradov, and I. I. Sobel'man
P. N. Lebedev Physical Institute of the Academy of Sciences
Moscow, USSR

I. INTRODUCTION

The radiation of an aggregate of N oscillators (atoms and molecules) localized in a volume with linear dimensions $a \ll \lambdabar$ (where $\lambdabar = \lambda/2\pi = c/\omega$ and λ is the wavelength) is characterized by a number of specific features connected with the so-called cooperative effect, i.e., with the existence of collective oscillations due to the interaction of the oscillators via their common radiation field.

The nature of this effect is easiest to explain with classical oscillators as an example (charge e, mass m, natural frequency ω_0). It is well known that the radiation of electromagnetic waves is accompanied by a reaction of the radiation field on the charges, i.e., by the appearance (occurrence) of radiation reaction forces. In the case when $a \ll \lambdabar$ the radiation reaction force acting on each oscillator of the system is [1]

$$\vec{F} = (2e/3c^3)\,\dddot{\vec{D}}, \quad \vec{D} = e \sum_i \vec{r}_i \tag{1}$$

---

*The talk given by Professor Sobel'man was based on this article, which appeared in Uspekhi Fizicheskikh Nauk 102, 43 (1970). The English translation [Soviet Physics Uspekhi 13, 576 (1971)] is reprinted here by permission of the publisher.

where $\vec{D}$ is the total dipole moment of the system. Therefore the equation of motion of each of the oscillators in the external field $\vec{E}^0$ is

$$\ddot{\vec{r}}_i - (2e/3mc^3)\,\dddot{\vec{D}} + w_0^2\vec{r}_i = (e/m)\,\vec{E}^0 \tag{2}$$

Multiplying each of these equations by e and summing over all the oscillators of the system, we obtain an equation of motion for $\vec{D}$:

$$\ddot{\vec{D}} - N(2e^2/3mc^3)\,\dddot{\vec{D}} + w_0^2\vec{D} = N(e^2/m)\,\vec{E}^0 \tag{3}$$

From this it is easy to find the radiation spectrum of the system

$$I(w) \sim |D_w|^2 \sim |w_0^2 - w^2 - iw\,N\gamma_0|^{-2}, \quad \gamma_0 = 2e^2w^2/3mc^3 \tag{4}$$

For one isolated oscillator, as is well known, we have

$$I(w) \sim |w_0^2 - w^2 - iw\gamma_0|^{-2} \tag{5}$$

here γ_0 is the radiative damping constant of the oscillator.

It is seen from (3) and (4) that the oscillations of the total dipole moment of the system in question, owing to the interaction of the oscillators via their common radiation field, attenuate by a factor N faster than the oscillations of one isolated oscillator. Accordingly, (4) contains the collective radiative width $N\gamma_0$.

It is easy to show that the same result is obtained by an elementary analysis of quantum radiators.

Let us consider a system of N identical atoms. The ground state of such a system is described by the wave function

$$\Psi_0 = \varphi_0(1)\varphi_0(2) \ldots \varphi_0(N) \tag{6}$$

where $\varphi(i)$ is the wave function of the i-th atom. The first excited level of the system, corresponding to excitation of one of the atoms, is N-fold degenerate (we assume for simplicity that the levels of one atom are not degenerate), and the states belonging to this level can be described either by wave functions of the type

$$\psi_1^{(i)} = \varphi_0(1)\,\varphi_0(2) \ldots \varphi_1(i) \ldots \varphi_0(N) \tag{7}$$

or by symmetrized linear combinations of these functions. If all the atoms are localized in a volume $V \ll \lambda^3$, then the interaction of the system with the electromagnetic field can be written in the form

$$\mathcal{H}' = (e/mc) \sum_i \vec{p}_i \cdot \vec{A}(r_i) \approx (e/mc) \vec{A} \cdot \sum_i \vec{p}_i$$

where $\vec{p}_i$ is the momentum of the atomic electron and $\vec{A}$ is the vector potential of the field. Since the operator $\mathcal{H}'$ is completely symmetrical with respect to permutations of the arguments i (the number of the atom) and the wave function of the ground state (6) is also fully symmetrical, the matrix element $\mathcal{H}'$ differs from zero only for transitions to a fully symmetrical excited state

$$\Psi_1 = (1/\sqrt{N}) \sum_i \psi_1^{(i)} \tag{8}$$

All the remaining linear combinations of the functions (7), orthogonal to the function (8), have a different symmetry. Therefore the matrix elements $\mathcal{H}''$ for the corresponding transitions are equal to zero. In other words, only the fully symmetrical state (8) is connected with Ψ_0 by a radiative transition.
The matrix element for the transition $\Psi_0 \rightarrow \Psi_1$ is

$$\langle \Psi_1 | \mathcal{H}' | \Psi_0 \rangle = (1/\sqrt{N}) \sum_i \langle \Psi_1^{(i)} | (e/mc) \vec{A} \cdot \sum_j \vec{p}_j | \Psi_0 \rangle = \sqrt{N}(e/mc) \vec{A} \cdot \vec{p}_{10} \tag{9}$$

where $\vec{p}_{10}$ is the matrix element of $\vec{p}$ for one isolated atom. It is seen even from (9) that the radiative decay of the state (8) is N times faster than in the case of a single atom. The width of the emission spectrum, which can be easily determined by using the well-known Weisskopf-Wigner method [2] (see [3] concerning this subject) is also N times larger than for the isolated atom.

The fact that radiating systems manifest collective properties is well known, and a number of concrete examples where this effect plays an important role can be cited. Thus, the effective cross section for the scattering of electromagnetic waves by a cluster of N electrons or by an atom containing N electrons is proportional to N in the case when $\lambda \ll a$ and proportional to N^2 if $\lambda \gg a$ [1].

The question of collective radiative damping has also been discussed many times in the literature in connection with a variety of problems (see, for example, [3-8]). There are, however, a number of problems connected with this effect, which either have not been discussed at all in the literature, or have not been considered quite fully. These include, in particular, the question of the radiative damping in the case of an infinite medium (in practice, in the case of volume $V \gg \lambdabar^3$).

Recently, careful investigations were made of the spectral-line broadening of noble-gas atoms at low pressures [9-13]. The results of these investigations were interpreted by the authors as evidence of the appearance of collective radiative damping. In fact, there are not sufficient grounds for such an interpretation of the experimental data obtained in Refs. 9-13. This is easiest to show by resorting to the known properties of the dielectric constant of rarefied gases $\varepsilon(\omega)$, i.e., by solving the problem of radiative damping within the framework of the macroscopic approach. As will be explained in Sec. II, no increase of the radiative width is connected with the cooperative effect in infinite media.

The macroscopic approach to the problem of interest to us, used in Sec. II for the case of large volumes ($a \gg \lambdabar$), is perfectly natural. In the case of small volumes ($a \ll \lambdabar$), on the other hand, the situation is not so simple. The point is that in practically all the cited papers devoted to the cooperative effect the analysis was based on the microscopic approach, i.e., on allowance for the interaction via the radiative field directly in the equations of motion of the oscillators. This naturally raises the question as to whether the collective radiative effect is present in the general scheme of the macroscopic description of the processes of radiation and scattering of electromagnetic waves or whether the microscopic analysis is essential in principle. In other words, is it possible to take into account the cooperative effect starting from the equations of macroscopic electrodynamics and ascribing to the medium in a small volume ($a \ll \lambdabar$) the same values of the dielectric constant and the magnetic permeability $\varepsilon(\omega)$ and $\mu(\omega)$ as in the case of an infinite medium?

This question is of interest for the following reasons. First of all, a negative answer to it would mean that number of the results obtained by exactly solving the equations of macroscopic electrodynamics, such as the Mie theory for the scattering of electromagnetic waves by small particles, reflection from a layer of thickness $a \ll \lambdabar$, etc., must be revised.

On the other hand, the microscopic approach is inevitably connected with the consideration of concrete models. Thus, for example, all the known published results concerning the influence of the cooperative effect on the emission spectrum of small volumes under thermal excitation pertain, in essence, to the case of a single isolated spectral line, corresponding to the transitions between nondegenerate levels. The foregoing pertains also to Dicke's well-known paper [4], which contains the most complete analysis of different manifestations of radiative coupling.

The main content of the present article is a comparison between the microscopic and macroscopic descriptions of the effect of radiative coupling in small and large volumes. The entire analysis is limited to conditions such that the substance is in the states of thermodynamic equilibrium, or when the radiation is due to thermal excitation. We therefore do not consider at all other aspects of the theory of the cooperative effect, such as radiation echo (see 4-14). As applied to the conditions in question, it is shown that the macroscopic approach permits a complete description of the cooperative effects. Moreover, within the framework of this approach, the limits of applicability of the approximations that must be made in the microscopic analysis become clear.

II.   RADIATIVE DAMPING IN UNBOUNDED MEDIA

The theory of collective radiative damping, developed in Ref. 4 for systems with linear dimensions $a \ll \lambdabar$, cannot be generalized directly to the case of unbounded media. At the same time, the question of the possible manifestation of this effect arises in a large number of physical problems. Thus, for example, in a series of recent studies [9-13] Kuhn, Vaughan, and Lewis investigated in detail the broadening of a number of spectral lines of noble-gas atoms. The density dependence of the width turned out to be in very good agreement with the theory of impact broadening. At the same time, for very small densities these investigations led to a somewhat unexpected result. The radiative width of a number of lines, obtained by extrapolating the measured widths to zero density, turned out to differ from the theoretical ones or from those determined by other methods. The authors point to the cooperative effect as a possible cause of the observed discrepancy [15]. This conclusion is based on reasons analogous to those discussed in Sec. I, as applied to the case of small volumes ($V \ll \lambdabar^3$), and on the results of Ref. 7, where the radiative damping constant γ, corresponding to the $p \rightarrow s$ transition in a system of two atoms separated by an arbitrary interatomic distance R was calculated. At the initial instant $t = 0$, one atom is in the p state and the second in the s state. The damping constant γ depends on the projection m of the momentum of the excited atom on the z axis which is directed along the vector $\vec{R}$. In the case when $m = 0$ we have

$$\gamma_\sigma = \gamma_0 \{1 + [3/(kR)^3](\sin kR - kR \cos kR)\}$$

For $m = \pm 1$,

$$\gamma_\pi = \gamma_0 \{1 + [3/2(kR)^3](- \sin kR + kR \cos kR + (kR)^2 \sin kR)\}$$

With decreasing R we get $\gamma_{\sigma, \pi} \rightarrow 2\gamma_0$. When R increases, γ_π differs from γ_0 by a quantity on the order of $\gamma_\pi - \gamma_0 \sim (\sin kR/kR)\gamma_0$, i.e., this difference decreases very slowly. This circumstance is the primary cause of the difficulty of generalizing the results of Ref. 7 to the case of a gas in a large volume, since one cannot limit oneself to allowing for nearest-neighbor interactions only.

The most general approach to our problem is to consider the connection of such characteristics of the medium as the absorption and emission spectra with the properties of the dielectric constant of the medium $\varepsilon(\omega)$.

As is well known, both the dissipation of electromagnetic energy in a medium and the thermal radiation of a medium can be expressed directly in terms of $\varepsilon(\omega)$. Knowing $\varepsilon(\omega)$, we can also determine the Einstein spectral coefficients for spontaneous emission $a_{ik}(\omega)$ per atom of the medium (with allowance for the interaction between the atoms):

$$\mathrm{Im}\,\{[\varepsilon(\omega) - 1]/[\varepsilon(\omega) + 2]\} = (\pi^2/3)(c/\omega)^3 \sum_{i,k} a_{ik}(\omega)[(g_i/g_k)n_k - n_i] \tag{10}$$

where g_i, g_k and n_i, n_k are the statistical weights and the populations of the levels i and k $(E_i > E_k)$,

$$\int a_{ik}(\omega)\,d\omega = A_{ik} = \frac{2\omega^2 e^2}{mc^3} \frac{g_k}{g_i} f_{ki} \tag{11}$$

A_{ik} is the Einstein integral coefficient, and f_{ki} is the oscillator strength of the transition $k \to i$.

The summation on the right-hand side of (10) extends over all possible transitions capable of making a contribution to the considered spectral region. For an isolated spectral line (in the case of sufficiently low densities) we can retain on the right-hand side of (10) only one term. In this case the frequency dependence of the coefficient $a_{ik}(\omega)$ can be obtained if the function $\varepsilon(\omega)$ is known.

Thus, the problem of calculating the emission spectrum, and by the same token of clarifying the role of the cooperative effect, reduces to the calculation of the dielectric constant. This problem has been discussed many times in the literature (see, for example, [16]), and this enables us to employ known results. However, before we proceed to discuss these results, let us show that by calculating $\varepsilon(\omega)$ in the microscopic theory of dispersion, the radiative interaction of the atoms, which is responsible for the cooperative effect in the case of small volumes, is completely taken into account.

Let us consider again the model of classical oscillators. This model makes it possible to establish all the main features of the manifestation of the radiative coupling in small and large volumes. All the results can be generalized without difficulty to the case of quantum systems. The equations of motion of the i-th oscillator (i. e., one of the oscillators of the medium), with allowance for the reaction of its own radiation field and the fields of all the remaining oscillators, written for the Fourier component of the dipole moment $\vec{d}_\omega^{(i)} = e\vec{r}_\omega^{(i)}$, is of the form

$$\vec{d}_\omega^{(i)}[-\omega^2 - i\omega(\gamma_0 + \Gamma) + \omega_0^2] = \frac{e^2}{m} \sum_{i \neq j} \text{curl curl} \frac{\exp[i(\omega/c)|\vec{R}_i - \vec{R}_j|]}{|\vec{R}_i - \vec{R}_j|} \vec{d}_\omega^{(j)} + \frac{e^2}{m} \vec{E}_\omega^0 \qquad (12)$$

where $\vec{R}_i$ and $\vec{R}_j$ are the coordinates of the centers of inertia of the oscillators, the constant Γ characterizes the possible electromagnetic-energy dissipation processes due to any other factor not represented in the right-hand side of the system (12).

If the oscillators are localized in a volume $V \ll \lambdabar^3$, and therefore the conditions $(\omega/c)|\vec{R}_i - \vec{R}_j| \ll 1$, are satisfied for all values of i and j, then the right-hand side of (12) can be expanded in a series of powers of $(\omega/c)|\vec{R}_i - \vec{R}_j|$. Retaining in this expansion only the first nonvanishing imaginary term, which, as it turns out, corresponds to allowance for the radiative reaction forces in the approximation (1),

$$\text{curl curl} \frac{\exp[i(\omega/c)|\vec{R}_i - \vec{R}_j|]}{|\vec{R}_i - \vec{R}_j|} \vec{d}_\omega^{(j)} \simeq i\frac{2}{3}\left(\frac{\omega}{c}\right)^3 \vec{d}_\omega^{(j)} \qquad (13)$$

we obtain

$$\vec{d}_\omega^{(i)}[-\omega^2 - i\omega\Gamma + \omega_0^2] = i\omega\gamma_0\vec{D}_\omega + \frac{e^2}{m}\vec{E}_\omega^0 \qquad (14)$$

where $\vec{D}_\omega$ is the Fourier component of the total dipole moment of the system. Summing (14) over all N oscillators of the system, we get

$$\vec{D}_\omega[-\omega^2 - i\omega(N\gamma_0 + \Gamma) + \omega_0^2] = N(e^2/m)\vec{E}_\omega^0 \qquad (15)$$

This equation, like (3), contains the collective radiative width.

Let us recall now the method used to calculate the dielectric constant $\varepsilon(\omega)$ with the aid of the system (12), (see Ref. 16 on this subject). Let us consider Eq. (12) without an external field. We average the right-hand side of (12) over the coordinates of the oscillators $\vec{R}_j$, assuming that each of the oscillators has an equal probability of being at any point in space, regardless of the locations of the other oscillators, including the oscillator i. Then

$$\sum_{j \neq i} = \lim_{V \to \infty}(N-1)\frac{1}{V}\int_\delta d\vec{R}_j \, \text{curl}_{\vec{R}_i} \text{curl}_{\vec{R}_i} \frac{\exp[i(\omega/c)|\vec{R}_i - \vec{R}_j|]}{|\vec{R}_i - \vec{R}_j|} \vec{d}_\omega(\vec{R}_j) \qquad (16)$$

where $N = nV$, and n is the oscillator concentration; the symbol δ at the integral sign denotes that the integration should be carried out over the region $|\vec{R}_i - \vec{R}_j| > \delta \to 0$. Using, further, the relation (see Ref. 16)

$$\int_\delta d\vec{R'} \, \text{curl}_{\vec{R}} \, \text{curl}_{\vec{R}} \, \frac{\exp[i(\omega/c)|\vec{R}-\vec{R'}|]}{|\vec{R}-\vec{R'}|} \, \vec{d}_\omega(\vec{R'}) =$$

$$\text{curl}_{\vec{R}} \, \text{curl}_{\vec{R}} \int d\vec{R'} \, \frac{\exp[i(\omega/c)|\vec{R}-\vec{R'}|]}{|\vec{R}-\vec{R'}|} \, \vec{d}_\omega(\vec{R'}) - \frac{8\pi}{3} \vec{d}_\omega(\vec{R})$$

we obtain an integral equation for $\vec{d}_\omega(\vec{R})$:

$$\vec{d}_\omega(\vec{R}) = n\alpha_0 \text{curl curl} \int d\vec{R'} \, \frac{\exp[i(\omega/c)|\vec{R}-\vec{R'}|]}{|\vec{R}-\vec{R'}|} \, \vec{d}_\omega(\vec{R'}) - \frac{8\pi}{3} n\alpha_0 \vec{d}_\omega(\vec{R}) \qquad (17)$$

Here α_0 is the polarizability of one isolated oscillator

$$\alpha_0 = (e^2/m)[\omega_0^2 - \omega^2 - i\omega(\gamma_0 + \Gamma)]^{-1} \qquad (18)$$

If we seek the solution of (17) in the form of plane waves $\vec{d}_\omega(\vec{R}) \sim \exp i\kappa R$, then it is easy to obtain a dispersion equation connecting κ with ω:

$$1 = n\alpha_0 \frac{4\pi\kappa^2}{\kappa^2 - (\omega/c)^2} - \frac{8\pi}{3} n\alpha_0 \qquad (19)$$

Putting $\kappa^2 = \epsilon(\omega)\omega^2/c^2 = \epsilon(\omega)k^2$ in (19), we obtain the well-known Lorentz-Lorenz formula for $\epsilon(\omega)$:

$$(\epsilon - 1)/(\epsilon+2) - (4\pi/3)n\alpha_0 = n(e^2/m)[\omega_0^2 - \omega^2 - i\omega(\gamma_0 + \Gamma)]^{-1} \qquad (20)$$

Formula (20) was obtained for the case of an ideal gas. As was shown by Klimontovich and Fursov [17] in the case of an arbitrary medium the radiative damping is determined by the constant

$$\gamma_{rad} = \gamma_0 \overline{\Delta N^2}/N$$

where $\overline{\Delta N^2}$ and N are respectively the mean square fluctuation and the average number of particles in a definite element of the body volume. For an ideal gas, as is well known, $\overline{\Delta N^2}/N = 1$ and $\gamma_{rad} = \gamma_0$. If we eliminate completely the possibility of fluctuations, by putting $\Delta N^2 = 0$, then, as can be seen from the foregoing formula, there is no radiative damping at all:

$$\frac{3}{4\pi} \frac{\varepsilon-1}{\varepsilon+2} = n\frac{e^2}{m} \frac{1}{\omega_0^2 - \omega^2 - i\omega\Gamma} \tag{21}$$

Thus, in a homogeneous medium of isotropic oscillators (without fluctuations) the interaction of oscillators via the radiation field leads not to an increase of the radiative width, as is in the case of a small volume, but to complete cancellation of the radiative damping. This result, first obtained by Mandel'shtam [18], is closely connected with the well-known fact that a homogeneous medium does not scatter electromagnetic waves.

The presence of fluctuations leads to scattering, and in an ideal gas the intensity of the scattering by the density fluctuations is equal to the sum of the intensities of scattering by each of the isolated oscillators. Simultaneously, $\varepsilon(\omega)$ is determined by formula (20), i.e., it has the same form as if there were no interaction of the oscillators via the radiation field at all.

In (12) above we did not take into account the motion of the oscillators. Generalization of these equations to the case of moving oscillators does not change the results of interest to us in any way.

From the foregoing statements concerning the connection between the Einstein spectral coefficients $a(\omega)$ and the function $\varepsilon(\omega)$, it follows that in an unbounded medium there is no increase of the radiative width as a result of the interaction of the oscillators via the common field of radiation. The radiative width remains exactly the same as in the case of an isolated oscillator. Thus, the effect whereby the radiative damping increases as a result of the coupling of the oscillators is peculiar only to small volumes $V \ll \lambda^3$, and, as will be shown below, is connected with peculiarities of the scattering of light by small volumes.

As to the anomalous behavior of the line width in the spectra of noble gases at low densities, observed in Refs. 9–13, it is apparently due to some other causes which have no bearing on the cooperative effect (see [19]).

III. SCATTERING OF ELECTROMAGNETIC WAVES BY SMALL PARTICLES. THERMAL RADIATION

Let us consider a spherical volume V whose radius a satisfies the condition $a \ll \lambda$, filled with classical isotropic oscillators.[*] In accordance with the

---

[*]Here and throughout we confine ourselves for simplicity to nonmagnetic substances with $\mu = 1$.

statements made in Sec. II, the dipole oscillations of such a volume should
be characterized by a radiative damping constant proportional to the total
number of oscillators $N=nV$ in the volume; this in principle can be manifest
in various radiative processes. It is convenient to start the analysis of such
effects with the problem of the scattering of electromagnetic waves.

If an external monochromatic field with amplitude $\vec{E}_\omega^0$ is incident on the
system, then Eq. (15) for the Fourier component of the total dipole moment
D_ω takes the form

$$\vec{D}_\omega[-\omega^2 - i\omega(N\gamma_0 + \Gamma) + \omega_0^2] - (e^2/m)N\vec{E}_\omega^0 \tag{22}$$

From this we can readily find the polarizability α_V of the volume:

$$\alpha_V = \frac{e^2}{m} \frac{nV}{\omega_0^2 - \omega^2 - i\omega(\Gamma + nV\gamma_0)} \tag{23}$$

Knowing the polarizability α_V, we can find the total effective cross section
σ_V (usually called the attenuation cross section) and the effective scattering
cross section σ_V':

$$\sigma_V = 4\pi k\, \mathrm{Im}\,\alpha_V = 6\pi\lambda^2 \frac{nV\gamma_0(nV\gamma_0 + \Gamma)\omega^2}{(\omega_0^2 - \omega^2)^2 + \omega^2(\Gamma + nV\gamma_0)^2} \tag{24}$$

$$\sigma_V' = \frac{8\pi}{3}k^4|\alpha_V|^2 = 6\pi\lambda^2 \frac{(nV\gamma_0)^2\omega^2}{(\omega_0^2 - \omega^2)^2 + \omega^2(\Gamma + nV\gamma_0)^2} \tag{25}$$

Let us express now the polarizability α_V in terms of the dielectric
constant ϵ from (21), using, as is customarily done in the theory of light
scattering by small particles, the same formula as in the case of a static
field [20]:[*]

$$\alpha_V^{st} = \frac{3}{4\pi} \frac{\epsilon-1}{\epsilon+2} V = \frac{e^2}{m} \frac{nV}{\omega_0^2 - \omega^2 - i\omega\Gamma} \tag{26}$$

---

[*]It should be recalled that the motion of particles localized in the volume
$V < \lambda^3$ does not lead to a Doppler broadening of the spectral lines, and
therefore $\epsilon(\omega)$ should not take the Doppler effect into account.

The difference between formulas (26) and (23) lies precisely in the fact that (23) contains the collective radiative width $nV\gamma_0$. From the fundamental point of view, this difference is quite significant. The point is that the polarizability α_V should satisfy the inequality

$$\text{Im}\,\alpha_V \geq 2k^3\,|\alpha_V|^2/3 \tag{27}$$

which is the consequence of the so-called optical theorem for dipole scattering, which relates the total cross section with the forward scattering amplitude. The physical meaning of (27) lies in the fact that the total cross section cannot be smaller than the scattering cross section σ'_V [see (24) and (25)]. It is to verify that the polarizability (23) does satisfy the condition (27), whereas (26) can lead to a contradiction.

Further, inasmuch as $\text{Im}\,\alpha_V \leq |\alpha'_V|$, it follows also from (27) that $|\alpha_V| \leq 2k^{-3}/3$. Therefore

$$\sigma'_V < 6\pi\lambda^2, \quad \sigma_V \leq 6\pi\lambda^2 \tag{28}$$

The quantity on the right-hand side of (28) is the well-known theoretical limit for the dipole-scattering cross section. It is easy to verify that the scattering cross section σ'_V calculated with the aid of (26) likewise does not satisfy this condition in the general case. On the other hand no difficulties arise in the case (23).

Let us consider therefore the problem of calculating α_V within the framework of macroscopic electrodynamics in somewhat greater detail.

Let the oscillator density, and consequently ϵ, depend on the coordinate R, and in such a way that $\epsilon(R) = \text{const} \neq 1$ in the volume V, and outside this volume $\epsilon(R)$ tends smoothly but quite rapidly to unity. We solve the scattering problem using the Maxwell equation

$$\text{curl curl}\,\vec{E}_\omega + (\omega^2/c^2)\epsilon(\vec{R})\,\vec{E}_\omega = 0 \tag{29}$$

It is convenient to change over from the differential equation (29) to the equivalent integral equation

$$\frac{\epsilon(\vec{R})+2}{3}\vec{E}_\omega(\vec{R}) = \int_\delta d\vec{R}'\,\frac{\epsilon(\vec{R}')-1}{4\pi}\,\text{curl}_{\vec{R}}\,\text{curl}_{\vec{R}}\,\frac{\exp[i(\omega/c)|\vec{R}-\vec{R}'|]}{|\vec{R}-\vec{R}'|}\vec{E}_\omega(\vec{R}') + \vec{E}^0_\omega(\vec{R}) \tag{30}$$

where $\vec{E}^0_\omega(\vec{R})$ is the incident wave. We denote the quantity on the left-hand side of (29) by $\vec{G}_\omega(\vec{R})$:

$$\vec{G}_\omega(\vec{R}) = [\epsilon(\vec{R}) + 2]\,\vec{E}_\omega(\vec{R})/3 \tag{31}$$

The vector $\vec{G}_\omega(\vec{R})$ satisfies an equation that follows directly from (30):

$$\vec{G}_\omega(\vec{R}) = \int\limits_\delta d\vec{R'}\, \frac{3}{4\pi}\frac{\epsilon(\vec{R'})-1}{\epsilon(\vec{R'})+2}\,\text{curl}_{\vec{R}}\,\text{curl}_{\vec{R}}\,\frac{\exp[i(\omega/c)|\vec{R}-\vec{R'}|]}{|\vec{R}-\vec{R'}|}\vec{G}_\omega(\vec{R'})+\vec{E}^0_\omega(\vec{R}) \qquad (32)$$

In solving this equation, we use precisely the same approximation as in the solution of the system (12), carrying out an expansion, similar to (13), of the integrand

$$\text{curl curl}\left[\frac{\exp[i(\omega/c)|\vec{R}-\vec{R'}|]}{|\vec{R}-\vec{R'}|}\vec{G}_\omega(\vec{R'})\right] \approx i\frac{2}{3}\left(\frac{\omega}{c}\right)^3\vec{G}_\omega(\vec{R}) \qquad (33)$$

In addition, we assume that inside the volume in question the field $\vec{G}_\omega(\vec{R'})$ does not change significantly. Then

$$\vec{G}_\omega(\vec{R}) = \frac{\vec{E}^0_\omega}{1 - i(3/4\pi)[(\epsilon-1)/(\epsilon+2)](2/3)(\omega/c)^3 V} \qquad (34)$$

Using the relation $\vec{d}_\omega = (\epsilon-1)\vec{E}_\omega/4\pi$, and also (31) and (34), we can express d_ω and $\vec{D}_\omega = V\vec{d}_\omega$ in terms of $\vec{E}^0_\omega$. Further, putting $\vec{D}_\omega = \alpha_V\vec{E}^0_\omega$, we obtain the polarizability of α_V:

$$\alpha_V = \frac{(3/4\pi)[(\epsilon-1)/(\epsilon+2)]\,V}{1-i(3/4\pi)[(\epsilon-1)/(\epsilon+2)](2/3)(\omega/c)^3\,V} = \frac{\alpha^{st}_V}{1-i(2/3)(\omega/c)^3\alpha^{st}_V} \qquad (35)$$

If we substitute in this formula ϵ from (21), then the resulting expression coincides exactly with (23), i.e., with the polarizability of the volume α_V, calculated from the microscopic equation (12) with allowance for the effects of collective radiative damping.

Thus, this effect falls entirely within the usual macroscopic approach. The fact that the second term in the denominator of (35) determines the correction precisely for the effect of the collective radiation damping is seen also from the following simple reasoning. After determining in the first approximation the dipole moment of the system with the aid of the quasistatic approximation $\vec{D}_\omega = \alpha^{st}_V\vec{E}^0_\omega$, we can refine this expression by adding to $\vec{E}$ the radiation field of the dipole $\vec{E}^{rad}_\omega$ calculated with the aid of (1):

$$\vec{D}_\omega = \alpha^{st}_V[\vec{E}_\omega + (2/3c^3)\,\dddot{\vec{D}}_\omega] \qquad (36)$$

From this we get

$$\vec{D}_\omega = \frac{\alpha_V^{st}}{1 - i(2/3)(\omega/c)^3 \alpha_V^{st}} \vec{E}_\omega \tag{37}$$

i. e. , we again arrive at formula (35). [*]

Formulas (35) are, evidently, much more general than (23), inasmuch as the derivation of these formulas does not involve any model of the oscillators (atoms or molecules) of the medium. Thus, in the case of an atomic gas it is possible to use in lieu of (21) the well-known expression for

$$\frac{3}{4\pi} \frac{\epsilon - 1}{\epsilon + 2} = \frac{e^2}{m} \sum_{i,k} f_{ik}\left(n_k - \frac{g_k}{g_i} n_i\right) \int \frac{F_{ik}(\omega')d\omega'}{(\omega_{ik} - \omega')^2 - \omega'^2 - i\omega' \Gamma} \tag{38}$$

The function $F_{ik}(\omega')$ describes the line broadening not connected with energy dissipation, i. e. , with inelastic collisions. For example, the function $F_{ik}(\omega')$ can specify the intensity distribution in a line broadened as a result of the Holtsmark or Weisskopf broadening mechanisms.

Let us proceed to clarify the limits of applicability of the formulas obtained. The main question that arises here is whether it is at all possible to obtain a situation whereby the effect of the collective radiative broadening plays an essential role and at the same time the approximation used in the derivation of (35) remains valid. The point is that in deriving these formulas we virtually had to assume [the field $\vec{G}_\omega(\vec{R})$ was taken outside the integral sign] that the dimensions of the volume in question are small not only compared with the wavelength in vacuum $\lambda = c/\omega$, but also compared with the wavelength in the medium or with the depth of penetration $\delta \sim (c/\omega)/\sqrt{|\epsilon|}$. This therefore raises the question as to whether the condition $(2/3)k^3 \alpha_V^{st} \gtrsim 1$ does not lead automatically to the inequality $\delta < a$, when the dipole approximation itself becomes meaningless.

It is seen from (35) that the second term in the denominator becomes of the order of or larger than unity, and simultaneously $\delta > a$ at such values of ϵ which satisfy the conditions

$$|(\epsilon-1)/(\epsilon+2)| \gtrsim 3(\lambda/a)^3/2, \quad \sqrt{|\epsilon|} < \lambda/a \tag{39}$$

It is easy to see that these conditions are compatible. Let us consider by way of an example ϵ from (21). The first of these conditions is realized

---

[*]G. A. Askar'yan called our attention to the possibility of deriving formula (37) in this manner.

when the inequalities $|\omega-\omega_0| \lesssim nV\gamma_0$ and $\Gamma < nV\gamma_0$ are satisfied. It is easy to verify that in this case $\delta \approx \chi \gg a$. It should be noted, however, that when $nV\gamma_0 > \Gamma$ the depth of penetration δ in the frequency region $|\omega - \omega_0 + \pi n\chi^3\gamma_0| \lesssim \Gamma$ becomes smaller than a. Then, as is well known [20-24], the principal role is assumed not by electric dipole scattering but by magnetic dipole scattering.

The question of applicability of the approximation used in the derivation of (35) will be discussed again after we consider the problem of reflection of electromagnetic waves from a thin layer. In this case the microscopic equation (12) also leads to the appearance of a collective radiative width, and the corresponding macroscopic Maxwell's equations admit of an exact solution in a simple analytic form. Consequently, the estimates of the limits of applicability become clearer.

It follows from the foregoing that the necessary condition for the appearance of the cooperative effect is smallness of the dissipative width compared with $nV\gamma_0$. This condition is not sufficient. Nowhere in the foregoing did we take into account the possibility of broadening of the spectral lines as a result of elastic collisions and of collisions accompanied by exchange of excitation quanta (the so-called resonant broadening, or broadening due to its own pressure). It is evident that the cooperative effect can appear only in the case when the corresponding widths Γ' [in the general case the width of the distribution $F(\omega')$ in (38)] are smaller than $nV\gamma_0$.

Let us estimate first the broadening due to its own pressure. For resonant lines this broadening is connected with the dipole–dipole interaction of the atoms of the same species. In essence we were forced to take into account the corresponding terms in the interaction of the oscillators in the system (12), without confining ourselves only to the first imaginary term in the expansion (13). It was also necessary to take into account the motion of the oscillators. The role of the dipole–dipole interaction, however, has been thoroughly investigated. The most complete analysis is contained in Ref. 23. In gases of not too high density, for an atomic transition with an oscillator strength f, this interaction leads to a broadening

$$\Gamma' \approx (e^2/m\omega_0)\, f_n = (3/2)n\gamma_0 f\chi^3 \tag{40}$$

For the case in question $f \approx 1$ and $V \ll \chi^3$. From this we get $\Gamma' \gg nV\gamma_0$. It can also be shown that for all atomic and molecular transitions, whether electric dipole, quadrupole, or magnetic dipole, the broadening due to its own pressure is larger than the possible value of the collective radiative width. Therefore in gases in this density region where the estimate (40) is applicable, the cooperative effect cannot play any role whatever. With increasing density, the situation may be different. In any case, formula (40) is certainly not applicable to condensed media. Therefore the inequality

$\Gamma' < n V \gamma_0$ may be satisfied in principle.

In concluding this section, let us stop to discuss briefly the process of thermal radiation by small volumes. Inasmuch as it was shown above that the macroscopic approach to the problem of scattering of electromagnetic waves by small volumes includes the description of the cooperative effect, there is no need for taking special account of this effect when calculating the intensity and the spectrum of the thermal radiation. It suffices to use the known results based on the Mie theory, and the general theory of equilibrium electric fluctuations [24, 25].

IV. REFLECTION FROM A THIN LAYER

Let us consider normal incidence (along the z axis) of a wave on a homogeneous layer of thickness a $\ll \lambda$. Replacing the summation over j in the system (12) by integration, we can transform this system without any simplifications into (see also [26])

$$
\vec{d}_\omega(z) \left[\omega_0^2 - \omega^2 - \frac{4\pi}{3} n \frac{e^2}{m} - i\omega\Gamma \right] =
$$

$$
i2\pi n \frac{e^2}{m} k \int_0^a e^{ik|z-z'|} \vec{d}_\omega(z') dz' + \frac{e^2}{m} \vec{E}_\omega^0 e^{ikz} \tag{41}
$$

Solving this equation in the approximation $\exp(ikz) \approx 1$, $\exp(ik|z-z'|) \approx 1$ and assuming that $\vec{d}_\omega(z')$ does not change significantly in the interval $(0, a)$, we obtain for $0 \leq z \leq a$

$$
\vec{d}_\omega \left[\omega_0^2 - \omega^2 - (4\pi/3) n \frac{e^2}{m} - i\omega\Gamma - i\omega 2\pi n a k(e^2/m) \right] = (e^2/m) \vec{E}_\omega^0 \tag{42}
$$

Knowing the induced dipole moments of each of the oscillators of the medium (42), we can find the field produced by them at large distances from the layer (in the wave zone)

$$
\vec{E}_\omega(z) = i2\pi k n a \vec{d}_\omega \begin{cases} e^{ikz} & z > a \\ e^{-ikz} & z < a \end{cases} \tag{43}
$$

From this we can readily obtain the following expressions for the amplitude reflection and transmission coefficients R and D, respectively:

$$R = \frac{2\pi i nake^2/m}{\omega_0^2 - \omega^2 - (4\pi/3)n(e^2/m) - i\omega[\Gamma + (3/4\pi)n\gamma_0 a \lambda^2]} \tag{44}$$

$$D = \frac{\omega_0^2 - \omega^2 - (4\pi/3)\,n\,(e^2/m) - i\omega\Gamma}{\omega_0^2 - \omega^2 - (4\pi/3)n(e^2/m) - i\omega[\Gamma + (3/4\pi)\,n\gamma_0 a \lambda^2]} \tag{45}$$

Thus, in this case the collective radiative width turns out to be equal to $3n\gamma_0 a \lambda^2/4\pi$, i.e., it is determined by the number of particles in the effective volume $3a\lambda^2/4\pi$.

Let us consider now the same problem, starting from the macroscopic Maxwell equations. Instead of the integral equation (30) we obtain for the one-dimensional problem being considered by us

$$\vec{E}_\omega(z) = \vec{E}_\omega^0 e^{ikz} + \frac{k}{2i} \int_0^a \exp(ik|z-z'|)\vec{E}_\omega(z')[1-\epsilon(z')]dz' \tag{46}$$

Solving this equation in the same approximation as (41), we obtain the field $\vec{E}_\omega(z)$ inside the layer

$$\vec{E}_\omega = \vec{E}_\omega^0[1 + i(ak/2)(1 - \epsilon)]^{-1} \tag{47}$$

For the reflected and transmitted waves we have

$$\vec{E}_\omega^{refl} = e^{-ikz}\frac{k}{2i} \int_0^a e^{ikz'}\vec{E}_\omega(z')[1-\epsilon(z')]dz' \tag{48}$$

$$\vec{E}_\omega^{trans} = e^{ikz}\left(\vec{E}_\omega^0 + \frac{k}{2i} \int_0^a e^{-ikz'}\vec{E}_\omega(z')[1-\epsilon(z')]\,dz'\right) \tag{49}$$

Hence

$$R = - i(ak/2)(1-\epsilon)[1+i(ak/2)(1-\epsilon)]^{-1} \tag{50}$$

$$D = [1 + i(ak/2)(1-\epsilon)]^{-1} \tag{51}$$

For ϵ from (21), formulas (50) and (51) coincide with (44) and (45), i.e., the macroscopic approach again makes it possible to take full account of the effect of collective radiative damping.

It is easy to see that expressions (50) and (51) satisfy the exact normali-zation condition: the sum of the reflected flux, the transmitted flux, and the energy absorbed in the layer is equal to the incident flux

$$\frac{c}{4\pi} |\vec{E}_\omega^0|^2 = \frac{c}{4\pi} |\vec{E}_\omega^0|^2 (|R|^2 + |D|^2) + \frac{\omega}{4\pi} \int |\vec{E}_\omega(z)|^2 \mathrm{Im}\, \epsilon(z) dz \qquad (52)$$

Let us ascertain now the relation of formulas (50) and (51) to the exact formulas for the reflection and transmission coefficients. For a homoge-neous layer (plane-parallel plate), as is well known [20–22]

$$R = r[1 - \exp(2ika\sqrt{\epsilon})] / [1 - r^2 \exp(2ika\sqrt{\epsilon})] \qquad (53)$$

$$D = (1 - r^2) / [\exp(-ika\sqrt{\epsilon}) - r^2 \exp(ika\sqrt{\epsilon})] \qquad (54)$$

where
$$r = (1 - \sqrt{\epsilon}) / (1 + \sqrt{\epsilon}) \qquad (55)$$

When $ka \ll 1$ and $ka\sqrt{|\epsilon|} \ll 1$ expressions (53) and (54) go over into (50) and (51). The second of these conditions means that the depth of penetration $\delta = \lambda/2\pi\sqrt{|\epsilon|}$ is large compared with the layer thickness a. On the other hand, the term $ak(1-\epsilon)/2$ in the denominators of (50) and (51) ceases to be negligibly small compared with unity if $|(ak/2)(1-\epsilon)| \gtrsim 1$.

Thus, the cooperative radiative effect plays an important role and is correctly described by formulas (50) and (51) in the region

$$(\lambda/2\pi a) \lesssim |\epsilon| \ll (\lambda/2\pi a)^2 \qquad (56)$$

The authors are grateful to V. L. Ginzburg for a discussion of the work and for a number of remarks.

REFERENCES

1. See, for example, L. D. Landau and E. M. Lifshitz, Teoriya polya, Fizmatgiz, 1960 [Classical Theory of Fields (Addison-Wesley, Reading, Mass., 1965)].
2. W. Heitler, The Quantum Theory of Radiation (Oxford University Press, Oxford, England, 1954).
3. V. M. Fain and Ya. I. Khanin, Kvantovaya radiofizika (Quantum Radio-physics)(Sov. Radio, Moscow, 1965).
4. R. H. Dicke, Phys. Rev. 93, 99 (1954).

834   V. A. Alekseev, A. V. Vinogradov, and I. I. Sobel'man

5. V. M. Fain, Usp. Fiz. Nauk 64, 273 (1958).
6. G. A. Askar'yan, Atomnaya energiya 4, 71 (1958).
7. D. A. Hutchinson and H. F. Hameka, J. Chem. Phys. 41, 2006 (1964).
8. E. A. Power, J. Chem. Phys. 46, 4297 (1967).
9. H. G. Kuhn and J. M. Vaughan, Proc. Roy. Soc. A277, 297 (1964).
10. H. G. Kuhn, Acta Phys. Pol. 26, 315 (1964).
11. J. M. Vaughan, Phys. Letters 21, 153 (1966); Proc. Roy. Soc. A295, 164 (1966).
12. H. G. Kuhn and E. L. Lewis, Proc. Roy. Soc. A299, 423 (1967).
13. J. M. Vaughan, Phys. Rev. 166, 13 (1968).
14. A. N. Oraevskii, Usp. Fiz. Nauk 91, 181 (1967)[Sov. Phys. Usp. 10, 45 (1967)].
15. H. G. Kuhn, E. L. Lewis, and J. M. Vaughan, Phys. Rev. Letters 15, 687 (1965).
16. M. Born and E. Wolf, Principles of Optics (Pergamon Press, New York, 1959).
17. Yu. L. Klimontovich and V. S. Fursov, Zh. Eksp. Teor. Fiz. 19, 819 (1949).
18. L. I. Mandel'shtam, Polnoe sobranie trudov (Collected Works), Vol. 1 (Akad. Nauk SSSR, Moscow,1957).
19. V. A. Alekseev, A. V. Vinogradov, and I. I. Sobel'man, J. Quant. Spectr. Radiat. Transfer 10, 55 (1970).
20. L. D. Landau and E. M. Lifshitz, Elektrodinamika sploshnykh sred (Gostekhizdat, Moscow, 1957) [Electrodynamics of Continuous Media (Addison-Wesley, Reading, Mass., 1959)].
21. I. E. Tamm, Osnovy teorii elektrichestva (Fundamentals of Electricity Theory) (Gostekhizdat, Moscow, 1957).
22. V. L. Ginzburg, Rasprostranenie elektromagnitnykh voln v plazme (Propagation of Electromagnetic Waves in a Plasma) (Nauka, Moscow, 1967).
23. Yu. A. Vdovin and V. M. Galitskii, Zh. Eksp. Teor. Fiz. 52, 1345 (1967)[Sov. Phys. JETP 25, 894 (1967)].
24. S. M. Rytov, Teoriya elektricheskikh fluktuatsii i teplovogo izlucheniya (Theory of Electric Fluctuations and Thermal Radiation), Akad. Nauk SSSR, Moscow, 1953.
25. M. L. Levin and S. M. Rytov, Teoriya ravnovesnykh teplovykh fluktuatsii v elektrodinamike (Theory of Equilibrium Thermal Processes) (Nauka, Moscow, 1967).
26. V. L. Lyuvoshitz, Zh. Eksp. Teor. Fiz. 52, 926 (1967)[Sov. Phys. JETP 25, 612 (1967)].

ATOMIC COHERENT STATES IN QUANTUM OPTICS

F. T. Arecchi
Universita di Pavia and C. I. S. E.
Milano, Italy

E. Courtens
IBM Research Laboratory
8803 Rüschlikon, Zurich, Switzerland

R. Gilmore*
Department of Physics
Massachusetts Institute of Technology
Cambridge, Massachusetts

H. Thomas
University of Frankfurt
Frankfurt, Germany

ABSTRACT

For the description of an assembly of two-level atoms, atomic coherent states can be defined which have properties analogous to those of the field coherent states. The derivation of the properties of the atomic coherent

*Present address: Department of Physics, University of South Florida, Tampa, Florida.

states is made easier by the use of a powerful disentangling theorem for exponential angular momentum operators. A complete labelling of the atomic states is developed and many of their properties are studied. In particular it is shown that the atomic coherent states are the quantum ana-logs of classical dipoles, and that they can be produced by classical fields.

I.   INTRODUCTION

Many problems in quantum optics can be dealt with in terms of the inter-action of an assembly of two-level atoms with a transverse electromagnetic field. In these problems a particular set of quantum states has to be selected for the description of both field and atoms. The choice of a parti-cular representation is always motivated by convenience rather than by necessity. A good example is given by the free field. Early treatments have made large use of Fock states, i. e., photon number states which are eigenstates of the free-field Hamiltonian. Although they form a perfectly valid basis for the corresponding Hilbert space, these states are poorly suited for the description of laser fields which contain a large and intrinsic-ally uncertain number of photons. To this effect another set of states, the so-called coherent states of the radiation field, sometimes called Glauber states, have been introduced and extensively studied [1, 2]. A single term of that representation is a good lowest-order solution to important dynamical problems associated with laser emission.

The coherent states of the radiation field have attractive properties. They are obtained from the vacuum state by a unitary shift operator, and are minimum uncertainty states, i.e., products of mean square deviations of conjugated variables are minimum in these states, e.g., $\langle \Delta p^2 \rangle \langle \Delta q^2 \rangle = \hbar^2/4$. Though not orthogonal, they obey a completeness relation, and hence form a good set of basis states. In fact the overcompleteness of coherent states allows one to expand many important field operators as a single integral over diagonal projectors on these states. Finally, these states correspond to the field radiated by classical currents, i. e., currents produced by moving charges for which the field reaction is neglected. In this sense these states provide a quantum description of classical fields.

One purpose of the present paper is to show that states with completely analogous properties can be defined for the free-atom assembly. In fact to each property of the atomic coherent states there exist a corresponding property of the field coherent states. This duality, far from being acciden-tal, will be shown to be deeply rooted, and related to the contraction of the rotation group, describing motions on a sphere, onto the translation group, describing motions on a plane.

For a single two-level system, that of atom n, the ground-state ket will be labelled $|\psi_2^n\rangle$ and the upper-state ket $|\psi_1^n\rangle$. Any operator acting on this system can be expanded in the set of Pauli matrices σ_x^n, σ_y^n, σ_z^n, plus the identity matrix I_2^n, associated with this particular atom. The two-level system is thus identical to a spin 1/2 system for which spin-up and spin-down operators are defined by

$$\sigma_\pm^n = \frac{1}{2}\,(\sigma_x^n \pm i\sigma_y^n) \tag{1.1}$$

For recollection, the commutation rules of these operators are

$$[\sigma_z, \sigma_\pm] = \pm 2\sigma_\pm, \qquad [\sigma_+, \sigma_-] = \sigma_z \tag{1.2}$$

The states $|\psi_1^n\rangle$ and $|\psi_2^n\rangle$ are eigenstates of σ_z^n. Such a choice of basis is convenient but by no means unique. Any other linear combination

$$|\psi_i^{n'}\rangle \equiv \sum_{j=1}^{2} U_{ij}\,|\psi_j^n\rangle \qquad (i = 1, 2) \tag{1.3}$$

may be chosen which preserves orthogonality and normalization. The most general transformations U_{ij} with these properties are the collection of 2×2 unitary matrices, which form the group U(2). The subgroup of transformations with determinant $+1$ forms the group SU(2), familiar from angular momentum analysis. U(2) and SU(2) differ by a trivial phase factor.

Turning to the assembly of N atoms, the corresponding Hilbert space is spanned by the set of 2^N product states

$$|\Phi_{i_1 i_2 \ldots i_N}\rangle \equiv \prod_{n=1}^{N} |\psi_{i_n}^n\rangle \qquad (i_n = 1, 2) \tag{1.4}$$

Collective angular momentum operators are defined by

$$J_\mu = \frac{1}{2}\sum_n \sigma_\mu^n \qquad\qquad (\mu = x, y, z) \tag{1.5a}$$

$$J_\pm = \sum_n \sigma_\pm^n \tag{1.5b}$$

$$J^2 = J_x^2 + J_y^2 + J_z^2 \tag{1.5c}$$

For the moment the effect of the different spatial positions of atoms 1, 2, ..., N is ignored. Following the historical development of quantum mechanics one could choose as another suitable basis, in place of (1.4), the set of eigenstates of the energy operator J_z. In this case, symmetry requirements usually indicate an appropriate complete set of commuting observables to which J_z belongs and whose simultaneous eigenstates form the basis. In analogy to angular momentum eigenstates these orthonormal states will be labelled

$$\left| \begin{array}{cc} J & \vec{\lambda} \\ M & ; \quad i \end{array} \right\rangle \tag{1.6}$$

where $J(J+1)$ and M are the eigenvalues of J^2 and J_z respectively. The quantum numbers $\vec{\lambda}$ and i are those additional eigenvalues which are required to provide a complete set of labels. They are related to the permutation properties of the free-atom Hamiltonian as explained in a longer paper [3]. The energy eigenstates (1.6) have been used in the study of superradiance [4, 5] and will be called Dicke states. They will be shown to have a close relationship to the Fock states of the free-field problem.

Another natural way to describe the N atoms is through the overcomplete set of product states

$$\left| \Phi(a_1, b_1; \ldots ; a_N, b_N) \right\rangle \equiv \prod_{n=1}^{N} \left(a_n | \psi_1^n \rangle + b_n | \psi_2^n \rangle \right) \tag{1.7}$$

with $|a_n|^2 + |b_n|^2 = 1$. These states display no correlation between different atoms. For any normalized state $|\Psi\rangle$ of the N atom assembly a degree of correlation can be defined in the following manner: one forms the overlap integral $|\langle \Psi | \Phi(a_1, b_1; \ldots ; a_N, b_N) \rangle |^2$ and maximizes the result with respect to the set (a_i, b_i), $i = 1$ to N. The complement to one of this maximized overlap integral is defined as the degree of atomic correlation of the state $|\Psi\rangle$. It can easily be seen that all states of the form (1.7) have zero correlation, whereas the Dicke states (1.6) of maximum J $(J = N/2)$ and small M $(M \simeq 0)$ have a correlation which approaches unity for large N values.

Another set of overcomplete states can be obtained by rotating the Dicke states $|J, -J; \vec{\lambda}, i \rangle$ through an angle $|\theta, \varphi\rangle$ in angular momentum space. These states, which can be labelled

$$\left| \begin{array}{cc} J & \vec{\lambda} \\ & ; \\ \theta, \varphi & i \end{array} \right\rangle \tag{1.8}$$

are the atomic coherent states. They will be named Bloch states in view of their resemblance with the spin states common in nuclear induction problems [6]. Only for $J = N/2$ are these states a subset of (1.7).

The remainder of the paper is divided as follows: Section II deals with the set of equations describing the atoms-field interaction. There the self-consistent approximation to the quantum equations of motion is shown to be equivalent to a semiclassical approach. The most natural representations for this description are those whose basis vectors are coherent states, both for field and atoms. In Sec. III we recall the main properties of the coherent states for a single-field mode. Section IV describes the symmetry properties of the atomic states. The atomic coherent states representation is fully described and compared with the Dicke representation. Both in Secs. III and IV we show the virtues of the coherent state representation in dealing with statistical averages. For sake of brevity the arguments will be presented in a half-heuristic way, leaving more rigorous proofs to a longer paper [3]. Furthermore, most considerations here are limited to the inter-action of a single-field mode with a collection of atoms corresponding to a single member of the set $(\vec{\lambda}, i)$. The notation in Sec. IV is therefore simplified, $|J, M\rangle$, or $|M\rangle$, replacing (1.6) and $|J; \theta, \varphi\rangle$, or $|\theta, \varphi\rangle$, replacing (1.8). Section V explains the group contraction procedure which allows one to derive all the properties of the field states from the corres-ponding properties of the atomic states in Sec. IV. Appendix I gives a disentangling theorem for exponential angular momentum operators, and some properties which result, such as formulas for the coupling of rota-tions. These disentangling properties should find great use in many other fields of physics where rotations are considered and expectation values have to be calculated. Finally Appendix II shows an example of the application of the disentangling theorem to the calculation of a generating function for expectation values of any product of angular momentum operators in Bloch states.

II.   FIELD-ATOMS INTERACTION

The interaction between the transverse electromagnetic field confined in a cavity of volume V and a set of two-level atoms uniformly distributed within the cavity is described by the following Hamiltonian (in frequency units)

$$\mathcal{K} = \sum_k \omega_k a_k^\dagger a_k + \frac{\omega_0}{2} \sum_{i=1}^N \sigma_z^i + \sum_{k,i} g_k [a_k \sigma_+^i e^{i\vec{k}\cdot\vec{r}_i} + a_k^\dagger \sigma_-^i e^{-i\vec{k}\cdot\vec{r}_i}] \qquad (2.1)$$

where $a_k^\dagger$, a_k are the Bose operators describing k, the field mode; σ_z^i and $\sigma_\pm^i$ are the Pauli operators for the i-th atom at position $\vec{r}_i$; and g_k is the follow-ing coupling constant (in mks units)

$$g_k = \frac{\omega_o}{\sqrt{2\varepsilon_o \hbar \omega_k V}} p_o \qquad (2.2)$$

p_o being the dipole matrix element between upper and lower states.   The detailed assumptions made in writing (2.1) and (2.2) are reviewed in Ref. 7.

The Hamiltonian (2.1) can be rewritten in terms of the collective atomic operators

$$J_{\pm}(k) = \sum_i \sigma_{\pm}^i e^{\pm i \vec{k} \cdot \vec{r}_i} \qquad (2.3a)$$

$$J_z = \frac{1}{2} \sum_i \sigma_z^i \qquad (2.3b)$$

(2.3a) reduces to (1.5b) when all distances $\vec{r}_i - \vec{r}_j$ are much smaller than a wavelength, so that the common phase factors $\exp(\pm i \vec{k} \cdot \vec{r}_i)$ can be taken out of the sum.

Writing the Heisenberg equations of motion for the operators leads to a k-k' coupling, and furthermore to the uselessness of the collective operators (2.3), as seen from the following commutation relation

$$[J_+(k),\ J_-(k')] = \sum_i \sigma_z^i \exp[i(\vec{k}-\vec{k}') \cdot \vec{r}_i]$$

Leaving to Ref. 3 the more general treatment of an extended medium with many k values, here we deal with a simplified Hamiltonian in terms of the operators, as

$$\mathcal{K} = \omega a^\dagger a + \omega_0 J_z + g(aJ_+ + a^\dagger J_-) \qquad (2.4)$$

This covers the following three cases: (i) point laser (interaction volume $V \ll \lambda^3$), (ii) single-mode running-wave laser, (iii) traveling-wave field in an amplifying or absorbing medium. The associated equations of motion are

$$\dot{a} = -i\omega a - igJ_-$$
$$\dot{J}_- = -i\omega_0 J_- + 2igaJ_z \qquad (2.5)$$
$$\dot{J}_z = -ig(aJ_+ - a^\dagger J_-)$$

and similar for conjugated operators $a^\dagger$ and J_+. If we take the expectation values over an initial field-atoms state $|t = 0\rangle$, we see that (2.5) form an unclosed set, in so far as the unknowns $\langle a \rangle$, $\langle a^\dagger \rangle$, $\langle J_\pm \rangle$ depend on expectation values of binary products such as $\langle aJ_+ \rangle$. Solving equations for binary quantities we would need knowledge of ternary products, and so on.

The self-consistent field approximation (SCFA) consists in assuming that binary products factor out as, e. g. ,

$$\langle aJ_+ \rangle \simeq \langle a \rangle \langle J_+ \rangle$$

This amounts to introducing "fluctuation" operators with zero expectation value such as

$$\delta a \equiv a - \langle a \rangle$$

and assuming that their correlations $\langle \delta a \, \delta J_+ \rangle$ are zero. In this approximation, Eqs. (2.5) reduce to the vector equation for the atomic vector operator

$$\langle \dot{\vec{J}} \rangle = \vec{\Omega} \times \langle \vec{J} \rangle, \quad \vec{\Omega} \equiv (2g\langle a \rangle, 0, \omega_0) \tag{2.7}$$

(we have assumed for simplicity perfect resonance $\omega = \omega_0$), and to the coupled equation

$$\langle \dot{a} \rangle = -i\omega \langle a \rangle - ig\langle J_- \rangle \tag{2.8}$$

for the field. Equations (2.7) and (2.8) form a closed nonlinear set of equations.

For an assigned source field $\langle a \rangle$, Eq. (2.7) describes the well-known Rabi-Bloch precession [6], responsible for displacing an atomic system from the ground state to a rotated state as (1.8). In the particular case we are considering ("point laser"), the rotated states (1.8) coincide with the uncorrelated states (1.7) as shown later in Sec. IV. Thus the Bloch states are the solution to the semiclassical problem of an atomic system illuminated by a classical field.

Vice versa, Eq. (2.8) describes the evolution of the field radiated by an assigned atomic current, proportional to $\langle J_- \rangle$. This semiclassical problem is also well known in connection with the so-called infrared catastrophe [8,1,2].

The states reached by a quantum field generated from the ground state by a classical current are those states that will be considered in detail in Sec. III, and that we have named Glauber states.

By joining the two semiclassical considerations, it results that a suitable (even though not necessarily unique) choice for Schrödinger states of a field-atom system starting from a "vacuum" (no photons in the field, atoms all excited) at t=0 are the product states of Glauber times Bloch states as suggested by Eqs. (2.7) and (2.8). This heuristic consideration shows the kinship between Glauber and Bloch states. Such a kinship will appear more dramatically in Secs. III and IV.

III.   DESCRIPTION OF THE FREE FIELD

A.   The Harmonic Oscillator States

In order to point out with maximum clarity the analogies between the free-field description and the free-atom description we start by listing here, in simple terms, the properties of the single harmonic oscillator. The equation numbering in this section and in Sec. IV is done in parallel.

The single harmonic oscillator is described by its canonically conjugated coordinates q, p with the commutation relation

$$[q, p] = i\hbar \tag{3.1}$$

One forms the usual lowering and raising operators

$$a = (2\hbar\omega m)^{-1/2}(\omega m q + ip) \tag{3.2a}$$

$$a^\dagger = (2\hbar\omega m)^{-1/2}(\omega m q - ip) \tag{3.2b}$$

Where $\omega m > 0$ is characteristic of the oscillator. These operators satisfy

$$[a, a^\dagger] = 1 \tag{3.3a}$$

from which one obtains

$$[a, a^\dagger a] = a \tag{3.3b}$$

$$[a^\dagger, a^\dagger a] = -a^\dagger \tag{3.3c}$$

The harmonic oscillator states, or Fock states, are the eigenstates of

$$N = a^\dagger a \tag{3.4}$$

and are given by [9]

$$|n\rangle = \frac{(a^\dagger)^n}{\sqrt{n!}} |0\rangle \qquad (n = 0, 1, 2 \ldots) \tag{3.5}$$

with eigenvalue n. The vacuum state $|0\rangle$ is the harmonic oscillator ground state defined by

$$a|0\rangle = 0 \tag{3.6}$$

B.   Coherent States of the Field

Let us consider the translation operator which produces a shift ξ in q and η in p:

$$T_\alpha \;=\; \exp[-(i/\hbar)(\xi p - \eta q)] \;=\; \exp(\alpha a^\dagger - \alpha^* a) \qquad (3.7)$$

where
$$\alpha \;=\; (2\hbar\omega m)^{-1/2}(\omega m \xi + i\eta) \qquad (3.7b)$$

A coherent state $|\alpha\rangle$ is obtained by translation of the ground state [1, 2]

$$|\alpha\rangle \;\equiv\; T_\alpha |0\rangle \qquad (3.8)$$

We shall name these states Glauber states, as they have been used extensively by Glauber in quantum optics [2].   Since

$$T_\alpha a T_\alpha^{-1} \;=\; a - \alpha , \qquad (3.9)$$

the state $|\alpha\rangle$ satisfies the eigenvalue equation

$$(a - \alpha)|\alpha\rangle \;=\; 0 \qquad (3.10)$$

Using the Baker–Campbell–Hausdorf theorem [10] or Feynman disentangling techniques [11] the translation operator can be written in the following forms:

$$T_\alpha \;=\; e^{|\alpha|^2/2}\, e^{-\alpha^* a}\, e^{\alpha a^\dagger} \;=\; e^{-|\alpha|^2/2}\, e^{\alpha a^\dagger}\, e^{-\alpha^* a} \qquad (3.11)$$

The second of these forms, which is known as the normally ordered form, gives immediately the expansion of $|\alpha\rangle$ in terms of Fock states

$$|\alpha\rangle \;=\; T_\alpha |0\rangle \;=\; e^{-|\alpha|^2/2}\, e^{\alpha a^\dagger} |0\rangle , \qquad (3.12)$$

from which, expanding the exponential and using (3.5), one obtains

$$\langle n|\alpha\rangle \;=\; e^{-|\alpha|^2/2}\, \frac{\alpha^n}{\sqrt{n!}} \qquad (3.13)$$

The scalar product of Glauber states can be obtained either from (3.12), using the disentangling theorem (3.11), or from (3.13), using the completeness property of Fock states, $\Sigma |n\rangle\langle n| = 1$.   One gets

$$\langle \alpha | \beta \rangle = \exp[-(1/2)(|\alpha|^2 - 2\alpha^*\beta + |\beta|^2)] \qquad (3.14a)$$

from which one obtains

$$|\langle \alpha | \beta \rangle|^2 = \exp(-|\alpha - \beta|^2) \qquad (3.14b)$$

The coherent states are minimum uncertainty packets. For three observables A, B, C, which obey a commutation relation $[A, B] = iC$, it is easy to show [9] that $\langle A^2 \rangle \langle B^2 \rangle \geq \langle C \rangle^2/4$. In particular, with $A = q-\xi$, $B = p-\eta$ and $C = \hbar$ one has

$$\langle (q - \xi)^2 \rangle \langle (p - \eta)^2 \rangle \geq \hbar^2/4 \qquad (3.15)$$

for any state. It is easy to show [2] that the equality sign holds for the coherent state $|\alpha\rangle$, where α is related to ξ and η by (3.7b). This establishes the minimum uncertainty property.

C.   The Coherent States as a Basis

We now consider the completeness properties of the coherent states. Using (3.13), and the completeness of Fock states $\sum_n |n\rangle\langle n| = 1$, one obtains straightforwardly

$$\int \frac{d^2\alpha}{\pi} |\alpha\rangle\langle\alpha| = 1 \qquad (3.16)$$

The expansion of an arbitrary state in Glauber states follows

$$|c\rangle \equiv \sum_n C_n |n\rangle = \int \frac{d^2\alpha}{\pi} \sum_n C_n |\alpha\rangle\langle\alpha|n\rangle$$

$$= \int \frac{d^2\alpha}{\pi} \exp[-(1/2)|\alpha|^2] f(\alpha^*) |\alpha\rangle \qquad (3.17a)$$

where

$$f(\alpha^*) \equiv \sum_n C_n \frac{(\alpha^*)^n}{\sqrt{n!}} = e^{-|\alpha|^2/2} \langle\alpha|c\rangle \qquad (3.17b)$$

Using (3.5) one sees that $|c\rangle$ can also be written as

$$|c\rangle = f(a^\dagger)|0\rangle \qquad (3.18)$$

where $f(a^\dagger)$ is defined by its expansion (3.17b). The scalar product of any two states $|c'\rangle$ and $|c\rangle$ is obtained from (3.16) and (3.17b)

$$\langle c'|c\rangle = \int \frac{d^2\alpha}{\pi} \langle c'|\alpha\rangle\langle\alpha|c\rangle$$

$$= \int \frac{d^2\alpha}{\pi} e^{-|\alpha|^2} [f'(\alpha^*)]^* f(\alpha^*) \tag{3.19}$$

In view of the completeness relations, operators F acting on this Hilbert space can be expanded as

$$F = \sum_{m,n} |m\rangle\langle m|F|n\rangle\langle n| \tag{3.20a}$$

or

$$F = \iint \frac{d^2\alpha\, d^2\beta}{\pi^2} |\beta\rangle\langle\beta|F|\alpha\rangle\langle\alpha| \tag{3.20b}$$

Due to the overcompleteness of the $|\alpha\rangle$ states, the expansion (3.20b) is in general not unique. This expansion is especially useful if it can be written in the diagonal form

$$F = \int d^2\alpha\, f(\alpha) |\alpha\rangle\langle\alpha| \tag{3.20c}$$

This will be further discussed for the case of the density matrix.

D. Statistical Operator for the Field

Up to now we have considered pure quantum states. Since a field in thermal equilibrium with matter at ordinary temperatures is essentially in the ground state ($\hbar\omega \gg k_B T$), this is an adequate description for any field obtained from thermal equilibrium in response to a classical current. However, the field radiated by an incoherently pumped medium is a statistical mixture described by a statistical operator ρ, which we assume normalized to unity,

$$\mathrm{Tr}\,\rho = 1 \tag{3.21}$$

With the help of this operator, the statistical average of any observable $F(a, a^\dagger)$ is obtained as

$$\langle F\rangle = \mathrm{Tr}\,\rho\, F \tag{3.22}$$

Of particular interest are statistical ensembles described by a statistical operator which is diagonal in the Glauber representation [2]

$$\rho = \int P(\alpha) \, |\alpha\rangle\langle\alpha| \, d^2\alpha \qquad (3.23)$$

where the normalization (3.21) requires

$$\int P(\alpha) \, d^2\alpha = 1 \qquad (3.24)$$

The statistical average of an observable F is then given by an average over the diagonal elements $\langle\alpha|F|\alpha\rangle$:

$$\langle F \rangle = \int P(\alpha)\langle\alpha|F|\alpha\rangle \, d^2\alpha \qquad (3.25)$$

The weight function $P(\alpha)$ has thus the properties of a distribution function in α-space, except that it is not necessarily positive.

Let us define a set of operators $\hat{X}(\lambda)$ such that their expectation values for coherent states

$$b^\alpha(\lambda) = \langle\alpha|\hat{X}(\lambda)|\alpha\rangle \qquad (3.26)$$

form a basis in the function space of functions of α. If the statistical ensemble has a diagonal representation (3.23), then the statistical averages of the operators form a kind of "characteristic function" of $P(\alpha)$:

$$X(\lambda) \equiv \langle\hat{X}(\lambda)\rangle = \int d^2\alpha \, P(\alpha)b^\alpha(\lambda) \qquad (3.27)$$

The weight function $P(\alpha)$ can be expressed in terms of $X(\lambda)$ with the help of the reciprocal basis $\bar{b}^\lambda(\alpha)$,

$$P(\alpha) = \int d^2\lambda X(\lambda) \, \bar{b}^\lambda(\alpha) \qquad (3.28)$$

A convenient basis is the Fourier basis

$$b^\alpha(\lambda) = \exp(\lambda\alpha^* - \lambda^*\alpha) \qquad (3.29a)$$

$$\bar{b}^\lambda(\alpha) = (2\pi)^{-2} \exp(-\lambda\alpha^* + \lambda^*\alpha) \qquad (3.29b)$$

which is generated by the normally ordered operators

$$\hat{X}_N(\lambda) = \exp(\lambda a^\dagger) \exp(-\lambda^*a) \qquad (3.29c)$$

The question of the existence of the P representation is a complicated one

[2,12]. Using the Fourier basis (3.29) it can be shown however that the mere existence of the inverse transformation (3.28) guarantees that the resulting function $P(\alpha)$ can be used to calculate the statistical average of any moment $\langle a^{\dagger m} a^n \rangle$ as if $P(\alpha)$ was the weight function defined in (3.23). This is due to the fact that the characteristic function $X_N(\lambda)$ plays the role of a generating function for $\langle a^{\dagger m} a^n \rangle$:

$$\langle a^{\dagger m} a^n \rangle = \left(\frac{\partial}{\partial \lambda}\right)^n \left(-\frac{\partial}{\partial \lambda^*}\right)^m X_N(\lambda)\Big|_{\lambda=0}$$

From which, by derivation of (3.27), one obtains

$$\langle a^{\dagger m} a^n \rangle = \int d^2\alpha \, P(\alpha) \, \langle \alpha | a^{\dagger m} a^n | \alpha \rangle$$

which is a particular case of (3.25) and proves the above statement. One could moreover introduce, in addition to (3.29c), symmetrically ordered $\hat{X}_S(\lambda)$, and antinormally ordered $\hat{X}_A(\lambda)$, exponential operators [2,10]. The Fourier transform of their statistical averages are the Wigner distribution, and the matrix element $\langle \alpha | \rho | \alpha \rangle / \pi$, respectively. We will not develop these aspects further as the corresponding expressions for atomic coherent states are rather involved, and of no clear use as yet.

IV. DESCRIPTION OF THE FREE ATOMS

A. The Angular Momentum States

As shown in Sec. I, angular momentum operators can be defined which act on the N-atom Hilbert space. In particular we can consider a subspace of degenerate eigenstates of J^2 with eigenvalues $J(J+1)$. Since J^2 commutes with J_x, J_y, J_z, these operators only connect states within the same subspace. In general, J^2 and J_z do not form a complete set of commuting observables. As explained in Ref. 3, such a complete set is formed by adding to J^2 and J_z some operators of the permutation group of N objects P_N. These operators play with respect to P_N the same role that J^2 and J_z have with respect to the three-dimensional rotation group. We shall assume that the subspace considered here has also been made invariant under these permutation operations, but for simplicity we shall omit for the time being to indicate this in the labelling of the states. The subspace we are dealing with is identical to a constant angular momentum Hilbert space. The Dicke states, which are the analog of the Fock states (3.5), and the Bloch states, which correspond to the Glauber state (3.8), are most easily defined within such a subspace. The equation numbering is in parallel with that of Sec. III. From the angular momentum operators J_x and J_y, which satisfy the commutation relation

$$[J_x, J_y] = i J_z \tag{4.1}$$

the lowering and raising operators are formed

$$J_- = J_x - i J_y \tag{4.2a}$$

$$J_+ = J_x + i J_y \tag{4.2b}$$

which obey

$$[J_-, J_+] = -2J_z \tag{4.3a}$$

$$[J_-, J_z] = J_- \tag{4.3b}$$

$$[J_+, J_z] = -J_+ \tag{4.3c}$$

The Dicke states, which are simply the usual angular momentum states, are defined as the eigenstates of

$$J_z = \frac{1}{2} (J_+ J_- - J_- J_+) \tag{4.4}$$

They are given by [9,13]

$$|M\rangle = \frac{1}{(M+J)!} \binom{2J}{M+J}^{-1/2} J_+^{M+J} |-J\rangle \tag{4.5}$$

$$(M = -J, -J+1, \ldots, J)$$

with eigenvalue M. They span the space of angular momentum quantum number J. The ground state $|-J\rangle$ is defined by

$$J_- |-J\rangle = 0 \tag{4.6}$$

B.   Coherent Atomic States

Let us consider the rotation operator which produces a rotation through an angle θ about an axis $\hat{n} \equiv (\sin\varphi, \cos\varphi, 0)$:

$$R_{\theta,\varphi} = e^{-i\theta J_n} = \exp[-i\theta(J_x \sin\varphi - J_y \cos\varphi)]$$

$$= \exp(\zeta J_+ - \zeta^* J_-) \tag{4.7a}$$

where

$$\zeta = \frac{\theta}{2} e^{-i\varphi} \tag{4.7b}$$

A coherent atomic state, or Bloch state, $|\theta,\varphi\rangle$ is obtained by rotation of the ground state $|-J\rangle$:

$$|\theta,\varphi\rangle \equiv R_{\theta,\varphi} |-J\rangle \tag{4.8}$$

Furthermore

$$R_{\theta,\varphi} J_n R_{\theta,\varphi}^{-1} = J_n$$

$$R_{\theta,\varphi} J_k R_{\theta,\varphi}^{-1} = J_k \cos\theta + J_z \sin\theta$$

$$R_{\theta,\varphi} J_z R_{\theta,\varphi}^{-1} = -J_k \cos\theta + J_z \sin\theta$$

where

$$J_n = J_x \sin\varphi - J_y \cos\varphi$$

$$J_k = J_x \cos\varphi + J_y \sin\varphi$$

which gives

$$J_+ = (J_k - i J_n) e^{i\varphi}$$

$$J_- = (J_k + i J_n) e^{-i\varphi}$$

Using these relations one obtains

$$R_{\theta,\varphi} J_- R_{\theta,\varphi}^{-1} = e^{-i\varphi} [J_- e^{i\varphi} \cos^2(\theta/2) - J_+ e^{-i\varphi} \sin^2(\theta/2) + J_z \sin\theta] \tag{4.9a}$$

and similar relations for J_+ and J_z:

$$R_{\theta,\varphi} J_+ R_{\theta,\varphi}^{-1} = e^{i\varphi} [J_+ e^{-i\varphi} \cos^2(\theta/2) - J_- e^{-i\varphi} \sin^2(\theta/2) + J_z \sin\theta] \tag{4.9b}$$

$$R_{\theta,\varphi} \, J_z \, R_{\theta,\varphi}^{-1} = J_z \cos\theta - J_- e^{i\varphi} \sin(\theta/2)\cos(\theta/2) - J_+ e^{-i\varphi} \sin(\theta/2)\cos(\theta/2)$$

$$(4.9c)$$

From (4.9a), and definition (4.8), one obtains the eigenvalue equation

$$[J_- e^{i\varphi} \cos^2(\theta/2) - J_+ e^{-i\varphi} \sin^2(\theta/2) + J_z \sin\theta] \, |\theta,\varphi\rangle = 0 \qquad (4.10a)$$

This equation, together with

$$J^2 \, |\theta,\varphi\rangle = J(J+1) \, |\theta,\varphi\rangle \qquad (4.10b)$$

specifies uniquely the Bloch state $|\theta,\varphi\rangle$. Note that the harmonic oscillator analog of (4.10b) would have been the trivial relation $(a^+ - \alpha^*)(a-\alpha)|\alpha\rangle = 0$. Other forms of the eigenvalue equation can be obtained using the relation

$$R_{\theta,\varphi} \, J_z \, R_{\theta,\varphi}^{-1} \, |\theta,\varphi\rangle = -J \, |\theta,\varphi\rangle$$

and (4.9c). The resulting equation can be combined with (4.10a) to eliminate one of the operators J_z, J_+, or J_-, giving

$$[J_- e^{i\varphi} \cos^2(\theta/2) + J_+ e^{-i\varphi} \sin^2(\theta/2)] \, |\theta,\varphi\rangle = J \sin\theta \, |\theta,\varphi\rangle \qquad (4.10c)$$

$$[J_- e^{i\varphi} \cos(\theta/2) + J_z \sin(\theta/2)] \, |\theta,\varphi\rangle = J \sin(\theta/2) \, |\theta,\varphi\rangle \qquad (4.10d)$$

$$[J_+ e^{-i\varphi} \sin(\theta/2) - J_z \cos(\theta/2)] \, |\theta,\varphi\rangle = J \cos(\theta/2) \, |\theta,\varphi\rangle \qquad (4.10e)$$

These additional relations are not independent of (4.10a) and (4.10b). One notes that these eigenvalue equations are more complicated than their counterpart (3.10). In particular they involve at least two of the three operators J_-, J_+, J_z. This feature is required by the more complicated commutation relation (4.1) which applies here.

Using the disentangling theorem for angular momentum operators (Appendix I), the rotation $R_{\theta,\varphi}$ given by (4.7a) becomes

$$R_{\theta,\varphi} = \exp(-\tau^* J_-) \exp[-\ln(1 + |\tau|^2) J_z] \exp(\tau J_+)$$

$$= \exp(\tau J_+) \exp[\ln(1 + |\tau|^2) J_z] \exp(-\tau^* J_-) \qquad (4.11a)$$

where

$$\tau \equiv e^{-j\varphi} \tan(\theta/2) \qquad (4.11b)$$

Let us point out that these expressions are singular for $\theta = \pi$, i.e., for the uppermost state. We may have to exclude from some of the following considerations the states contained within an infinitesimally small circle around $\theta = \pi$. The validity of expressions such as (4.13) for $\theta = \pi$ is usually not affected and can be checked directly. The last form of (4.11a), which we call the normally ordered form, gives immediately the expansion of $|\theta,\varphi\rangle$ in terms of Dicke states:

$$|\theta,\varphi\rangle = R_{\theta,\varphi} |-J\rangle = [1/(1 + |\tau|^2)]^J \, e^{\tau J_+} \, |-J\rangle \qquad (4.12)$$

from which, expanding the exponential and using (4.5) one obtains

$$\langle M|\theta,\varphi\rangle = \binom{2J}{M+J}^{1/2} \frac{\tau^{M+J}}{(1+|\tau|^2)^J} = \binom{2J}{M+J}^{1/2} \sin^{J+M}(\theta/2) \cos^{J-M}(\theta/2) \, e^{-i(J+M)\varphi}$$

$$(4.13)$$

Since the Dicke states form a basis for a well-known irreducible representation of the rotation group, these results could have been derived using the appropriate Wigner $\mathfrak{D}^{(J)}$ matrix [13]. The same remark applies to Eqs. (3.12) and (3.13): these could have been obtained without using the Baker-Campbell-Hausdorf formula, from the transformation properties of an irreducible representation of the group of operations T_α.

The overlap of two Bloch states is obtained either from (4.12), using the disentangling theorem for exponential angular momentum operators, or from (4.13), using the completeness property of Dicke states $\Sigma_M |M\rangle\langle M| = 1$. One obtains

$$\langle \theta,\varphi|\theta',\varphi'\rangle = \left(\frac{(1 + \tau^*\tau')^2}{(1 + |\tau|^2)(1 + |\tau'|^2)} \right)^J$$

$$= e^{iJ(\varphi-\varphi')} \left(\cos\frac{\theta-\theta'}{2} \cos\frac{\varphi-\varphi'}{2} - i \cos\frac{\theta+\theta'}{2} \sin\frac{\varphi-\varphi'}{2} \right)^{2J} \qquad (4.14a)$$

from which one obtains

$$|\langle \theta,\varphi|\theta',\varphi'\rangle|^2 = \cos^{4J}(\Theta/2) \qquad (4.14b)$$

where τ is given by (4.11b), τ' is given by the same equation written with the primed quantities, and Θ is the angle between the (θ,φ) and (θ',φ') directions, as given by

$$\cos\Theta = \cos\theta \cos\theta' + \sin\theta \sin\theta' \cos(\varphi-\varphi')$$

The Bloch states form minimum uncertainty packets. The uncertainty relation can be defined in terms of the set of rotated operators $(J_\xi, J_\eta, J_\zeta) = R_{\theta,\varphi} (J_x, J_y, J_z) R^{-1}_{\theta,\varphi}$. These three observables obey a commutation relation of the type $[A, B] = iC$ with $A = J_\xi$, $B = J_\eta$, $C = J_\zeta$, from which they have the uncertainty property

$$\langle J_\xi^2 \rangle \langle J_\eta^2 \rangle \geq \tfrac{1}{4} \langle J_\zeta \rangle^2 \tag{4.15}$$

for any states. It is easy to show that the equality sign holds for the Bloch state $|\theta,\varphi\rangle$, which is therefore a minimum uncertainty state.

C.   The Bloch States as a Basis

Let us now consider the completeness properties of the Bloch states. Using (4.13), and the completeness of Dicke states $\Sigma_M |M\rangle\langle M| = 1$, one obtains

$$(2J+1)\int \frac{d\Omega}{4\pi} |\theta,\varphi\rangle\langle\theta,\varphi|$$

$$= (2J+1)\int \frac{d\Omega}{4\pi} \sum_{M, M'} \binom{2J}{M+J}^{\frac{1}{2}} \binom{2J}{M'+J}^{\frac{1}{2}} e^{i(M'-M)\varphi}$$

$$\times \left(\cos\frac{\theta}{2}\right)^{2J-M-M'} \left(\sin\frac{\theta}{2}\right)^{2J+M+M'} |M\rangle\langle M'|$$

$$= (2J+1)\int_0^\pi d\theta \frac{\sin\theta}{2} \sum_M \binom{2J}{M+J} \left(\cos\frac{\theta}{2}\right)^{2J-2M} \left(\sin\frac{\theta}{2}\right)^{2J+2M} |M\rangle\langle M|$$

$$= \sum_M |M\rangle\langle M| = 1 \tag{4.16}$$

The expansion of an arbitrary state in Bloch states follows:

$$|c\rangle = \sum_M C_M |M\rangle = (2J + 1) \int \frac{d\Omega}{4\pi} \sum_M C_M |\theta,\varphi\rangle\langle\theta,\varphi|M\rangle$$

$$= (2J + 1) \int \frac{d\Omega}{4\pi} \frac{f(\tau^*)}{(1 + |\tau|^2)^J} |\theta,\varphi\rangle \tag{4.17a}$$

where

$$f(\tau^*) \equiv \sum_M C_M \left(\frac{2J}{J+M}\right)^{1/2} (\tau^*)^{J+M} = (1 + |\tau|^2)^J \langle \theta,\varphi|c\rangle \qquad (4.17b)$$

Using (4.5) one sees that $|c\rangle$ can also be written as

$$|c\rangle = f\left(\frac{1}{J+1-J_z} J_+\right) |-J\rangle \qquad (4.18)$$

The amplitude function $f(\tau^*)$ is, by its definition (4.17b), a polynomial of degree 2J. However any function which has a Maclaurin expansion can be taken as a suitable amplitude function in (4.17a) or (4.18). Indeed the powers of τ^* higher than 2J gives zero contribution in (4.17a) and (4.18). The coefficients C_M are then obtained from the first $(2J + 1)$ terms of the Maclaurin series, using (4.17b).

The scalar product of two states characterized by their amplitude function is, from (4.16) and (4.17b)

$$\langle c'|c\rangle = (2J + 1) \int \frac{d\Omega}{4\pi} \langle c'|\theta,\varphi\rangle\langle\theta,\varphi|c\rangle$$

$$= (2J + 1) \int \frac{d\Omega}{4\pi} \frac{1}{(1 + |\tau|^2)^{2J}} [f'(\tau^*)]^* f(\tau^*) \qquad (4.19)$$

Since (4.17b) was used to derive this equation, its validity is restricted to amplitude functions which are polynomials of degree 2J.

In view of the completeness relations, operators G acting on this Hilbert space can be expanded as

$$G = \sum_{M,M'} |M\rangle\langle M|G|M'\rangle\langle M'| \qquad (4.20a)$$

or

$$G = \frac{(2J + 1)^2}{(4\pi)^2} \iint d\Omega \, d\Omega' \, |\theta,\varphi\rangle\langle\theta,\varphi|G|\theta',\varphi'\rangle\langle\theta',\varphi'| \qquad (4.20b)$$

However, G is completely defined by the $(2J + 1)^2$ matrix elements $\langle M|G|M'\rangle$. It results that, except for pathological cases, an operator can always be written in the diagonal form

$$G = \int d\Omega \, g(\theta,\varphi) \, |\theta,\varphi\rangle\langle\theta,\varphi| \qquad (4.20c)$$

where $g(\theta,\varphi)$ is given by a series expansion

$$g(\theta,\varphi) = \sum_{\ell,m} G_{\ell,m} \, Y_\ell^m (\theta,\varphi)$$

In accordance with Appendix IV of Ref. 3 only the $(2J + 1)^2$ first terms of this sum contribute to (4.20c). These are the terms for which $0 \leq \ell \leq 2J$. The corresponding coefficients $G_{\ell,m}$ can be expressed as a function of the matrix element $\langle M | G | M' \rangle$.

D.   Statistical Operators for the Atoms

In order to describe an incoherently pumped system of atoms we introduce a statistical operator ρ with the properties

$$\mathrm{Tr}\, \rho = 1 \qquad\qquad (4.21)$$

$$\langle G \rangle = \mathrm{Tr}\, \rho\, G \qquad\qquad (4.22)$$

As before, the considerations are restricted to states belonging to a single constant angular momentum subspace, and therefore the statistical operator described here does not allow for the most general mixing of atomic states. Of particular interest is the expression of ρ in a diagonal Bloch representation

$$\rho = \int P(\theta,\varphi) \, |\theta,\varphi\rangle\langle\theta,\varphi| \, d\Omega \qquad\qquad (4.23)$$

with the normalization

$$\int P(\theta,\varphi) \, d\Omega = 1 \qquad\qquad (4.24)$$

The statistical average of an observable G is then given by

$$\langle G \rangle = \int P(\theta,\varphi) \, \langle \theta,\varphi | G | \theta,\varphi \rangle \qquad\qquad (4.25)$$

The weight function $P(\theta,\varphi)$ has thus the properties of a distribution function on the unit sphere, except that it is not necessarily positive.

Let us define a set of operators $\hat{X}_\lambda$ such that their expectation values for Bloch states

$$b_\lambda^{(\theta,\varphi)} = \langle \theta,\varphi | \hat{X}_\lambda | \theta,\varphi \rangle \qquad\qquad (4.26)$$

form a basis in the space of functions on the unit sphere. Since in this space a discrete basis can be chosen, the parameter λ can be restricted to discrete values $\lambda = 1, 2, \ldots$. For a statistical ensemble described by (4.23), the statistical averages of the operators $\hat{X}_\lambda$ form a set of "characteristic coefficients" of $P(\theta, \varphi)$:

$$X_\lambda \equiv \mathrm{Tr}\, \rho\, \hat{X}_\lambda = \int d\Omega\, P(\theta,\varphi)\, b_\lambda^{(\theta,\varphi)} \tag{4.27}$$

The weight function can be expressed as a series with the help of the reciprocal basis $\bar{b}^\lambda(\theta, \varphi)$,

$$P(\theta,\varphi) = \sum_\lambda X_\lambda\, \bar{b}^\lambda(\theta,\varphi) \tag{4.28}$$

A convenient basis is given by the spherical harmonics

$$b_\lambda^{(\theta,\varphi)} = Y_\ell^m(\theta,\varphi) \qquad\qquad (\lambda \equiv \ell, m) \tag{4.29a}$$

$$\bar{b}^\lambda(\theta,\varphi) = Y_\ell^{-m}(\theta,\varphi) \tag{4.29b}$$

which are generated by the spherical harmonic operators [14]

$$\hat{X}_\lambda = y_\ell^m(\vec{J}) \tag{4.29c}$$

The already mentioned fact that a diagonal representation always exists in the atomic case also corresponds to the fact that for a given J only the $(2J + 1)^2$ operators y_ℓ^m with $\ell \le 2J$ are different from zero. The finite dimensionality of the basis is required, since ρ is completely determined by its $(2J + 1)^2$ matrix elements $\langle M|\rho|M'\rangle$ in the Dicke representation.

Other differences with the field case should also be noted. First, the spherical harmonic operators are usually written in a fully symmetrized form, whereas the operators (3.29c) are normally ordered. This is only a formal difficulty as it should be possible to write normally ordered and anti-normally ordered multipole operators with properties similar to the $y_\ell^m(\vec{J})$. A second, and more fundamental, difference is that the expectation values X_λ are not generating functions for products of the type $\langle J_+^m J_z^n J_-^p \rangle$, in view of the discreteness of the set. This does not cause much difficulty, as generating functions can be defined from exponential operators (Appendix II) whose expectation values can be calculated with the help of the disentangling theorem (Appendix I). It is tempting to use for the $\hat{X}_\lambda$'s of Eq. (4.29c) these exponential operators themselves. Though the parallel with the field case then seems more transparent, the use of the discrete set $\hat{X}_\lambda$ may be of more fundamental significance as it takes into account symmetry properties of the states.

A final comment should be made about the difficulty of dealing with

creation and annihilation operations in a finite Hilbert space. The existence
of two terminal states, $|J\rangle$ and $|-J\rangle$, requires the presence of a third
operator with the properties of J_z, and prevents the writing of an eigenvalue
equation in terms of one compound operator alone. For instance, the
comparison of (4.18) and (3.18) suggests that $J_-(J + 1 - J_z)^{-1}$ could be a
"good" annihilation operator. Using (4.13) one finds immediately

$$J_-(J + 1 - J_z)^{-1} |\theta,\varphi\rangle = \tau|\theta,\varphi\rangle - \tau \sin^{2J}(\theta/2) e^{-2iJ\varphi} |J\rangle$$

which for small θ and large J is almost an eigenvalue equation: so the
application of the operator reproduces $|\theta,\varphi\rangle$ except for the uppermost Dicke
state. There is no doubt that a theory could be developed in terms of more
complicated annihilation and creation operators of such type, but the
advantages are not clear.

V.   CONTRACTION, OR THE RELATION BETWEEN ATOMIC STATES AND FIELD STATES

The extreme similarity between the treatments of Secs. IV and III suggests
a close connection between atomic and field states. This connection is
made here through a process known as group contraction. The time evolution of a single two-level atom is governed by a 2×2 unitary transformation
matrix. The commutation relations for the generators of the group U(2) are
rewritten here:

$$[J_z, J_\pm] = \pm J_\pm$$

$$[J_+, J_-] = 2 J_z \tag{5.1}$$

$$[\vec{J}, J_0] = 0$$

where J_0 in the third relation is essentially the identity. An arbitrary 2×2
unitary transformation matrix is given by

$$U(2) = \exp(i \sum_\mu \lambda_\mu J_\mu) \tag{5.2}$$

where the summation is over all four indices and the λ_μ's are c-number
parameters which characterize the group operation.

    If another set of generators h_+, h_-, h_z, h_0, is related to J_+, J_-, J_z, J_0
by a nonsingular transformation $A_{\nu\mu}$

$$h_\nu = \sum_\mu A_{\nu\mu} J_\mu \tag{5.3a}$$

then the group operation (5.2) may be written

$$\exp(i \sum_\mu \lambda_\mu J_\mu) = \exp(i \sum_\nu \alpha_\nu h_\nu) \tag{5.4}$$

with

$$\alpha_\nu = \sum_\mu \lambda_\mu (A^{-1})_{\mu\nu} \tag{5.3b}$$

We select the following transformation A, which depends on a real parameter c:

$$
\begin{bmatrix} h_+ \\ h_- \\ h_z \\ h_0 \end{bmatrix}
=
\begin{bmatrix} c & 0 & 0 & 0 \\ 0 & c & 0 & 0 \\ 0 & 0 & 1 & 1/2c^2 \\ 0 & 0 & 0 & 1 \end{bmatrix}
\begin{bmatrix} J_+ \\ J_- \\ J_z \\ J_0 \end{bmatrix}
\tag{5.5}
$$

It is easily verified that the h_ν's satisfy the commutation relations

$$[h_z, h_\pm] = \pm h_\pm$$

$$[h_+, h_-] = 2c^2 h_z - h_0 \tag{5.6}$$

$$[\vec{h}, h_0] = 0$$

In the limit $c \to 0$ the transformation A becomes singular and A^{-1} fails to exist. Nevertheless, the commutation relations (5.6) are well defined, and in fact identical to the commutation relations (3.6) under the identification

$$\lim_{c \to 0} h_z = n = a^\dagger a; \quad \lim_{c \to 0} h_+ = a^\dagger; \quad \lim_{c \to 0} h_- = a \tag{5.7}$$

Although the inverse A^{-1} (5.3b) does not exist as $c \to 0$, the parameter α_ν may approach a well-defined limit if we demand all the parameters λ_μ to shrink ("contract") to zero in the limit $c \to 0$, in such a way that the following ratios are well defined:

$$\lim i\lambda_+/c = \lim e^{-i\varphi}\theta/2c = \alpha$$

$$\lim i\lambda_-/c = \lim \left(-e^{i\varphi}\theta/2c\right) = -\alpha^*$$

$$\lim \lambda_z/c = \lim c\,\lambda_z/c^2 = 0 \qquad (5.8)$$

Within any $(2J+1)$-dimensional representation of the group U(2) the eigenvalue of the diagonal operator h_z is

$$h_z\,|J,M\rangle = (J_z + 1/2c^2)|J,M\rangle = |J,M\rangle(M + 1/c^2) \qquad (5.9)$$

We demand this have a definite limit as $c \to 0$. Physically, for both Fock and Dicke states we progress upward from the ground or vacuum state. It is convenient to demand that the (energy) eigenvalue in (5.9) be zero for the ground state $M = -J$

$$\lim_{c \to 0} (-J + 1/2c^2) = 0 \qquad (5.10)$$

In the limit $c \to 0$, $2Jc^2 = 1$, the unitary irreducible representation $D^J[U(2)]$ goes over into the unitary irreducible representation for the contracted group with generators (5.6). In simple words, the contraction procedure amounts to letting the radius of the Bloch sphere tend to infinity as $1/c^2$, while considering smaller and smaller rotations on the sphere. The motion on the sphere then becomes identical to the motion on the bottom tangent plane which goes over into the phase plane of the harmonic oscillator.

This procedure for contracting groups, commutation relations, and representations will now be used to show the similarity between Dicke and Fock states. We define

$$|\infty,n\rangle = \lim_{c \to 0} |J,M\rangle \qquad (J+M=n \ \text{fixed}) \qquad (5.11)$$

Then

$$a^\dagger a|\infty,n\rangle = \lim (J_z + 1/c^2)|J,M\rangle$$

$$= \lim |J,M\rangle[J+M + (-J + 1/2c^2)] \qquad (5.12a)$$

$$= n\,|\infty,n\rangle$$

The computations for $a^\dagger$ and a are handled in an entirely analogous way:

$$a^\dagger|\infty,n\rangle = \lim h_+|J,M\rangle = \sqrt{n+1}\,|\infty,n+1\rangle \qquad (5.12b)$$

$$a|\infty,n\rangle = \lim h_-|J,M\rangle = \sqrt{n}\,|\infty,n-1\rangle \qquad (5.12c)$$

These equations provide a straightforward connection between Dicke and Fock states. The operators h_+, h_- and h_z contract to $a^\dagger$, a, and $a^\dagger a$ with the proper commutation relations (3.3), and with the proper matrix elements between contracted Dicke states as shown in (5.12). The contracted Dicke states (5.11) can thus be identified with the Fock states, and we conclude that every property of Dicke and Bloch states listed in Sec. IV must contract to a corresponding property of Fock and Glauber states listed in Sec. III. The contraction procedure is summarized in Table I.

TABLE I. RULES FOR CONTRACTION OF THE ANGULAR MOMENTUM ALGEBRA TO THE HARMONIC OSCILLATOR ALGEBRA.
[The limit of the angular momentum quantities (1st line) for $c \to 0$ are the corresponding harmonic oscillator quantities (2nd line).]

Operators	Coordinates	Eigen-values	Eigen-states	Coherent states
Angular momentum				
cJ_+, cJ_-, $J_z + \dfrac{1}{2c^2}$	$\dfrac{\theta}{2c} e^{-i\varphi}$	$2c^2 J$, $J + M$	$\|J, M\rangle$ (Dicke)	$\|\theta, \varphi\rangle$ (Bloch)
Harmonic oscillator				
$a^\dagger$, a, $a^\dagger a$	α	1, n	$\|\infty, n\rangle$ (Fock)	$\|\alpha\rangle$ (Glauber)

We demonstrate this correspondence in some particular cases:

Example 1: Just as the angular momentum eigenstates $|J, M\rangle$ are obtained from the ground state $|J, -J\rangle$ by $(J+M)$ successive applications of the shift-up operator J_+, the Fock state $|n\rangle$ is obtained by n successive applications of $a^\dagger$. By contraction of (4.5) we get

$$|\infty, n\rangle = \lim \frac{(J_+)^n}{[2J!n!/(2J-n)!]^{1/2}} |J, -J\rangle$$

$$= \lim \frac{(cJ_+)^n}{[(2Jc^2)^n n!]^{1/2}} |J, -J\rangle$$

$$= \frac{(a^\dagger)^n}{\sqrt{n!}} \; |\infty, 0\rangle \tag{5.13}$$

which is nothing but (3.5).

Example 2: Let us now contract Bloch states to Glauber states, using equation (4.12) and Table I:

$$|\alpha\rangle = \lim |\theta, \varphi\rangle = \lim \left(\frac{1}{1 + |\tau|^2} \right)^J e^{\tau J_+} |-J\rangle$$

$$= \lim \, [1 - 2c^2(\alpha\alpha^*/2)]^{1/2c^2} \, e^{\,\alpha a^\dagger} \, |0\rangle$$

$$= e^{-\alpha\alpha^*/2} \, e^{\alpha a^\dagger} \, |0\rangle \tag{5.14}$$

which is nothing but (3.12).

We leave it to the reader to verify that every equation of Sec. IV goes over to the corresponding equation of Sec. III under contraction. This is true in particular of the disentangling theorem (4.11) whose contracted limit is the Baker-Campbell-Hausdorf formula (3.11) [cf. Appendix I, Eq. (A1.4)]. In general all properties related to angular momentum have as a counterpart a harmonic oscillator property. Thus, the total angular momentum contracts to the harmonic oscillator Hamiltonian, with the spherical harmonics (and their properties) contracting to the harmonic oscillator eigenfunctions (and corresponding properties).

APPENDIX I.   DISENTANGLING THEOREM FOR ANGULAR MOMENTUM OPERATORS

In dealing with noncommuting exponential operators it is very useful to be able to change a symmetrized exponential operator into an ordered product of exponential operators. The well-known Baker-Campbell-Hausdorf formula (3.11) is of this type. Similar expressions can be obtained for angular momentum operators. We proceed to the derivation of these expressions by first considering the 2 x 2 matrix representation of the rotation group,

$$J_+ = \begin{pmatrix} 0 & 1 \\ 0 & 0 \end{pmatrix} ; \qquad J_- = \begin{pmatrix} 0 & 0 \\ 1 & 0 \end{pmatrix} ; \qquad J_z = \begin{pmatrix} \frac{1}{2} & 0 \\ 0 & -\frac{1}{2} \end{pmatrix}$$

which is the faithful group representation of smallest dimension.

By Maclaurin series expansion one finds

$$\exp(w_+J_+ + w_-J_- + w_zJ_z) = \begin{pmatrix} \cosh k + \dfrac{w_z}{2}\dfrac{\sinh k}{k} & w_+ \dfrac{\sinh k}{k} \\ \\ w_- \dfrac{\sinh k}{k} & \cosh k - \dfrac{w_z}{2}\dfrac{\sinh k}{k} \end{pmatrix}$$

$$(AI.1)$$

with

$$k = (w_+w_- + w_z^2/4)^{1/2},$$

and similarly

$$\exp(x_+J_+)\exp[(\ln x_z)J_z]\exp(x_-J_-) = \begin{pmatrix} \sqrt{x_z} + \dfrac{x_+x_-}{\sqrt{x_z}} & \dfrac{x_+}{\sqrt{x_z}} \\ \\ \dfrac{x_-}{\sqrt{x_z}} & \dfrac{1}{\sqrt{x_z}} \end{pmatrix} \qquad (AI.2)$$

$$\exp(y_-J_-)\exp[(\ln y_z)J_z]\exp(y_+J_+) = \begin{pmatrix} \sqrt{y_z} & y + \sqrt{y_z} \\ \\ y - \sqrt{y_z} & \dfrac{1}{\sqrt{y_z}} + y_+y_-\sqrt{y_z} \end{pmatrix}$$

$$(AI.3)$$

Equating these three matrices element by element gives expressions for each set of coefficients in terms of the others. This procedure gives four equations for three variables, but since the J matrices are traceless, the determinant of each group operation [(AI.1)-(AI.3)] is unity; therefore only three of these equations are independent. The applicability of the resulting operator equation

$$\exp(w_+J_+ + w_-J_- + w_zJ_z) = \exp(x_+J_+)\exp[(\ln x_z)J_z]\exp(x_-J_-)$$

$$= \exp(y_-J_-)\exp[(\ln y_z)J_z]\exp(y_+J_+) \qquad (AI.4)$$

is not restricted to the 2 x 2 matrix representation. The algebra of infinitesimal rotation operators maps onto the rotation group, which is represented by the exponential operators. Any relation between exponential operators, i.e., between group operations, which is valid for one particular faithful representation of the group remains valid for all others. Therefore the equalities (AI.4) are general.

Using these relations, expressions for the rotation operator (4.7) are obtained:

$$R_{\theta,\varphi} = \exp(\zeta J_+ - \zeta^* J_-) = \exp(\tau J_+) \exp[\ln(1 + |\tau|^2) J_z] \exp(-\tau^* J_-)$$

$$= \exp(-\tau^* J_-) \exp[-\ln(1 + |\tau|^2) J_z] \exp(\tau J_+) \qquad \text{(AI.5)}$$

where ζ and τ are given in (4.7b) and (4.11b). If we let $\zeta = c\alpha$, $\zeta^* = c\alpha^*$, $a^\dagger = cJ_+$, $a = cJ_-$, and $J_z = a^\dagger a - 1/2c^2$, following the contraction procedure of Sec. V, Eq. (AI.5) gives, in the limit $c \to 0$,

$$T_\alpha = \exp(\alpha a^\dagger - \alpha^* a) = \exp(\alpha a^\dagger) \exp(- |\alpha|^2/2) \exp(-\alpha^* a)$$

$$= \exp(-\alpha^* a) \exp(|\alpha|^2/2) \exp(\alpha a^\dagger) \qquad \text{(AI.6)}$$

which is the Baker-Campbell-Hausdorf formula (3.11). A more general expression can be obtained by the contraction of (AI.4). If we let $w_+ = c\alpha$, $w_- = c\beta^*$, and $w_z = 0$, we obtain

$$\exp(\alpha a^\dagger - \beta^* a) = \exp(\alpha a^\dagger) \exp(- \alpha\beta^*/2) \exp(-\beta^* a)$$

$$= \exp(-\beta^* a) \exp(\alpha\beta^*/2) \exp(\alpha a^\dagger) \qquad \text{(AI.7)}$$

Using this relation, together with (AI.6), one obtains, after some manipulations

$$T_\alpha T_\beta = \exp[(\alpha\beta^* - \alpha^*\beta)/2] \, T_{\alpha+\beta} \qquad \text{(AI.8)}$$

which describes the composition of translations T_α. The use of this equation allows one to derive (3.14a) very simply. Of course, Eq. (AI.8) can also be obtained by contraction of a similar equation for the composition of rotations:

$$R_{\theta',\varphi'} R_{\theta,\varphi} = R_{\Theta,\Phi} \exp(-i\psi J_z) \qquad \text{(AI.9)}$$

where Θ, Φ, and ψ are to be determined. We note that the $R_{\theta,\varphi}$'s do not form a group, since we have restricted ourselves to rotations around an axis in the (x, y) plane. It is therefore necessary to allow for a rotation around the z axis on the right-hand side of (AI.9). This rotation simply amounts to changing the phase factor of the single atom eigenstates $|\Psi_i^n\rangle$. The angles Θ, Φ, and ψ could be obtained by manipulating (AI. 4) and (AI. 5). A simpler procedure is to use the 2×2 matrix representation as in (AI. 1)–(AI. 3). By application of (AI. 1) one has

$$R_{\theta, \varphi} = \frac{1}{(1 + |\tau|^2)^{1/2}} \begin{pmatrix} 1 & \tau \\ -\tau^* & 1 \end{pmatrix} \tag{AI.10}$$

APPENDIX II.   GENERATING FUNCTIONS FOR EXPECTATION VALUES WITHIN BLOCH STATES

Using the disentangling theorem of Appendix I together with the definition of Bloch states by the rotation (4. 8), or equivalently (4. 12), it is easy to construct generating functions for normally ordered, antinormally ordered, and symmetrized expectation values of products of powers of the operators J_+, J_z, J_- within Bloch states.

We define the following expectation values:

$$X_N(\alpha, \beta, \gamma) = \langle \theta, \varphi | \, e^{\alpha J_+} e^{\beta J_z} e^{\gamma J_-} | \theta, \varphi \rangle \tag{AII.1a}$$

$$X_A(\alpha, \beta, \gamma) = \langle \theta, \varphi | \, e^{\gamma J_-} e^{\beta J_z} e^{\alpha J_+} | \theta, \varphi \rangle \tag{AII.1b}$$

$$X_S(\alpha, \beta, \gamma) = \langle \theta, \varphi | \, \exp(\alpha J_+ + \beta J_z + \gamma J_-) | \theta, \varphi \rangle \tag{AII.1c}$$

and will show that these functions can easily be calculated. These functions are generating functions since one has

$$\left[\left(\frac{\partial}{\partial \alpha} \right)^a \left(\frac{\partial}{\partial \beta} \right)^b \left(\frac{\partial}{\partial \gamma} \right)^c X_N \right]_{\alpha = \beta = \gamma = 0} = \langle \theta, \varphi | \, J_+^a J_z^b J_-^c \, | \theta, \varphi \rangle \tag{AII.2a}$$

$$\left[\left(\frac{\partial}{\partial \alpha} \right)^a \left(\frac{\partial}{\partial \beta} \right)^b \left(\frac{\partial}{\partial \gamma} \right)^c X_A \right]_{\alpha = \beta = \gamma = 0} = \langle \theta, \varphi | \, J_-^c J_z^b J_+^a \, | \theta, \varphi \rangle \tag{AII.2b}$$

$$\left[\left(\frac{\partial}{\partial \alpha} \right)^a \left(\frac{\partial}{\partial \beta} \right)^b \left(\frac{\partial}{\partial \gamma} \right)^c X_S \right]_{\alpha = \beta = \gamma = 0} = \langle \theta, \varphi | \, S\{J_+^a J_z^b J_-^c\} \, | \theta, \varphi \rangle \tag{AII.2c}$$

where $S\{ \ \}$ means the fully symmetrized sum of products, which is equal to the sum of all distinct permutations of the factors within the bracket divided by the number $(a+b+c)!/(a!b!c!)$ of these permutations.

It is sufficient to compute one of the generating functions, as the other two are then given by using the disentangling theorem of Appendix I. It is simplest to do this for X_A. Using (4.12) one has

$$X_A = \frac{1}{(1 + |\tau|^2)^{2J}} \langle -J | e^{(\tau^* + \gamma)J_-} e^{\beta J_z} e^{(\tau + \alpha)J_+} | -J \rangle \qquad \text{(AII.3)}$$

This expression is then put in normally ordered form (AI.2), in which case only the term $\exp[(\ln x_z)J_z]$ contributes to the expectation value in the ground state: $\langle -J | \exp[(\ln x_z)J_z] | -J \rangle = 1/x_z^J$. One obtains

$$\frac{1}{\sqrt{x_z}} = \frac{1}{\sqrt{y_z}} + y_+ y_- \sqrt{y_z} = e^{-\beta/2} + (\tau + \alpha)(\tau^* + \gamma) e^{\beta/2} \qquad \text{(AII.4)}$$

which gives immediately

$$X_A = \left(\frac{e^{-\beta/2} + e^{\beta/2}(\tau + \alpha)(\tau^* + \gamma)}{1 + |\tau|^2} \right)^{2J}$$

$$= \left(e^{-\beta/2} \cos^2(\theta/2) + e^{\beta/2} [\sin(\theta/2) e^{-i\varphi} + \alpha \cos(\theta/2)] \right.$$

$$\left. \times [\sin(\theta/2) e^{i\varphi} + \gamma \cos(\theta/2)] \right)^{2J} \qquad \text{(AII.5)}$$

REFERENCES

1.   J. Schwinger, Phys. Rev. **91**, 728 (1953).
2.   R. J. Glauber, Phys. Rev. **131**, 2766 (1963); R. J. Glauber, in Quantum Optics and Electronics, edited by C. De Witt et al. (Gordon and Breach, New York, 1965).
3.   F. T. Arecchi, E. Courtens, R. Gilmore, and H. Thomas, Phys. Rev. A **6**, 2211 (1972).
4.   R. H. Dicke, Phys. Rev. **93**, 99 (1954).
5.   F. T. Arecchi and E. Courtens, Phys. Rev. A **2**, 1730 (1970).
6.   F. Bloch, Phys. Rev. **70**, 460 (1946).
7.   F. T. Arecchi, G. Masserini, and P. Schwendimann, Riv. Nouvo Cimento **1**, 181 (1969).
8.   F. Bloch and A. Nordsieck, Phys. Rev. **52**, 54 (1937).
9.   A. Messiah, Mécanique Quantique (Dunod, Paris, 1959).

10.   H. Haken, in Encyclopedia of Physics edited by S. Flügge (Springer-
      Verlag, Berlin, 1970), Vol. XXV, Chap. 2.
11.   R. P. Feynman, Phys. Rev. 84, 108 (1951).
12.   R. J. Glauber in Physics of Quantum Electronics, edited by P. L.
      Kelley, B. Lax and P. E. Tannenwald (Columbia, New York, 1966);
      J. R. Klauder and E. C. G. Sudarshan, Fundamentals of Quantum
      Optics (Benjamin, New York, 1968).
13.   E. P. Wigner, Group Theory and Its Application to the Quantum
      Mechanics of Atomic Spectra (Academic, New York, 1959).
14.   E. Callen and H. Callen, Phys. Rev. 129, 578 (1963).

THEORY OF A GAS LASER AMPLIFIER FOR BROADBAND FIELDS

J. H. Parks
Department of Physics, University of Southern California
Los Angeles, California

ABSTRACT

This paper presents a semiclassical theory of the amplification of a broad-band optical-frequency field which propagates through a high-gain gas discharge. When the broadband field originates as spontaneous emission noise, the time dependence will characteristically include rapid amplitude and phase fluctuations. However, it is shown that a consistent perturbational treatment can be formulated when the Fourier components of the field are considered to be slowly varying quantities compared to the rate of change of the induced polarization. In this case the electromagnetic field and the polarization are represented as Fourier integrals and the propagation equations are derived for the amplitude and phase of the quasi-Fourier component at frequency ω. The induced polarization is calculated to third order in the field using the density matrix equations of motion for a gas of two-level atoms. The linear polarization involves each field component independently and leads to a gain-induced narrowing of the field bandwidth. As the field strength increases, the nonlinear interaction reduces the gain available for each field component and, in addition, introduces amplitude and phase coupling among these components. The extent of the interaction between field components is determined by the atomic radiative linewidth and the theory is developed for both inhomogeneously broadened and homogeneously broadened amplifying transitions. When the amplifying transition has a Doppler-broadened linewidth, computer solutions indicate that the power spectrum broadens in the presence of these saturation effects even when the phases are assumed to be randomly distributed. The amplitude coupling, which still remains in this approximation, reduces the saturation of individual broadband field components by diffusing the energy throughout the field bandwidth. An amplifier with a homogeneous collision-broadened linewidth continues to gain-narrow the field bandwidth in the saturation region, when the phases remain randomly distributed. In this case, however, the general expression for the nonlinear polarization including phase terms contains a frequency dependence related to saturation rebroadening. These nonlinear effects are also considered in the presence of two oppositely travelling broadband fields. Quantitative results are obtained using parameters for the 3.508-μ transition of Xe, and are shown to be consistent with linewidth studies of the high-gain Xe^{136} noise amplifier by Gamo and Chuang.

I. INTRODUCTION

This paper presents a theory [1,2] which describes the amplification and saturation behavior of a broadband electromagnetic field as it propagates through a high-gain gas amplifier. This case is of particular interest because an adequate theoretical description has not been developed which treats the nonlinear coupling among the components of a broadband field as it builds up from radiative noise. Previous analyses [3,4] of broadband amplification have concentrated on the variation of the field bandwidth with intensity and neglected the detailed behavior of these coupling processes. The present perturbational treatment of the induced polarization includes terms up to the third power of the field which allows the subtle coupling effects among field components to be considered analytically. This third-order theory is also a useful formulation in which to consider the random phase approximation of the field. The validity of ignoring the coupling between the phase and amplitude equations which results in this approximation is questionable for arbitrarily high field strengths, although this has been done in the previous treatments.

In this paper we will consider an optical-frequency amplifier characterized by an atomic linewidth which is inhomogeneously broadened by the Doppler effect. A homogeneously broadened amplifying transition is also examined. The frequency distribution of the electric field is represented by a Fourier integral and it is important to emphasize that in this treatment the total ensemble of broadband field components is interacting with the atoms at the same time. In a nonlinear medium this leads to results which are quite different from sweeping a monochromatic field through the atomic linewidth of the amplifier. In the latter case, the monochromatic fields cannot interact with the atoms simultaneously, so that each field component is constrained to propagate <u>independently</u>. In contrast, the nonlinear broadband field equations predict an interesting coupling among the amplitudes and phases of the Fourier components which changes the saturation behavior considerably.

The characteristics of the amplified field follow when the equations of motion for the induced polarization are combined with the electromagnetic field equations. A solution to the field equations is found for the important special case in which the phases are randomly distributed. Although the phase equations separate in this approximation, the coupling of Fourier amplitudes is still retained for a Doppler-broadened amplifying transition. When the initial field is chosen to characterize spontaneous emission in the amplifying trnasition, this model provides a qualitative description of the processes occurring in high-gain noise amplifiers, such as those reported [5] in Xe, N_2, and Pb vapor discharges. In the case of a homogeneously broadened amplifying transition, it has been possible to compare the characteristics of the general form for the induced polarization with solutions found in the random phase approximation.

II. THE ELECTROMAGNETIC FIELD EQUATIONS

We consider the amplification of a classical electromagnetic wave as it travels through a long cylindrical gas discharge which maintains a population inversion between two energy levels separated by $E = \hbar w_0$. These levels are radiatively coupled by the propagating electromagnetic field which is assumed to have a broad frequency spectrum centered about w_0. The macroscopic polarization induced in the medium by the field is included in Maxwell's equations and appears as a self-consistent source term in the wave equation

$$\frac{\partial^2}{\partial z^2} \vec{E} - \frac{1}{c^2} \frac{\partial^2}{\partial t^2} \vec{E} = \frac{4\pi}{c^2} \frac{\partial^2}{\partial t^2} \vec{P} \tag{1}$$

The field is assumed to be uniformly distributed over the cross section of the discharge tube, and $\vec{P}$ will be treated as a perturbation for which $\vec{\nabla} \cdot \vec{P} \approx 0$. Radiative losses which might arise from scattering have been neglected in Eq. (1) since we are considering amplifier lengths for which the saturated gain will still greatly exceed these losses. The solutions obtained in this case will not represent a steady state in the sense that the field energy is continually increasing, although at a slower rate in the presence of saturation. Clearly, if arbitrarily long amplifying lengths are of interest,

e. g., interstellar amplification [6], these damping terms are necessary to reach steady-state propagation.

To concentrate on the frequency characteristics of the field, we will treat propagation only along the $\hat{z}$ direction, which ignores the nonlinear coupling which may arise between waves travelling off axis. Here z is the coordinate measured along the axis of the discharge tube and the transverse optical field, of inital value E_0, is linearly polarized along the $\hat{x}$ direction. The field is specified by the integral form

$$\vec{E}(z,t) = \hat{x} \int_{-\infty}^{\infty} \frac{d\omega}{2\pi} \, E_\omega(z_0,t_0) \, \exp\{i[\omega(t-t_0)-k(z-z_0)]\} \tag{2}$$

representing a superposition of travelling waves of different frequencies ω. The complex field amplitude is defined by $E_\omega \equiv (E_0/2)a_\omega(z_0,t_0)\exp i\varphi_\omega(z_0,t_0)$ for $\omega > 0$, and $E_\omega^* = E_{-\omega}$ ensures a real field. The amplitudes a_ω and phases φ_ω are functions of the frequency ω, denoted by the subscript, and also depend on the space and time parameters (z_0,t_0). The macroscopic polarization can be similarly written

$$\vec{P}(z,t) = \hat{x} \int_{-\infty}^{\infty} \frac{d\omega}{2\pi} \, P_\omega(z_0,t_0) \, \exp\{i[\omega(t-t_0)-k(z-z_0)]\} \tag{3}$$

for an isotropic gas medium. In this complex notation we define $P_\omega \equiv \frac{1}{2}(p_\omega - ip_\omega')\exp(i\varphi_\omega)$ for $\omega > 0$ and $P_\omega^* = P_{-\omega}$, where p_ω and p_ω' are the in-phase and out-of-phase real polarization amplitudes.

It is important to point out that the field E(z,t) defined in Eq. (2) is independent of the parameters (z_0,t_0). This is rigorously satisfied when E_ω is given by $E_\omega(z_0,t_0) = E(\omega)\exp[i(\omega t_0-kz_0)]$ which is a solution to the equations $\partial E/\partial t_0 = 0$ and $\partial E/\partial z_0 = 0$. In this case the representation for E(z,t) becomes a Fourier integral in which the complex amplitude $E(\omega)$ is a function only of the frequency ω. Instead of rigorously removing the time dependence of the field and polarization via complete Fourier integrals as in Ref. (3), we choose to study the propagation behavior by deriving the equation of motion for each quasi-Fourier component $E_\omega(z_0,t_0)$. For this purpose, the representation [7] given by Eq. (2) suggests the propagation of field components between the points (z_0,t_0) and (z,t) under the assumption that the amplitude and phase can be specified by the initial values at (z_0,t_0). Although this description motivates the choice of this representation, the parameters (z_0,t_0) are not restricted to represent a physical space-time point in the sense that a field measured at (z,t) was at the position z_0 at an earlier time t_0.

When the induced polarization can be reliably calculated using third-order perturbation theory, it will be shown that to a good approximation we may consider E_ω and P_ω as slowly varying quantities within a certain time interval $(t-t_0)$. In this slowly varying amplitude and phase approximation (SVAP), $E_\omega(z_0,t_0)$ may be considered as a propagating Fourier transform which describes the evolution of the field spectrum as it travels through the amplifier. This treatment will indicate conditions under which a field at the position of measurement (z,t) may be accurately characterized by the amplitude E_ω at the time t_0. Although the field may contain rapid time fluctuations such as a spontaneous noise field, this formalism identifies the spectral component as the slowly varying physical entity.

Since $\vec{E}(z,t)$ does not depend on the parameters (z_0,t_0), it follows from the conditions $\partial E/\partial t_0 = 0$ and $\partial E/\partial z_0 = 0$ that the space and time derivatives of the field can be written as

$$\frac{\partial \vec{E}}{\partial z} = \hat{x} \int_{-\infty}^{\infty} \frac{d\omega}{2\pi} \frac{\partial E_\omega}{\partial z_0} \exp\{i[\omega(t-t_0)-k(z-z_0)]\} \tag{4a}$$

and

$$\frac{\partial \vec{E}}{\partial t} = \hat{x} \int_{-\infty}^{\infty} \frac{d\omega}{2\pi} \frac{\partial E_\omega}{\partial t_0} \exp\{i[\omega(t-t_0)-k(z-z_0)]\} \tag{4b}$$

Noting that the second space and time derivatives can operate on the exponential factor alone, the wave equation reduces to

$$\int_{-\infty}^{\infty} \frac{d\omega}{2\pi} \left(-ik\frac{\partial E_\omega}{\partial z_0} - i\frac{\omega}{c^2}\frac{\partial E_\omega}{\partial t_0} + 4\pi\frac{\omega^2}{c^2}P_\omega\right)\exp\{i[\omega(t-t_0)-k(z-z_0)]\} = 0 \tag{5}$$

The macroscopic polarization will be calculated in the following sections as a quantum-mechanical response of the atomic medium to the field $E(z,t)$. In this case $P_\omega(z_0,t_0)$ does not rigorously represent the Fourier component of this response but merely the amplitude of the polarization associated with the frequency ω in the SVAP approximation. The nonlinear contribution of P_ω will be seen to be independent of the time t only in this approximation. In this restricted sense, we apply the usual orthogonality relations to Eq. (5) to obtain a first-order partial differential equation for the field component

$$\frac{\partial E_\omega}{\partial t_0} + c\frac{\partial E_\omega}{\partial z_0} = -4\pi i\omega_0 P_\omega \tag{6}$$

In Eq. (6) the presence of dispersion has been neglected by taking $k = \omega/c$, and the frequency factors have been evaluated at ω_0, the atomic resonance.

For the field representation given by Eq. (2), the wave equation reduces to the first-order differential form which usually arises as the result of a slowly varying envelope approximation. To display the equations determining the amplitude a_w and the phase φ_w, we write

$$E_o \left(\frac{\partial a_w}{\partial t_o} + c \frac{\partial a_w}{\partial z_o} \right) = -8\pi w_o \, \text{Re} \left\{ i[\exp(-i\varphi_w) \, P_w] \right\} \tag{7a}$$

$$E_o a_w \left(\frac{\partial \varphi_w}{\partial t_o} + c \frac{\partial \varphi_w}{\partial z_o} \right) = -8\pi w_o \, \text{Im} \left\{ i[\exp(-i\varphi_w) \, P_w] \right\} \tag{7b}$$

III.   THE INDUCED POLARIZATION

A.   Integration of the Density Matrix Equations

The macroscopic polarization $P(z, t)$ (electric dipole moment density) is the resultant of an ensemble of atomic dipoles, $\vec{p} = e\vec{x}$, induced in the gas medium through the interaction with an electromagnetic field. This polarization is then considered to act as a self-consistent source for the field in accordance with Maxwell's equations. The following quantum-mechanical calculation of this polarization will utilize the density matrix formalism to derive an expression for the component, $P_w(z_o, t_o)$. This treatment follows the theory of Lamb [8] which was originally applied to the atomic response of a gas laser medium. Consider an ensemble of two-level atoms interacting with a travelling electromagnetic field. The eigenfunctions $|a\rangle$ and $|b\rangle$ of the unperturbed Hamiltonian, H_0, represent the upper and lower atomic levels respectively, and determine the atomic resonance frequency $w_o = (E_a - E_b)/\hbar$. In the presence of the field, the perturbed Hamiltonian, $H = H_o - \vec{p} \cdot \vec{E}(z, t)$, leads to the differential equations of motion for the elements of the density matrix operator, $\rho(z, t)$

$$\dot{\rho}_{ab} + (\gamma_{ab} + iw_o) \rho_{ab} = iV(z, t)(\rho_{aa} - \rho_{bb}) \tag{8a}$$

$$\dot{\rho}_{aa} + \gamma_a \rho_{aa} = iV(z, t)(\rho_{ab} - \rho_{ba}) + \Lambda_a \tag{8b}$$

$$\dot{\rho}_{bb} + \gamma_b \rho_{bb} = -iV(z, t)(\rho_{ab} - \rho_{ba}) + \Lambda_b \tag{8c}$$

$$\rho_{ba} = \rho_{ab}^* \tag{8d}$$

where $V(z,t) = -\mu E(z,t)/\hbar$ and $\mu = \langle a \,|ex|\, b \rangle$. The radiative and collision processes depopulate the two atomic levels at the rates γ_a and γ_b, respectively, and in the absence of the field, the polarization decays with the rate $\gamma_{ab} = \frac{1}{2}(\gamma_a + \gamma_b)$, referred to as the homogeneous radiative linewidth. When the effect of elastic collision processes [9] are included, we generally find $\gamma_{ab} \gtrsim (\gamma_a + \gamma_b)$. This can be taken into account by introducing γ_{ab} as an independent parameter; however we will not emphasize the treatment of atomic collisions in this paper. In these equations, the terms Λ_a and Λ_b are the probabilities per unit time per unit volume that an atom is excited into states $|a\rangle$ and $|b\rangle$ by some external mechanism responsible for sustaining the population inversion. These pumping rates will be considered to vary slowly over times comparable to the atomic lifetimes.

The formal solutions to these first order differential equations can be combined to yield an expression proportional to the polarization

$$(\rho_{ab} + \rho_{ba}) = i \int_{-\infty}^{t} dt' \exp[-(i\omega_0 + \gamma_{ab})(t-t')]V(t')\{\rho_{aa} - \rho_{bb}\}_{t'} + c.c. \tag{9}$$

and the population inversion

$$(\rho_{aa} - \rho_{bb}) = \left(\frac{\Lambda_a}{\gamma_a} - \frac{\Lambda_b}{\gamma_b}\right) - \int_{-\infty}^{t} dt' \int_{-\infty}^{t'} dt'' \{\exp[-\gamma_a(t-t')] + \exp[-\gamma_b(t-t')]\} V(t')V(t'')$$

$$\times \{\rho_{aa} - \rho_{bb}\}_{t''} \{\exp[-(i\omega_0 + \gamma_{ab})(t'-t'')] + c.c.\} \tag{10}$$

An expansion of the polarization in powers of $E(t)$ is obtained by iterating the formal solution for $(\rho_{aa} - \rho_{bb})$ in Eq. (10). When $(\mu E/\hbar)^2 \ll 1$, we can safely truncate the expansion at the second-order term $\sim E^2$. In this case, the population inversion has changed only slightly from its initial value and we have

$$\left(\rho_{aa} - \rho_{bb}\right)_t = \left(\frac{\Lambda_a}{\gamma_a} - \frac{\Lambda_b}{\gamma_b}\right) - \left(\frac{\Lambda_a}{\gamma_a} - \frac{\Lambda_b}{\gamma_b}\right) \int_{-\infty}^{t} dt' \int_{-\infty}^{t'} dt'' \, V(t')V(t'')$$

$$\times \{\exp[-\gamma_a(t - t')] + \exp[-\gamma_b(t - t')]\} \{\exp[-(i\omega_0 + \gamma_{ab})(t' - t'')] + c.c.\} \tag{11}$$

Finally, using Eqs. (3) and (9), we extract the polarization component at frequency ω

$$P_w(z_o, t_o) = \frac{iN\mu^2}{\hbar} E_w(z_o, t_o) \int_{-\infty}^{t} dt' (\rho_{aa} - \rho_{bb})_{t'}$$

$$\times (\exp\{[i(w_o - w) - \gamma_{ab}](t-t')\} - \exp\{[-i(w_o + w) - \gamma_{ab}](t - t')\}) \tag{12}$$

Since $P_{-w} = P_w^*$, we need only consider the polarization for $w > 0$ and the first exponential term in Eq. (12) containes the important resonance behavior for positive w. In the following derivation the spatial parameter z_o is suppressed to simplify the notation.

B. The Polarization Including Atomic Motion

To describe a Doppler-broadened amplifying transition, we consider the polarization of a particular atomic velocity ensemble. The motion of an atom moving with velocity component v in the z direction, parallel to the wave vector $\vec{k}$, can be included by considering the field to interact with the atom at points (z, t) described by a classical trajectory [8]. If an atom is observed at the point (z, t) after following the classical path $z = z_i + v(t - t_i)$, the field which interacted with this atom at some earlier point (z', t') can be expressed as $E(z', t') = E(z_i + v(t' - t_i, t'))$. Inserting $z_i = z - v(t - t_i)$, we have $E(z', t') = E(z - v(t - t'), t')$ which is independent of the time t_i when the atom was initially excited.

The component, $P_{wv}(t_o)$, for this atomic ensemble follows from Eq. (12) by transforming $E(z', t')$ as described above:

$$P_{wv}(t_o) = \frac{i\mu^2}{\hbar} E_w(t_o) \int_{-\infty}^{t} dt' (\rho_{aa} - \rho_{bb})_{t'v}$$

$$\times \left(\exp\{[i(w_o - w + kv) - \gamma_{ab}](t - t')\} - \exp\{[-i(w_o + w - kv) - \gamma_{ab}](t - t')\} \right) \tag{13}$$

The sum of these contributions from all velocity ensembles gives the component of the macroscopic polarization

$$P_w(t_o) = \int_{-\infty}^{\infty} P_{wv}(t_o) \, dN_v \tag{14}$$

where $dN_v = NW(v) \, dv$ is the number of atoms having a velocity between v and $v + dv$. We let $W(v)$ describe a Maxwell–Boltzmann velocity distribution,

$W(v) = (1/u\sqrt{\pi})\exp[-(v/u)^2]$, in which $u = \sqrt{2}\,v_{rms}$ is the most probable speed.

C. Linear and Nonlinear Polarization

The linear polarization follows by taking $(\rho_{aa} - \rho_{bb})_{t',v} = (\Lambda_a/\gamma_a - \Lambda_b/\gamma_b)$ in Eq. (13) to obtain

$$P_\omega^L = i\mu n_0 \frac{\mu E_\omega}{\hbar} \int_{-\infty}^{\infty} \frac{W(v)\,dv}{\gamma_{ab} + i(\omega - \omega_0 - kv)} = i\mu n_0 \frac{\mu E_\omega}{\hbar} \frac{1}{ku} H(x,\eta) \qquad (15)$$

in which the initial population inversion density is given by $n_0 = (n_a - n_b) = N(\Lambda_a/\gamma_a - \Lambda_b/\gamma_b)$. The velocity integral has been expressed in terms of $H(x,\eta)$, a complex function of $\varphi = x + i\eta = [(\omega - \omega_0)/ku] + i\gamma_{ab}/ku$, whose properties are summarized in Ref. (10).

As the electric field is amplified by the high-gain medium, it is necessary to account for the decrease in population inversion by considering subsequent iterations for $(\rho_{aa} - \rho_{bb})$. In this perturbative treatment, we retain only the second iteration expressed by Eq. (11) and write the resonant contribution of $(\rho_{aa} - \rho_{bb})_{t',v}$ with the change of variables $\tau' = t' - t'$, $\tau'' = t'' - t'''$

$$\left(\rho_{aa} - \rho_{bb}\right)_{t',v} = \left(\frac{\Lambda_a}{\gamma_a} - \frac{\Lambda_b}{\gamma_b}\right) - \left(\frac{\Lambda_a}{\gamma_a} - \frac{\Lambda_b}{\gamma_b}\right)\left(\frac{\mu}{\hbar}\right)^2$$

$$\times \int_0^\infty \frac{d\omega'}{2\pi} \int_0^\infty \frac{d\omega''}{2\pi} \left\{ \left[E_{\omega'} E_{\omega''}^* \exp\{i[(\omega' - \omega'')(t' - t_0) - (k' - k'')(z - z_0)]\} \right. \right.$$

$$\times \int_0^\infty d\tau' \int_0^\infty d\tau'' \left(\exp\{-[i(\omega' - \omega'') + \gamma_a]\tau'\} + \exp\{-[i(\omega' - \omega'') + \gamma_b]\tau'\} \right)$$

$$\times \exp\{-[(\omega_0 - \omega'' + kv) + \gamma_{ab}]\tau''\} \Big] + c.c. \Big\} \qquad (16)$$

in which we have taken $k'v \approx k''v \approx kv$. In the presence of a broadband field, the saturation of the population inversion occurs through the interaction of an atom with those fields $E_{\omega''}$ having frequencies lying within the homogeneous width, γ_{ab}. We also observe that $(\rho_{aa} - \rho_{bb})$ is modulated at the difference frequency $(\omega' - \omega'')$ by the nonlinear response of the atomic medium to the field components $E_{\omega'}$ and $E_{\omega''}$. This population modulation is neglected in a rate equation approximation [11], and in such a treatment the amplitude and phase coupling cannot be accurately preserved in the nonlinear contribution. In fact a derivation of the nonlinear response based on the rate equation approximation would result in equations similar to those presented in Sec. IV which follow from a random phase representation for the broadband field. The modulation of ρ_{aa} and ρ_{bb}, sometimes referred to as population pulsations, has also been shown to be important in formulating gas laser theories [8, 12].

Using the field-dependent part of $(\rho_{aa} - \rho_{bb})_{t'v'}$, we obtain the following expression for the <u>nonlinear</u> contribution to the polarization

$$P_\omega^{NL}(z_0, t_0) = -i\mu n_0 \frac{\mu E_\omega(z_0, t_0)}{\hbar} \left(\frac{\mu}{\hbar}\right)^2 \int_0^\infty \frac{d\omega'}{2\pi} \int_0^\infty \frac{d\omega''}{2\pi} \left(\frac{1}{ku}\right)^2$$

$$\times \left\{\left[E_{\omega'}E_{\omega''}^* \exp\{i[(\omega'-\omega'')(t-t_0) - (k'-k'')(z-z_0)]\}\left(\frac{1}{\gamma_a + i(\omega'-\omega'')} + \frac{1}{\gamma_b + i(\omega'-\omega'')}\right)\right.\right.$$

$$\times \frac{1}{2\eta + i(x+x'-2z'')} \left(H(x+x'-x'', \eta) + H^*(x'', \eta)\right)\Bigg]$$

$$+ \left[E_{\omega'}^* E_{\omega''} \exp\{-i[(\omega'-\omega'')(t-t_0) - (k'-k'')(z-z_0)]\}\left(\frac{1}{\gamma_a - i(\omega'-\omega'')} + \frac{1}{\gamma_b - i(\omega'-\omega'')}\right)\right.$$

$$\left.\left.\times \frac{i}{(x-x')}\left(H(x-x'+x'', \eta) - H(x'', \eta)\right)\right]\right\} \tag{17}$$

where $x = (\omega - \omega_0)/ku$ and $\eta = \gamma_{ab}/ku$. Although the inversion $(\rho_{aa} - \rho_{bb})$ given by Eq. (16) exhibits a nonlinear dependence on field components within the interval γ_{ab}, observe that the nonlinear polarization couples fields within an interval $\sim 2\gamma_{ab}$.

As discussed briefly in Sec. II, the polarization component P_ω does not represent the Fourier transform of $P(z, t)$ in the nonlinear region. It is clear from Eq. (17) that the polarization component P_ω^{NL} depends explicitly on the time interval $(t-t_0)$ and as a result will produce sidebands on the field component E_ω appearing at $\omega \pm (\omega'-\omega'')$. In this sense E_ω does not uniquely determine the energy of the spectral component of the field $E(z, t)$ at frequency ω. This ambiguity can be removed by restricting the interval $(t-t_0)$ so that within this time the exponential factor is slowly varying and can be approximated by taking $\exp[i(\omega'-\omega'')(t-t_0)] \approx 1$. The complex Lorentzian factors in Eq. (17) involving γ_a and γ_b indicate that sizeable contributions to P_ω^{NL} occur when the frequencies ω' and ω'' are within $(\omega' - \omega'') \approx \gamma_a$ or γ_b. Since these decay rates are comparable to the homogeneous linewidth γ_{ab}, we find that this approximation for the exponential is is valid for times $(t - t_0) \ll (\gamma_{ab})^{-1}$. The SVAP approximation is then appropriate to describe the field component E_ω when it is varying slowly compared to the rate γ_{ab}. These conditions are necessary to derive an internally consistent treatment of broadband amplification, and are generally required in perturbative calculations of the polarization to obtain a slowly varying field solution.

The nonlinear polarization does not exhibit an explicit dependence on the time $(t - t_0)$ in this approximation so that Eq. (5) is rigorously satisfied. In this case, the time parameter is not essential to the treatment and we remove the implicit dependence on t_0 by taking $z_0 = ct_0$ and evaluate E_ω at $z = z_0$. The polarization can be expressed in this SVAP approximation by

$$
P_\omega^{NL}(z) = -i\mu n_0 \frac{\mu E_\omega(z)}{\hbar} \left(\frac{\mu}{\hbar}\right)^2 \int_0^\infty \frac{d\omega'}{2\pi} \int_0^\infty \frac{d\omega''}{2\pi} \left(\frac{1}{ku}\right)^2
$$

$$
\times \Bigg\{ \Bigg[E_{\omega'} E_{\omega''}^* \left(\frac{1}{\gamma_a + i(\omega' - \omega'')} + \frac{1}{\gamma_b + i(\omega' - \omega'')} \right)
$$

$$
\times \frac{1}{2\eta + i(x + x' - 2x'')} \left(H(x + x' - x'', \eta) + H^*(x'', \eta) \right) \Bigg]
$$

$$
+ \Bigg[E_{\omega'}^* E_{\omega''} \left(\frac{1}{\gamma_a - i(\omega' - \omega'')} + \frac{1}{\gamma_b - i(\omega' - \omega'')} \right)
$$

$$
\times \frac{1}{(x - x')} \left(H(x - x' + x'', \eta) - H(x'', \eta) \right) \Bigg] \Bigg\} \tag{18}
$$

The nonlinear polarization still retains the effects of population modulation and these Raman or parametric terms may be considered to be included to the lowest order in the SVAP approximation. However, the third-order polarization provides an insight into the initial processes which couple the field components E_ω and these effects will be considered in Sec. IV.

IV. BROADBAND NOISE AMPLIFIER

The propagation of the field component E_ω is described in the SVAP approximation by the equation

$$\frac{dE_\omega}{dz} = -2\pi i \frac{\omega_o}{c} \left\{ P_\omega^L + P_\omega^{NL} \right\} \tag{19}$$

The coupling between amplitudes and phases introduced by P_ω^{NL} suggests the possibility that the SVAP formalism might be applied to study the spectral energy distribution of the amplified field and also the development of phase coherence among the components. We shall limit the scope of the present work to a consideration of the power spectrum when the phases are randomly distributed, considering both inhomogeneously and homogeneously broadened amplifying transitions.

In this section we obtain the nonlinear equations for the amplification of a random field by assuming the phases remain randomly distributed over the interval $(0, 2\pi)$. This important limiting case is a useful model for the noise amplifier, in which the initial broadband field originates as spontaneous emission and is rapidly amplified as it propagates through the active medium. In this approximation, the field equations follow more directly by using the series representation for the field

$$E(t) = E_o \sum_{n=1}^{N} c_n \cos(\omega_n t - k_n z + \varphi_n) \tag{20}$$

to calculate the polarization and then taking the appropriate limit to obtain these equations for a continuum frequency spectrum.

A. Power Spectrum for Inhomogeneous Broadening

In the discrete formalism each set of phases $\{\varphi_n\}$ represents a particular ensemble of field components and an average over many such ensembles reduces the contribution of the nonlinear polarization to a single sum over a discrete frequency variable. This follows directly from the assumption of randomly distributed phases since in this case the average of the phase factor $\exp[i(\varphi_{\omega'} - \varphi_{\omega''})]$ results in a Kronecker delta, $\langle \exp[i(\varphi_p - \varphi_m)] \rangle = \delta_{pm}$. The appropriate quantity to represent by an ensemble average is the increased power in the amplified field given by

$$\langle i\omega_o E_n^* P_n \rangle_\varphi = -\frac{\mu^2 n_o}{4\hbar} \frac{\omega_o}{ku} E_o^2 c_n^2 H(x_n, \eta)$$

$$+ \mu n_o \omega_o \frac{\mu E_o^2}{4\hbar} c_n^2 \left(\frac{\mu E_o}{2\hbar}\right)^2 \left(\frac{1}{ku}\right)^2 \left(\frac{\gamma_a + \gamma_b}{\gamma_a \gamma_b}\right) \quad \text{(cont'd)}$$

$$\times \sum_{p=0}^{N} c_p^2 \left(\frac{1}{2\eta + i(x_n - x_p)} \left[H(x_n, \eta) + H^*(x_p, \eta) \right] \right.$$

$$\left. + \frac{i}{x_n - x_p} \left[H(x_n, \eta) - H(x_p, \eta) \right] \right) \tag{21}$$

where $E_n = \frac{1}{2} E_0 c_n \exp(i\varphi_n)$. The square of the amplitude, c_n^2, is related to the power spectrum component u_n at frequency ω_n by $c_n^2 = 2u_n \Delta\nu$ and $E_0^2 u_n$ has the units of power per unit frequency.

In the limit $\Delta\nu \to 0$ and $N \to \infty$, the sum in Eq. (21) approaches an integral and we derive the following equation for the power spectrum

$$\frac{d}{dz} E_0^2 u_\omega \, d\omega = -\frac{8\pi}{c} \, \text{Re}\left(\lim_{\Delta\nu, N} \langle i\omega_0 E_n^* P_n \rangle_\varphi \right) \tag{22}$$

where

$$\lim_{\Delta\nu, N} \langle i\omega_0 E_n^* P_n \rangle_\varphi \to -\frac{\mu^2 n_0}{2\hbar} \frac{\omega_0}{ku} E_0^2 u_\omega \, d\omega H(x, \eta)$$

$$+ \frac{\sqrt{\pi}\,\mu^2 n_0}{2\hbar ku} \frac{\omega_0}{\hbar} \left(\frac{\mu E_0}{\hbar} \right)^2 \frac{1}{\gamma_a \gamma_b} E_0^2 u_\omega \, d\omega$$

$$\times \int_0^\infty d\omega' \, u_{\omega'} \frac{1}{2\sqrt{\pi}} \left(\frac{2\eta}{2\eta + i(x-x')} \left[H(x, \eta) + H^*(x', \eta) \right] \right.$$

$$\left. + \frac{i2\eta}{x - x'} \left[H(x, \eta) - H(x', \eta) \right] \right) \tag{23}$$

The equation for the power spectrum of the amplified field can then be reexpressed in the form

$$\frac{du_x}{dz} = g_0 u_x \exp(-x^2) - g_0 u_x S \int_{-\infty}^{\infty} dx' \, u_{x'} A_{xx'} \tag{24}$$

where g_0 is the linear gain and S is the saturation parameter given by

$$g_0 = \frac{4\pi^{3/2} \mu^2 n_0}{\hbar cku}, \quad S = \left(\frac{\mu E_0}{\hbar} \right)^2 \frac{1}{\gamma_a \gamma_b} \tag{25}$$

The real and imaginary parts of $H(x, \eta) \equiv H_R - iH_I$ are approximated [10] in the Doppler limit ($\eta \ll 1$) and used to define the field coupling function $A_{xx'}$ given by

$$A_{xx'} = \frac{1}{2\sqrt{\pi}} \frac{(2\eta)^2}{(2\eta)^2 + (x - x')^2}$$

$$\times \left(H_R(x, \eta) + H_R(x', \eta) + \frac{2\eta}{x' - x} [H_I(x', \eta) - H_I(x, \eta)] \right) \qquad (26)$$

The integration variable in Eq. (24) has been transformed to the normalized frequency $x' = (\omega' - \omega_0)/ku$ by noting $u_{x'} dx' = u_{\omega'} d\omega'$; and the lower integration limit extended from $x' = -\omega_0/ku \to -\infty$ without appreciable error.

u_ω Solution for Linear Polarization. In the linear regime ($S \to 0$) the amplification of each component of a broadband field occurs independently in the sense that P_ω^L depends only on the field component at frequency ω. In this limit, Eq. (24) is identical to the field equation describing the amplification of a monochromatic field at frequency ω, and for this reason we will refer to the monochromatic field treatment as the independent field model when comparing it with the broadband field results. When $S = 0$, we find a solution for the power spectrum

$$u_\omega(z) = u_\omega^o \exp\left\{ g_o z \exp\left[-\left(\frac{\omega - \omega_o}{ku} \right)^2 \right] \right\} \qquad (27)$$

which exhibits an exponential growth of the input power spectrum u_ω^o governed by the linear incremental gain coefficient g_o. When $g_o z$ is sufficiently large, the exponential in Eq. (27) cannot be expanded and in this case the amplification factor has a frequency width narrower than the Doppler width $\Delta\omega_D = ku \sqrt{\ln 2}$. For example, if the input spectrum is flat, $u_\omega^o = u_o$, the half-width of the amplified radiation which has propagated a distance L is given by

$$\Delta\omega_{1/2} = \Delta\omega_D \left[\ln\left(\frac{g_o L}{g_o L - \ln 2} \right) \left(\ln 2 \right)^{-1} \right]^{1/2}$$

$$\simeq \Delta\omega_D / \sqrt{g_o L} \quad \text{for } g_o L \gg 1 \qquad (28)$$

The linewidth of radiation emitted from an N_2 noise amplifier 100 cm in length has been measured [13] to be a factor of 3 to 5 narrower than the Doppler width. This spectral narrowing effect in the presence of a linear polarization has been considered previously [14] and has also been observed in several other high-gain amplifiers [15] having unsaturated exponential gains in the range $g_0 \approx 0.1$ cm^{-1} to 1.5 cm^{-1}. The spectral narrowing which results for the input $u_\omega^0 = 1/\sqrt{\pi}\, \exp\{-[(\omega-\omega_0)/ku]^2\}$ is shown in Fig. 2 as the curve for S = 0.

$\underline{u_\omega \text{ Solution for a Saturated Amplifier.}}$ When the atomic medium is being saturated by the intense amplified field, the nonlinear contribution in Eq. (24) must be included. The principal features of the nonlinear coupling term are given by the distribution of $A_{xx'}$ which is shown in Fig. 1 for $\eta = 0.01$. We note the narrow resonance of width 4η about each value $x = x'$ corresponding to a frequency interval $\sim 4\gamma_{ab}$ within which the field components are strongly

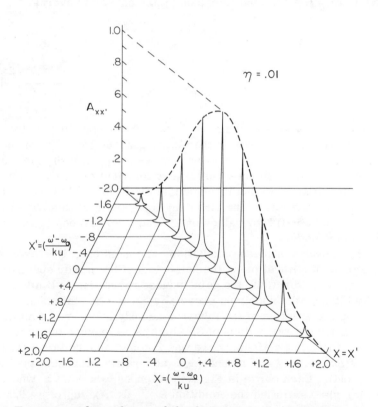

Figure 1. Frequency dependence of the function $A_{xx'}$. The dashed curve shows the distribution of diagonal terms, $x = x'$. Each peak represents a resonance region of $A_{xx'}$ within which the Fourier components are strongly coupled. This resonance width is $\delta x \approx 4\eta = 4\gamma_{ab}/ku$ and these curves have been calculated for $\eta = 0.01$. A finite set of peaks have been shown for clarity, but $A_{xx'}$ includes a continuum of these resonances distributed along the diagonal.

coupled. The relative strength of each nonlinear contribution is determined by the diagonal values of $A_{xx'}$ given by $A_{xx} = \exp(-x^2)$. We may interpret the integral in Eq. (24) as representing a diffusion of field energy at frequency ω to nearby components at ω' which will tend to reduce the saturation of the field E_ω. Considering the set of equations for all the components, the net diffusion of field energy will extend toward the wings of the Doppler-broadened linewidth. Aside from this subtle coupling behavior, the general feature of the nonlinear term results from a depletion of the inversion in each atomic velocity ensemble similar to the case of a monochromatic field.

Computer solutions which are appropriate to describe a saturated noise amplifier have been obtained using IBM 360 scientific subroutines. The initial noise field is characterized by the ensemble averages

$$\langle E_\omega^o \rangle_\varphi = 0$$

and

$$\langle (E_\omega^o)^2 \rangle_\varphi = E_o^2 u_\omega^o = E_o^2 \; \frac{1}{ku\sqrt{\pi}} \; \exp\left[-\left(\frac{\omega - \omega_o}{ku}\right)^2\right] \tag{29}$$

in which the power spectrum, u_ω^o, at $z = 0$ has been chosen to represent spontaneous emission within the inhomogeneous linewidth of the amplifying transition. This initial condition ignores the spatial distribution of spontaneous emission and assumes the largest contribution in the output of a long cylindrical high-gain amplifier originates from fields generated in a small volume near one end. The classical description of this initial noise field is assumed to be valid for values of E_o^2 which correspond to an average of one photon per mode.

The results of these calculations are most clearly presented by noting the effects of saturation on the half-width of the power spectrum. In the presence of saturation, the gain-induced narrowing is limited to a minimum half-width and the spectrum eventually becomes rebroadened. These principal characteristics of the inhomogeneously broadened amplifier are shown in Fig. 2 for several values of S, which is a measure of the input power level for constant η. Particular care was taken to assure that a frequency grid size of $\Delta x = 0.01$ was fine enough to accurately represent the integral term in Eq. (24). Each curve in Fig. 2 is extended to a point z' where the saturation for $x = 0$ satisfies the condition $S \int_{-\infty}^{\infty} dx' \, A_{0x'} u_{x'}(z') = 0.5$, which serves as a reasonable limit for the perturbation calculation.

As η increases, the region over which field components are strongly coupled will expand and change the contribution of the saturation integral. Since $\gamma_{ab} = (\gamma_a + \gamma_b)/2$, the parameters S and η cannot be specified independently and the most useful independent parameters for u_x are the initial power, $P_o = (c/8\pi)E_o^2 A$, and η. In Figs. 3 and 4 we use values of μ, γ_a and γ_b given below for the high-gain Xe transition and arbitrarily take $A = 1 \text{ cm}^2$

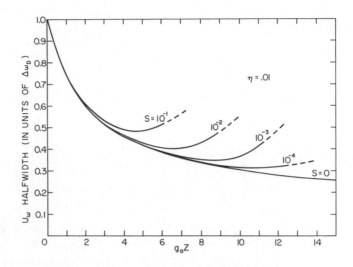

Figure 2.   Half-width of the power spectrum as a function of amplifier length. The normalized half-width $\Delta\omega_h/\Delta\omega_D$ at the length z is obtained from the solutions to Eq. (24) by calculating the values x_h for which $u_x(z) = 0.5u_0$ and taking $\Delta\omega_h/\Delta\omega_D = x_h/\sqrt{\ln 2}$.   The length is scaled by the unsaturated gain g_0. Curves are shown for several values of the saturation parameter S, determined by the input power level when the linewidth parameter is held constant, $\eta = 0.01$.   The dashed portion of each curve roughly indicates the limit of the perturbation calculation.   The linear solution (S = 0) given by Eq. (27) is shown for comparison.

in order to calculate S for several values of P_0.   A variation of the linewidth parameter, η, implies a change in γ_{ab} resulting from similar changes in both γ_a and γ_b.   For example, this can occur at high pressures when collision effects dominate the relaxation rates, and in this case $\eta \sim p$ and $S \sim 1/p^2$ for a pressure p.   Note that a change of the parameter S accompanies either a variation in η or P_0.   In Fig. 3 we show the half-width saturation for three sets of parameters (η, P_0).   These have been chosen to compare the saturation behavior resulting from an increase in η with a case in which P_0 decreases but the parameter S remains unchanged.   We note that the saturation effects are reduced more effectively by changing the power from 10^{-8} to 10^{-9} W than by increasing η by a factor of 3.   This is expected if we interpret the frequency coupling as a mechanism which diffuses the increasing field energy throughout the field bandwidth.   An increase in η extends this diffusion and reduces the broadening of the power spectrum; however, when P_0 is decreased, the saturation is reduced more

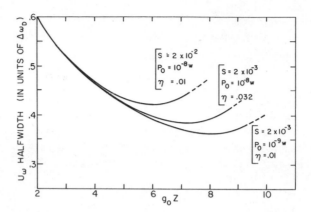

Figure 3. Half-width of the power spectrum versus $g_0 z$ for various sets of parameters (P_0, η). These sets were chosen to compare the change in saturation behavior of solutions to Eq. (24) when (a) η increases from 0.01 to 0.032 for fixed $P_0 = 10^{-9}$ W, and (b) P_0 decreases from 10^{-8} to 10^{-9} W for fixed $\eta = 0.01$. The change in S for both cases (a) and (b) is the same.

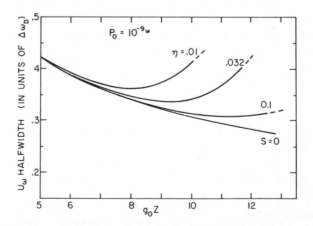

Figure 4. Half-width of the power spectrum versus $g_0 z$ is shown for solutions to Eq. (24) using several values of the linewidth parameter η. The input power level is held constant, $P_0 = 10^{-9}$ W, and the value of the parameter S is calculated as described in Sec. IV A. The linear solution (S = 0) given by Eq. (27) is shown for comparison.
884

effectively by starting with less energy in each field component in addition to the nonlinear coupling effects. In Fig. 4 the saturation of the power spectrum is shown for different η, holding P_0 constant.

The N_2[1] and Xe noise amplifiers have shown direct evidence of saturation broadening and a detailed study of the 3.508 μ high-gain transition in Xe, $5d[7/2]_3^o \to 6p[5/2]_2$ by Gamo and Chuang [15], provides an opportunity to compare the broadband amplifier theory with experimental data. Gamo has measured the linewidth of 3.508 μ radiation emitted from a He-Xe^{136} mixture as a function of discharge length using a high-resolution scanning interferometer. The input power parameter, P_0, is calculated from Gamo's output power measurements for several amplifier lengths which exhibit unsaturated exponential gain ($g_0 = 0.177$ cm^{-1}). Assuming this input originates as spontaneous emission near one end of the discharge region, we find $P_0 \approx 2 \times 10^{-9}$ W and use this value for the computer solution. Previous theoretical calculations provide the Xe transition matrix element [16], $\mu^2 = 73(ea_0)^2$, and the radiative lifetimes [17] $\tau_a = 1.35$ μsec, $\tau_b = 44$ nsec. Taking the full Doppler width to be 117 MHz, we find the linewidth parameter $\eta = 0.032$. The linewidth measurements were made for a Xe/He mixture at partial pressures of 75 μ Xe/0.8 mm He, however the effects of collision broadening on the atomic linewidth are not included in this calculation. In Fig. 5 the half-width variation with $g_0 z$ is shown for these parameters by the solid curve and Gamo's measurements in the saturation region are indicated by the data points. Clearly this model is not precise enough to account for the details of the experimental results. In addition to the pressure broadening of γ_{ab}, a spatially distributed input field and the presence of a saturating field traveling in the opposite direction will all tend to decrease the power spectrum broadening in the direction indicated by the data.

It is important to point out the improvement of the broadband theory over an independent field model shown by the dashed curve in Fig. 5. The independent field spectrum is given by

$$u_x = u_x^o \frac{\exp\left[g_o z \exp(-x^2)\right]}{1 + u_x^o S\left\{\exp\left[g_o z \exp(-x^2)\right] - 1\right\}} \tag{30}$$

for $u_x^o S \left\{\exp[g_o z \exp(-x^2)] - 1\right\} \ll 1$, and describes the result of sweeping a monochromatic field through the linewidth of the amplifying transition. The bandwidth of an ensemble of amplified monochromatic fields measured in this way will not show any effects of frequency coupling and in this sense represents an ensemble of independent field components. In Fig. 5 we observe that the half-width for the independent fields saturates at a smaller value of $g_o z$ than the broadband solution, which is to be expected since the interaction

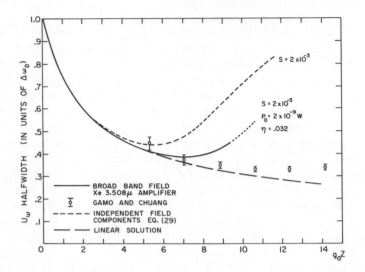

Figure 5.   Half-width of the power spectrum versus $g_0 z$ for the xenon 3.508-μ amplifier parameters.   Calculated values of $\tau_a = 1.35$ μsec and $\tau_b = 44$ nsec from Ref. 17 and $\mu^2 = 73(ea_0)^2$ from Ref. 16 were used to obtain $\eta = 0.032$ and $S = 2 \times 10^{-3}$ for $P_0 = 10^{-9}$ W.   The solution to Eq. (24) using these parameters is given by the solid curve.   Gamo's measurements of the radiative linewidth are indicated by the data points.   The dashed curve represents the half-width of an ensemble of independent field components described by Eq. (30) for $S = 2 \times 10^{-3}$.   The linear solution ($S = 0$) given by Eq. (27) is shown to emphasize the slow rebroadening of the experimental linewidth.

between different components in the broadband field slows down the amplification and thus the saturation of any single component.   The saturation of individual spectral components shown in Fig. 6 emphasizes this effect.   Note that the nonlinear behavior of the frequency component at $\omega = \omega_0 + \Delta\omega_D$ is quite small, but still shows a distinct difference between the monochromatic and broadband solutions which indicates the sensitivity of the coupling process.

B.   Power Spectrum for Homogeneous Broadening

The amplification of a broadband field is fundamentally different when the amplifying transition is homogeneously broadened, for example, by atomic

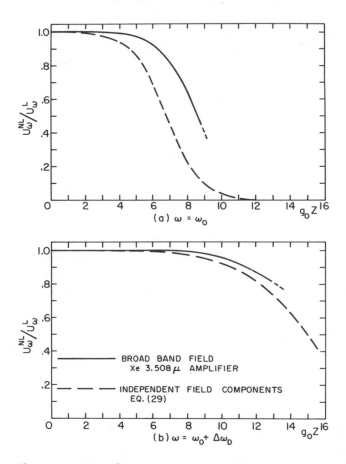

Figure 6.   The saturation of Fourier components at (a) $\omega = \omega_0$ and (b) $\omega = \omega_0 + \Delta\omega_D$ versus $g_0 z$.   The ratio of the power spectrum component u_ω^{NL}, calculated from Eq. (24), and the linear solution u_ω^{L}, from Eq. (27), shows the effects of saturation in the solid curves.   This ratio is obtained for an ensemble of independent fields from Eq. (30) indicated by the dashed curve.

collisions at high pressures.   In this section we examine the results of a random phase approximation, and also consider the implications for a case in which we do not impose this constraint on the field.   Note that all the amplified field components will now lie <u>within</u> the linewidth γ_{ab}, so that the results in this section can only be considered in the SVAP approximation.   Since the nonlinear behavior will become important after the field bandwidth has been gain narrowed to less than γ_{ab}, the major contributions to P_ω^{NL} will occur for $E_{\omega'}$ and $E_{\omega''}$ near ω_0 such that $\omega' - \omega'' \ll \gamma_{ab}$. This indicates that the SVAP approximation provides a consistent description of the nonlinear polarization which is obtained from Eq. (14) by taking $W(v) = \delta(v)$.

When the phases of the field components are assumed to be randomly distributed, the following equation is derived in the discrete representation

$$
\mathrm{Re}\left(\lim_{\Delta\nu,\,N}\langle i\omega_0 E_n^* P_n\rangle_\varphi\right) \rightarrow -\frac{\mu^2 n_0}{2\hbar}\frac{\omega_0}{\gamma_{ab}}E_0^2 u_\omega\,d\omega\left[\frac{\gamma_{ab}^2}{\gamma_{ab}^2 + (\omega - \omega_0)^2}\right.
$$

$$
\left. +\left(\frac{\mu E_0}{\hbar}\right)^2\frac{1}{\gamma_a\gamma_b}\frac{\gamma_{ab}^2}{\gamma_{ab}^2+(\omega-\omega_0)^2}\int_0^\infty d\omega'\,2u_{\omega'}\frac{\gamma_{ab}^2}{\gamma_{ab}^2+(\omega'-\omega_0)^2}\right] \tag{31}
$$

The nonlinear contribution has been reduced to a single frequency integral similar to the case of inhomogeneous broadening; however, the integral is now <u>independent</u> of ω. This indicates that for homogeneous broadening, the random phase approximation eliminates amplitude coupling as well as phase coupling. The equation for the power spectrum follows from Eqs. (22) and (31).

$$
\frac{du_x}{dz} = g_0\frac{1}{1+x^2}u_x\left(1 - S\int_{-\infty}^\infty dx'\,u_{x'}\frac{2}{1+x'^2}\right) \tag{32}
$$

in which

$$
g_0 = \frac{4\pi\omega_0}{c}\frac{\mu^2}{\hbar}\frac{n_0}{\gamma_{ab}}\quad cm^{-1} \tag{33}
$$

The saturation parameter, S, is expressed by Eq. (25) and the normalized frequency is now $x = (\omega - \omega_0)/\gamma_{ab}$. The computer solutions to Eq. (32) result in the curves shown in Fig. 7 which give the power spectrum half-width versus g_0z for a range of values of S. In these calculations, the initial field is assumed to originate as spontaneous emission from the homogeneously broadened amplifier transition, and the corresponding initial power spectrum is taken in the form

$$
u_x^0 = \frac{1}{\pi}\frac{1}{1+x^2} \tag{34}
$$

Since the saturation integral in Eq. (32) exhibits a dependence on z but not on the frequency ω, the frequency behavior of u_x is identical to that of the

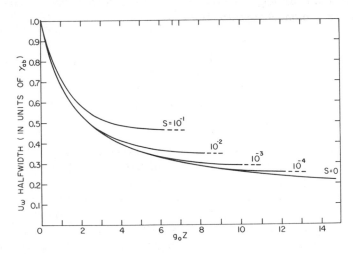

Figure 7.   Half-width of the power spectrum versus $g_0 z$ for a homogeneously broadened amplifying transition.   The half-width is calculated for solutions to Eq. (32) for different values of the saturation parameter S, which in this case completely determines the characteristics of the nonlinear solution.

linear contribution.   For this reason the saturated amplifier solutions predict a continual narrowing of the power spectrum with increasing z as shown in Fig. 7, although the bandwidth narrows less rapidly in the nonlinear region.   In this case the integral represents a weighted average of the power over the homogeneous linewidth rather than expressing coupling between field components.

It is interesting to examine the physical basis for the different saturation behavior of the homogeneous and inhomogeneous broadband amplifiers.   In the saturation region of an inhomogeneously broadened amplifier, each field component can interact with only a small ensemble of the number of moving atoms.   Those atoms having $v \approx 0$ can be heavily saturated by strong fields near ω_0; however, a less-intense off-resonant field component at ω may be saturating the atoms with finite v very weakly.   For this reason, a field at ω will eventually grow more rapidly with distance than fields near ω_0 which will rebroaden the field bandwidth.   Although the presence of amplitude coupling will modify these saturation effects, this basic description is valid.   In contrast, every atom in a homogeneously broadened amplifier can interact with the total distribution of field components.   In this case, the most intense fields will always present the most probable interaction even when these components become saturated.   An atom may absorb a photon $\hbar\omega$, but when interacting with the total field distribution, the atom will most probably emit a photon near $\hbar\omega_o$. To emphasize this interpretation, consider the solution for a monochromatic field given by

$$u_x = u_x^0 \frac{\exp[g_o z/(1+x^2)]}{1 + u_x^0[2S/(1+x^2)](\exp[g_o z/(1+x^2)]-1)} \tag{35}$$

which is the power spectrum observed by sweeping a monochromatic field through the homogeneous amplifier linewidth. In this case the atom is constrained to interact only by the absorption and emission of radiation at a frequency ω. The probability of an atom interacting with a field at ω is proportional to $[\gamma_{ab}^2 + (\omega - \omega_o)^2]^{-1}$, the homogeneous line shape, and the fields near ω_o will grow and saturate more rapidly since they interact with a larger proportion of the inversion density. The continuing amplification of off-resonant fields produces the broadening observed in Fig. 8, similar to the inhomogeneous case. These results for a monochromatic field can also be interpreted in terms of a power broadening of the spectral linewidth; however, the concept of power broadening is not applicable to describe the saturation of a broadband field spectrum.

The more general form of the nonlinear polarization, P_ω^{NL}, depends on ω explicitly in the double-frequency integral through the complex Lorentzian $[\gamma_{ab} - i(\omega_o - \omega - \omega' + \omega'')]^{-1}$. This indicates that the linear and nonlinear contributions will exhibit different frequency characteristics when the polarization includes phase terms. To examine the consequences of this frequency dependence, the complex Lorentzian can be expanded in a power series in powers of $(\omega' - \omega'')$. If we consider the initial nonlinear behavior after the gain-narrowing effect has reduced the power spectrum to a width much less than γ_{ab}, we may take $\omega' - \omega'' \ll \gamma_{ab}$ and truncate this expansion after a few terms given by

$$\frac{1}{\gamma_{ab} - i(\omega_o - \omega - \omega' + \omega'')} \approx \frac{\gamma_{ab} + i(\omega_o - \omega)}{\gamma_{ab}^2 + (\omega - \omega_o)^2}$$

$$-i(\omega' - \omega'')\left(\frac{\gamma_{ab} + i(\omega_o - \omega)}{\gamma_{ab}^2 + (\omega_o - \omega)^2}\right)^2 \tag{36}$$

The first nonlinear term shows the same frequency dependence on ω as the

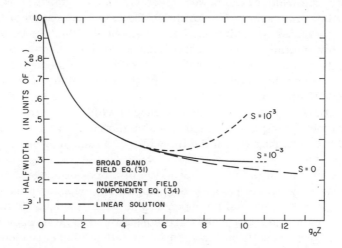

Figure 8. Half-width of the power spectrum versus $g_0 z$ for a homogeneously broadened amplifying transition. The broadband solution is compared with the results of an ensemble of independent fields given by Eq. (35). The spectral gain narrowing indicated by the linear solution (S = 0) emphasizes that the spectrum narrows at a slower rate in the presence of saturation.

linear polarization, similar to the random phase result. However, the ω dependence of the second nonlinear term indicates a different saturation behavior for field components closer to line center. This term is peaked within a width significantly <u>narrower</u> than the linear contribution and changes sign for $|\omega - \omega_0| > \gamma_{ab}$. It is interesting to point out that this functional dependence is similar to the nonlinear polarization for a monochromatic field which leads to saturation rebroadening. The higher terms of the power series expansion will generally be more sharply peaked, and show more frequent sign changes, however it is beyond the intent of this analysis to examine if this frequency behavior is a sufficient condition for rebroadening. At the very least, these considerations suggest that the possibility of saturation broadening in homogeneous amplifiers may be related to a change in the statistics of the propagating field.

V. CONCLUDING REMARKS

A theoretical description of the propagation of a broadband field in a high-gain amplifier has been developed to include the initial nonlinear behavior arising for intense amplification. When the amplifying transition is Doppler broadened, the width of the power spectrum in the saturation region is determined by two processes. In addition to the usual saturation of the atomic velocity ensembles by individual field components which tends to rebroaden the field bandwidth, we observe the presence of coupling among Fourier components which reduces the saturation effect. This coupling resembles a diffusion process by which the energy of the most intense components near ω_0 is redistributed throughout the field bandwidth. The sensitivity of this coupling process to the atomic homogeneous width γ_{ab} suggests another method of measuring this parameter using the nonlinear behavior of radiative interactions occurring within the linewidth of an inhomogeneously broadened transition.

The evolution of the power spectrum of a broadband field propagating in a _homogeneously_ broadened amplifier appears to depend on the extent to which the phases of the field components become coupled. In the random phase approximation, the phases are constrained to be independent random variables, which results in the absence of saturation rebroadening. The general expression for the nonlinear polarization was examined more closely for the case of homogeneous broadening. Although the frequency dependence of the polarization does suggest the possibility of spectral rebroadening, it is not conclusive. The analytical form of the general equations are rather unwieldy to treat the phase coupling problem in the Fourier integral representation; however, the characteristics indicated by these amplifier solutions are a useful guide for a time-domain computer treatment presently underway [18]. Aside from the frequency behavior, another interesting question concerns the photon statistics of the amplified field, particularly in the case of homogeneous broadening. By treating the initial noise source as a quantized radiation field, spatially distributed throughout the amplifier, it may be possible to investigate the change in statistics as the nonlinear interaction develops.

ADDENDUM (SEPTEMBER, 1972)

During this past year, the theory of broadband amplification given above has been extended to include the effects of population pulsations more completely. The present formulation [19] avoids the need for the approximation of slowly varying exponentials in Eq. (17) and treats these modulation terms rigorously. For the case of inhomogeneous broadening, this more complete treatment enhances the coupling among Fourier components by a factor of 2. When the amplifier is homogeneously broadened, the field equations now show evidence of _coupling_ within the linewidth in sharp contrast to the form given by Eq. (32). The expanded theory including these results is being readied for publication.

In addition to the perturbation theory treatment of broadband amplification, a much more inclusive time-domain analysis [20] has been completed. This treatment includes the effects of strong saturation through higher–order terms, allows for changes in the field statistics, and provides computer calculations for several cases of interest. Among the more interesting results, it was found that an input of randomly spaced pulses characteristic of spontaneous noise can develop into a train of well–separated pulses having separations greater than the pulse width. This occurs for inhomogeneous broadening and is accompanied by changes in the higher–order moments of the field. It was also found that the field bandwidth in a homogeneously broadened amplifier may undergo rebroadening in the nonlinear regime. This rebroadening is associated with the onset of extreme power broadening and is thus sensitive to the unsaturated loss coefficient as well as atomic parameters which determine the steady–state field strength. These results are also in preparation for publication.

REFERENCES

1. J. H. Parks, Ph.D. thesis (MIT, 1968).
2. J. H. Parks and A. Javan, J. Opt. Soc. Am. $\underline{61}$, 658 (1971).
3. M. M. Litvak, Phys. Rev. A $\underline{2}$, 2107 (1970).
4. L. Allen and G. I. Peters, J. Phys. A $\underline{4}$, 564 (1971).
5. Xe: J. W. Kluver, J. Appl. Phys. $\underline{37}$, 2987 (1966). N_2: D. A. Leonard, Appl. Phys. Letters $\underline{7}$, 4 (1965); E. T. Gerry, Appl. Phys. Letters $\underline{7}$, 6 (1965). Pb: G. R. Fowles and W. T. Silfvast, Appl. Phys. Letters $\underline{6}$, 236 (1965).
6. M. M. Litvak, A. L. McWhorter, M. L. Meeks, and H. J. Zeiger, Phys. Rev. Letters $\underline{17}$, 821 (1966).
7. This representation was suggested by A. Szöke to avoid an ambiguous form for the field used in an earlier version of the theory, in which $E_{(t)}$ was considered an explicit function of the time t. Although the original formalism, Ref.(1), arrived at identical physical results, the use of the parameters (z_0, t_0) leads to a more consistent treatment of the field propagation in the slowly varying amplitude and phase approximation.
8. W. E. Lamb, Jr., Phys. Rev. $\underline{134}$, A1429 (1964).
9. A. Szöke and A. Javan, Phys. Rev. $\underline{145}$, 137 (1965).
10. B. Fried and S. Conte, The Plasma Dispersion Function (Academic Press, New York, 1961). Briefly, $H(x, \eta)$ is related to the plasma dispersion function $Z(\varphi)$ by $H(x, \eta) = iZ^*(\varphi)$. The expansion for $\eta \ll 1$ follows from $Z(\varphi) = i\sqrt{\pi} \exp(-\varphi^2)[1 + erf(i\varphi)]$, having the leading terms

$$H(x, \eta) \approx \sqrt{\pi} \exp(-x^2) - 2\eta[1 - 2x \exp(-x^2) \int_0^x \exp(y^2)\, dy]$$

$$-i[2 \exp(-x^2) \int_0^x \exp(y^2)\, dy + 2\sqrt{\pi}\eta x \exp(-x^2)]$$

11. The rate equation approximation is discussed in Ref. 8, Sec. 18.

12. B. J. Feldman and M. S. Feld, Phys. Rev. A 1, 1375 (1970).

13. J. H. Parks, D. R. Rao and A. Javan, Appl. Phys. Letters 13, 142 (1968).

14. A. Yariv and J. P. Gordon, Proc. IEEE 51, 4 (1963); A. Yariv and H. Kogelnik, Proc. IEEE 52, 165 (1964).

15. In addition to Ref. 13, gain-induced spectral narrowing has been measured in the following amplifiers. GaAs: A. Yariv and R. Leite, J. Appl. Phys. 34, 3410 (1963); Pb: W. T. Silfvast and J. S. Deech, Appl. Phys. Letters 11, 97 (1967); Th: A. A. Isaev, P. I. Ishchenko, and G. G. Petrash, ZhETF Pis. Red. 6, 619 (1967), [JETP Letters, 6, 118 (1967)]; Ne: D. A. Leonard and W. R. Zinky, Appl. Phys. Letters 12, 113 (1968); HF and DF: J. Goldhar, R. M. Osgood, Jr., and A. Javan, Appl. Phys. Letters 18, 167 (1971); Xe: S. Chuang and H. Gamo, Appl. Phys. Letters 19, 150 (1971).

16. W. L. Faust and R. A. McFarlane, J. Appl. Phys. 35, 2010 (1964).

17. P. O. Clark et al., Investigation of the DC-Excited Xenon Laser, Final Report, J. P. L. Contract No. 950803 (1965).

18. F. A. Hopf and J. H. Parks, J. Opt. Soc. Am. 61, 659 (1971). This time-domain computer study includes phase coupling effects and extends the solutions for the amplified field beyond the limitations imposed by perturbation theory.

19. J. H. Parks, A. Szőke, and A. Javan, (to be published).

20. F. Hopf, J. H. Parks and A. Szőke, (to be published).

OPTICS AND OTHER APPLICATIONS

ACTIVE INTEGRATED OPTICS*

Amnon Yariv
California Institute of Technology
Pasadena, California

I. INTRODUCTION

Observation of light guiding in thin dielectric films was first made in 1961 in optical fibers by Snitzer and Osterberg [1] and by Kapany and Burke [2] and in p-n junctions in 1963 by Bond, Cohen, Leite, and Yariv [3, 4]. Shubert and Harris [5] (1968) recognized the importance of structures incorporating thin films for "Integrated Data Processors". Miller [6] (1969) added the word "optics" thus giving rise to the term "integrated optics" by which this field has come to be known.

This paper is devoted primarily to what one may call "active integrated optics", i.e., the performance of a variety of active tasks including light generation, modulation, switching, and correlation in thin-film circuits. This particular area of activity traces its beginning to experiments by Nelson and Rinehart [7] (1964) in which the guiding of light in GaP p-n junctions was accompanied by electro-optic modulation. In 1965 we started at Caltech on experiments designed to demonstrate guiding and electro-optic

---

*Work supported by the Advanced Research Projects Agency of the Department of Defense and Monitored by the Army Research Office-Durham under Contract DAHCO4 68 C 0041.

switching in thin films. The demonstration of these effects [8, 9] had to wait until 1970 for the maturation of the epitaxial film technology. Work on switching and mode coupling in thin films by sound waves has been reported recently by two IBM groups [10, 11].

The ability to design active thin-film components requires a separate understanding of the electromagnetic properties of the modes which can propagate in thin films and their interplay with the photoelastic and electro-optic phenomena used in active coupling and switching.

This article reviews the theoretical background of the different building-block disciplines with special emphasis on the theoretical aspects of combining them to perform various optical functions. The recent experimental progress and a number of potential applications are considered in the process.

II. REVIEW OF THIN-FILM PROPAGATION THEORY

The basic thin-film waveguide geometry is shown in Fig. 1. A guiding layer with a dielectric constant ϵ_2 and thickness t is sandwiched between a substrate with a dielectric constant* $\epsilon_1 < \epsilon_2$ and an overlayer with a dielectric constant ϵ_3. The width of the guide in the y direction is assumed large compared to the wavelength so that its effect on the mode characteristics can be neglected. If $\epsilon_1 \neq \epsilon_3$ the propagating electromagnetic modes of the structure do not possess odd or even symmetry, however, and can be written in general as

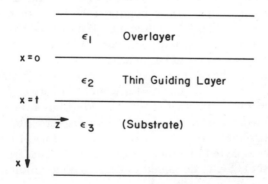

Figure 1. A three-layer dielectric waveguide.

*In this paper we use the values of the dielectric constant relative to the volume value ϵ_0 so that $\epsilon = n^2$, where n is the index of refraction.

$$E_y(TE) \propto \begin{cases} \exp(qx - i\beta z) & x < 0 \\[2mm] [\sin(hx) + \dfrac{h}{q}\cos(hx)]\exp(-i\beta z) & 0 < x < t \\[2mm] \exp[-p(x-t) - i\beta z] & x > t \end{cases} \qquad (1)$$

$$H_y(TM) \propto \begin{cases} \exp(qx - i\beta z) & x < 0 \\[2mm] \sin[(hx) + \dfrac{\epsilon_1}{\epsilon_2}\dfrac{h}{q}\cos(hx)]\exp(-i\beta z) & 0 < x < t \\[2mm] \exp[-p(x-t) - i\beta z] & x > t \end{cases} \qquad (2)$$

It is an extremely fortunate circumstance that in most of the structures of practical interest the condition

$$|\epsilon_2 - \epsilon_3| \ll |\epsilon_2 - \epsilon_1| \qquad (3)$$

i.e., the dielectric discontinuity at one boundary designated here as the 1-2 interface, is much larger than at the other one. This happens for example when medium 1 is metal or air while medium 3 is some dielectric with $\epsilon_3 \simeq \epsilon_2$. When condition (3) is fulfilled it is a simple matter to show [12] that $q \gg h$ so that the field solutions for both the TE and TM waves can be approximated by

$$H_y(TM), \ E_y(TE) \simeq \sin(hx) \qquad (4)$$

It also follows directly [12] that in this case the dispersion relations for the various modes, the mode cutoff conditions and the solutions for h, β, and p are nearly the same as those of the odd modes in the symmetric waveguide made up of media 3-2-3 with the thickness of layer 2 being 2t. Condition (3), which leads to the inequality $h \ll q$, causes the exponential tail in region 1 to extend over a distance which is very short compared to t. If region 1 is a metallic conductor, this small but finite penetration can be the main cause for the mode attenuation and cannot, consequently, be ignored in the solution of the propagation β. This point is discussed further below.

The propagation constants, p, h, and β, are determined by solving the wave equation along with the proper boundary conditions at $x = 0$ and $x = t$ [13]. In the limit of negligible penetration into the overlayer ($q \to \infty$) i.e., when condition (3) is fulfilled, we need to solve simultaneously the equations

$$\text{TE(m)} \quad ht = \tan^{-1} \frac{p}{h} + m\frac{\pi}{2}$$

$$m = 1,3,5,\ldots \tag{5}$$

$$\text{TM(m)} \quad ht = \tan^{-1} \frac{\epsilon_2}{\epsilon_3}\frac{p}{h} + m\frac{\pi}{2}$$

and the equation

$$(pt)^2 + (ht)^2 = (\epsilon_2 - \epsilon_3)(2\pi t/\lambda_o)^2 \tag{6}$$

The $\tan^{-1}$ in (5) refers to the principal value, between $-\pi/2$ and $\pi/2$. $\epsilon_i \equiv n_i^2$ refers to the relative dielectric constants, while λ_o is the wavelength of light in vacuum. The cutoff condition for the m-th (TE or TM) mode occurs when the corresponding p is zero so that from (5)

$$(h^{(m)}t)_{\text{cutoff}} = m\pi/2, \quad m = 1,3,5,\ldots \tag{7}$$

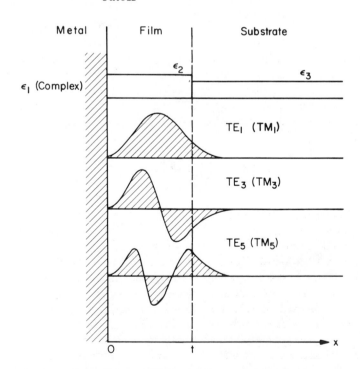

Figure 2.   Some low-order modes (m = 1, 3, 5) in the odd symmetry dielectric waveguide.

The conditions that must be satisfied so that the m-th mode is confined (i.e., above cutoff) is derived by putting $p = 0$ and using the cutoff value of $h^{(m)}$ from (7) in (6). The result is

$$\sqrt{\epsilon_2 - \epsilon_3}\ (2\pi t/\lambda_0) > m\pi/2, \qquad m = 1, 3, 5, \ldots \qquad (8)$$

As the parameter $(\epsilon_2 - \epsilon_3)\ 2\pi t/\lambda_0$ increases, $h^{(m)}$ increases asymptotically from its cutoff value of $m\pi/2t$ to $(m+1)\pi/2t$. It follows that the m-th mode undergoes $(m-1)/2$ zero crossing in $0 < x < t$. Some low-order profiles are shown in Fig. 2.

The graphical procedure for obtaining p and h from the intersection of the curves represented by (5) and (6) is shown in Fig. 3. The values of h thus obtained can then be used in the relation

$$\beta^2 = (2\pi/\lambda_0)^2\,\epsilon_2 - h^2$$

which is the wave equation $(\nabla^2 + k\ )E = 0$ in region 2, to obtain the dispersion $(\beta$ versus $\omega)$ curves. The result is illustrated in Fig. 4.

Two features of these waves will be needed in our subsequent discussion and should be emphasized here. One is that a given mode near its cutoff condition propagates with a velocity c_0/n_3 which is consistent with the fact that near cutoff most of the mode energy is in region 3 since $p \simeq 0$. As the cutoff condition is exceeded, i.e., as the difference

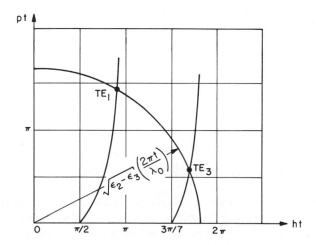

Figure 3.   Procedure for determining the mode constants p and h.

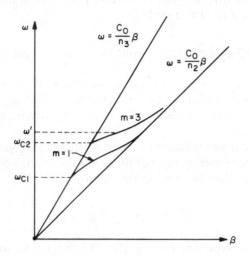

Figure 4.   The dispersion curves for m = 1 and m = 3 modes.

$$\sqrt{\epsilon_2 - \epsilon_3} \left(\frac{2\pi t}{\lambda_o}\right) - \frac{m\pi}{2}$$

increases, $p \to \infty$ and the mode becomes confined to the thin film ($0 < x < t$) region. Its phase velocity, appropriately, approaches the value of c_o/n_2. Also of interest, especially in nonlinear mode interactions to be discussed below, is the fact that well above cutoff the constant $h^{(m)}$ of the m-th mode approaches its limiting value $(m+1)\pi/2t$ and the modes become orthogonal in the sense

$$\int_{-\infty}^{\infty} u_m(x)u_n(x)\, dx \simeq \int_0^t u_m(x)u_n(x)\, dx \to 0, \quad m \neq n \tag{9}$$

III.   MATERIAL CONSIDERATIONS

The choice of materials for dielectric waveguides will depend on the specific applications envisaged. But just as silicon and germnium have come to play a key role in integrated electronics, it is possible already to develop approximate criteria which point toward certain materials as candidates for active integrated optics applications. Some of the more important requisite properties are

1. transparency and good optical quality for light in the visible and near-visible regions of the spectrum;

2. material should lend itself easily to interfacing with electronic circuits;

3. the material should be capable of light generation and detection;

4. the material should be capable of performing light switching and modulation functions. More specifically, it should possess large electro-optic and photoelastic figures of merit so that modulation and switching of light by either of these two techniques can be used;

5. the material should be suitable for thin-film dielectric waveguide fabrication.

There are many materials that can satisfy reasonably well one or two of these requirements and it is conceivable that future integrated circuits will combine a number of them for specific applications. It is interesting to note, however, that at least one class of known materials already comes close to fulfilling all of these requirements. This is the semiconductor GaAs and its related alloys, such as $Ga_{1-x}Al_xAs$ and $GaAs_{1-x}P_x$.

It may prove useful to go down our list of critcria and see how they apply, for example, to the GaAs alloy system.

1. GaAs single crystals have a ($300°K$) energy gap of 1.45 eV (~ 0.85 μ) and have a useful transparency range of $1-12$ μ. The energy gap and hence the optical cutoff frequency can be "pushed" into the visible in the Al or P alloys.

2. The ability to easily dope GaAs, to fabricate p-n junctions, and make Ohmic contacts should prove useful in electrical interfacing. An example of such an application is described in Sec. VII.

3. GaAs is the material used most widely for injection semiconductor diodes. Such diodes emitting coherently (lasers) or incoherently (LED) can thus be built into the circuit. GaAs p-n junctions, reverse biased or un-biased (photovoltaic), can also be used for efficient light detection. A thin film circuit incorporating an injection laser is shown in Fig. 5.

4. Both the electro-optic figure of merit [14] ($n_1^3 r_{41} = 6 \times 10^{-11}$ m/V) and the photoelectic figure of merit [15] ($n^6 p^2 / \rho v_s^3 \simeq 10^{-13}$) in GaAs are among the largest.

5. Epitaxial techniques for growing $Ga_{1-x}Al_xAs$ are well developed and have received a strong impetus because of their importance in the fields of solid state microwave oscillators and the heterojunction injection lasers [16]. The presence of a fraction x of aluminum causes the index to change by approximately

$$\Delta n \sim -0.4x$$

904   A. Yariv

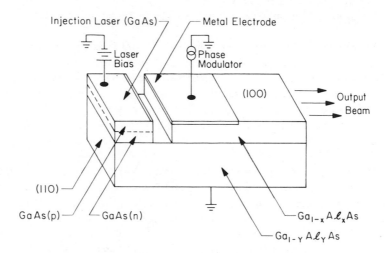

Figure 5.   A Ga$_{1-x}$Al$_x$As dielectric waveguide incorporating an injection laser and an electro-optic phase modulator.

so that a simple dielectric waveguide can be fabricated by growing epitaxially a Ga$_{1-x}$Al$_x$As layer on a substrate containing a larger aluminum concentration.   Such a waveguide is shown in Fig. 6.

The techniques of epitaxy are easily adaptable to growing a large number of layers with different indices of refraction so that we can envisage bulk volume integrated circuits stacked on top of each other and interconnected or coupled together.

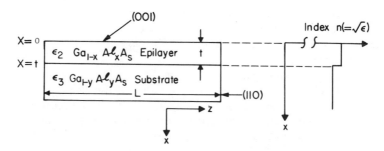

Figure 6.   The index profile of air-Ga$_{1-x}$Al$_x$As – Ga$_{1-y}$Al$_y$As waveguide.

IV.   PHOTOELASTIC BRAGG SWITCHING

The two main approaches so far to active switching used (1) the photoelastic effect and (2) the electro-optic effect.   The first photoelastic switching experiment performed at IBM laboratories [10] is illustrated in Fig. 7.   An optical mode coupled into a glass thin-film guide crosses a surface acoustic wave at the Bragg angle [15]

$$\theta_B = \sin^{-1}(\lambda_g/2\lambda_s) \tag{10}$$

where λ_s is the wavelength of the acoustic wave, while $\lambda_g = 2\pi/\beta$ is the mode (guide) wavelength.   A fraction

$$\eta = \sin^2 \frac{\omega\ell}{2c_o} \Delta n$$

is scattered into a beam at $2\theta_B$ as shown (ω is the optical radian frequency, ℓ the interaction path length, and Δn the index modulation amplitude due to the acoustic strain). Δn can be related to the acoustic intensity $I(W/m^2)$ by means of the acoustic figure of merit [15] $M \equiv n^6 p^2/\rho v_s^3$ where p is the appropriate photoelastic tensor element and ρ the mass density.   This results in

$$\Delta n = (MI/2)^{1/2}$$

Since the acoustic energy is limited to a depth $\sim \lambda_s/4$ we have $I \simeq 4P_a/W\lambda_s$

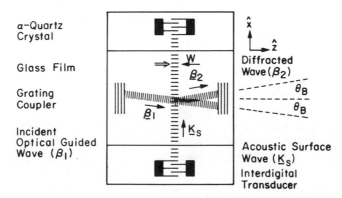

Figure 7.   Bragg diffraction of dielectric waveguide modes by a surface acoustic wave (after Ref. 10).

where W is the width of the acoustic beam, and P_a is the total acoustic power. Taking $\ell = W/\cos\theta_B \simeq W$, we get

$$\eta \;=\; \sin^2\!\left[\, \frac{\pi}{\lambda_o} \left(\frac{2MW}{\lambda_s}\, P_a\right)^{\!1/2}\, \right] \tag{11}$$

The acoustic power needed to effect complete switching ($\eta = 1$) is thus

$$(P_a)_{\eta=1} \;=\; \lambda_o^2 \lambda_s / 8MW \tag{12}$$

As an estimate of the typical switching power in GaAs we use the following data:

$$W \;=\; 0.5 \text{ mm}, \qquad \lambda_o \;=\; 1\ \mu, \qquad v_s \;=\; 5 \times 10^5 \text{ cm/sec}$$

$$f_s \;=\; 3 \times 10^8 \text{ Hz}, \qquad\qquad M \;=\; 10^{-13} \text{ (mks)}$$

This gives

$$(P_a)_{\eta=1} \;\simeq\; 40 \text{ mW}$$

This is a modest amount of power and serves to emphasize the attractiveness of acoustic switching. Attempts to reduce it much further by increasing the sound frequency will probably be thwarted by the increasing attenuation of the acoustic beam.

A basic limitation of the acoustic Bragg deflection is the switching time. This is due to the finite bandwidth of the acoustic transducer and the highest usable frequency. Assuming that acoustic propagation losses will prevent the use of frequencies far in excess of 3×10^8 Hz and that the bandwidth is ~10% we find that the modulation bandwidth is ~ 3×10^7Hz.

V.   PHOTOELASTIC MODE COUPLING

Another elegant application of sound surface waves in thin-film optical waveguides was demonstrated by an IBM group [11] who used the photoelastic interaction to couple energy between two thin film TE_1 and TE_3 modes. The surface sound wave in this case is launched in a direction parallel to the optical modes. Energy is coupled photoelastically from the TE_1 mode to the TE_3 mode and vice versa by a sound wave whose propagation constant is K_s. The experimental situation is demonstrated in Fig. 8. For efficient coupling between two modes, say i and j, it is necessary that

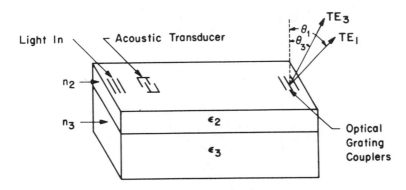

Figure 8.   Coupling between dielectric waveguide modes TE_1 and TE_3 by means of an acoustic surface.

the phase-matching condition

$$\beta_i - \beta_j = \pm K_S \tag{13}$$

be satisfied. Since $K_S = \omega_S/v_S$, efficient coupling between a given pair of modes occurs at a particular sound frequency. The graphical procedure for determining ω_S for coupling a TE_3 mode at ω' and a TE_1 mode at $\omega' \pm \omega_S$, for example, is illustrated in Fig. 9.

A.   Theory of Photoelastic Mode Coupling

Nonlinear coupling between a number of modes is expected to play an important role in switching and correlation applications [17] so that an understanding of the basic limitations and possibilities is in order.
    We can start with the basic equations governing the coupling between two optical plane waves by an acoustic wave. For collinear propagation along the z direction these equations become [15]

$$\frac{dE_i}{dz} = -i\eta\, E_i \exp[-i(k_i - k_j \mp K_S)z]$$

$$\frac{dE_i}{dz} = -i\eta\, E_j \exp[i(k_i - k_j \mp K_S)z] \tag{14}$$

where i and j correspond to the polarization directions of the optical waves;

908   A. Yariv

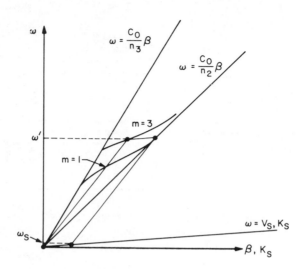

Figure 9.   The graphical procedure for determining the acoustic frequency ω_s for phase-matched ($\beta_1 - \beta_3 + K_s$) coupling between two dielectric waveguide modes by a surface sound wave.

K_s appears with a (−) sign for sound propagating in the +z direction and a (+) sign for the −z direction, and E represents the electric field.   The coupling constant η is given by

$$\eta = \frac{\pi n^3}{2\lambda_0} \, p_{ijkl} \, S_{kl} \tag{15}$$

where S_{kl} is the strain amplitude due to an acoustic wave

$$s_{kl}(z, t) = S_{kl} \, \exp[i(\omega_s t \mp K_s z)]$$

and p_{ijkl} is the photoelastic tensor element.
    In the case of interaction between localized modes, rather than plane waves, we replace the field amplitudes E and s by normalized mode amplitudes according to

$$E_{i,j}(x, z) = \epsilon_{i,j}(z) \, u_{i,j}(x)$$

$$s_{kl}(x, z) = S_{kl}(z) \, u_s(x) \tag{16}$$

where the u(x) functions describe the transverse (x) mode variation and are normalized so that

$$\int_{-\infty}^{\infty} u_{i,j,s}^2(x)\, dx = 1 \tag{17}$$

It follows from (16) and (17) that the total optical power in a mode is

$$P_i = c_0 n_i \varepsilon_0 W |e_i|^2$$

where W is the width of the mode in the y direction. The total acoustic mode power (due to s_{kl}) is

$$P_a = \frac{1}{2} \rho v_s^3 W |S_{kl}|^2 \tag{18}$$

$$\frac{de_j}{dz} = -ig\varepsilon_i \exp[-i(\beta_i - \beta_j \mp K_s)z]$$

$$\frac{de_i}{dz} = -ig\varepsilon_j \exp[i(\beta_i - \beta_j \mp K_s)z] \tag{19}$$

where

$$g = \frac{\pi n^3}{2\lambda_0} p_{ijkl}\, S_{kl} \int_{-\infty}^{\infty} u_i(x)u_j(x)u_s(x)\, dx \tag{20}$$

and the β's refer to mode propagation constants. Note that the overlap integral (20) has its maximum value when the two optical modes are of the same order, say m, while the sound wave is uniform in a region $0 < x < t$ and is zero elsewhere. If the optical modes are assumed well above their threshold value, we have

$$u_i(x) = u_j(x) = \begin{cases} \sqrt{2/t}\, \sin\frac{m\pi}{t} x & 0 < x < t \\ \\ 0 & \text{elsewhere} \end{cases}$$

$$u_s(x) = \begin{cases} \sqrt{1/t} & 0 < x < t \\ \\ 0 & \text{elsewhere} \end{cases}$$

The overlap integral becomes

$$J_{ijs} \equiv \int_{-\infty}^{\infty} u_i(x)u_j(x)u_s(x)\, dx = \frac{1}{t^{1/2}} \tag{20a}$$

An inspection of (19) reveals that a prerequisite for cumulative energy exchange is the so-called phase-matching condition

$$\beta_i - \beta_j \mp K_s = 0 \tag{13}$$

and in addition a nonvanishing overlap integral. This last point will be discussed further below. When condition (13) is satisfied the solution of (19) is

$$\epsilon_i(z) = \epsilon_i(0)\cos gz - i\epsilon_j(0)\sin gz$$
$$\epsilon_j(z) = \epsilon_j(0)\cos gz - i\epsilon_i(0)\sin gz \tag{21}$$

In mode switching applications we usually deal with a single input, say $\epsilon_i(0)$, so that

$$\epsilon_i(0) = \epsilon_i(0)\cos gz$$
$$\epsilon_j(0) = -i\epsilon_i(0)\sin gz \tag{22}$$

Complete switching in a distance $z = \ell$ occurs when $g\ell = \pi/2$ which, using (18) and (20), requires an acoustic power

$$P_{a(switching)} = \frac{\rho v_s^3 W \lambda_o^2}{2\ell^2 n^6 p_{ijkl}^2 J_{ijs}^2} \tag{23}$$
$$= \frac{W \lambda_o^2}{2\ell^2 M J_{ijs}^2}$$

where M is the acoustic figure of merit [15]

$$M \equiv \frac{n^6 p_{ijkl}^2}{\rho v_s^3}$$

The nature of the mode profiles is thus very important in determining the acoustic switching power. If as an example the two modes are different ($m \neq n$) TE or TM modes and the sound intensity is uniform (i.e.,

$u_s(x)$ = const.) then well above threshold we have according to (9) $J_{ijs} = 0$ and the switching power is infinite. The lowest threshold results from coupling two similar modes with a nearly uniform sound wave where all three proiles occupy the same region $0 < x < t$. For the case of coupling of modes with index m so that $ht \simeq m\pi$ and a sound wave limited to a depth t, we get, using (20a),

$$P_{a(\text{switching})} = \frac{W\lambda_0^2 t}{4\ell^2 M} \tag{24}$$

For a numerical estimate consider the case of switching in GaAs where $M \simeq 10^{-13}$ (mks). Using: $W = 0.5$ mm, $\lambda_0 = 1$ μ, $\ell = 5$ mm, and $t = 1$ μ we get

$$P \simeq 10^{-4} \text{ W}$$

This should be compared to the switching power for Bragg deflection ($\sim 4 \times 10^{-2}$ W) calculated in Sec. IV. The difference is due to the longer interaction path in the collinear case.

To couple together, as an example, the orthogonally polarized TE_1 and TM_1 modes we need a strain wave s_{kl} in a crystalline medium possessing a nonvanishing p_{ijkl} where i and j are the polarizations of the TE and TM modes. If the crystalline medium is optically anisotropic, the polarization directions of the TE and TM modes may correspond to two different indices of refraction and thus possess different values of β. In this case the β mismatch will be made up, as in (13), by using the appropriate acoustic frequency.

VI.   SIGNAL CORRELATION WITH PHOTOELASTIC COUPLING

The photoelastic mode coupling described above can provide a convenient means for performing correlation functions [17]. To understand how this works consider the case where the acoustic strain is in the form of a pulse whose amplitude envelope is $S(t - z/v_s)$, i.e., a pulse propagating with a velocity v_s. Similarly, let the optical input be a pulse $\mathcal{E}_i(t - z/c_i)$. Since $c_i = 3 \times 10^4 v_s$, the optical pulse will overtake the acoustic pulse and cross past it. The overlap region acts as a giant phase-matched dipole radiating into the $\mathcal{E}_j$ mode. At any one time t the output field at $\omega_j = \omega_i \pm \omega_s$ is proportional to the instantaneous integral over z of the product of both fields. Since the pulses move relative to each other, the output field $\mathcal{E}_j(t)$ corresponds to the correlation function. This result follows directly from (19) which for $(\beta_i - \beta_j \mp K_s) = 0$ and $g\ell \ll \pi$ give

$$\mathcal{E}_j(\ell, t) \propto \int_0^\ell S(z - v_s t) \, \mathcal{E}_i(z - c_i t) \, dz$$

where c_i is the phase velocity of the i-th optical mode. Because of the huge disparity between the sound velocity v_s and c_i we can consider the acoustic pulse length to be much shorter than the interaction distance, so that the integration limits can be taken as $-\infty$ to ∞. Since the last integral depends only on the difference of the arguments we can write it as

$$\mathcal{E}_j(\ell, t) \propto \int_{-\infty}^{\infty} S(z) \, \mathcal{E}_i[z - (c_i - v_s)t] \, dz \tag{25}$$

which is in the form of a correlation integral. The correlation signal is thus recovered by detection of the envelope of the output $\mathcal{E}_j$.

VII.   ELECTRO-OPTIC SWITCHING

Another phenomenon used recently [8, 9] in thin-film switching is the electro-optic effect. This requires the choise of a thin-film guide material which is electro-optic and the ability to impose an electric field on the thin film. Both of these requirements were met by using high-resistivity GaAs which is an excellent electro-optic material. The electric field was provided by a reverse-biased Schottky junction at the metal-GaAs junction as shown in Fig. 10.

In one of the experiments the application of a voltage to the guide region resulted in a differential phase shift (retardation) between the TM_1 and the TE_1 modes given, in the particular geometry depicted in Fig. 11, by

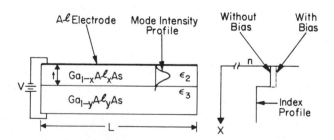

Figure 10.   Electric field control of the dielectric discontinuity $\epsilon_2 - \epsilon_3$ in an electro-optic thin-film guide. The application of a field causes a change in the index of refraction "seen" by a guided mode. The magnitude of the change depends on the crystal orientation, field direction, and mode polarization, and can be different for different modes.

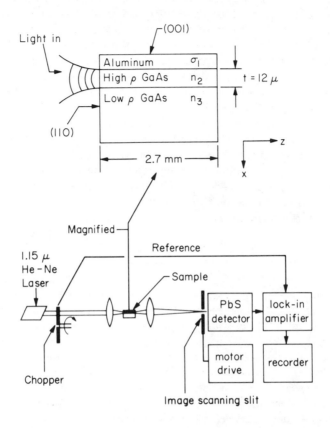

Figure 11.   Electro-optic thin-film polarization modulator (after Ref. 8).

$$\Gamma = (\beta_{TM} - \beta_{TE})\ell = \frac{\pi \ell n_o^3 r_{41}}{\lambda_o} \left(\frac{V}{t}\right) \qquad (26)$$

where n_o is the index of refraction and r_{41} the electro-optic coefficient of GaAs. V is the reverse bias voltage. This retardation causes a power transfer of a fraction

$$\eta = \sin^2 \Gamma/2$$

from the input polarization (which in the experiment was at $45°$ to the TE and TM polarizations) to one at right angles to it.

A different orientation of the crystal [input face $(\bar{1}10)$ and electric field normal to (110)] would cause mode coupling in the sense that a fraction η of the power in an input TE (or TM) will be transferred in a distance ℓ to the TM (or TE) mode. Either form of power transfer can be used to perform modulation or switching. This requires the incorporation of a polarizer which would discriminate against one polarization component. No practical solution to the thin-film polarizer problem has been advanced yet. One possibility is suggested by recent calculations [18] of the effect of a metallic overlay (see Fig. 1) on the TE and TM mode losses. It is found that the TM modes are always losser and under certain circumstances can undergo an attenuation which is two orders of magnitude larger than that of the TE mode. A section of the thin-film guide with a metal layer on top should thus absorp most of the TM mode power. One of the main attractions of using the electro-optic effect is in its promise for low power and high-speed switching and modulation. Conventional transverse electro-optic modulators require a minimum modulation power of [19]

$$P_{min} = \frac{\epsilon_m \epsilon_o \lambda_o^3}{n^7 r^2} \, m^2 B \tag{27}$$

where ϵ_m is the relative dielectric constant at the modulation frequency, r the appropriate electro-optic coefficient, m the modulation index, and B the modulation bandwidth. The lack of any geometrical factors in (27) is due to the fact that the factor t^2/ℓ which would normally appear has been set to $4\lambda_o/n\pi$ [19] which is near the optimum value consistent with the diffraction of a fundamental Gaussian beam. This limitation does not exist when the modulation is done on a guided thin-film mode. This is due to the fact that the mode does not diffract, since the natural tendency to do so is counterbalanced by the refocusing due to the dielectric discontinuity. This results in a modulation power of

$$\frac{\pi \lambda_o^2 \epsilon_m \epsilon_o}{4 n^6 r^2} \left(\frac{Wt}{\ell}\right) m^2 B \tag{28}$$

The difference between a bulk modulator and a thin-film modulator is illustrated in Table I which compares a conventional modulator using $LiTaO_3$ and a thin-film modulator using $Ga_{1-x}Al_xAs$.

When one of the principal dielectric axes in the presence of an electric field is parallel to the polarization direction of the TE or TM mode, the application of a voltage causes a pure phase modulation. In the geometry illustrated in Fig. 5, for example, the applied voltage causes the TE mode to "see" an index of refraction

TABLE I.   MODULATOR COMPARISON

Modulator Type	$Ga_{1-x}Al_xAs$ Thin Film	$LiTaO_3$ Bulk Modulator
t--thickness	1 μ	0.25 mm
w--width	0.1 mm	0.25 mm
L--length	1 cm	1 cm
ϵ_m--dielectric constant	11	43
λ_o--optical wavelength	8500 Å	6328 Å
B--bandwidth	100 MHZ	100 MHZ
V--modulator voltage for m = 40%	0.1 V	9 V
P--power for 40% AM	0.2 mW	60 mW

$$n_{TE} = n_o - \frac{1}{2} n_o^3 r_{41} E \tag{29}$$

while the index "seen" by the TM mode is unchanged and remains n_o.   A sinusoidal voltage will thus cause a phase modulation of the TE mode only with a modulation index

$$\delta = \frac{\pi \ell n^3 r_{41}}{\lambda_o} \left(\frac{V_m}{t} \right) \tag{30}$$

where V_m is the amplitude of the applied voltage.   Using $\ell = 1$ cm, $\lambda_o, t = 1$ μ and $n^3 r_{41} = 6 \times 10^{-11}$ (GaAs) we get $\delta \simeq 1.9/V$.   This plus the example of Table I illustrates that wide band modulation in thin films can be achieved at voltages of ~1 V and power levels in the milliwatt region.

VIII.   ELECTRO-OPTIC FIELD PENETRATION CONTROL

Another way in which the electro-optic effect in thin films can be used to control the flow of mode energy is by varying electrically the penetration of the mode energy profile into the adjacent lower index medium.

The condition for confined propagation of the mode m is that $p^{(m)} > 0$ which according to (8) happens when

$$\sqrt{\epsilon_2 - \epsilon_3} \,(2\pi t/\lambda_0) \;>\; m\pi/2$$

Near cutoff (i.e., $p \gtrsim 0$) we can obtain by manipulating (5) and (6)

$$p \;\simeq\; \frac{m\pi}{4t} \left(\frac{(\epsilon_2 - \epsilon_3)(2\pi t/\lambda_0)^2}{(m\pi/2)^2} - 1 \right) \tag{31}$$

so that the penetration depth p^{-1} of the field into region 3 can be controlled strongly by varying the dielectric discontinuity $(\epsilon_2 - \epsilon_3)$. An example of this field penetration control is shown in Fig. 12 in which the dependence of the index of refraction "seen" by the TE_1 mode on the electric field Eq. (29), is used to control p near cutoff.

There are numerous ways in which this penetration control can be utilized. One application is suggested directly by the experimental data of Fig. 12. By changing p from a large positive value to one slightly above or below zero, we cause the power density in region 2 to go down drastically, thus obtaining on-off switching.

Another potential application of this effect is in controlling the degree of coupling between two adjacent waveguides as illustrated in Fig. 13. A voltage applied between the two guides changes $\epsilon_2 - \epsilon_3$ in the transverse (y)

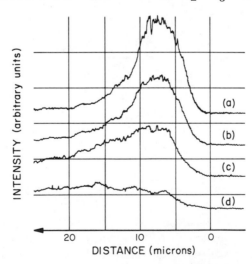

Figure 12.   Electro-optic control of the field penetration. By varying the applied voltage the field penetration and confinement in a GaAs thin-film guide are controlled. (a) TE_1 mode V = 100 V; (b) TE_1 mode V = 70 V; (c) TE_1 mode V = 30 V; (d) TM_1 mode V = 100 V (after Ref. 9).

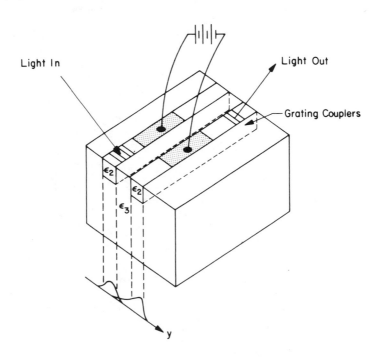

Figure 13.   Coupling between two guides by means of field-controlled penetration.

direction, thus controlling the coupling, as represented by the overlap of the interpenetrating tails between the guides. By controlling the voltage we may thus transfer all or part of the power launched into one mode to the other mode and thus use it for switching, modulation, or coupling.

IX.   APERTURE PHASE CONTROL

Another potential application of the electro-optic effect in thin films is in the ability to synthesize an arbitrary phase variation in the cross section of an optical beam. A two-dimensional version of such a device is illustrated in Fig. 14. The input beam is divided among a large number of guides. The output phase of each component beam is controlled by an applied voltage. A linear voltage profile leads to beam deflection, while a quadratic variation, for example, causes focusing with a focal length which depends on the quadratic constant. The present example illustrates the possibility of

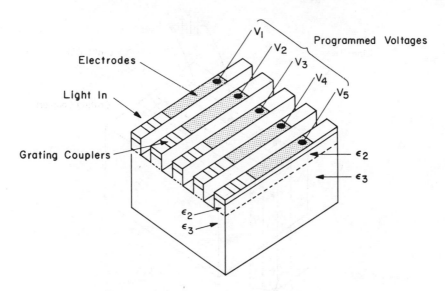

Figure 14.   Electro-optic control of aperture phase for programmable lens and deflector applications.

programmable beam deflection, focusing, and shaping in one plane. A complete control of the output beam will be made possible if the thin-film guides can be fabricated not only in parallel rows as in Fig. 14 but stacked on top of each other.

X.   CONCLUSION

We have considered the problem of guiding light in thin dielectric films and that of actively interfering with the flow of power in the modes by means of the electro-optic and photoelastic effects. A review of the current state of the art and the potentialities suggested by theoretical considerations suggest that the basic functions of light generation, switching, guiding, and detection can be performed in a single monolithic material. It seems reasonable to assume that these techniques will find important applications in the newly developing field of integrated optics.

REFERENCES

1. J. E. Snitzer and H. Osterberg, J. Opt. Soc. Am. $\underline{51}$, 499 (1961).
2. N. S. Kapany and J. J. Burke, J. Opt. Soc. Am. $\underline{51}$, 1067 (1961).
3. A. Yariv and R. C. Leite, Appl. Phys. Letters $\underline{2}$, 55 (1963).
4. W. L. Bond, B. C. Cohen, R. C. Leite, and A. Yariv, Appl. Phys. Letters $\underline{2}$, 57 (1963).
5. R. Shubert and J. H. Harris, IEEE Trans. Microwave Theory Tech. $\underline{MTT-16}$, 1048 (1968).
6. S. E. Miller, Bell Syst. Tech. J. $\underline{48}$, 2059 (1969).
7. D. F. Nelson and F. K. Rinehart, Appl. Phys. Letters $\underline{5}$, 148 (1964).
8. D. Hall, A. Yariv, and E. Garmire, Opt. Comm. $\underline{1}$, 403 (1970).
9. D. Hall, A. Yariv, and E. Garmire, Appl. Phys. Letters $\underline{17}$, 127 (1970).
10. L. Kuhn, M. L. Dakss, P. F. Heidrick, and B. A. Scott, Appl. Phys. Letters $\underline{17}$, 265 (1970).
11. L. Kuhn, P. F. Heidrick, and E. G. Lean, Appl. Phys. Letters $\underline{19}$, 428 (1971).
12. D. Hall, Ph.D. thesis (California Institute of Technology, 1971).
13. R. E. Collin, Field Theory of Guided Waves (McGraw-Hill, New York, 1960).
14. A. Yariv, Introduction to Quantum Electronics (Holt, Rinehart, and Winston, New York, 1971), pp. 228–230.
15. R. W. Dixon, J. Appl. Phys. $\underline{38}$, 5149 (1967). See also, Ref. 14, Chap. 12.
16. H. Kressel, H. Nelson, and F. Z. Hawrglo, J. Appl. Phys. $\underline{41}$, 2019 (1970).
17. The use of acoustic surface waves only for this application was demonstrated by C. F. Quate and R. B. Thompson, Appl. Phys. Letters $\underline{16}$, 494 (1970), and also by M. Luukkala and G. S. Kino, Appl. Phys. Letters $\underline{18}$, 393 (1971).
18. D. Hall, E. Garmire, and A. Yariv (unpublished).
19. See Ref. 14, p. 326.

GEOPHYSICAL STRAIN MEASUREMENT BY

OPTICAL INTERFEROMETRY

R. H. Lovberg* and J. Berger
University of California, San Diego

ABSTRACT

A Michelson interferometer having a 800-m arm has been employed for observations of earth crustal strains resulting from lunar tides, earthquakes, and slow motions of large crustal plates. The optical path is above the ground, but protected by an evacuated pipe. Connection of the optical components to the ground is made by massive granite piers. The laser light source is stabilized by reference to a passive optical resonant cavity, and is apparently constant in wavelength to within a few parts in 10^{10} over periods of weeks. Normal mode oscillations of the whole earth excited by earthquakes have been observed.

I. GEOPHYSICAL BACKGROUND

If the outer crust of our planet were viewed by an observer placed far enough away to see the globe as a whole, and if time were, for him, compressed so that the age of the earth occupied only a few minutes, then this observer

---

*On leave of absence at the UKAEA Culham Laboratory, Abingdon, Berks, England.

would, according to modern evidence, see a number of large semirigid crustal plates moving about with respect to one another. They would experience some sliding encounters, occasional direct frontal collisions that would produce wrinkling at the locality of contact, and, occasionally, one plate could be seen to slip beneath another. On the whole, however, the motions would appear rather smooth and continuous, albeit with an overall randomness in organization.

Suppose further that our hypothetical viewer could, through some cosmic instrumental means, expand the time scale of his observation so that a minute to him might represent perhaps one million years rather than the original thousand million. This thousand-fold slowing of events would render the relative plate motions nearly imperceptible except on a much magnified distance scale. But now, an important new property of the motion would appear: the sliding of plates past one another would be seen as a somewhat irregular, "chattering" motion, and the smoothness of the whole process would be lost. If the cosmic watcher were particularly fortunate during a certain half second or so of the developing panorama before him, he might notice, at a few places on the globe, the flickering appearance of small spots or features that had not been there before; they probably would soon be gone, however, and the geological drama would proceed on its way. These otherwise unimportant transient appearances, however, represent the whole history of civilized man upon the earth, and so, because of our quite parochial interest in this instant of time, we must once again expand the viewer's scale around it until, let us say, he would see a hundred earth years pass by during a second of his own time. By now, geological motions on a large scale would seem completely to have stopped, although the works of men would spread and contract like amoebae upon the land. But, viewed with a fine enough microscope, say over a few meters distance, the junction between two crustal plates would still be very active, only now the motion would be highly irregular, consisting of abrupt jumps, separated by motionless intervals. These jumps, or sudden, local displacements along crustal boundaries, release strain energy built up during quiescent periods by the long-range average plate motions, and radiate shock waves that we know as earthquakes.

The intervals between these fault slippages vary greatly from one region to another, depending upon the character of the adjoining materials, the direction of the applied stress, and the average (over geologic times) relative rate of motion of the two sides. For example, the famous San Andreas fault in California, which is the boundary between the Pacific and North American crustal plates, is known to slip at an average rate of about 2 cm/year. Some portions of it actually move about this much each year, the motion occurring as a series of quite frequent but weak earthquakes. Conversely, some sections have not moved at all in over 100 years, and it is estimated that in portions not too far from such populated areas as Los

Angeles, we owe an effective debt of about 5 m of displacement. In these matters, nature is a totally merciless creditor; the debt must be repaid, and if the repayment must occur all at once, the reckoning will be terrible, since it will cause an earthquake somewhat in excess of magnitude 8.

II. STRAIN MEASUREMENT

Since earthquakes are so frequently destructive of life and property, and also occur without prior warning, it is natural that a most urgent goal of applied science should be the invention of some means for their prediction. In this context, we would be content to define prediction as merely a significant degree of improvement in our presently poor ability to define times and places of maximum hazard. Recently, however, there have been suggestions from theoretical geophysicists that it is reasonable to expect certain observable precursory events, particularly changes in the rate of strain accumulation in the neighborhood of a fault, at some time (so far unspecified) before the actual break occurs [1].

Seismographs, which have been almost the only instruments employed in earthquake studies, are essentially useless for the observation of strain or even of fairly rapid strain rates, since they characteristically respond only to the very-high-frequency disturbances associated with earthquakes themselves and their aftershocks. Any search for possible strain precursors must be done with instruments responding to the slowest movements; the essential distinction in our requirements is then that we require a displacement, rather than a velocity or acceleration transducer. By definition, this means that a strain gauge must be employed.

The measurement of strain (and here we discuss linear strain for simplicity) is, in its most basic sense, the determination of the distance between two reference points imbedded in a medium under study. Two important requirements are implied by this simple definition. First, since a length measurement is always a comparison between the object being measured and some standard object, the standard itself must be adequately constant and stable. (The measurement of fine machine parts with a rubber ruler is obviously bad practice). Second, the medium between the two reference points must be homogeneous, and typical of the overall region under examination.

Both of these seemingly simple requirements have, nonetheless, been so difficult to meet in the case of earth strain studies, that until very recently, little reliable data have been obtained. For many years, geophysicists have attempted to monitor strain by using long built-up rods of quartz, some as long as 100 m, suspended between piers attached to the bedrock within deep mines or tunnels [2]. Typically, the rod is connected to one pier, and approaches the second pier closely enough so that a sensitive displacement transducer (e. g., capacitor) can be used to measure the remaining gap. Then, to the extent that the quartz rod is dimensionally invariant, changes in this gap represent earth-strain variations averaged over the quartz rod length. The difficulty in this method arises because the magnitudes of strain which are important geophysically are so very small. For example, regional tectonic strain rates in the Southern California area are estimated to be about 10^{-8} per day, and when one considers that quartz has a thermal expansion coefficient of 5×10^{-7}/deg, it follows that a fraction of a degree diurnal fluctuation in the local air temperature would mask any data having a one-day period.

More serious than this problem, however, is the severe limitation on the size of crustal sample available to a solid strainmeter. Crustal rocks are almost always fractured to some degree, and for depths less than a kilometer or so, where lithostatic pressures bind the fractures, various blocks tend to superimpose some individual motions upon the average associated with area-wide strain. We can say, equivalently, that there is a great deal of noise in the spatial wave-number spectrum of strain for values higher than the order of 1 km^{-1}. Clearly, then, a very long measurement baseline is advantageous since, for a given strain, the absolute change in separation of its ends is proportional to its length for wavelengths much longer than the baseline, whereas for highly local motions, relative displacement of the two ends is uncorrelated, and hence does not increase with base length. In short, the signal-to-noise ratio of an earth strainmeter designed for tectonic studies is proportional to its length.

III. THE APPLICATION OF LASER INTERFEROMETRY

With the advent of the laser, it has become reasonable to consider optical interferometry as a means for geophysical strain measurement. Coherence lengths at least as great as desirable strain baselines can be obtained, and means are available for stabilizing laser wavelengths to the necessary degree.

Several groups have initiated work along these lines [3-5]. While the objectives of these experiments have tended to differ, particularly in the portion of the strain frequency spectrum under study, all have had two characteristics in common. First, it is necessary that the long test arm be evacuated, since the inhomogeneity and turbulence in even the quietest air causes intolerable fluctuations in the optical path length. Second, the laser wavelength is stabilized, either by active feedback circuitry, saturation techniques, or by careful stabilization of the environment of an otherwise free-running laser.

The particular experiment we describe here differs from all others, however, in that we have deliberately chosen to do the strain measurement at the surface of the ground rather than in a deep mine or tunnel, as is usual [6,7]. The contending arguments for each of these placements might be summarized as follows:

Location of the strainmeter in a deep site such as a mine provides a more reliable coupling to the bedrock, eliminates diurnal and seasonal heating and cooling as well as weathering as perturbing factors, and usually also provides an extremely stable atmospheric environment for the instrument.

Location on the surface allows a freer choice of sites. One cannot expect suitable caverns to exist at just those places where, on geophysical grounds, it would be most important to make measurements. While the surface itself is subject to rapid environmental changes, the ground only a meter or two below is quite well insulated, and one can easily anchor the reference piers in holes dug to such depths. Also, since most of the residual strain noise present near the surface is of quite local character, one can hope to achieve a useful signal-to-noise ratio (as mentioned earlier) by making the baseline long enough. Our present instruments, now two in number, were designed and built in the hope that the second of these two arguments is valid.

IV. DETAILS OF INTERFEROMETER DESIGN

A. Configuration

We employ, for our experiments, an interferometer of the Michelson configuration, with the long arm 800 m long, and the short arm 30 cm long. The main optical components are all in the atmosphere within highly insulated houses whose temperature is maintained constant within two degrees. The main portion of the long baseline arm is a metal tube of 15 cm diameter, evacuated by rotary pumps to a pressure of about 1 mTorr. The residual air path outside the vacuum windows of the pipe is made equal in length to that of the reference arm, so that barometric pressure variations produce no spurious strain readings. A single-frequency He–Ne laser of about 100 μW power at 6328 Å illuminates the instrument. The beam, after passing through an isolator, is expanded to 2.5 cm diameter for transmission down the pipe; after 800 m, it has been expanded by diffraction to about 5 cm diameter. A focusable "cat's eye" retro-reflector of 5 cm aperture returns the beam to the sending point, where it is recombined with the local reference beam to form fringes. Since the local reflector is also a cat's eye, focused at infinity, the interference occurs with uniform phase over the whole field, i.e., we employ the central fringe.

Ambiguity in the direction of strain changes is resolved by insertion of a phase-shifting plate into half of the reference beam, so that in the fringe field, one-half of the beam reaches maximum intensity a quarter-cycle before the other half. Each half is detected separately, and the signals fed to a reversible counter that adds or subtracts fringe counts depending upon the lag-lead relationship of the two channels.

There is no apparent loss of fringe visibility due to lack of temporal beam coherence over the 1600-m path difference. Even after incurring beam losses by reflection at the vacuum windows, some loss of power by vignetting at the retro-reflector aperture, and a loss deliberately inserted by slight defocusing of the return beam in order to reduce sensitivity to steering drift, we retain a ratio I_{max}/I_{min} in the fringes of between 2 and 3.

B. Stabilization

The laser originally employed (a Spectra-Physics No. 119) is stabilized by means of frequency modulation of its output around the central Lamb dip, and correction of the center frequency, by an ac servo loop, toward the bottom of the dip. By such a means, long-term stability of the order of 10^{-9} is reportedly attained. We quickly discovered, however, that because of the greatly different interferometer arm lengths and resulting retardation effects, the mixed beams differed in frequency at the detector by hundreds of kilohertz, forcing the counters to operate at this rate in order not to confuse the data. As a result, the stabilizer was changed altogether to a design in which the laser output is referred to the resonant frequency of a very stable passive optical cavity.

The reference cavity is a 2.5-cm-diam 30-cm-long tube of annealed quartz, against which two mirrors, one a flat and the other a sphere of 3 m radius, are held by spring retainers. This cavity has a finesse of about 100 for the TM_{00} mode, and operates in about the millionth order; thus, a relative frequency shift of 10^{-8} sweeps the light through an entire resonance. In order to develop a dc correction signal for the piezoelectric mirror mount in the laser, an ac servo loop is still employed, except that the frequency of the reference cavity itself is modulated, rather than that of the laser. We accomplish this modulation by acoustically vibrating the quartz tube at the frequency of its lowest axial compressional mode, which is 5 kHz when the loading of the mirrors is taken into account. Coils are fastened around the tube near both ends and radial magnetic fields from external permanent magnets pass through them. A power of about 1 mW at 5 kHz in each coil stretches and contracts the whole structure axially by about 20 Å. This otherwise small excursion is enough to produce a large ac modulation in the cavity optical transmission, and allows the use of a conventional ac servo loop, i.e., a circuit in which the output of the reference cavity detector and the 5-kHz driving oscillator are compared in a phase detector whose output (quasi-dc corrects the laser frequency toward the center of the reference cavity passband. The overall loop gain is 150 dB at 5×10^{-6} Hz falling off at 6 dB per octave to unity gain at 130 Hz.

The reference cavity, together with its detector, preamplifier, support structure, and magnets, is sealed within a heavy aluminum cannister in a dry-argon atmosphere. The cannister is wound with heater wires, and enclosed in insulation consisting of styrene foam and aluminum foil radiation shielding. Signals from a thermistor bridge circuit on the container are fed through high-gain amplifiers to the heater supply, providing final temperature stability of better than 10^{-3} °C for the reference cell. The corresponding frequency drift, due to thermal fluctuations, is thus no greater than 5×10^{-10}. There are, indeed, materials superior to quartz in their thermal coefficients which might have been employed as the reference cavity spacer. However, no available material is superior to quartz in terms of its dimensional stability over long periods. Quartz has not, as far as we can determine, ever been observed to "creep", or flow under mechanical stress. This property is much more important than low thermal coefficient, since, while one may detect temperature fluctuations and correct for them, correction for dimensional changes arising from stress is not possible.

C. Mechanical Design

Coupling to the earth itself is accomplished by mounting the main interferometer and the retro-reflector upon massive granite columns, each about 3 m long and 1 m square in cross section. These piers are cemented into holes of a little over 2 m depth, with cement surrounding only the bottom 1 m of the pier. Thus, perturabtions caused by personnel walking about in the shelter houses are not transmitted detectably to the piers.

The vacuum pipe enclosing the long optical arm is held approximately 1 m above ground level by a series of support frames spaced at 8-m intervals. A special problem is caused by the fact that, being in the open atmosphere, the pipe expands and contracts considerably from night to day and with changes in season. A typical diurnal variation is 50 cm in total length. An actual change of this magnitude in the residual air path in the long arm would be very serious, since each millimeter of added air path produces about one fringe of indicated strain change. We have, therefore, constructed at each end of the pipe a servo-actuated sliding expansion section that maintains the constancy of distance between each window and its corresponding pier to within about 10 μ. The pipe is free to move on rollers at each support point, with the exception of its midpoint, where it is anchored to the ground.

D. Isolator Design

The means by which isolation of the laser from the external resonant systems set up by the interferometer geometry is accomplished here is perhaps unusual in some respects. Ordinarily, one may achieve isolation between two optical resonators by use of a polarizer and a quarter-wave plate, and indeed, we employ such a device at the laser output. However, since the beam splitter exhibits a different transmission (and reflection) coefficient for wave components polarized parallel and perpendicular to the plane of incidence, the beams passing through the two interferometer arms will be in different states of polarization: we may say, technically, that two different Jones matrices transform the beams for transmission and reflection. When they return together to the isolator, the difference in their polarization ellipses is doubled again, and while it is possible to block the return of one beam by suitable adjustment of the isolator, it cannot be done for both at the same time.

In order to resolve this problem, we have introduced into the local short Michelson arm a component called an "auxiliary beam splitter". Its function is only to alter the polarization ellipse of this beam in such a way that when it returns to the isolator it is polarized identically to the long-arm beam, thus allowing full isolation. Such a procedure is workable because modern optical technology has made available beam splitters having various ratios of parallel-to-perpendicular component transmission; this availability, in addition to fine tuning of the ratio which is accomplished simply by adjusting the angle of the auxiliary beamsplitter with respect to the short-arm axis, has enabled us to achieve, at least in an operational sense, perfect isolation of the laser.

V. ILLUSTRATIONS AND FURTHER DETAILS

We present, in this section, several figures that illustrate the actual construction of the equipment so far described, and elaborate on some of the design features.

A schematic representation of the first instrument constructed is shown in Fig. 1. This strainmeter is located within the city limits of San Diego, California, on a piece of land selected not for its intrinsic geophysical interest, but simply because of its proximity to the University campus, its relative flatness, and its availability. We have, therefore, viewed the installation primarily as a test facility for the interferometer, and not as a geophysical experiment. A heavily travelled motorway passes within 1 km of the site, and a nearby air base generates considerable acoustic noise. It is remarkable, in retrospect, how little noise in the interferometer output has appeared from these sources. The original vacuum pipe is stainless

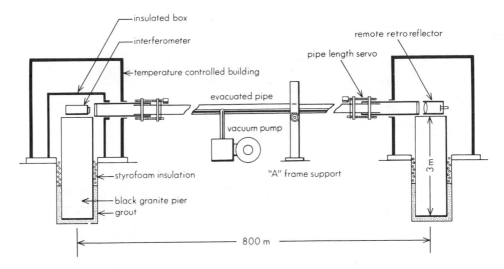

Figure 1. Layout of interferometer.

steel of 1.5 mm thickness, and welded at 8-m intervals into a single unit.
The support crossarms within the frames can be adjusted to hold the tube at
any height from just over ground level, up to a height of nearly 2 m; thus,
the ground beneath the tube does not have to be perfectly level, but should
not vary by much more than 1 m from a level mean. In a kilometer span,
however, this is still a very exacting requirement, and indeed, we have had
to do small amounts of excavation in setting up the line.

A schematic diagram of the interferometer chassis itself is shown in
Fig. 2. Near the center of the table the laser, consisting of plasma tube (S),
frame (A), and mirror translator (T), rests upon a coarse steering platform
(E) together with isolator (B), relay mirror (C), and expanding telescope (D).
Micrometers (F) and (G) steer the beam vertically and horizontally by turning the platform (H). The main beamplitter (I) is placed immediately ahead
of the expander, with the auxiliary beamsplitter (J) between it and the local
retro-reflector (K). Detection of fringes is done by two photomultipliers in
housing (L), the pairing being necessary because, as mentioned earlier, the
beam must be broken into two phase-shifted components in order to remove
directional ambiguity in strain. The phase plate itself is mounted over the
local cat's-eye lens. Fine directional steering of the beams is done at
mirror (C), advantage being taken of the fact that any directional change here
is reduced, in the output beam, by the magnification ratio of the expanding
telescope. While provision was included originally for a servomechanism to
correct possible steering drifts in the beam, its use was never found

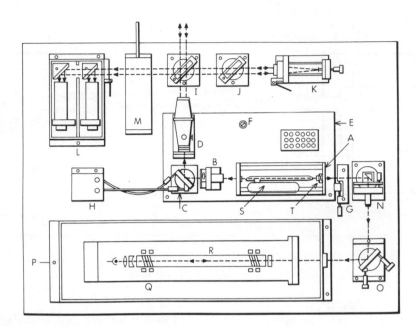

Figure 2. Optical and stabilizing components on main pier.

necessary.

Through the rear mirror of the laser, a weak beam is directed by way of two relay mirrors into reference cavity (R). Since the mirror spacing of the resonator is far less than confocal, there is negligible degeneracy among the multitude of available modes, and extreme care must be taken in illumination of the cavity if one wishes to excite, as is desirable, only the TM_{00} mode. In particular, the direction and position of entry of the beam into the cavity must be correct, as well as the wavefront curvature. Since there are two independently steerable relay mirrors in the beam and also a small adjustable collimating telescope (N), these criteria can all be met, although not without the expenditure of considerable time and patience in the adjustment.

Component (M), finally, is a linear potentiometer mounted upon the granite interferometer table, with its sliding arm connected mechanically to the end of the vacuum pipe. This resistor is one arm of a bridge whose output is used to control the pipe extension mechanism responsible for holding the residual air path constant.

The reference cavity cell is shown in some additional detail in the

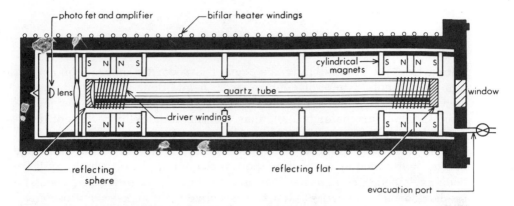

Figure 3.   Fabry-Perot stabilizing reference cavity.

schematic diagram of Fig. 3.   The radial magnetic field through the acoustic driver windings is set up from pairs of axially magnetized cylindrical Alnico magnets facing each other in opposing polarity.   Detection of cavity transmission is done by a photo-field effect transistor.

A block schematic diagram of the stabilizer electronic system is shown in Fig. 4.   The reader should find it self-explanatory.

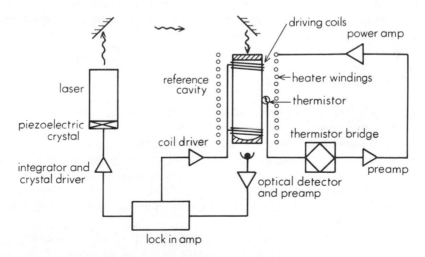

Figure 4.   Block diagram of stabilizer electronics.

VI. OBSERVATIONAL RESULTS

A. Data Presentation and Logging

Output from the reversible fringe counter is recorded digitally on magnetic tape at a sample rate of one per second. In addition, a digital-to-analog converter feeds a chart recorder at the data logging station (a few meters from the interferometer house), thus allowing a quick inspection of instrument operation.

An extremely useful visual representation of the instantaneous fringe output is provided by connecting the two photomultipliers (which, as has been explained, are in phase quadrature with respect to fringe passage) to the horizontal and vertical plates, respectively, of an x-y oscilloscope. Here one observes a spot moving about in a circular path: each clockwise revolution represents one fringe of baseline lengthening, i.e., $\lambda/2$ length change, and conversely, a counterclockwise turn means $\lambda/2$ shortening. High-frequency noise components are observed as oscillation of the spot back and forth along the circumference. This presentation also allows instant recognition of (a) loss of laser isolation, in which case the spot shows a preference for one azimuth, (b) drop in laser power, indicated by shrinkage of the circle, or (c) steering errors in the long beam, which usually results in distortion of the circle into an eccentric ellipse.

The magnitude of high-frequency noise in both of the present 800-m instruments is usually less than one-half fringe. Most of it appears to be of acoustic origin from such components as air conditioners, etc. When jet aircraft take off from the nearby station, the noise (in the frequency range of several tens of hertz) can reach a magnitude of two fringes.

B. Geophysical Data

The major strain component of geological origin that can be recognized easily within periods of a few days is the tidal strain. Its origin is identical to that of the ocean tides, and so, its appearance, as a record in the time domain is nearly identical. It consists of two frequency components, one diurnal and one semidiurnal with their relative phases shifting according to the phase of the moon. Since the elastic properties of the Earth are fairly well known, the amplitude of these earth tides can be predicted with good accuracy.

In Fig. 5, we show the raw output of the interferometer over a period of 18 h. Each of the fine steps observable in the curve is one fringe shift, or a strain change of 4×10^{-10}.

Figure 6 is a superposition of two curves: the solid line is the observed

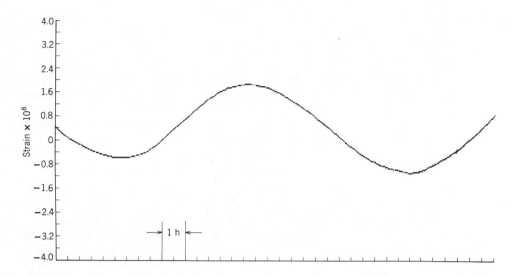

Figure 5. Raw output of strainmeter over one day. Steps indicate discrete fringe counts.

tidal signal over 10 days, and the broken curve is the theoretically predicted tidal record. While the agreement is roughly satisfactory, it turns out that the discrepancy can be almost perfectly removed by correcting for the effects of ocean tidal loading (we are only 15 km from the Pacific coast [8]. A phasor diagram of the semidiurnal components, both observed

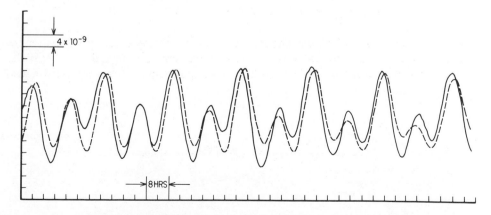

Figure 6. Ten-day record of tides (solid curve) and theoretically predicted tides for ideal earth (dashed curve).

and theoretical, is shown in Fig. 7. The difference vector is seen to agree extremely well both in amplitude and phase, with a strain phasor computed on the basis of the known ocean tidal amplitudes in the San Diego region [9, 10].

The main aim of this research, however, is the observation of trends in the long-term or secular strain rate. In a record of 42 days' duration, it was plain that a fairly steady increase in strain was occurring. The average was 7.7×10^{-9} per day, which agrees in order of magnitude with what can be guessed on the basis of geological history. However, the constancy of this particular record is not typical of later data; rates have varied between nearly zero, and double the above value. Certainly, there is no a priori reason to expect a steady strain rate over periods as short as a few months.

One must, of course, consider seriously the possibility that this "secular strain" record contains a component of, or at worst a preponderance of, instrumental drift. We find it difficult, however, to find ways in which the laser frequency could have shifted by a factor of 10^{-6}, which by now is the net strain accumulation over the baseline of our older instrument. The reference cavity temperature is monitored by an independent pair of thermistors located at different places in the oven, so that a failure of the controller would be easily seen. Any jump of orders of axial resonance in the stabilizer would be recognizable by a corresponding discontinuity in the strain record. Pier drifts through settling or curing of the cement remain as a serious possibility, although motion of the interferometer pier, if it occurs as a tipping about an axis near its bottom, would be accompanied by a serious and observable drift in beam steering. This has not occurred.

VII.   THE INSTALLATION AT PIÑON FLAT

The first installation of one of these interferometers at a site specifically chosen for its geophysical interest was made during February and March, 1971, at Piñon Flat in the San Jacinto Mountains about 200 km northwest of

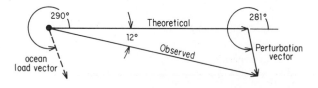

Figure 7.   Phasor diagram of semidiurnal component of tides. The dashed vector is the strain computed from the tides at the seashore, 15 km away.

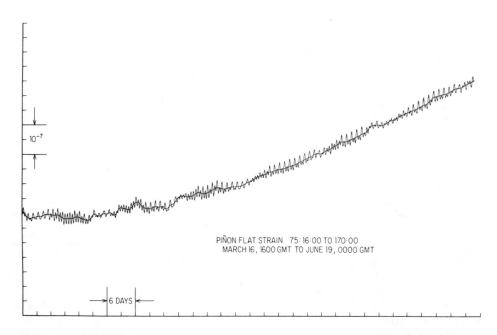

Figure 8.   95 days' record from the new installation at Piñon Flat.   The uniform secular strain rate in the last half of the record is 7×10^{-9} per day.

San Diego.   This location is near a section of the San Andreas fault where a transition from a "locked" condition (i.e., no observed slippage for over 100 years) to an active condition (frequent minor jumps) occurs.   The site is remote enough so that man-caused noise is very low.

Special features of the new instrument are (1) a modification of the counter electronics, producing a count for every one-fourth fringe, or 10^{-10} strain, has been installed.   (2) A gentle curvature of the otherwise smooth terrain compelled us to insert a 1.5° bend in the pipe at its midpoint, and to bend the beam, accordingly, at this point by means of a pair of small-angle prisms.   The instrument has performed very well from the first weeks of its being turned on.   In Fig. 8 we display a 95-day record, starting a few days from the beginning of its operation.   It is interesting to note that the maximum secular strain rate (about 7×10^{-9}/day) was observed after a month or so of running, rather than at the beginning of the run, as might have been the case had pier movement due to cement curing been a serious factor.

Finally, as an example of data obtainable by a least-count sensitivity of 10^{-10}, we present two records obtained from the earthquake near Valparaiso in Chile on July 9, 1971.   It had magnitude 7.6 (Richter), and produced a

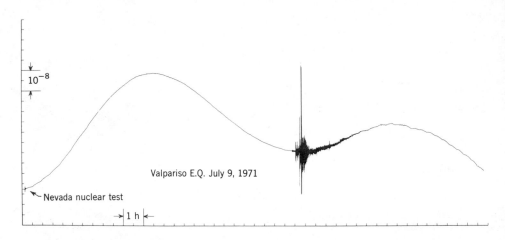

Figure 9.   Output of Piñon Flat strainmeter during Chile earthquake.   Note the presence of low-frequency strain components for many hours afterwards.

peak-to-peak strain of about 5×10^{-8}, or 500 quarter-fringe counts.   The real-time record, shown in Fig. 9 is not in itself exceptional, since the signal was very large.   However, a Fourier analysis was performed upon two segments of this record, the first being over the 12 h preceding the earthquake, and the second the 12 h after its occurrence.   These two spectra are drawn together in Fig. 10.   The most significant feature of this plot is that the majority of peaks in the post-earthquake spectrum can be associated with well-established normal modes of oscillation of the whole earth.   The lowest modes, having frequencies not much over one cycle per hour, required many hours of integration for their determination;   they probably account for the visible irregularity in the last part of the record.

VIII.   CONCLUSION

The exceedingly difficult and confusing science of tectonic strain measurement has been made much easier since the laser has made very long beams of coherent light available to us.   We may suppose that with some development in the simplification and cheapening of such instruments as we have described here, one may look forward toward building arrays of strainmeters, thus making strain field mapping a possibility.   By such means, it is not unrealistic to hope that we may some day succeed in predicting with useful accuracy the time and location of earthquakes.

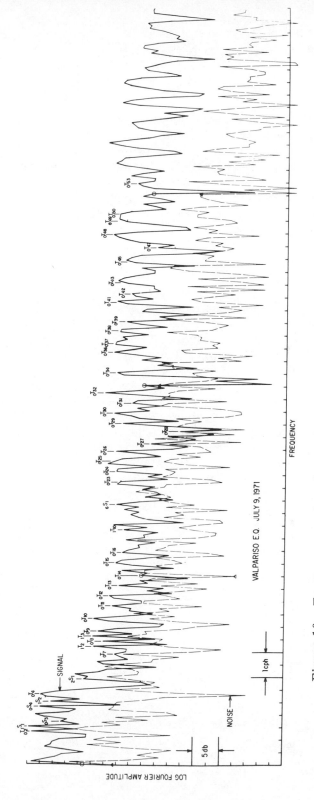

Figure 10. Frequency spectra of record of Fig. 9, taken for 12 h preceding Valparaiso earthquake, and 12 h after. Frequencies of earth normal modes are indicated.

937

ACKNOWLEDGMENTS

This work has been supported by grants from the National Atmospheric and Oceanic Administration and the National Science Foundation.
Frank Wyatt has designed and constructed most of the electronic circuitry, and has been responsible for much of the design and organization of mechanical components.
Finally, we wish to express our sincere thanks to the staff of the UKAEA Culham Laboratory for their generosity in helping us prepare this manuscript at relatively short notice.

REFERENCES

1. F. Press (private communication).
2. H. Benioff, Bull. Seismol. Soc. Amer. 25, 283 (1935).
3. V. Vali and R. C. Bostrum, Rev. Sci. Inst. 39, 1304 (1968).
4. H. S. Boyne et al., in Laser Applications in the Geosciences (Western Periodicals, North Hollywood, Calif., 1970), p. 215
5. G. C. P. King and D. Davis, The Application of Modern Physics to the Earth and Planetary Interiors (Wiley, London, 1969), p. 513.
6. J. Berger and R. H. Lovberg, Rev. Sci. Inst. 40, 1569 (1969).
7. J. Berger and R. H. Lovberg, Science 170, 296 (1970).
8. J. Berger, thesis (University of California at San Diego, 1970).
9. W. E. Farrell, thesis (University of California at San Diego, 1970).
10. W. Munk, F. Snodgrass and M. Wimbush, Geophys. Fluid Dyn. 1, 161 (1970).

LASER APPLICATIONS, IMPORTANT AND DELIGHTFUL

Paul H. Lee
Department of Physics, University of California
Santa Barbara, California

ABSTRACT

Major uses of lasers are discussed by type and in the order of their commercial economic importance. A few applications are mentioned which have unusual technological or human appeal. The economic information presented was freshly gathered during visits the author made to many of the larger manufacturers of lasers. Conjectures are made about laser applications in the future.

I. INTRODUCTION

There is now a voluminous informal literature on the subject of laser applications. It is not the purpose of this brief review to catalog these applications completely, nor to analyze them from a purely scientific point of view. Instead, the plan is to discuss only a few laser applications which have an unusual human, technological, or social importance.

The concept of importance has many facets. For example, I learned recently that the rubber nipples for baby nursing bottles are now pierced using laser light. A physicist might consider this trivial indeed, but for parents who are struggling with a hot needle at 2:00 am, this could be the all

939

important laser application.

Raw economic statistics are one useful measure of importance, and therefore, throughout this discussion are the answers to such questions as: How big is the total yearly laser market? Which are the economically important types of lasers? For what applications are these lasers sold?

The economic statistics given here pertain to the year 1970. They were not derived from library search, but entirely from information gathered during a visit to each of the major laser manufacturers in California [1]. In combination, these companies make more than half of the lasers sold throughout the world.

During these visits, I felt welcome enough to ask for detailed figures about company sales and for estimates of the present world market for lasers. These companies seemed without exception to be well-informed about their own business posture and also about how well their competition was doing. These reports on business performance were remarkably inter-consistent. There was a standard request by most of these companies not to ascribe detailed information directly to its individual or corporate source. For this reason, the edited and compiled data only will be presented here.

The plants I visited were well equipped and had a businesslike atmos-phere. All were the survivors of recent lean years during which they had found it necessary to run a "tight ship." I learned that about 50 companies are engaged in the manufacture of lasers but only five of these have annual sales in this field of more than $1 million. Together these five account for about 45% of the total dollar sales in the U.S.A.

I also learned that, in the lingo of laser manufacturing and sales, one talks about laser devices and laser systems. In these terms, a laser device is the laser oscillator itself together with such parts as a protective case and power supply which are considered necessary to make a minimum work-ing product. Any instrument dependent on an embodied laser, but more complex or expensive than a laser device, is called a laser system. There is also a market for accessories which is small compared with that for the two major classifications. Accessories include any optical, electro-optical, and mechanical components manufactured especially for use in conjunction with laser devices.

The major manufacturers of laser devices feel strongly that the future growth of their enterprises will depend on their success in developing and selling laser systems as well. The reason for this is clear from the follow-ing figures:

For the year 1970, the world sales for laser devices was just under $18 million, with an annual rate of increase of about 10%. The total world commercial sales for laser systems together with laser devices was $55 million. The annual rate of increase in these overall sales is about 30%. $40 million dollars in laser systems and devices were made and sold commercially in the U.S.A.

These numbers do not include sums spent to procure operational military equipment. They also do not include salaries, overhead, and operating expenses for laser research and development (whether in government laboratories, universities, or industry), except when these expenses appear as purchases of commercially catalogued laser devices or systems. There has been considerable speculation about the total size of all such laser expenditures with estimates ranging between 100 and 200 million dollars annually. I have found no new financial information to contribute to these estimates. There is also a difference of opinion about how much weight one should attach to military or R and D expenditures when assessing the economic and social importance of our new laser industry.

Those interested in the growth of the laser industry are disappointed to note that as yet there are no popular consumer products containing a laser. Few have been seriously proposed. Nevertheless, the recent increase in the sale of laser systems for industrial, military, and medical purposes is encouraging. Growth into a far larger, more broadly based laser industry may still be "just around the corner."

At the present time, and for every classification of lasers, one-half or more of those commercially manufactured are bought to do research, development, or teaching in the basic and applied sciences. These lasers are usually used for "one-of-a-kind" applications which cover an interest range from trivial to fascinating. Thus, the laser industry still has a strong feedback loop to scientific free-thinkers who may contribute ideas for new fields of industrial growth.

Some examples of the major ways lasers are now being used will be discussed briefly in order of laser type in the following sections.

II. HE-NE LASER APPLICATIONS

Approximately 19,000 He-Ne laser devices are manufactured yearly and are sold for a total of $6.3 million. More than half of these are used in all kinds of scientific laboratories as general purpose coherent light sources. The most frequent use is for the alignment of other laboratory apparatus.

Perhaps the most spectacular example of this kind of use occurs at the Stanford Linear Accelerator Center [2]. By the use of 297 all-metal Fresnel zone plates distributed along the 3-km length of the accelerator, its evacuated tube can be aligned both horizontally and vertically within ± 0.1 mm, or about 1 part in 3 times 10^7 of the overall length.

A few relatively expensive single-mode frequency-stabilized 6328-Å lasers are built for the illumination of precision interferometers. Some of these are used for secondary length standards, for geophysical measurements, and for machine tool control. Laser systems for machine tool control have growing annual sales of several million dollars a year.

The remaining financially important He-Ne laser systems are used for alignment purposes in the construction industry. Of these, laser transits account for about $1 million annual sales. But the surprisingly larger sales in this field, $6 million annually, are chalked up by some remarkably simple systems for laying sewer pipe at a controlled down-hill grade of about 1.5°. Suitcase kits for assuring the uniform flow of "effluent" are available complete with convenient auxilliary hardware including pencil and paper. These little laser systems can easily save a construction contractor $2000 per week in the costs of labor and open trench time.

III. ION LASER APPLICATIONS

Nearly 500 ion laser devices are manufactured yearly, and sold for a total sum approaching $5 million. Again more than half of these units are used in R and D laboratories in diverse small-quantity applications which include holography, scattering, data storage, display, writing, and facsimile applications. The major use of ion laser systems is in Raman spectroscopy apparatus. Sales of instruments for this purpose total about $5 million annually.

The surprise in the field of argon ion laser systems is the new $2.5 million annual sales level of instruments for retinal coagulation. These systems are self-contained in fairly elaborate cases. In one model, the laser beam is reflected down an enclosed articulated arm to a moveable opthalmascope head boresighted and focused to coincide with the beam.

Retinal coagulation has been very successful for the periodic treatments of diabetic retinopathy [3]. It prevents or greatly retards blindness which otherwise results from retinal hemorraging. Such hemorraging is often a concomitant of severe diabetes.

IV. OTHER LASER APPLICATIONS

Table I shows annual sales figures for Ruby/Glass, CO_2, YAG, and other laser devices. None of these seriously challenge the unit or the dollar sales levels of the He-Ne or ion laser types.

Ruby or glass lasers continue to be most effective devices for generating high power and high peak power pulses with a slow repetition rate. Applications are made to high-speed photography, holography, and range finding. A few remarkably large units have been built to generate very-high-temperature plasmas in nuclear research laboratories.

Although sales of YAG laser devices are relatively low, a number of quite expensive systems are being sold for cutting and trimming in manufacturing plants. Some have complex automatic controls. Quite a few YAG

TABLE I.   SALES OF LASER DEVICES IN 1970[a]

Laser Type	Units/Yr.	$ Million/Yr.
He-Ne	19,000	6.3
Ion	500 max	5.0 max
Ruby/Glass		2.5
YAG		1.5
CO_2	130 max	1.5
All others		1.0
Total		17.8[b]

[a]Compiled from information furnished by major
laser manufacturers in California.
[b]Estimated accuracy ± 10%.

laser systems are used in the electronics industry, where resistor trimming
is the most common application.

Most CO_2 laser systems likewise are many times more expensive than
the laser devices they use.   The mass common applications are in industry
for the cutting and welding of metals, for the drilling of ceramics, and for
the burning of wood, paper, cloth, and plastics.   Continuous output power
levels over 250 W are available at better than 10% overall efficiency, and
maintainance problems are low.

V.   LASERS IN INDUSTRY

The output light from several different lasers is employed in industrial tools.
A number of laser devices have shown adequate reliability in industrial
service.   Chosen carefully, they can sometimes offer unique advantages over
other methods of manufacture.   Probably in part because of its close association with the Bell Laboratories, Western Electric Company is unusually
well informed and creative about the use of lasers in its industrial operations.
Table II shows twelve different applications of lasers reported in use at
Western Electric factories.   Some of these applications are intellectually so
intriguing they would make great adult toys, but every one is done in full
financial earnest.

Perhaps the most delightful application of a laser system to a production
industry is the clothing pattern cutting machine being developed by Hughes
for Genesco [4].   Conceptually, with this device one can have clothing

TABLE II.   LASER APPLICATIONS IN WESTERN ELECTRIC
            MANUFACTURE[a]

Application	Example	Laser
Measurement	Mask defect identification	He-Ne
	Continuous monitoring of wire drawing	
	Interferometer for step and repeat	
Heating and melting	Welding metal relay springs	Ruby
Material removal:		
Bulk	Resize diamond wire drawing dies	Ruby
	Alumina and silicon substrate separation	CO_2, Nd:YAG
	Hole drilling in substrates	
Film	Carbon resistor trimming	
	Thin-film resistor trimming	Nd:YAG
	Dynamic tuning of quartz filters	
	Direct pattern generation	
Photographic exposure	Mask artwork generation	Ar

[a]Information from J. E. Geusic, Bell Telephone Laboratories, Holmdel, N. J.

economically cut to his personal specifications by a CO_2 laser beam under robot control.

VI.   FUTURE LASER APPLICATIONS

From the discussion above, it seems likely that the present major applications of lasers will live and grow in the future.   These present applications are

   1.   as general research laboratory tools, especially for alignment;
   2.   as tools for surveying and alignment in the construction industries;
   3.   for medical instruments--especially for treatments and surgery of the eyes;
   4.   as a fixed optical frequency source for interferometer illumination,

and for Raman spectroscopy;

    5. as a narrow-beam source for machining operations such as cutting, trimming, drilling, and welding;

    6. as a tool for mining and tunneling through rock.

It seems to me that applications such as these are likely to continue to grow until they become many times more important economically and socially than they now are. They will grow by reason of increased confidence and refined technology simply because there is no evidence at this time of market saturation.

One can only conjecture about how the field of laser applications might change as new science evolves. For instance, what will be the practical impact on science and technology if an x-ray laser is discovered soon?

One can be more confident in predicting how existing knowledge may be extrapolated to future applications [5]. Within this scope it seems to me that the impending development of practical tunable coherent light sources will soon revolutionize not only the scientific control and measurement of light but also a large peripheral field of general instrumentation. Spectroscopy resolution will increase dramatically and spectrophotometry will become simple, practical, and accurate.

It also seems to me that lasers will be widely used within the next 20 years for the handling of information: I think that lasers will become important components in communication systems. They are very likely to be used with short-range medium-band wave guides for video telephones. They may also be used to produce long-range broad-band beams for communication in outer space. I also think that lasers will be used to process information in digital computers in many differing ways. Diode lasers may become especially useful for these latter applications.

There has been some expression of disappointment that many dramatically widespread and practical applications have not been realized during this past decade of laser history. Nevertheless the prognosis remains strongly positive for the growth of important new laser applications in the years to come.

REFERENCES

1. The author acknowledges the very kind cooperation of Herbert Dwight and Robert Mortenson, Spectra-Physics, Inc., Mountain View, Calif.; M. W. Dowley and George Stephans, Coherent Radiation, Palo Alto, Calif.; M. L. Stitch, Korad Department, Union Carbide Corp., Santa Monica, Calif.; R. C. Rempel, Chromatix, Mountain View, Calif.

2. W. B. Herrmannsfeldt, M. J. Lee, J. J. Spranza, and K. R. Trigger, Appl. Opt. **7**, 995 (1968); J. J. Spranza, Stanford Linear Accelerator

Center, Stanford, Calif. (private communication).

3. H. L. Little, H. C. Zweng, and R. R. Peabody, Trans. Am. Acad.
   Ophth. $\underline{74}$, 85 (1970).
4. Hughes Industrial Systems development for Genesco, Fredricksburg, Va.
5. The author feels indebted to P. L. Kelley, Lincoln Laboratory,
   Lexington, Mass., for unusually helpful discussions about present and
   future applications of lasers.

PARTICIPANTS AND INDEX OF CONTRIBUTIONS

947

INDEX

949